T0324846

Machine Learning for Engineers

This self-contained introduction to machine learning, designed from the start with engineers in mind, will equip students and researchers with everything they need to start applying machine learning principles and algorithms to real-world engineering problems. With a consistent emphasis on the connections between estimation, detection, information theory, and optimization, it includes: an accessible overview of the relationships between machine learning and signal processing, providing a solid foundation for further study; clear explanations of the differences between state-of-the-art machine learning techniques and conventional model-driven methods, equipping students and researchers with the tools necessary to make informed technique choices; demonstration of the links between information-theoretical concepts and their practical engineering relevance; and reproducible examples using MATLAB®, enabling hands-on experimentation. Assuming only a basic understanding of probability and linear algebra, and accompanied by lecture slides and solutions for instructors, this is the ideal introduction to machine learning for engineering students of all disciplines

Osvaldo Simeone is Professor of Information Engineering at King's College London, where he directs the King's Communications, Learning & Information Processing (KCLIP) lab. He is a Fellow of the IET and of the IEEE.

Machine Learning for Engineers

Osvaldo Simeone
King's College London

CAMBRIDGE
UNIVERSITY PRESS

University Printing House, Cambridge CB2 8BS, United Kingdom

One Liberty Plaza, 20th Floor, New York, NY 10006, USA

477 Williamstown Road, Port Melbourne, VIC 3207, Australia

314–321, 3rd Floor, Plot 3, Splendor Forum, Jasola District Centre, New Delhi – 110025, India

103 Penang Road, #05–06/07, Visioncrest Commercial, Singapore 238467

Cambridge University Press is part of the University of Cambridge.

It furthers the University's mission by disseminating knowledge in the pursuit of education, learning, and research at the highest international levels of excellence.

www.cambridge.org
Information on this title: www.cambridge.org/highereducation/isbn/9781316512821
DOI: 10.1017/9781009072205

© Cambridge University Press 2023

This publication is in copyright. Subject to statutory exception and to the provisions of relevant collective licensing agreements, no reproduction of any part may take place without the written permission of Cambridge University Press.

First published 2023

A catalogue record for this publication is available from the British Library.

Library of Congress Cataloging-in-Publication Data
Names: Simeone, Osvaldo, author.
Title: Machine learning for engineers : principles and algorithms through signal processing and information theory / Osvaldo Simeone, King's College London.
Description: First edition. | Cambridge ; New York, NY : Cambridge University Press, 2022. | Includes bibliographical references and index.
Identifiers: LCCN 2022010244 | ISBN 9781316512821 (hardback)
Subjects: LCSH: Engineering–Data processing. | Machine learning. |
BISAC: TECHNOLOGY & ENGINEERING / Signals & Signal Processing
Classification: LCC TA345 .S5724 2022 | DDC 620.00285–dc23/eng/20220504
LC record available at https://lccn.loc.gov/2022010244

ISBN 978-1-316-51282-1 Hardback

Additional resources for this publication at www.cambridge.org/simeone

Cambridge University Press has no responsibility for the persistence or accuracy of URLs for external or third-party internet websites referred to in this publication and does not guarantee that any content on such websites is, or will remain, accurate or appropriate.

Contents

Preface

Overview

This book provides a self-contained introduction to the field of machine learning through the lens of signal processing and information theory for an audience with an engineering background. This preface explains why the book was written and what you will find in it.

Background and Motivation

Advances in machine learning and artificial intelligence (AI) have made available new tools that are revolutionizing science, engineering, and society at large. Modern machine learning techniques build on conceptual and mathematical ideas from stochastic optimization, linear algebra, signal processing, Bayesian inference, as well as information theory and statistical learning theory. Students and researchers working in different fields of engineering are now expected to have a general grasp of machine learning principles and algorithms, and to be able to assess the relative relevance of available design solutions spanning the space between model- and data-based methodologies. This book is written with this audience in mind.

In approaching the field of machine learning, students of signal processing and information theory may at first be ill at ease in reconciling the similarity between the techniques used in machine learning – least squares, gradient descent, maximum likelihood – with differences in terminology and emphasis (and hype?). Seasoned signal processing and information-theory researchers may in turn find the resurgence of machine learning somewhat puzzling ("didn't we write off that technique three decades ago?"), while still being awed by the scale of current applications and by the efficiency of state-of-the-art methods. They may also pride themselves on seeing many of the ideas originating in their communities underpin machine learning solutions that have wide societal and economic repercussions.

Existing books on the subject of machine learning come in different flavors: Some are compilations of algorithms mostly intended for computer scientists; and others focus on specific aspects, such as optimization, Bayesian reasoning, or theoretical principles. Books that have been used for many years as references, while still relevant, appear to be partly outdated and superseded by more recent research papers.

In this context, what seems to be missing is a textbook aimed at engineering students and researchers that can be used for self-study, as well as for undergraduate and graduate courses alongside modules on statistical signal processing, information theory, and optimization. An ideal text should provide a principled introduction to machine learning that highlights connections with estimation, detection, information theory, and optimization, while offering a concise but extensive coverage of state-of-the-art topics and simple, reproducible examples. Filling this gap in the bookshelves of engineering libraries is the ambition of this book.

As I end this first section without mentioning "deep learning", some readers may start to worry. Indeed, deep learning is seen in this book "merely" as a particularly effective way to define and train a specific class of models. I have attempted to place it in the context of related methods, and I have often preferred to focus on simpler models that are better suited to illustrate the underlying ideas. While running counter to the general trend on the computer science side of machine learning, the focus on simple, reproducible examples is intended as a means to strip away aspects of scale to reveal concepts, intuition, and key techniques.

Intended Audience

This book is intended for a general audience of students, engineers, and researchers with a background in probability and signal processing. (I also hope that it will be of some interest to students and researchers working in information theory, who may not be aware of some of the connections explored here.)

To offer a self-contained introduction to these intended readers, the text introduces supervised and unsupervised learning in a systematic fashion – including the necessary background on linear algebra, probability, and optimization – taking the reader from basic tools to state-of-the-art methods within a unified, coherent presentation. Information-theoretic concepts and metrics are used throughout the text to serve as training criteria and performance measures. Later chapters explore topics that are subject to intense research at the time of writing, in the hope that they can serve as launching pads for original research and new contributions.

Why This Book?

I am often asked by colleagues and students with a background in engineering to suggest "the best place to start" to get into the field of machine learning. I typically respond with a list of books: For a general, but slightly outdated introduction, read this book; for a detailed survey of methods based on Bayesian signal processing, check this other reference; to learn about optimization for machine learning, I found this text useful; for theoretical aspects, here is another text; and, for more recent developments, a list of papers is attached. Unfortunately, the size and number of these references may be intimidating.

I started writing this book in the form of lectures notes for a course taught at the New Jersey Institute of Technology (NJIT). My motivation was, and remains, that of distilling the array of references mentioned above into a single text that offers a balanced presentation connecting to other courses on statistical signal processing, information theory, and optimization. This initial effort led to the monograph [1], which contains some of the material covered here.

Years later, across the ocean, now with a family and forced to work from home, I have had a chance to revise the original text [1] for a module on machine learning I am currently teaching at King's College London (KCL). The current book was born as a result of this activity. It is meant to serve as a full textbook, including background material; expanded discussions on principles and algorithms, such as automatic differentiation, contrastive learning, and energy-based models; completely new chapters on optimization, variational inference and learning, information-theoretic learning, Bayesian learning, transfer learning, meta-learning, and federated learning, as well as on further topics for research; new examples and figures; and end-of-chapter problems.

The number of machine learning books is quite extensive and growing. There are classical books, such as the texts by Bishop [2] and Hastie, Tibshirani, and Friedman [3], which provide general introductions in the form of an extensive review of techniques, not including more recent developments. In this book, I have taken a different approach, striving for unification within the framework of information theory and probabilistic models, with the aim of also covering more advanced material. The text by Murphy [4] focuses on probabilistic models, and can serve as a complementary reference to fill in the details of specific techniques that are not detailed here. Theoretical books, such as the excellent reference by Shalev-Shwartz and Ben-David [5], may not be the most natural choice for researchers interested in general principles, intuition, and applications, rather than detailed theoretical derivations. The text by Watt, Borhani, and Katsaggelos [6] takes an optimization perspective, and can also be used as a complement for this book. The book by Theodoridis [7] is an extensive reference on Bayesian learning. For readers interested in programming for deep learning, a useful reference is the book by Zhang, Lipton, Li, and Smola [8]. There are also various books that focus on applications, but mostly in the form of edited contributions (e.g., [9]).

Using This Book

This book can be used for self-study, as a reference for researchers seeking as an entry point in the field of machine learning, as well as a textbook. Courses on machine learning are primarily taught in computer science departments, but they are also increasingly part of the portfolio of engineering departments. This text can be adopted for a senior undergraduate course or for a more advanced graduate course on machine learning within an engineering curriculum.

An undergraduate course, like the one I currently teach at KCL, can cover Part I and Part II by discussing Chapter 1 to Chapter 6 in full, while including a selection of topics from Chapter 7 (see the Organization section below). A more advanced course could be taught by selecting topics from Chapter 7, as well as Part III and Part IV.

A note of warning for the more formally minded readers: While I have tried to be precise in the use of notation and in the formulation of statements, I have purposely avoided a formal – theorem, proof, remark – writing style, attempting instead to follow a more "narrative" and intuitive approach. I have endeavored to include only results that are easy to interpret and prove, and the text includes short proofs for the main theoretical results. I have also excluded any explicit mention of measure-theoretic aspects, and some results are provided without detailing all the underlying technical assumptions.

Throughout the text, I have made an effort to detail simple, reproducible examples that can be programmed in MATLAB® or Python without making use of specialized libraries in a relatively short amount of time. In particular, all examples in the book have been programmed in MATLAB® (with one exception for Fig. 13.17). Complex, large-scale experiments can be easily found in the literature and in books such as [8].

End-of-chapter problems are provided for Chapter 2, Part II, and Part III. (For the topics covered in Part IV, the reader is invited to reproduce the experiments provided in the text and to search the literature for open problems and more complex experiments.) The problems offer a mix of analytical and programming tasks, and they can be solved using the material and tools covered in the text. Solutions for instructors can be found at the book's website. Questions requiring programming are indicated with an asterisk (*).

When teaching, I find it useful to discuss some of end-of-chapter problems as part of tutorial sessions that encompass both theoretical and programming questions. I have also found it easier to use MATLAB®, particularly with students having little expertise in programming.

A list of recommended resources is included at the end of each chapter. The main purpose of these notes is to offer pointers for further reading. I have not attempted to assign credit for the techniques presented in each chapter to the original authors, preferring instead to refer to textbooks and review papers.

In writing, editing, and re-editing the text, I was reminded of the famous spider (*Cyclosa octotuberculata* if you wish to look it up) that weaves information about the positions of previously observed preys into its web. Like the spider, we as engineers use mathematics as a form of extended cognition, outsourcing mental load to the environment. In making various passes through the text, I have picked up my previous trails on the web, discovering new ideas and correcting old ones. I realize that the process would be never-ending, weaving and unweaving, and I offer my apologies to the reader for any mistakes or omissions in the current text. A list of errors will be maintained at the book's website.

Organization

The text is divided into four main parts. The first part includes a general introduction to the field of machine learning and to the role of engineers in it, as well as a background chapter used to set the notation and review the necessary tools from probability and linear algebra. The second part introduces the basics of supervised and unsupervised learning, including algorithmic principles and theory. The third part covers more advanced material, encompassing statistical learning theory, exponential family of distributions, approximate Bayesian inference, information maximization and estimation, and Bayesian learning. Finally, the fourth part moves beyond conventional centralized learning to address topics such as transfer learning, meta-learning, and federated learning. Following is a detailed description of the chapters.

Part I: Introduction and Background

Chapter 1. When and How to Use Machine Learning. This chapter motivates the use of machine learning – a data-driven inductive bias-based design methodology – as opposed to the conventional model-driven domain knowledge-based design engineers are more accustomed to; it provides general guidelines for the use of machine learning; and it introduces the landscape of machine learning frameworks (supervised, unsupervised, and reinforcement learning).

Chapter 2. Background. Chapter 2 reviews the necessary key concepts and tools from probability and linear algebra.

Part II: Fundamental Concepts and Algorithms

Chapter 3. Inference, or Model-Driven Prediction. As a benchmark for machine learning, this chapter reviews optimal Bayesian inference, which refers to the ideal scenario in which the statistical model underlying data generation is known. The chapter focuses on detection and estimation under both hard and soft predictors. It provides an introduction to key information-theoretic metrics as performance measures adopted for optimal inference, namely cross entropy, Kullback–Liebler (KL) divergence, entropy, mutual information, and free energy.

Chapter 4. Supervised Learning: Getting Started. Chapter 4 covers the principles of supervised learning in the frequentist framework by introducing key definitions, such as inductive bias, model class, loss function, population loss, generalization, approximation and estimation errors, and overfitting and underfitting. It distinguishes between the problems of training hard and soft predictors, demonstrating the connections between the two formulations.

Chapter 5. Optimization for Machine Learning. The previous chapter considers only problems for which closed-form, or simple numerical, solutions to the learning problem are available. Chapter 5 presents background on optimization that is needed to move beyond the simple settings studied in Chapter 4, by detailing gradient descent, stochastic gradient descent, backpropagation, as well as their properties.

Chapter 6. Supervised Learning: Beyond Least Squares. This chapter builds on the techniques introduced in Chapters 4 and 5 to cover linear models and neural networks, as well as generative models, for binary and multi-class classification. The chapter also introduces mixture models and non-parametric techniques, as well as extensions to regression.

Chapter 7. Unsupervised Learning. Chapter 7 introduces unsupervised learning tasks within a unified framework based on latent-variable models, distinguishing directed discriminative and generative models, autoencoders, and undirected models. Specific algorithms such as K-means clustering, expectation maximization (EM), energy model-based training, and contrastive representation learning, are detailed. An underlying theme of the chapter is the use of supervised learning as a subroutine to solve unsupervised learning problems.

Part III: Advanced Tools and Algorithms

Chapter 8. Statistical Learning Theory. The previous chapters have left open an important question: How many samples are needed to obtain desirable generalization performance? This chapter addresses this question by reviewing the basics of statistical learning theory with an emphasis on probably approximately correct (PAC) learning theory. Limitations of the theory and open problems are also discussed.

Chapter 9. Exponential Family of Distributions. Another limitation of the previous chapters is their reliance on Gaussian and Bernoulli distributions. This chapter provides a general framework to instantiate probabilistic models based on the exponential family of distributions (which includes as special cases Gaussian and Bernoulli distributions). Generalized linear models (GLMs) are also introduced as conditional extensions of exponential-family models.

Chapter 10. Variational Inference and Variational Expectation Maximization. As discussed in Chapter 7, Bayesian inference is a key subroutine in enabling learning for models with latent variables, including most unsupervised learning techniques and mixture models for supervised learning. Exact Bayesian inference becomes quickly intractable for model sizes of practical interest, and, in order to scale up such methods, one needs to develop approximate Bayesian inference strategies. This chapter elaborates on such techniques by focusing on variational inference (VI) and variational EM (VEM). As an application of VEM, the chapter describes variational autoencoders (VAE).

Chapter 11. Information-Theoretic Inference and Learning. All the inference and learning problems studied in the previous chapters can be interpreted as optimizing specific information-theoretic metrics – most notably the free energy in the general framework of VI and VEM. This chapter explores this topic in more detail by covering both likelihood-based and likelihood-free learning problems. For likelihood-based models, problem formulations, such as maximum entropy, are introduced as generalized forms of VI and VEM. For likelihood-free models, two-sample estimators of information-theoretic metrics – most generally f-divergences – are described and leveraged to define generative adversarial networks (GANs).

Chapter 12. Bayesian Learning. This chapter provides an introduction to Bayesian learning, including motivation, examples, and main techniques. State-of-the-art parametric methods, such as stochastic gradient Langevin dynamics (SGLD), are covered, along with non-parametric methods such as Gaussian processes (GPs).

Part IV: Beyond Centralized Single-Task Learning

Chapter 13. Transfer Learning, Multi-task Learning, Continual Learning, and Meta-learning. As discussed in Part II and Part III, machine learning typically assumes that the statistical conditions during training match those during testing, and that training is carried out separately for each learning problem. This chapter introduces formulations of learning problems that move beyond this standard setting, including transfer learning, multi-task learning, continual learning, and meta-learning. Among the specific algorithms covered by the chapter are likelihood- and regularization-based continual learning, as well as model agnostic meta-learning (MAML). Bayesian perspectives on these formulations are also presented.

Chapter 14. Federated Learning. Previous chapters assume the standard computing architecture of centralized data processing. A scenario that is of increasing interest involves separate learners, each with its own local data set. Training algorithms that operate in a decentralized way without the exchange of local data sets are often labeled as "federated". This chapter provides an introduction to federated learning, presenting basic algorithms such as federated averaging (FedAvg) and generalizations thereof, covering privacy aspects via differential privacy (DP), and Bayesian solutions based on federated VI.

Part V: Epilogue

Chapter 15. Beyond This Book. This chapter provides a brief look at topics not covered in the main text, including probabilistic graphical models, causality, quantum machine learning, machine unlearning, and general AI.

Bibliography

[1] O. Simeone, "A brief introduction to machine learning for engineers," *Foundations and Trends in Signal Processing*, vol. 12, no. 3–4, pp. 200–431, 2018.

[2] C. M. Bishop, *Pattern Recognition and Machine Learning*. Springer, 2006.

[3] T. Hastie, R. Tibshirani, and J. Friedman, *The Elements of Statistical Learning*. Springer, 2001.

[4] K. P. Murphy, *Machine Learning: A Probabilistic Perspective*. The MIT Press, 2012.

[5] S. Shalev-Shwartz and S. Ben-David, *Understanding Machine Learning: From Theory to Algorithms*. Cambridge University Press, 2014.

[6] J. Watt, R. Borhani, and A. Katsaggelos, *Machine Learning Refined: Foundations, Algorithms, and Applications*. Cambridge University Press, 2020.

[7] S. Theodoridis, *Machine Learning: A Bayesian and Optimization Perspective*. Academic Press, 2015.

[8] A. Zhang, Z. C. Lipton, M. Li, and A. J. Smola, *Dive into Deep Learning*. https://d2l.ai, 2020.

[9] M. R. Rodrigues and Y. C. Eldar, *Information-Theoretic Methods in Data Science*. Cambridge University Press, 2021.

Acknowledgements

I have read somewhere that writing is all about compressing time – the time, that is, spent by the writer in researching, selecting and organizing, and finally committing to page. A reader may need only five minutes to read a paragraph that the writer spent hours laboring on. Compression – writing – is successful if is nearly lossless, allowing the reader to reproduce the ideas in the mind of the writer with limited effort. Calibrating the level of compression has been an ongoing concern in writing this text, and I have been helped in making choices by readers of the original monograph [1] and by students at NJIT and KCL who have attended my courses on machine learning. A special mention goes to Kfir Cohen, who has patiently sifted through hundreds of slides for typos and imperfections (and prepared Fig. 7.2).

I have learned many of the topics covered in this book alongside colleagues and students: variational inference with Dr. Hyeryung Jang and Dr. Nicolas Skatchkovsky; Bayesian learning with Dr. Rahif Kassab; meta-learning with Dr. Sangwoo Park and Dr. Sharu Theresa Jose; differential privacy with Dr. Dongzhu Liu; information-theoretic learning with Dr. Jingjing Zhang; and the list goes on. Professor Bipin Rajendran has motivated me to look into probabilistic machine learning, and I am grateful for his illuminating explanations; Professor Petar Popovski has been inspirational as ever throughout the long period of gestation of this book and has provided very useful and detailed comments; Dr. Jakob Hoydis has provided useful initial feedback; and Dr. Onur Sahin has offered much needed reminders not to forget to use cases. I am grateful to all my colleagues at NJIT and at KCL, as well as my collaborators around the world, for making my work life so fulfilling. Thank you also to Dr. Hari Chittoor for spotting a number of typos, to Dr. Yusha Liu for preparing Fig. 12.5, and to Dr. Sangwoo Park for preparing Fig. 13.17.

I would like to gratefully acknowledge support by the European Research Council (ERC) under the European Union's Horizon 2020 Research and Innovation Programme (grant agreement No. 725732).

At Cambridge University Press, I have been lucky to work with Helen Shannon, Elizabeth Horne, and Jane Adams. Special thanks also to Julie Lancashire for believing in this project from the beginning and to Khaled Makhshoush for graciously agreeing to design the cover of this book. Finally, I am grateful to John King for his excellent and thoughtful work on editing the text.

This work is dedicated to my parents, to Lisa, Noah, Lena, and to my extended family across two continents.

Notation

General Conventions

- Random variables or random vectors – both abbreviated as rvs – are represented using Roman typeface, while their values and realizations are indicated by the corresponding standard font. For instance, the equality $\mathrm{x} = x$ indicates that rv x takes value x.
- An exception is made for rvs denoted by Greek letters, for which the same symbol is used for both rvs and realizations, with their use made clear by the context.
- All vectors are taken to be in column form.
- Matrices are indicated using uppercase fonts, with Roman typeface used for random matrices.
- Calligraphic fonts are used for sets.
- The distribution of an rv x, which may be either a probability mass function (pmf) for a discrete rvs or a probability density function (pdf) for continuous rvs, is denoted as $p(x)$. To specify a given numerical value x for rv x, we will also write $p(\mathrm{x} = x)$.
- For the Kullback–Liebler (KL) divergence, which involves two distributions $p(x)$ and $q(x)$, we use both notations $\mathrm{KL}(p||q)$ and $\mathrm{KL}(p(x)||q(x))$. The latter notation is particularly useful for conditional distributions $p(x|y)$ and $q(x|y)$, in which case the KL divergence $\mathrm{KL}(p(x|y)||q(x|y))$ is evaluated for a fixed value y. The same approach is applied to the other information-theoretic metrics listed below.
- In numerical calculations, we will often only keep the first two decimal digits by rounding up the second digit at each intermediate step.

Notations

General

$\mathbb{R}$	set of real numbers
$\mathbb{R}^N$	set of $N \times 1$ vectors of real numbers
$\mathbb{R}^+$	set of non-negative real numbers
$(a, b]$	interval between two real numbers a and b with $a < b$, excluding a and including b
(a, b)	interval between two real numbers a and b with $a < b$, excluding a and b
$[a, b]$	interval between two real numbers a and b with $a < b$, including a and b
$\{1, \ldots, K\}$	set including all integer numbers from 1 to K
$\{(a_k)_{k=1}^{K}\}$ or $\{a_k\}_{k=1}^{K}$	set $\{a_1, \ldots, a_K\}$
$x_{\mathcal{S}}$	set of elements x_k indexed by the integers $k \in \mathcal{S}$
$\mathcal{S}^N$	set of all $N \times 1$ vectors with entries taking values in set $\mathcal{S}$
$|\mathcal{S}|$	cardinality of a set $\mathcal{S}$
$\propto$	proportional to
$\exists$	there exists
$\sum_x$	sum over all values of variable x

Functions

$f(\cdot)$	a function
$f(x)$	output of function $f(x)$ for input x, or the function itself
$\nabla f(x)$	gradient vector of function $f(x)$
$\nabla^2 f(x)$	Hessian matrix of function $f(x)$
$\log(\cdot)$	natural logarithm
$\log_2(\cdot)$	logarithm in base 2
$\vert \cdot \vert$	absolute value (magnitude)
$\mathbb{1}(\cdot)$	indicator function: it equals 1 if the argument is true and 0 otherwise
$\delta(\cdot)$	Dirac delta function or Kronecker delta function ($\delta(x - x') = \mathbb{1}(x = x')$)
$\sigma(\cdot)$	sigmoid function: $\sigma(x) = 1/(1 + \exp(-x))$
$\lceil \cdot \rceil$	ceiling function (smallest larger integer)
$\text{step}(\cdot)$	step function: $\text{step}(x) = 1$ if $x > 0$, and $\text{step}(x) = 0$ if $x < 0$
$\min_x f(x)$	minimization problem for function $f(\cdot)$ over x
$\arg\min_x f(x)$	an element x in the set of minimizers of function $f(x)$
$\mathcal{O}(f(x))$	a function of the form $af(x) + b$ for some constants a and b

Linear Algebra

$[x]_i$	ith element of vector x
$[A]_{ij}$ or $[A]_{i,j}$	(i,j)th element of matrix A
$[A]_{r:}$ or $[A]_{:c}$	rth row and cth column of matrix A, respectively
x^T and X^T	transpose of vector x and matrix X
$\Vert a \Vert^2 = \sum_{i=1}^N a_i^2$	quadratic, or ℓ_2, norm of a vector $a = [a_1, \ldots, a_N]^T$
$\Vert a \Vert_1 = \sum_{i=1}^N \vert a_i \vert$	ℓ_1 norm
$\Vert a \Vert_0$	ℓ_0 pseudo-norm, which returns the number of non-zero entries of vector a
I_L	$L \times L$ identity matrix
I	identity matrix when dimension is clear from context
0_L	$L \times L$ all-zero matrix or $L \times 1$ all-zero vector, as clear from the context
1_L	$L \times 1$ all-one vector
$\det(\cdot)$	determinant of a square matrix
$\text{tr}(\cdot)$	trace of a square matrix
$\text{diag}(\cdot)$	column vector of elements on the main diagonal of the argument matrix
$\text{Diag}(\cdot)$	square diagonal matrix with main diagonal given by the argument vector
$\odot$	element-wise product

Probability

$x \sim p(x)$	rv x is distributed according to distribution $p(x)$
$x_n \underset{\text{i.i.d.}}{\sim} p(x)$	rvs $x_n \sim p(x)$ are independent and identically distributed (i.i.d.)
$p(x\vert y)$ or $p(x\vert y = y)$	conditional distribution of x given the observation of rv y $= y$
$(x\vert y = y) \sim p(x\vert y)$	rv x is drawn according to the conditional distribution $p(x\vert y = y)$
$E_{x \sim p(x)}[\cdot]$	expectation of the argument over the distribution of the rv x $\sim p(x)$
$E_{x \sim p(x\vert y)}[\cdot]$	conditional expectation of the argument over the distribution $p(x\vert y)$

$\Pr_{x \sim p(x)}[\cdot]$	probability of the argument over the distribution of the rv x $\sim p(x)$
$\Pr[\cdot]$	probability of the argument when the distribution is clear from the context
$\text{Bern}(x|q)$	Bernoulli pmf with parameter $q \in [0, 1]$
$\text{Cat}(x|q)$	categorical pmf with parameter vector q
$\mathcal{N}(x|\mu, \Sigma)$	multivariate Gaussian pdf with mean vector μ and covariance matrix Σ
$\text{Beta}(z|a, b)$	beta distribution with parameters a and b
$\mathcal{U}(x|a, b)$	uniform pdf in the interval $[a, b]$
$\text{Var}(p(x))$ or $\text{Var}(p)$	variance of rv x $\sim p(x)$
$\text{Var}(x)$	variance of rv x $\sim p(x)$ when the distribution $p(x)$ is clear from the context

Information-Theoretic Metrics

$\text{H}(p(x))$ or $\text{H}(p)$	entropy of rv x $\sim p(x)$
$\text{H}(x)$	entropy of rv x $\sim p(x)$ when the distribution $p(x)$ is clear from the context
$\text{H}(x|y)$	conditional entropy of rv x given rv y
$\text{I}(x; y)$	mutual information between rvs x and y
$\text{H}(p||q)$	cross entropy between distributions p and q
$\text{KL}(p||q)$	KL divergence between distributions p and q
$\text{F}(p||\tilde{p})$	free energy for distribution p and unnormalized distribution $\tilde{p}$
$\text{JS}(p||q)$	Jensen–Shannon divergence between distributions p and q
$\text{D}_f(p||q)$	f-divergence between distributions p and q
$\text{IPM}(p||q)$	integral probability metric between distributions p and q

Learning-Related Quantities

$\mathcal{D}$	training set
$L_p(\theta)$	population loss as a function of the model parameter θ
$L_{\mathcal{D}}(\theta)$	training loss as a function of the model parameter θ
$\theta_{\mathcal{D}}^{ERM}$	model parameter obtained via empirical risk minimization (ERM)
$\theta_{\mathcal{D}}^{ML}$	model parameter obtained via maximum likelihood (ML)
$\theta_{\mathcal{D}}^{MAP}$	model parameter obtained via maximum a posteriori (MAP)
$\text{ExpFam}(x|\eta)$	distribution in the exponential family with natural parameter vector η
$\text{ExpFam}(x|\mu)$	distribution in the exponential family with mean parameter vector μ

Acronyms

ADF	Assumed Density Filtering
AI	Artificial Intelligence
BFL	Bayesian Federated Learning
BN	Bayesian Network
CDF	Cumulative Distribution Function
CDL	Contrastive Density Learning
CRL	Contrastive Representation Learning
DAG	Directed Acyclic Graph
DP	Differential Privacy
DV	Donsker–Varadhan
ELBO	Evidence Lower BOund
EM	Expectation Maximization
EP	Expectation Propagation
ERM	Empirical Risk Minimization
EWC	Elastic Weight Consolidation
FIM	Fisher Information Matrix
FL	Federated Learning
FVI	Federated Variational Inference
GAN	Generative Adversarial Network
GD	Gradient Descent
GGN	Generalized Gauss–Newton (GGN)
GLM	Generalized Linear Model
GOFAI	Good Old Fashioned AI
GP	Gaussian Process
GVEM	Generalized Variational Expectation Maximization
GVI	Generalized Variational Inference
i.i.d.	independent identically distributed
IPM	Integral Probability Metric
ITM	Information-Theoretic Measure
JS	Jensen–Shannon
KDE	Kernel Density Estimation
KL	Kullback–Leibler
K-NN	K-Nearest Neighbors
LDR	Log-Distribution Ratio
LLR	Log-Likelihood Ratio
LS	Least Squares
MAML	Model Agnostic Meta-Learning
MAP	Maximum A Posteriori
MC	Monte Carlo
MCMC	Markov Chain Monte Carlo

MDL	Minimum Description Length
ML	Maximum Likelihood
MMD	Maximum Mean Discrepancy
MRF	Markov Random Field
OMP	Orthogonal Matching Pursuit
PAC	Probably Approximately Correct
PCA	Principal Component Analysis
pdf	probability density function
PGM	Probabilistic Graphical Model
pmf	probability mass function
QDA	Quadratic Discriminant Analysis
RBM	Restricted Boltzmann Machine
RKHS	Reproducing Kernel Hilbert Space
rv	random variable, or random vector
SGD	Stochastic Gradient Descent
SGLD	Stochastic Gradient Langevin Dynamics
SG-MCMC	Stochastic Gradient Markov Chain Monte Carlo
SVGD	Stein Variational Gradient Descent
TV	Total Variation
VAE	Variational AutoEncoder
VC	Vapnik–Chervonenkis
VCL	Variational Continual Learning
VEM	Variational Expectation Maximization
VI	Variational Inference

Part I

Introduction and Background

1   When and How to Use Machine Learning

1.1   Overview

This chapter aims to motivate the study of machine learning, having in mind as the intended audience students and researchers with an engineering background.

Learning Objectives and Organization of the Chapter. By the end of this chapter, the reader should be able to addresses the following basic questions:

- What is machine learning for?
- Why are machine learning and "AI" so popular right now?
- What can engineers contribute to machine learning?
- What is machine learning?
- When to use machine learning?

Each of these questions will be addressed in a separate section in the rest of this chapter.

1.2   What is Machine Learning For?

Machine learning is currently the dominant form of **artificial intelligence (AI)** – so much so that the label "AI" has by now become synonymous with the data-driven pattern recognition methods that are the hallmark of machine learning. Machine learning algorithms underlie a vast array of applications, with real-life implications for individuals, companies, and governments. Here are some examples:

- Governments use it to decide on visa applications.
- Courts deploy it to rule on requests for bail.
- Police departments rely on it to identify suspects.
- Schools and universities apply it to assign places.
- Banks deploy it to grant or deny credit.
- Financial institutions run it to operate in the stock market.
- Companies apply it to optimize their hiring decisions.
- Online vendors leverage it to provide recommendations and track users' preferences.
- Individuals interact with it when "conversing" with virtual personal assistants and finding the best route to a destination. (Quotation marks seem necessary when using verbs that imply intentionality to algorithms.[1])

Not all cases of machine learning or AI use are equally legitimate, successful, or even morally justifiable. The "algorithm" (another necessary use of quotation marks) relied on

---

[1] The story of the ELIZA computer program provides an interesting cautionary tale.

by the UK government to predict students' marks in 2020 was actually based on a single, poorly thought-out, formula. Machine learning-based face-recognition tools are notorious for being inaccurate and biased. So are many suspect-identifying, credit-scoring, and candidate-selecting tools based on machine learning that have been widely reported to produce unfair outcomes.

This discussion points to an important distinction: It is one thing to design technically sound data-driven solutions, and it is another thing altogether to ensure that these solutions are deployed to benefit society. While this book will focus on the technical problem, there is clearly a need for engineers to work alongside regulators and social scientists to ensure that proper legal and ethical guidelines are applied. In the words of the American statistician Nate Silver, "The numbers have no way of speaking for themselves. We speak for them, we imbue them with meaning".

1.3   Why Study Machine Learning Now?

In 1950, **Claude Shannon**, the father of information theory, published a seminal paper entitled *Programming a Computer for Playing Chess*. In the paper, he envisions

(1) Machines for designing filters, equalizers, etc.
(2) Machines for designing relay and switching circuits.
(3) Machines which will handle routing of telephone calls based on the individual circumstances rather than by fixed patterns.
(4) Machines for performing symbolic (non-numerical) mathematical operations.
(5) Machines capable of translating from one language to another.
(6) Machines for making strategic decisions in simplified military operations.
(7) Machines capable of orchestrating a melody.
(8) Machines capable of logical deduction.

Despite being over 70 years old, this passage reads almost like a summary of recent breakthroughs in machine learning. In the same year, **Alan Turing**, writing for the Conference on Information Theory, explained the idea of learning machines as follows: "If, as is usual with computing machines, the operations of the machine itself could alter its instructions, there is the possibility that a learning processing could by this means completely alter the programme in the machine." Using language that has hardly aged, Turing outlines what computer scientists would describe today as **"Programming 2.0"**: Programmers no longer need to explicitly code algorithmic steps to carry out a given task; instead, it is the machine that automatically infers the algorithmic steps from data. Most commonly, data are in the form of examples that "supervise" the machine by illustrating the desired relationship between input and some target variables or actions; think, for instance, of a data set of emails, each labeled as either "spam" or "not spam".

Work by pioneers such as Shannon and Turing ushered in the **cybernetic age** and, with it, the first wave of AI. **First-wave AI systems** – known today as **good old fashioned AI (GOFAI)** – were built on **deduction** and on handcrafted collections of facts. Within two decades, this logic-based approach was found to be inadequate vis-à-vis the complexity of perception in the real world, which can hardly be accommodated within GOFAI's world of well-defined distinct objects. By the late 1980s, GOFAI had fallen under the weight of its own exaggerated promises.

The **second, current, wave of AI** is dominated by machine learning, which is built on **induction**, that is, generalization from examples, and statistical pattern recognition. Linear methods, such as support vector machines, were the techniques of choice in the earlier years of the second wave, while more recent years, stretching to today, have witnessed the rise of **deep learning**. (Accordingly, some partition this second wave into two separate waves.[2])

The breakthroughs during the latter part of the second wave have met many of the goals set out in Shannon's paper from 1950, thanks mostly to the convergence of **"big data"** and **"big compute"**. For example, AlphaGo Zero – a more powerful version of the AI that famously beat a human world champion at the game of Go – required training more than 64 GPU workers and 19 CPU parameter servers for weeks, with an estimated hardware cost of US$25 million. OpenAI's highly publicized video game-playing program needed training for an equivalent of 45,000 years of game play, costing millions of dollars in rent access for cloud computing. The focus on scale has also led to concerns about the **accessibility** of the technology, with large multinationals scooping up most of the world's data in large data centers. The trend is expected to continue for the foreseeable future, thanks also to the rise of the **Internet of Things (IoT)**.

This book focuses on second-wave AI ideas and principles with the ambition of helping rebuild the bridge between machine learning and the fields of information theory and signal processing that existed at the time of Shannon's and Turing's work.

As a final note, it is useful to emphasize that the scope of machine learning should not be conflated with that of AI. The field of AI is far broader, including also first-wave AI methods based on logic, symbolic reasoning, and deduction, as well as other areas such as planning. The road to achieving **"general AI"** appears to be long and to stretch beyond the boundaries of machine learning: In the words of the American computer scientist Fei-Fei Li, with today's AI, i.e., with machine learning, "A machine … can make a perfect chess move while the room is on fire." An outlook on the topic of general AI is provided in Chapter 15.

1.4 What is Machine Learning?

To introduce the data-driven statistical pattern recognition approach at the core of machine learning, let us first review the standard **domain knowledge-based model-driven design methodology** that should be familiar to all engineers.

1.4.1 Domain Knowledge-Based Model-Driven Design

Engineers are often faced with the problem of designing algorithms to be deployed in a given real-world setting to meet certain performance requirements. To choose examples close to Shannon's work, a group of engineers may be tasked with devising an algorithm to compress images in such a way that they can be reconstructed with limited distortion; or to invent algorithms capable of communicating reliably over a wireless channel.

To address this type of problem, the standard domain knowledge-based model-driven design methodology complies with the block diagram shown in Fig. 1.1, carrying out the following steps:

---

[2] Besides deduction and induction, there is a third form of reasoning, or inference, that AI has yet to master: abduction. Abductive inferences involve intuition, guesswork, and the ability to form hypotheses to be revised via observation.

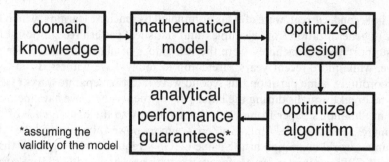

Figure 1.1 Conventional domain knowledge-based model-driven design methodology.

1. **Definition of a mathematical model based on domain knowledge**. Based on domain knowledge, i.e., information about the "physics" of the problem, one constructs a mathematical model of the problem of interest that involves all the relevant variables. Importantly, the mathematical model is **white-box**, in the sense that the relationship between inputs and outputs variables is mediated by quantities and mechanisms that have physical counterparts and interpretations.
2. **Model-based optimization**. Given the mathematical model, one formulates an optimization problem with the aim of identifying an optimal algorithm. Importantly, the optimized algorithm offers analytical performance guarantees under the assumption that the model is correct.

Example 1.1

To illustrate this methodology, consider the mentioned problem of designing a reliable wireless digital communication system. The outlined conventional methodology starts by establishing, or retrieving, domain knowledge about the physics of radio propagation from a transmitter to a receiver in the environment of interest. This step may entail consulting with experts, relying on textbooks or research papers, and/or carrying out measurement campaigns. As a result of this first step, one obtains a **mathematical model** relating input variables, e.g., input currents at the transmitter's antennas, and output variables, e.g., output currents at the receiver's antennas. The resulting white-box model is inherently **interpretable**. For instance, the relationship between the signals at the transmitter's and receiver's antennas may depend on the number and type of obstacles along the way.

The second step is to leverage this model to define, and hopefully solve, a **model-based optimization problem**. For the example at hand, one should optimize over two algorithms – the first, the encoder, mapping the information bits to the input currents of the transmit antennas, and the second, the decoder, mapping the output currents at the receiver to the decoded bits. As an optimization criterion, we can adopt the average number of incorrectly decoded bits, which we wish to minimize while ensuring a power or energy constraint at the transmitter. Driven by the model, the problem can be mathematically stated and ideally solved, yielding provably optimal algorithms, whose performance can be backed by analytical guarantees. Crucially, any such guarantee is only reliable and trustworthy to the extent that one can trust the starting mathematical model.

1.4.2   Inductive Bias-Based Data-Driven Design: Machine Learning

Machine learning follows an **inductive bias-based** – rather than domain knowledge-based – **data-driven** – rather than model-driven – design methodology that is illustrated by the block diagram in Fig. 1.2. This methodology proceeds as follows:

1. **Definition of an inductive bias**. As discussed, the conventional design methodology assumes the availability of a sufficiently detailed and precise knowledge about the problem domain, enabling the definition of a trustworthy, white-box, mathematical model of the "physics" of the variables of interest. In contrast, when adopting a machine learning methodology, one implicitly assumes that such knowledge is incomplete, or at least that the available models are too complex to be used in the subsequent step of model-based optimization. To account for this **model deficit** or **algorithmic deficit**, the machine learning methodology starts off by restricting the type of algorithms that may work well for the problem at hand. Broadly speaking, the selected class of algorithms encompasses **black-box** mappings between input and output variable; think of polynomial functions with a given degree. This class of mappings, which we will refer to as **model class**, constitutes (part of) the **inductive bias** that the methodology builds upon to extract information from data.[3]

2. **Training**. Machine learning aims to extract patterns and regularities from data that may generalize outside the available data set, i.e., to the "real world". This process, known as **training**, operates within the confines set by the inductive bias: Based on data, a specific (black-box) model is selected within the model class posited by the inductive bias, and no models outside it are allowed as the outputs of training. As opposed to the analytical performance guarantees provided by the conventional methodology, only statistical metrics can generally be evaluated on the trained model. Their validity and relevance rests on the extent to which the available data are deemed to be representative of the conditions one would encounter when deploying the system.

A popular cartoon[4] summarizes – and somewhat simplifies! – this process in the form of Algorithm 1.1.

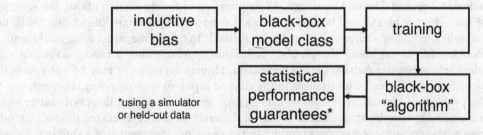

Figure 1.2 Inductive bias-based data-driven design methodology adopted by machine learning.

---

[3] The definition of an inductive bias may be the result of abductive inference.
[4] https://xkcd.com/1838/

Algorithm 1.1: Machine learning as an algorithm (according to an xkcd cartoon)

---

while *numbers don't look right* **do**
> pour data into pile of linear algebra
> collect answers

end

Example 1.2

To elaborate on the methodology followed by machine learning, consider again the problem of designing the transmitter and receiver of the wireless communication link described in Example 1.1. Assume that a model for the propagation environment is not available; perhaps we are faced with a new communication setting, and we do not have the resources to carry out an extensive measurement campaign or to consult with experts to obtain an accurate "physics"-based model. What we assume to be true, instead, is that a general class of **black-box models**, such as linear functions or neural networks, can implement effective transmitting and receiving algorithms. We may know this because of prior experience with similar problems, or through some trial-and-error process. This choice defines the **inductive bias**.

We also assume that we can collect data by transmitting over the channel. **Training** leverages these data to select a specific model within the given model class posited as part of the inductive bias. In the problem at hand, training may output two neural networks, one to be used by the transmitter and one by the receiver. The performance obtained upon deployment can be estimated using (additional) available data, as long as the latter are deemed to be sufficiently rich to accurately reflect real-world conditions.

Concluding this section, a couple of remarks are in order.

- **Inductive bias and domain knowledge**. As emphasized in the discussion above, the selection of the inductive bias should be guided as much as possible by domain knowledge. While not enough domain knowledge is assumed to be available to determine an accurate mathematical model, information about the problem is invaluable in selecting a model class that may effectively generalize outside the available data. Domain knowledge may be as basic as the identification of specific **invariances** in the desired input–output mapping. For instance, in the problem of classifying images of cats against images of dogs, it does not matter *where* the cat or dog is in the image. This translational invariance can be leveraged to select a model class whose constituent classifiers produce the same output irrespective of a shift in the input. A recent example of the importance of choosing a suitable inductive bias is provided by the success of DeepMind's AlphaFold system in predicting the structure of a protein from its primary sequence. A key element of AlphaFold is, in fact, the way in which its predictive model class encodes symmetry principles that facilitate reasoning over protein structures in three dimensions.

- **White-box vs. black-box models**. Care was taken in the discussion above to distinguish between white-box models – the type of mathematical models assumed by the conventional engineering methodology – and black-box models and model classes – the type of input–output mappings assumed as part of the inductive bias by machine learning. While we use in both cases we use the term "models", their application and significance are different, and are key to understanding the distinction between the two methodologies.

1.5   Taxonomy of Machine Learning Methods

There are three main broad classes of machine learning algorithms, which differ in the way data are structured and presented to the learning agent: supervised, unsupervised, and reinforcement learning. This section provides a short introduction.

1.5.1   Supervised Learning

In supervised learning, the learning agent is given a **training data set** including multiple samples. Each data sample consists of a pair **"(input, desired output)"**: Each data point hence "supervises" the learner as to the best output to assign to a given input. This is illustrated by the example in Fig. 1.3, where the input variable is given by a point in the two-dimensional plane, while the target variable is binary, with the two possible values represented as a circle and a cross.

In supervised learning, the goal is to generalize the input–output relationship – which is partially exemplified by the training data – outside the training set. In the example in Fig. 1.3, the problem amounts to **classifying** all points in the plane as being part of the "circle class" or of the "cross class". The trained model may be expressed as a line separating the regions of inputs belonging to either class.

A practical example is the problem of classifying emails as spam or not spam, for which we may have access to a corpus of emails, each labeled as "spam" or "not spam".

Figure 1.3 Supervised vs. unsupervised learning.

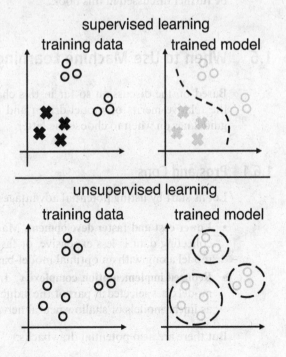

1.5.2 Unsupervised Learning

In unsupervised learning, the learning agent is again given a training set containing multiple samples. However, each sample only contains the "input", without specifying what the desired output should be. For instance, in Fig. 1.3, the learner is given several points on the plane without any label indicating the target variable that should be assigned to each point. Unsupervised learning generally leverages **similarities** and **dissimilarities** among training points in order to draw conclusions about the data.

A specific example of an unsupervised task – but by no means not the only one! – is **clustering**, whereby the data set is partitioned into groups of similar inputs. For the corpus of emails mentioned above, in the absence of the "spam"/"not spam" labels, clustering may aim to group together emails that are about similar topics.

1.5.3 Reinforcement Learning

In reinforcement learning, the learning agent is not given a fixed data set; instead, data are collected as the agent interacts with the environment over time. Specifically, based on its current "understanding" of the environment, at each given time step the agent takes an action; observes its impact on the evolution of the system; and adjusts its strategy for the following step. To enable adaptation, the agent is given a reward signal that can be used to reinforce useful behavior and discard actions that are harmful to the agent's performance.

With its focus on feedback-based dynamic adaptation, reinforcement learning is closely related to control theory. Unlike standard control theory, however, in reinforcement learning one does not assume the availability of a (white-box) model for the evolution of the system. The agent can only rely on data and on the rewards to optimize its operation. Reinforcement learning is quite different in nature from supervised and unsupervised learning, and it will not be further discussed in this book.

1.6 When to Use Machine Learning?

Based on the discussion so far in this chapter, we can draw some general conclusions about the relative merits of model-driven and data-driven design methodologies, as well as some guidelines on when to choose the latter.

1.6.1 Pros and Cons

Let us start by listing potential advantages of the machine learning methodology:

- **Lower cost and faster development**. Machine learning can reduce the time to deployment if collecting data is less expensive, or faster, than acquiring a "physics"-based mathematical model along with an optimal model-based algorithm.
- **Reduced implementation complexity**. Trained models can be efficiently implemented if the model class selected as part of the inductive bias contains low-complexity "algorithms", such as linear models or shallow neural networks.

But there are also potential drawbacks:

- **Limited performance guarantees**. While model-driven designs can be certified with analytical performance guarantees that are valid under the assumed model, the same is not true for data-driven solutions, for which only statistical guarantees can be provided.
- **Limited interpretability**. Machine learning models are black boxes that implement generic mappings between input and output variables. As such, there is no direct way to interpret their operation by means of mechanisms such as **counterfactual queries** ("what would have happened if … ?"). This issue can be partially mitigated if the selected model class contains structured models that encode a specific inductive bias. For instance, if a model class relies on given hand-crafted features of the data, one may try to tease apart the contribution of different features to the output.

1.6.2 Checklist

Given these pros and cons, when should we use machine learning? The following check list provides a useful roadmap.

1. **Model or algorithm deficit?** First off, one should rule out the use of domain knowledge-based model-driven methods for the problem at hand. There are two main cases in which this is well justified:
   - **model deficit**: a "physics"-based mathematical model for the variables of interest is not available.
   - **algorithm deficit**: a mathematical model is available, but optimal or near-optimal algorithms based on it are not known or too complex to implement.

2. **Availability of data?** Second, one should check that enough data are available, or can be collected.
3. **Stationarity?** Third, the time over which the collected data are expected to provide an accurate description of the phenomenon of interest should be sufficiently long to accommodate data collection, training, and deployment.
4. **Performance guarantees?** Fourth, one should be clear about the type of performance guarantees that are considered to be acceptable for the problem at hand.
5. **Interpretability?** Finally, one should assess to what degree "black-box" decisions are valid answers for the setting under study.

The checklist is summarized in Algorithm 1.2.

1.7 Summary

- Machine learning – often conflated with AI – is extensively used in the real world by individuals, companies, and governments, and its application requires both technical knowledge and ethical guidelines.
- The field of AI has gone through various waves of funding and hype, and its basic tenets are decades old, dating back to work done during the cybernetic age. The current, second, wave revolves around statistical pattern recognition – also known as machine learning.
- Machine learning is distinct from the domain knowledge-based model-driven design methodology that is most familiar to engineers. Rather than relying on optimization over

---

Algorithm 1.2: Should I use machine learning?

---

if *there is a model or algorithm deficit* **then**
    if *data are available or can be generated* **then**
        if *environment is stationary* **then**
            if *model-based performance guarantees are not needed* **then**
                if *interpretability is not a requirement* **then**
                    | apply machine learning
                end
            end
        end
    end
end

---

a "physics"-based analytical model, machine learning adopts an inductive bias-based data-driven methodology.

- The inductive bias consists of a model class and a training algorithm. Based on training data, the training algorithm selects one model from the model class. The inductive bias should ideally be selected based on domain knowledge.

- Machine learning methods can be broadly partitioned into supervised learning, unsupervised learning, and reinforcement learning. In supervised learning, the training algorithm is given examples of an input–output relationship; in unsupervised learning, only "inputs" are available; while in reinforcement learning the agent collects data as it interacts with the environment.

- Machine learning can help lower cost, deployment time, and implementation complexity, at the cost of offering more limited performance guarantees and interpretability as compared to the standard model-based methodology.

1.8   Recommended Resources

For further reading on the cybernetic age and on the early days of AI, I recommend the books by Kline [1] and by Wooldridge [2] (the latter includes the story of ELIZA), as well as Lem's "Cyberiad" [3]. The classical books by Pagels [4] and by Hofstadter [5] are also still well worth reading. For discussions on the role of AI and machine learning in the "real world", interesting references are the books by Levesque [6], Russell [7], and Christian [8]. The philosophical relationship between GOFAI and machine learning is covered by Cantwell Smith [9]. If you wish to learn more about the "extractive" aspects of AI – from natural resources, communities, individuals – Crawford's book [10] provides an informative introduction.

Dozens of papers on machine learning are posted every day on repositories such as arXiv.org. Incentives to publish are such that one should generally question results that appear exaggerated. Are the experiments reproducible? Are the data representative of real-world scenarios or instead biased in favor of the use of a specific method? At the time of writing, a particularly evident failure of machine learning is the lack of effective AI tools to diagnose COVID-19 cases [11], despite hundreds of (published) claims of success stories.

Bibliography

[1] R. R. Kline, *The Cybernetics Moment: Or Why We Call Our Age the Information Age*. The Johns Hopkins University Press, 2015.

[2] M. Wooldridge, *The Road to Conscious Machines: The Story of AI*. Penguin Books UK, 2020.

[3] S. Lem, *The Cyberiad: Fables for the Cybernetic Age*. Penguin Books UK, 2014.

[4] H. R. Pagels, *The Dreams of Reason: The Computer and the Rise of the Sciences of Complexity*. Bantam Dell Publishing Group, 1988.

[5] D. R. Hofstadter, *Gödel, Escher, Bach: An Eternal Golden Braid*. Basic Books, 1979.

[6] H. J. Levesque, *Common Sense, the Turing Test, and the Quest for Real AI*. The MIT Press, 2017.

[7] S. Russell, *Human Compatible: Artificial Intelligence and the Problem of Control*. Penguin Books, 2019.

[8] B. Christian, *The Alignment Problem: How Can Machines Learn Human Values?* Atlantic Books, 2021.

[9] B. Cantwell Smith, *The Promise of Artificial Intelligence: Reckoning and Judgment*. The MIT Press, 2019.

[10] K. Crawford, *The Atlas of AI*. Yale University Press, 2021.

[11] M. Roberts, D. Driggs, M. Thorpe, *et al.*, "Common pitfalls and recommendations for using machine learning to detect and prognosticate for covid-19 using chest radiographs and CT scans," *Nature Machine Intelligence*, vol. 3, no. 3, pp. 199–217, 2021.

2 Background

2.1 Overview

This chapter provides a refresher on probability and linear algebra with the aim of reviewing the necessary background for the rest of the book. Readers not familiar with probability and linear algebra are invited to first consult one of the standard textbooks mentioned in Recommended Resources, Sec. 2.14. Readers well versed on these topics may briefly skim through this chapter to get a sense of the notation used in the book.

Learning Objectives and Organization of the Chapter. By the end of this chapter, the reader should be able to:

- understand and manipulate random variables (Sec. 2.2), expectations (Sec. 2.3), and variance (Sec. 2.4);
- work with vectors and matrices (Secs. 2.5 and 2.6); and
- understand and use random vectors, marginal and conditional distributions, independence and chain rule, Bayes' theorem, and the law of iterated expectations (Sec. 2.7 to Sec. 2.12).

2.2 Random Variables

Random variables are the key building blocks from which machine learning models are constructed. Randomness, in fact, is necessary to account for the inherent "noisiness" of the data generation process, as well as for the uncertainty caused by the availability of limited data at the learning agent.

A **random variable** is a numerical quantity that takes values probabilistically in a set $\mathcal{X}$ of possible outcomes. This is in contrast to **deterministic variables**, which have well-defined, fixed values. Before measuring a random variable, one only knows how likely it is for the observation of the random variable to take any of the values in the set $\mathcal{X}$. After the measurement, the random variable takes a fixed value in $\mathcal{X}$, which is referred to as the **realization** of the random variable.

Throughout this book, with the exception of Greek letters, we use Roman fonts to denote random variables. We specifically distinguish between a random variable x and a realization x of the random variable. Accordingly, we write x $= x$ for the event that random variable x produces a realization x, that is, that random variable x takes value x.

The domain $\mathcal{X}$ of a **discrete random variable** x contains a discrete finite number of values, e.g., values 0 and 1 in set $\mathcal{X} = \{0, 1\}$. For discrete random variables, the domain is also known as the **alphabet**. In contrast, a **continuous random variable** x can take values on a continuum, e.g., on the real line $\mathcal{X} = \mathbb{R}$.

2.2.1 Discrete Random Variables

Let $\Pr[\mathcal{E}]$ be the probability of an event $\mathcal{E}$. A discrete random variable x is described by a **probability mass function (pmf)** $p(x)$. A pmf is defined as the function

$$p(x) = \Pr[\mathrm{x} = x], \tag{2.1}$$

that is, the value of the pmf $p(x)$ corresponds to the probability that a draw of the random variable x results in the realization $\mathrm{x} = x$. Being a probability, by (2.1), a pmf satisfies the inequalities $p(x) \geq 0$ for all values $x \in \mathcal{X}$ and the equality $\sum_{x \in \mathcal{X}} p(x) = 1$. Accordingly, the sum of the pmf across all values in the domain (or alphabet) $\mathcal{X}$ must equal one. In order to indicate that realizations of random variable x are drawn according to pmf $p(x)$, we use the notation $\mathrm{x} \sim p(x)$.

Frequentist Interpretation of Probability. It is sometimes useful to think of probabilities in terms of their **frequentist interpretation**. To describe this perspective, imagine repeating the experiment of measuring a discrete random variable x multiple times. We can then compute the frequency at which each value x is observed by evaluating the fraction of times $\mathrm{x} = x$ is measured over the total number of measurements. The frequentist interpretation of probability is based on the fact that, if the number of measurements is large and the measurements are independent (see Sec. 2.10), this ratio tends to the probability $p(x)$. (This is the **law of large numbers**, which is detailed in Sec. 8.3.)

Accordingly, we can think of the value $p(x)$ of the pmf for a given x as the fraction of times that we would observe $\mathrm{x} = x$ if we were to repeat the measurement of random variable x many times. We can express this condition using the approximate equality

$$p(x) \simeq \frac{\text{number of samples equal to value } x}{\text{total number of samples}}, \tag{2.2}$$

which becomes increasingly accurate as the number of samples grows larger. It is emphasized that the frequentist implementation of probability is not always applicable, since certain measurements and observations cannot be repeated. (What is the probability that it will rain on February 22, 2036? What is the probability that neutrinos move faster than light?)

Bernoulli Random Variable. A Bernoulli random variable $\mathrm{x} \sim \mathrm{Bern}(q)$ takes values in the set $\mathcal{X} = \{0, 1\}$, i.e., Bernoulli random variables are binary. Its pmf is denoted as

$$p(x) = \mathrm{Bern}(x|q), \tag{2.3}$$

where $p(1) = \mathrm{Bern}(1|q) = \Pr[\mathrm{x} = 1] = q \in [0, 1]$ represents the probability of the event $\mathrm{x} = 1$, which we also write as $p(\mathrm{x} = 1)$. The complement of this probability is the probability of the event $\mathrm{x} = 0$, namely $p(0) = \mathrm{Bern}(0|q) = \Pr[\mathrm{x} = 0] = 1 - q$, which we also write as $p(\mathrm{x} = 0)$. Note that a Bernoulli pmf depends on the single parameter $q \in [0, 1]$.

For a Bernoulli random variable, it is also useful to define the **odds** of it being equal to 1 as

$$\mathrm{odds} = \frac{q}{1 - q}. \tag{2.4}$$

The odds (2.4) is a *relative* measure of the probability of the rv being 1 as opposed to taking value 0. They are related to the probability q by (2.4) as $q = \mathrm{odds}/(1 + \mathrm{odds})$.

Categorical (or Multinoulli) Random Variable. As an important generalization, a categorical (or multinoulli) random variable $\mathrm{x} \sim \mathrm{Cat}(q)$ takes values in the set $\mathcal{X} = \{0, 1, \ldots, C - 1\}$ for some integer $C \geq 1$. Note that, with $C = 2$, we obtain a Bernoulli random variable. The pmf of a categorical random variable x is denoted as

$$p(x) = \mathrm{Cat}(x|q), \tag{2.5}$$

Figure 2.1 A categorical pmf.

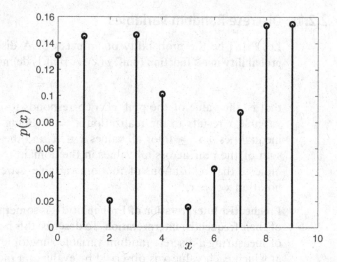

which depends on the probability vector (a vector is an ordered collection of numbers; see Sec. 2.7)

$$q = \begin{bmatrix} q_0 \\ q_1 \\ \vdots \\ q_{C-1} \end{bmatrix}. \tag{2.6}$$

Specifically, the pmf defines the probabilities $p(x) = \mathrm{Cat}(x|q) = \Pr[\mathrm{x} = x] = q_x$, which we also write as $p(\mathrm{x} = x)$, for $x \in \{0, 1, \ldots, C - 1\}$. Note that the probability vector q must satisfy the conditions $q_k \geq 0$ for all $k = 0, 1, \ldots, C - 1$ and $\sum_{k=0}^{C-1} q_k = q_0 + q_1 + \cdots + q_{C-1} = 1$. An example of a categorical pmf is shown in Fig. 2.1.

Categorical random variables are often represented in machine learning using a **one-hot vector** representation. Given a categorical random variable x, we define its one-hot representation as the $C \times 1$ vector x^{OH} that contains all zeros except for a single 1 in the $(k + 1)$th position when $\mathrm{x} = k$. For example, if $C = 4$, $\mathrm{x} = 0$ can be represented by the one-hot vector

$$\mathrm{x}^{OH} = \begin{bmatrix} 1 \\ 0 \\ 0 \\ 0 \end{bmatrix} \tag{2.7}$$

and $\mathrm{x} = 2$ by the one-hot vector

$$\mathrm{x}^{OH} = \begin{bmatrix} 0 \\ 0 \\ 1 \\ 0 \end{bmatrix}. \tag{2.8}$$

It may be useful to note that the one-hot vector x^{OH} is not *efficient* as a binary representation of a categorical variable x. This is in the sense that the one-hot vector x^{OH} includes C binary digits (bits), while $\log_2(C)$ bits would be sufficient to represent x (assuming $\log_2(C)$ is an integer). Nevertheless, the one-hot representation turns out to be very useful when considered in conjunction with pmfs. To see this, note that the probability vector (2.6) is of dimension $C \times 1$ as the one-hot vector, and there is a one-to-one correspondence between the entries of the two vectors: The probability of the kth entry being equal to 1 in the one-hot vector x^{OH} is given by the corresponding entry q_k in vector q for $k = 0, 1, \ldots, C - 1$.

2.2.2 Continuous Random Variables

A continuous random variable x is described by a **probability density function (pdf)** $p(x)$, which we denote as x $\sim p(x)$. The use of the same notation $p(x)$ for both pmfs and pdfs is intentional, and it will allow us to discuss both discrete and continuous random variables in a unified fashion. Unlike pmfs, a pdf $p(x)$ does not represent a probability, but instead a probability *density*.

To explain this point, it is useful to consider the **frequentist interpretation** of a pdf. Accordingly, the product $p(x)dx$ between the value of the pdf $p(x)$ at some point $x \in \mathbb{R}$ and a very small interval length dx is interpreted as the fraction of times that we observe a realization of the random variable x in an interval of length dx around value x if we repeat the measurement of x many times (assuming that it is possible):

$$p(x)dx \simeq \frac{\text{number of samples in the interval } \left(x - \frac{dx}{2}, x + \frac{dx}{2}\right)}{\text{total number of samples}}. \tag{2.9}$$

So, in order to obtain a probability, one must multiply a pdf by an interval length. It is also worth noting that, since $p(x)dx$ represents a probability and not $p(x)$, the value of a pdf $p(x)$ is not constrained to be smaller than 1, unlike for pmfs.

By (2.9), we can compute the probability of an interval $[a, b]$ with $a \leq b$ via the integral

$$\Pr[x \in (a,b)] = \int_a^b p(x)dx. \tag{2.10}$$

Furthermore, a pdf must satisfy the conditions $p(x) \geq 0$ and $\int_{-\infty}^{+\infty} p(x)dx = 1$.

Gaussian Random Variable. A Gaussian random variable x $\sim \mathcal{N}(\mu, \sigma^2)$ is characterized by a pdf

$$p(x) = \mathcal{N}(x|\mu, \sigma^2) = \frac{1}{\sqrt{2\pi\sigma^2}} \exp\left(-\frac{(x-\mu)^2}{2\sigma^2}\right) \tag{2.11}$$

that depends on two parameters – **mean** μ and **variance** σ^2. Observe that, as mentioned, the pdf $\mathcal{N}(x|\mu, \sigma^2)$ can take values larger than 1. Gaussian random variables are "light-tailed" in the sense that the probability of points away from the mean decreases quickly. An illustration of a Gaussian pdf, along with three intervals (a, b) and associated probabilities (2.10) can be found in Fig. 2.2. Note that the mean defines the peak, or the **mode**, of the pdf, and that the probability of finding a Gaussian random variable within the interval $[\mu - 3\sigma, \mu + 3\sigma]$ around the mean μ is 0.997.

Reparametrization of a Gaussian Random Variable. A Gaussian random variable x $\sim \mathcal{N}(\mu, \sigma^2)$ has the useful property that it can be reparametrized as a function of a **standard Gaussian variable** z $\sim \mathcal{N}(0, 1)$, which has zero mean and variance equal to 1. In particular, the random variable x can be expressed as

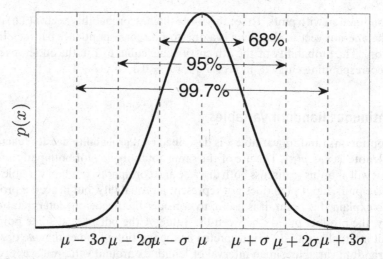

Figure 2.2 A Gaussian pdf. The intervals $[\mu - \sigma, \mu + \sigma]$, $[\mu - 2\sigma, \mu + 2\sigma]$, and $[\mu - 3\sigma, \mu + 3\sigma]$ contain the realization of random variable $x \sim \mathcal{N}(\mu, \sigma^2)$ with probabilities approximately equal to 0.68, 0.95, and 0.997, respectively.

$$x = \mu + \sigma z, \tag{2.12}$$

where $z \sim \mathcal{N}(0, 1)$. This is in the sense that the distribution of the random variable x in (2.12) is the Gaussian pdf $\mathcal{N}(x|\mu, \sigma^2)$. This reparametrization property will allow us in Part III of the book to simplify the optimization of parameters (μ, σ^2) in some learning problems.

Other pdfs. Other distributions for continuous random variables may be more useful if one expects to observe values away from the mean with a higher chance: "Heavy-tailed" distributions include the Laplace pdf $p(x) = \frac{1}{2b} \exp\left(-\frac{|x-\mu|}{b}\right)$ with mean μ and variance $2b^2$, the Pareto distribution, and the Student's t-distribution.

2.3   Expectation

Expectations play an essential role in machine learning, most notably in the definition of training criteria. In fact, as we will see, a learning agent typically wishes to minimize the average, or expected, value of some loss function, with the average taken over the distribution of the random variables of interest. From a frequentist perspective, the expectation captures the average value of a random variable when many measurements of it are available.

Expectation of a Discrete Random Variable. The expectation of a discrete random variable is defined as

$$E_{x \sim p(x)}[x] = \sum_{x \in \mathcal{X}} p(x) \cdot x. \tag{2.13}$$

Under a frequentist viewpoint, by (2.2), we can think of the expectation as the average observed value of random variable x if we repeat the measurement of x many times:

$$E_{x \sim p(x)}[x] \simeq \sum_{x \in \mathcal{X}} \frac{\text{number of samples equal to value } x}{\text{total number of samples}} \cdot x. \tag{2.14}$$

More generally, the expectation of a function of a discrete random variable is given as

$$E_{x \sim p(x)}[f(x)] = \sum_{x \in \mathcal{X}} p(x) \cdot f(x). \tag{2.15}$$

This is also known as the *law of the unconscious statistician*. Generalizing (2.14), we can hence think of the expectation as the average observed value of $f(x)$ if we repeat the measurement of x, and hence of $f(x)$, many times:

$$E_{x \sim p(x)}[f(x)] \simeq \sum_{x \in \mathcal{X}} \frac{\text{number of samples equal to value } x}{\text{total number of samples}} \cdot f(x). \tag{2.16}$$

Applying the definition of expectations to Bernoulli and categorical variables, we obtain the following general expressions:

- For **Bernoulli random variables**, the expectation of any function $f(\cdot)$ is given by

$$E_{x \sim \text{Bern}(q)}[f(x)] = (1 - q) \cdot f(0) + q \cdot f(1); \text{and} \tag{2.17}$$

- for **categorical random variables**, we have

$$E_{x \sim \text{Cat}(q)}[f(x)] = \sum_{x=0}^{C-1} q_x f(x)$$

$$= q_0 f(0) + q_1 f(1) + q_2 f(2) + \cdots + q_{C-1} f(C - 1). \tag{2.18}$$

Example 2.1

Consider a random variable x $\sim$ Cat(q) with $q_0 = 0.1$, $q_1 = 0.2$, and $q_2 = 0.7$, along with function $f(x) = x^2$. The expectation of this function is given as

$$E_{x \sim \text{Cat}(q)}[x^2] = 0.1 \cdot 0^2 + 0.2 \cdot 1^2 + 0.7 \cdot 2^2 = 3. \tag{2.19}$$

It is useful to keep in mind that, for discrete random variables, expectations generalize probabilities. To see this, define the **indicator function**

$$\mathbb{1}(a) = \begin{cases} 1 & \text{if } a = \text{true} \\ 0 & \text{if } a = \text{false.} \end{cases} \tag{2.20}$$

The indicator (2.20) function equals 1 if the condition $x = a$ is true and it equals zero otherwise. Then, we have the equality

$$E_{x \sim p(x)}[\mathbb{1}(x = a)] = \Pr[x = a] = p(x = a). \tag{2.21}$$

Expectation of a Continuous Random Variable. The expectation of a function of a continuous random variable is defined as

$$E_{x \sim p(x)}[f(x)] = \int_{-\infty}^{+\infty} p(x) f(x) dx. \tag{2.22}$$

Accordingly, in a manner similar to discrete random variables, one can think of the expectation as the average observed value of $f(x)$ if we repeat the measurement of x many times.

For **Gaussian random variables**, we have the notable expectations

$$E_{x \sim \mathcal{N}(\mu,\sigma^2)}[x] = \mu$$
$$\text{and } E_{x \sim \mathcal{N}(\mu,\sigma^2)}[x^2] = \mu^2 + \sigma^2. \tag{2.23}$$

Linearity of the Expectation. For both discrete and continuous random variables, the expectation is linear. This is in the sense that, given any two functions $f(\cdot)$ and $g(\cdot)$, we have the equality

$$E_{x \sim p(x)}[af(x) + bg(x)] = aE_{x \sim p(x)}[f(x)] + bE_{x \sim p(x)}[g(x)] \tag{2.24}$$

for any constants a and b. Note that, by induction, this property extends to weighted sums including any number of terms.

2.4 Variance

Suppose that you have access only to the distribution of a random variable x – which value x would you guess for x before you observe it? Given the interpretation (2.14) of the expectation as the average value of repeated experiments, it may be reasonable to choose $E_{x \sim p(x)}[x]$ as the predicted value. How accurate would this guess be?

One way to measure the accuracy of this prediction is to consider the expected value of the **squared error** $(x - E_{x \sim p(x)}[x])^2$. This expectation defines the variance of a random variable. Specifically, the **variance** of a pmf or pdf $p(x)$ measures the average squared spread of the distribution around its mean, i.e.,

$$\text{Var}(p(x)) = E_{x \sim p(x)}[(x - E_{x \sim p(x)}[x])^2] = E_{x \sim p(x)}[x^2] - (E_{x \sim p(x)}[x])^2. \tag{2.25}$$

We will also write Var(x) in lieu of Var($p(x)$) when the distribution of random variable x is clear from the context. Intuitively, the variance Var(x) measures how difficult – in terms of squared error – it is to predict the value of the random variable x. In other words, it gauges the uncertainty in the value of random variable x prior to its measurement.

For Bernoulli, categorical, and Gaussian random variables, we have the following:

- for a **Bernoulli random variable** x $\sim$ Bern(q), the variance is given as

$$\text{Var}(x) = q - q^2 = q(1 - q); \tag{2.26}$$

- for a **categorical random variable** x $\sim$ Cat(q), we have

$$\text{Var}(x) = \sum_{x=0}^{C-1} q_x \left(x - \left(\sum_{x=0}^{C-1} q_x x \right) \right)^2 ; \text{ and} \tag{2.27}$$

- for a **Gaussian random variable** x $\sim \mathcal{N}(\mu,\sigma^2)$, the variance is

$$\text{Var}(x) = \sigma^2. \tag{2.28}$$

Example 2.2

For a categorical variable with $C = 4$, $q_0 = q_2 = q_3 = 0$, and $q_1 = 1$, the mean equals $\sum_{x=0}^{C-1} q_x x = 1$, and the variance is

$$\text{Var(x)} = (1 - 1)^2 = 0. \tag{2.29}$$

The pmf is concentrated at $x = 1$, and the variance of the random variable is zero. This variable is *deterministic* in the sense that there is no uncertainty in its realization.

Example 2.3

For a categorical variable with $C = 4$ and $q_0 = q_1 = q_2 = q_3 = 1/4$, the mean equals $\sum_{x=0}^{C-1} q_x x = 1/4 \cdot (1 + 2 + 3) = 3/2$, and the variance is

$$\text{Var(x)} = \frac{1}{4}\left(\left(0 - \frac{3}{2}\right)^2 + \left(1 - \frac{3}{2}\right)^2 + \left(2 - \frac{3}{2}\right)^2 + \left(3 - \frac{3}{2}\right)^2\right) = \frac{5}{4}. \tag{2.30}$$

This categorical random variable is "maximally random" in the sense that it produces all possible realizations with equal probabilities. Its variance is accordingly larger than Example 2.2.

2.5 Vectors

Vectors represent the most basic data format used in machine learning. A vector is an array of scalars, in which the position of each scalar in the array matters. We have already encountered a few vectors, such as the one-hot representation (cf. (2.7)) of a categorical random variable. In this section, we recall basic definitions and some relevant operations for machine learning.

2.5.1 Definitions

We will use lower-case fonts for vectors, which will be assumed to be in **column form**. For example, an $L \times 1$ (column) vector x will be denoted as

$$x = \begin{bmatrix} x_1 \\ x_2 \\ \vdots \\ x_L \end{bmatrix}, \tag{2.31}$$

where the dimension L will be clear from the context. Note that each scalar x_i has a specific position, i, in the array. The space of all L dimensional vectors with real entries is denoted as $\mathbb{R}^L$. Special vectors are the all-zero vector 0_L and the all-one vector 1_L, both of dimension $L \times 1$.

A vector $x \in \mathbb{R}^L$ can be thought of in any of the following ways:

- as a point in the L-dimensional space $\mathbb{R}^L$, whose coordinates along each axis are given by the respective entries $x_1, x_2, \ldots$ and x_L;
- as an "arrow" starting from the origin (the origin is the all-zero vector) and pointing to (i.e., ending at) x in $\mathbb{R}^L$; and

Figure 2.3 Illustrations of key definitions regarding vectors in $\mathbb{R}^2$ ($L = 2$ and $a > 0$ in the figure). The circle represents vectors with ℓ_2 norm equal to 1.

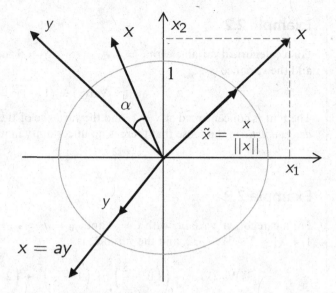

- as an "arrow" starting from any point y in $\mathbb{R}^L$ and pointing to (i.e., ending at) $y + x$.

Figure 2.3 illustrates the first two interpretations for a vector $x = [x_1, x_2]^T$ in $\mathbb{R}^2$, which is shown as a point and as an arrow ending at a point with coordinates x_1 and x_2 along the two axes.

The **transpose** of vector x in (2.31) is the $1 \times L$ row vector

$$x^T = [x_1 x_2 \cdots x_L]. \tag{2.32}$$

Note that we have $(x^T)^T = x$.

2.5.2   Inner Product

The inner product is a fundamental operation in machine learning, and it can be interpreted as providing a measure of "similarity" between two vectors with the same dimension L. More precisely, from a geometric perspective, the inner product between two vectors x and y in $\mathbb{R}^L$ measures how aligned the two vectors are. Mathematically, the inner product between two vectors x and y is defined as

$$x^T y = \sum_{i=1}^{L} x_i y_i, \tag{2.33}$$

which amounts to the sum of the element-wise products of the elements of the two vectors. We will see below that the inner product is related to the angle α between the two vectors x and y, which is illustrated in Fig. 2.3.

Example 2.4

Consider the 49×1 vectors x and y displayed in Fig. 2.4 as 7×7 matrices (obtained by placing successive disjoint 7×1 subvectors from x and y side by side as columns of the respective matrices). The inner products $x^T y$ for the vectors in both rows can be easily seen to equal 6. According to the inner product, the pairs of vectors x and y in the two rows are hence equally similar.

Figure 2.4 Two examples of vectors x and y, displayed as matrices.

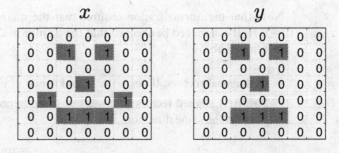

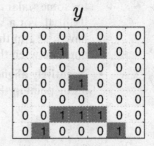

2.5.3 Squared ℓ_2 Norm

The squared ℓ_2 norm of vector x is defined as the inner product of a vector with itself, i.e.,

$$||x||^2 = x^T x = \sum_{i=1}^{L} x_i^2. \tag{2.34}$$

Geometrically, the ℓ_2 norm $||x||$ of vector x coincides with the length of the arrow representing it. One can always **normalize** a vector (with non-zero norm) so that it has ℓ_2 norm equal to 1 as

$$\tilde{x} = \frac{x}{||x||}. \tag{2.35}$$

In fact, it can be easily checked that we have $||\tilde{x}||^2 = 1$. An example is illustrated in Fig. 2.3.

2.5.4 Cosine Similarity

The **cosine** of the angle $\alpha \in (-\pi, \pi]$ between two vectors x and y in $\mathbb{R}^L$ can provide a useful measure of the similarity of the two vectors (see Fig. 2.3). In fact, from standard trigonometry, the cosine is zero, $\cos(\alpha) = 0$, if the two vectors are orthogonal, i.e., if $\alpha = \pi/2$ or $\alpha = -\pi/2$; it takes its maximum positive value $\cos(\alpha) = 1$ when $\alpha = 0$ and hence the vectors are aligned; and it takes its maximum negative value $\cos(\alpha) = -1$ when $\alpha = \pi$ and hence the vectors point in diametrically opposite directions. For these reasons, the cosine $\cos(\alpha)$ is also referred to as the **cosine similarity** between vectors x and y.

Given two vectors x and y in $\mathbb{R}^L$, the cosine, also known as the cosine similarity, is defined as the following normalized version of the inner product $x^T y$:

$$\cos(\alpha) = \frac{x^T y}{||x|| ||y||} \in [-1, 1]. \tag{2.36}$$

Note that the normalization ensures that the quantity in (2.36) takes values in the interval $[-1, 1]$. It can indeed be checked that the definition (2.36) satisfies the trigonometric conditions mentioned above:

- **Orthogonal vectors.** If $\cos(\alpha) = 0$, and hence $x^T y = 0$, the two vectors are orthogonal.
- **Positively aligned vectors.** If $\cos(\alpha) = 1$ and hence $x^T y = ||x||||y||$, the two vectors x and y point in the same direction, i.e., we have

$$x = ay \tag{2.37}$$

for some scalar $a > 0$. This can be directly verified by the chain of equalities $x^T y = a||y||^2 = ||x||||y||$. For a vector y, Fig. 2.3 displays an aligned vector $x = ay$ with $a > 0$. (In fact, we have $a > 1$ in the figure since x has a larger norm.) More generally, if $0 < \cos(\alpha) \leq 1$, the angle between the two vectors is acute (i.e., $|\alpha| < \pi/2$ for $\alpha \in (-\pi, \pi)$).

- **Negatively aligned vectors.** If $\cos(\alpha) = -1$ and hence $x^T y = -||x||||y||$, the two vectors x and y point in the opposite direction, i.e., we have

$$x = -ay \tag{2.38}$$

for some scalar $a > 0$. This can be similarly verified and illustrated. More generally, if $-1 \leq \cos(\alpha) < 0$, the angle between the two vectors is obtuse (i.e., $|\alpha| > \pi/2$ for $\alpha \in (-\pi, \pi)$).

That the cosine similarity (2.36) is bounded in the interval $[-1, 1]$ follows from the **Cauchy–Schwartz inequality**

$$|x^T y| \leq ||x||||y||, \tag{2.39}$$

which says that the absolute value of the inner product, $|x^T y|$, can never be larger than the product of the ℓ_2 norms $||x||||y||$.

The cosine similarity is used in many machine learning algorithms, such as for **recommendation systems**: Two users with closely aligned "profile" vectors x and y, obtained from their respective ratings of items, are likely to provide recommendations that are useful for one another. As a measure of relatedness, the cosine similarity (2.36) has the useful property of being **invariant** to positive scalings of the vectors x and y: If we multiply vectors x and/or y by, possibly different, positive constants, the cosine of the angle between the two vectors does not change.

Example 2.5

Consider again the 49×1 vectors x and y displayed in Fig. 2.4 (see Example 2.4). While the inner products for the pairs of vectors in the two rows are the same, their cosine similarities are different. In fact, for the top-row vectors, we have $\cos(\alpha) = 6/(\sqrt{8} \cdot \sqrt{6}) = 0.87$ (since the norms of the two vectors are $||x|| = \sqrt{8}$ and $||y|| = \sqrt{6}$); and for the bottom-row vectors, we have $\cos(\alpha) = 6/(\sqrt{8} \cdot \sqrt{8}) = 0.75$. The fact that, when using the cosine similarity metric, the top-row vectors are deemed to be more similar than the bottom-row ones complies with intuition, although this is not a general fact. Another interesting observation is that, as mentioned, if we multiply any of the vectors by a positive scalar, the cosine similarity remains invariant, while this is not the case for the inner product.

2.5.5 Linearly Independent Vectors

A set of vectors $\{x_1, \ldots, x_K\}$, with $x_k \in \mathbb{R}^L$ for all $k = 1, \ldots, K$, is said to be **linearly independent** if none of the vectors can be written as a linear combination of the others, i.e., if there are no coefficients $\{a_k\}_{k=1}^K$ such that $x_i = \sum_{k=1, k \neq i}^K a_k x_k$ for some $i \in \{1, 2, \ldots, K\}$. Note that, throughout the book, the context will help clarify whether the notation x_i represents a vector or a scalar. As an example, in Fig. 2.3, the vectors x and y for which the mutual angle α is shown are linearly independent since they are not (positively or negatively) aligned.

2.6 Matrices

2.6.1 Definitions

Matrices are collections of vectors. We denote them using upper-case fonts. Accordingly, an $L \times M$ matrix is written as

$$
A = \begin{bmatrix}
a_{11} & a_{12} & \cdots & a_{1M} \\
a_{21} & & & a_{2L} \\
\vdots & & & \vdots \\
a_{L1} & a_{L2} & \cdots & a_{LM}
\end{bmatrix},
\tag{2.40}
$$

where $[A]_{ij} = a_{ij}$ is the (i,j)th (ith row and jth column) element of matrix A. By (2.40), one can think of a matrix as a collection of row vectors, stacked on top of each other; or as a collection of column vectors, set side by side. Note that the scalar elements a_{ij} of a matrix are indexed by two integers, i and j, while a single index is sufficient for a vector. Vectors and matrices can be further generalized by **tensors**, whose scalar entries can depend on more than two indices. (Vectors are tensors of order 1, and matrices are tensors of order 2.)

The **transpose** of the $L \times M$ matrix A is given by the $M \times L$ matrix

$$
A^T = \begin{bmatrix}
a_{11} & a_{21} & \cdots & a_{L1} \\
a_{12} & & & a_{L2} \\
\vdots & & & \vdots \\
a_{1M} & a_{2M} & \cdots & a_{LM}
\end{bmatrix},
\tag{2.41}
$$

which is obtained by arranging the transpose of the columns of A on top of each other.

For an $L \times 1$ vector x, the $L \times L$ **diagonal matrix** $\mathrm{Diag}(a)$ is defined as

$$
\mathrm{Diag}(a) = \begin{bmatrix}
a_1 & 0 & \cdots & 0 \\
0 & a_2 & & 0 \\
\vdots & & \ddots & \vdots \\
0 & 0 & \cdots & a_L
\end{bmatrix}.
\tag{2.42}
$$

We also write as

$$
\mathrm{diag}(A) = \begin{bmatrix}
a_{11} \\
a_{22} \\
\vdots \\
a_{LL}
\end{bmatrix}
\tag{2.43}
$$

the vector of elements on the **diagonal** of any square $L \times L$ matrix A. A special diagonal matrix is given by the $L \times L$ **identity** matrix

$$
I_L = \mathrm{Diag}(1_L).
\tag{2.44}
$$

2.6.2   Products

Outer Product. The outer product of two $L \times 1$ vectors x and y is given by the $L \times L$ matrix

$$xy^T = \begin{bmatrix} x_1y_1 & x_1y_2 & \cdots & x_1y_L \\ x_2y_1 & x_2y_2 & & x_2y_L \\ \vdots & & & \vdots \\ x_Ly_1 & x_Ly_2 & \cdots & x_Ly_L \end{bmatrix}, \tag{2.45}$$

which contains all pair-wise products of elements from the two vectors.

Matrix–Vector Multiplication. Denote as

$$[A]_{r:} = [a_{r1}, a_{r2}, \ldots, a_{rM}] \tag{2.46}$$

the $1 \times M$ vector containing the rth row of A. The **right multiplication** of an $L \times M$ matrix A with an $M \times 1$ vector x yields the $L \times 1$ vector

$$Ax = \begin{bmatrix} [A]_{1:}x \\ [A]_{2:}x \\ \vdots \\ [A]_{L:}x \end{bmatrix}. \tag{2.47}$$

Therefore, the matrix–vector product is a collection of inner products between the rows of A and x.

The **left multiplication** of an $L \times M$ matrix A with the transpose of an $L \times 1$ vector x yields the $1 \times M$ vector

$$x^T A = (A^T x)^T. \tag{2.48}$$

As an example of a matrix multiplication, for $L \times 1$ vectors x and y, we have

$$\mathrm{Diag}(x)y = \mathrm{Diag}(y)x = \begin{bmatrix} x_1y_1 \\ x_2y_2 \\ \vdots \\ x_Ly_L \end{bmatrix} = x \odot y, \tag{2.49}$$

where we have used the **element-wise product** notation $x \odot y$.

Matrix–Matrix Multiplication. Generalizing the matrix–vector product, for $L \times M$ and $M \times D$ matrices A and B, respectively, the product AB is given by the $L \times D$ matrix

$$AB = \begin{bmatrix} A[B]_{:1} & A[B]_{:2} & \cdots & A[B]_{:D} \end{bmatrix}$$

$$= \begin{bmatrix} [A]_{1:}B \\ [A]_{2:}B \\ \vdots \\ [A]_{L:}B \end{bmatrix}, \tag{2.50}$$

where $[B]_{:c}$ represents the $M \times 1$ cth column vector of matrix B. Therefore, matrix–matrix multiplications can be obtained as collections of matrix–vector products.

As an example, for $L \times 1$ vectors x and y, we have

$$\mathrm{Diag}(x)\mathrm{Diag}(y) = \mathrm{Diag}(x \odot y). \tag{2.51}$$

2.6.3   Symmetric and Positive (Semi-)Definite Matrices

An $L \times L$ **square matrix** A has an equal number of rows and columns. A **symmetric matrix** A is a square matrix that satisfies the equality $A^T = A$. Equivalently, symmetric matrices are such that the equality $[A]_{ij} = [A]_{ji}$ holds for all elements outside the main diagonal, i.e., for all $i \neq j$. Diagonal matrices are symmetric.

Eigendecomposition. An $L \times L$ symmetric matrix can be written using its **eigendecomposition** as

$$A = U \Lambda U^T, \tag{2.52}$$

where $U = [u_1, \ldots, u_L]$ is an $L \times L$ matrix with orthogonal unitary-norm columns $\{u_i\}_{i=1}^L$ and $\Lambda = \text{Diag}(\lambda)$ is a diagonal matrix. The columns of matrix U are known as the **eigenvectors** of matrix A, and the elements $\lambda = [\lambda_1, \ldots, \lambda_L]$ on the diagonal of Λ are referred to as the **eigenvalues** of matrix A.

The eigendecomposition (2.52) can also be expressed as the linear combination of outer products

$$A = \sum_{i=1}^{L} \lambda_i u_i u_i^T. \tag{2.53}$$

This form of the eigendecomposition highlights the fact that each eigenvector u_i is associated with the corresponding eigenvalue λ_i.

Gershgorin Theorem. Computing the eigenvalues and eigenvectors of a symmetric matrix generally requires specialized numerical routines. There is, however, a special case in which this calculation is trivial: diagonal matrices. For a diagonal matrix A, we trivially have $A = \Lambda$, i.e., the eigenvalues coincide with the diagonal elements of A. Interestingly, if a matrix is *close* to being diagonal, one can get an approximation of the eigenvalues via the Gershgorin theorem.

The Gershgorin theorem says that, if the elements on the diagonal of a symmetric matrix are larger than the rest – in some specific sense to be detailed next – then the eigenvalues are close to the elements on the diagonal. Note that this is clearly true for diagonal matrices.

To elaborate, for some symmetric matrix A, define as $a_{-i} = \sum_{j=1, j \neq i}^{L} |a_{ij}|$ the sum of the absolute values of the elements outside the main diagonal for row i. With this definition, the Gershgorin theorem says that all the eigenvalues of matrix A are in the union of the intervals $[a_i - a_{-i}, a_i + a_{-i}]$ for $i = 1, \ldots, L$. That is, every eigenvalue must be in an interval $[a_i - a_{-i}, a_i + a_{-i}]$ around some diagonal element a_i.

Positive (Semi-)Definite Matrices. A symmetric matrix A with all non-negative eigenvalues is said to be **positive semi-definite**, and denoted as

$$A \succeq 0. \tag{2.54}$$

If all the eigenvalues of A are strictly positive, we say that the symmetric matrix A is **positive definite**, and we denote it as

$$A \succ 0. \tag{2.55}$$

For an $L \times L$ symmetric matrix A, we can define the associated **quadratic form** as

$$x^T A x = \sum_{i=1}^{L} \sum_{j=1}^{L} a_{ij} x_i x_j = \sum_{i=1}^{L} a_{ii} x_i^2 + 2 \sum_{i=1}^{L-1} \sum_{j=i+1}^{L} a_{ij} x_i x_j, \tag{2.56}$$

which is a function of the vector $x \in \mathbb{R}^L$. An $L \times L$ matrix is positive semi-definite if and only if we have the inequality

$$x^T A x \geq 0 \tag{2.57}$$

for all $x \in \mathbb{R}^L$, that is, if the associated quadratic form is non-negative everywhere. Furthermore, it is positive define if and only if we have the inequality

$$x^T A x > 0 \tag{2.58}$$

for all $x \in \mathbb{R}^L$ with $x \neq 0_L$.

Let us see two applications of this result. First, for a diagonal matrix, we have

$$x^T \mathrm{Diag}(a) x = \sum_{i=1}^{L} a_i x_i^2, \tag{2.59}$$

and hence a diagonal matrix is positive semi-definite or definite if all the elements on the diagonal are, respectively, non-negative or positive. Second, given any $L \times M$ matrix B, the $L \times L$ matrix

$$A = B B^T \tag{2.60}$$

is always symmetric and positive semi-definite. To see this, note that we have the inequality

$$x^T A x = x^T B B^T x = (B^T x)^T (B^T x) = ||B^T x||^2 \geq 0. \tag{2.61}$$

2.6.4 Invertible Matrices

A square $L \times L$ matrix is **invertible** if its columns are linearly independent, or equivalently if its rows are linearly independent. If matrix A is invertible, there exists an $L \times L$ **inverse matrix** A^{-1} that satisfies the equality $A^{-1} A = I_L$.

As an example, a matrix $A = \mathrm{Diag}(a)$ is invertible if all the diagonal elements are non-zero. In this case, we have the inverse matrix

$$A^{-1} = \mathrm{Diag} \left(\begin{bmatrix} \frac{1}{a_1} \\ \vdots \\ \frac{1}{a_L} \end{bmatrix} \right). \tag{2.62}$$

As a generalization, a symmetric matrix $A = U \Lambda U^T$ is invertible if all its eigenvalues are different from zero, and its inverse is given as $A^{-1} = U \Lambda^{-1} U^T$.

If matrix A is invertible, the solution of the linear system $Ax = y$, with fixed matrix A and vector y, yields $x = A^{-1} y$.

2.6.5 Trace and Determinant

The **trace** of a square matrix is equal to the sum of the elements on its diagonal, i.e.,

$$\mathrm{tr}(A) = \sum_{i=1}^{L} a_{ii}. \tag{2.63}$$

The **determinant** of a symmetric matrix is equal to the product of its eigenvalues

$$\det(A) = \prod_{i=1}^{L} \lambda_i. \tag{2.64}$$

For a symmetric matrix, the trace can also be written as the sum of the eigenvalues

$$\mathrm{tr}(A) = \sum_{i=1}^{L} \lambda_i. \tag{2.65}$$

2.7 Random Vectors

An array

$$\mathbf{x} = \begin{bmatrix} x_1 \\ x_2 \\ \vdots \\ x_L \end{bmatrix} \tag{2.66}$$

of L jointly distributed random variables defines a random vector. A random vector may contain a mix of discrete or continuous random variables $x_1, x_1, \ldots, x_L$.

2.7.1 Discrete Random Vectors

When all constituent random variables are discrete, the joint distribution of a random vector is defined by a **joint pmf** $p(x)$. In this case, each random variable x_i with $i = 1, \ldots, L$ belongs to a discrete finite set $\mathcal{X}$. We can also write this condition in vector form as $x \in \mathcal{X}^L$ to indicate that all L entries are from alphabet $\mathcal{X}$. The joint pmf $p(x)$ defines the probability of random vector x being equal to some (deterministic) vector $x \in \mathcal{X}^L$, which is the probability that random variable x_1 equals the first entry x_1 of vector x, *and* random variable x_2 equals the second entry x_2 of vector x, and so on. We can therefore write the equalities

$$\begin{aligned} p(x) &= p(x_1, \ldots, x_L) \\ &= \Pr[x = x] \\ &= \Pr[x_1 = x_1 \text{ and } x_2 = x_2 \ldots \text{ and } x_L = x_L]. \end{aligned} \tag{2.67}$$

Being a probability, the joint pmf satisfies the conditions $p(x) \geq 0$ and

$$\sum_{x_1 \in \mathcal{X}} \sum_{x_2 \in \mathcal{X}} \cdots \sum_{x_L \in \mathcal{X}} p(x) = 1. \tag{2.68}$$

Note that the sum in (2.68) is over all $|\mathcal{X}|^L$ values vector x can take.

Using a frequentist interpretation, we can think of the value $p(x)$ assumed by the joint pmf at a point $x = [x_1, \ldots, x_L]^T$ as the fraction of times that we observe the configuration $(x_1 = x_1, x_2 = x_2, \ldots, x_L = x_L)$ if we repeat the measurement of random vector x many times, i.e.,

$$p(x_1, \ldots, x_L) \simeq \frac{\text{number of samples equal to } (x_1, \ldots, x_L)}{\text{total number of samples}}, \tag{2.69}$$

Table 2.1 The joint pmf for a jointly Bernoulli random vector with $L = 2$ as a table of probability values

$x_1 \backslash x_2$	0	1
0	$p(0,0)$	$p(0,1)$
1	$p(1,0)$	$p(1,1)$

where, as usual, the approximation is accurate when the number of observations is large.

Jointly Bernoulli Random Vector. A two-dimensional jointly Bernoulli random vector $x = [x_1, x_2]^T$ is defined by a joint pmf $p(x_1, x_2)$ that can be specified by a table of the form in Table 2.1. In the table, the row index specifies the value of random variable x_1, while the column index describes the value of random variable x_2. The term $p(0,0)$ represents the probability of the event $(x_1 = 0, x_2 = 0)$, and a similar definition applies to all other entries. A general L-dimensional jointly Bernoulli random vector is similarly specified by a multi-dimensional table that describes the joint pmf $p(x)$ for all 2^L possible configuration of the binary vector.

2.7.2   Continuous Random Vectors

When all constituent random variables are continuous, the joint distribution of the random vector is defined by a joint pdf $p(x)$. In this case, all entries are real, i.e., $x_i \in \mathbb{R}$ for $i = 1, \ldots, L$, and hence we have $x \in \mathbb{R}^L$. Generalizing the one-dimensional case, the value $p(x)$ of a joint pdf does not directly define a probability; instead, the product $p(x)dx$ equals the probability of observing the random vector x within a hypercube of (small) volume dx around vector $x \in \mathbb{R}^L$. Probabilities of arbitrary regions in $\mathbb{R}^L$ can accordingly be obtained from the joint pdf via integration. For instance, the probability of a hypercube with the ith side along variable x_i spanning the interval $[a_i, b_i]$, for $a_i < b_i$, is given as

$$\Pr\left[x_1 \in [a_1, b_1] \text{ and } x_2 \in [a_2, b_2] \ldots \text{ and } x_L \in [a_L, b_L]\right] = \int_{a_1}^{b_1} \int_{a_2}^{b_2} \cdots \int_{a_L}^{b_L} p(x) dx_1 dx_2 \cdots dx_L.$$

(2.70)

The joint pdf satisfies the conditions $p(x) \geq 0$ and $\int_{-\infty}^{+\infty} \int_{-\infty}^{+\infty} \cdots \int_{-\infty}^{+\infty} p(x) dx_1 dx_2 \cdots dx_L = 1$.

Jointly (or Multivariate) Gaussian Random Vector. Jointly Gaussian random vectors, also know as multivariate Gaussian random vectors, are a key example of continuous random vectors. A jointly Gaussian pdf $\mathcal{N}(x|\mu, \Sigma)$ is defined by an $L \times 1$ **mean vector** μ and an $L \times L$ **covariance matrix** $\Sigma \succeq 0$. Note that the covariance matrix is constrained to be positive semi-definite. As we will see, the covariance describes the statistical dependence among the variables in the random vector $x \sim \mathcal{N}(\mu, \Sigma)$. In the special case in which the covariance matrix is positive definite, i.e., when $\Sigma \succ 0$, the joint pdf has the general expression

$$\mathcal{N}(x|\mu, \Sigma) = \frac{1}{2\pi\sqrt{\det(\Sigma)}} \exp\left(-\frac{1}{2}(x - \mu)^T \Sigma^{-1} (x - \mu)\right).$$

(2.71)

To understand this expression and the role of the covariance in it, let us focus now on the case with two variables ($L = 2$). A two-dimensional jointly Gaussian random vector can be written as

$$x = \begin{bmatrix} x_1 \\ x_2 \end{bmatrix} \sim \mathcal{N}(\mu, \Sigma), \text{ with } \mu = \begin{bmatrix} \mu_1 \\ \mu_2 \end{bmatrix} \text{ and } \Sigma = \begin{bmatrix} \sigma_1^2 & \sigma_{12} \\ \sigma_{12} & \sigma_2^2 \end{bmatrix}, \tag{2.72}$$

where we have introduced the **covariance** between random variables x_1 and x_2,

$$\sigma_{12} = \text{Cov}(x_1, x_2). \tag{2.73}$$

The covariance matrix Σ in (2.72) can be seen to be positive semi-definite if the condition

$$|\sigma_{12}| \le \sigma_1 \sigma_2 \tag{2.74}$$

holds, and it is positive definite if

$$|\sigma_{12}| < \sigma_1 \sigma_2. \tag{2.75}$$

When the inequality (2.75) holds, the Gaussian joint pdf (2.71) can be written for the special case $L = 2$ as

$$\mathcal{N}(x|\mu, \Sigma) = \frac{1}{2\pi \sigma_1 \sigma_2 \sqrt{1 - \rho^2}} \exp\left(-\frac{1}{2(1 - \rho^2)}\left[\sum_{k=1}^{2}\left(\frac{(x_k - \mu_k)^2}{\sigma_k^2}\right) - \frac{2\rho(x_1 - \mu_1)(x_2 - \mu_2)}{\sigma_1 \sigma_2}\right]\right), \tag{2.76}$$

where we have introduced the **covariance coefficient**

$$\rho = \frac{\sigma_{12}}{\sigma_1 \sigma_2}. \tag{2.77}$$

Note that, when expressed in terms of the covariance coefficient, the inequalities (2.74) and (2.75) are equivalent to $|\rho| \le 1$ and $|\rho| < 1$, respectively.

The mean vector and the diagonal elements of the covariance in (2.72) correspond to the respective means and variances of the constituent random variables x_1 and x_2, i.e., we have $\mu_1 = E_{x \sim \mathcal{N}(\mu, \Sigma)}[x_1]$ and $\mu_2 = E_{x \sim \mathcal{N}(\mu, \Sigma)}[x_2]$, as well as $\sigma_1^2 = \text{Var}(x_1)$ and $\sigma_2^2 = \text{Var}(x_2)$. The novel element here is the presence of the covariance σ_{12} as the off-diagonal term of the covariance matrix. As we will discuss, the covariance models the degree of **statistical dependence** of the two constituent random variables x_1 and x_2.

Example 2.6

The joint pdf (2.76) is illustrated in Figs. 2.5 and 2.6 using two different representations, namely a three-dimensional plot and a contour-line plot. The latter can be thought of as a bird's-eye view of the former. The left-hand and right-hand figures differ in the value of the covariance coefficient ρ. As in the single-variable Gaussian pdf, the mean vector μ is seen to specify the position of the peak, or **mode**, of the distribution, while the covariance matrix – and specifically the covariance coefficient ρ in the figures – defines the shape of the Gaussian "hat". Intuitively, when the pdf is circularly symmetric, as with $\rho = 0$ in the left-hand figures, no statistical dependence exists between the two variables; while if $\rho \ne 0$, as with $\rho = 0.9$ in the right-hand figures, plausible values for the two random variables follow a clear pattern that can be leveraged for prediction of one variable from the other. We will return to this point in the next subsection.

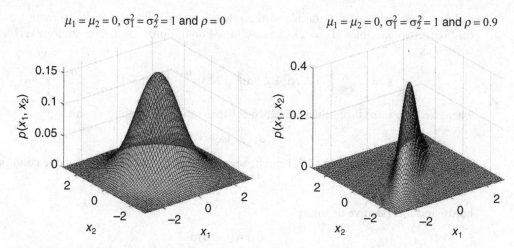

Figure 2.5 Jointly Gaussian pdfs with $L = 2$ represented as a three-dimensional plot. The left figure corresponds to the case of uncorrelated random variables, $\rho = 0$, while the right figure represents positively correlated variables with $\rho = 0.9$.

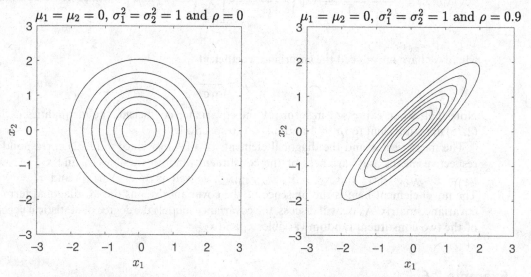

Figure 2.6 Jointly Gaussian pdfs with $L = 2$ represented as contour lines. Each line represent the values of x that have the same density value $p(x)$ for a selected subset of such values. The left figure corresponds to the case of uncorrelated random variables, $\rho = 0$, while the right figure represents positively correlated variables with $\rho = 0.9$.

Reparametrization of a Jointly Gaussian Random Vector. Generalizing the reparametrization property introduced in Sec. 2.2.2, an L-dimensional jointly Gaussian random vector $x \sim \mathcal{N}(\mu, \Sigma)$ has the useful property that it can be reparametrized as a function of the standard Gaussian vector $z \sim \mathcal{N}(0_L, I_L)$, which has all-zero mean vector 0_L and covariance matrix equal to the identity I_L. In particular, the random vector x can be expressed as

$$x = \mu + \Sigma^{1/2} z, \tag{2.78}$$

where $z \sim \mathcal{N}(0_L, I_L)$, in the sense that the distribution of random vector x in (2.78) is the jointly Gaussian pdf $\mathcal{N}(x|\mu, \Sigma)$. The $L \times L$ square root of the covariance matrix, $\Sigma^{1/2}$, can be obtained as $\Sigma^{1/2} = U \Lambda^{1/2} U^T$, where $\Sigma = U \Lambda U^T$ is the eigendecomposition of the covariance matrix and $\Lambda^{1/2}$ is a diagonal matrix with each element equal to the square root of the corresponding eigenvalue in matrix Λ.

2.7.3 Covariance and Covariance Coefficient

Consider any two random variables with joint pmf or pdf $p(x) = p(x_1, x_2)$, *not necessarily jointly Gaussian*, and denote their respective expectations as μ_1 and μ_2. Then, the **covariance** of the random variables x_1 and x_2 is defined as

$$\begin{aligned}
\text{Cov}(x_1, x_2) = \text{Cov}(x_2, x_1) &= \text{E}_{x \sim p(x_1, x_2)}[(x_1 - \mu_1)(x_2 - \mu_2)] \\
&= \text{E}_{x \sim p(x_1, x_2)}[x_1 x_2] - \mu_1 \mu_2,
\end{aligned} \tag{2.79}$$

where the equivalence between the two expectations can be easily proved. Note that the covariance generalizes the definition of variance in the sense that we have the equalities $\text{Cov}(x_1, x_1) = \text{Var}(x_1)$ and $\text{Cov}(x_2, x_2) = \text{Var}(x_2)$. Furthermore, for a two-dimensional Gaussian random vector (2.72), the covariance is given by σ_{12} in (2.73).

Using its definition (2.79), we can interpret the covariance as a metric of statistical dependence of the two random variables:

- **Uncorrelated variables**. If $\text{Cov}(x_1, x_2) = 0$, deviations from the respective means of the two variables, i.e., $(x_1 - \mu_1)$ and $(x_2 - \mu_2)$, tend to cancel each other out on average. This implies that there is no linear relationship between the two random variables.
- **Positively correlated variables**. If the covariance is positive, deviations from the respective means of the two variables tend to have the same sign in order for the expectation of the product in (2.79) to be positive. Hence, if x_1 is larger than its mean, i.e., if $x_1 > \mu_1$, we should expect that x_2 is also larger than its mean, i.e., $x_2 > \mu_2$; and, similarly, if $x_1 < \mu_1$, we should expect that $x_2 < \mu_2$. Overall, this implies that there exists a positive linear trend relating the two deviations.
- **Negatively correlated variables**. If the covariance is negative, deviations from the respective means of the two variables tend to have opposite signs. Accordingly, if $x_1 > \mu_1$, we should expect that $x_2 < \mu_2$; and if $x_1 < \mu_1$, that $x_2 > \mu_2$, implying a negative linear trend for the two deviations.

This analysis suggests that the covariance plays a similar role to the inner product between two vectors in the context of random variables. Like the inner product (cf. (2.39)), the covariance satisfies its own version of the **Cauchy–Schwartz inequality**, namely

$$|\text{Cov}(x_1, x_2)|^2 \leq \text{Var}(x_1)\text{Var}(x_2). \tag{2.80}$$

Therefore, in a manner that parallels the cosine similarity for vectors, the **covariance coefficient** can be used as a normalized measure of "alignment" of two random variables, since it satisfies the normalization condition

$$\rho = \frac{\text{Cov}(x_1, x_2)}{\sqrt{\text{Var}(x_1)\text{Var}(x_2)}} \in [-1, 1]. \tag{2.81}$$

Accordingly, we can distinguish the following situations:

- **Uncorrelated variables**. If $\rho = 0$, the covariance is zero, and the two variables are uncorrelated.
- **Positively correlated variables**. When $\rho = 1$, the two variables satisfy an exact positive linear relationship in the sense that we have

$$(x_1 - \mu_1) = a(x_2 - \mu_2) \tag{2.82}$$

with $a > 0$ – the two variables are fully, and positively, correlated. More generally if $0 < \rho \leq 1$, as in Figs. 2.5 and 2.6, the two variables are positively correlated and there exists a "noisy" positive linear relationship between the two.
- **Negatively correlated variables**. When $\rho = -1$, we have

$$(x_1 - \mu_1) = -a(x_2 - \mu_2) \tag{2.83}$$

with $a > 0$, in which case the two variables are fully, and negatively, correlated. More generally if $-1 \leq \rho < 0$, the two variables are negatively correlated and there exists a "noisy" negative linear relationship between the two.

A necessary word of caution: Based on this discussion, one may be tempted to say that a positive covariance implies that, when one variable increases, the other does too, while, for a negative covariance, when one variable increases, the other decreases. This statement, however, should be interpreted carefully, lest it is misconstrued to imply an actual **causality** relationship: Famously, *correlation (or covariance) does not imply causation*. The reader is referred to Chapter 15 for further discussion on causality.

Multivariate Gaussian Random Vector. The expression for the jointly Gaussian pdf (2.76) with $L = 2$ reflects the role highlighted above of the covariance and covariance coefficient. In particular, when $\rho > 0$, the term $\rho(x_1 - \mu_1)(x_2 - \mu_2)$ in (2.76) ensures that the pdf value is larger when the deviations $(x_1 - \mu_1)$ and $(x_2 - \mu_2)$ have the same sign, while, for $\rho < 0$, the pdf value is larger when they have the opposite signs. No preferred direction exists when $\rho = 0$.

Example 2.7

To illustrate the impact of the covariance coefficient ρ on the shape of the multivariate Gaussian pdf with $L = 2$, let us return to Figs. 2.5 and 2.6. As seen, these figures plot two examples of zero-mean jointly Gaussian pdfs, one with $\rho = 0$ (left figures) and the other with $\rho = 0.9$ (right figures), assuming equal variances for the two random variables. For $\rho = 0$, given a value x_1, positive and negative values x_2 with the same magnitude have the same density $\mathcal{N}(x|0, \Sigma)$. In contrast, when the covariance coefficient ρ is positive and large (here $\rho = 0.9$), for a positive value x_1, the probability density $\mathcal{N}(x|0, \Sigma)$ is much larger for positive values x_2 than it is for negative values. This highlights the positive linear relationship that exists between the two variables.

The role of the covariance in shaping and rotating the hat-shaped Gaussian distribution extends more generally to any number L of dimensions. In particular, the eigenvectors of the covariance matrix determine orthogonal directions in $\mathbb{R}^L$ along which the pdf is arranged, and the corresponding eigenvalues indicate the spread of the pdf along each direction. The reader is referred to standard textbooks on probability (see Recommended Resources, Sec. 2.14), as well as to Sec. 5.7, which discusses the role of the Hessian in defining the shape of quadratic functions. (The exponent of the Gaussian pdf (2.71) is a quadratic function with Hessian given by $-\Sigma^{-1}$.)

2.8   Marginal Distributions

Given a joint distribution $p(x_1, x_2)$, what is the distribution $p(x_1)$ of random variable x_1? More generally, given a joint distribution $p(x_1, x_2, \ldots, x_L)$, what is the distribution $p(x_i)$ of one of the random variables? This is known as the **marginal distribution** of x_i.

To address this question, consider first the case of two random variables and discuss the problem of obtaining the marginal $p(x_1)$ given a joint distribution $p(x_1, x_2)$. (The same discussion applies, *mutatis mutandis*, to $p(x_2)$.) The marginal of random variable x_1 is obtained as

$$p(x_1) = \sum_{x_2 \in \mathcal{X}} p(x_1, x_2) \text{ for discrete random variables, and}$$

$$= \int_{-\infty}^{+\infty} p(x_1, x_2) dx_2 \text{ for continuous random variables.} \tag{2.84}$$

In words, the marginal distribution of random variable x_1 is obtained by summing, or integrating, over all the values of the other variable x_2. This operation indicates, in a sense, that all possible values x_2 should be considered when evaluating the probability of random variable x_1 – i.e., we "don't care" about the value of random variable x_2.

This point is easier to understand if one considers the frequentist interpretation of probability. In fact, for discrete random variables, by summing over all values x_2 in (2.69), the marginal distribution $p(x_1)$ measures the fraction of times that we observe $x_1 = x_1$ in a sample (x_1, x_2), irrespective of what the value of random variable x_2 is. We can express this interpretation with the approximate equality

$$p(x_1) \simeq \frac{\text{number of samples } (x_1, *)}{\text{total number of samples}}, \tag{2.85}$$

where the "don't care" notation "$*$" indicates that the random variable x_2 can take any value. Therefore, the marginal distribution $p(x_1)$ can be estimated by neglecting x_2 from every sample (x_1, x_2). An analogous discussion applies to continuous random variables, with the usual caveat that one estimates probabilities of hypercubes in $\mathbb{R}^L$ and not of individual values.

The definition of marginal distribution extends to any entry x_i of a random vector x of arbitrary dimension L by summing or integrating – by **marginalizing** – over all the values of all the constituent random variables except x_i. Denoting as x_{-i} the vector x that excludes the entry x_i, the marginal of random variable x_i is hence computed as

$$p(x_i) = \sum_{x_{-i} \in \mathcal{X}^{L-1}} p(x) \text{ for discrete random variables, and}$$

$$= \int p(x) dx_{-i} \text{ for continuous random variables,} \tag{2.86}$$

where the sum and the integral are over the domain of vector x_{-i}.

Table 2.2 A joint pmf $p(x_1, x_2)$

$x_1\backslash x_2$	0	1
0	0.45	0.05
1	0.1	0.4

Figure 2.7 Marginal pdfs for a two-dimensional jointly Gaussian pdf, which is illustrated through its contour lines.

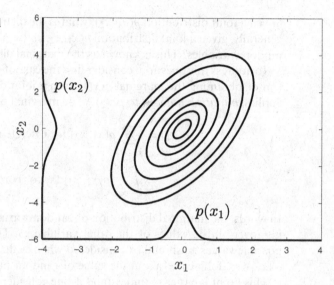

Example 2.8

Let us compute the marginal $p(x_1 = 0)$ for the jointly Bernoulli random vector with joint pmf in Table 2.2. Using the formula (2.84), we have $p(x_1 = 0) = 0.45 + 0.05 = 0.5$, i.e., $x_1 \sim \text{Bern}(0.5)$.

Example 2.9

For a two-dimensional jointly Gaussian random vector $x \sim \mathcal{N}(\mu, \Sigma)$ with joint pdf as in (2.76), the marginals can be obtained as $x_1 \sim \mathcal{N}(\mu_1, \sigma_1^2)$ and $x_2 \sim \mathcal{N}(\mu_2, \sigma_2^2)$. Therefore, the marginals of a multivariate Gaussian pdf are single-variate Gaussian pdfs. This is illustrated in Fig. 2.7.

2.9   Conditional Distributions

As we have seen, marginal distributions disregard the variables that are not of interest: To compute the marginal $p(x_i = x_i)$ from the joint distribution $p(x)$ of an L-dimensional random vector, we sum or integrate over all configurations with $x_i = x_i$, irrespective of the value of the other variables x_{-i}, via (2.86). The marginal distribution can hence be estimated, following a frequentist interpretation, by dropping the vector x_{-i} from all samples x. To introduce the starkly different concept of conditional distribution, let us focus again first on the case $L = 2$.

The **conditional distribution** of x_1 given $\mathrm{x}_2 = x_2$, denoted as $p(\mathrm{x}_1 = x_1 | \mathrm{x}_2 = x_2)$, provides the answer to the question: If $\mathrm{x}_2 = x_2$, what is the probability distribution of x_1 at value x_1? Let us understand this question using the frequentist interpretation of probability. Assume as usual that we have access to many samples (x_1, x_2). Now, since we condition on the event $\mathrm{x}_2 = x_2$, we should only keep samples (x_1, x_2) with $\mathrm{x}_2 = x_2$. All other samples that do not comply with the assumed condition $\mathrm{x}_2 = x_2$ are thus irrelevant to the aim of addressing this question, so that the kept samples are of the form $(*, x_2)$ with any ("don't care") value of x_1 and $\mathrm{x}_2 = x_2$. The conditional pmf $p(x_1 | x_2)$ measures the fraction of times that we observe only $\mathrm{x}_1 = x_1$ among the samples we have retained. We can write this as the approximation

$$p(x_1 | x_2) \simeq \frac{\text{number of samples } (x_1, x_2)}{\text{number of samples } (*, x_2)}. \tag{2.87}$$

So, the value of x_2 is not at all disregarded as when computing marginal distributions. On the contrary, it determines which configurations are relevant to the evaluation of the conditional distribution of interest.

Example 2.10

As of January 2021, the fraction of people that were vaccinated against COVID-19 in the 10 countries with the largest number of vaccinations was around 1.5 in 100. We can therefore write the conditional probability

$$p(\text{vaccinated} | \text{top-ten countries}) \simeq 0.015. \tag{2.88}$$

In contrast, if we condition on specific countries, the fractions of vaccinated citizens vary widely, e.g.,

$$p(\text{vaccinated} | \text{Israel}) \simeq 0.38 \tag{2.89}$$

and

$$p(\text{vaccinated} | \text{UK}) \simeq 0.08. \tag{2.90}$$

(Editor's note: things have changed somewhat in subsequent months.)

Mathematically, the conditional distribution is defined as

$$
\begin{aligned}
p(x_1 | x_2) &= \frac{p(x_1, x_2)}{p(x_2)} \\
&= \frac{p(x_1, x_2)}{\sum_{x_1 \in \mathcal{X}} p(x_1, x_2)} \text{ for discrete random variables, and} \\
&= \frac{p(x_1, x_2)}{\int_{-\infty}^{\infty} p(x_1', x_2) dx_1'} \text{ for continuous random variables.}
\end{aligned}
\tag{2.91}
$$

We write

$$(\mathrm{x}_1 | \mathrm{x}_2 = x_2) \sim p(x_1 | x_2) \tag{2.92}$$

to indicate that random variable x_1 follows distribution $p(x_1|x_2)$ when conditioned on the event $x_2 = x_2$.

The definition (2.91) can be readily understood by using the frequentist interpretation of probability. In fact, for discrete random variables, we have the approximation (2.69) for the joint pmf and (2.85) for the marginal pmf, and hence we can write

$$p(x_1|x_2) = \frac{p(x_1, x_2)}{p(x_2)} \simeq \frac{\text{number of samples } (x_1, x_2)}{\text{number of samples } (*, x_2)}, \tag{2.93}$$

which matches the approximation (2.87) we have used to introduce the notion of conditional distribution.

Example 2.11

For the jointly Bernoulli random vector with pmf $p(x_1, x_2)$ in Table 2.2, what is the conditional distribution $p(x_1 = 0|x_2 = 1)$? Using the formula (2.91), we get

$$p(x_1 = 0|x_2 = 1) = \frac{p(x_1 = 0, x_2 = 1)}{p(x_2 = 1)} = \frac{0.05}{0.05 + 0.4} = 0.11, \tag{2.94}$$

i.e., $(x_1|x_2 = 1) \sim \text{Bern}(0.89)$.

Conditional vs. Marginal Distribution. By (2.91), computing a conditional distribution $p(x_1|x_2)$ involves evaluating the marginal $p(x_2)$ of the conditioning variable. So, evaluating the marginal distribution of a random variable requires us to sum over all values of the *other* variable; while computing a conditional distribution requires us to sum over all values of the variable itself (for a fixed value of the other variable).

Multivariate Gaussian Random Vector. For a two-dimensional jointly Gaussian random vector $x \sim \mathcal{N}(\mu, \Sigma)$, the conditional distribution can be obtained from (2.91) as

$$(x_1|x_2 = x_2) \sim \mathcal{N}\left(\mu_1 + \frac{\sigma_{12}}{\sigma_2^2}(x_2 - \mu_2), \sigma_1^2 - \frac{\sigma_{12}^2}{\sigma_2^2}\right). \tag{2.95}$$

Therefore, for Gaussian random vectors, the conditional distributions are also Gaussian. This is illustrated in Fig. 2.8, which shows the contour lines of the joint pdf (gray), along with the conditional distributions $p(x_2|x_1)$ for $x_1 = -1, 0,$ and 1.

It is useful to interpret the conditional distribution in the context of the problem of **predicting** the value of random variable x_1 given the observation $x_2 = x_2$ of random variable x_2. Prediction is a central concern of machine learning, and it will be the focus of much of Part II of this book.

Given $x_2 = x_2$, a natural prediction for x_1, assuming that the joint distribution $p(x_1, x_2)$ is known, is given by the mean of the conditional distribution (2.95), also known as the **conditional mean**. From (2.95), the conditional mean is given as

$$\mathrm{E}_{x_1 \sim p(x_1|x_2)}[x_1] = \mu_1 + \frac{\sigma_{12}}{\sigma_2^2}(x_2 - \mu_2)$$

$$= \mu_1 + \rho\frac{\sigma_1}{\sigma_2}(x_2 - \mu_2), \tag{2.96}$$

Figure 2.8 Conditional pdfs $p(x_2|x_1)$ for $x_1 = -1, 0$, and 1 for a two-dimensional jointly Gaussian pdf, whose contour lines are shown in gray.

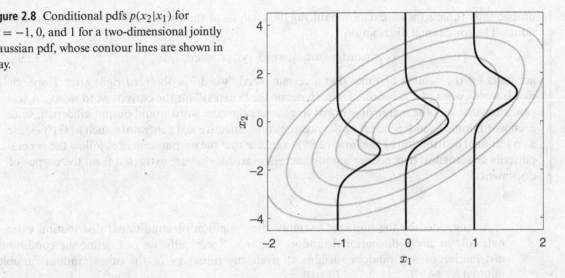

where in the second line we have used the definition of the covariance coefficient ρ in (2.81). The prediction of x_1 via the conditional mean (2.96) is in line with the discussion in Sec. 2.7.3 regarding the significance of the covariance coefficient:

- **Uncorrelated variables.** If the two variables are uncorrelated, i.e., $\rho = 0$, the predictor (2.96) equals the mean μ_1 of random variable x_1, indicating that the random variable x_2 does not provide useful information about x_1.

- **Positively correlated variables.** If the two variables are positively correlated, i.e., $\rho > 0$, the prediction of random variable x_1 deviates from its mean μ_1 to a degree that increases proportionally to the corresponding deviation $x_2 - \mu_2$ of random variable x_2.

- **Negatively correlated variables.** If the two variables are negatively correlated, i.e., $\rho < 0$, the prediction of random variable x_1 deviates from its mean μ_1 to a degree that decreases proportionally to the corresponding deviation $x_2 - \mu_2$ of random variable x_2.

The **conditional variance** – i.e., the variance of the conditional distribution (2.95) –

$$E_{x_1 \sim p(x_1|x_2)}[(x_1 - E_{x_1 \sim p(x_1|x_2)}[x_1])^2] = \sigma_1^2 - \frac{\sigma_{12}^2}{\sigma_2^2} = \sigma_1^2(1 - \rho^2) \qquad (2.97)$$

decreases as the square of the covariance coefficient ρ increases. This indicates that the prediction (2.96) is more accurate when the covariance coefficient is larger in absolute value.

Example 2.12

As a final example of the difference between conditional and marginal distributions, consider a natural language generator such as GPT-3. The marginal distribution $p(\text{word})$ of a certain word, evaluated over a corpus of documents, accounts for the frequency with which the word appears in the documents. For instance, words such as "and" and "the" tend to have the largest marginal probabilities (unless they are removed during the analysis of the corpus of documents, as is often done). The conditional distribution

$$p(\text{word}_i | \text{word}_{i-1}, \text{word}_{i-2}, \dots, \text{word}_{i-T}) \tag{2.98}$$

accounts for the number of times that a certain word "word$_i$" is observed right after T specific words word$_i$, word$_{i-1}$, ..., word$_{i-T}$, which define the **"context"** for the current word word$_i$. A text generator that uses the marginal $p(\text{word})$ to generate the next word would output gibberish, since each word would be independent of the previous one. Effective text generators, such as GPT-3, use a conditional distribution of the form (2.98) to ensure that the output sentences follow the general patterns of statistical dependence among successive words that are extracted from the corpus of documents.

As suggested by the previous example, the definition of conditional distribution extends naturally to any L-dimensional random vector x. Specifically, we can define the conditional distribution of any random variable x$_i$ given any subset x$_{\mathcal{I}}$ of the other random variables indexed by a set $\mathcal{I} \subseteq \{1, 2, \dots, L\} \setminus \{i\}$, as

$$p(x_i | x_{\mathcal{I}}) = \frac{p(x, x_{\mathcal{I}})}{p(x_{\mathcal{I}})}. \tag{2.99}$$

2.10 Independence of Random Variables and Chain Rule of Probability

Independent Random Variables. Random variables in a jointly distributed random vector are independent if their joint distribution can be written as a product of the marginal distributions as

$$p(x_1, x_2, x_3, \dots, x_N) = p(x_1)p(x_2)p(x_3) \cdots p(x_N) = \prod_{i=1}^{N} p(x_i). \tag{2.100}$$

Equivalently, given any subset $\mathcal{I} \subseteq \{1, 2, \dots, N\} \setminus \{i\}$ of variables other than variable x$_i$, the conditional distribution of x$_i$ equals the marginal, i.e.,

$$p(x_i | x_{\mathcal{I}}) = p(x_i), \tag{2.101}$$

where $x_{\mathcal{I}}$ represents the vector that includes the entries x_i with $i \in \mathcal{I}$. This can again be interpreted in terms of **prediction**: Knowing the value of one or more random variables is not useful for predicting a random variable that is independent of them.

Using this definition, jointly Gaussian random variables can easily be seen to be independent if their covariance matrix is diagonal, that is, if they are mutually uncorrelated. Note that zero correlation does not imply independence in general, but this implication holds true for jointly Gaussian random variables.

Example 2.13

The jointly Bernoulli random variables with pmf in Table 2.3 are independent as can be directly checked by applying the definition (2.100).

Table 2.3 A joint pmf $p(x_1, x_2)$

$x_1 \backslash x_2$	0	1
0	0.09	0.81
1	0.01	0.09

Table 2.4 A joint pmf $p(x_1, x_2)$

$x_1 \backslash x_2$	0	1
0	0.25	0.25
1	0.25	0.25

Independent and Identically Distributed (i.i.d.) Random Variables. Random variables are said to be independent and identically distributed (i.i.d.) if they are independent (cf. (2.100)) and the marginal distributions $p(x_i)$ are identical.

Example 2.14

If x_1 and x_2 are i.i.d. Bern(0.5) random variables, what is the joint distribution? Using (2.100), we get $p(x_1, x_2) = p(x_1)p(x_2) = \text{Bern}(x_1|0.5) \cdot \text{Bern}(x_2|0.5)$, which is described in Table 2.4.

Chain Rule of Probability. Generalizing (2.100) to any random vector, the **chain rule of probability** stipulates that the joint distribution can be factorized as

$$p(x_1, x_2, x_3, \ldots, x_N) = p(x_1)p(x_2|x_1)p(x_3|x_1, x_2) \cdots p(x_N|x_1, x_2, \ldots, x_{N-1})$$

$$= \prod_{i=1}^{N} p(x_i|x_1, \ldots, x_{i-1}). \tag{2.102}$$

This relationship can be easily derived by using recursively the definition of conditional distribution (2.91), which can be rewritten as

$$p(x_1, x_2) = p(x_1)p(x_2|x_1). \tag{2.103}$$

The chain rule (2.102) indicates that one can describe arbitrary statistical dependencies among random variables by considering one random variable at a time, and including for each random variable x_i the conditional distribution $p(x_i|x_1, \ldots, x_{i-1})$. The rule (2.102) can be applied with any ordering of the random variables. The following example helps to make sense of this equality.

Example 2.15

Consider again Example 2.12. The joint distribution of the words in a text, with each word defined by a categorical random variable x_i (the number of words is finite for all practical purposes), can be described via the chain rule (2.102). In it, the conditional distribution $p(x_i|x_1, \ldots, x_{i-1})$ specifies

Table 2.5 A conditional pmf $p(x_2|x_1)$

$x_1\backslash x_2$	0	1
0	0.9	0.1
1	0.2	0.8

how word x_i depends on *all* prior words $x_1, \ldots, x_{i-1}$. In practical models, as seen in Example 2.12 (cf. (2.98)), one generally limits the memory of the model to a context of T prior words.

Example 2.16

If $x_1 \sim \text{Bern}(0.5)$ and $p(x_2|x_1)$ is given as in Table 2.5, with each row representing the corresponding conditional distribution, what is the joint distribution? Using the chain rule, we get $p(x_1, x_2) = p(x_1)p(x_2|x_1)$, which yields the joint distribution in Table 2.2.

2.11 Bayes' Theorem

Bayes' theorem is one of the central tools in machine learning, underlying the problem of **inference** (Chapter 3) – that is, of predicting one random variable from another given their joint distribution – as well as the framework of **Bayesian learning** (Chapter 12).[1]

The fundamental underlying problem is that of predicting random variable x_2 given an observation $x_1 = x_1$ of a statistically dependent random variable x_1 when their joint distribution $p(x_1, x_2)$ is known. The joint distribution $p(x_1, x_2) = p(x_2)p(x_1|x_2)$ is determined by two probability distributions:

- **Prior distribution**: The marginal $p(x_2)$ of the target variable x_2 can be thought of as representing the state of knowledge about x_2 *prior* to the observation of x_1. A more spread-out marginal $p(x_2)$ indicates a larger **prior uncertainty** about the true value of x_2, while a more concentrated marginal indicates a smaller prior uncertainty.
- **Likelihood function**. The conditional distribution $p(x_1|x_2)$ of the observation x_1 given the target variable x_2 can be used to model the measurement process: It indicates how *likely* it is to observe $x_1 = x_1$ when $x_2 = x_2$. For instance, if the observation x_1 is a noisy version of the target variable x_2, the relationship between x_1 and x_2 can be modeled by choosing a conditional $p(x_1|x_2)$ that accounts for measurement noise (see Sec. 3.4).

Before stating Bayes' theorem, it is useful to observe that the prior distribution does not typically have a frequentist interpretation. This is in the sense that it typically reflects a state of knowledge – that of the predictor of x_2 before observing x_1. In such situations, the prior is said to account for the **epistemic uncertainty** of the predictor prior to the observation of x_1. In contrast, the likelihood often has a frequentist interpretation, according to which it captures the fraction of times that an observation x_2 is made when $x_1 = x_1$ holds over a large run of measurements of the two variables.

Bayes' theorem relates the prior and the likelihood to the **posterior distribution** $p(x_2|x_1)$, which is the distribution of the target variable conditioned on the observation. The posterior

---

[1] Bayes' theorem was actually formulated in its current form by Pierre Simon Laplace. Laplace's first essay on the subject was entitled "Mémoire on the Probability of the Causes Given Events", which gives a good idea of what Bayes' theorem is about.

distribution $p(x_2|x_1)$ describes the state of uncertainty about x_2 of the predictor after observing $x_1 = x_1$. Like the prior, the posterior is typically interpreted in terms of epistemic uncertainty, capturing the state of the knowledge of the predictor of x_2 after observing $x_1 = x_1$.

Bayes' theorem amounts to the fundamental equality

$$\underbrace{p(x_2|x_1)}_{\text{posterior distribution}} = \underbrace{p(x_2)}_{\text{prior distribution}} \cdot \underbrace{\frac{p(x_1|x_2)}{p(x_1)}}_{\text{likelihood ratio}}, \qquad (2.104)$$

which can be easily proved by using the definition of conditional distribution twice (this is left as an exercise).

Bayes' theorem (2.104) describes how the prior of random variable x_2 is modified as a result of the observation of random variable x_1:

- If x_1 is independent of x_2, the likelihood ratio $p(x_1|x_2)/p(x_1)$ equals 1, and the posterior equals the prior: The prediction of x_2 cannot make use of the observation of an independent variable x_1.
- If the likelihood ratio is larger than 1, i.e., if $p(x_1|x_2) > p(x_1)$, observing $x_1 = x_1$ is more likely when $x_2 = x_2$ than it is a priori, when we have no information about x_2. Therefore, the observation $x_1 = x_1$ lends evidence to the possibility that we have $x_2 = x_2$. Accordingly, the posterior distribution of the value x_2 is increased by a factor equal to the likelihood ratio.
- Finally, if $p(x_1|x_2) < p(x_1)$, the opposite is true, and the observation $x_1 = x_1$ reduces the predictor's belief that we have $x_2 = x_2$. Accordingly, the posterior distribution of the value x_2 is decreased by a factor equal to the likelihood ratio.

Overall, we can think of the **likelihood ratio** as quantifying the evidence that $x_2 = x_2$ based on the observation of $x_1 = x_1$. Bayes' theorem says that the probability of the event $x_2 = x_2$ given the observation of $x_1 = x_1$ is larger or smaller than its corresponding prior by a multiplicative factor given by the likelihood ratio. Broadly speaking, Bayes' theorem provides a tool to extend logical arguments regarding the inference of causes from effects to a probabilistic setting that accounts for the uncertainty associated with prior knowledge of the causes and with the available data.

Example 2.17

As a simple intuitive example, consider the problem of detecting whether a certain event is happening or not based on the observation of a binary alarm signal. For instance, the event being monitored may be the presence of a burglar in the premises monitored by an alarm system. To set up the problem, we introduce a target variable x_2 as a Bernoulli variable indicating whether the event being detected is actually on ($x_2 = 1$) or not ($x_2 = 0$). The observation is modeled as a jointly distributed Bernoulli random variable x_1 that equals 1 if the alarm is active or 0 if the alarm is not reporting any anomaly.

The prior $p(x_2)$ represents the probability, or belief, that the event being monitored occurs at any given time in the absence of additional information. The presence of a burglar in one's home should be rather unlikely, and so one would assign a low prior probability $p(x_2 = 1)$. Say that the alarm goes off – we observe $x_1 = 1$. How has the probability that a burglar is in the premises ($x_2 = 1$) changed?

This change depends on the alarm, and specifically on the likelihood ratio $p(x_1 = 1|x_2 = 1)/p(x_1 = 1)$. If the alarm tends to go off a lot – irrespective of whether the monitored event is actually happening – the marginal $p(x_1 = 1)$ will be large and close to the conditional probability $p(x_1 = 1|x_2 = 1)$ of the alarm sounding for the legitimate reason of informing about an actual ongoing event. Accordingly, the likelihood ratio $p(x_1 = 1|x_2 = 1)/p(x_1 = 1)$ will be close to 1.

Therefore, by Bayes' theorem, hearing the alarm (again!) does not change our belief on x_2 by much since the alarm is not reliable (it "cries wolf" too often to be believable).

Example 2.18

For a numerical example in the setting of Example 2.17, let us specify prior $x_2 \sim \text{Bern}(0.5)$ – there is a 50% chance of the monitored event being on; and likelihood defined by $(x_1|x_2 = 0) \sim \text{Bern}(0.1)$ – the alarm is not too likely to activate when the event does not occur – and $(x_1|x_2 = 1) \sim \text{Bern}(0.8)$ – the alarm is more likely to activate when the event is on. With these choices, using Bayes' theorem, we have the posterior

$$p(x_2 = 1|x_1 = 1) = p(x_2 = 1)\frac{p(x_1 = 1|x_2 = 1)}{p(x_1 = 1)}$$
$$= 0.5 \cdot \frac{0.8}{0.5 \cdot 0.1 + 0.5 \cdot 0.8} = 0.5 \cdot \frac{0.8}{0.45} = 0.89. \qquad (2.105)$$

The observation of $x_1 = 1$ has thus increased the probability of the event $x_2 = 1$, since the likelihood ratio $0.8/0.45 = 1.78$ is larger than 1.

Example 2.19

According to Public Health England (PHE), the best antibody tests for COVID-19 are 70% **sensitive** (if taken at least 14 days after infection) and 98% **specific**. Based on this, if I get a positive result, how sure can I be that it is correct?

First of all, let us translate the sensitivity and specificity of the test into conditional probabilities in order to use Bayes' theorem. The fact that the test is 70% sensitive indicates that we have the conditional probability $\text{Pr}(\text{positive}|\text{COVID}) = 0.7$ of obtaining a positive test when infected by the virus; while the 98% specificity means that we have $\text{Pr}(\text{negative}|\text{no COVID}) = 0.98$. To answer the question, we need to evaluate the posterior probability

$$\text{Pr}(\text{COVID}|\text{positive}) = \underbrace{\text{Pr}(\text{COVID})}_{\text{prior}} \cdot \underbrace{\frac{\text{Pr}(\text{positive}|\text{COVID})}{\text{Pr}(\text{positive})}}_{\text{likelihood ratio}}. \qquad (2.106)$$

To do this, we need to specify the prior. Note that the prior also affects the likelihood ratio through the denominator $\text{Pr}(\text{positive})$.

To this end, suppose first that you are in London and that you have all the symptoms. Then, it is likely, even before you take the test, that you have COVID. For this case, let us set the prior as $\text{Pr}(\text{COVID}) = 0.5$. Then, by Bayes' theorem, we have

$$\text{Pr}(\text{COVID}|\text{positive}) = \underbrace{0.5}_{\text{prior}} \cdot \underbrace{\frac{0.7}{0.5 \cdot 0.7 + 0.5 \cdot 0.02}}_{\text{likelihood ratio}}$$
$$= 0.5 \cdot 1.94 = 0.97. \qquad (2.107)$$

Suppose instead that you have been shielding in Cornwall and have no symptoms. For this case, let us set the prior as $\text{Pr}(\text{COVID}) = 0.05$. Then, by Bayes' theorem, we have

$$\text{Pr}(\text{COVID}|\text{positive}) = \underbrace{0.05}_{\text{prior}} \cdot \underbrace{\frac{0.7}{0.05 \cdot 0.7 + 0.95 \cdot 0.02}}_{\text{likelihood ratio}}$$

$$= 0.05 \cdot 12.96 = 0.65. \tag{2.108}$$

The answer hence depends strongly on the state of prior knowledge.

2.12   Law of Iterated Expectations

The law of iterated expectations provides a useful way to compute an expectation over multiple random variables. The idea is that the expectation is computed over one variable at a time in a manner that follows the chain rule of probability.

Consider first the case of two jointly distributed random variables. According to the law of iterated expectations, in order to compute the expectation $\mathrm{E}_{(x_1,x_2) \sim p(x_1,x_2)}[f(x_1,x_2)]$ of some function $f(x_1,x_2)$ over a distribution $p(x_1,x_2) = p(x_2)p(x_1|x_2)$, we can proceed in two steps:

1. First, we compute the conditional average

$$\mathrm{E}_{x_1 \sim p(x_1|x_2)}[f(x_1,x_2)] = \sum_{x_1 \in \mathcal{X}_1} p(x_1|x_2)f(x_1,x_2) = F(x_2) \tag{2.109}$$

   for all values x_2 – we denote this conditional average as the function $F(x_2)$. In (2.109), the expectation is expressed by assuming a discrete rv x_1, but a similar expression applies for continuous rvs by replacing the sum with an integral.

2. Then, we average the conditional expectation over $x_2 \sim p(x_2)$, obtaining

$$\mathrm{E}_{(x_1,x_2) \sim p(x_1,x_2)}[f(x_1,x_2)] = \mathrm{E}_{x_2 \sim p(x_2)}[F(x_2)]. \tag{2.110}$$

The order of the random variables can also be inverted (see the following example), and the procedure naturally extends to an arbitrary number of variables by leveraging in a similar manner the chain rule of probability.

Example 2.20

For the joint distribution $p(x_1,x_2)$ with $x_1 \sim \text{Bern}(0.5)$, $(x_2|x_1 = 0) \sim \text{Bern}(0.1)$, and $(x_2|x_1 = 1) \sim \text{Bern}(0.8)$, let us compute the average $\mathrm{E}_{(x_1,x_2) \sim p(x_1,x_2)}[x_1 x_2]$ by using the law of iterated expectations. Note that, in this simple example, we can easily compute the expectation without using the law of iterated expectations, as

$$\mathrm{E}_{(x_1,x_2) \sim p(x_1,x_2)}[x_1 x_2] = 0.45 \cdot 0 + 0.1 \cdot 0 + 0.05 \cdot 0 + 0.4 \cdot 1 = 0.4. \tag{2.111}$$

We apply the law of iterated expectations by averaging first over x_2 and then over x_1. Accordingly, in the first step, we compute the conditional expectations

$$F(0) = \mathrm{E}_{x_2 \sim p(x_2|x_1=0)}[x_1 x_2] = \mathrm{E}_{x_2 \sim p(x_2|x_1=0)}[0 \cdot x_2] = 0,$$
$$F(1) = \mathrm{E}_{x_2 \sim p(x_2|x_1=1)}[x_1 x_2] = \mathrm{E}_{x_2 \sim p(x_2|x_1=1)}[1 \cdot x_2] = 0.8, \tag{2.112}$$

and in the second step we finally obtain

$$\mathrm{E}_{x_1 \sim p(x_1)}[F(x_1)] = 0.5 \cdot 0 + 0.5 \cdot 0.8 = 0.4, \tag{2.113}$$

as in (2.111).

2.13   Summary

- In this chapter, we have reviewed basic notions of probability and linear algebra. The tools and concepts developed here will find application in the rest of the book.
- Jointly distributed random vectors are described by joint distributions – pmfs or pdfs – from which one can evaluate expectations, as well as marginal and conditional distributions.
- When a frequentist interpretation is applicable, probabilities, and more generally expectations, can be estimated as empirical averages over independent random draws of rvs.
- The variance captures the spread of a random variable when measured in terms of the average quadratic distance to its mean.
- Linear algebra operations – most notably, inner products and matrix–vector multiplications – play a key role in machine learning. For instance, inner products can be used to quantify the "similarity" between two vectors.
- Symmetric matrices can be described in terms of their eigendecomposition. All positive semi-definite matrices have only non-negative eigenvalues, while all positive definite matrices have only positive eigenvalues.
- Marginal distributions only reflect the distribution of a subset of variables when the others are neglected, while conditional distributions are obtained by fixing the value of a subset of variables.
- Bayes' theorem provides a principled way to describe the effect of observations on the uncertainty of a random variable by incorporating prior knowledge. Bayes' theorem is typically employed in inferential problems, in which probabilities account for states of epistemic uncertainty.

2.14   Recommended Resources

There are many excellent textbooks on both linear algebra and probability. For an entry-level book on probability, good choices include [1] and [2], while for linear algebra one can recommend [3], as well as the classical text [4]. A historical account of the many lives of Bayes' theorem can be found in [5].

Problems

2.1   Consider a Bernoulli random variable x $\sim$ Bern(0.2). (a) Draw the pmf. (b*) Generate $N = 10$ independent realizations of random variable x. (c*) Use these samples to estimate the probability $p(x = 1)$. (d*) Repeat the previous two points using $N = 1000$ samples. (e) Comment on your results.

2.2   Consider a categorical (or multinoulli) random variable x $\sim$ Cat($[0.2, 0.1, 0.3, 0.4]^T$). (a) Draw the pmf. (b*) Generate $N = 10$ independent realizations of random variable x. (c*) Using these samples, estimate the probabilities $q_k = p(x = k)$ for $k = 0, 1, 2, 3$. (d*) Repeat the previous two points using $N = 1000$ samples. (e) Comment on your results.

2.3   Consider a Gaussian random variable x $\sim$ $\mathcal{N}(-3, 4)$. (a) Draw the pdf. (b*) Generate $N = 10$ independent realizations of random variable x. (c*) Using these samples estimate the probability $\Pr[x \in (-3, 3)]$. (d*) Repeat the previous two points using $N = 1000$ samples. (e) Comment on your results.

2.4 (a) Given a random variable x $\sim$ Cat($q = [0.2, 0.1, 0.3, 0.4]^T$), compute the expectation

$$E_{x \sim \text{Cat}(q)}[x^2 + 3 \exp(x)].$$

(b*) Verify your calculation using an empirical estimate obtained by drawing random samples.

2.5 (a) Given a random variable x $\sim \mathcal{N}(-3, 4)$, compute the expectation

$$E_{x \sim \mathcal{N}(-3, 4)}[x + 3x^2].$$

(b*) Verify your calculation using an empirical estimate obtained by drawing random samples.

2.6 Plot the variance of Bernoulli random variable Bern(p) as a function of p. What is the value of p that maximizes uncertainty?

2.7 Compute the variance of random variable x $\sim$ Cat($q = [0.2, 0.1, 0.3, 0.4]^T$).

2.8 Given a categorical random variable x $\sim$ Cat(q) with $q = [0.2, 0.1, 0.3, 0.4]^T$, what is the expectation $E_{x \sim \text{Cat}(q)}[\mathbb{1}(x = 0)]$? For a Gaussian random variable x $\sim \mathcal{N}(x|0, 1)$, what is the expectation $E_{x \sim \mathcal{N}(x|0, 1)}[\mathbb{1}(x = 0)]$?

2.9 Consider vectors $x = [2, 1]^T$ and $y = [-1, 3]^T$. (a) Represent the vectors in the two-dimensional plane $\mathbb{R}^2$. (b) Compute their inner product. (c) Compute the squared ℓ_2 norms of the two vectors. (d) Compute the cosine of the angle between the two vectors. (e) Are the two vectors linearly independent? (f) Determine a vector that is orthogonal to x. (g) Normalize vector y so that it has unitary norm. (h) Find a vector such that the cosine of the angle with x equals 1, and a vector with cosine equal to -1. (i) Compute the element-wise product $x \odot y$. (l) Plot all the vectors determined in the previous points.

2.10 Consider matrices

$$A = \begin{bmatrix} 1 & -1 \\ -1 & 2 \end{bmatrix} \text{ and } B = \begin{bmatrix} 2 & 3 & -1 \\ -1 & 2 & 3 \end{bmatrix}.$$

(a) Compute the product AB. (b) Compute the product $B^T A^T$. (c) Compute the product Diag($[1, 2]^T$)B. (d) Is A symmetric? (e*) If it is symmetric, evaluate eigenvectors and eigenvalues of A. (f*) Is A positive definite, i.e., is $A \succ 0$? (g*) Plot the quadratic form $x^T A x$ as a function of vector $x = [x_1, x_2]^T$ for $x_1 \in [-2, 2]$ and $x_2 \in [-2, 2]$. (h) Is BB^T positive definite? Is it invertible?

2.11 Consider the jointly Bernoulli random vector x $= [x_1, x_1]^T$ with the joint pmf $p(x_1, x_2)$ given by Table 2.6. (a) Compute the marginal $p(x_2)$ and the conditional distribution $p(x_2|x_1 = 1)$. (b*) Generate $N = 100$ realizations of the random vector x in the previous problem and estimate the probabilities of each of the four configurations of the outputs. (c*) Repeat with $N = 10,000$ and discuss your results.

2.12 For a jointly Gaussian random vector x $= [x_1, x_2]^T$ with mean vector $\mu = [\mu_1, \mu_2]^T$ and covariance

$$\Sigma = \begin{bmatrix} \sigma_1^2 & \sigma_{12} \\ \sigma_{12} & \sigma_2^2 \end{bmatrix},$$

Table 2.6 A joint pmf $p(x_1, x_2)$

$x_1 \backslash x_2$	0	1
0	0.45	0.05
1	0.1	0.4

prove the second equality in

$$\sigma_{12} = E_{x \sim \mathcal{N}(\mu, \sigma^2)}[(x_1 - \mu_1)(x_2 - \mu_2)] = E_{x \sim \mathcal{N}(\mu, \sigma^2)}[x_1 x_2] - \mu_1 \mu_2.$$

2.13 Consider a jointly Gaussian random vector with an all-zero mean vector and covariance

$$\Sigma = \begin{bmatrix} 2 & -1 \\ -1 & 2 \end{bmatrix}.$$

(a) Compute the covariance coefficient ρ. (b) Verify that the covariance matrix is positive definite. (c) Evaluate the expectation $E_{x \sim \mathcal{N}(0, \Sigma)}[(x_1 + x_2)^2]$. (d) Evaluate the expectation $E_{x \sim \mathcal{N}(0, \Sigma)}[(x_1 - x_2)^2]$ and compare with your result in the previous point. (e) Modify the covariance σ_{12} so that $E_{x \sim \mathcal{N}(0, \Sigma)}[(x_1 + x_2)^2] = 0$. (f) Modify the covariance σ_{12} so that $E_{x \sim \mathcal{N}(0, \Sigma)}[(x_1 - x_2)^2] = 0$.

2.14 (a*) Produce 3D plots for a jointly Gaussian pdf with mean vector $\mu = [0, 0]^T$ and the same covariance in Problem 2.13. (b*) Repeat for covariance

$$\Sigma = \begin{bmatrix} 2 & -1.9 \\ -1.9 & 2 \end{bmatrix}.$$

(c*) Repeat for mean vector $\mu = [5, 7]^T$ and either one of the two covariances above.

2.15 (a*) Produce contour plots for the multivariate Gaussian pdfs considered in Problem 2.14.

2.16 For a jointly Gaussian random vector, interpret the formula for the conditional distribution (2.95) in terms of prediction of x_1 given an observation $x_2 = x_2$. How is the formula simplified when $\sigma_1 = \sigma_2$ and $\mu_1 = \mu_2 = 0$?

2.17 Consider a jointly Gaussian vector with all-zero mean vector and covariance matrix Σ defined by $\sigma_1 = \sigma_2 = 1$ and $\sigma_{12} = -0.1$. (a) For the linear predictor $\hat{x}_2 = ax_1$ given some real number a, compute the mean squared error

$$E_{x \sim \mathcal{N}(0, \Sigma)}[(\hat{x}_2 - x_2)^2]$$

as a function of a. (b) Then, optimize over a by equating the derivative with respect to a to zero. (c) How does this solution compare with the conditional mean ρx_2 considered in Problem 2.16?

2.18 Using the formula for the joint pdf of a two-dimensional jointly Gaussian vector, show that, if the covariance is zero, i.e., if $\sigma_{12} = 0$, then the two variables are independent.

2.19 Suppose that 1% of women suffer from breast cancer, and that a test for the condition is 90% accurate but reports false positives 9% of the time. If a woman tests positive, what is the chance that she has the disease? The most common answer among a sample of doctors (as reported by Steven Pinker) is 80–90%, but the true answer is just 9%. (a) Argue that the correct answer is indeed 9% by picturing 1000 women, 10 of whom (1%) suffer from breast cancer. Reflecting the rates of error of the test, 9 of the 10 women with the condition are correctly diagnosed, while around 89 healthy individuals are incorrectly diagnosed. Considering now only the individuals with a positive test, what is the fraction that has the disease? (It should be around 9%). (b) Confirm your results by applying Bayes' theorem.

Table 2.7 A joint pmf $p(x_1, x_2)$

$x_1 \backslash x_2$	0	1
0	0.6	0.1
1	0.1	0.2

2.20 We have a test for screening cancer that is 90% sensitive (i.e., Pr(positive|cancer) = 0.9) and 90% specific (i.e., Pr(negative|no cancer) = 0.9). (a) Assuming that 1% of the population has cancer, what is the fraction of positive tests that correctly detects cancer? (b) What happens if the test is 100% specific? (c) For the case in part (b), would your answer change if the test were less sensitive?

2.21 For the joint distribution $p(x_1, x_2)$ in Table 2.7, compute the average

$$\mathrm{E}_{(x_1, x_2) \sim p(x_1, x_2)}[x_1 x_2 + x_2^2]$$

using the law of iterated expectations.

Bibliography

[1] S. Kay, *Intuitive Probability and Random Processes Using MATLAB*. Springer Science+Business Media, 2006.

[2] R. D. Yates and D. J. Goodman, *Probability and Stochastic Processes: A Friendly Introduction for Electrical and Computer Engineers*. John Wiley & Sons, 2014.

[3] G. Strang, *Linear Algebra and Learning from Data*. Wellesley-Cambridge Press, 2019.

[4] G. H. Golub and C. F. Van Loan, *Matrix Computations*. The Johns Hopkins University Press, 2013.

[5] S. McGrayne, *The Theory That Would Not Die: How Bayes' Rule Cracked the Enigma Code, Hunted Down Russian Submarines, and Emerged Triumphant from Two Centuries of Controversy*. Yale University Press, 2011.

Part II

Fundamental Concepts and Algorithms

3 Inference, or Model-Driven Prediction

3.1 Overview

As discussed in Chapter 2, learning is needed when a "physics"-based mathematical model for the data generation mechanism is not available or is too complex to use for design purposes. As an essential benchmark setting, this chapter discusses the ideal case in which an accurate mathematical model *is* known, and hence learning is not necessary. As in large part of machine learning, we specifically focus on the problem of **prediction**. The goal is to predict a target variable given the observation of an input variable based on a mathematical model that describes the joint generation of both variables. Model-based prediction is also known as **inference**.

Along the way, this chapter will introduce key concepts such as loss function, population loss, optimal hard and soft predictors, log-loss, cross-entropy loss, and cross entropy, as well as the information-theoretic metrics of entropy, conditional entropy, KL divergence, mutual information, and free energy.

Learning Objectives and Organization of the Chapter. By the end of this chapter, the reader should be able to:

- understand the definition of inference as the optimization of hard and soft predictors (Sec. 3.2);
- formulate the optimization of hard predictors, and understand the concepts of loss function and population loss (Sec. 3.3);
- evaluate optimal inference for jointly Gaussian random variables (Sec. 3.4);
- understand and interpret the information-theoretic metrics of KL divergence and cross entropy, which are functionals of two distributions (Sec. 3.5);
- formulate the problem of optimizing soft prediction via the minimization of the population log-loss, and understand the definitions of entropy and conditional entropy – two important information-theoretic measures related to the performance of optimal soft predictors (Sec. 3.6);
- understand and interpret the related metric of mutual information (Sec. 3.7);
- interpret the log-loss as a "universal" loss function (Sec. 3.8); and
- understand the free energy as an information-theoretic metric for the definition of optimal inference problems (Sec. 3.9).

About the Notation. Before we start, two remarks about the notation used in this book:

- **Random variables and random vectors**. Throughout this book, we will use the abbreviation **rv** to refer to both random vectors and random variables. Random variables are special cases of random vectors with a single element, and this notation will allow us to keep the presentation general.

- **Optimization**. Given a function $f(z)$ of some (discrete or continuous) vector (or scalar) quantity z, we write $\min_z f(z)$ to indicate the problem of minimizing the function over z; and $\arg\min_z f(z)$ to denote the set of optimal solutions of the minimization problem $\min_z f(z)$. Note that we will often keep the optimization domain implicit when it encompasses the entire domain of variable z. We will also write $z^* = \arg\min_z f(z)$ to indicate that z^* is one of the minimizers of function $f(z)$. (A more precise notation would be $z^* \in \arg\min_z f(z)$.) We refer to Chapter 5 for more discussion on optimization.

3.2   Defining Inference

3.2.1   Input, Output, and Population Distribution

In inference problems, we have two types of variables:

- an **input** rv x, and
- an **output**, or **target**, rv t.

Both input and output are generally vectors with multiple elements. The general problem of inference consists in designing a predictor of the target rv t given input rv x based on a model of the relationship between these two rvs.

The **model** is specified by the **true joint distribution** $p(x, t)$ of the two rvs, which is also referred to as the **population distribution**, for reasons that will be made clear in the next chapter (see Sec. 4.2.3). The population distribution $p(x, t)$ fully describes the mechanism that generates joint realizations of rvs x and t.

In learning problems, the population distribution is not known, and only data in the form of input–output realizations are available; while in inference problems the population distribution is known and data are not required. (Note, in fact, that, based on the model $p(x, t)$, one can potentially generate an arbitrary amount of data samples for rvs x and t.) The material in this chapter hence provides the necessary background for the rest of the book by dealing with the ideal, reference, case in which an accurate model – the population distribution $p(x, t)$ – for the variables of interest is available.

3.2.2   Detection and Estimation

Inference can be classified into two distinct, but related, problems:

- **Detection**. In detection problems, the target t takes values in a discrete and finite set.
- **Estimation**. In estimation problems, the target rv t is continuous.

In both cases, the input rv x can be discrete, continuous, or a mixture of both.

Example 3.1

In this chapter, we will often consider the following bare-bones version of a **weather prediction** problem. In it, the input variable x represents the weather on the current day, and the target variable t represents the weather on the following day. Both rvs are assumed to be binary, with value "0" indicating a rainy day and value "1" a sunny day. Since the target is discrete, this is a **detection problem**.

Table 3.1 Population distribution for the weather prediction problem (Example 3.1)

$x \backslash t$	0	1
0	0.45	0.05
1	0.1	0.4

This being an instance of inference, we have access to the population distribution $p(x, t)$ for the two variables. We may have obtained it, for instance, from textbooks, experts, or measurements. For this example, the population distribution is given by the joint pmf specified by Table 3.1. According to this model, for any two consecutive days, we have a probability of 45% that they are both rainy; 40% that they are both sunny; 5% that a rainy day is followed by a sunny day; and 10% that a sunny day is followed by a rainy day.

3.2.3 Hard and Soft Predictors

The predictor, for both detection and estimation problems, may be one of two different types:

- **Hard predictors**. A hard predictor specifies a single predicted value $\hat{t}$ of the target variable t for every observation x $= x$. We will write $\hat{t}(x)$ for the predictive function mapping input x to prediction $\hat{t}(x)$. Hard predictors are also known as **point predictors**.
- **Soft predictors**. A soft predictor specifies a "score" for each possible value of the target t given input x $= x$. We will write $q(t|x)$ for the function returning the score of value t for a given value x. This is assumed to be a probability distribution for rv t given any fixed values x, i.e., $q(t|x)$ is a conditional distribution. Specifically, the soft predictor $q(t|x)$ is a conditional pmf in detection problems (where rv t is discrete), and a conditional pdf in estimation problem (where rv t is continuous).The problem of optimizing a soft predictor $q(t|x)$ based on the joint distribution $p(x, t)$ is known as **Bayesian inference.**

Example 3.1 (continued)

For the weather prediction example, a possible **hard predictor** is defined as follows: When $x = 0$, predict $\hat{t} = 0$; and when $x = 1$, predict $\hat{t} = 1$. This corresponds to the hard predictive function $\hat{t}(x)$ defined as $\hat{t}(0) = 0$ and $\hat{t}(1) = 1$.

Moreover, a possible **soft predictor** is as follows: When $x = 0$, output the scores 0.8 for $t = 0$ and 0.2 for $t = 1$; and when $x = 1$, the scores are 0.3 for $t = 0$ and 0.7 for $t = 1$. This corresponds to the conditional $q(t|x)$ defined as $q(1|0) = 0.2$ (and hence $q(0|0) = 0.8$) and $q(1|1) = 0.7$ (and hence $q(0|1) = 0.3$). These hard and soft predictors appear reasonable in light of the fact that the weather conditions across successive days are likely to be the same under the model in Table 3.1. We will later see that the described soft predictor, as may be guessed from Table 3.1, is not optimal.

The scores $q(t|x)$ given an input x quantify the **uncertainty** associated with any possible (hard) prediction t. For instance, if we have $q(1|1) = 0.7$ as in Example 3.1, the soft predictor assigns

Table 3.2 Posterior distribution $p(t|x)$ **for Example 3.1**

$x \backslash t$	0	1
0	0.9	$0.05/(0.45 + 0.05) = 0.1$
1	0.2	$0.4/(0.1 + 0.4) = 0.8$

a 30% chance that the prediction $\hat{t} = 1$ for $x = 1$ is incorrect. Therefore, a soft predictor offers more information than a hard predictor by providing **"error bars"** for any given possible prediction. An optimal soft predictor should specifically report a level of uncertainty that depends on the randomness of the relationship between rvs x and t that is encoded by the model $p(x, t)$.

One can always turn a soft predictor $q(t|x)$ into a hard predictor $\hat{t}(x)$ – but not vice versa. This can be done, for instance, by predicting the value t that has the largest score according to the soft predictor, i.e.,

$$\hat{t}(x) = \arg\max_t q(t|x). \tag{3.1}$$

Note that there may be multiple values of t that are assigned the maximum score, in which case this notation is intended as identifying any one of the maximizers (see Sec. 3.1).

A natural choice for the soft predictor $q(t|x)$ of t given x $= x$ is the **posterior distribution** $p(t|x)$. We will indeed see later in this chapter (Sec. 3.6) that setting $q(t|x) = p(t|x)$ is optimal in a specific mathematical sense. For this reason, the soft predictor given by the posterior, i.e., $q(t|x) = p(t|x)$, is also known as the **optimal Bayesian predictor** or **optimal soft predictor**.

Example 3.1 (continued)

For the joint pmf in Table 3.1, the optimal soft predictor $p(t|x)$ is computed by using the definition of posterior distribution as in Table 3.2; i.e., we have

$$(t|x = 0) \sim q(t|0) = p(t|0) = \text{Bern}(t|0.1)$$
$$\text{and } (t|x = 1) \sim q(t|1) = p(t|1) = \text{Bern}(t|0.8). \tag{3.2}$$

This soft predictor can be used to draw the following conclusions: If x $= 0$, we can offer the hard prediction $\hat{t}(0) = 0$ with associated probability of error of 0.1; and if x $= 1$, we can offer the hard prediction $\hat{t}(1) = 1$ with associated probability of error of 0.2. Note that we have chosen the hard predictor (3.1) that maximizes the posterior probability, and hence minimizes the probability of error.

3.3   Optimal Hard Prediction

In the preceding example, given the population distribution, we have chosen the hard predictor (3.1), which was argued to minimize the probability of error. How do we formalize and generalize the optimal design of hard predictors? This is the subject of this section, which will cover the key concepts of loss function, population loss, and optimal hard prediction.

3.3.1 Loss Function

In order to define the problem of designing a hard predictor, it is necessary to measure the quality of a prediction $\hat{t}$ when the correct value for the target variable is t. This is done by introducing a **loss function**

$$\ell(t, \hat{t}) \tag{3.3}$$

that quantifies the "loss" accrued when one predicts $\hat{t}$ and the true target variable is t. Note that the loss function is fixed and not subject to optimization.

We will assume without loss of generality that the loss function satisfies the two conditions

zero loss for exact prediction: $\ell(t, \hat{t}) = 0$ if $t = \hat{t}$

and **non-negativity**: $\ell(t, \hat{t}) \geq 0$. $\tag{3.4}$

That is, the loss is non-negative, and it equals zero if the prediction is exact (it may also equal zero if the prediction is not exact).

The following choices of loss functions are standard and will be used throughout this book:

- **Estimation**. For estimation, one often adopts the ℓ_k loss for some integer $k \geq 1$, which is defined as

$$\ell(t, \hat{t}) = \ell_k(t, \hat{t}) = |t - \hat{t}|^k. \tag{3.5}$$

  The most common choice for the exponent is $k = 2$, which yields the **quadratic loss** $\ell_2(t, \hat{t}) = (t - \hat{t})^2$.

- **Detection**. For detection, the typical choice is the **detection-error loss**, which is defined as

$$\ell(t, \hat{t}) = \mathbb{1}(t \neq \hat{t}), \tag{3.6}$$

where we have used the indicator function (2.20). That is, the detection-error loss returns 1 in case of a prediction error, and 0 otherwise. Moving beyond the detection-error loss, one can more generally give different weights to distinct error events $\{t \neq \hat{t}\}$, as illustrated by the following example.

Example 3.1 (continued)

For the weather prediction problem in Example 3.1, one would use the detection-error loss if the two errors – predicting sun for a rainy day and rain for a sunny day – were deemed to be equally problematic. Suppose, instead, that predicting sunshine ($\hat{t} = 1$) on a day that turns out to be rainy ($t = 0$) is deemed to be 10 times worse than predicting rain ($\hat{t} = 0$) on a sunny day ($t = 1$). We can represent these preferences in the loss function $\ell(t, \hat{t})$ represented in Table 3.3, in which the first error produces a loss equal to 1, while the second yields a loss of 0.1.

3.3.2 Population Loss

The loss function $\ell(t, \hat{t})$ measures how well a prediction $\hat{t}$ reflects the true value t. In order to design a hard predictor $\hat{t}(\cdot)$, we need a way to gauge how well the *entire* predictive function $\hat{t}(\cdot)$

Table 3.3 Example of an alternative loss function $\ell(t, \hat{t})$ for a binary target variable t

$t\backslash\hat{t}$	0	1
0	0	1
1	0.1	0

fits the relationship between input and target variables described by the population distribution $p(x, t)$. A natural idea is to consider the average loss over the joint distribution $p(x, t)$ of input and target. This yields the **population loss** as

$$L_p(\hat{t}(\cdot)) = \mathrm{E}_{(x,t)\sim p(x,t)}[\ell(t, \hat{t}(x))]. \tag{3.7}$$

Observe that in (3.7) we have used the notation $\hat{t}(\cdot)$ to emphasize that the population loss depends on the entire function $\hat{t}(x)$ obtained by varying x, and that the subscript "p" stands for "population".

As we will formalize in the next subsection, the goal of optimal hard prediction is minimizing the population loss over the predictive function $\hat{t}(\cdot)$.

The Population Detection-Error Loss is the Probability of Detection Error. For detection problems, the population detection-error loss equals the **probability of error** $\mathrm{Pr}_{(x,t)\sim p(x,t)}[t \neq \hat{t}(x)]$. Therefore, a hard predictor that minimizes the population detection-error loss also minimizes the probability of error. To see this, let us for simplicity consider the case in which rv x is also discrete, although the result applies also for continuous inputs. Under this assumption, the probability of error can be written as

$$\mathrm{Pr}_{(x,t)\sim p(x,t)}[t \neq \hat{t}(x)] = 1 - \mathrm{Pr}_{(x,t)\sim p(x,t)}[t = \hat{t}(x)] = 1 - \sum_x p(x, \hat{t}(x)), \tag{3.8}$$

where the term $\sum_x p(x, \hat{t}(x))$ is the **probability of correct detection** for predictor $\hat{t}(x)$. Now, under the detection loss, the population loss (3.7) is computed as

$$L_p(\hat{t}(\cdot)) = \mathrm{E}_{(x,t)\sim p(x,t)}[\mathbb{1}(\hat{t}(x) \neq t)] = \sum_{x,t} p(x, t)\mathbb{1}(\hat{t}(x) \neq t)$$

$$= \sum_{x, t\neq\hat{t}(x)} p(x, t) = 1 - \sum_x p(x, \hat{t}(x)), \tag{3.9}$$

which coincides with the probability of error (3.8), as claimed.

3.3.3 Optimal Hard Prediction

For a given loss function, a hard predictor is said to be optimal if it minimizes the population loss. Mathematically, an **optimal hard predictor** $\hat{t}^*(\cdot)$ satisfies the condition

$$\hat{t}^*(\cdot) = \arg\min_{\hat{t}(\cdot)} L_p(\hat{t}(\cdot)). \tag{3.10}$$

Note that there may be multiple optimal hard predictors, all achieving the same **minimum population loss** $L_p(\hat{t}^*(\cdot))$.

Example 3.1 (continued)

Let us return again to the weather prediction problem and adopt the detection-error loss. What is the population loss for the (not very reasonable in light of Table 3.1) hard predictor $\hat{t}(0) = 1$ and $\hat{t}(1) = 1$? As discussed, the population loss is equal to the probability of error (3.8), which is computed as

$$L_p(\hat{t}(\cdot)) = 0.45 \times \underbrace{\mathbb{1}(\hat{t}(0) \neq 0)}_{=1} + 0.05 \times \underbrace{\mathbb{1}(\hat{t}(0) \neq 1)}_{=0}$$

$$+ 0.1 \times \underbrace{\mathbb{1}(\hat{t}(1) \neq 0)}_{=1} + 0.4 \times \underbrace{\mathbb{1}(\hat{t}(1) \neq 1)}_{=0} = 0.55. \tag{3.11}$$

Is this the optimal hard predictor? For any predictor $\hat{t}(\cdot)$, we can write the population loss as

$$L_p(\hat{t}(\cdot)) = 0.45 \times \mathbb{1}(\hat{t}(0) = 1) + 0.05 \times \mathbb{1}(\hat{t}(0) = 0)$$

$$+ 0.1 \times \mathbb{1}(\hat{t}(1) = 1) + 0.4 \times \mathbb{1}(\hat{t}(1) = 0), \tag{3.12}$$

which we need to optimize over the predictions $\hat{t}(0)$ and $\hat{t}(1)$ in order to find the optimal hard predictor (3.10). Specifically, to obtain the optimal prediction $\hat{t}^*(0)$, we need to minimize the first two terms in (3.12). This yields $\hat{t}^*(0) = 0$, since we have $p(x = 0, t = 0) = 0.45 > p(x = 0, t = 1) = 0.05$. The optimal prediction $\hat{t}^*(1)$ is computed by minimizing the last two terms, which yields $\hat{t}^*(1) = 1$, since $p(x = 1, t = 1) = 0.4 > p(x = 1, t = 0) = 0.1$. The corresponding minimum population loss is

$$L_p(\hat{t}^*(\cdot)) = 0.05 + 0.1 = 0.15. \tag{3.13}$$

In the preceding example, the optimal hard predictor was obtained by minimizing the population loss separately for each value x of the input. We will now see that this procedure can be generalized to any loss function $\ell(t, \hat{t})$, for both estimation and detection problems. Specifically, we will conclude that the optimal hard prediction $t^*(x)$ can be directly computed from the optimal soft predictor, i.e., the posterior

$$p(t|x) = \frac{p(x, t)}{p(x)}, \tag{3.14}$$

separately for each value x.

Using the law of iterated expectations (see Sec. 2.12), the population loss (3.7) can be written as

$$L_p(\hat{t}(\cdot)) = \mathrm{E}_{x \sim p(x)}[\mathrm{E}_{t \sim p(t|x)}[\ell(t, \hat{t}(x))]] \tag{3.15}$$

by taking first the inner expectation with respect to the optimal soft predictor $p(t|x)$ and then the outer expectation with respect to the input marginal $p(x)$. It follows that the optimal hard prediction $\hat{t}^*(x)$ for any given value x can be obtained by minimizing the inner expectation as

$$\hat{t}^*(x) = \arg\min_{\hat{t}} \mathrm{E}_{t \sim p(t|x)}[\ell(t, \hat{t})]. \tag{3.16}$$

So, as anticipated, the optimal hard prediction can be computed separately for each value x, and it is a function of the corresponding optimal soft predictor $p(t|x)$.

Example 3.1 (continued)

Let us apply the general formula (3.16) to this recurring example. By (3.16), under the detection-error loss, the optimal prediction $\hat{t}^*(0)$ is given as

$$\hat{t}^*(0) = \arg\min_{\hat{t} \in \{0,1\}} \mathrm{E}_{t \sim p(t|0)}[\mathbb{1}(t \neq \hat{t})], \tag{3.17}$$

where we have

$$\mathrm{E}_{t \sim p(t|0)}[\mathbb{1}(t \neq \hat{t})] = 0.9 \times \mathbb{1}(\hat{t} \neq 0) + 0.1 \times \mathbb{1}(\hat{t} \neq 1). \tag{3.18}$$

Therefore, the optimal prediction is $\hat{t}^*(0) = 0$, which yields the conditional probability of error $\mathrm{E}_{t \sim p(t|0)}[\mathbb{1}(t \neq 0)] = 0.1$, when conditioned on x = 0. In a similar way, we have $\hat{t}^*(1) = 1$ (this is left as an exercise for the reader).

We now detail the solution of the optimal hard prediction problem (3.16) for three important special cases. The first is the baseline scenario when x and t are independent, while the second and third settings correspond to the use of the ℓ_2 loss and the detection-error loss, respectively.

3.3.4 Optimal Hard Prediction When x and t Are Independent

As a special case, suppose that x and t are independent, i.e., $p(x,t) = p(x)p(t)$ (see Sec. 2.10). What is the optimal hard predictor of t given x? In this case, using x to predict t is not useful, since the posterior distribution $p(t|x)$ equals the marginal $p(t)$:

$$p(t|x) = \frac{p(x,t)}{p(x)} = \frac{p(x)p(t)}{p(x)} = p(t). \tag{3.19}$$

Therefore, the only information we can use to predict t is the distribution $p(t)$, and the optimal predictor is a constant $\hat{t}^*$ that does not depend on the value of x, i.e.,

$$\hat{t}^* = \arg\min_{\hat{t}} \left\{ L_p(\hat{t}) = \mathrm{E}_{t \sim p(t)}[\ell(t, \hat{t})] \right\}. \tag{3.20}$$

We will see examples of the application of (3.20) to the ℓ_2 loss and the detection-error loss in the next two subsections.

3.3.5 Optimal Hard Prediction under the Quadratic Loss

In this subsection, we study the important case of the ℓ_2 loss (quadratic loss), and start by assuming that x and t are independent. We will then generalize the design of the optimal hard predictor to any joint distribution $p(x,t)$.

When x and t are Independent. When x and t are independent, using (3.20), the optimal hard predictor is the constant

$$\hat{t}^* = \arg\min_{\hat{t}} \left\{ L_p(\hat{t}) = \mathrm{E}_{t \sim p(t)}[(t - \hat{t})^2] \right\} = \mathrm{E}_{t \sim p(t)}[t], \tag{3.21}$$

where the minimization can be carried out by taking the derivative of the population loss and equating it to zero (see Chapter 5 for further details). By (3.21), the optimal hard prediction under the ℓ_2 loss is the **mean** of the distribution $p(t)$. This result is in line with intuition: If we only know the distribution $p(t)$, the constant predictor that minimizes the squared error is the expectation of rv $t \sim p(t)$.

The corresponding **minimum population loss** – that is, the population loss of the optimal predictor – is

$$
\begin{aligned}
L_p(\hat{t}^*) &= E_{t\sim p(t)}[(t - \hat{t}^*)^2] \\
&= E_{t\sim p(t)}[(t - E_{t\sim p(t)}[t])^2] \\
&= \text{Var(t)},
\end{aligned}
\tag{3.22}
$$

which is the **variance** of distribution $p(t)$.

General Case. In the more general case in which x and t are statistically dependent, problem (3.16) can be solved in an analogous way for each fixed value x, yielding the **posterior mean**

$$
\hat{t}^*(x) = E_{t\sim p(t|x)}[t]
\tag{3.23}
$$

as the optimal hard predictor. Note that, if x and t are independent, the posterior mean reduces to the mean in (3.21).

The corresponding minimum population loss is the **average posterior variance**

$$
\begin{aligned}
L_p(\hat{t}^*(\cdot)) &= E_{(x,t)\sim p(x,t)}[(t - \hat{t}^*(x))^2] \\
&= E_{(x,t)\sim p(x,t)}[(t - E_{t\sim p(t|x)}[t])^2] \\
&= E_{x\sim p(x)}[\underbrace{E_{t\sim p(t|x)}[(t - E_{t\sim p(t|x)}[t])^2]}_{=\text{Var(t|x)}}] \\
&= E_{x\sim p(x)}[\text{Var(t|x)}],
\end{aligned}
\tag{3.24}
$$

where we have used the law of iterated expectations and we have defined the **conditional variance**, or **posterior variance**, Var(t|x).

3.3.6 Optimal Hard Prediction under the Detection-Error Loss

We now turn to studying optimal hard prediction under the detection-error loss.

When x and t are Independent. Under the detection-error loss, when x and t are independent, using (3.20), the optimal hard predictor is the constant

$$
\hat{t}^* = \arg\min_{\hat{t}} \left\{ L_p(\hat{t}) = E_{t\sim p(t)}[\mathbb{1}(t \neq \hat{t})] \right\} = \arg\max_t p(t),
\tag{3.25}
$$

as can be easily proved in a manner similar to the derivation of the optimal hard predictor in Example 3.1. So, the optimal hard prediction is the **mode** of the distribution $p(t)$. Note that the mode is generally distinct from the mean, which is the optimal prediction under the ℓ_2 loss.

General Case. Generalizing the case of independent input and target variables, the optimal hard predictor (3.16) is given by the **posterior mode**

$$
\hat{t}^*(x) = \arg\max_t p(t|x).
\tag{3.26}
$$

This is referred to as the **maximum a posterior (MAP) predictor**, and is also written as

$$
\hat{t}^{MAP}(x) = \arg\max_t p(t|x).
\tag{3.27}
$$

Following Sec. 3.3.2, the corresponding **minimum population loss** is the **minimum probability of error**, i.e.,

$$L_p(\hat{t}^*(\cdot)) = \mathrm{E}_{(\mathrm{x,t})\sim p(x,t)}[\mathbb{1}(\mathrm{t} \neq \hat{t}^*(\mathrm{x}))] = \mathrm{E}_{\mathrm{x}\sim p(x)}[\mathrm{E}_{\mathrm{t}\sim p(t|\mathrm{x})}[\mathbb{1}(\mathrm{t} \neq \hat{t}^*(\mathrm{x}))]]$$

$$= \mathrm{E}_{\mathrm{x}\sim p(x)}\left[\sum_{t\neq \hat{t}^*(x)} p(t|\mathrm{x})\right] = 1 - \mathrm{E}_{\mathrm{x}\sim p(x)}\left[p(\hat{t}^*(\mathrm{x})|\mathrm{x})\right]$$

$$= 1 - \mathrm{E}_{\mathrm{x}\sim p(x)}\left[\max_t p(t|\mathrm{x})\right]. \tag{3.28}$$

Example 3.1 (continued)

For the joint pmf in Table 3.1, under the quadratic loss, the optimal hard predictor for input $x = 0$ is given by the posterior mean

$$\hat{t}^*(0) = 0.9 \times 0 + 0.1 \times 1 = 0.1, \tag{3.29}$$

while under the detection-error loss, the optimal hard predictor is the MAP predictor

$$\hat{t}^*(0) = 0. \tag{3.30}$$

We leave the optimal hard predictor for $x = 1$ as an exercise.

3.4   Optimal Prediction for Jointly Gaussian Random Vectors

In this section, we study optimal soft and hard prediction for the important special case of Gaussian rvs. We specifically consider the standard estimation problem in which one would like to predict a Gaussian target rv t based on an observation x subject to additive Gaussian noise. We first investigate the case in which both variables are scalar quantities, and then we extend the analysis to Gaussian vectors.

3.4.1   The Case of Jointly Gaussian Random Variables

Adopting the terminology first introduced in the context of Bayes' theorem in Sec. 2.11, we view the joint distribution $p(x,t) = p(t)p(x|t)$ as the combination of a prior distribution $p(t)$ and a likelihood $p(x|t)$. In this subsection, we specifically assume:

- the **Gaussian prior** $\mathrm{t} \sim \mathcal{N}(v, \alpha^{-1})$, where v is the prior mean and α represents the **precision**, i.e., the inverse of the variance, of the prior distribution; and
- the **Gaussian likelihood** $(\mathrm{x}|\mathrm{t} = t) \sim \mathcal{N}(t, \beta^{-1})$ for the noisy observation x, which can be equivalently described as

$$\mathrm{x} = \mathrm{t} + \mathrm{z}, \text{ with } \mathrm{z} \sim \mathcal{N}(0, \beta^{-1}), \tag{3.31}$$

where the observation noise z is independent of t, and β is the precision of the observation. (The equivalence follows from the reparametrization property (2.12).)

This situation is quite common. It arises when one has prior knowledge, with precision α, about a value v that is most probable for the target t, as well as a measurement x of the target variable t that is affected by Gaussian additive noise as per (3.31).

In a manner similar to (2.95), the optimal soft predictor – the posterior distribution – can be computed as

$$p(t|x) = \mathcal{N}\left(t \left| \frac{\alpha v + \beta x}{\alpha + \beta}, \frac{1}{\alpha + \beta} \right. \right). \tag{3.32}$$

Interestingly, the precision $\alpha + \beta$ of the posterior (3.32) is the sum of the precisions of prior and likelihood – an intuitive result. Furthermore, by (3.32), if α is much larger than β, i.e., if the precision of the prior is much larger than that of the observation, we have the approximation $p(t|x) \simeq \mathcal{N}(t|v, \alpha^{-1}) = p(t)$, and hence the observation x is effectively irrelevant and can be discarded. On the other hand, if the precision of the observation, β, is much larger than that of the prior, α, the approximation $p(t|x) \simeq \mathcal{N}(t|x, \beta^{-1})$ holds, and hence the prior becomes irrelevant.

An illustration of prior, likelihood (as contour lines), and posterior can be found in Fig. 3.1. Note that the likelihood $p(x|t)$ should be read for any fixed value t as a function of x. The contour lines clearly illustrate the mean t, as well as the spread dictated by the variance β^{-1} of the likelihood $p(x|t)$. Figure 3.2 displays the effect of the choice of precisions of prior and likelihood on the posterior. It is seen that, when the precision of the prior, α, increases, the posterior mean tends to move towards the prior mean and the variance decreases. Furthermore, when the precision of the likelihood, β, increases, the posterior mean tends to move towards the observation x and the variance also decreases.

By (3.32), under the ℓ_2 loss, the optimal hard predictor is given by the posterior mean

$$\hat{t}^*(x) = \frac{\alpha v + \beta x}{\alpha + \beta}, \tag{3.33}$$

and the minimum population loss is given by the average posterior variance

$$E_{x \sim p(x)}[\text{Var}(t|x)] = \text{Var}(t|x) = \frac{1}{\alpha + \beta}. \tag{3.34}$$

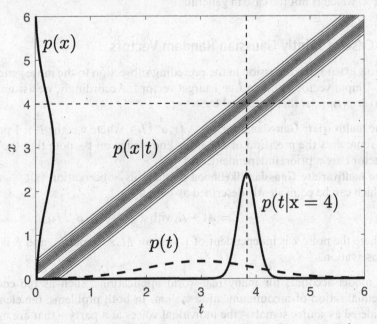

Figure 3.1 Prior $p(t)$, contour lines of the likelihood $p(x|t)$, and posterior $p(t|\text{x} = 4)$ for two-dimensional jointly Gaussian variables. (The marginal $p(x)$ is also shown.)

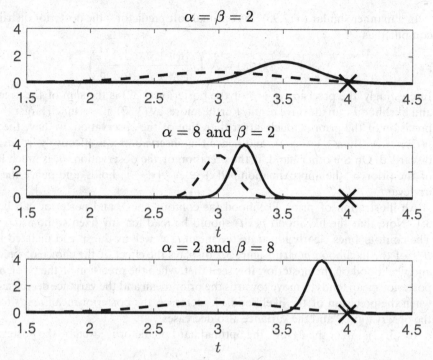

Figure 3.2 Illustration of the effect of the precision of the prior (α) and the likelihood (β) on the posterior distribution for jointly Gaussian variables. The prior is shown as a dashed line, the posterior is a solid line, and the observation ($x = 4$) is represented as a cross.

Note that, for this example, as highlighted in (3.34), the posterior variance is the same for every value x, which is not the case in general.

3.4.2 The Case of Jointly Gaussian Random Vectors

We now extend the discussion in the preceding subsection to the more general case involving an $M \times 1$ input vector x and an $L \times 1$ target vector t. Accordingly, we assume a joint distribution $p(x, t) = p(t)p(x|t)$ that is defined by:

- the **multivariate Gaussian prior** t $\sim \mathcal{N}(v, \alpha^{-1}I_L)$, where v is the $L \times 1$ prior mean vector and α represents the precision of the prior knowledge on t – note that the entries of the target vector t are a priori independent; and
- the **multivariate Gaussian likelihood** of the noisy observation (x|t $= t$) $\sim \mathcal{N}(At, \beta^{-1}I_M)$, which can be equivalently described as

$$x = At + z, \text{ with } z \sim \mathcal{N}(0_M, \beta^{-1}I_M), \tag{3.35}$$

where the noise z is independent of t, A is an $M \times L$ matrix, and β is the precision of the observation.

This model accounts for many real-world applications, such as the **cocktail party problem** and equalization of a communication system. In both problems, the elements of vector t are considered as source signals – the individual voices at a party – that are mixed through matrix

A before being observed as vector x subject to Gaussian noise as per (3.35). The source signals t are assumed to be a priori independent with mean ν and precision α. The aim is to extract the source signals t based solely on the noisy mixed signals x.

Generalizing (3.32), the optimal soft predictor can be computed as

$$p(t|x) = \mathcal{N}(\Theta^{-1}(\alpha\nu + \beta A^T x), \Theta^{-1}), \qquad (3.36)$$

where

$$\Theta = \alpha I_L + \beta A^T A \qquad (3.37)$$

is the precision matrix of the posterior. Note that this distribution reduces to the posterior (3.32) if all quantities are scalar ($L = M = 1$) and we set $A = 1$.

Example 3.1

As a numerical example, we set

$$\nu = \begin{bmatrix} 1 \\ 1 \end{bmatrix}, \alpha = 1, A = \begin{bmatrix} 10 & 0 \\ 0 & 1 \end{bmatrix}, \beta = 2, \text{ and x} = \begin{bmatrix} 25 \\ 2 \end{bmatrix}, \qquad (3.38)$$

and we compute the optimal soft predictor, as well as the optimal hard predictor, under the ℓ_2 loss. To start, let us evaluate the product

$$A^T A = \begin{bmatrix} 100 & 0 \\ 0 & 1 \end{bmatrix}, \qquad (3.39)$$

and the posterior (3.36) as

$$\left(t \middle| \text{x} = \begin{bmatrix} 25 \\ 2 \end{bmatrix}\right) \sim \mathcal{N}\left(\underbrace{\begin{bmatrix} 1/201 & 0 \\ 0 & 1/3 \end{bmatrix}\left(\begin{bmatrix} 1 \\ 1 \end{bmatrix} + \begin{bmatrix} 500 \\ 4 \end{bmatrix}\right)}_{\begin{bmatrix} 501/201 \\ 5/3 \end{bmatrix}}, \begin{bmatrix} 1/201 & 0 \\ 0 & 1/3 \end{bmatrix}\right). \qquad (3.40)$$

The optimal hard predictor is given by the posterior mean vector

$$\hat{t}^*\left(\begin{bmatrix} 25 \\ 2 \end{bmatrix}\right) = \begin{bmatrix} 501/201 \\ 5/3 \end{bmatrix}. \qquad (3.41)$$

3.5   KL Divergence and Cross Entropy

We now pause our investigation of optimal model-based prediction to introduce the most basic, and arguably most important, information-theoretic metric: the **Kullback–Liebler (KL) divergence**. The KL divergence will be used in the next section to formalize the problem of optimal Bayesian inference, and it will also be encountered throughout the book as a key criterion for the definition of learning problems.

3.5.1 Log-Distribution Ratio

To start, consider two distributions $p(t)$ and $q(t)$. How different are they? For a fixed value t, a reasonable way to quantify their dissimilarity – which will turn out to be very useful – is the **log-distribution ratio (LDR)**

$$\log\left(\frac{p(t)}{q(t)}\right), \tag{3.42}$$

i.e., the natural logarithm of the ratio $p(t)/q(t)$. The LDR provides a signed measure of the dissimilarity of the two distribution at t: The LDR is positive if $p(t) > q(t)$; negative if $p(t) < q(t)$; and zero if $p(t) = q(t)$. Furthermore, the magnitude of the LDR is large for values t at which the two distributions differ more significantly. Note that the LDR is well defined only for values t at which $q(t) \neq 0$, although, when both $p(t)$ and $q(t)$ equal zero, we can set the LDR conventionally to 1.

   The LDR is also referred to as the **log-likelihood ratio (LLR)** when emphasizing the connection to classification problems (see Sec. 11.8), as well as the **log-density ratio** when $p(t)$ and $q(t)$ are pdfs.

Example 3.2

For $p(t) = \mathcal{N}(t\,|-1,1)$ and $q(t) = \mathcal{N}(t\,|1,1)$, the two distributions and the LDR (3.42) (rescaled for clarity) are illustrated in Fig. 3.3. As discussed, the LDR is positive for values t that are more probable under distribution $p(t)$; it is negative for values t that are more probable under $q(t)$; and it is zero otherwise.

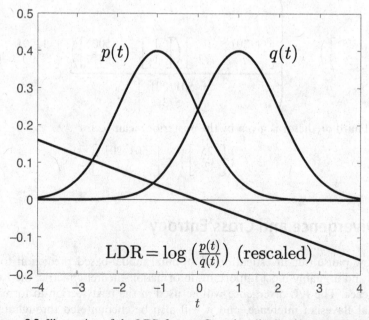

Figure 3.3 Illustration of the LDR for two Gaussian distributions $p(t)$ and $q(t)$.

3.5.2   KL Divergence

As discussed, the LDR (3.42) measures the discrepancy between the two distributions $p(t)$ and $q(t)$ at a specific value t. We would now like to define a single metric that quantifies the overall dissimilarity between the two distributions. The KL divergence provides a standard solution to this problem, and is defined as the average value of the LDR with respect to distribution $p(t)$. In formulas, the KL divergence between two distributions $p(t)$ and $q(t)$ is defined as

$$\mathrm{KL}(p\|q) = \mathrm{E}_{t\sim p(t)}\left[\log\left(\frac{p(\mathrm{t})}{q(\mathrm{t})}\right)\right]. \tag{3.43}$$

We will also write $\mathrm{KL}(p(t)\|q(t))$ when we want to emphasize the notation – here t – for the underlying rvs. Intuitively, if the LDR is small on average – and hence the KL divergence (3.43) is also small – the two distributions can be deemed to be similar in terms of KL divergence. On the other hand, if the LDR is large on average – and hence the KL divergence (3.43) is also large – the discrepancy between the two distributions may be considered as large in terms of KL divergence.

As a measure of distance between two distributions, the KL divergence enjoys some useful properties, but also some peculiarities. Before we delve into these aspects, let us make two technical points on the definition (3.43). (Readers not interested in technicalities can skip this paragraph.) First, when $p(t) = 0$, we take the corresponding term $p(t)\log(p(t)/q(t))$ that appears in the KL divergence to be zero, even when $q(t) = 0$. This can be justified by taking the limit $p(t) \to 0$. Second, the KL divergence is well defined only when, for all values of t for which $q(t) = 0$, we also have $p(t) = 0$. This condition is known as **absolute continuity** of $p(t)$ with respect to $q(t)$. When this condition is violated, one takes the KL divergence to be infinite, i.e., to equal ∞.

Having clarified these technical aspects, let us return to the use of the KL divergence as a measure of discrepancy between two distributions $p(t)$ and $q(t)$.

- **Unit of measure (nats and bits).** The KL divergence is measured in **natural units of information (nats)**. If we were to substitute the natural log with the log in base 2, the KL divergence would be measured in **bits**. We can convert the KL divergence from nats to bits as

$$\mathrm{KL}(p\|q)[\mathrm{bits}] = \log_2(e) \cdot (\mathrm{KL}(p\|q)[\mathrm{nats}])$$
$$= 1.44 \cdot (\mathrm{KL}(p\|q)[\mathrm{nats}]). \tag{3.44}$$

Unless stated otherwise, we will use the natural logarithm and measure all information-theoretic metrics in nats.

- **Non-negativity (Gibbs' inequality).** The KL divergence (3.43) averages the LDR over $p(t)$. Since the LDR tends to be positive when $p(t)$ is large, one may expect $\mathrm{KL}(p\|q)$ to be non-negative – a useful property for a similarity metric. This turns out to be true, and is known as **Gibbs' inequality**: Given two distributions $p(t)$ and $q(t)$, the KL divergence satisfies the inequality

$$\mathrm{KL}(p\|q) \geq 0. \tag{3.45}$$

Furthermore, the equality $\mathrm{KL}(p\|q) = 0$ holds if and only if the two distributions $p(t)$ and $q(t)$ are identical. A proof of this inequality can be found in Appendix 3.A.

- **Lack of symmetry.** Distance measures are symmetric: Swapping the roles of p and q should not change their mutual distance. In contrast, the KL divergence is generally asymmetric, i.e., we have the inequality

$$\mathrm{KL}(p\|q) \neq \mathrm{KL}(q\|p). \tag{3.46}$$

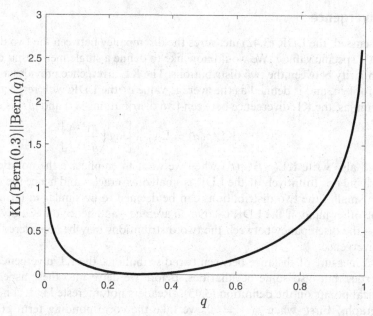

Figure 3.4 KL divergence KL(Bern(t|0.3)||Bern(t|q)) as a function of probability q.

So, in terms of KL divergence, p may be close to q but not vice versa. (For a practical example of an asymmetric "distance" measure, think about the time needed to go from one location to another, as compared to the time required to go back, on an inclined road.)

Example 3.3

For the distributions $p(t) = \text{Bern}(t|p)$ and $q(t) = \text{Bern}(t|q)$, the KL divergence is

$$\text{KL}(p\|q) = p\log\left(\frac{p}{q}\right) + (1-p)\log\left(\frac{1-p}{1-q}\right). \tag{3.47}$$

As seen in Fig. 3.4, the KL divergence KL(Bern(t|0.3)||Bern(t|q)) is zero when $q = 0.3$, and hence the two distribution are identical; and is larger than zero otherwise, growing larger as q moves away from 0.3. Here, and throughout the text, we use both notations KL(Bern(t|p)||Bern(t|q)), emphasizing the pmfs being compared, and KL(Bern(p)||Bern(q)), the latter having the benefit of brevity. We will also apply this convention to other distributions (see the next example).

Example 3.4

For $p(t) = \mathcal{N}(t|\mu_1, \sigma_1^2)$ and $q(t) = \mathcal{N}(t|\mu_2, \sigma_2^2)$, the KL divergence is

$$\text{KL}(p\|q) = \frac{1}{2}\left(\frac{\sigma_1^2}{\sigma_2^2} + \frac{(\mu_1 - \mu_2)^2}{\sigma_2^2} - 1 + \log\left(\frac{\sigma_2^2}{\sigma_1^2}\right)\right). \tag{3.48}$$

Therefore, the KL divergence is positive when either or both means and variances of the two distributions are different. In the special case $\sigma_1^2 = \sigma_2^2 = \sigma^2$, we have

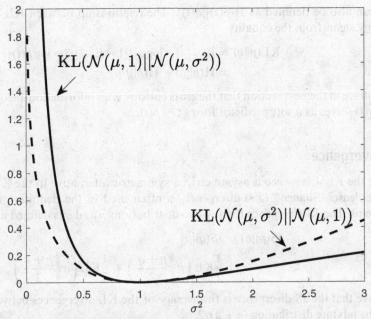

Figure 3.5 KL divergence between Gaussian distributions $KL(\mathcal{N}(\mu,1)\|\mathcal{N}(\mu,\sigma^2))$ and $KL(\mathcal{N}(\mu,\sigma^2)\|\mathcal{N}(\mu,1))$ as a function of σ^2.

$$KL(p\|q) = \frac{1}{2}\frac{(\mu_1 - \mu_2)^2}{\sigma^2}. \tag{3.49}$$

Figure 3.5 shows the two KL divergences $KL(\mathcal{N}(\mu,1)\|\mathcal{N}(\mu,\sigma^2))$ and $KL(\mathcal{N}(\mu,\sigma^2)\|\mathcal{N}(\mu,1))$ for any mean μ as a function of the variance σ^2. Note that the two KL divergences are different because of the lack of symmetry of the KL divergence. Furthermore, thinking of the distribution $\mathcal{N}(\mu,1)$ as being fixed and the distribution $\mathcal{N}(\mu,\sigma^2)$ as being an approximation of $\mathcal{N}(\mu,1)$, we can make the following observations from the figure:

- The KL divergence $KL(\mathcal{N}(\mu,1)\|\mathcal{N}(\mu,\sigma^2))$ penalizes more significantly approximations $\mathcal{N}(\mu,\sigma^2)$ that do not **cover** the given distribution $\mathcal{N}(\mu,1)$, i.e., approximations $\mathcal{N}(\mu,\sigma^2)$ with small values of the variance σ^2.
- In contrast, the KL divergence $KL(\mathcal{N}(\mu,\sigma^2)\|\mathcal{N}(\mu,1))$ "prefers" approximations $\mathcal{N}(\mu,\sigma^2)$ that capture the **mode** of the given distribution $\mathcal{N}(\mu,1)$, while not being too spread out.

This discussion suggests that using either divergence in machine learning problems leads to solutions with different properties in terms of their capacity to match the data distribution.

3.5.3 Cross Entropy

The KL divergence can be expressed as a function of another important information-theoretic quantity, namely the **cross entropy** between distributions p and q, which is defined as

$$H(p\|q) = \mathrm{E}_{t\sim p(t)}[-\log q(t)], \tag{3.50}$$

and can also be denoted as $H(p(t)||q(t))$. The relationship between KL divergence and cross entropy stems from the equality

$$\begin{aligned} KL(p||q) &= E_{t \sim p(t)}[-\log q(t)] - E_{t \sim p(t)}[-\log p(t)] \\ &= H(p||q) - H(p||p). \end{aligned} \tag{3.51}$$

We will see in the next section that the cross entropy is an information-theoretic measure of how well $q(t)$ serves as a soft predictor for rv $t \sim p(t)$.

3.5.4   JS Divergence

While the KL divergence is asymmetric, a symmetric alternative to the KL divergence, known as the **Jensen–Shannon (JS) divergence**, is often used in the definition of machine learning problems. The JS divergence for any two distributions p and q is defined as

$$\begin{aligned} JS(p||q) &= JS(q||p) \\ &= \frac{1}{2} KL\left(p \left\| \frac{p+q}{2} \right.\right) + \frac{1}{2} KL\left(q \left\| \frac{p+q}{2} \right.\right). \end{aligned} \tag{3.52}$$

Observe that the JS divergence is the average of the KL divergences between each distribution and the **mixture distribution** $(p + q)/2$.

By Gibbs' inequality (3.45), the JS divergence satisfies the inequality $JS(p||q) \geq 0$, with equality if and only if p and q are identical. It can also be proved that, unlike the KL divergence, the JS divergence is upper bounded as

$$JS(p||q) \leq \log(2). \tag{3.53}$$

Therefore, the JS divergence does not need the absolute continuity condition required by the KL divergence to be well defined (i.e., to take finite values). The JS divergence can be related to the minimal population loss of a specific detection problem, as detailed in Appendix 3.B.

3.6   Optimal Soft Prediction

Earlier in this chapter, in Sec. 3.2, it was stated that the posterior distribution $p(t|x)$ is the optimal soft predictor $q(t|x)$. A first way to formalize this claim was discussed in Sec. 3.3, where it was shown that the optimal hard predictor can be obtained from the posterior for any loss function. In this section, we provide a different perspective on the optimality of the posterior distribution by demonstrating that the posterior minimizes the population loss for a specific loss function – the log-loss.

3.6.1   Log-Loss, or Cross-Entropy Loss

To start, consider the use of a generic predictive distribution $q(t|x)$ – not necessarily the posterior $p(t|x)$. Can we optimize over $q(t|x)$ to define an optimal soft predictor? To this end, we need to measure the loss accrued by the soft predictor $q(t|x)$ when applied on an input x for which the true value of the target variable is t.

This can be done by using a **scoring rule**, the most popular of which is the **log-loss**, also known as **cross-entropy loss**. A scoring rule should produce a large loss when the soft predictor $q(t|x)$

assigns a low score to the true target value t, and a small loss when the soft predictor $q(t|x)$ assigns a large score to the true target value t. To this end, it should be a decreasing, or at least a non-increasing, function of the soft prediction $q(t|x)$ (for any fixed x and t). The log-loss for a soft predictor $q(t|x)$ on a pair (x, t) is defined as

$$-\log q(t|x). \tag{3.54}$$

Note that we will often, as in (3.54), omit the parentheses that identify the argument of the logarithm when this does not cause confusion. As desired, the log-loss is a strictly decreasing function of the soft prediction $q(t|x)$.

Discrete Target. If the target t is a discrete rv, i.e., if we have a detection problem, the soft predictor $q(t|x)$ defines a conditional pmf and is hence constrained to lie in the interval $[0, 1]$. Therefore, the log-loss $-\log q(t|x)$ is always non-negative ($-\log q(t|x) \geq 0$), and it equals zero ($-\log q(t|x) = 0$) if and only if we have $q(t|x) = 1$. Therefore, a zero log-loss is only obtained when the correct value t is assigned the maximum possible score $q(t|x) = 1$. This is illustrated in Fig. 3.6.

Continuous Target. In contrast, for a continuous target rv t, i.e., for an estimation problem, the soft prediction $q(t|x)$ is a conditional pdf, and is hence non-negative and possibly larger than 1. Think for instance of a Gaussian soft predictor – the Gaussian pdf can take values larger than 1. Therefore, the log-loss $-\log q(t|x)$ can be negative when t is continuous. In practice this is not an issue, since the log-loss is used to compare different predictors, and hence what matters is the relative value between log-losses obtained by the predictors under comparison. The log-loss for an estimation problem is illustrated in Fig. 3.7.

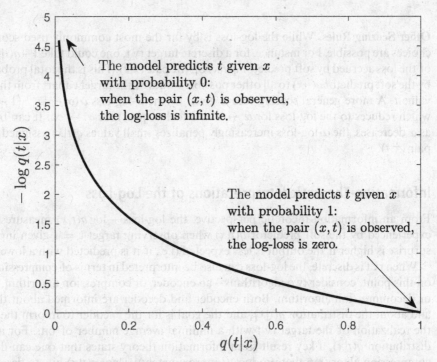

The model predicts t given x with probability 0: when the pair (x, t) is observed, the log-loss is infinite.

The model predicts t given x with probability 1: when the pair (x, t) is observed, the log-loss is zero.

Figure 3.6 Log-loss for a discrete target variable t (detection problem). The soft prediction $q(t|x)$ is in the interval $[0, 1]$, and the log-loss is non-negative.

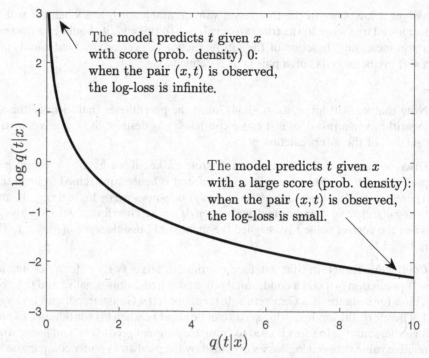

Figure 3.7 Log-loss for a continuous target variable t (estimation problem). The soft prediction $q(t|x)$ is non-negative and it can be larger than 1, and the log-loss can be negative.

Other Scoring Rules. While the log-loss is by far the most commonly used scoring rule, other choices are possible. For instance, for a discrete target rv t, one could take $1 - q(t|x)$ as a measure of the loss accrued by soft predictor $q(t|x)$ on the pair (x, t). This is the total probability assigned by the soft predictor $q(\cdot|x)$ to all other possible values of the target t apart from the true observed value t. A more general scoring rule is given by the α-**log-loss** $\alpha(\alpha - 1)^{-1} \left(1 - q(t|x)^{(1-1/\alpha)}\right)$, which reduces to the log-loss for $\alpha \to 1$ and to $1 - q(t|x)$ for $\alpha \to \infty$. It can be checked that, as α decreases, the α-log-loss increasingly penalizes small values $q(t|x)$ assigned to an observed pair (x, t).

3.6.2 Information-Theoretic Interpretations of the Log-Loss

From an information-theoretic perspective, the log-loss $- \log q(t|x)$ measures the **"surprise"** experienced by the soft predictor $q(t|x)$ when observing target t $= x$ given input x $= x$: The surprise is higher if the output is less expected (i.e., if it is predicted with a lower score).

When rv t is discrete, the log-loss can also be interpreted in terms of **compression**. To elaborate on this point, consider two algorithms – an encoder, or compression algorithm, and a decoder, or decompression algorithm. Both encoder and decoder are informed about the value x $= x$ and about the distribution $q(t|x)$, and the goal is for the encoder to inform the decoder about the realization of the target rv t with a minimal average number of bits. For any conditional distribution $q(t|x)$, a key result from information theory states that one can design a **lossless** compression algorithm that produces (approximately) $- \log_2 q(t|x)$ bits to describe each value t of the target. Values of the target rv t that are less surprising under the soft predictor $q(t|x)$

are assigned fewer bits, while more surprising outputs are assigned more bits. In fact, it turns out that it is not possible to design lossless compressors that produce a smaller number of bits on average that this class of algorithms. For a real-life counterpart of this approach, think of the Morse code, which assigns longer descriptions to letters that occur less frequently and are hence more surprising.

For continuous rvs t, a similar interpretation can be obtained for the compression of a **quantized** version of the target t. See Recommended Resources, Sec. 3.11, for more discussion on the relationship between compression and log-loss.

3.6.3 Population Log-Loss and Optimal Soft Prediction

Following the approach taken in Sec. 3.3 to define optimal hard predictors as the minimizers of the population loss (cf. (3.10)), in the rest of this section we explore the definition of optimal soft predictors through the minimization of the **population log-loss**, or **population cross-entropy loss**. The population log-loss is the average of the log-loss over the population distribution, in a manner akin to the population loss (3.7), and is defined as

$$L_p(q(\cdot|\cdot)) = \mathrm{E}_{(\mathrm{x,t})\sim p(x,t)}[-\log q(\mathrm{t|x})]. \tag{3.55}$$

In (3.55), the notation $q(\cdot|\cdot)$ is used to indicate that the population log-loss depends on the soft predictions $q(t|x)$ for all values of x and t.

A soft predictor $q(t|x)$ is said to be optimal if it minimizes the population log-loss, i.e., an **optimal soft predictor** solves the problem

$$\min_{q(\cdot|\cdot)} L_p(q(\cdot|\cdot)). \tag{3.56}$$

3.6.4 Optimal Soft Prediction When x and t are Independent

To make progress in addressing problem (3.56), in this subsection we first consider the simpler case in which rvs x and t are independent, and hence the only information we have about the target variable t is the marginal distribution $p(t)$. We are therefore interested in optimizing over a soft predictor $q(t)$ that does not depend on x.

Omitting the dependence on x from (3.56), the problem of interest is the minimization of the population log-loss

$$\min_{q(\cdot)} \left\{ L_p(q(\cdot)) = \mathrm{E}_{\mathrm{t}\sim p(t)}[-\log q(\mathrm{t})] \right\}. \tag{3.57}$$

By (3.50), the population log-loss in (3.57) coincides with the cross entropy between $p(t)$ and $q(t)$, i.e.,

$$L_p(q(\cdot)) = \mathrm{H}(p||q). \tag{3.58}$$

This justifies the name "cross-entropy loss" for the log-loss. So, minimizing the population log-loss is equivalent to finding the distribution $q(t)$ that yields the minimal value for the cross entropy $\mathrm{H}(p||q)$.

By Gibbs' inequality (3.45), we have the inequality

$$\mathrm{KL}(p||q) = \mathrm{H}(p||q) - \mathrm{H}(p,p) \geq 0, \tag{3.59}$$

or equivalently $\mathrm{H}(p||q) \geq \mathrm{H}(p||p)$, with equality if and only if the two distributions are identical, i.e., if $p(t) = q(t)$. This demonstrates that the minimum population log-loss – that is, the minimum cross entropy H($p\|q$) – is attained when $q(t) = p(t)$, and hence the marginal $q^*(t) = p(t)$ is the optimal soft predictor.

This result should not come as a surprise: If we have access to the true distribution $p(t)$, the best way to assign scores to values t is to follow the distribution $p(t)$, giving higher scores to values that are more likely to be observed.

Example 3.5

Consider the distribution $t \sim p(t) = \text{Bern}(t|0.7)$ and assume that the soft predictor $q(t) = \text{Bern}(t|0.5)$ is used. The population log-loss $L_p(q(\cdot))$, i.e., the cross entropy H($p\|q$), is given as

$$\text{H}(p\|q) = 0.7 \cdot (-\log(0.5)) + 0.3 \cdot (-\log(0.5)) = 0.69 \text{ (nats)}. \tag{3.60}$$

In contrast, the cross entropy of the optimal soft predictor $q(t) = p(t)$ is

$$\text{H}(p\|p) = 0.7 \cdot (-\log(0.7)) + 0.3 \cdot (-\log(0.3)) = 0.61 \text{ (nats)}, \tag{3.61}$$

which is smaller than the population log-loss (equal to 0.69 nats) obtained by the first predictor. The cross entropy (3.61) is in fact the minimum possible value of the population log-loss, and hence no soft predictor can obtain a smaller value.

3.6.5   Entropy

As concluded in the preceding subsection, when rvs x and t are independent, the minimum population log-loss is obtained when the soft predictor $q(t)$ equals the population distribution $p(t)$. Therefore, the minimum value of the population log-loss is given as

$$\min_{q(\cdot)} \text{H}(p\|q) = \text{H}(p\|p) = \text{H}(p), \tag{3.62}$$

which is known as the **entropy** of rv t $\sim p(t)$. In other words, the entropy

$$\text{H}(p) = \text{E}_{t \sim p(t)} \left[-\log p(t) \right] \tag{3.63}$$

measures the population log-loss obtained with the optimal soft predictor. We will also write the entropy (3.63) as H($p(t)$); or as H(t) if the distribution of t is clear from the context.

Based on (3.63), the entropy can be interpreted as a measure of *irreducible uncertainty* for rv t $\sim p(t)$: No soft predictor can decrease the population log-loss below the entropy. Target variables t with larger entropy can hence be considered to be more "random", and more difficult to predict.

Information-Theoretic Interpretation of the Entropy. Following the information-theoretic interpretation of the log-loss for discrete rvs given in Sec. 3.6.2, the entropy H(t) represents (roughly speaking) the minimum average number of bits required to describe in a lossless fashion the outcome of a random draw of rv t $\sim p(t)$. More "random" variables, with a larger entropy, tend to have more surprising outcomes, and hence they require a larger number of bits to be compressed on average.

KL Divergence as Regret. By (3.51), the KL divergence is given as

$$\text{KL}(p\|q) = \text{H}(p\|q) - \text{H}(p). \tag{3.64}$$

Given that the entropy represents the minimum population log-loss, the KL divergence measures the average *excess* population log-loss incurred when using a generally suboptimal soft predictor q as compared to the optimal predictor p. Unlike the cross entropy, the KL divergence is hence a measure not of loss but of "regret", or excess loss.

Entropy and Differential Entropy. Experts on information theory will have by now noted that the entropy has been introduced without making the necessary differentiation between discrete and continuous rvs. In fact, for continuous rvs, one should more correctly talk about **differential entropy**, rather than entropy. Among the main distinctions between entropy and differential entropy is the fact that, unlike the entropy – i.e., $H(p)$ in (3.63) for a pmf $p(t)$ – the differential entropy – i.e., $H(p)$ in (3.63) for a pdf $p(t)$ – can be negative. This is a direct consequence of the fact that the log-loss for continuous rvs can be negative (see Fig. 3.7). Appendix 3.C discusses some of the other key differences between entropy and differential entropy.

Having made this important distinction, in this book we will adopt the convention of not distinguishing explicitly between entropy and differential entropy, identifying both with the term "entropy". This will allow concepts such as optimal soft prediction to be presented in a unified way for both discrete and continuous rvs.

Example 3.6

The (differential) entropy of a Gaussian rv $t \sim \mathcal{N}(\mu, \sigma^2)$ is

$$H(t) = H(\mathcal{N}(\mu, \sigma^2)) = \frac{1}{2} \log(2\pi e \sigma^2), \tag{3.65}$$

when measured in nats. The entropy of a Gaussian rv is hence independent of the mean, and it increases *logarithmically* with the variance σ^2. Note that, when the variance σ^2 is sufficiently small, the differential entropy (3.65) is negative.

3.6.6 Optimal Soft Prediction

To Recap. To summarize the discussion so far, for the special case in which we only know the marginal distribution $p(t)$ (i.e., when x and t are independent):

- the soft optimal predictor $q(t)$ that minimizes the population log-loss, i.e., the cross entropy $H(p||q)$, is given by $q^*(t) = p(t)$; and
- the minimum cross entropy is given by the entropy $H(p) = H(p||p)$; and
- the difference between cross entropy and entropy is the KL divergence

$$KL(p||q) = H(p||q) - H(p) \geq 0, \tag{3.66}$$

which is an asymmetric measure of distance between true distribution $p(t)$ and soft predictor $q(t)$.

In this subsection, we generalize these conclusions to the original problem (3.56) of evaluating the optimal soft predictor for a general population distribution $p(x, t)$.

Generalizing Optimal Soft Prediction. In a manner similar to what we have done for hard predictors in (3.15), using the law of iterated expectations, we can write the population log-loss as

$$L_p(q(\cdot|\cdot)) = \mathrm{E}_{\mathrm{x}\sim p(x)}[\underbrace{\mathrm{E}_{\mathrm{t}\sim p(t|\mathrm{x})}[-\log q(\mathrm{t}|\mathrm{x})]}_{\mathrm{H}(p(t|\mathrm{x})||q(t|\mathrm{x}))}]. \tag{3.67}$$

This expresses the population log-loss as the average over $p(x)$ of the cross entropy $\mathrm{H}(p(t|x)||q(t|x))$ between the conditional distributions $p(t|x)$ and $q(t|x)$ for a fixed value of x. Hence, the optimal soft predictor can be obtained separately for each value of x by minimizing the cross entropy as

$$q^*(\cdot|x) = \arg\min_{q(\cdot|x)} \mathrm{H}(p(t|x)||q(t|x)). \tag{3.68}$$

Note that the only difference with respect to problem (3.57) is the conditioning on a given value of x. Therefore, from Gibbs' inequality, following the analysis in Sec. 3.6.4, the optimal soft predictor is given by

$$q^*(t|x) = p(t|x). \tag{3.69}$$

As anticipated, we can finally conclude that the posterior distribution $p(t|x)$ is the optimal soft predictor under the log-loss.

3.6.7 Conditional Entropy

In the preceding subsection, we have seen that the posterior distribution is the optimal soft predictor under the log-loss. This implies that the minimum population log-loss is given as

$$L_p(p(\cdot|\cdot)) = \mathrm{E}_{\mathrm{x}\sim p(x)}[\underbrace{\mathrm{E}_{\mathrm{t}\sim p(t|\mathrm{x})}[-\log p(\mathrm{t}|\mathrm{x})]}_{\mathrm{H}(p(t|\mathrm{x}))}], \tag{3.70}$$

where $\mathrm{H}(p(t|x))$ is the entropy of the conditional distribution $p(t|x)$ for a fixed value x. The quantity $L_p(p(\cdot|\cdot))$, which averages the entropies $\mathrm{H}(p(t|\mathrm{x}))$ over rv $\mathrm{x} \sim p(x)$, is known as the **conditional entropy** of t given x under the joint distribution $p(x,t)$. When the joint distribution $p(x,t)$ is clear from the context, we will also write the conditional entropy as

$$\mathrm{H}(\mathrm{t}|\mathrm{x}) = \mathrm{E}_{(\mathrm{x},\mathrm{t})\sim p(x,t)}[-\log p(\mathrm{t}|\mathrm{x})]. \tag{3.71}$$

Generalizing the discussion on the entropy in Sec. 3.6.5, the conditional entropy can be interpreted as a measure of **irreducible unpredictability** of the target rv t given the observation of rv x. It quantifies how difficult – in terms of log-loss – it is to predict t from x under the joint distribution $p(x,t)$. Equivalently, it provides a measure of uncertainty about t when x is known under the model $p(x,t)$. When x and t are independent, the conditional entropy reduces to the entropy, i.e., we have the equality $\mathrm{H}(\mathrm{t}|\mathrm{x}) = \mathrm{H}(\mathrm{t})$.

As in the case of the entropy, we will not distinguish between conditional differential entropy – the correct terminology for continuous rvs t – and conditional entropy for discrete rvs t, using the expression "conditional entropy" throughout.

Example 3.1 (continued)

Consider again the joint distribution $p(x,t)$ in Table 3.1, and assume the soft predictor $q(t|x)$ given in Table 3.4. What is the population log-loss of this predictor?

Let us first compute the cross entropies $\mathrm{H}(p(t|\mathrm{x}=0), q(t|\mathrm{x}=0))$ and $\mathrm{H}(p(t|\mathrm{x}=1), q(t|\mathrm{x}=1))$ for the two possible values of rv x, and then evaluate the population log-loss via (3.67). Using the

Table 3.4 Example of a soft predictor $q(t|x)$

$x \backslash t$	0	1
0	0.6	0.4
1	0.6	0.4

conditional distribution $p(t|x)$ in Table 3.2, the cross entropies are given as

$$H(p(t|x=0)||q(t|x=0)) = 0.9 \cdot (-\log(0.6)) + 0.1 \cdot (-\log(0.4)) = 0.55 \text{ (nats)} \quad (3.72)$$

$$H(p(t|x=1)||q(t|x=1)) = 0.2 \cdot (-\log(0.6)) + 0.8 \cdot (-\log(0.4)) = 0.83 \text{ (nats)}. \quad (3.73)$$

Therefore, by (3.67), the population log-loss is the average

$$E_{x \sim p(x)}[H(p(t|x)||q(t|x))] = 0.5 \cdot 0.55 + 0.5 \cdot 0.83 = 0.69 \text{ (nats)}. \quad (3.74)$$

The minimum population log-loss is the conditional entropy $H(t|x)$, which can be computed as

$$E_{(x,t) \sim p(x,t)}[-\log p(t|x)] = 0.45 \cdot (-\log(0.9)) + 0.05 \cdot (-\log(0.1))$$
$$+ 0.1 \cdot (-\log(0.2)) + 0.4 \cdot (-\log(0.8))$$
$$= 0.41 \text{ (nats)}. \quad (3.75)$$

The conditional entropy is smaller than the population log-loss (equal to 0.69 nats) obtained by the first soft predictor, indicating that the latter is suboptimal.

The concept of conditional entropy as the minimum of the population loss can also be extended to hard predictors and arbitrary loss functions, as detailed in Appendix 3.D.

3.6.8 On Soft and Hard Predictors

It may at first glance be surprising that, even when choosing the optimal soft predictor $q^*(t|x)$ that matches the true conditional distribution $p(t|x)$, the population log-loss $L_p(q^*(\cdot|\cdot))$, i.e., the conditional entropy $H(t|x)$, is not zero. This surprise may stem from a parallel with the case of hard predictors: If one chooses a hard predictor $\hat{t}$ that always produces the correct value t, the population loss is zero (under assumption (3.4)). That this parallel is incorrect points to a conceptual difference between soft and hard prediction, which is elaborated on in this subsection by focusing on a discrete rv t.

A soft predictor must assign a score $q(t|x)$ to *every* value of rv t. Given a realization t = t, the log-loss resulting from this choice is $-\log q(t|x)$ when x = x. The log-loss is zero only if the score assigned to the actual realized value t takes the maximum possible value $q(t|x) = 1$. Since the score function is a probability, we have the sum $\sum_{t'} q(t'|x) = 1$, which implies that, with this assignment, all other values $t' \neq t$ must be assigned a zero score, i.e., $q(t'|x) = 0$. Therefore, this soft predictor would incur an infinite log-loss $-\log q(t'|x) = -\log(0) = \infty$ on any target value $t' \neq t$.

The discussion in the preceding paragraph demonstrates that the only situation in which we can guarantee a zero population log-loss is if, for every value of x, the conditional population distribution $p(t|x)$ assigns probability 1 to one value t and 0 probability to all other values.

This happens if the target rv t is a (deterministic) function of the observation rv x. In fact, in this case, the soft predictor can assign the maximum score to the only possible value t for any value of x, incurring a zero population log-loss. Therefore, leaving aside the very special case in which a functional relationship exists between t and x, the minimum population log-loss is typically positive (for discrete rvs), even for an optimal soft predictor.

3.6.9   Optimal Soft Predictor As Minimizer of the KL Divergence

As discussed throughout Sec. 3.6, the optimal soft predictor is obtained as the minimizer of the average cross entropy (3.70). This final subsection demonstrates that optimal soft prediction can be equivalently formulated as the minimization of the average KL divergence between true posterior $p(t|x)$ and soft predictor $q(t|x)$, where the average is taken with respect to the marginal $p(x)$. This result hinges on a generalization of the relationship (3.51) among KL divergence, cross entropy, and conditional entropy.

By (3.51), for any fixed value x, we can write the equality

$$\mathrm{KL}(p(t|x)||q(t|x)) = \mathrm{H}(p(t|x)||q(t|x)) - \mathrm{H}(p(t|x)). \tag{3.76}$$

Therefore, by averaging over x $\sim p(x)$, we obtain the key equality

$$\begin{aligned}
\mathrm{E}_{\mathrm{x}\sim p(x)}[\mathrm{KL}(p(t|\mathrm{x})||q(t|\mathrm{x}))] &= \mathrm{E}_{\mathrm{x}\sim p(x)}[\mathrm{H}(p(t|\mathrm{x})||q(t|\mathrm{x}))] - \mathrm{E}_{\mathrm{x}\sim p(x)}[\mathrm{H}(p(t|\mathrm{x}))] \\
&= L_p(q(\cdot|\cdot)) - \mathrm{E}_{\mathrm{x}\sim p(x)}[\mathrm{H}(p(t|\mathrm{x}))] \\
&= L_p(q(\cdot|\cdot)) - \mathrm{H}(\mathrm{t}|\mathrm{x})
\end{aligned} \tag{3.77}$$

involving KL divergence, cross entropy, and conditional entropy. As anticipated, since the conditional entropy $\mathrm{H}(\mathrm{t}|\mathrm{x})$ does not depend on the soft predictor $q(t|x)$, the minimizer of the population log-loss $L_p(q(\cdot|\cdot))$ also minimizes the average KL divergence (3.77).

Furthermore, by optimizing the average KL divergence separately for each value x, the optimal soft predictor can be obtained as the minimizer

$$q^*(\cdot|x) = \arg\min_{q(\cdot|x)} \mathrm{KL}(p(t|x)||q(t|x)) = p(\cdot|x) \tag{3.78}$$

of the KL divergence between the conditional distributions $p(t|x)$ and $q(t|x)$. This result follows directly from Gibbs' inequality (3.45).

3.7   Mutual Information

The mutual information $\mathrm{I}(\mathrm{x};\mathrm{t})$ is an important measure of statistical dependence between two rvs x and t that are distributed according to a joint distribution $p(x,t)$. As such, the mutual information plays a role similar to the covariance as a metric that gauges the degree of predictability of one variable given the other. As seen in Sec. 2.7.3, the covariance is a measure of *quadratic* loss, and it inherently refers to *linear* predictability: It quantifies how well, in terms of average ℓ_2 loss, the target t can be predicted using a *hard* predictor given by a linear function of x. In contrast, as we discuss in this section, the mutual information measures predictability in terms of *log-loss* without constraining the form of the *soft* predictor.

3.7.1 Definition of the Mutual Information

To introduce the mutual information, recall from the previous sections that:

- the entropy H(t) represents the minimum population log-loss when the only information available about rv t is the marginal distribution $p(t)$; and
- the conditional entropy H(t|x) represents the minimum population log-loss when we also have access to the observation x to predict t.

The **mutual information** I(x; t) between rvs x and t is the difference between entropy and conditional entropy:

$$I(x; t) = H(t) - H(t|x). \tag{3.79}$$

Therefore, the mutual information measures by how much one can decrease the population log-loss via the observation of rv x as compared to the case when only the distribution $p(t)$ of the target is known. As such, it provides a measure of predictability of rv t given the observation of rv t.

Example 3.7

Consider the jointly Gaussian distribution $p(x, t) = p(t)p(x|t)$ studied in Sec. 3.4.1, with t $\sim$ $\mathcal{N}(v, \alpha^{-1})$ and $(x|t = t) \sim \mathcal{N}(t, \beta^{-1})$. What is the mutual information I(x; t)? From (3.65), the entropy is

$$H(t) = H(\mathcal{N}(v, \alpha^{-1})) = \frac{1}{2} \log(2\pi e \alpha^{-1}), \tag{3.80}$$

and, using (3.32), the conditional entropy is

$$H(t|x) = \frac{1}{2} \log(2\pi e(\alpha + \beta)^{-1}). \tag{3.81}$$

Therefore, the mutual information is

$$I(x; t) = H(t) - H(t|x) = \frac{1}{2} \log\left(1 + \frac{\beta}{\alpha}\right). \tag{3.82}$$

The mutual information (3.82) is easily seen to be non-negative; to increase with the ratio β/α between the precision of the observation and the precision of the prior; and to equal zero only if $\beta = 0$. Therefore, the higher the precision of the observation, β, is as compared to the precision of the prior, α, the more the prediction of t can be improved, on average, by observing x.

3.7.2 Some Properties of the Mutual Information

The mutual information enjoys some useful properties, which will be proved in the next subsection:

- **Non-negativity**. The mutual information is non-negative:

$$I(x; t) \geq 0, \tag{3.83}$$

and we have I(x; t) = 0 only if x and t are independent. Therefore, having access to x improves the population log-loss for the prediction of t (by I(x; t)) unless the two rvs are independent.

Note that, for continuous rvs, both H(t) and H(t|x) can be negative, but their difference, i.e., the mutual information, is always non-negative.

- **Symmetry**. The mutual information is symmetric in the sense that we have the equality

$$I(x; t) = I(t; x) = H(x) - H(x|t). \tag{3.84}$$

Therefore, the improvement in terms of average log-loss accrued on the prediction of t given x is equal to the corresponding improvement for the prediction of x given t.

3.7.3 Mutual Information and KL Divergence

To prove the non-negativity and symmetry properties of the mutual information, it is sufficient to note that the mutual information can be written in terms of the KL divergence as

$$
\begin{aligned}
I(x; t) &= H(t) - H(t|x) \\
&= E_{(x,t) \sim p(x,t)} \left[\log \left(\frac{p(t|x)}{p(t)} \right) \right] \\
&= E_{(x,t) \sim p(x,t)} \left[\log \left(\frac{p(x,t)}{p(x)p(t)} \right) \right] \\
&= KL(p(x,t) \| p(x)p(t)).
\end{aligned} \tag{3.85}
$$

Accordingly, the mutual information can be interpreted as measuring by how much the joint distribution $p(x,t)$ differs from the product distribution $p(x)p(t)$, under which x and t are independent. This offers another way of thinking of the mutual information $I(x; t)$ as a measure of the statistical dependence of x and t: The "further" the joint distribution $p(x,t)$ is from being equal to the product of marginals, the more the two rvs are dependent.

By (3.85), the non-negativity of the mutual information (cf. (3.83)) then follows directly from Gibbs' inequality (3.45), and its symmetry (cf. (3.84)) is a direct consequence of the fact that the expression $KL(p(x,t) \| p(x)p(t))$ depends only on the joint and on the product of the two marginals.

3.7.4 Mutual Information and Explained Uncertainty

In line with the interpretation of entropy in terms of uncertainty and degree of randomness of a rv, another useful way to think about the mutual information is as **"explained uncertainty"**. To formalize this concept, we can rewrite the mutual information as

$$\underbrace{H(t)}_{\text{total uncertainty}} = \underbrace{H(t|x)}_{\text{residual, i.e., unexplained, uncertainty}} + \underbrace{I(x; t)}_{\text{explained uncertainty}}. \tag{3.86}$$

In (3.86), the total a priori uncertainty about t – its entropy H(t) – is decomposed into the sum of the residual uncertainty that remains even after observing x – the conditional entropy H(t|x) – and the part of the entropy that is explained by the observation of x. The latter is measured by the mutual information, which can accordingly be interpreted as explained uncertainty.

In passing, we note that the decomposition (3.86) is related to, but distinct from, the decomposition of the variance, also known as the **law of total variance**,

$$\underbrace{\text{Var}(t)}_{\text{total variance}} = \underbrace{E_{x \sim p(x)}[\text{Var}(t|x)]}_{\text{residual, i.e., unexplained, variance}} + \underbrace{\text{Var}(E_{t \sim p(t|x)}[t|x])}_{\text{explained variance}}, \tag{3.87}$$

where the conditional variance $\mathrm{Var}(t|x)$ is defined in (3.24), and the outer variance in the second term is evaluated with respect to rv $x \sim p(x)$.

3.7.5 Data Processing Inequality

The mutual information satisfies the intuitive property that **post-processing** of data cannot create information. To elaborate, consider again an input x and a target variable t. Assume now that the input x is processed via a random transformation $p(y|x)$ producing a rv $y \sim p(y|x)$ for $x = x$. One can think of y as a (generally random) function of x. It can be proved that the mutual information $I(y; t)$ cannot be larger than the mutual information $I(x; t)$, i.e.,

$$I(y; t) \leq I(x; t), \tag{3.88}$$

which is known as the **data processing inequality**. The data processing inequality (3.88) says that the information about a target variable t that is provided by rv x cannot be increased by processing x. When equality holds in (3.88), we say that y is a **sufficient statistic** of x for the prediction of t. A sufficient statistic "summarizes" all the information in x that is relevant for the prediction of t.

3.7.6 Variational Bounds on the Mutual Information

For later reference, we finally state two useful bounds on the mutual information that follow directly from Gibbs' inequality (3.45). These bounds will play a particularly important role in Part III of the book. The bounds are derived in Appendix 3.E, and are given as follows:

- **Upper bound on the mutual information.** For any distribution $q(t)$, we have the inequality

$$
\begin{aligned}
I(x; t) &\leq H(p||q) - H(t|x) \\
&= E_{(x,t) \sim p(x,t)} \left[\log \left(\frac{p(t|x)}{q(t)} \right) \right],
\end{aligned}
\tag{3.89}
$$

with equality if $q(t) = p(t)$.

- **Lower bound on the mutual information.** For any conditional distribution $q(t|x)$, we have the inequality

$$
\begin{aligned}
I(x; t) &\geq H(t) - E_{x \sim p(x)} \left[H(p(t|x)||q(t|x)) \right] \\
&= E_{(x,t) \sim p(x,t)} \left[\log \left(\frac{q(t|x)}{p(t)} \right) \right],
\end{aligned}
\tag{3.90}
$$

with equality if $q(t|x) = p(t|x)$.

As a first application, using the lower bound (3.90), we can conclude that the optimal soft predictor $q(t|x)$, which by (3.68) minimizes the average cross entropy (the negative of the second term in the first line of (3.90)), also maximizes a lower bound on the mutual information.

3.8   Log-Loss As a "Universal" Loss Function

Notwithstanding the distinctions between the optimal design of hard and soft predictors discussed in Sec. 3.6.8, the problem of designing soft predictors essentially includes that of designing hard predictors as a special case. To frame the design of a hard predictor $\hat{t}(x)$ in terms of a soft predictor $q(t|x)$, one needs to define the latter as a function of $\hat{t}(x)$ in such a way that the resulting log-loss $-\log q(t|x)$ equals the desired loss function $\ell(t, \hat{t}(x))$. This can be done successfully for many cases of interest.

To elaborate on this point, let us define a Gaussian soft predictor $q(t|x) = \mathcal{N}(t|\hat{t}(x), \sigma^2)$, where $\hat{t}(x)$ is any hard predictor. The corresponding log-loss is

$$-\log q(t|x) = \frac{1}{2\sigma^2}(t - f(x))^2 + \text{constant}$$

$$= \frac{1}{2\sigma^2}\ell(t, f(x)) + \text{constant}, \tag{3.91}$$

where the constant term does not depend on input x, and we have used the quadratic loss $\ell(t, \hat{t}) = (t - \hat{t})^2$. Therefore, apart from inessential multiplicative and additive constants, the log-loss is equivalent to the ℓ_2 loss for the hard predictor $\hat{t}(x)$.

More generally, for a hard predictor $\hat{t}(x)$ and loss function $\ell(t, \hat{t}(x))$, we can define the soft predictor

$$q(t|x) = \exp(-a \cdot \ell(t, \hat{t}(x)) + b), \tag{3.92}$$

where $a > 0$ is some parameter and b is selected so as to ensure that $q(t|x)$ is a conditional distribution (i.e., that it integrates or sums to 1). If the constant b exists and it does not depend on the hard predictor $\hat{t}(x)$, the choice (3.92) guarantees the log-loss $-\log q(t|x)$ equals the desired loss function $\ell(t, \hat{t}(x))$, apart from inessential multiplicative and additive constants. This is, for instance, the case for the ℓ_2 loss as per (3.91).

We can conclude that the log-loss is a "universal" measure of loss in the sense that it recovers other loss functions $\ell(t, \hat{t})$ through a suitable choice of the soft predictor. The centrality of probabilistic thinking in machine learning calls to mind **James Clerk Maxwell**'s famous quote, "The true logic of this world is in the calculus of probabilities."

3.9   Free Energy

In the previous sections, we have used the cross entropy $\text{H}(p(t|x)||q(t|x))$ and the KL divergence $\text{KL}(p(t|x)||q(t|x))$ to evaluate the quality of a soft predictor $q(t|x)$ under conditional population distribution $p(t|x)$ for any fixed value x of the input. In this section, we introduce an alternative information-theoretic metric that quantifies the performance of a soft predictor – the **free energy**. Apart from being central to many advanced machine learning algorithms such as expectation maximization (discussed in Chapter 7), the free energy provides a useful interpretation of the optimal soft predictor as balancing predictive accuracy and predictive uncertainty (or, conversely, predictive confidence). The free energy, also known as *variational* free energy in the literature, is the subject of this section.

3.9.1 Minimizing the "Reverse" KL Divergence

In Sec. 3.6, we have seen that the optimal soft predictor $q^*(t|x)$, minimizing the population log-loss, is the posterior distribution $q^*(t|x) = p(t|x)$, which can be obtained as the minimizer of the KL divergence as in (3.78). To introduce the free energy, we start with a simple observation: By Gibbs' inequality (3.45), we can equivalently formulate the problem of optimizing a soft predictor as the minimization of the **"reverse" KL divergence**, i.e.,

$$q^*(\cdot|x) = \arg \min_{q(\cdot|x)} \text{KL}(q(t|x)||p(t|x)) = p(t|x). \tag{3.93}$$

The equivalence between the two formulations (3.78) and (3.93) is in the sense that the solutions of both problems are equal to the posterior $q^*(t|x) = p(t|x)$. Note that, as discussed in Sec. 3.5 (see, e.g., Fig. 3.5), the equality between the minimizers does not imply that the two KL divergences have the same values for any given soft predictor $q(\cdot|x)$.

3.9.2 Introducing the Free Energy

For any fixed x, let us add the term $-\log p(x)$ to the objective $\text{KL}(q(t|x)||p(t|x))$ in (3.93). This operation does not change the minimizer $q^*(t|x)$ since the added term, $-\log p(x)$, is a constant that does not depend on the soft predictor $q(t|x)$. With this addition, the objective in (3.93) becomes

$$\text{KL}(q(t|x)||p(t|x)) - \log p(x) = \text{E}_{\text{t}\sim q(t|x)}\left[-\log(p(\text{t}|x) \cdot p(x))\right] - \text{E}_{\text{t}\sim q(t|x)}\left[-\log q(t|x)\right]$$
$$= \text{E}_{\text{t}\sim q(t|x)}\left[-\log p(x,\text{t})\right] - \text{H}(q(t|x)). \tag{3.94}$$

Note that we have used the trivial equality $-\log p(x) = \text{E}_{\text{t}\sim q(t|x)}[-\log p(x)]$, which holds for any fixed x. This modified objective is defined as the **free energy**

$$\text{F}(q(t|x)||p(x,t)) = \text{E}_{\text{t}\sim q(t|x)}\left[-\log p(x,\text{t})\right] - \text{H}(q(t|x)), \tag{3.95}$$

where x is fixed. As we will clarify later in this section, the free energy differs from the cross entropy and KL divergence in that it depends on a probability distribution, $q(t|x)$, and on an **unnormalized** probability distribution, $p(x,t)$, of rv t (where x is fixed).

A note on terminology: the expression "free energy" comes from **thermodynamics**. In it, the amount of work that can be extracted from a system – the free energy – equals the overall energy of the system discounted by the entropy, i.e.,

$$\text{free energy = overall energy − entropy.} \tag{3.96}$$

Accordingly, the entropy represents the part of the overall energy that cannot be turned into work. Observe the parallel between (3.96) and (3.95). In order to distinguish between (3.95) and the thermodynamic free energy, it is common to refer to (3.95) as *variational* free energy. In this book, in order to simplify the terminology we will not add this qualifier.

For our purposes, the significance of the free energy is that, as mentioned, the optimal soft prediction problem (3.93) can be equivalently restated as the minimization

$$q^*(\cdot|x) = \arg \min_{q(\cdot|x)} \text{F}(q(t|x)||p(x,t)) = p(t|x) \tag{3.97}$$

for the given fixed value x. In words, *optimal Bayesian inference can be formulated as the minimization of the free energy*.

3.9.3 Interpreting the Free Energy

In this subsection, we argue that the free energy $\mathrm{F}(q(t|x)\|p(x,t))$ balances two conflicting requirements in evaluating the quality of a soft predictor $q(t|x)$ – predictive accuracy and confidence.

- **Predictive loss.** The first term in (3.95), namely $\mathrm{E}_{t\sim q(t|x)}[-\log p(x,t)]$, can be considered as a measure of predictive loss. Its presence in (3.95) pushes the minimizer of the free energy in (3.97) towards a loss-minimizing *hard* predictor. To see this, note that this term tends to be small if the soft predictor $q(t|x)$ assigns a high score to values t that yield a large probability $p(x,t)$, or equivalently a large conditional probability $p(t|x)$. In fact, for a discrete rv t, the first term is maximized by the deterministic predictor

$$q(t|x) = \mathbb{1}(t - \hat{t}^{MAP}(x)), \tag{3.98}$$

where $\hat{t}^{MAP}(x)$ is the MAP predictor (3.27). This soft predictor assigns maximum score to the MAP prediction and a zero score to all other values $t \neq \hat{t}^{MAP}(x)$. It is a very "confident" predictor.

- **Confidence.** The second term, namely the negative entropy $-\mathrm{H}(q(t|x))$, quantifies the uncertainty associated with the prediction, penalizing soft predictors $q(t|x)$ that are too confident. In fact, this term tends to the small when the entropy is large, and hence the predictor maintains some level of uncertainty. Its role in problem (3.97) is to ensure that the minimizer of the free entropy does not collapse to an overconfident hard predictor as in (3.98).

3.9.4 Alternative Views on the Free Energy

The free energy (3.95) can also be viewed and interpreted in the following two useful ways, to which we will return several times throughout the book:

- **Free energy as "loss – KL".** The free energy in (3.96) can be equivalently written, for a fixed x, as

$$\mathrm{F}(q(t|x)\|p(x,t)) = \mathrm{E}_{t\sim q(t|x)}\left[-\log p(x|t)\right] + \mathrm{KL}(q(t|x)\|p(t)). \tag{3.99}$$

This alternative formulation can again be interpreted as the difference between a loss term – the first – and a penalty term – the second. The loss term has a similar interpretation to the original definition (3.95), while the second term penalizes soft predictors that assign scores differing significantly from the marginal $p(t)$. In a manner analogous to the entropy penalty in (3.95), the second term in (3.99) promotes soft predictors whose score assignment $q(t|x)$ does not depend too much on the input x, deviating from the marginal $p(t)$.

- **Free energy as KL divergence with an unnormalized distribution.** The free energy can be equivalently thought of as a KL divergence in which the second distribution is unnormalized. This is in the sense that we can write

$$\mathrm{F}(q(t|x)\|p(x,t)) = \mathrm{E}_{t\sim q(t|x)}\left[\log\left(\frac{q(t|x)}{p(x,t)}\right)\right], \tag{3.100}$$

for any fixed x. Comparing (3.100) with the definition (3.43) of the KL divergence, we observe that the free energy replaces a distribution over t ($q(t)$ in (3.100)) with the *unnormalized* distribution $p(x,t)$, which, when viewed as a distribution over t for a fixed x, need not sum or integrate to 1.

3.9.5 General Definition of the Free Energy Involving an Unnormalized Distribution

As seen in the preceding subsection, the free energy is an information-theoretic metric involving a distribution and an unnormalized distribution. To elaborate, fix a distribution $q(t)$ – possibly conditioned on a fixed value of another rv x – and an unnormalized distribution $\tilde{p}(t)$. An unnormalized distribution $\tilde{p}(t)$ is a non-negative function, i.e., $\tilde{p}(t) > 0$, that can be normalized to produce a distribution $p(t)$. This is in the sense that there exists a finite constant Z such that $p(t) = \tilde{p}(t)/Z$ is a distribution (a pmf or a pdf). Specifically, for discrete rvs, the normalizing constant is $Z = \sum_t \tilde{p}(t)$, while for continuous rvs the sum over all possible values t is replaced by the corresponding integral over the domain of rv t.

Given a distribution $q(t)$ and an unnormalized distribution $\tilde{p}(t)$ defined over the same domain, we generalize the definition (3.100) of free energy as

$$F(q(t)\|\tilde{p}(t)) = E_{t \sim q(t)}\left[\log\left(\frac{q(t)}{\tilde{p}(t)}\right)\right]$$
$$= E_{t \sim q(t)}[-\log\tilde{p}(t) + \log q(t)]$$
$$= E_{t \sim q(t)}[-\log\tilde{p}(t)] - H(q(t)). \tag{3.101}$$

Informally, following the second bullet in the preceding subsection, we can think of the free energy $F(q(t)\|\tilde{p}(t))$ as a generalization of the KL divergence $KL(q(t)\|p(t))$ that allows for the second argument function to be unnormalized.

A key result that we will turn to many times in this book is that the minimizer of the free energy $F(q(t)\|\tilde{p}(t))$ is the normalized distribution $q^*(t) = \tilde{p}(t)/Z = p(t)$, i.e.,

$$q^*(\cdot) = \arg\min_{q(\cdot)} F(q(t)\|\tilde{p}(t)) = \frac{\tilde{p}(t)}{Z}. \tag{3.102}$$

This is proved in Appendix 3.F. As a special case, if we set $\tilde{p}(t) = p(t,x)$ for a fixed value x, this result recovers the equality (3.97).

3.10 Summary

- Inference – also known as model-based prediction – refers to the optimal prediction of a target variable t from an input variable x when the joint distribution $p(x,t)$ of the two rvs is known. We distinguish between detection problems, in which the target t is discrete (e.g., predicting rainy vs. sunny weather), and estimation problems, in which the target t is continuous (e.g., predicting the temperature).
- The optimal hard predictor minimizes the population loss, and the optimization can be done separately for each input x via (3.16), based on the posterior distribution $p(t|x)$.
- A soft predictor $q(t|x)$ assigns a score for each value of rv t, in the form of a conditional probability distribution given observation x. The optimal soft predictor solves the problem of minimizing the population log-loss, or cross-entropy loss, and is given by the posterior distribution $q^*(t|x) = p(t|x)$.
- The minimum population log-loss is known as conditional entropy of rv t given x, and is defined as $H(t|x) = E_{(x,t) \sim p(x,t)}[-\log p(t|x)]$. The conditional entropy is a measure of uncertainty about variable t when variable x is known. When the only information available about t is its distribution, the minimum population log-loss is given by the entropy $H(t) = E_{t \sim p(t)}[-\log p(t)]$. The entropy is a measure of uncertainty about variable t when no additional information is available apart from the marginal distribution $p(t)$.

- The KL divergence between distributions p and q is defined as $\mathrm{KL}(p\|q) = \mathrm{H}(p\|q) - \mathrm{H}(p) \geq 0$, where equality holds if and only if q is identical to p. The inequality is known as Gibbs' inequality. The KL divergence is an asymmetric measure of distance between two distributions.
- The mutual information $\mathrm{I}(x;t) = \mathrm{H}(t) - \mathrm{H}(t|x)$ measures by how much we can decrease the population log-loss for the prediction of rv t by measuring rv x, and is accordingly a measure of statistical dependence between x and t. The mutual information can be written in terms of the KL divergence as $\mathrm{I}(x;t) = \mathrm{KL}(p(x,t)\|p(x)p(t))$.
- The optimal soft predictor $p(t|x)$ can also be obtained as the minimizer of the free energy $\mathrm{F}(q(t|x)\|p(x,t)) = \mathrm{E}_{t \sim q(t|x)}\left[-\log p(x,t)\right] - \mathrm{H}(q(t|x))$.
- More generally, the free energy $\mathrm{F}(q(t)\|\tilde{p}(t))$ between a distribution $q(t)$ and an unnormalized distribution $\tilde{p}(t)$ is minimized by the corresponding normalized distribution $p(t) = \tilde{p}(t)/Z$, where the constant Z ensures the normalization of $p(t)$.

3.11 Recommended Resources

The material covered in this chapter partly overlaps with a number of standard textbooks on estimation and detection, such as [1, 2, 3], and on information theory, such as [4, 5]. Relevant material can also be found in textbooks on signal processing, such as [6, 7], and on Bayesian machine learning [8]. A well-argued and entertaining historical perspective on frequentist and Bayesian approaches to statistics is offered by [9]. The focus on the free energy is aligned with recent formulations of learning problems to be covered in Part III (as well as to some theories of intelligence [10]; see Sec. 7.10.12). A recent history of thermodynamics can be found in [11], and some aspects related to the "physics" of information are also presented in [12]. An interesting, and still relevant, early work on the connection between log-loss and loss functions for hard predictors is [13].

Problems

3.1 (a) For the joint distribution $p(x,t)$ in Table 3.5, compute the optimal soft predictor, the optimal hard predictor for the detection-error loss, and the corresponding minimum population detection-error loss. (b) For the same joint distribution, how does the optimal hard predictor change when we adopt the loss function illustrated by Table 3.3?

3.2 For the joint distribution $p(x,t)$ in Table 3.6, compute the optimal soft predictor, the optimal hard predictor for the detection-error loss, and the corresponding minimum population detection-error loss.

3.3 Given the posterior distribution $p(t|x)$ in Table 3.7, obtain the optimal point predictors under ℓ_2 and detection-error losses, and evaluate the corresponding minimum population losses when $p(x = 1) = 0.5$.

Table 3.5 Example of joint pmf $p(x,t)$

$x\backslash t$	0	1
0	0.4	0.3
1	0.1	0.2

Table 3.6 Example of joint pmf $p(x,t)$

$x\backslash t$	0	1
0	0.16	0.24
1	0.24	0.36

Table 3.7 Example of conditional pmf $p(t|x)$

$x\backslash t$	0	1
0	0.9	0.1
1	0.2	0.8

3.4 If the posterior distribution is given as

$$(t|x = x) \sim \mathcal{N}(\sin(2\pi x), 0.1),$$

what is the optimal hard predictor under the ℓ_2 loss? What is the MAP predictor?

3.5 Consider the posterior distribution

$$p(t|x) = 0.5\mathcal{N}(t|3, 1) + 0.5\mathcal{N}(t| - 3, 1).$$

(a*) Plot this posterior distribution. (b) What is the optimal hard predictor under the quadratic, or ℓ_2, loss? (c) What is the MAP hard predictor?

3.6 Consider the joint distribution $p(x,t) = p(t)p(x|t)$ defined by prior $t \sim \mathcal{N}(2,1)$ and likelihood $(x|t = t) \sim \mathcal{N}(t, 0.1)$. (a) Derive the optimal soft predictor. (b) Derive the optimal hard predictor under the ℓ_2 loss. (c*) Plot prior and optimal soft predictor for $x = 1$. (d) Repeat the previous points for prior $t \sim \mathcal{N}(2, 0.01)$ and comment on the result.

3.7 Consider the joint distribution $p(x,t) = p(t)p(x|t)$ defined by prior $t \sim \mathcal{N}(0_2, I_2)$ and likelihood given as $x = [2,1]t + z$, with $z \sim \mathcal{N}(0, 0.1I_2)$. (a*) Derive the optimal soft predictor. (b) Derive the optimal hard predictor under the ℓ_2 loss. (c*) Plot contour lines of prior and optimal soft predictor for $x = 1$. (d) Comment on the result.

3.8 (a*) Plot the KL divergence $\text{KL}(p\|q)$ between $p(t) = \text{Bern}(t|0.4)$ and $q(t) = \text{Bern}(t|q)$ as a function of $q \in [0, 1]$. Prepare two plots, one in nats and one in bits. Repeat for $\text{KL}(q\|p)$.

3.9 Compute $\text{KL}(p\|q)$ between $p(t) = \text{Cat}(t|[0.4, 0.1, 0.5]^T)$ and $q(t) = \text{Cat}(t|[0.1, 0.5, 0.4]^T)$ in nats.

3.10 (a*) Plot the KL divergence between $p(t) = \mathcal{N}(t| - 1, 1)$ and $q(t) = \mathcal{N}(t|\mu, 1)$ as a function of μ (in nats). (b*) Plot the KL divergence between $p(t) = \mathcal{N}(t| - 1, 1)$ and $q(t) = \mathcal{N}(t| - 1, \sigma^2)$ as a function of σ^2 (in nats). (c*) Plot the KL divergence $\text{KL}(q\|p)$ with the same distributions as in point (b*) (in nats). Discuss your results.

3.11 Consider a soft predictor $q(t) = \text{Bern}(t|0.6)$. Recall that a soft predictor assigns score $q(t)$ to all values of $t \in \{0, 1\}$. (a) Calculate the log-loss, in nats, for all possible values of t and comment on your result. (b) With this soft predictor, which value of t has the smallest log-loss? Interpret your conclusions in terms of information-theoretic surprise.

3.12 Consider a rv with distribution $p(t) = \text{Bern}(t|0.4)$, and a soft predictor $q(t) = \text{Bern}(t|0.6)$. (a) Calculate the population log-loss, i.e., the cross entropy between $p(t)$ and $q(t)$, in nats. (b) Repeat with the soft predictor $q(t) = \text{Bern}(t|0.4)$, and compare with the result obtained in point (a).

Table 3.8 Example of joint pmf $p(x,t)$

$x\backslash t$	0	1
0	0.3	0.4
1	0.1	0.2

Table 3.9 Example of joint pmf $p(x,t)$

$x\backslash t$	0	1
0	0	0.5
1	0.5	0

3.13 Consider distributions $p(t) = \text{Bern}(t|0.4)$, and soft predictor $q(t) = \text{Bern}(t|0.6)$. Compute the KL divergence $\text{KL}(\text{Bern}(0.4)||\text{Bern}(0.6))$ in two different ways: (a) through direct calculation by using the definition of KL divergence, and (b) using the relationship of the KL divergence with cross entropy and entropy. Recall that the latter relationship views the KL divergence as a measure of regret for not having used the correct predictor.

3.14 Consider a soft predictor $q(t) = \mathcal{N}(t|-1, 3)$. (a) Plot the log-loss, in nats, across a suitable range of values of t and comment on your result. (b) With this soft predictor, which value of t has the smallest log-loss? Interpret your conclusions in terms of information-theoretic surprise.

3.15 Consider a rv with distribution $p(t) = \mathcal{N}(t|-2, 1)$, and a soft predictor $q(t) = \mathcal{N}(t|1, 3)$. (a) Calculate the population log-loss, i.e., the cross entropy between $p(t)$ and $q(t)$, in nats. (b) Repeat with the soft predictor $q(t) = \mathcal{N}(t|-2, 1)$, and compare with the result obtained in point (a).

3.16 Consider a soft predictor $q(t) = \mathcal{N}(t|\mu, 1)$ for some value of μ. The population distribution of the rv to be predicted is $p(t) = \mathcal{N}(t|-1, 1)$. (a*) Plot the cross entropy $\text{H}(p||q)$, the entropy $\text{H}(p)$, and the KL divergence $\text{KL}(p||q)$ in the same figure and discuss their relationship.

3.17 For the joint distribution $p(x,t)$ in Table 3.8, compute the population log-loss obtained by the soft predictor $q(t = 1|x = 0) = 0.4$ and $q(t = 1|x = 1) = 0.6$ in two ways: (a) using directly the definition of population log-loss; and (b) using the expression in terms of the cross entropy. (c) Derive the optimal soft predictor and the corresponding population log-loss. (d) What are the conditional entropy $\text{H}(t|x)$ and the entropy $\text{H}(t)$?

3.18 Compute the conditional entropy $\text{H}(t|x)$ and the entropy $\text{H}(t)$ for the joint distribution in Table 3.6.

3.19 (a) Compute the mutual information $\text{I}(x; t)$ for the joint distribution in Table 3.9. (b) Repeat for the joint distribution $p(x,t)$ in Table 3.6, and discuss your results.

3.20 Consider the population distribution $p(x,t) = p(t)p(x|t)$ with $t \sim \mathcal{N}(0, 1)$, and $(x|t = t) \sim \mathcal{N}(t, 1)$, i.e., $x = t + z$, with independent noise $z \sim \mathcal{N}(0, 1)$. (a) For the soft predictor $q(t|x) = \mathcal{N}(t|-x, 0.5)$, compute the population log-loss. (b) What is the optimal soft predictor? (c) What is the corresponding minimum population log-loss?

3.21 We are given the population distribution $p(x,t) = p(t)p(x|t)$ with $t \sim \mathcal{N}(0, 1)$, and $(x|t = t) \sim \mathcal{N}(t, \beta^{-1})$, i.e., $x = t + z$, with independent noise $z \sim \mathcal{N}(0, \beta^{-1})$. (a*) Plot the mutual information $\text{I}(x; t)$ as a function of β and comment on your results.

3.22 Prove that the entropy H(x) of a random vector $x = [x_1, \ldots, x_M]^T$ with independent entries $\{x_1, \ldots, x_M\}$ is given by the sum $H(x) = \sum_{m=1}^{M} H(x_m)$. Use this result to compute the (differential) entropy of vector $x \sim \mathcal{N}(0_M, I_M)$.

3.23 Prove that the optimal constant predictor of a rv $t \sim \mathcal{N}(\mu, \sigma^2)$ under the ℓ_2 loss is the mean μ.

Appendices

Appendix 3.A: Proof of Gibbs' Inequality

To prove Gibbs' inequality (3.45), we will make use of **Jensen's inequality**. Jensen's inequality applies to convex functions. As we will detail in Sec. 5.6, a twice differentiable function is convex if its second derivative is positive throughout its domain. Given a convex function $f(x)$, Jensen's inequality is stated as

$$E_{x \sim p(x)}\left[f(x)\right] \geq f\left(E_{x \sim p(x)}[x]\right) \tag{3.103}$$

with equality if and only if $f(x)$ is constant for all values x with $p(x) > 0$.

Since $f(x) = -\log(x)$ is convex (see Sec. 5.6), we can apply Jensen's inequality to it. Considering first a discrete rv t, we have

$$\begin{aligned} KL(p||q) = E_{t \sim p(t)}\left[-\log\left(\frac{q(t)}{p(t)}\right)\right] &\geq -\log\left(E_{t \sim p(t)}\left[\frac{q(t)}{p(t)}\right]\right) \\ &= -\log\left(\sum_t p(t)\frac{q(t)}{p(t)}\right) = -\log\left(\sum_t q(t)\right) \\ &= -\log(1) = 0, \end{aligned} \tag{3.104}$$

with equality if and only if $p(t) = q(t)$. The sum in (3.104) is replaced by an integral for continuous rvs.

Appendix 3.B: On the JS Divergence

Consider a binary detection problem in which the target rv t is Bern(0.5). In formulas, we have the joint distribution

$$\begin{aligned} t &\sim Bern(0.5) \\ (x|t=0) &\sim p(x|t=0) \\ (x|t=1) &\sim p(x|t=1), \end{aligned} \tag{3.105}$$

for some arbitrary conditional distributions $p(x|t=0)$ and $p(x|t=1)$. Note that the marginal of rv x is given as

$$p(x) = \frac{p(x|t=1) + p(x|t=0)}{2}. \tag{3.106}$$

The minimum population log-loss for the prediction of t given the observation of rv x must intuitively depend on how dissimilar the two conditional distributions $p(x|t=0)$ and $p(x|t=1)$ are. If the input x follows significantly distinct distributions when conditioned on $t = 0$ and $t = 1$, values x that are likely to be observed when $t = 0$ may not be as probable when $t = 1$, yielding evidence in favor of t = 0; while values x for which $p(x|t = 1) > p(x|t = 0)$ bring evidence in favor of the choice t = 1. In accordance with this intuition, we will demonstrate in this appendix that the minimum population log-loss for the problem at hand is given by the JS divergence between the conditional distributions $p(x|t = 0)$ and $p(x|t = 1)$ (apart from an additive constant).

As seen in Sec. 3.6.7, the minimum population log-loss, obtained by minimizing over all possible soft predictors, equals the conditional entropy H(t|x). We will show next that the conditional entropy for the joint distribution (3.105) satisfies

$$H(t|x) = \log(2) + JS(p(x|t = 0)||p(x|t = 1)). \tag{3.107}$$

Hence, a larger JS divergence implies that it is more difficult to predict the value of the binary rv t from x under the joint distribution (3.105).

Using the definition of conditional entropy (3.71), this equality can be proved as follows:

$$H(t|x) = E_{(x,t) \sim p(x,t)}[-\log p(t|x)]$$

$$\overset{\text{LIE}}{=} \frac{1}{2} \sum_{t=0}^{1} E_{x \sim p(x|t=t)}[-\log p(t|x)]$$

$$= \frac{1}{2} \sum_{t=0}^{1} E_{x \sim p(x|t=t)}\left[-\log\left(\frac{0.5 \cdot p(x|t = t)}{p(x)}\right)\right]$$

$$= \log(2) + \frac{1}{2} \sum_{t=0}^{1} \underbrace{E_{x \sim p(x|t=t)}\left[-\log\left(\frac{p(x|t = t)}{p(x)}\right)\right]}_{\text{KL}\left(p(x|t) \,\middle\|\, \frac{p(x|t=0)+p(x|t=1)}{2}\right)}$$

$$= \log(2) + JS(p(x|t = 0)||p(x|t = 1)), \tag{3.108}$$

where in the second equality we used the law of iterated expectations (LIE) from Sec. 2.12; the third equality follows from Bayes' theorem (see Sec. 2.11); in the fourth, we have plugged in the marginal $p(x) = 0.5(p(x|t = 0) + p(x|t = 1))$; and in the last equality we have used the definition of the JS divergence (3.52).

Appendix 3.C: Some Differences between Entropy and Differential Entropy

In this section, we briefly summarize some differences between the entropy of discrete rvs and the differential entropy of continuous rvs. In both cases, the definition is given by the expectation (3.63). For all proofs and details, see Recommended Resources, Sec. 3.11.

Discrete rvs. For a discrete rv t $\sim p(t)$, one can think of the entropy H(t) as the logarithm of the "effective" alphabet size as determined by the pmf $p(t)$. This interpretation is substantiated by the inequalities

$$0 \le H(t) \le \log|\mathcal{T}|, \tag{3.109}$$

where $|\mathcal{T}|$ denotes the cardinality of the alphabet $\mathcal{T}$ of rv t, and by the following facts:

- By (3.109), the entropy is non-negative, and it is minimized, taking value equal to zero, when the rv t is deterministic. In this case, the effective alphabet size is 1, as rv t takes a single value with probability 1 (i.e., $p(t) = \mathbb{1}(t = t_0)$ for some $t_0 \in \mathcal{T}$).

- It is maximized, taking value $\log |\mathcal{T}|$, when the distribution is uniform over its alphabet ($p(t) = 1/|\mathcal{T}|$), and hence the effective alphabet size is $|\mathcal{T}|$.

Continuous rvs. For a continuous rv $t \sim p(t)$, the differential entropy can be thought of as the logarithm of the "effective" support of the distribution. Accordingly, the differential entropy does not satisfy the inequalities (3.109), and it can generally take any real value, i.e.,

$$-\infty \leq H(t) \leq +\infty. \tag{3.110}$$

The differential entropy tends to $-\infty$ when the support of the pdf $p(t)$ becomes increasingly small, while it tends to $+\infty$ when the support of the pdf $p(t)$ becomes increasingly large.

As an example, for a Gaussian rv, the differential entropy (3.65) tends to $-\infty$ as the variance σ^2 goes to zero; and to $+\infty$ as the variance goes to infinity. A zero variance implies a distribution with vanishingly small support, while an infinite variance corresponds, in the limit, to a uniform distribution on the entire real line.

It is finally emphasized that, unlike for discrete rvs, a continuous rv with zero differential entropy is not a deterministic variable. For instance, by (3.65), a Gaussian rv with variance $\sigma^2 = 1/(2\pi e)$ has differential entropy $H(\mathcal{N}(\mu, \sigma^2)) = 0$.

Appendix 3.D: Generalized Entropy and Generalized Conditional Entropy

The concepts of entropy and conditional entropy can be generalized so as to apply also to hard predictors under any loss function. This appendix briefly elaborates on this extension.

To start, recall that the entropy is the minimum population log-loss for soft prediction when one only knows the distribution $p(t)$ of the target variable t; and that, more generally, the conditional entropy is the minimum population log-loss when we have access to a statistically dependent input variable x and to the joint distribution $p(x, t)$. Furthermore, their difference defines the mutual information.

For a hard predictor $\hat{t}(\cdot)$ and any loss function $\ell(t, \hat{t})$, we can similarly define the following quantities:

- **Generalized entropy.** The generalized entropy $H_\ell(t)$ is the minimum population loss when one knows only the distribution $p(t)$ of the target variable t, i.e.,

$$H_\ell(t) = L_p(\hat{t}^*) = E_{t \sim p(t)}[\ell(t, \hat{t}^*)], \tag{3.111}$$

where $\hat{t}^*$ is the optimal hard predictor in (3.20). Therefore, the generalized entropy measures how much uncertainty there is about rv t when the uncertainty is measured by the average loss function $\ell(t, \hat{t})$. Note that the definition of generalized entropy is loss function dependent.

- **Generalized conditional entropy.** The generalized conditional entropy $H_\ell(t|x)$ is the minimum population loss when the predictor has access to the observation of an input variable x, i.e.,

$$H_\ell(t|x) = L_p(\hat{t}^*(\cdot)) = E_{(x, t) \sim p(x, t)}[\ell(t, \hat{t}^*(x))], \tag{3.112}$$

where the optimal hard predictor is defined in (3.10). The generalized conditional entropy hence measures how much uncertainty there is about t given observation x, on average over x, when the uncertainty is measured by the average loss function $\ell(t, \hat{t})$.

- **Generalized mutual information.** The generalized mutual information can be analogously defined as the difference

$$I_\ell(x; t) = H_\ell(t) - H_\ell(t|x), \tag{3.113}$$

which inherits the same interpretation of the conventional mutual information in terms of improvement in the predictability of t upon the observation of rv x. Note that the generalized mutual information need not be symmetric in x and t since the loss function is defined on the space of the target variable t.

Appendix 3.E: Proof of Mutual Information Bounds

In this section, we derive the bounds on the mutual information introduced in Sec. 3.7.6. Both bounds follow directly from Gibbs' inequality (3.45), which can be restated as

$$H(p||q) \geq H(p), \tag{3.114}$$

with equality if and only if the two distributions p and q are identical. The inequalities in Sec. 3.7.6 follow by applying (3.114) to the mutual information

$$
\begin{aligned}
I(x;t) &= H(t) - H(t|x) \\
&= H(p(t)) - E_{x \sim p(x)}\left[H(p(t|x)) \right].
\end{aligned}
\tag{3.115}
$$

Specifically, upper bounding the entropy in (3.115) with the cross entropy via (3.114), we have the following upper bound:

$$
\begin{aligned}
I(x;t) &\leq H(p(t)||q(t)) - H(t|x) \\
&= E_{(x,t) \sim p(x,t)}\left[\log\left(\frac{p(t|x)}{q(t)} \right) \right],
\end{aligned}
\tag{3.116}
$$

which holds for any distribution $q(t)$ and recovers the bound (3.89). Upper bounding the conditional entropy in (3.115) with the cross entropy via (3.114), we have the lower bound

$$
\begin{aligned}
I(x;t) &\geq H(t) - E_{x \sim p(x)}\left[H(p(t|x)||q(t|x)) \right] \\
&= E_{(x,t) \sim p(x,t)}\left[\log\left(\frac{q(t|x)}{p(t)} \right) \right],
\end{aligned}
\tag{3.117}
$$

which holds for any conditional distribution $q(t|x)$ and recovers the bound (3.90).

Appendix 3.F: Minimizing the Free Energy over the First Term

In this appendix, we prove the equality (3.102), which shows that the minimizer of the free energy $F(q(t)||\tilde{p}(t))$ with respect to the first term is the normalized distribution $p(t) = \tilde{p}(t)/Z$. To this end, note that we have the equalities

$$
\begin{aligned}
KL(q(t)||p(t)) &= F(q(t)||\tilde{p}(t)) + E_{t \sim q(t)}[\log(Z)] \\
&= F(q(t)||\tilde{p}(t)) + \log(Z),
\end{aligned}
\tag{3.118}
$$

and that the normalizing constant Z does not depend on $q(t)$. Therefore, the minimizer of the free energy is the same as the minimizer of the KL divergence $KL(q(t)||p(t))$, yielding the optimal distribution $q^*(t) = p(t)$ and concluding the proof.

Bibliography

[1] S. M. Kay, *Fundamentals of Statistical Signal Processing*. Prentice Hall PTR, 1993.

[2] H. V. Poor, *An Introduction to Signal Detection and Estimation*. Springer Science+Business Media, 2013.

[3] H. L. Van Trees, *Detection, Estimation, and Modulation Theory, Part I: Detection, Estimation, and Linear Modulation Theory*. John Wiley & Sons, 2004.

[4] T. M. Cover and J. A. Thomas, *Elements of Information Theory*. John Wiley & Sons, 2006.

[5] D. J. MacKay, *Information Theory, Inference and Learning Algorithms*. Cambridge University Press, 2003.

[6] M. Vetterli, J. Kovačević, and V. K. Goyal, *Foundations of Signal Processing*. Cambridge University Press, 2014.

[7] U. Spagnolini, *Statistical Signal Processing in Engineering*. Wiley Online Library, 2018.

[8] S. Theodoridis, *Machine Learning: A Bayesian and Optimization Perspective*. Academic Press, 2015.

[9] A. Clayton, *Bernoulli's Fallacy: Statistical Illogic and the Crisis of Modern Science*. Columbia University Press, 2022.

[10] K. Friston, "The free-energy principle: A unified brain theory?" *Nature Reviews Neuroscience*, vol. 11, no. 2, pp. 127–138, 2010.

[11] P. Sen, *Einstein's fridge*. William Collins, 2021.

[12] C. Scharf, *The Ascent of Information*. Riverhead Books, 2021.

[13] E. Levin, N. Tishby, and S. A. Solla, "A statistical approach to learning and generalization in layered neural networks," *Proceedings of the IEEE*, vol. 78, no. 10, pp. 1568–1574, 1990.

4 Supervised Learning: Getting Started

4.1 Overview

As seen in the preceding chapter, when a reliable model $p(x, t)$ is available to describe the probabilistic relationship between input variable x and target variable t, one is faced with a model-based prediction problem, also known as inference. Inference can in principle be optimally addressed by evaluating functions of the posterior distribution $p(t|x) = p(x, t)/p(x)$ of the output t given the input x. In fact, this distribution provides the optimal soft predictor; and, from it, one can compute the optimal hard predictor for any loss function. What can one do if an accurate model $p(x, t)$ is not known, or when computing the posterior $p(t|x)$ – or the optimal hard predictor directly – from $p(x, t)$ is too computationally demanding? This chapter provides an answer to this question by introducing the inductive bias-based data-driven design methodology of supervised learning.

Supervised learning assumes the availability of data in the form of examples of pairs (x, t) of inputs x and desired outputs t. As we will discuss in this chapter, using these data, one can design both hard and soft predictors. In order to focus on basic concepts, we will study simple training problems that can be solved in closed form, relegating the discussion of more complex optimization strategies to the next chapter.

Learning Objectives and Organization of the Chapter. By the end of this chapter, the reader should be able to:

- understand the problems of regression and classification within a frequentist learning framework, which is the most commonly used and is, adopted for the entire Part II of this book (Sec. 4.2);
- formulate and address the problem of training hard predictors (Sec. 4.3);
- carry out inductive bias selection via validation (Sec. 4.4);
- understand and evaluate the trade-off between bias and estimation error (Sec. 4.5);
- appreciate more modern views on generalization beyond the standard bias-estimation error trade-off (Sec. 4.6);
- understand and apply regularization (Sec. 4.7);
- formulate the problem of training soft predictors via maximum likelihood (ML) learning (Sec. 4.9);
- formulate the problem of training soft predictors via maximum a posteriori (MAP) learning (Sec. 4.10); and
- draw the connection between parametric and non-parametric methods for linear models, and understand the use of kernel methods for regression (Sec. 4.11).

4.2 Defining Supervised Learning

Supervised learning can be formulated within different algorithmic frameworks, most notably by adopting either a frequentist or a Bayesian viewpoint. In this chapter, and for the entire Part II of this book, we focus on frequentist learning, while Part III, starting with Chapter 12, introduces Bayesian learning. This section starts by defining key concepts that are common to the problems of training hard and soft predictors within a frequentist framework.

4.2.1 Regression and Classification

In a manner similar to inference (see the previous chapter), we can distinguish two different types of supervised learning problems:

- **regression** problems, in which the target variables t are continuous-valued; and
- **classification** problems, in which the target variables t take values in a discrete and finite set.

The correspondence between supervised learning and inference problems is illustrated in Table 4.1. We now introduce regression and classification in turn.

Regression. In the problem of regression, we are given a **training set** $\mathcal{D}$ of N **training data points** $\mathcal{D} = \{(x_n, t_n)_{n=1}^N\}$, where

- x_n represents the nth **input**, also known as **covariate** or **explanatory vector**; and
- t_n represents the corresponding nth **desired output**, also known as the **target variable** or **response**. Note that we will take the target variable t to be a scalar, although there are problems in which the target quantity can be a vector (see, e.g., semantic segmentation).

The goal is to **generalize** the relationship between input and output variables *outside* the training set $\mathcal{D}$ in the following sense: Given a **test input** x, which is as yet unobserved, we would like to extrapolate from the training set a predicted output t. This is illustrated in Fig. 4.1, where training data are represented as circles, and the position of a possible test value of the input x is marked as a dashed line.

Classification. In classification, we are also given a training set of N training data points $\mathcal{D} = \{(x_n, t_n)_{n=1}^N\}$, but the scalar target variables t_n take values in a discrete and finite set. In the case of classification, the target variables are also referred to as **labels** (the term is sometimes also applied to regression). The goal is again that of generalizing the relationship between input and label outside the training set. An illustration can be found in Fig. 4.2, in which the inputs are points on the plane and the two possible values of the binary label are represented as circles and crosses, respectively.

Table 4.1 Inference vs. supervised learning

domain of the target	inference	supervised learning
discrete and finite	detection	classification
continuous	estimation	regression

Figure 4.1 Illustration of the problem of regression.

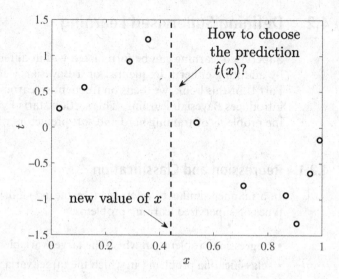

Figure 4.2 Illustration of the problem of classification.

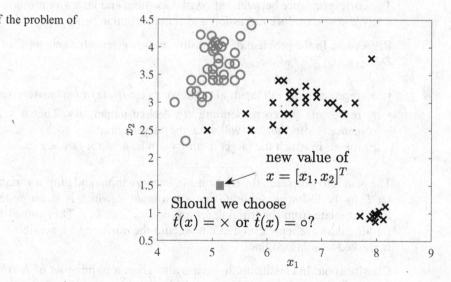

4.2.2   Generalization

As discussed, the main goal of learning is **generalization**. In this subsection, we elaborate on this concept by contrasting it with memorization and by describing the central role of the inductive bias in it.

Generalization vs. Memorization. While memorizing refers to the retrieval of a value t_n corresponding to an already observed pair (x_n, t_n) in the training set $\mathcal{D}$, generalization refers to the extension of the relationship between input and output, which is partially revealed by $\mathcal{D}$, to values of the input that are outside $\mathcal{D}$. Generalization is also known as **induction**, which is the process of deriving general rules from examples. A learning procedure that generalizes well enables the prediction of the output t for an an input x that was not previously encountered.

More poetically, in the (often quoted) words of **Jorge Luis Borges**, "To think is to forget details, generalize, make abstractions."

Inductive Bias. In order to generalize, and hence in order to learn, it is *necessary* to make prior assumptions about the relationship between input and output.

Take another look at the regression problem of Fig. 4.1: Why should some values of the target variable t be more probable for the indicated test input value x? After all, the training data are consistent with infinitely many functions relating x and t that behave in any arbitrary way outside the training set. Stated differently, we have no reason to predict t at a new value x – unless, that is, we are willing to make some assumptions about the mechanism generating pairs (x, t).

As another example, consider again the classification problem in Fig. 4.2. Without some additional assumptions, the test input x can be legitimately classified both as a circle and as a cross.

The set of all assumptions made on the relationship between input and output to enable learning is known as **inductive bias**, since it "biases" the induction process of deriving general rules from data. For example, for the regression problem in Fig. 4.1, we could decide to assume that the function relating output t and input x is a polynomial of some degree; and, in the classification problem of Fig. 4.2, we could assume that clusters of close points, in terms of Euclidean distance, should have the same label.

Without an inductive bias, generalization is not possible. If we are not willing to make any assumption about the relationship between input and output, the predicted value for the test input x in Fig. 4.1 could legitimately be anywhere on the dashed line; and the test point in Fig. 4.2 could just as well be assigned to the circle class or to the cross class. An inductive bias is required to constrain the relationship between x and t, which can otherwise be arbitrary outside the training set.

The choice of the inductive bias is critical. In fact, the performance of the learned predictor can only be as good as the inductive bias allows it to be. For instance, in Fig. 4.1, if we constrain the relationship between x and t to be linear, the trained predictor will have to describe a straight line on the plane, no matter how much training data one has access to. An incorrect choice of the inductive bias can thus cause a large "bias", making it impossible for the learning process to identify an effective predictor within the constraints imposed on the inductive process.

Based on this discussion, one may be tempted to conclude that the right approach is to make the broadest possible assumptions so as to ensure limited bias. However, an excessively broad inductive bias would make it more difficult for the learning agent to estimate a suitable predictor based on the available, limited, training data. As a result, a broad inductive bias would yield a low **"bias"**, but also a large **"estimation error"** caused by the availability of limited data. These terms will be elucidated and clarified in the rest of this chapter.

No Free Lunch. Formalizing the intuitive discussion in this subsection about the need for an inductive bias, the **no-free-lunch theorem** states that there is no **universal** procedure that can map training data to a predictor in such a way that it outperforms all other predictors for all possible training and test data. Since such a universal procedure does not exist, learning must rely on specialized methods that constrain the set of predictors through the selection of an inductive bias.

Transductive vs. Inductive Learning. This chapter, as well as the rest of the book, adopts the **inductive** learning framework described above, whereby the goal is to generalize to new inputs outside the training set. An alternative formulation of the learning problem would require the learner to use training data to label *specific* target data which are available at the time of training. In this **transductive** learning setting, the learner is given not only the training set $\mathcal{D}$ but also unlabeled test data – in the form of inputs x_n for $n = N + 1, \ldots, N + N'$ – and the goal is to assign labels to the N' test samples.

4.2.3 Training and Test Data

In the frequentist formulation of learning, each data point (x_n, t_n) is assumed to be generated independently of the ground-truth distribution $p(x, t)$, which is known as the **population distribution**. The **training set** $\mathcal{D}$ is hence assumed to be distributed as

$$(\mathrm{x}_n, \mathrm{t}_n) \underset{\text{i.i.d.}}{\sim} p(x, t), \text{for } n = 1, \ldots, N, \tag{4.1}$$

where the notation " $\underset{\text{i.i.d.}}{\sim}$ " highlights that the samples are drawn in an i.i.d. manner.

The **test sample** (x, t) is similarly assumed to be drawn independently of the training set $\mathcal{D}$ from the population distribution as

$$(\mathrm{x}, \mathrm{t}) \underset{\text{independent of } \mathcal{D}}{\sim} p(x, t). \tag{4.2}$$

Based on the training set $\mathcal{D}$, the outcome of a learning procedure is a hard predictor $\hat{t}_{\mathcal{D}}(x)$ or a soft predictor $q_{\mathcal{D}}(t|x)$, where the subscript $\mathcal{D}$ is used to emphasize the dependence on the training data $\mathcal{D}$.

Inference vs. Learning. As we have discussed in the preceding chapter, when the population distribution $p(x, t)$ is known, we do not need training data $\mathcal{D}$ – we can generate as many data points as we wish! – and we are faced with a standard inference problem. When the distribution $p(x, t)$ is not known or not tractable, we are confronted with a learning problem, which is the subject of this chapter.

4.2.4 Population Distribution

As seen in the preceding subsection, the population distribution describes the mechanism generating both training and test data. Borrowing the taxonomy proposed by David Spiegelhalter (see Recommended Resources, Sec. 4.13), the population distribution $p(x, t)$ may be interpreted in one of the following ways:

- as a distribution over a **literal** population of individuals, e.g., the distribution of vote t in an election for a voter with features x, such as gender and income level – in this case, the population is finite, though potentially very large, and randomness is introduced by sampling from the population;
- as a distribution over a **virtual** population, e.g., the distribution of air pollution level t in a geographical location x subject to measurement errors – in this case, the population is potentially infinite and random measurements can be taken at will; or
- as a distribution over a **metaphorical** population, e.g., the distribution of the number t of murders in a given year and place, with both features collected in a vector x – in this case, there is no randomness, but we can act as if the observation is one of a set of possible counterfactual histories that may have happened by chance.

Importantly, in all of these cases, the population distribution should also account for any source of **distortion and bias** that is inherent in the data generation process. A particularly timely example involves classification problems in which labels are assigned by hand, and hence label generation reflects specific viewpoints or perspectives, e.g., in defining the gender or profession of individuals depicted in input images.

4.2.5 Recurring Examples

To provide concrete, simple examples, this chapter will study two recurring examples.

Example 4.1

The first setting is the weather prediction problem introduced in Example 3.1. In it, the input x represents the current day's weather – x = 0 indicating a rainy day and x = 1 a sunny day – and the target t represents the next day's weather, which is encoded in the same way. As seen in Chapter 3, if the joint distribution $p(x, t)$ is known, we have a detection problem: Based on the posterior $p(t|x)$ we can obtain the optimal hard predictor $\hat{t}^*(x)$, which depends on the loss function being minimized; as well as the optimal soft predictor $q^*(t|x) = p(t|x)$, which minimizes the population log-loss.

Suppose now that the population distribution $p(x, t)$ is not known, and hence we are faced with a classification problem. To address it, we are given a training data set $\mathcal{D}$ of $N = 1000$ measurements (x_n, t_n). Supervised learning aims to use data set $\mathcal{D}$ to train a hard predictor $\hat{t}_\mathcal{D}(x)$ for $x \in \{0, 1\}$ or a soft predictor $q_\mathcal{D}(t|x)$ for $x, t \in \{0, 1\}$.

Given that we only have four possible configurations for each observation, namely $(0, 0)$, $(0, 1)$, $(1, 0)$, and $(1, 1)$, the data set $\mathcal{D}$ can be described by specifying how many observation of each type are included in $\mathcal{D}$. To this end, we introduce the function

$$N[x, t] = \sum_{n=1}^{N} \mathbb{1}(x_n = x, t_n = t), \tag{4.3}$$

which counts the number of observations $(x_n = x, t_n = t)$ with configuration (x, t). For this recurring example, we will assume that the count function $N[x, t]$ is given by Table 4.2. That is, we have $N[0, 0] = \sum_{n=1}^{N} \mathbb{1}(x_n = 0, t_n = 0) = 507$ observations of pairs $(x_n = 0, t_n = 0)$; and so on.

By dividing all entries in the count table by the number of observations, N, we obtain the **empirical distribution**

$$p_\mathcal{D}(x, t) = \frac{N[x, t]}{N} \tag{4.4}$$

associated with the data set $\mathcal{D}$. From Table 4.2, the empirical distribution is given by Table 4.3.

Intuitively, if the number of data points, N, is very large, the empirical distribution $p_\mathcal{D}(x, t)$ is close to the (unknown) population distribution $p(x, t)$, and supervised learning is therefore able to approximate well the optimal predictors obtained in Chapter 3.

Table 4.2 Count table for Example 4.1
($N = 1000$)

$x\backslash t$	0	1
0	507	39
1	76	378

Table 4.3 Empirical distribution $p_{\mathcal{D}}(x,t)$ for Example 4.1

$x \backslash t$	0	1
0	0.507	0.039
1	0.076	0.378

Example 4.2

In the second recurring example, we wish to design a predictor of the continuous-valued temperature t $\in \mathbb{R}$ (measured on some scale) as a function of the (scalar) position x along a corridor. We assume the population distribution $p(x,t) = p(x)p(t|x)$ given as x $\sim p(x) = \mathcal{U}(x|0,1)$ – a uniformly distributed position along a corridor of unitary length – and

$$(\text{t}|\text{x} = x) \sim p(t|x) = \mathcal{N}(t|\sin(2\pi x), 0.1). \tag{4.5}$$

Accordingly, for a given position x, the observed temperature is $\sin(2\pi x) \pm \sqrt{0.1} = \sin(2\pi x) \pm 0.32$ with 95% probability (see Fig. 2.2).

If the joint distribution $p(x,t)$ is known, we have an estimation problem, as detailed in Chapter 3. In this case, the optimal soft predictor is given by the posterior $p(t|x)$, while the optimal hard predictor under the quadratic loss is the posterior mean $\hat{t}(x) = \sin(2\pi x)$. In this chapter, we will suppose that the population distribution is unknown, and hence we are faced with a regression problem. To address it, we are given some training data set $\mathcal{D}$, an example of which is shown in Fig. 4.1. Supervised learning aims to use this data set to obtain a hard predictor $\hat{t}_{\mathcal{D}}(x)$ for $x \in [0,1]$ or a soft predictor $q_{\mathcal{D}}(t|x)$ for $x \in [0,1]$ and $t \in \mathbb{R}$.

4.3   Training Hard Predictors

In this section, we introduce the problem of supervised learning for the design of hard predictors.

4.3.1   Model Class

As discussed in the preceding section, in order to enable learning, we must first make some assumptions about the relationship between input and output as part of the **inductive bias**. To this end, the learner constrains the hard predictor to belong to a **model class** $\mathcal{H}$ of hard predictors of the form $\hat{t}(\cdot|\theta)$ that are parametrized by a **model parameter vector** $\theta \in \Theta$ in some set Θ, i.e.,

$$\mathcal{H} = \{\hat{t}(\cdot|\theta) : \theta \in \Theta\}. \tag{4.6}$$

Training a hard predictor amounts to identifying a model parameter $\theta \in \Theta$ based on the training set $\mathcal{D}$. As a result of training, the learner hence obtains a hard predictor $\hat{t}(\cdot|\theta)$ in model class $\mathcal{H}$.

Example 4.1 (continued)

For Example 4.1, the set of all possible hard predictors can be defined as a model class $\mathcal{H}$ with predictors defined as $\hat{t}(0|\theta) = \theta_1$ and $\hat{t}(1|\theta) = \theta_2$, with $\theta_1 \in \{0, 1\}$ and $\theta_2 \in \{0, 1\}$ representing the respective predictions when the input is $x = 0$ and $x = 1$. The model parameter vector is $\theta = [\theta_1, \theta_2]^T$, which lies in the discrete set $\Theta = \{[0, 0]^T, [0, 1]^T, [1, 0]^T, [1, 1]^T\}$.

As an example of a constrained model class, one could impose that the same prediction $\theta \in \{0, 1\}$ be made for both values of the input $x \in \{0, 1\}$, i.e., $\hat{t}(0|\theta) = \theta$ and $\hat{t}(1|\theta) = \theta$ with model parameter $\theta \in \Theta = \{0, 1\}$. This inductive bias may be reasonable if we expect the weather to be either always sunny or always rainy.

Example 4.2 (continued)

For the regression problem in Example 4.2, the learner may choose a model class $\mathcal{H}$ consisting of polynomial functions with some fixed degree M, i.e.,

$$\hat{t}(x|\theta) = \theta_0 + \theta_1 x + \theta_2 x^2 + \cdots + \theta_M x^M$$
$$= \sum_{i=0}^{M} \theta_i x^i = \theta^T u(x). \tag{4.7}$$

In (4.7), we have defined the **feature vector** $u(x) = [1, x, x^2, \cdots, x^M]^T$ and the model parameter vector $\theta = [\theta_0, \theta_1, \cdots, \theta_M]^T \in \mathbb{R}^{M+1}$ that determines the coefficients of the polynomial.

Predictors of the form (4.7) are referred to as **linear** because they are linear functions of the model parameter θ. Note that the dependence on x is not linear, as the prediction $\hat{t}(x|\theta)$ depends on the input x through a vector of features $u(x)$ that includes monomials of x.

One possible justification for the choice of the inductive bias defined by the polynomial regression function (4.7) is the **Weierstrass approximation theorem**, which states that any continuous function in a finite interval can be approximated by a polynomial function of degree M with an arbitrarily small error by choosing the degree M to be sufficiently large. The selection of the degree M is, however, critical: As M increases, the model becomes richer, and able to approximate a larger class of functions, but also more demanding on the training process given the larger number of parameters to optimize in vector θ.

The degree M is an example of a **hyperparameter**. Hyperparameters are numerical quantities that identify the inductive bias – in this case, the "capacity" of the model class.

4.3.2 Training Loss

As discussed in Sec. 3.3, we can measure the quality of a hard predictor by adopting a loss function $\ell(t, \hat{t})$ such as the ℓ_2 loss or the detection-error loss. Accordingly, the loss of a hard predictor $\hat{t}(\cdot|\theta)$, and hence of a model parameter θ, on a pair (x, t) can be obtained as

$$\ell(t, \hat{t}(x|\theta)). \tag{4.8}$$

In a frequentist formulation, for a fixed model class $\mathcal{H}$, the general goal of supervised learning is to minimize the **population loss**

$$L_p(\theta) = \mathrm{E}_{(x, t) \sim p(x, t)}[\ell(t, \hat{t}(x|\theta))], \tag{4.9}$$

over the model parameter vector $\theta \in \Theta$, i.e., to address the problem

$$\min_{\theta \in \Theta} L_p(\theta). \tag{4.10}$$

The population loss is also known as **generalization loss** or **out-of-sample loss**, since it can be interpreted as the average loss measured on the test pair $(x, t) \sim p(x, t)$ (see (4.2)). Therefore, by minimizing the population loss, a hard predictor also minimizes the average loss over a randomly generated test point (4.2).

However, unlike for inference, in a learning problem the population loss $L_p(\theta)$ cannot be computed since the learner does not know the population distribution $p(x, t)$. Based on the training data $\mathcal{D}$, the learner can, however, evaluate the **training loss**

$$L_{\mathcal{D}}(\theta) = \frac{1}{N} \sum_{n=1}^{N} \ell(t_n, \hat{t}(x_n | \theta)). \tag{4.11}$$

The training loss measures the *empirical* average of the loss accrued by the predictor $\hat{t}(\cdot | \theta)$ on the examples (x_n, t_n) in the training set $\mathcal{D}$. As such, the training loss $L_{\mathcal{D}}(\theta)$ is an estimate of the population loss $L_p(\theta)$ based on the training data set. Note that this estimate is just that, and training and population losses are generally different, i.e., $L_{\mathcal{D}}(\theta) \neq L_p(\theta)$.

The relationship between training and population losses is characterized by the following important properties.

- **The training loss replaces the population distribution with the empirical distribution.** Let us first consider discrete rvs x and t, for which the empirical distribution $p_{\mathcal{D}}(x, t)$ of a given data set $\mathcal{D}$ is defined as in (4.4). Recall that $N[x, t]$ is the number of observations in $\mathcal{D}$ that are equal to (x, t). In this case, the training loss can be equivalently written as the average of the loss with respect to the empirical distribution , i.e.,

$$L_{\mathcal{D}}(\theta) = E_{(x,t) \sim p_{\mathcal{D}}(x,t)} \left[\ell(t, \hat{t}(x|\theta)) \right]. \tag{4.12}$$

Therefore, the training loss is obtained by substituting the population distribution in (4.9) with the empirical distribution (4.4). This result is proved via the following series of equalities:

$$\begin{aligned} L_{\mathcal{D}}(\theta) &= \frac{1}{N} \sum_{n=1}^{N} \ell(t_n, \hat{t}(x_n|\theta)) \\ &= \frac{1}{N} \sum_{x,t} N[x, t] \ell(t, \hat{t}(x|\theta)) \\ &= \sum_{x,t} p_{\mathcal{D}}(x, t) \ell(t, \hat{t}(x|\theta)), \end{aligned} \tag{4.13}$$

where the sums are over all possible values (x, t). The relationship (4.12) can also be obtained for continuous rvs by defining the empirical distribution as a sum of Dirac delta functions on the data points. This is detailed in Appendix 4.A.

- **The training loss is asymptotically accurate for any fixed model parameter θ.** The training loss is an estimate of the population loss that becomes increasingly accurate as the number of data points, N, increases. Informally, for any fixed θ, we have the limit

$$L_{\mathcal{D}}(\theta) \to L_p(\theta) \tag{4.14}$$

"with high probability" as $N \to \infty$. The "with high probability" condition refers to the probability that the rv $L_\mathcal{D}(\theta)$ – where the randomness stems from the stochastic generation of the training data – tends to the (deterministic) population loss $L_p(\theta)$ as N grows large. This result follows from the **law of large numbers**, which can be, again informally, stated as follows: Given i.i.d. rvs $z_1, z_2, \ldots, z_N \sim p(z)$ with a finite variance, the empirical average satisfies the limit

$$\frac{1}{N} \sum_{n=1}^{N} z_n \to \mathrm{E}_{z \sim p(z)}[z] \tag{4.15}$$

"with high probability" as $N \to \infty$. In the limit (4.14), the role of each rv z_n in (4.15) is played by the loss $\ell(t_n, \hat{t}(x_n|\theta))$ for the nth data point. A more precise statement of the law of large numbers can be found in Sec. 8.3.

4.3.3 Empirical Risk Minimization

The asymptotic correctness of the estimate provided by the training loss for the population loss suggests **empirical risk minimization (ERM)** as a principle for learning. ERM consists of the problem of minimizing the training loss, i.e.,

$$\min_{\theta \in \Theta} L_\mathcal{D}(\theta) \tag{4.16}$$

over the model parameter θ, obtaining the ERM solution

$$\theta_\mathcal{D}^{ERM} = \arg \min_{\theta \in \Theta} L_\mathcal{D}(\theta). \tag{4.17}$$

As in Chapter 3, we take the notation arg min to denote one of the optimal solutions, and we refer to Sec. 5.3 for more discussion on this point. By the limit (4.14), as N grows large, one expects that the ERM solution should tend to the ideal model parameter θ that minimizes the population loss as per (4.10).

Example 4.1 (continued)

For Example 4.1, let us assume the unconstrained model class $\mathcal{H}$ that is defined by the model parameter vector $\theta = [\theta_1, \theta_2]^T$ with $\theta_1, \theta_2 \in \{0, 1\}$. With this choice, using the empirical distribution $p_\mathcal{D}(x, t)$ in Table 4.3, the training loss (4.13) under the detection-error loss is given as

$$L_\mathcal{D}(\theta) = 0.507 \cdot \mathbb{1}(\theta_1 \neq 0) + 0.039 \cdot \mathbb{1}(\theta_1 \neq 1)$$
$$+ 0.076 \cdot \mathbb{1}(\theta_2 \neq 0) + 0.378 \cdot \mathbb{1}(\theta_2 \neq 1). \tag{4.18}$$

By minimizing the training loss $L_\mathcal{D}(\theta)$, the ERM solution is given as $\theta_\mathcal{D}^{ERM} = [0, 1]^T$.

Generalizing this example, when the predictor is unconstrained and the target variable t is discrete, under the detection-error loss, the ERM predictor can be computed as the maximization of the empirical distribution

$$\hat{t}_\mathcal{D}^{ERM}(x) = \arg \max_t p_\mathcal{D}(x, t). \tag{4.19}$$

The proof of this claim is left as an exercise.

Example 4.2 (continued)

For Example 4.2, assuming the class of linear models (4.7), under the ℓ_2 loss, ERM tackles the problem

$$\min_{\theta \in \mathbb{R}^D} \left\{ L_{\mathcal{D}}(\theta) = \frac{1}{N} \sum_{n=1}^{N} (t_n - \theta^T u(x_n))^2 \right\}, \tag{4.20}$$

where $D = M + 1$. We will see in the next subsection that this optimization is an example of a least squares (LS) problem, and that it can be solved in closed form.

4.3.4 Solving a Least Squares Problem

Least Squares (LS) Problem. The ERM problem (4.20) is one of the most common in machine learning. It is encountered every time the ℓ_2 loss is adopted and the model class consists of linear predictive models of the form

$$\mathcal{H} = \left\{ \hat{t}(x|\theta) = \theta^T u(x) = \sum_{i=1}^{D} \theta_i u_i(x) \right\}, \tag{4.21}$$

where $u(x) = [u_1(x), \ldots, u_D(x)]^T$ is a $D \times 1$ vector of features. For instance, for the polynomial regression model assumed in Example 4.2, we have $D = M + 1$ features, which are given as $u_d(x) = x^{d-1}$ for $d = 1, \ldots, D$. In this subsection, we will derive and describe a closed-form solution for problem (4.20).

To proceed, define the $N \times D$ **data matrix** obtained by stacking data features for each data point by rows:

$$X_{\mathcal{D}} = \begin{bmatrix} u(x_1)^T \\ u(x_2)^T \\ \vdots \\ u(x_N)^T \end{bmatrix}. \tag{4.22}$$

Note that the nth row of the data matrix is the feature vector corresponding to one input data point x_n; and that the dth column corresponds to a different feature $u_d(x)$. In fact, we can write the data matrix (4.22) as

$$X_{\mathcal{D}} = \begin{bmatrix} u_1(x_1) & u_2(x_1) & \cdots & u_D(x_1) \\ u_1(x_2) & u_2(x_2) & \cdots & u_D(x_2) \\ \vdots & \vdots & \vdots & \vdots \\ u_1(x_N) & u_2(x_N) & \cdots & u_D(x_N) \end{bmatrix}. \tag{4.23}$$

Defining also the **target vector**

$$t_{\mathcal{D}} = \begin{bmatrix} t_1 \\ t_2 \\ \vdots \\ t_N \end{bmatrix}, \tag{4.24}$$

the ERM problem (4.20) can be expressed as

$$\min_{\theta \in \mathbb{R}^D} \left\{ L_{\mathcal{D}}(\theta) = \frac{1}{N} ||t_{\mathcal{D}} - X_{\mathcal{D}}\theta||^2 \right\}. \tag{4.25}$$

This can easily be verified by using the definition of ℓ_2 norm (see Problem 4.3). Optimizations of the form (4.25) are known as **least squares (LS) problems**.

Solution of the LS Problem. To address the LS problem (4.25), let us study first the conventional **overdetermined** case with no fewer data points than features, i.e., with $N \geq D$. In this case, assuming that the matrix $X_{\mathcal{D}}^T X_{\mathcal{D}}$ is invertible, the problem (4.25) has the closed-form solution

$$\theta_{\mathcal{D}}^{ERM} = (X_{\mathcal{D}}^T X_{\mathcal{D}})^{-1} X_{\mathcal{D}}^T t_{\mathcal{D}}. \tag{4.26}$$

(If $N < D$ or, more generally, if matrix $X_{\mathcal{D}}^T X_{\mathcal{D}}$ is not invertible, i.e., it is rank-deficient, a solution is given as $\theta_{\mathcal{D}}^{ERM} = X_{\mathcal{D}}^\dagger t_{\mathcal{D}}$, with $X_{\mathcal{D}}^\dagger$ being the pseudo-inverse of the data matrix.)

Interpreting the LS Solution. Denote as $u_d = [u_d(x_1), u_d(x_2), \ldots, u_d(x_N)]^T$ the dth column of the data matrix $X_{\mathcal{D}}$ in (4.23). This is an $N \times 1$ vector that collects the values taken by dth feature $u_d(\cdot)$ across the data points in the training set. We can refer to it as the **per-feature sample vector** for data set $\mathcal{D}$.

To facilitate the interpretation of the LS solution (4.26), assume temporarily that the data matrix satisfies the equality $X_{\mathcal{D}}^T X_{\mathcal{D}} = I_D$. This hypothetical equality would imply that all per-feature sample vectors u_d have unitary average energy, i.e., $[X_{\mathcal{D}}^T X_{\mathcal{D}}]_{(d,d)} = ||u_d||^2 = 1$; and that they are orthogonal, i.e., $[X_{\mathcal{D}}^T X_{\mathcal{D}}]_{(d,d')} = u_d^T u_{d'} = 0$ for $d' \neq d$. So, all features are equally "large" and they are "uncorrelated". The more general case is discussed in Sec. 4.11.

Under this assumption, the LS solution (4.26) can be simplified as

$$\theta_{\mathcal{D}}^{ERM} = X_{\mathcal{D}}^T t_{\mathcal{D}} = \sum_{n=1}^{N} t_n \cdot u(x_n). \tag{4.27}$$

This expression shows that the ERM parameter vector $\theta_{\mathcal{D}}^{ERM}$, which solves the LS problem (4.20), "summarizes" the training set via a weighted sum of the feature vectors $u(x_n)$ corresponding to the data points in the training set. In this average, the weight of each feature vector $u(x_n)$ is given by the corresponding target t_n.

Furthermore, the ERM hard predictor is obtained as

$$\hat{t}(x|\theta_{\mathcal{D}}^{ERM}) = (\theta_{\mathcal{D}}^{ERM})^T u(x) = u(x)^T \theta_{\mathcal{D}}^{ERM}$$

$$= \sum_{n=1}^{N} t_n \cdot \underbrace{(u(x)^T u(x_n))}_{\text{inner product between } u(x) \text{ and } u(x_n)}. \tag{4.28}$$

This result, in turn, demonstrates that the optimal hard predictor for a new input x is a weighted sum of contributions from all training points. In it, the contribution of each training point x_n – given by the corresponding target t_n – is weighted by how much the feature vector $u(x_n)$ for data point n is "correlated" with the feature vector $u(x)$ of the new input x. A way to think about (4.28) is that the inner product $u(x)^T u(x_n)$ quantifies the amount of "**attention**" that the predictor should pay to example (x_n, t_n) when predicting the output of input x.

4.4   Inductive Bias Selection and Validation

How should we select an inductive bias? This section elaborates on this question by studying the problem of selecting a model class $\mathcal{H}$. Based on the discussion so far, ideally, the model class should be rich enough to include predictors that reflect the behavior of the variables of interest *outside* the training set well, while also enabling optimization of the model parameter θ with tractable computational and sample complexity. The **sample complexity** of a model class $\mathcal{H}$ refers to the amount of data that is needed to identify an effective predictor – through the model parameter θ – within $\mathcal{H}$. These two requirements – "richness" and sample complexity – are generally in conflict with each other, with the former calling for a larger model class $\mathcal{H}$ and the latter for a smaller $\mathcal{H}$. This section will introduce validation as a solution to the problem of resolving this tension. In Chapter 13, we will discuss a more sophisticated approach based on **meta-learning**, whereby data from related learning tasks are used for inductive bias optimization.

Example 4.2 (continued)

For Example 4.2, let us apply the ERM predictor (4.26) to the data set in Fig. 4.3 (represented as circles). The figure shows examples of ERM predictors obtained with different values of the degree M of the feature vector $u(x)$. The **hyperparameter** M determines the capacity of the model class $\mathcal{H}$, and is hence part of the assumed inductive bias. The figure shows that widely different behaviors of the predictor are obtained with different values of M outside the training set. How should we choose M?

Throughout this section, we will use Example 4.2 to elaborate on the problem of optimizing the degree M – as an instance of inductive bias selection. To proceed, let us define

Figure 4.3 Examples of hard predictors trained via ERM with different values of hyperparameter M given the training set $\mathcal{D}$ shown in the figure as circles. The dashed line illustrates the population-optimal unconstrained predictor $t^*(x)$.

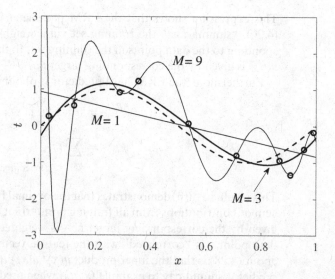

$$\mathcal{H}_M = \left\{ \hat{t}(x|\theta) = \theta^T u(x) = \sum_{i=0}^{M} \theta_i x^i \right\}, \qquad (4.29)$$

the model class of polynomials (4.7) of degree – or model order – M.

Benchmark Predictors. The ideal goal of learning is to minimize the population loss over all possible hard predictors, while in practice training is limited to the minimization of the training loss (or variations thereof) within a given model class. Accordingly, the problem of selecting the model class involves comparisons among the following predictors:

- **Population-optimal unconstrained predictor**. The population-optimal *unconstrained* predictor $\hat{t}^*(\cdot)$ minimizes the population loss without any constraint on the model class, i.e.,

$$\hat{t}^*(\cdot) = \arg\min_{\hat{t}(\cdot)} L_p(\hat{t}(\cdot)). \qquad (4.30)$$

  The population loss $L_p(\hat{t}^*(\cdot))$ obtained by this predictor is hence the minimum possible value of the population loss.

- **Population-optimal within-class predictor**. For the model class $\mathcal{H}_M$ in (4.29) of order M, the population-optimal *within-class* predictor is given as $\hat{t}(\cdot|\theta_M^*)$, where the **population-optimal model parameter** is

$$\theta_M^* = \arg\min_{\theta \in \mathbb{R}^{M+1}} L_p(\theta). \qquad (4.31)$$

  Unlike the unconstrained predictor $t^*(\cdot)$, the population-optimal within-class predictor $\hat{t}(\cdot|\theta_M^*)$ is constrained to belong to the model class $\mathcal{H}_M$. By definition, no other predictor within this class can improve over the population loss $L_p(\theta_M^*)$.

- **Trained ERM predictor**. For the model class $\mathcal{H}_M$ of order M in (4.29), the ERM predictor is given as $\hat{t}(\cdot|\theta_M^{ERM})$, where the ERM solution (4.17)

$$\theta_M^{ERM} = \arg\min_{\theta \in \mathbb{R}^{M+1}} L_{\mathcal{D}}(\theta) \qquad (4.32)$$

  is as in (4.42). Note that, in this section, as per (4.32), we will drop the dependence of the ERM solution on $\mathcal{D}$ in order to emphasize its dependence on M.

Performance Metrics. We will compare the performance of these predictors inside and outside the training set in order to obtain insights into the optimal selection of M. To this end, we will make use of two important metrics related to the choice of the model class, which will be formalized in the next section.

- **Estimation error**. The estimation error measures how well the ERM solution θ_M^{ERM} estimates the population-optimal within-class model parameter θ_M^*. A small estimation error indicates that the ERM solution is almost as good as it can be, given the constraint imposed by the selected degree M; while a large estimation error suggests that the available data are insufficient to identify an effective predictor with the model class.

- **Bias**. The bias measures how well the population-optimal within-class predictor $\hat{t}(\cdot|\theta_M^*)$ approximates the population-optimal unconstrained predictor $\hat{t}^*(\cdot)$. A small bias indicates that the model class is large enough to include good approximations of the population-optimal predictor, while a large bias implies that there are no good approximations of the population-optimal unconstrained predictor within the model class.

Figure 4.4 ERM predictor $\hat{t}(x|\theta_1^{ERM})$ (solid line) and population-optimal within-class predictor $\hat{t}(x|\theta_1^*)$ (dashed line), both with $M = 1$, along with the population-optimal unconstrained predictor $t^*(x)$ (gray line), for Example 4.2. Circles represent the data points in the training set $\mathcal{D}$.

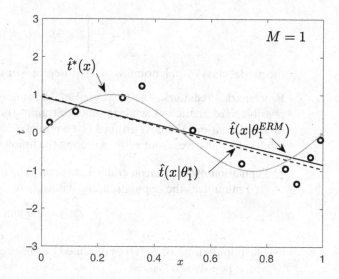

4.4.1   Underfitting

With $M = 1$, the ERM predictor $\hat{t}(x|\theta_1^{ERM})$, the population-optimal within-class predictor $\hat{t}(x|\theta_1^*)$, and the population-optimal unconstrained predictor $t^*(x)$ are shown in Fig. 4.4, along with the data points in the training set $\mathcal{D}$. As illustrated in the figure, with $M = 1$ the ERM predictor $\hat{t}(x|\theta_1^{ERM})$ **underfits** the training data, in the sense that the predictor does not predict well even the target variables in the training set. Underfitting is caused by the fact that model class $\mathcal{H}_1$ is not rich enough to capture the variations present in the training examples.

Underfitting yields both a large training loss $L_{\mathcal{D}}(\theta_{\mathcal{D}}^{ERM})$ and a large population loss $L_p(\theta_{\mathcal{D}}^{ERM})$. Therefore, it can generally be detected by computing the training loss $L_{\mathcal{D}}(\theta_M^{ERM})$ for different values of M.

Overall, with a small M, we have:

- a small **estimation error**, since the ERM solution is a good approximation of the population-optimal within-class parameter, i.e., $\theta_1^{ERM} \simeq \theta_1^*$; and
- a large **bias**, since, even with the population-optimal within-class parameter θ_1^*, the predictor $\hat{t}(x|\theta_1^*)$ is very different from the population-optimal unconstrained predictor $\hat{t}^*(x)$.

The bias is caused by the choice of a model class with insufficient capacity: The model class does not include any predictor that can represent the training data well, let alone generalize outside it. Therefore, the bias cannot be reduced with more data. Even with unlimited data, we would only have an accurate estimate of the population-optimal within-class predictor, i.e., we would have $\theta_1^{ERM} = \theta_1^*$.

4.4.2   Overfitting

With $M = 9$, the hard predictors $\hat{t}(x|\theta_9^{ERM})$ and $\hat{t}(x|\theta_9^*)$ are shown in Fig. 4.5. As illustrated by the figure, with $M = 9$ the ERM predictor $\hat{t}(x|\theta_9^{ERM})$ **overfits** the data: The model is too rich

Figure 4.5 ERM predictor $\hat{t}(x|\theta_9^{ERM})$ (solid line) and population-optimal within-class predictor $\hat{t}(x|\theta_9^*)$ (dashed line), both with $M = 9$, along with the population-optimal unconstrained predictor $t^*(x)$ (gray line), for Example 4.2. Circles represent the data points in the training set $\mathcal{D}$.

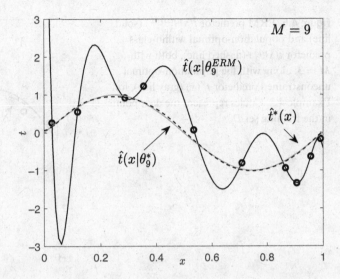

and, in order to account for the observations in the training set, it yields inaccurate predictions outside it. With overfitting, the model is **memorizing** the training set, rather than **learning** how to generalize to new examples. (Note that memorization does not, by itself, imply the failure of generalization, as further discussed in Sec. 4.6.)

Using the bias-estimation error terminology introduced above, with $M = 9$ we have:

- a large **estimation error**, since the ERM solution θ_9^{ERM} is not a good approximation of the population-optimal within-class parameter θ_9^*; and
- a small **bias**, since, with the population-optimal within-class parameter vector θ_9^*, the predictor $\hat{t}(x|\theta_9^*)$ would be a good approximation of the population-optimal unconstrained predictor $\hat{t}^*(x)$, i.e., $\hat{t}(x|\theta_9^*) \simeq \hat{t}^*(x)$.

Therefore, if we had more data, the problem of overfitting would be solved: With a larger N, we would be able to ensure that the ERM solution is a good approximation of the population-optimal within-class model, i.e., $\theta_9^{ERM} \simeq \theta_9^*$, with the latter providing a predictor $\hat{t}(x|\theta_9^*) \simeq \hat{t}^*(x)$ close to the population-optimal unconstrained predictor $t^*(x)$.

With the given amount of data shown in Fig. 4.6, the choice $M = 3$, illustrated in Fig. 4.6, seems to be a more reasonable one, as it trades some small bias – since $\hat{t}(x|\theta_3^*) \neq \hat{t}^*(x)$ – with a much reduced estimation error – since $\theta_3^{ERM} \simeq \theta_3^*$. But reaching this conclusion hinges on knowledge of the population loss (to obtain the population-optimal predictors), which is not available at the learner. This issue is addressed by validation, as discussed next.

4.4.3 Validation

As we have argued in the preceding subsection, an optimized selection of the model order requires a means to estimate the population loss so as to detect overfitting. The population loss (4.9) is the expectation over a test sample $(x,t) \sim p(x,t)$ that is drawn independently of the training set. Therefore, in order to estimate this expectation, one should use samples from the

Figure 4.6 ERM predictor $\hat{t}(x|\theta_3^{ERM})$ (solid line) and population-optimal within-class predictor $\hat{t}(x|\theta_3^*)$ (dashed line), both with $M = 3$, along with the population-optimal unconstrained predictor $t^*(x)$ (gray line), for Example 4.2. Circles represent the data points in the training set $\mathcal{D}$.

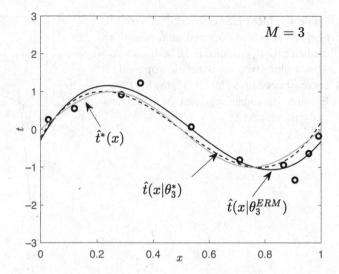

population distribution that are independent of the samples in the training set. Accordingly, **validation** estimates the population loss by using a data set, known as **validation set**, that is distinct from the training set.

To elaborate, define as $\mathcal{D}^v = \{(x_n^v, t_n^v)_{n=1}^{N^v}\}$ the validation data set of N^v data points. Importantly, the validation data set should not include data points from the training set $\mathcal{D}$. The validation estimate $\hat{L}_p(\theta)$ of the population loss for a model parameter vector θ is obtained as the empirical loss on the validation set, i.e.,

$$\hat{L}_p(\theta) = L_{\mathcal{D}^v}(\theta) = \frac{1}{N^v} \sum_{n=1}^{N^v} \ell(t_n^v, \hat{t}(x_n^v|\theta)). \tag{4.33}$$

In particular, given the ERM solution $\theta_{\mathcal{D}}^{ERM}$, the estimate of the population loss via validation is

$$\hat{L}_p(\theta_{\mathcal{D}}^{ERM}) = L_{\mathcal{D}^v}(\theta_{\mathcal{D}}^{ERM}). \tag{4.34}$$

The validation loss (4.34) is an **unbiased** and **consistent** estimate of the population loss $L_p(\theta_{\mathcal{D}}^{ERM})$. Specifically, the estimate is unbiased because the expectation of the estimate coincides with the population loss. Furthermore, it is consistent since, by the law of large numbers, as the number of validation points, N^v, increases, the estimate tends (with high probability) to the population loss.

What would happen if we reused the training set for validation? To address this question, assume, for the sake of argument, that the ERM solution $\theta_{\tilde{\mathcal{D}}}^{ERM}$ is obtained based on a data set $\tilde{\mathcal{D}} = \mathcal{D} \cup \mathcal{D}^v$ that also includes the training set $\mathcal{D}$. Then, the validation-based estimate $\hat{L}_p(\theta_{\tilde{\mathcal{D}}}^{ERM}) = L_{\mathcal{D}^v}(\theta_{\tilde{\mathcal{D}}}^{ERM})$ would no longer be an unbiased estimate of the population loss of the ERM predictor. In fact, this estimate would tend to be smaller than the true value of the population loss: The ERM solution $\theta_{\tilde{\mathcal{D}}}^{ERM}$ minimizes the empirical loss on the data set $\tilde{\mathcal{D}}$, hence making the loss on the samples in set $\mathcal{D}^v$ smaller, on average, than it would have been on an independent test sample.

Figure 4.7 Training loss and population loss, estimated via validation, as a function of the hyperparameter M, for Example 4.2. (The figure shows the square root of the loss.)

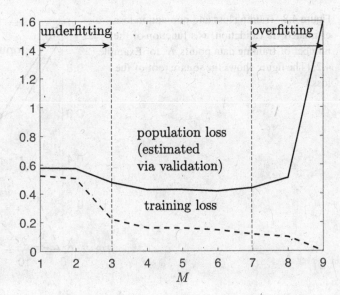

Example 4.2 (continued)

Returning to Example 4.2, we plot the training loss $L_{\mathcal{D}}(\theta_M^{ERM})$ and the validation-based estimate $\hat{L}_p(\theta_M^{ERM})$ of the population loss vs. the model order M in order to draw some conclusions about the best choice for M. These curves are displayed in Fig. 4.7, which plots the square root of the loss, i.e., the root mean squared error. The figure identifies the underfitting regime in which training loss – as well as the population loss – are large, and the overfitting regime in which there is a large gap between training and population losses. As we will discuss in Chapter 8, this gap is referred to as **generalization error**. The figure allows us to choose a suitable value of M that minimizes the estimate of the population loss. In this regard, the plot confirms that the choice $M = 3$ is a reasonable one.

The optimal choice of the model order M depends on the amount of available data. In fact, as the number N of training points increases, the estimation error decreases, and the performance becomes bounded by the bias. This in turn calls for the selection of a larger value of M in order to reduce the bias. This is illustrated in Fig. 4.8. As the figure also shows, when data are limited, it is generally preferable to choose a smaller model order M. This selection reduces the estimation error, which dominates the population loss in the data-limited regime.

4.4.4 K-Fold Cross-Validation

Validation requires setting aside some part of the data set, and is hence inefficient when the availability of data is limited. A common way to improve data efficiency for model selection is **K-fold cross-validation**, in which training and validation sets are pooled together in a single data set $\tilde{\mathcal{D}}$, while avoiding the bias problem described in the preceding subsection. As illustrated in Fig. 4.9, for any given model class $\mathcal{H}$, K-fold cross-validation follows these steps:

Figure 4.8 Training loss and population loss, estimated via validation, as a function of the number of training data points, N, for Example 4.2. (The figure shows the square root of the loss.)

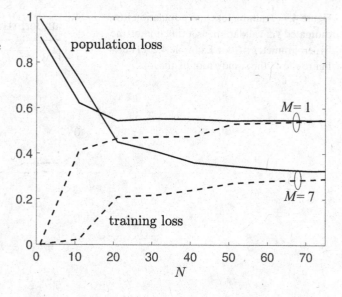

Figure 4.9 Illustration of K-fold cross-validation.

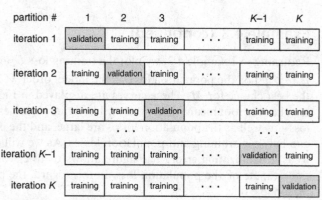

1. Divide the data set $\tilde{\mathcal{D}}$ into K partitions as $\tilde{\mathcal{D}} = \{\tilde{\mathcal{D}}_1, \ldots, \tilde{\mathcal{D}}_K\}$.
2. For each iteration $k = 1, \ldots, K$, train model $\theta^{ERM}_{\tilde{\mathcal{D}}_{-k}}$ on data set $\tilde{\mathcal{D}}_{-k} = \tilde{\mathcal{D}} \backslash \tilde{\mathcal{D}}_k$, which excludes the kth partition.
3. Obtain the validation estimate $L_{\mathcal{D}_k}(\theta^{ERM}_{\tilde{\mathcal{D}}_{-k}})$, where the kth partition, which was not used for training, is used as validation data.
4. Compute the final estimate of the population loss for model class $\mathcal{H}$ as

$$\hat{L}_p = \frac{1}{K} \sum_{k=1}^{K} L_{\mathcal{D}_k}(\theta^{ERM}_{\tilde{\mathcal{D}}_{-k}}). \tag{4.35}$$

5. Finally, choose the model $\mathcal{H}$ with the lowest estimated population loss.

Cross-validation provides an estimate of the minimum population loss for a given model class $\mathcal{H}$, and it can hence be used for model and hyperparameter selection. Importantly, this estimate is averaged over different model parameters $\theta^{ERM}_{\tilde{\mathcal{D}}_{-k}}$ that are trained using different subsets of the original data set $\tilde{\mathcal{D}}$. Therefore, unlike validation, cross-validation does not produce an estimate of the population loss for a single trained model obtained from the training set. Once the model class $\mathcal{H}$ is selected via cross-validation, conventional training should be carried out so as to obtain a single model parameter. It is emphasized that the statistical properties of the cross-validation estimate are more complex than for validation, with the latter providing easy-to-prove unbiasedness and consistency.

4.5   Bias and Estimation Error

Generalizing the discussion in the preceding section around Example 4.2, how can we formalize the trade-off between bias and estimation error? This section provides an answer to this question by introducing the important decomposition of the population loss in terms of minimum unconstrained population loss, bias, and estimation error; as well as by presenting the concepts of epistemic uncertainty and aleatoric uncertainty.

4.5.1   Population Loss = Minimum Population Loss + Bias + Estimation Error

Given a model class $\mathcal{H} = \{t(x|\theta) : \theta \in \Theta\}$, a training data set $\mathcal{D}$, and any training algorithm, e.g., ERM, we start by generalizing the definitions of the benchmark predictors given in the preceding section as follows:

- **Population-optimal unconstrained predictor**. The population-optimal unconstrained predictor $\hat{t}^*(\cdot)$ is defined as in (4.30).
- **Population-optimal within-class predictor**. Generalizing (4.31) to any model class, the population-optimal within-class predictor is defined as $\hat{t}(\cdot|\theta_{\mathcal{H}}^*)$ with **population-optimal model parameter vector**

$$\theta_{\mathcal{H}}^* = \underset{\theta \in \Theta}{\operatorname{argmin}}\, L_p(\theta). \tag{4.36}$$

- **Trained predictor**. The trained predictor is given as $\hat{t}(\cdot|\theta_{\mathcal{D}})$, with trained model parameter $\theta_{\mathcal{D}} \in \Theta$, e.g., the ERM solution in (4.17).

With these definitions at hand, the population loss obtained by any training algorithm producing model parameter $\theta_{\mathcal{D}}$ can be decomposed as

$$L_p(\theta_{\mathcal{D}}) = \underbrace{L_p(\hat{t}^*(\cdot))}_{\text{minimum unconstrained population loss}} + \underbrace{(L_p(\theta_{\mathcal{H}}^*) - L_p(\hat{t}^*(\cdot)))}_{\text{bias}}$$
$$+ \underbrace{(L_p(\theta_{\mathcal{D}}) - L_p(\theta_{\mathcal{H}}^*))}_{\text{estimation error}}.$$

Note that this equality can be trivially verified. In words, this corresponds to the identity

$$\text{population loss } = \text{minimum population loss } + \text{bias } + \text{estimation error}. \tag{4.37}$$

Accordingly, the actual performance of a training algorithm – its population loss $L_p(\theta_{\mathcal{D}})$ – is kept away from the ideal, minimum possible value $L_p(\hat{t}^*(\cdot))$ by the sum of bias and estimation error.

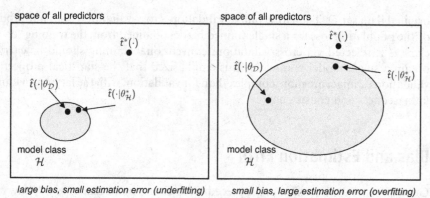

Figure 4.10 An illustration of the impact of the capacity of the model class on bias and estimation error.

On the Impact of the Capacity of a Model Class. This decomposition is useful in understanding the impact of the choice of the model class $\mathcal{H}$ on the performance of a trained model $\theta_{\mathcal{D}}$. On the one hand, increasing the capacity of the model class $\mathcal{H}$ (e.g., the degree M) decreases the bias, since the optimum in-class predictor $\hat{t}(\cdot|\theta_{\mathcal{H}}^*)$ will tend to be a better approximation of the population-optimal unconstrained predictor $\hat{t}^*(\cdot)$. On the other hand, it increases the estimation error, since the trained model $\hat{t}(\cdot|\theta_{\mathcal{D}})$ will tend to be further from the population-optimal within-class model $\hat{t}(\cdot|\theta_{\mathcal{H}}^*)$. This situation is sketched in the right-hand part of Fig. 4.10. Conversely, as illustrated in the left-hand part of Fig. 4.10, a small capacity of the model class $\mathcal{H}$ tends to decrease the estimation error – as $\hat{t}(\cdot|\theta_{\mathcal{D}})$ gets closer to $\hat{t}(\cdot|\theta_{\mathcal{H}}^*)$ – and increase the bias – as $t(\cdot|\theta_{\mathcal{H}}^*)$ tends to be further from $\hat{t}^*(\cdot)$.

On the Impact of the Number of Data Points. The optimal balance between bias and estimation error depends on the number of data points, N, in the training set. In fact, increasing N does not modify the bias, since the model parameter $\theta_{\mathcal{H}}^*$ and the predictor $\hat{t}^*(\cdot)$ do not depend on N; but it decreases the estimation error, as the estimated model parameter $\theta_{\mathcal{D}}$ tends to be closer to $\theta_{\mathcal{H}}^*$.

This behavior is illustrated in Fig. 4.11, which is obtained from Example 4.2 with ERM used as the training procedure. The population loss $L_p(\hat{t}^*(\cdot))$ of the optimal unconstrained predictor is the smallest possible population loss. The choice of the model class $\mathcal{H}$ implies an irreducible gap with respect to this optimal performance, which is quantified by the bias, i.e., by the difference $L_p(\theta_{\mathcal{H}}^*) - L_p(\hat{t}^*(\cdot))$. The population loss obtained by the trained model, $L_p(\theta_{\mathcal{D}})$, tends to the population-optimal within-class performance $L_p(\theta_{\mathcal{H}}^*)$ as N grows, yielding a vanishing estimation error $L_p(\theta_{\mathcal{D}}) - L_p(\theta_{\mathcal{H}}^*)$.

For reference, the figure also shows the training loss $L_{\mathcal{D}}(\theta_{\mathcal{D}})$. Unlike the population losses, the training loss increases with N, since more data points typically make it more difficult to obtain a good fit. As N increases, by the law of large numbers, the training loss tends, from below, to the corresponding population loss $L_p(\theta_{\mathcal{D}})$.

4.5.2 Epistemic and Aleatoric Uncertainty

The "population loss = minimum population loss + bias + estimation error" decomposition in (4.37) applies to the overall data set $\mathcal{D}$. We now consider a more granular version of this decomposition that applies separately to each value x of the input. This will allow us to

Figure 4.11 Population losses of the optimal unconstrained predictor $\hat{t}^*(\cdot)$, the optimal within-class model parameter $\theta_{\mathcal{H}}^*$, and the trained model $\theta_{\mathcal{D}}$, along with the training loss $L_{\mathcal{D}}(\theta_{\mathcal{D}})$ for Example 4.2.

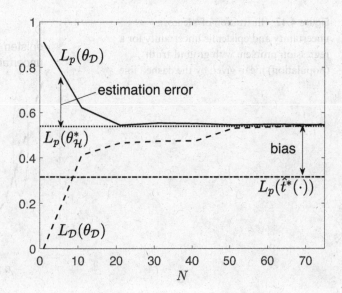

decompose the average loss obtained by a trained predictor $\hat{t}(x|\theta_{\mathcal{D}})$ on a given input x – which can be thought of as a measure of uncertainty in the target t given the observation x – in terms of two different types of uncertainty – aleatoric and epistemic. This decomposition provides a useful conceptual framework in which to interpret the performance of a trained predictor, and it plays a particularly important role in Bayesian learning (see Chapter 12).

To elaborate on this point, let us write as

$$L_p(\theta|x) = E_{t\sim p(t|x)}[\ell(t, \hat{t}(x|\theta))] \tag{4.38}$$

the population loss of a predictor $\hat{t}(x|\theta)$ conditioned on a given value x of the input. We similarly write

$$L_p(\hat{t}^*(x)|x) = E_{t\sim p(t|x)}[\ell(t, \hat{t}^*(x))] \tag{4.39}$$

for the corresponding conditional population loss of the population-optimal unconstrained predictor $\hat{t}^*(x)$.

In a manner similar to (4.37), for a trained predictor $\hat{t}(x|\theta_{\mathcal{D}})$, we then have the decomposition

$$L_p(\theta_{\mathcal{D}}|x) = \underbrace{L_p(\hat{t}^*(x)|x)}_{\text{aleatoric uncertainty}} + \underbrace{(L_p(\theta_{\mathcal{H}}^*|x) - L_p(\hat{t}^*(x)|x))}_{\text{bias}} + \underbrace{(L_p(\theta_{\mathcal{D}}|x) - L_p(\theta_{\mathcal{H}}^*|x))}_{\text{epistemic uncertainty}}. \tag{4.40}$$

Accordingly, the overall predictive uncertainty $L_p(\theta_{\mathcal{D}}|x)$ associated with trained predictor $\hat{t}(x|\theta_{\mathcal{D}})$ on input x can be decomposed into:

- the **bias** contribution $L_p(\theta_{\mathcal{H}}^*|x) - L_p(\hat{t}^*(x)|x)$, which is dictated by the choice of the model class;
- the **aleatoric uncertainty** contribution $L_p(\hat{t}^*(x)|x)$, which measures the inherent randomness of the mechanism $p(t|x)$ generating output t for input x; and
- the **epistemic uncertainty** contribution $L_p(\theta_{\mathcal{D}}|x) - L_p(\theta_{\mathcal{H}}^*|x)$, which instead quantifies the uncertainty about the optimal choice of the model parameter within the model class $\mathcal{H}$ given the access to a limited data set $\mathcal{D}$.

Figure 4.12 Illustration of aleatoric uncertainty and epistemic uncertainty for a regression problem with ground-truth (population) mean given by the dashed line.

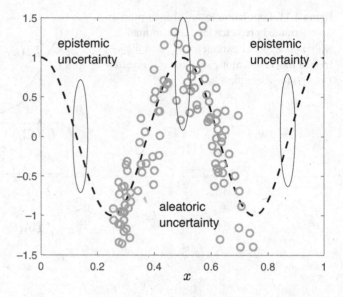

Importantly, no amount of data can reduce aleatoric uncertainty, which is inherent in the nature of the relationship between input and output variables, as it corresponds to the performance obtained by the population-optimal unconstrained predictor. In stark contrast, epistemic uncertainty exists only if access to data is limited: When more and more data are collected, the trained model parameter $\theta_{\mathcal{D}}$ will tend to the population-optimal within-class model $\theta_{\mathcal{H}}^*$, yielding a vanishing epistemic uncertainty.

An illustration of epistemic and aleatoric uncertainty can be found in Fig. 4.12. The figure shows data generated from a population distribution with mean equal to the dashed line. For values of x around which a significant number of data points (circles) are available, the observed uncertainty is to be ascribed to the inherent randomness of the data generation mechanism – this is an example of aleatoric uncertainty. In contrast, in parts of the input space in which data are not available, uncertainty is, at least partly, epistemic, since it can be reduced by collecting more data.

As a final note on this topic, the decomposition (4.40) is particularly useful for **active learning** protocols. In active learning, rather than being faced with a fixed data set, the learner is allowed to choose the next input value x at which to measure the output $t \sim p(t|x)$. The decomposition above suggests that one should ideally select values of x with a large epistemic uncertainty, since the latter can be reduced when collecting more data. In other words, epistemic uncertainty quantifies potential knowledge gains that can be accrued with additional data. More discussion on active learning can be found in Chapter 12.

4.6   Beyond the Bias vs. Estimation Error Trade-Off

While providing a theoretical framework in which to understand the performance of data-driven systems, the bias vs. estimation trade-off outlined in the preceding section is not without its limitations. These have become particularly clear in recent years with the success

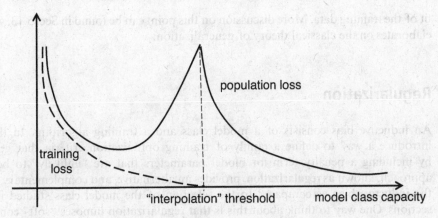

Figure 4.13 The "double descent" phenomenon.

of large-capacity machine learning models like deep neural networks (see Chapter 6). In fact, the classical view on generalization outlined in the preceding section predicts that an increase in the model class capacity eventually leads to a growing gap between training and population losses due to overfitting (see, e.g., Fig. 4.7). But this is not what is observed when training large models like deep neural networks: As illustrated in Fig. 4.13, instead of continuously growing, the population loss is often seen to decrease after the **"interpolation point"** at which the training loss is close to zero and hence the model has "memorized" the training set. This is known as the **"double descent"** behavior of the population loss.

The key issue appears to be – and this is a topic of current research at the time of writing – that the classical bias vs. estimation error framework described in the preceding section conflates the choice of the inductive bias with that of the capacity of a model. But this is an oversimplification that omits critical elements such as the suitability of the specific model class to the given population distribution, as well as the choice of the training algorithm.

Regarding the **interplay between population distribution and choice of the model class**, while a large-capacity model may fit the training data almost perfectly, this "overfitting" does not necessarily imply a degradation of performance on the test set for *every* possible population distribution. For instance, model classes such as neural networks appear to be particularly well suited to operate on natural signals such as images, making overfitting to training data less problematic.

As for the role of the **training algorithm**, as will be detailed in the next chapter, practical optimization algorithms used to train machine learning models operate via local search steps. As such, they are generally not guaranteed to obtain optimal solutions to the ERM problem (or variations thereof). Larger models, when coupled with suboptimal training solutions, may provide a better inductive bias than smaller models. In fact, overparametrized, i.e., large-capacity, models may make it easier to find good model parameter vectors, since there may be more such solutions in a larger model space. For instance, it has been shown that local-search training algorithms may fail to train a model even when data have been generated from the model itself; while they succeed when the model class is larger than the model used to generate training data.

Finally, some results suggest that local optimization algorithms act as **implicit regularizers** (see the next section) that can obtain solutions that generalize well even when yielding an excellent fit of the training data. More discussion on this point can be found in Sec. 5.13, while Chapter 8 elaborates on the classical theory of generalization.

4.7   Regularization

An inductive bias consists of a model class and a training algorithm. In this section, we introduce a way to define a family of training optimization criteria that generalize ERM by including a penalty term for model parameters that are "unlikely" to be correct. This approach, known as **regularization**, provides an alternative, and complementary, way to control the inductive bias as compared to the selection of the model class studied in the previous sections. One way to think about this is that regularization imposes "soft" constraints on the model class that are enforced during training.

Example 4.2 (continued)

For Example 4.2, given the data set in Fig. 4.1, evaluating the ERM solution θ_M^{ERM} for different values of M yields:

- for $M = 1$, $\theta_1^{ERM} = [0.93, \, -1.76]^T$;
- for $M = 3$, $\theta_3^{ERM} = [-0.28, 13.32, \, -35.92, 22.56]^T$; and
- for $M = 9$, $\theta_9^{ERM} = [13.86, -780.33, 12.99 \cdot 10^3, \, -99.27 \cdot 10^3, 416.49 \cdot 10^3, \, -1.03 \cdot 10^6, 1.56 \cdot 10^6,$ $1.40 \cdot 10^6, 0.69 \cdot 10^6, -1.44 \cdot 10^6]^T$.

Based on this evidence, overfitting, which we know from Fig. 4.5 to occur when $M = 9$, appears to be manifested in a large value of the squared ℓ_2 norm $||\theta_M^{ERM}||^2$ of the trained model parameter vector. This suggests that we should ensure that the norm of the trained vector is small enough if we want to avoid overfitting.

4.7.1   Ridge Regression

The condition identified in the preceding example can be enforced by adding a term to the training loss that penalizes model parameter vectors with a large squared ℓ_2 norm. This is known as ℓ_2 **regularization**. The resulting learning problem is given by the optimization

$$\min_{\theta \in \Theta} \left\{ L_{\mathcal{D}}(\theta) + \frac{\lambda}{N} ||\theta||^2 \right\}, \tag{4.41}$$

where $\lambda > 0$ is a **hyperparameter** that defines the desired trade-off between minimization of the training loss and regularizing penalty. The learning criterion $L_{\mathcal{D}}(\theta) + \lambda/N \cdot ||\theta||^2$ is known as **regularized training loss** with an ℓ_2 regularization. The division of the hyperparameter λ by N in (4.41) ensures that, as N increases, the relative weight of the regularizing term in problem (4.41) decreases as compared to the training loss. This is a useful property since, as N grows, the training loss becomes an increasingly accurate estimate of the population loss, and regularization is no longer required to avoid overfitting. That said, for a fixed N, one can practically remove the division by N and redefine the hyperparameter as $\tilde{\lambda} = \lambda/N$.

Figure 4.14 Training and population losses as a function of the regularization hyperparameter λ for Example 4.2, with $M = 9$.

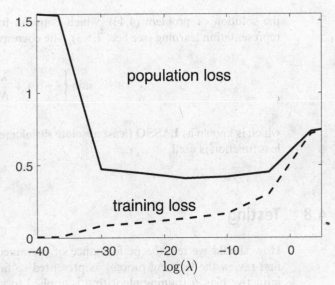

When the loss function is quadratic, as for in Example 4.2, the optimization (4.41) is referred to as **ridge regression**. As will be detailed in the next chapter, this is an LS problem, which has the closed-form solution

$$\theta_{\mathcal{D}}^{R-ERM} = (\lambda I_D + X_{\mathcal{D}}^T X_{\mathcal{D}})^{-1} X_{\mathcal{D}}^T t_{\mathcal{D}}. \tag{4.42}$$

Each value of the hyperparameter λ in (4.41) yields a different training criterion and correspondingly a distinct inductive bias. To select the hyperparameter λ, one can use validation with the aim of identifying a value λ that minimizes an estimate of the population loss.

Example 4.2 (continued)

The training and population losses are shown as a function of the regularization hyperparameter λ for Example 4.2 in Fig. 4.14, with $M = 9$. Increasing λ has a similar effect as decreasing the capacity of the model, moving the operating regime closer to underfitting, while decreasing λ has a similar effect as increasing the capacity of the model class, eventually causing overfitting.

4.7.2 Regularized ERM

Generalizing ridge regression, we define as **regularized ERM** the learning problem

$$\theta_{\mathcal{D}}^{R-ERM} = \arg\min_{\theta \in \Theta} \left\{ L_{\mathcal{D}}(\theta) + \frac{\lambda}{N} R(\theta) \right\}, \tag{4.43}$$

where $R(\theta)$ is a **regularization function**. The inclusion of the regularization function in (4.43) penalizes model parameter vectors θ with larger values $R(\theta)$. It is emphasized that regularization is not applied to the population loss, which is defined in the usual way as in (4.9). Regularization only modifies the training algorithm through the learning criterion in (4.43).

Apart from the squared ℓ_2 regularization function $R(\theta) = \|\theta\|^2$, other standard choices include the ℓ_1 norm $R(\theta) = \|\theta\|_1 = \sum_{i=0}^{M} |\theta_i|$. This penalty term promotes the sparsity of the solution of problem (4.43), which is useful in many signal recovery algorithms and in representation learning (see Sec. 7.5.3). The corresponding optimization problem is given as

$$\min_{\theta} \left\{ L_{\mathcal{D}}(\theta) + \frac{\lambda}{N} \|\theta\|_1 \right\}, \tag{4.44}$$

which is known as **LASSO (least absolute shrinkage and selection operator)** when the quadratic loss function is used.

4.8   Testing

How should we test the performance of a trained model? To address this question, let us first review the learning process as presented so far. As illustrated in Fig. 4.15, based on an inductive bias, a training algorithm is applied to a training set in order to produce a model parameter vector whose population loss is estimated via validation. In practice, one typically iterates several times between training and validation in order to obtain the final trained model. When several iterations of training and validation are performed, the validation loss cannot be used as an estimate of the population loss for the trained model. In fact, when the validation set is reused across the mentioned iterations, the validation loss becomes a *biased* estimate of the population loss that tends to underestimate the population loss. This conclusion follows the same arguments provided in Sec. 4.4 in support of validation: After more than one iteration, the trained model is effectively optimized to fit the validation data, which can no longer be used to estimate the population loss.

In light of this, as seen in Fig. 4.15, the final estimate of the population loss should be done on a separate set of data points, referred to as the **test set**, that are set aside for this purpose and not used at any stage of inductive bias selection or learning. This estimate of the population loss is known as **test loss**. In competitions among different machine learning algorithms, the test set is kept by a judge to evaluate the submissions, and is never shared with the contestants.

Figure 4.15 An overview of training, validation, and testing, along with the corresponding data sets.

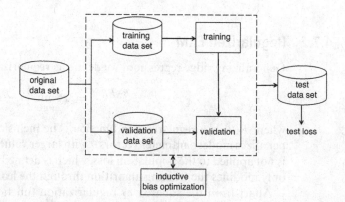

4.9 Training Soft Predictors

In this section, we introduce the problem of supervised learning for the design of soft predictors.

4.9.1 Model Class

A model class of soft predictors is given by a family of conditional distributions $p(t|x,\theta)$ parametrized by a **model parameter vector** $\theta \in \Theta$ as

$$\mathcal{H} = \{p(t|x,\theta) : \theta \in \Theta\}. \tag{4.45}$$

Each distribution $p(t|x,\theta)$, specified by a value of model parameter vector θ, represents a soft predictor. We remark that the notation $p(t|x,\theta)$ is different from that used in Chapter 3 to describe a soft predictor, namely $q(t|x)$. The rationale for this notational choice is that the dependence on the model parameter θ should provide sufficient context to identify $p(t|x,\theta)$ as a trainable soft predictor. Furthermore, the use of this notation has the advantage of emphasizing that the general goal of the predictor $p(t|x,\theta)$ is to be a good approximation of the true posterior $p(t|x)$.

Example 4.2 (continued)

For the regression problem in Example 4.2, we can define a probabilistic model class that includes soft predictors of the form

$$p(t|x,\theta) = \mathcal{N}(t|\mu(x|\theta),\beta^{-1}), \tag{4.46}$$

for some fixed precision parameter $\beta > 0$ (see Sec. 3.4), where the mean is given by the polynomial function

$$\mu(x|\theta) = \sum_{j=0}^{M} \theta_j x^j = \theta^T u(x), \tag{4.47}$$

where the $(M + 1) \times 1$ feature vector $u(x)$ is defined as in (4.7). Note that this model class assumes that the target has mean equal to the hard predictor (4.7) and fixed variance β^{-1}. The randomness assumed by this model class allows us to account, to some extent, for the inherent aleatoric uncertainty related to the relationship between x and t under the population distribution.

Discriminative vs. Generative Models. In this chapter, we cover **discriminative** probabilistic models that take the form of parametric conditional distributions $p(t|x,\theta)$. An alternative type of model class is given by **generative** probabilistic models that include parametric joint distributions of the form $p(x,t|\theta)$. Writing $p(x,t|\theta) = p(x|\theta)p(t|x,\theta)$ (by the chain rule of probability) reveals that, unlike discriminative models, generative models also account for the marginal distribution $p(x|\theta)$ – possibly dependent on model parameters θ – of the input x. This type of inductive bias can be useful if one can more naturally model the distribution $p(x|t,\theta)$ of input x given the target variable t. Discriminative models can be obtained from generative models, since one can compute (at least in principle) $p(t|x,\theta) = p(x,t|\theta)/p(x|\theta)$. But the converse is not true, as discriminative models do not account for the input marginal $p(x|\theta)$.

We refer to Chapter 6 for further discussion on the relationship between discriminative and generative models.

4.9.2 Population Log-Loss

As discussed in Sec. 3.6, the quality of a soft prediction $p(t|x, \theta)$ for a pair (x, t) can be evaluated via the **log-loss**, or **cross-entropy loss**,

$$- \log p(t|x, \theta). \tag{4.48}$$

Accordingly, for a fixed model class $\mathcal{H}$, the general goal of supervised learning for soft predictors is to minimize the **population log-loss**

$$L_p(\theta) = \mathrm{E}_{(\mathrm{x,t}) \sim p(x, t)}[- \log p(\mathrm{t}|\mathrm{x}, \theta)], \tag{4.49}$$

which represents the average log-loss measured on an independently generated test pair $(\mathrm{x}, \mathrm{t}) \sim p(x, t)$.

The population log-loss can be expressed as the average **cross entropy** between the population posterior distribution $p(t|x)$ and the soft predictor $p(t|x, \theta)$, as in

$$L_p(\theta) = \mathrm{E}_{\mathrm{x} \sim p(x)}[\mathrm{H}(\underbrace{p(t|\mathrm{x})}_{\text{true posterior distribution}} \| \underbrace{p(t|\mathrm{x}, \theta)}_{\text{soft predictor}})], \tag{4.50}$$

where we have used the definition of the cross entropy (see Sec. 3.6.1),

$$\mathrm{H}(p(t|x) \| p(t|x, \theta)) = \mathrm{E}_{\mathrm{t} \sim p(t|x)}[- \log p(\mathrm{t}|x, \theta)]. \tag{4.51}$$

4.9.3 Training Log-Loss and Maximum Likelihood Learning

Based on the training data $\mathcal{D}$, the learner can compute the **training log-loss** for a given model parameter vector θ as

$$L_{\mathcal{D}}(\theta) = \frac{1}{N} \sum_{n=1}^{N} (- \log p(t_n|x_n, \theta)). \tag{4.52}$$

The training log-loss measures the empirical average of the log-loss accrued by the predictor on the examples in the training set $\mathcal{D}$. The negative log-loss, $\log p(t_n|x_n, \theta)$, is also referred to as the **log-likelihood** of model parameter θ for data point (x_n, t_n).

The ERM problem applied to the log-loss is given as the minimization

$$\theta_{\mathcal{D}}^{ML} = \arg \min_{\theta \in \Theta} L_{\mathcal{D}}(\theta). \tag{4.53}$$

Since the negative log-loss is known as log-likelihood, this problem is referred to as **maximum likelihood (ML) learning**.

Example 4.2 (continued)

Let us formulate ML learning for Example 4.2 using the class of predictors $p(t|x, \theta) = \mathcal{N}(t|\mu(x|\theta), \beta^{-1})$ with mean function (4.47). The log-loss for any example (x, t) is given as

$$- \log p(t|x, \theta) = \frac{\beta}{2}(t - \mu(x|\theta))^2 - \frac{1}{2}\log(2\pi\beta). \tag{4.54}$$

Omitting multiplicative and additive constants that do not depend on the parameter vector θ, the ML problem (4.53) can hence be equivalently stated as

$$\min_{\theta \in \Theta} \frac{1}{N} \sum_{n=1}^{N} (t_n - \mu(x_n|\theta))^2. \qquad (4.55)$$

Therefore, ML learning of a Gaussian probabilistic model class with linear mean (4.47) is equivalent to ERM problem (4.20) for the training of the model class of linear hard predictors (4.7) under the ℓ_2 loss. Therefore, the same closed-form solution (4.26) applies. This equivalence between ML and ERM problems for the given settings can also be directly concluded based on the relationship between log-loss with Gaussian soft predictors and the ℓ_2 loss that was highlighted in Sec. 3.8.

4.9.4 Interpretations of ML Learning

We now discuss a number of useful *interpretations* of ML learning, which provide conceptual frameworks in which to think about the properties of ML learning.

- **ML learning chooses a model parameter θ that maximizes the probability of the training data set under the model**. Given that the data points are assumed to be i.i.d., the log-probability of observing the outputs $t_{\mathcal{D}} = \{t_1, \ldots, t_N\}$ in the training set given the inputs $x_{\mathcal{D}} = \{x_1, \ldots, x_N\}$ can be written as

$$\log p(t_{\mathcal{D}}|x_{\mathcal{D}}, \theta) = \log \left(\prod_{n=1}^{N} p(t_n|x_n, \theta) \right)$$

$$= \sum_{n=1}^{N} \log p(t_n|x_n, \theta) = -N \cdot L_{\mathcal{D}}(\theta). \qquad (4.56)$$

Therefore, minimizing the training log-loss $L_{\mathcal{D}}(\theta)$ is equivalent to maximizing the log-probability $p(t_{\mathcal{D}}|x_{\mathcal{D}}, \theta)$ of the data under the model. It is useful to remark that this probability is assigned by the specific model $p(t_{\mathcal{D}}|x_{\mathcal{D}}, \theta)$, and is generally distinct from the ground-truth conditional distribution $p(t|x)$ obtained from the population distribution (see Sec. 9.7 for further discussion on this point).

- **ML minimizes the surprise of the predictor when observing the training data**. As discussed in Sec. 3.6, the information-theoretic surprise of the predictor $p(t|x, \theta)$ when observing a pair (x, t) is given by the log-loss $-\log p(t|x, \theta)$. More generally, the surprise of the predictor in observing the entire training data set $\mathcal{D}$ is quantified by $\log p(t_{\mathcal{D}}|x_{\mathcal{D}}, \theta)$ in (4.56). Therefore, by the previous interpretation, ML minimizes the information-theoretic surprise of the predictor in observing the target variables in the training set $\mathcal{D}$ (for the given inputs $x_{\mathcal{D}}$). This interpretation is leveraged in some models of biological behavior to explain how organisms optimize their internal model (predictor) of their sensory inputs (see Sec. 7.10.12). As a side note, in some problems, one may also be interested in *maximizing* the surprise, particularly when optimizing over the next observation to be made by an agent in active learning (see Sec. 4.5).

- **ML learning minimizes the average cross entropy between the empirical distribution of the data and the corresponding distribution under the model**. Using (4.12) with the log-loss allows us to write the training log-loss as

$$L_{\mathcal{D}}(\theta) = E_{(x,t) \sim p_{\mathcal{D}}(x,t)} \left[-\log p(t|x, \theta) \right]$$
$$= E_{x \sim p_{\mathcal{D}}(x)} \left[H(p_{\mathcal{D}}(t|x) || p(t|x, \theta)) \right]. \tag{4.57}$$

Therefore, ML learning minimizes the average cross entropy between the empirical conditional distribution of the data $p_{\mathcal{D}}(t|x) = p_{\mathcal{D}}(x, t)/p_{\mathcal{D}}(x)$ and the corresponding distribution $p(t|x, \theta)$ under the model.

- **ML learning minimizes the average KL divergence between the empirical distribution of the data and the corresponding distribution under the model.** Recalling the identity $\text{KL}(p||q) = H(p||q) - H(p)$ from the previous chapter, by (4.57), ML can be equivalently formulated as the minimization of the average KL divergence

$$\theta_{\mathcal{D}}^{ML} = \arg \min_{\theta \in \Theta} E_{x \sim p_{\mathcal{D}}(x)} \left[\text{KL}(p_{\mathcal{D}}(t|x) || p(t|x, \theta)) \right]. \tag{4.58}$$

Therefore, by the law of large numbers, as $N \to \infty$ we informally have the limit $p_{\mathcal{D}}(x, t) \to p(x, t)$ and ML minimizes the average KL divergence

$$E_{x \sim p(x)} [\text{KL}(p(t|x) || p(t|x, \theta))]. \tag{4.59}$$

With a high-capacity model class, the minimum of the resulting problem is attained when the model equals the population-optimal unconstrained soft predictor, i.e., for $p(t|x, \theta) \simeq p(t|x)$.

- **ML learning maximizes a lower bound on the mutual information $I(x; t)$ evaluated with respect to the empirical joint distribution** $p_{\mathcal{D}}(x, t)$. For a discussion and a proof of this interpretation, we refer to Appendix 4.B.

4.9.5   From Soft to Hard Predictors

As discussed in Sec. 3.3, once a soft predictor is trained, one can in principle obtain the corresponding optimal hard predictor for any loss function $\ell(t, \hat{t})$ by solving the problem

$$\hat{t}(x|\theta) = \arg \min_{\hat{t}} E_{t \sim p(t|x, \theta)} [\ell(t, \hat{t})] \tag{4.60}$$

separately for each value x. In this sense, probabilistic model learning is a more general problem than training hard predictors, since we can obtain an optimized hard predictor, for any loss function, from the optimal soft predictor. This conclusion is further compounded by the "universality" of the log-loss as loss function detailed in Sec. 3.8.

4.10   MAP Learning and Regularization

As will be seen in this section, regularization has a natural probabilistic interpretation when applied to the ML problem. Specifically, regularization encodes prior information about the model parameter vector θ in the form of a distribution $p(\theta)$ on the model parameter space. This distribution, known as the **prior**, is part of the inductive bias. The prior $p(\theta)$ provides a priori "scores" to all values of θ, assigning low scores to values that are undesirable or a priori unlikely. While the prior $p(\theta)$ also plays a central role in Bayesian learning (Chapter 12), it is important to emphasize that the techniques discussed in this subsection are still frequentist.

Example 4.2 (continued)

As we have discussed in Sec. 4.7, model parameter vectors with a large norm are unlikely to be optimal, since they are typically a hallmark of overfitting. This prior information on the problem implies that it is sensible to assign a low prior score to values of θ that have a large norm. To this end, one possible choice is to assume the Gaussian prior

$$\theta \sim p(\theta) = \mathcal{N}(\theta | 0_D, \alpha^{-1} I_D) \tag{4.61}$$

for some precision parameter α and $D = M + 1$. A larger prior precision α indicates a more confident prior, which makes it less likely that values with a large norm are selected during training.

4.10.1 MAP Learning

How do we integrate the prior distribution $p(\theta)$ with the likelihood

$$p(t_{\mathcal{D}} | x_{\mathcal{D}}, \theta) = \prod_{n=1}^{N} p(t_n | x_n, \theta) \tag{4.62}$$

evaluated on the training set $\mathcal{D}$? The answer lies in the chain rule of probability reviewed in Sec. 2.10. Accordingly, the joint distribution of model parameter vector θ and the target variables $t_{\mathcal{D}}$ in the data set $\mathcal{D}$ given the inputs $x_{\mathcal{D}}$ factorizes as

$$p(\theta, t_{\mathcal{D}} | x_{\mathcal{D}}) = p(\theta) p(t_{\mathcal{D}} | x_{\mathcal{D}}, \theta). \tag{4.63}$$

Note that the covariates $x_{\mathcal{D}}$ are to be considered as fixed in the joint distribution (4.63).

With the joint distribution (4.63), we can now consider the inference problem of predicting model parameter θ given the observation of the data set $\mathcal{D}$ within the framework developed in Chapter 3. By accounting for the prior distribution, we have hence turned the learning problem into one of inference. As we will discuss in Chapter 12, Bayesian learning obtains a soft predictor of the model parameter vector θ. In contrast, as detailed next, frequentist learning evaluates a hard predictor.

Specifically, one typically adopts the maximum a posteriori (MAP) hard predictor of the model parameter vector,

$$\theta_{\mathcal{D}}^{MAP} = \arg \min_{\theta \in \Theta} \left\{ \underbrace{-\frac{1}{N} \log p(t_{\mathcal{D}} | x_{\mathcal{D}}, \theta)}_{\text{training log-loss}} + \frac{1}{N} \underbrace{\left(-\log p(\theta) \right)}_{\text{regularization}} \right\}. \tag{4.64}$$

MAP learning tackles problem (4.64) to obtain the trained predictor $\hat{t}(\cdot | \theta_{\mathcal{D}}^{MAP})$.

The name "MAP learning" may seem at first misleading since problem (4.64) amounts to the maximization of the logarithm of the *joint* distribution of model parameter and data, and not of the posterior. However, the optimization (4.64) yields the same result as maximizing the posterior distribution $p(\theta | \mathcal{D})$ of the model parameters, since the posterior distribution is proportional to the joint distribution, i.e., we have

$$p(\theta | \mathcal{D}) \propto p(\theta, t_{\mathcal{D}} | x_{\mathcal{D}}), \tag{4.65}$$

with a normalization factor that does not depend on θ.

Example 4.2 (continued)

With the Gaussian prior (4.61), MAP learning (4.64) is easily seen to be equivalent to the ridge regression problem (4.41), where the hyperparameter λ is related to the precisions of prior and likelihood as $\lambda = \alpha/\beta$.

Laplace Prior and ℓ_1 Norm Regularization. Other standard examples of the prior distribution include the Laplace pdf, which yields the ℓ_1 norm regularization function $R(\theta) = \|\theta\|_1 = \sum_{i=0}^{M} |\theta_i|$. More discussion on the prior distribution can be found in Chapter 12 in the context of Bayesian learning.

4.10.2 MAP Learning As Minimum Description Length Inference

In this subsection, we revisit MAP learning through the lens of information theory by following the interpretation of the log-loss introduced in Sec. 3.6.1 in terms of information-theoretic surprise and compression. According to this perspective, as we will detail next, MAP learning aims to minimize the description length of the target variables $t_{\mathcal{D}}$, and is hence an instance of **minimum description length (MDL)** inference.

Let us elaborate on this point by first reviewing the discussion in Sec. 3.6.1. In information theory, for a discrete rv $z \sim p(z)$, the log-loss $-\log p(z)$, or $-\log_2 p(z)$, is referred to as **description length** and is measured in nats or bits, respectively. The terminology reflects the key result that an encoder can communicate the realization $z = z$ to a decoder that is aware of the distribution $p(z)$ by transmitting a message of (approximately) $-\log_2 p(z)$ nats, i.e., $-\log_2 p(z)$ bits. Accordingly, the more "surprising" a message is, the longer is its description.

With this interpretation in mind, the criterion optimized by MAP learning, namely

$$\underbrace{(-\log p(\theta))}_{\text{description length for } \theta} + \underbrace{(-\log p(t_{\mathcal{D}}|x_{\mathcal{D}},\theta))}_{\text{description length for the target rvs } t_{\mathcal{D}}} \tag{4.66}$$

has the following **"two-part description" interpretation**. Consider a communication setting in which both encoder and decoder have access to the prior $p(\theta)$ and the likelihood $p(t|x,\theta)$. They also both know the covariates $x_{\mathcal{D}}$ in the training set $\mathcal{D}$. The encoder first describes the parameter vector θ with $-\log p(\theta)$ nats; and then the target variables $t_{\mathcal{D}}$ are described for the given θ with $-\log p(t_{\mathcal{D}}|x_{\mathcal{D}},\theta)$ nats. MAP learning chooses the parameter θ that minimizes the length of this description. Therefore, MAP learning can be interpreted as minimizing the description length of the target variables $t_{\mathcal{D}}$ using a two-part code that first encodes the model parameter and then the data $t_{\mathcal{D}}$.

4.11 Kernel Methods

This chapter, and most of this book, focus on **parametric** models, for which the model class includes parametric functions describing hard or soft predictors. As a result of learning, parametric methods "summarize" the training data set $\mathcal{D}$ into a trained parameter vector $\theta_{\mathcal{D}}$, in the sense that the prediction function depends on the data set $\mathcal{D}$ only through $\theta_{\mathcal{D}}$. In contrast, **non-parametric** methods, to be introduced in this section, produce predictions for each test input x that depend directly on the training data (rather than on a parametric "summary" of the data set). The inductive bias adopted by non-parametric methods encodes the assumption that the prediction $\hat{t}(x)$ for an input x should be "similar" to the outputs t_n associated with inputs x_n in the training set that are "similar" to x.

In this section, we focus on a specific class of non-parametric techniques known as **kernel methods**, and we concentrate on regression. Section 6.3.5 will cover applications of kernel methods to classification.

4.11.1 Introducing Kernel Methods

Consider again the ERM regression problem (4.25) for the linear model class (4.21) under the ℓ_2 loss. To get started, assume that the features are uncorrelated and normalized as described in Sec. 4.3.4, so that, from (4.28), the ERM trained linear hard predictor $\hat{t}(x|\theta^{ERM}) = (\theta^{ERM})^T u(x)$ can be expressed as

$$\hat{t}(x|\theta_{\mathcal{D}}^{ERM}) = \sum_{n=1}^{N} t_n \cdot \kappa(x, x_n), \tag{4.67}$$

where

$$\kappa(x, x_n) = u(x)^T u(x_n) \tag{4.68}$$

is the inner product, or correlation, between the features of the input x and the features of training point x_n. As detailed in Sec. 2.5, we can think of the inner product as a measure of the similarity between x and x_n

The key observation from (4.67) is that, in order to compute the prediction $\hat{t}(x|\theta_{\mathcal{D}}^{ERM})$ on an input variable x, one does not need to explicitly specify a feature vector $u(x)$. All we need is to be able to compute correlations $\kappa(x, x_n)$ between input x and each data point x_n.

Another way to think about the predictor (4.67) is that the kernel value $\kappa(x, x_n)$ quantifies by how much the prediction $\hat{t}(x|\theta_{\mathcal{D}}^{ERM})$ on an input x changes if the label t_n of data point x_n is modified.

The resulting general form of a kernel-based predictor is illustrated in Fig. 4.16. In it, the prediction on an input x is obtained by first computing all the correlations $\kappa(x, x_n)$ between the input x and each training point x_n. These N correlations are combined, with each nth term being weighted by a model parameter ϑ_n. Unlike the direct implementation of the linear predictor $\hat{t}(x|\theta_{\mathcal{D}}^{ERM}) = (\theta_{\mathcal{D}}^{ERM})^T u(x)$, the architecture in Fig. 4.16 does not require the computation of a D-dimensional feature vector $u(x)$. As a result, the kernel approach offers an efficient way to implement a linear predictor when N < D, that is, for large feature vectors – which may even take an infinite number of dimensions.

Kernel methods can be formally studied within the mathematical framework of **reproducing kernel Hilbert spaces (RKHSs)**. An RKHS is defined by a kernel function $\kappa(x, x')$, and it encompasses functions of the form illustrated in Fig. 4.16, i.e.,

$$\hat{t}(x|\vartheta) = \sum_{n=1}^{N} \vartheta_n \cdot \kappa(x, x_n), \tag{4.69}$$

which depend on a vector of parameters $\vartheta = [\vartheta_1, \ldots, \vartheta_N]^T$. RKHSs prescribe a specific inner product operation that "reproduces" the kernel function. See Recommended Resources, Sec. 4.13, for details.

Figure 4.16 Unlike parametric methods, kernel methods define predictors that depend directly on the similarity between the current input x and the training samples x_n through kernel function $\kappa(x, x_n)$.

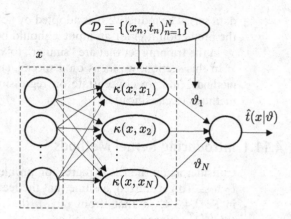

4.11.2 Kernel Functions

A kernel is a function of the form

$$\kappa(x, x') = u(x)^T u(x'), \tag{4.70}$$

where the feature vector $u(x)$ takes values in a D-dimensional space, with D being potentially infinite. Given a function $\kappa(x, x')$, how can we verify that it is a valid kernel, that is, that there exists a, possibly infinite-dimensional, feature vector $u(x)$ satisfying (4.70)?

Given a function $\kappa(x, x')$ and a training set $\mathcal{D} = \{(x_n)_{n=1}^N\}$, let us define $K_\mathcal{D}$ as the $N \times N$ **kernel matrix** with entries

$$[K_\mathcal{D}]_{n,m} = \kappa(x_n, x_m) \tag{4.71}$$

for $n, m \in \{1, \dots, N\}$. The kernel matrix contains the correlations among all pairs of data points. According to **Mercer's theorem**, the function $\kappa(x, x')$ is a kernel if and only if all kernel matrices that may be obtained from it are positive semi-definite, i.e., if we have $K_\mathcal{D} \succeq 0$ for all possible data sets $\mathcal{D}$. Accordingly, to avoid confusion with other types of kernel functions to be encountered later (see Chapter 7), we will refer to kernel functions of the form (4.70) as **positive semi-definite kernels**.

Positive semi-definite kernels satisfy properties akin to the inner product, including the following counterpart of the Cauchy–Schwartz inequality (2.39): $\kappa(x, x')^2 \leq \kappa(x, x) \cdot \kappa(x', x')$. Examples of positive semi-definite kernel functions include the **polynomial kernel**

$$\kappa(x, x') = (x^T x')^r \tag{4.72}$$

for some positive integer r, which can be proved to correspond to the inner product $\kappa(x, x') = u(x)^T u(x')$ with a feature vector $u(\cdot)$ of size growing exponentially with r; the **Gaussian radial basis function (RBF)** kernel

$$\kappa(x, x') = \exp\left(-\frac{||x - x'||^2}{2\sigma^2}\right), \tag{4.73}$$

which can be seen to correspond to the inner product between feature vectors of infinite dimension; and the **inverse multi-quadric (IMQ) kernel**

$$\kappa(x_n, x_m) = (\alpha + ||x_n - x_m||^2)^\beta,$$ (4.74)

for some $\alpha > 0$ and $\beta \in (-1, 0)$, which also corresponds to a feature vector with an infinite dimension.

The choice of the positive semi-definite kernel should generally encode any information one has about the problem as part of the inductive bias. For example, if one knows that the relationship between input and target variable has some periodicity properties, then a **periodic kernel** can be selected. This can be done, e.g., by considering a linear combination of sinusoidal functions of the difference $x_n - x_m$ for scalar inputs. Furthermore, positive semi-definite kernels can also be defined for objects other than numerical vectors. For instance, one can introduce kernels that quantify the similarity between sentences on the basis of the count of some common words.

4.11.3 "Kernelizing" Ridge Regression

The discussion in this section has so far built on the simplified form (4.67) of the trained predictor obtained via the ERM problem (4.20) for the linear model (4.21). We now derive more formally a "kernelized" version of the solution (4.42) of the more general ridge regression problem (4.41). The starting point is to note that the ridge regression solution (4.42) can also be written by the **matrix inversion lemma**[1] as

$$\theta_{\mathcal{D}}^{R-ERM} = \left(\lambda I_D + \underbrace{X_{\mathcal{D}}^T X_{\mathcal{D}}}_{\text{empirical correlation matrix } (D \times D)} \right)^{-1} X_{\mathcal{D}}^T t_{\mathcal{D}}$$

$$= X_{\mathcal{D}}^T \left(\lambda I_N + \underbrace{X_{\mathcal{D}} X_{\mathcal{D}}^T}_{K_{\mathcal{D}} = \text{kernel matrix } (N \times N)} \right)^{-1} t_{\mathcal{D}}.$$ (4.75)

From this alternative expression, we have

$$\theta_{\mathcal{D}}^{R-ERM} = \sum_{n=1}^{N} \tilde{t}_n \cdot u(x_n),$$ (4.76)

where the normalized target vector is $\tilde{t}_{\mathcal{D}} = (\lambda I_N + K_{\mathcal{D}})^{-1} t_{\mathcal{D}}$ with $\tilde{t}_{\mathcal{D}} = [\tilde{t}_1, \ldots, \tilde{t}_N]^T$. The optimal predictor is given as

$$\hat{t}(x | \theta^{R-ERM}) = u(x)^T \theta_{\mathcal{D}}^{R-ERM}$$

$$= \sum_{n=1}^{N} \tilde{t}_n \cdot \underbrace{u(x)^T u(x_n)}_{\kappa(x, x_n)},$$ (4.77)

which has the desired form illustrated in Fig. 4.16 with $\vartheta_n = \tilde{t}_n$ for $n = 1, \ldots, N$.

---

[1] The matrix inversion lemma, or Woodbury matrix identity, corresponds to the equality $(A + UCV)^{-1} = A^{-1} - A^{-1}U(C^{-1} + VA^{-1}U)^{-1}VA^{-1}$, where all matrix products and inverses are assumed to be well defined.

4.11.4 Reduced-Complexity Kernel Methods

As seen in Fig. 4.16, kernel methods generally require one to go over the entire data set in order to obtain a prediction. Several approaches can be used to alleviate this computational challenge. **Sparse kernel methods** limit the sum in (4.77) to a subset of training points (see also Sec. 6.3.5). Another related idea is to summarize the data into a smaller set of **prototypical vectors**. This is especially sensible in classification problems, for which a prototypical vector can be assigned to each class.

4.12 Summary

- When the population distribution is not known but one has access to a training set of examples of input and target variables, supervised learning can be used to design hard and soft predictors.
- Learning aims to generalize outside the training set, and its ultimate goal is to minimize the (unknown) population loss of the trained predictor over the set of all possible predictors.
- By the no-free-lunch theorem, generalization is only possible if an inductive bias – specifying model class and training algorithm – is selected prior to training.
- The learning process for parametric models proceeds by the following steps:

  1. **Inductive bias selection.** The inductive bias specifies:
     - a model class

$$\mathcal{H} = \{\hat{t}(x|\theta) : \theta \in \Theta\} \text{ for hard predictors} \tag{4.78}$$

$$\mathcal{H} = \{p(t|x, \theta) : \theta \in \Theta\} \text{ for (discriminative) probabilistic models;} \tag{4.79}$$

     - a per-sample loss function $f(x, t|\theta)$

$$f(x, t|\theta) = \ell(t, \hat{t}(x|\theta)) \text{ for hard predictors} \tag{4.80}$$

$$f(x, t|\theta) = -\log p(t|x, \theta) \text{ for (discriminative) probabilistic models;} \tag{4.81}$$

     - a regularization function $R(\theta)$, which is obtained from a prior distribution $p(\theta)$ for probabilistic models (as $R(\theta) \propto -\log p(\theta)$); and
     - a training algorithm to tackle the optimization problem

$$\min_{\theta} \underbrace{L_{\mathcal{D}}(\theta)}_{\frac{1}{N}\sum_{n=1}^{N} f(x_n, t_n|\theta)} + \frac{\lambda}{N} R(\theta), \tag{4.82}$$

  which is referred to as regularized ERM for hard predictors and MAP learning for probabilistic models.

  2. **Training.** Use the training algorithm to obtain θ by tackling the regularized ERM or MAP learning problems.
  3. **Validation.** Estimate the population loss obtained with the trained model parameter θ using data set aside for this purpose.
  4. **Revision of inductive bias.** Repeat steps 1–3 to optimize hyperparameters defining model class and training algorithm.
  5. **Test.** Use test data, which need to be separate from the training and validation data, to obtain the final estimate of the population loss.

- The population loss of a trained predictor can be decomposed into minimum population loss, bias, and estimation error. When applied separately to each value of x, the corresponding decomposition yields a quantification of the distinct contributions of bias, aleatoric uncertainty, and epistemic uncertainty to the overall predictive uncertainty.
- The bias can be controlled by choosing the model class, while the estimation error depends on both model class and amount of available data. As indicated by recent studies, the choice of the training algorithm and the "match" between data distribution and model class also play important roles in determining the generalization performance.
- Non-parametric methods do not rely on predictors that depend on finite-dimensional model parameter vectors, obtaining predictions as a function directly of the training data points. Kernel methods, as an important instance of non-parametric techniques, allow the efficient use of infinite-dimensional feature vectors.

4.13 Recommended Resources

The material in this chapter is partly inspired by the classical textbook by Bishop [1]. A lively and informative high-level introduction to statistics can be found in [2]. For discussions on modern views concerning the trade-off between bias and estimation error, useful readings at the time of writing include [3] and references therein. A comprehensive review of kernel methods is offered by the book [4]. Finally, a discussion about the inductive biases inherent in the way humans learn and interpret their sensory inputs can be found in [5].

Problems

4.1 We have a classification problem with true joint distribution $p(x, t)$ given in Table 4.4. (a) What is the optimal soft predictor? (b) What is the optimal hard predictor $\hat{t}^*(\cdot)$ under the detection-error loss, and what is the minimum population loss $L_p(\hat{t}^*(\cdot))$? (c*) Generate a training set $\mathcal{D}$ of $N = 10$ i.i.d. examples from this joint distribution. (d*) Compute the empirical distribution $p_{\mathcal{D}}(x, t)$. (e*) Repeat with $N = 10,000$ samples.

Table 4.4 Example of joint pmf $p(x, t)$

$x \backslash t$	0	1
0	0.1	0.2
1	0.2	0.1
2	0.1	0.3

4.2 For the joint distribution in Table 4.4, generate a data set $\mathcal{D}$ of $N = 10$ data points. (a*) For this data set, assuming the detection-error loss, compute the training loss $L_{\mathcal{D}}(\hat{t}(\cdot))$ for the predictors $\hat{t}(0) = 0$, $\hat{t}(1) = 1$, and $\hat{t}(2) = 0$. (b*) Identify the predictor $\hat{t}_{\mathcal{D}}(\cdot)$ that minimizes the training loss, i.e., the ERM predictor, assuming a model class $\mathcal{H}$ with $\hat{t}(0|\theta) = \theta_1$, $\hat{t}(1|\theta) = \theta_2$, and $\hat{t}(2|\theta) = \theta_3$ for $\theta_1, \theta_2, \theta_3 \in \{0, 1\}$, and compare its training loss with the predictor considered in the preceding point. (c*) Compute the population loss of the ERM predictor. (d*) Using your previous results, calculate the optimality gap $(L_p(\hat{t}_{\mathcal{D}}(\cdot)) - L_p(\hat{t}^*(\cdot)))$. (e*) Repeat the points above with $N = 10,000$ and comment on your result.

4.3   Show that the training loss for Example 4.2 in the text can be written as

$$L_{\mathcal{D}}(\theta) = \frac{1}{N} ||t_{\mathcal{D}} - X_{\mathcal{D}}\theta||^2.$$

4.4   Show that the condition $X_{\mathcal{D}}^T X_{\mathcal{D}} = I_D$ on the data matrix (4.22) imposes that, on average over the data set, all the features $u_d(x)$, $d = 1, \ldots, D$, have the same expected energy, and they are orthogonal.

4.5   (a*) Write a function `thERM=LSsolver(X,t)` that takes as input an $N \times D$ data matrix `X`, an $N \times 1$ desired output vector `t`, and outputs the LS solution `theta` in (4.26). (b*) Load the training data $\mathcal{D}$ from the file `sineregrdataset` on the book's website. The variables `x` and `t` contain $N = 20$ training inputs and outputs, respectively. (c*) Plot the values of `t` against the values of `x` in the training set. (d*) We now wish to implement regression with feature vector $u(x) = [1, x, x^2, \ldots, x^M]^T$. With $M = 5$, build the $N \times (M+1)$ data matrix $X_{\mathcal{D}}$ and assign it to a matrix `X`. (e*) Use the function `LSsolver` you have designed in point (a*) to obtain the ERM solution $\theta_{\mathcal{D}}^{ERM}$. (f*) Plot the hard predictor $\hat{t}(x|\theta_{\mathcal{D}}^{ERM})$ in the figure that contains the training set. (g*) Evaluate the training loss for the trained predictor. (h*) Repeat steps (d*)–(g*) with $M = 3$ and $M = 10$, and compare the resulting training errors.

4.6   (a*) Continuing Problem 4.5, for the data set `sineregrdataset`, compute the training loss $L_{\mathcal{D}}(\theta_M^{ERM})$ for values of the model capacity M between 1 and 6, and plot the training loss vs. M in this range. (b*) Explain the result in terms of the capacity of the model to fit the training data. (c*) Estimate the population loss using validation by computing the empirical loss on the held-out data `xval` and `tval` (which you can find in the data set `sineregrdataset`). Plot the validation error on the same figure. (d*) Which value of M would you choose? Explain by using the concepts of bias and estimation error.

4.7   In this problem, we continue Problem 4.6 by evaluating the impact of regularization. (a*) For $M = 6$, set $\lambda = \exp(-10)$ and compute the regularized ERM solution $\theta_{\mathcal{D}}^{R-ERM}$. Recall that this is also known as ridge regression. (b*) Plot the hard predictor $\hat{t}(x|\theta_{\mathcal{D}}^{R-ERM})$ as a function of x in the figure that contains the training set. Note that, when $\lambda = 0$, we obtain the ERM solution. (c*) Repeat for $\lambda = \exp(-20)$. (d*) Evaluate the training loss for the trained predictor for both choices of λ and comment on the comparison.

4.8   Continuing Problem 4.7, for $M = 6$, consider ten possible values for λ, namely $\lambda = \exp(v)$ with $v = \log(\lambda)$ taking one of ten equally spaced values. (a*) For each value of λ, compute the training loss $L_{\mathcal{D}}(\theta_{\mathcal{D}}^{R-ERM})$. Plot these as a function of $\log(\lambda)$. Comment on your results. (b*) Evaluate the estimate of the population loss on the held-out set variables `xval` and `tval`. Plot the validation-based estimate of the population loss in the same figure. Which value of λ would you choose?

4.9   Assume that we have a training data set with the empirical distribution $p_{\mathcal{D}}(x,t)$ given in Table 4.5. The population distribution is in Table 4.4. We consider the class of hard predictors

$$\mathcal{H} = \{\hat{t}(x|\theta) : \hat{t}(0|\theta) = \theta_1,$$
$$\hat{t}(1|\theta) = \hat{t}(2|\theta) = \theta_2 \text{ for } \theta_1, \theta_2 \in \{0,1\}\}.$$

Under the detection-error loss, evaluate (a) the population-optimal unconstrained predictor $\hat{t}^*(\cdot)$; (b) the population-optimal within-class model parameter $\theta_{\mathcal{H}}^*$; and (c) the ERM model $\theta_{\mathcal{D}}^{ERM} \in \Theta$. (d) Using these results, decompose the population loss of the ERM

Table 4.5 Example of joint pmf $p(x, t)$

$x \backslash t$	0	1
0	0.15	0.2
1	0.2	0.05
2	0.25	0.15

predictor in terms of the minimum unconstrained population loss, bias, and estimation error.

4.10 We have a classification problem with the joint distribution $p(x, t)$ in Table 4.4. We focus on the class of soft predictors

$$\mathcal{H} = \{p(t|x, \theta) : p(t = 1|x = 0, \theta) = \theta_1 \in [0, 1],$$
$$p(t = 1|x = i, \theta) = \theta_2 \in [0, 1] \text{ for } i = 1, 2 \text{ for } \theta = (\theta_1, \theta_2)\}.$$

(a) Evaluate the population-optimal within-class soft predictor $\theta_{\mathcal{H}}^*$ under the log-loss, and calculate the resulting population log-loss $L_p(\theta_{\mathcal{H}}^*)$. (b) Write the population log-loss $L_p(\theta_{\mathcal{H}}^*)$ in terms of the cross entropy between population distribution and soft predictor. (c) Given the within-class soft predictor $p(t|x, \theta_{\mathcal{H}}^*)$, obtain the optimal hard predictor under the detection-error loss.

4.11 In Problem 4.10, we considered the ideal situation in which the population distribution $p(x, t)$ is known. In this problem, we assume the same population distribution and the same model class $\mathcal{H}$ of soft predictors, and we study the learning problem. To this end, assume that we have a training data set of $N = 100$ data points with the empirical distribution $p_{\mathcal{D}}(x, t)$ in Table 4.5. (a) Obtain the ML model $\theta_{\mathcal{D}}^{ML}$. (b) Calculate the corresponding population log-loss $L_p(\theta_{\mathcal{D}}^{ML})$ and compare it with the population loss $L_p(\theta_{\mathcal{H}}^*)$ of the population-optimal within-class predictor obtained in Problem 4.10.

4.12 Continuing Problem 4.11, let us now consider the effect of regularization. To this end, assume that, based on prior knowledge, we have good reason to choose the following prior

$$p(\theta) = p(\theta_1, \theta_2) = p(\theta_1)p(\theta_2)$$

with marginals $p(\theta_i) = q$ if $0 \leq \theta_i < 0.5$ and $p(\theta_i) = 2 - q$ if $0.5 \leq \theta_i \leq 1$ for $i = 1, 2$ and some fixed $0 \leq q \leq 2$. (a) Verify that the marginals $p(\theta_i)$ are valid pdfs. (b) Write the conditions defining the MAP model parameters $\theta_{\mathcal{D}}^{MAP}$ for any N and prior parameter q. (c*) Plot the population log-loss $L_p(\theta_{\mathcal{D}}^{MAP})$ for the empirical distribution at hand (see Problem 4.9) as a function of q in the interval $[0, 2]$ (recall that we have $N = 100$). (d*) As a reference, plot the population log-loss obtained by ML and comment on the comparison between the two losses.

4.13 Consider again the data set `sineregrdataset` and the problem of ridge regression (see Problem 4.7). (a*) Implement the kernelized ridge regression solution by assuming the vector of features $u(x) = [1, x, x^2, \ldots, x^M]^T$ with $M = 5$ using (4.75). Verify that the solution is equivalent to that obtained in Problem 4.7 by plotting the trained hard predictor as a function of x. (b*) Substitute the kernel defined by the vector $u(x)$ with the RBF kernel and plot the resulting predictor by trying different values of σ^2. (c*) Repeat the previous point with the IMQ kernel by trying different values of α and β. Comment on your results.

4.14 Import data from the USPS data set `USPSdataset9296` from the book's website. It contains 9296 16×16 images of handwritten digits that are stacked into vectors of size 256×1 and placed, transposed, as rows in matrix X, along with the corresponding label vector t. (a*) Extract all the images corresponding to digits 0 and 1, and divide them into a training set $\mathcal{D}$ containing the first 70% of examples from each class, and a validation set containing the last 30% of examples from each class. (b) While this is a classification problem, we will treat it here as a regression problem. To this end, consider the 257×1 vector of features $u(x_n) = [1, x_n^T]^T$ and the linear model $\hat{t}(x|\theta) = \theta^T u(x)$, along with the quadratic loss, and formulate the ERM problem with quadratic loss using the training data. (c*) Plot the prediction $\hat{t}(x_n|\theta^{ERM}) = (\theta^{ERM})^T u(x_n)$ as a function of $n = 1, 2, \ldots, N$. In the same figure, plot the true labels t_n. (d*) Now consider a vector of features $u(x_n) = [1, \tilde{x}_n^T]^T$, where $\tilde{x}_n^T$ contains the first nine entries of the vector x_n. In the same figure, add the resulting prediction $\hat{t}(x|\theta^{ERM})$. (e) Comment on the comparison between the predictors obtained in points (c*) and (d*).

Appendices

Appendix 4.A: ML as Minimization of the Cross Entropy for Continuous rvs

To see how one can define ML as the minimization of the cross entropy for continuous rvs, consider the simpler case in which x is independent of t and hence the training log-loss is given by $L_{\mathcal{D}}(\theta) = 1/N \sum_{n=1}^{N} (-\log p(t_n|\theta))$. The extension to the more general case that also includes the input x requires a more cumbersome notation, but it follows in the same way. For a continuous rv t, given data set $\mathcal{D} = \{(t_n)_{n=1}^{N}\}$, we define the empirical distribution as

$$p_{\mathcal{D}}(t) = \frac{1}{N} \sum_{n=1}^{N} \delta(t - t_n), \tag{4.83}$$

where $\delta(\cdot)$ is the Dirac impulse function. Note that definition (4.83) is consistent with that given for discrete rvs as long as we substitute the Dirac delta function with a Kronecker delta function $\mathbb{1}(x = x_0)$.

With definition (4.83), using the standard "sifting" property of the Dirac impulse function, the cross entropy between empirical distribution and model distribution can be computed as

$$H(p_{\mathcal{D}}(t)\|p(t|\theta)) = \mathrm{E}_{t \sim p_{\mathcal{D}}(t)} \left[-\log p(t|\theta) \right]$$

$$= \frac{1}{N} \sum_{n=1}^{N} (-\log p(t_n|\theta)) = L_{\mathcal{D}}(\theta), \tag{4.84}$$

which is indeed the training log-loss as in the case of discrete rvs.

For data sets including both covariates and target variables, the empirical distribution is defined, generalizing (4.83), as a sum of N Dirac delta functions, each centered at the nth data point (x_n, t_n) for $n = 1, \ldots, N$.

Appendix 4.B: ML as Information Maximization

We have seen in the preceding appendix that ML can be interpreted as the minimization of the average cross entropy $E_{x \sim p_{\mathcal{D}}(x)} \left[H(p_{\mathcal{D}}(t|x) \| p(t|x, \theta)) \right]$. Consider now the mutual information $I(x; t)$ under the joint distribution $(x, t) \sim p_{\mathcal{D}}(x, t)$. Note that this mutual information is computable based only on the training set, since it depends on the empirical joint distribution $p_{\mathcal{D}}(x, t)$. We denote as $p_{\mathcal{D}}(t)$ the marginal of the empirical joint distribution $p_{\mathcal{D}}(x, t)$, and similarly for the marginal $p_{\mathcal{D}}(x)$ and for the conditional distribution $p_{\mathcal{D}}(t|x)$.

For any model distribution $p(t|x, \theta)$, recall from Sec. 3.7 that we have the bound (3.90), i.e.,

$$I(x; t) \geq E_{(x, t) \sim p_{\mathcal{D}}(x, t)} \left[\log \left(\frac{p(t|x, \theta)}{p_{\mathcal{D}}(t)} \right) \right]$$

$$= E_{x \sim p_{\mathcal{D}}(x)} \Bigg[\underbrace{E_{t \sim p_{\mathcal{D}}(t|x)} \left[\log \left(p(t|x, \theta) \right) \right]}_{-H(p_{\mathcal{D}}(t|x) \| p(t|x, \theta))} \Bigg] + H(t). \tag{4.85}$$

It follows that the ML solution $p(t|x, \theta)$, which minimizes the average cross entropy (the negative of the first term in (4.85)), also maximizes a lower bound on the mutual information.

Bibliography

[1] C. M. Bishop, *Pattern Recognition and Machine Learning*. Springer, 2006.

[2] D. Spiegelhalter, *The Art of Statistics: Learning from Data*. Penguin Books UK, 2019.

[3] P. L. Bartlett, A. Montanari, and A. Rakhlin, "Deep learning: A statistical viewpoint," arXiv:2103.09177, 2021.

[4] B. Schölkopf and A. Smola, *Learning with Kernels: Support Vector Machines, Regularization, Optimization, and Beyond*. The MIT Press, 2002.

[5] S. Gershman, *What Makes Us Smart*. Princeton University Press, 2021.

5 Optimization for Machine Learning

5.1 Overview

In the examples studied in Chapter 4, the exact optimization of the (regularized) training loss was feasible through simple numerical procedures or via closed-form analytical solutions. In practice, exact optimization is often computationally intractable, and scalable implementations must rely on approximate optimization methods that perform **local, iterative updates** in search of an optimized solution. This chapter provides an introduction to local optimization methods for machine learning.

Learning Objectives and Organization of the Chapter. By the end of this chapter, the reader should be able to:

- understand the role of optimization for training (Sec. 5.2);
- understand the goal of optimization (Sec. 5.3);
- apply optimality conditions for single-variable functions (Secs. 5.4 and 5.5);
- recognize convex functions, and apply optimality conditions for convex functions (Sec. 5.6);
- understand and apply optimality conditions for the more general case of functions of multiple variables (Sec. 5.7);
- implement gradient descent (GD) (Sec. 5.8);
- understand the convergence properties of GD (Sec. 5.9);
- appreciate the use of second-order methods (Sec. 5.10);
- implement stochastic GD (SGD), the workhorse of modern machine learning (Sec. 5.11);
- understand the convergence properties of SGD (Sec. 5.12);
- appreciate the role of optimization as part of the inductive bias (Sec. 5.13); and
- evaluate gradients based on symbolic differentiation, numerical differentiation, and back-prop (Secs. 5.14 and 5.15).

5.2 Optimization for Training

As we have seen in Chapter 4, training is typically formulated as the solution of problems of the form

$$\min_{\theta \in \Theta} \left\{ g(\theta) = \frac{1}{N} \sum_{n=1}^{N} f(x_n, t_n | \theta) + \frac{\lambda}{N} R(\theta) \right\}, \tag{5.1}$$

for loss function $f(x, t | \theta)$ and regularization function $R(\theta)$. In this chapter, we will study this problem by focusing on the typical **unconstrained** case in which the domain Θ is the entire space $\mathbb{R}^D$ of D-dimensional vectors θ for some dimension $D \geq 1$.

Problem (5.1) can be equivalently restated as the **finite-sum optimization**

$$\min_{\theta \in \mathbb{R}^D} \left\{ g(\theta) = \frac{1}{N} \sum_{n=1}^{N} g_n(\theta) \right\}, \tag{5.2}$$

where the regularized loss function associated with each data point (x_n, t_n) is defined as

$$g_n(\theta) = f(x_n, t_n | \theta) + \frac{\lambda}{N} R(\theta). \tag{5.3}$$

To keep the terminology general, we will refer to the function $g(\theta)$ under optimization in (5.2) as the **cost function** in this chapter. Furthermore, we will refer to θ as either a **vector** or a **point** in $\mathbb{R}^D$.

5.3 Solving an Optimization Problem

What does it mean to solve the optimization problem (5.2)?

5.3.1 Global Minima

Ideally, solving the optimization problem (5.2) entails either producing at least one **global minimum** of the cost function $g(\theta)$ or indicating that no global minimum exists. Globally optimal points, or **global minima**, are vectors $\theta^* \in \mathbb{R}^D$ such that the inequality

$$g(\theta) \geq g(\theta^*) \tag{5.4}$$

holds for all vectors $\theta \in \mathbb{R}^D$. This inequality states that there are no other parameter vectors in $\mathbb{R}^D$ that yield a strictly smaller value of the cost function than $g(\theta^*)$.

The set of global minima of the cost function $g(\theta)$ is denoted as $\arg\min_\theta g(\theta)$. Accordingly, the notation $\theta^* \in \arg\min_\theta g(\theta)$ indicates that a vector θ^* is a global minimum. As was done in the previous chapters, we will also use the simpler, but less accurate, notation $\theta^* = \arg\min_\theta g(\theta)$ to express the condition that θ^* is a global minimum. This notation is formally correct only in case there is a single global minimum θ^*.

Depending on the cost function $g(\theta)$, we may have one of the following situations:

- The set of global minima $\arg\min_\theta g(\theta)$ is **empty**. This happens if the function is **unbounded below**, i.e., if it tends to $-\infty$ by taking a suitable limit on θ; or if it is bounded by an **infimum** value, but there is no choice of θ that achieves the infimum. An example of the first case is the cost function $g(\theta) = \theta$ for $\theta \in \mathbb{R}$; and one for the second situation is $g(\theta) = \exp(\theta)$ for $\theta \in \mathbb{R}$, whose infimum value is 0.
- The set of global minima $\arg\min_\theta g(\theta)$ contains a **single minimum**. An example is the cost function $g(\theta) = \theta^2$ for $\theta \in \mathbb{R}$.
- The set of global minima $\arg\min_\theta g(\theta)$ contains **multiple minima**. Examples include all constant cost functions.

5.3.2 Local Minima

Obtaining global minima is typically infeasible when the model parameter dimension D is large. In this case, which is common in machine learning, one is conventionally satisfied with

Figure 5.1 Function $g(\theta) = \theta^4 - 4\theta^2 + 2\theta - 4$.

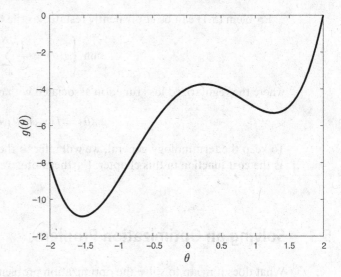

producing local minima as the solution of an optimization problem. Locally optimal points, or **local minima**, are values θ^* such that the inequality (5.4) holds only for all values of θ in a **neighborhood** of θ^*, and not for all possible vectors θ in $\mathbb{R}^D$. A neighborhood of point θ^* contains all vectors $\theta \in \mathbb{R}^D$ that are close to θ^* in terms of Euclidean distance, i.e., that satisfy the inequality

$$||\theta - \theta^*|| \leq \epsilon \tag{5.5}$$

for some sufficiently small $\epsilon > 0$.

Example 5.1

Consider the function $g(\theta) = \theta^4 - 4\theta^2 + 2\theta - 4$ shown in Fig. 5.1. (The curve continues its upward trajectory beyond the boundaries of the plot.) Point $\theta^* = -1.52$ is the only global minimum, and is hence also a local minimum, while $\theta^* = 1.27$ is a local minimum, but not a global minimum.

5.3.3 Stationary Points

In practice, when D is large, even computing a local minimum may be computationally challenging. In this case, a stationary point θ^* is generally considered to provide an acceptable solution for the optimization problem (5.2). A **stationary point**, also known as **critical point**, θ^* satisfies the condition that the cost function $g(\theta)$ is locally flat in a small neighborhood of θ^*. We will make this notion precise in the next section by using the first derivatives of the function $g(\theta)$. The next section will also describe how stationary points are related to local and global minima.

5.3.4 Stochastic Optimality Guarantees

In machine learning, the number of data points, N, is typically large too. In this case, as we will detail in Sec. 5.11, it is useful to use **stochastic** optimization algorithms that select points from the data set at random in order to carry out local updates. With randomized algorithms, optimality guarantees can only be provided in a probabilistic sense. For instance, an optimization algorithm may reach a stationary point with high probability or on average over the randomness of the procedure. This aspect will be covered in Sec. 5.12.

5.4 First-Order Necessary Optimality Condition for Single-Variable Functions

How can we obtain globally or locally optimal points? To elaborate, in this section we study cost functions that depend on a scalar parameter $\theta \in \mathbb{R}$, i.e., we set $D = 1$. We will assume here and throughout this chapter that the functions $\{g_n(\theta)\}_{n=1}^{N}$, and hence also the cost function $g(\theta)$, are differentiable as needed.

5.4.1 Local Linear Approximation

By the **first-order Taylor approximation**, a function $g(\theta)$ is approximated in the neighborhood of a value $\theta^0 \in \mathbb{R}$ by an affine function (i.e., a line) as

$$g(\theta) \simeq g(\theta^0) + \frac{dg(\theta^0)}{d\theta}(\theta - \theta^0). \tag{5.6}$$

The notation $dg(\theta^0)/d\theta$ indicates the first derivative of function $g(\theta)$ evaluated at point θ_0, i.e.,

$$\frac{dg(\theta^0)}{d\theta} = \frac{dg(\theta)}{d\theta}\bigg|_{\theta=\theta^0}. \tag{5.7}$$

Examples of linear approximations obtained via (5.6) are illustrated in Fig. 5.2.

The linear approximation (5.6) is clearly exact at $\theta = \theta^0$. To measure its quality as θ moves away from θ^0, it is useful to consider the Taylor expansion with **Lagrange remainder**:

$$g(\theta) = g(\theta^0) + \frac{dg(\theta^0)}{d\theta}(\theta - \theta^0) + \underbrace{\frac{1}{2}\frac{d^2g(\theta^m)}{d\theta^2}(\theta - \theta^0)^2}_{\text{Lagrange remainder}}, \tag{5.8}$$

which holds for some $\theta^m \in [\theta, \theta^0]$. Equality (5.8) shows that the error of the first-order Taylor approximation at a point θ scales as the square of the distance from θ to the point θ^0 at which the approximation is computed. Furthermore, the rate of growth of the error depends on the **second derivative**, which, as we will see, quantifies the **curvature** of the function. The more curved a function is, the more quickly the linear approximation in (5.6) becomes inaccurate as we move away from θ^0.

Figure 5.2 First-order, or linear, Taylor approximations around points $\theta^0 = -1$, $\theta^0 = -1$, and $\theta^0 = 2.5$ for the function $g(\theta) = \theta^4 - 3\theta^3 + 2$.

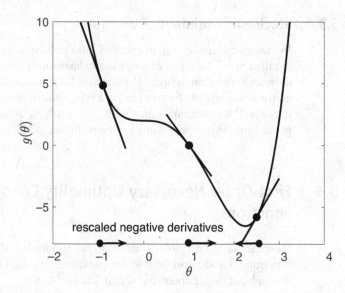

5.4.2   First-Order Necessary Optimality Condition

Substituting the linear approximation (5.6) into the condition (5.4), we obtain that, in order for a point $\theta^* \in \mathbb{R}$ to be a local minimum, the inequality

$$g(\theta) \simeq g(\theta^*) + \frac{dg(\theta^*)}{d\theta}(\theta - \theta^*) \geq g(\theta^*) \qquad (5.9)$$

must hold in a neighborhood of θ^*. If the derivative at θ^* is not zero, i.e., if $dg(\theta^*)/d\theta \neq 0$, the term

$$\frac{dg(\theta^*)}{d\theta}(\theta - \theta^*) \qquad (5.10)$$

can be negative, since $(\theta - \theta^*)$ takes both positive and negative values as θ varies in $\mathbb{R}$. Therefore, for a point θ^* to be a local minimum, it must be a **stationary point**, that is, it must satisfy the condition

$$\frac{dg(\theta^*)}{d\theta} = 0. \qquad (5.11)$$

   To restate the main conclusion, stationarity is a necessary condition for a point to be a local minimum. Stationarity is therefore known as a **first-order necessary optimality condition**, which we can summarize with the logical implication

$$\text{global minimum} \implies \text{local minimum} \implies \text{stationary point.} \qquad (5.12)$$

As indicated, since a global optimum is a local optimum point, stationarity is also a necessary condition for global optimality.

   It is important to reiterate that stationarity is only a *necessary* condition: Not all stationary points are local minima. In fact, stationary points can also be **local maxima** or **saddle points**. Local maxima are defined as for minima but with the sign in the inequality (5.4) reversed. Saddle points are stationary points that are neither local maximal nor local minima.

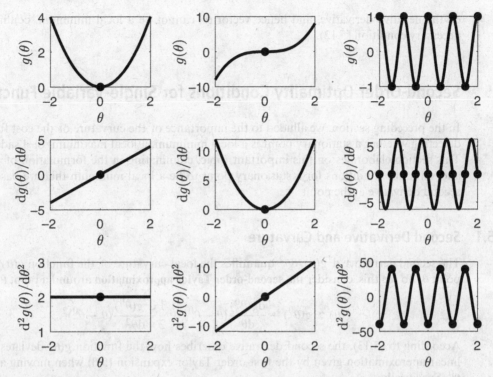

Figure 5.3 Examples of stationary points that are local minima (left column and odd markers in the right columns); saddle points (central column); and local maxima (even markers in the right column).

Example 5.2

Consider the three cost functions in Fig. 5.3. The leftmost function has a stationary point at $\theta = 0$ that is a local (and global) minimum; the function in the center has a stationary point at $\theta = 0$ that is a saddle point; and the rightmost function has multiple local (and global) minima as well as multiple local (and global) maxima. Note that the curvature of the central function changes as it goes through the saddle point: The curvature is negative (i.e., pointing down) to the left of the point and positive (i.e., pointing up) to the right. We will define the notion of curvature formally in the next section.

5.4.3 Derivative As Direction of Local Descent

Another way to think about the necessary optimality condition (5.11) is to regard the negative derivative $-dg(\theta^0)/d\theta$ as a "pointer" in the direction of decrease of the function around point θ^0. Specifically, at each point θ^0, imagine adding an arrow on the horizontal axis with length given by the magnitude $|dg(\theta^0)/d\theta|$ and direction given by the sign of the negative gradient $-dg(\theta^0)/d\theta$. As illustrated in Fig. 5.2, this arrow points in the direction of local descent for the function $g(\theta)$ around θ^0, and its magnitude indicates the extent to which the function can be decreased by moving by a small step along the direction of the negative derivative. If the derivative is not zero at the point θ^0, the function can be decreased by a small step in the direction of the negative derivative, and hence vector θ^0 cannot be a local minimum, confirming the necessary condition (5.12).

5.5   Second-Order Optimality Conditions for Single-Variable Functions

In the preceding section, we alluded to the importance of the curvature of the cost function in detecting whether a stationary point is a local minimum, a local maximum, or a saddle point. This section elaborates on this important topic, culminating in the formulation of necessary and sufficient conditions for a stationary point to be a local minimum that are based on the second derivative at the point.

5.5.1   Second Derivative and Curvature

The second derivative $d^2g(\theta)/d\theta^2$ quantifies the local curvature of the function $g(\theta)$ around a point θ. To see this, consider the **second-order Taylor approximation** around a point θ^0:

$$g(\theta) \simeq g(\theta^0) + \frac{dg(\theta^0)}{d\theta}(\theta - \theta^0) + \frac{1}{2}\frac{d^2g(\theta^0)}{d\theta^2}(\theta - \theta^0)^2. \tag{5.13}$$

According to (5.13), the second derivative describes how the function $g(\theta)$ deviates from the linear approximation given by the first-order Taylor expansion (5.6) when moving away from θ^0. Specifically:

- when $d^2g(\theta^0)/d\theta^2 = 0$, the function as a **zero curvature** around θ^0, i.e., it coincides with its local linear approximation around θ^0;
- when $d^2g(\theta^0)/d\theta^2 > 0$, the function has a **positive curvature** around θ^0, i.e., it has a "U shape" around θ^0, with a larger positive second derivative indicating a narrower "U"; and
- when $d^2g(\theta^0)/d\theta^2 < 0$, the function has a **negative curvature** around θ^0, i.e., it has an "upside-down U shape" around θ^0, with a more negative second derivative indicating a narrower upside-down "U".

The second derivative is, of course, the first derivative of the first derivative. That taking the derivative twice should quantify the curvature of the function around a point can be understood as follows. Consider as an example a stationary point θ^0 at which we have a positive second derivative, $d^2g(\theta^0)/d\theta^2 > 0$. This implies that the first derivative is locally increasing around θ^0. Moving from the left towards θ^0, the first derivative must hence be negative, then cross zero at θ^0, and finally take larger positive values as θ increases beyond θ^0. Therefore, the function has a "U shape" around θ^0, which is narrower if the slope of the first derivative increases, that is, if the second derivative $d^2g(\theta^0)/d\theta^2$ is larger.

Example 5.3

Figure 5.3 shows the second derivatives for the three functions described in Example 5.2. The leftmost function has a positive, and constant, curvature for all values θ (a hallmark of quadratic functions). The central function has a negative curvature on the left of $\theta = 0$ and a positive curvature on the right of this point. Finally, the rightmost function has a positive curvature around the local (and global) minima and a negative curvature around the local (and global) maxima.

This example confirms that the second derivative can be useful for distinguishing between minima, maxima, and saddle points.

5.5.2   Second-Order Sufficient and Necessary Optimality Conditions

We now get to the main messages of this section in the form of optimality conditions for a stationary point θ^*. Recall that the focus on stationary points comes with no loss of generality since, by the first-order necessary condition (5.12), only stationary points can be local minima.

The **second-order sufficient optimality condition** states that, if a stationary point θ^* has a positive second derivative, $d^2 g(\theta^*)/d\theta^2 > 0$, then it is a local minimum. We can express this condition with the implication

$$\text{stationary point } \theta^* \text{ with } \frac{d^2 g(\theta^*)}{d\theta^2} > 0 \implies \text{local minimum.} \tag{5.14}$$

To see this, note that a positive curvature implies that the function cannot decrease below $g(\theta^*)$ when moving in a small neighborhood around θ^*. Conversely, if the second derivative is negative at a stationary point, the point is a local maximum.

Example 5.4

The leftmost function in Fig. 5.3 has a positive curvature around the stationary point $\theta^* = 0$ (and, in fact, on the entire domain). Therefore, by the second-order sufficient optimality condition, the stationary point $\theta^* = 0$ is a local minimum. Note that the positive curvature can be inferred both from the positive value of the second derivative at $\theta^* = 0$ and from the local increase of the first derivative around this point.

The **second-order necessary optimality condition** states that a stationary point θ^* can be a local minimum only if its second derivative is non-negative, i.e., $d^2 g(\theta^*)/d\theta^2 \geq 0$. We can express this condition with the implication

$$\text{stationary point } \theta^* \text{ is a local minimum} \implies \frac{d^2 g(\theta^*)}{d\theta^2} \geq 0. \tag{5.15}$$

In fact, if the curvature is negative, i.e., if $d^2 g(\theta^*)/d\theta^2 < 0$, the stationary point θ^* is a local maximum and cannot be a local minimum.

Importantly, the second-order sufficient condition (5.14) and necessary condition (5.15) do not allow us to draw any definite conclusions about stationary points θ^* with zero curvature, i.e., with $d^2 g(\theta^*)/d\theta^2 = 0$. Such points *can* be local minima, but they can also be local maxima or saddle points. To find out, one needs to consider higher-order derivatives and hence more refined approximations than the second-order Taylor expansion (5.13).

Example 5.5

For the central function in Fig. 5.3, the curvature changes from negative to positive as θ varies in a neighborhood of the stationary point $\theta^* = 0$. The second derivative at $\theta^* = 0$ equals zero, and hence the second-order conditions cannot be used to determine whether the point $\theta^* = 0$ is a local minimum.

5.5.3 Summary of First- and Second-Order Optimality Conditions

Based on the first-order and second-order optimality conditions, to solve a minimization problem, one can in principle proceed as follows:

1. Apply the first-order necessary optimality condition to obtain all stationary points and evaluate the corresponding second derivatives.
2. All stationary points with positive curvature are local minima.
3. Stationary points with zero curvature must be individually checked to determine if they are local minima, saddle points, or local maxima,
4. Among the local minima, the points with the minimum value of the cost function are globally optimal if the set of global minima is not empty.

Example 5.6

For the function $g(\theta) = \theta^4 - 3\theta^3 + 2$ in Fig. 5.2, the first-order optimality condition $dg(\theta)/d\theta = 4\theta^3 - 9\theta^2 = 0$ yields the stationary points $\theta = 0$ and $\theta = 9/4$. Furthermore, we have the second derivative $d^2g(\theta)/d\theta^2 = 12\theta^2 - 18\theta$, which equals zero at $\theta = 0$, and is larger than zero at $\theta = 9/4$. By inspection of the function in Fig. 5.2, we see that the first point, with zero curvature, is a saddle point, as the curvature is positive on one side of the point and negative on the other side. This can be mathematically checked by evaluating the second derivative $d^2g(\epsilon)/d\theta^2$ for a small value $\epsilon > 0$, i.e., on the right of $\theta = 0$, to find that it is negative; while $d^2g(-\epsilon)/d\theta^2$ on the left of $\theta = 0$ is positive. In contrast, by the second-order sufficient optimality condition, $\theta^* = 9/4$ is a local minimum, and, being the only one, also the only global minimum since the set of global minima is not empty. (The latter condition can be verified by inspection of the function in this example.)

5.6   Optimality Conditions for Convex Single-Variable Functions

As discussed in the preceding section, in order to validate the local optimality of a stationary point, one must generally compute the second derivative at the point – and, if the second derivative is zero, also higher-order derivatives. However, there is a class of functions for which the first-order condition is **sufficient** to establish local optimality. Even more remarkably, the first-order condition not only determines local optimality but also **global** optimality. This is the class of **convex functions**, which plays an important role in optimization and machine learning. (We hasten to note that modern machine learning models based on neural networks yield non-convex cost functions, as we will see in Sec. 5.7.)

5.6.1   Definition of Convex Single-Variable Functions

A (twice differentiable) function $g(\theta)$ of a variable $\theta \in \mathbb{R}$ is **convex** if it has a *non-negative* curvature on its entire domain, i.e., if $d^2g(\theta)/d\theta^2 \geq 0$ for *all* $\theta \in \mathbb{R}$.

There are also several equivalent, alternative definitions that may be useful both to build an intuition about convex functions and as a tool to prove or disprove convexity:

- A function $g(\theta)$ is convex if its first derivative is non-decreasing.
- A function $g(\theta)$ is convex if it is never below the first-order Taylor approximation evaluated at any point $\theta^0 \in \mathbb{R}$, i.e., if

$$g(\theta) \geq g(\theta^0) + \frac{\mathrm{d}g(\theta^0)}{\mathrm{d}\theta}(\theta - \theta^0) \tag{5.16}$$

for all $\theta \in \mathbb{R}$.

- A function $g(\theta)$ is convex if the value $g(\theta)$ in the interval $\theta \in [\theta^1, \theta^2]$ is never above the segment drawn between points $g(\theta^1)$ and $g(\theta^2)$ for any two $\theta^1 \leq \theta^2$ in $\mathbb{R}$; more precisely, for all $\alpha \in [0,1]$, we have the inequality

$$g(\alpha\theta_1 + (1-\alpha)\theta_2) \leq \alpha g(\theta^1) + (1-\alpha)g(\theta^2). \tag{5.17}$$

Importantly, convexity does not require differentiability, and hence the last two definitions, which do not rely on derivatives, are in fact more general and more broadly applicable. An example of a convex but non-differentiable function $g(\theta)$ is $g(\theta) = |\theta|$, for which the first derivative is not defined at $\theta = 0$.

Some Convex Functions. Examples of convex functions in $\theta \in \mathbb{R}$ include:

- $g(\theta) = a\theta + b$, for any $a, b \in \mathbb{R}$;
- $g(\theta) = \theta^{2k}$, for any integer $k \geq 1$;
- $g(\theta) = \exp(\theta)$; and
- any linear combination of convex functions with non-negative coefficients.

5.6.2 Stationarity As a Sufficient Global Optimality Condition for Convex Functions

The key property of convex functions as it pertains to optimization is the following: If the cost function $g(\theta)$ is convex, any stationary point θ^* is a global minimum, i.e.,

cost function $g(\theta)$ is convex and θ^* is a stationary point $\Longrightarrow \theta^*$ is a global minimum

$$\Longrightarrow \theta^* \text{ is a local minimum.} \tag{5.18}$$

Therefore, when the function is convex, it is sufficient to apply the first-order optimality condition in order to obtain the set of global minima.

Example 5.7

Consider the **quadratic function**

$$g(\theta) = c + b\theta + \frac{1}{2}a\theta^2 \tag{5.19}$$

with $a > 0$. Since this is a linear combination of convex functions with non-negative coefficients (see the list in the preceding subsection), the function is convex. To see this, we can also simply note that the second derivative $\mathrm{d}^2 g(\theta)/\mathrm{d}\theta^2 = a$ is positive for all $\theta \in \mathbb{R}$. The first-order optimality condition $\mathrm{d}g(\theta)/\mathrm{d}\theta = b + a\theta = 0$ yields the only stationary point,

$$\theta^* = -b/a, \tag{5.20}$$

which is the only global minimum by property (5.18) of convex functions.

5.6.3 Strictly Convex Functions

As discussed, by (5.18), all local minima of a convex function are global. A convex function can have no minima, one minimum, or multiple minima. In the special case of **strictly convex functions**, the (global) minimum, if it exists, is unique. Therefore, a strictly convex function can have at most one stationary point, which is also the only global minimum.

A (twice differentiable) function $g(\theta)$ is strictly convex if it has a *positive* curvature on its entire domain, i.e., if $d^2g(\theta)/d\theta^2 > 0$ for *all* $\theta \in \mathbb{R}$. For example, a linear function $g(\theta) = a\theta + b$, for any $a, b \in \mathbb{R}$, is convex but not strictly convex (since $d^2g(\theta)/d\theta^2 = 0$); a quadratic function $g(\theta) = \theta^2$ is strictly convex (since $d^2g(\theta)/d\theta^2 = 2$); and $g(\theta) = \exp(\theta)$ is strictly convex (since $d^2g(\theta)/d\theta^2 = \exp(\theta)$). Note that $g(\theta) = \theta^2$ has a single global minimum $\theta^* = 0$; while $g(\theta) = \exp(\theta)$ has no stationary points and hence the set of global minima is empty. As another example, the quadratic function (5.19) with $a > 0$ is strictly convex, and has a unique global minimum (5.20).

5.7 Optimality Conditions for Multi-variable Functions

In this section, we extend the optimality conditions discussed in the last three sections to cost functions $g(\theta)$ of multiple variables. Accordingly, the optimization is over the parameter vector $\theta = [\theta_1, \ldots, \theta_D]^T \in \mathbb{R}^D$ with any integer $D \geq 1$. We will consider both first-order and second-order optimality conditions, as well as convex functions. The first step in this direction is to generalize the concept of derivative.

5.7.1 Gradient and First-Order Necessary Optimality Condition

The counterpart of the derivative for multi-variable functions is the $D \times 1$ **gradient vector**

$$\nabla g(\theta) = \begin{bmatrix} \frac{\partial g(\theta)}{\partial \theta_1} \\ \vdots \\ \frac{\partial g(\theta)}{\partial \theta_D} \end{bmatrix} \tag{5.21}$$

that collects the partial derivatives of the function $g(\theta)$ with respect to all variables θ_d with $d = 1, \ldots, D$. When we wish to emphasize that the partial derivatives are computed with respect to the elements of vector θ, we will also write the gradient (5.21) as $\nabla_\theta g(\theta)$.

The gradient defines the **first-order Taylor approximation** around any point $\theta^0 = [\theta_1^0, \ldots, \theta_D^0]^T$ as

$$g(\theta) \simeq g(\theta^0) + \nabla g(\theta^0)^T (\theta - \theta^0)$$

$$= g(\theta^0) + \sum_{i=1}^{D} \frac{\partial g(\theta^0)}{\partial \theta_i} (\theta_i - \theta_i^0). \tag{5.22}$$

Note that, when $D = 1$, the approximation (5.22) reduces to (5.6).

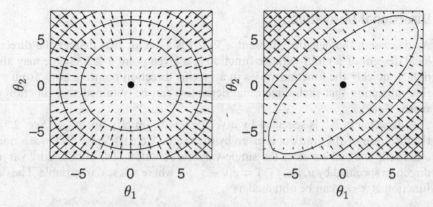

Figure 5.4 Contour lines and negative gradient vectors at a grid of equispaced points for cost functions $g(\theta) = 2\theta_1^2 + 2\theta_2^2$ (left) and $g(\theta) = 2\theta_1^2 + 2\theta_2^2 - 3\theta_1\theta_2$ (right). The negative gradient vectors point in the direction of steepest local decrease of the function, which does not generally coincide with a direction towards a minimum of the cost function. In particular, negative gradient and direction towards the global minimum match for the first function, which is characterized by a uniform curvature in all directions; but this is not so for the second function, which has different curvatures along distinct directions on the plane.

Generalizing the discussion on the derivative in Sec. 5.4.3, we can interpret the negative gradient $-\nabla g(\theta)$ as a vector in $\mathbb{R}^D$ starting at θ and pointing in the direction of **maximal local descent** of the objective function around point θ.

Example 5.8

Figure 5.4 displays the negative gradient vector $-\nabla g(\theta)$ at a grid of points θ on the plane $\mathbb{R}^D$ for cost functions $g(\theta) = 2\theta_1^2 + 2\theta_2^2$ and $g(\theta) = 2\theta_1^2 + 2\theta_2^2 - 3\theta_1\theta_2$, whose contour lines are also shown. The negative gradient $-\nabla g(\theta)$ indicates the direction along which the function can be decreased the most when making small steps around θ. The norm of the gradient – that is, the length of the arrow – quantifies the magnitude of this local decrease.

5.7.2   First-Order Necessary Optimality Condition

By (5.4), using the arguments presented in Sec. 5.4, in order for a point θ to be a local minimum, it must be **stationary**, i.e., it must have an all-zero gradient:

$$\nabla g(\theta) = 0_D. \tag{5.23}$$

In other words, as in the case of single-variable functions, stationarity is a necessary condition for local and global optimality as per (5.12). We will hence refer to the equality (5.23) as the **first-order necessary optimality condition**.

Example 5.9

For both functions in Fig. 5.4, the only stationary point is $\theta = 0_2$, as can be seen by imposing the condition (5.23). This is also the only global minimum for both functions.

5.7.3 Directional Derivative

As discussed, the negative gradient $-\nabla g(\theta)$ is a vector that defines the direction of steepest local descent of a multi-variable function at a given point $\theta \in \mathbb{R}^D$. One may also be interested in how quickly the function varies in a specific direction pointing away from θ. We will see in this subsection that such directional derivative can be obtained as a function of the gradient vector $\nabla g(\theta)$.

A **direction** in $\mathbb{R}^D$ is specified by a vector u with unitary norm (see Sec. 2.5). (Accordingly, the direction of the gradient is given by the vector $\nabla g(\theta)/\|\nabla g(\theta)\|$.) Once a normalized vector u is fixed, one can consider the single-variable function $f(x)$ obtained by varying θ along the direction specified by u, i.e., $f(x) = g(\theta + x \cdot u)$, where x is scalar variable. The derivative of this function at $x = 0$ can be obtained as

$$\nabla g(\theta)^T u, \tag{5.24}$$

and is known as the **directional derivative** of function $g(\theta)$ at θ in the direction u.

5.7.4 Hessian Matrix

Following the discussion in Sec. 5.5, the curvature of a function $g(\theta)$ at a point θ^0 quantifies how quickly the function deviates locally from the first-order Taylor approximation. In the case of multi-variable functions, the curvature hence depends on the direction u along which the vector θ is varied away from θ^0. In this subsection, we will see that the curvatures in all directions u around a point θ^0 can be obtained from a single matrix – the Hessian.

The **Hessian** of a function $g(\theta)$ at a point θ is the $D \times D$ matrix of second derivatives

$$\nabla^2 g(\theta) = \begin{bmatrix} \frac{\partial^2 g(\theta)}{\partial \theta_1^2} & \frac{\partial^2 g(\theta)}{\partial \theta_1 \partial \theta_2} & \frac{\partial^2 g(\theta)}{\partial \theta_1 \partial \theta_3} & \cdots & \frac{\partial^2 g(\theta)}{\partial \theta_1 \partial \theta_D} \\ \frac{\partial^2 g(\theta)}{\partial \theta_2 \partial \theta_1} & \frac{\partial^2 g(\theta)}{\partial \theta_2^2} & \frac{\partial^2 g(\theta)}{\partial \theta_2 \partial \theta_3} & \cdots & \frac{\partial^2 g(\theta)}{\partial \theta_2 \partial \theta_D} \\ \frac{\partial^2 g(\theta)}{\partial \theta_3 \partial \theta_1} & \frac{\partial^2 g(\theta)}{\partial \theta_3 \partial \theta_2} & \frac{\partial^2 g(\theta)}{\partial \theta_3^2} & \cdots & \frac{\partial^2 g(\theta)}{\partial \theta_3 \partial \theta_D} \\ \vdots & \vdots & \vdots & \ddots & \vdots \\ \frac{\partial^2 g(\theta)}{\partial \theta_D \partial \theta_1} & \frac{\partial^2 g(\theta)}{\partial \theta_D \partial \theta_2} & \frac{\partial^2 g(\theta)}{\partial \theta_D \partial \theta_3} & \cdots & \frac{\partial^2 g(\theta)}{\partial \theta_D^2} \end{bmatrix}. \tag{5.25}$$

Note that, for clarity, the Hessian (5.25) is shown for $D \geq 3$, and that the Hessian for $D = 2$ is given by the upper left 2×2 submatrix in (5.25).

The second derivative $\partial^2 g(\theta)/\partial \theta_i \partial \theta_j$ is the first derivative with respect to θ_j of the first derivative with respect to θ_i of the function $g(\theta)$. Therefore, when $i = j$, it measures the curvature of the function when moving along the axis θ_i; while, when $i \neq j$, it measures the degree to which the change of the function along axis θ_i varies as we move along axis θ_j. The order of differentiation is immaterial, and we have the equality

$$\frac{\partial^2 g(\theta)}{\partial \theta_i \partial \theta_j} = \frac{\partial^2 g(\theta)}{\partial \theta_j \partial \theta_i} \tag{5.26}$$

for all $i, j = 1, \ldots, D$, which implies that the Hessian is a symmetric matrix.

Armed with the Hessian, we can generalize (5.13) to obtain the **second-order Taylor approximation** around a point θ^0 as

$$g(\theta) \simeq g(\theta^0) + \nabla g(\theta^0)(\theta - \theta^0) + \frac{1}{2}(\theta - \theta^0)^T \nabla^2 g(\theta^0)(\theta - \theta^0). \tag{5.27}$$

This approximation shows that the deviation from the first-order approximation (5.22) around a point θ^0 depends on the quadratic form $(\theta - \theta^0)^T \nabla^2 g(\theta^0)(\theta - \theta^0)$ associated with the Hessian matrix $\nabla^2 g(\theta^0)$ (see Sec. 2.6.3 for the definition of quadratic form).

As reviewed in Sec. 2.6, a symmetric matrix can be decomposed into its eigenvalues and eigenvectors. Therefore, we can write the Hessian as

$$\nabla^2 g(\theta) = U(\theta) \Lambda(\theta) U(\theta)^T, \tag{5.28}$$

where $U(\theta)$ is a $D \times D$ matrix with orthogonal unitary-norm columns – the eigenvectors – and $\Lambda(\theta) = \text{Diag}(\lambda(\theta))$ is a $D \times D$ diagonal matrix with eigenvalues on the main diagonal given by the vector $\lambda(\theta) = [\lambda_1(\theta), \ldots, \lambda_D(\theta)]^T$. Each eigenvector $[U(\theta)]_{:i}$, which is the ith column of the eigenvector matrix $U(\theta)$, determines an orthogonal direction of variability of the function around point θ. The curvature in each direction $[U(\theta)]_{:i}$ is given by the corresponding eigenvalue $\lambda_i(\theta)$ for $i = 1, \ldots, D$.

Example 5.10

Consider again the cost function $g(\theta) = 2\theta_1^2 + 2\theta_2^2 - 3\theta_1\theta_2$ in Fig. 5.4 (right). The Hessian is given by the matrix

$$\nabla^2 g(\theta) = \begin{bmatrix} 4 & -3 \\ -3 & 4 \end{bmatrix}. \tag{5.29}$$

The eigenvectors can be computed as

$$U(\theta) = \frac{1}{\sqrt{2}} \begin{bmatrix} -1 & -1 \\ 1 & -1 \end{bmatrix}, \tag{5.30}$$

and the eigenvalues are $\lambda(\theta) = [7, 1]^T$. Note that the Hessian, for this particular function, is the same for every point θ – a property common to all quadratic functions (see also Sec. 5.7.6). The directions defined by the eigenvectors are illustrated in Fig. 5.5 at $\theta = 0$ (the same directions would apply to every point on the plane for this function). It is observed that the curvature in the direction of the first eigenvector $u_1 = [U(\theta)]_{:1}$, i.e., the first eigenvalue $\lambda_1 = 7$, is larger than the curvature in the direction of the second eigenvector $u_2 = [U(\theta)]_{:2}$, i.e., the second eigenvalue $\lambda_2 = 1$.

Figure 5.5 Contour lines and eigenvectors of the Hessian at $\theta = 0$, scaled by the corresponding eigenvalue, for the function $g(\theta) = 2\theta_1^2 + 2\theta_2^2 - 3\theta_1\theta_2$.

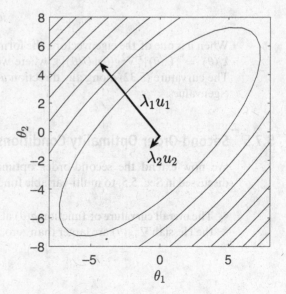

Figure 5.6 The function $g(\theta) = \theta_1^2 - \theta_2^2$ has a saddle point at $\theta = 0_2 = [0,0]^T$.

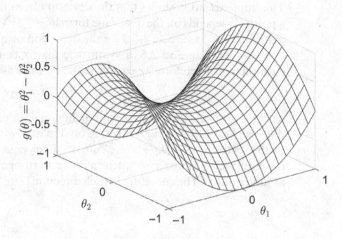

Example 5.11

For the quadratic function $g(\theta) = \theta_1^2 - \theta_2^2$, which is illustrated in Fig. 5.6, the Hessian is also constant for all $\theta \in \mathbb{R}^2$, and it is given as

$$\nabla^2 g(\theta) = \begin{bmatrix} 2 & 0 \\ .0 & -2 \end{bmatrix}. \tag{5.31}$$

The eigenvectors can be computed as $U(\theta) = I_2$ and the eigenvalues are $\lambda(\theta) = [2, -2]^T$. The two eigenvectors are hence aligned with the two axes. Furthermore, the curvature along axis θ_1 is positive, while that along θ_2 is negative, as described by the two eigenvalues. Since the curvature changes as θ moves through the stationary point $\theta = 0_2 = [0,0]^T$, the latter is a saddle point.

More generally, the curvature around point θ along a direction defined by a unitary-norm vector u can be obtained as

$$u^T \nabla^2 g(\theta) u. \tag{5.32}$$

When u is one of the eigenvectors, the formula (5.32) returns the corresponding eigenvalue, i.e., $\lambda_i(\theta) = [U(\theta)]_{:i}^T \nabla^2 g(\theta)[U(\theta)]_{:i}$, where we recall that $[U(\theta)]_{:i}$ represents the ith eigenvector. The curvature (5.32) along any direction u can be easily seen to be a linear combination of the eigenvalues.

5.7.5   Second-Order Optimality Conditions

We now extend the second-order optimality conditions from single-variable functions, as discussed in Sec. 5.5, to multi-variable functions. To this end, we need the following definitions:

- The **overall curvature** of function $g(\theta)$ at a point θ is said to be **positive** if *all* the eigenvalues of the Hessian $\nabla^2 g(\theta)$ are larger than zero, that is, if the Hessian is positive definite, $\nabla^2 g(\theta) \succ 0$.

- The **overall curvature** of function $g(\theta)$ at a point θ is said to be **non-negative** if *all* the eigenvalues of the Hessian $\nabla^2 g(\theta)$ are larger than, *or equal to*, zero, that is, if the Hessian is positive semi-definite, $\nabla^2 g(\theta) \succeq 0$.

Henceforth, when referring to the **curvature** of a function of multiple variables, we will mean the overall curvature as defined here.

Generalizing the **second-order sufficient optimality condition** in (5.14), if the curvature at a stationary point θ^* is positive – i.e., if $\nabla^2 g(\theta^*) \succ 0$ – then θ^* is a local minimum:

$$\text{stationary point } \theta^* \text{with } \nabla^2 g(\theta^*) \succ 0 \implies \text{local minimum.} \tag{5.33}$$

Conversely, if all eigenvalues are negative at a stationary point, the point is a local maximum.

Moreover, generalizing (5.15), the **second-order necessary optimality condition** states that, in order for a stationary point θ^* to be a local minimum, it must have a non-negative curvature, i.e., $\nabla^2 g(\theta^*) \succeq 0$:

$$\text{stationary point } \theta^* \text{is a local minimum} \implies \nabla^2 g(\theta^*) \succeq 0. \tag{5.34}$$

In fact, if the curvature is negative along any direction, the stationary point is either a local maximum or a saddle point. In particular, if some of the eigenvalues are positive and others negative (some may be equal to zero) at a stationary point, the stationary point is a saddle point. An example of a saddle point is illustrated in Fig. 5.6.

These conditions do not allow one to draw definite conclusions about the local optimality of a stationary point θ^* when the eigenvalues of the Hessian $\nabla^2 g(\theta^*)$ are non-negative – satisfying the necessary condition (5.34) – with some eigenvalues being equal to zero – making the sufficient condition (5.33) inapplicable. For such a stationary point, one should consider higher-dimensional derivatives in order to classify the point as local minimum, local maximum, or saddle point.

5.7.6 Optimality Conditions for Convex Functions

Generalizing the discussion in the preceding section, convex functions of multiple variables have the property that all stationary points are global minima (and hence also local minima). Therefore, when the function is convex, we do not need to verify second-order conditions, and it is sufficient to apply the first-order optimality condition.

A function $g(\theta)$ of vector $\theta \in \mathbb{R}^D$ is said to be **convex** if it has a *non-negative* curvature in its entire domain, i.e., if all the eigenvalues of the Hessian $\nabla^2 g(\theta)$ are non-negative, or $\nabla^2 g(\theta) \succeq 0$, for all $\theta \in \mathbb{R}^D$. We can also give equivalent definitions in a manner similar to single-valued functions, as detailed in Appendix 5.A.

A function $g(\theta)$ of vector $\theta \in \mathbb{R}^D$ is said to be **strictly convex** if it has a *positive* curvature in its entire domain, i.e., if all the eigenvalues of the Hessian $\nabla^2 g(\theta)$ are positive, or $\nabla^2 g(\theta) \succ 0$, for all $\theta \in \mathbb{R}^D$. A strictly convex function can have at most one (global) minimum point. Therefore, if it exists, the only stationary point of a strictly convex function is the sole global minimum.

Example 5.12

Quadratic functions have the general form

$$g(\theta) = c + b^T \theta + \frac{1}{2}\theta^T A\theta, \tag{5.35}$$

where b is a $D \times 1$ vector and A is a $D \times D$ symmetric matrix. The Hessian of a quadratic function is given as $\nabla^2 g(\theta) = A$ (this is why we multiply by $1/2$ in (5.35)). Therefore, if matrix A is positive semi-definite, i.e., if $A \succeq 0$, the quadratic function is convex; and, if matrix A is positive definite, i.e., if $A \succ 0$, the quadratic function is strictly convex.

Let us assume that we have $A \succ 0$, and hence the quadratic function (5.35) is **strictly convex**. Recall that, for strictly convex functions, there can be at most one stationary point – and hence a single global minimum. Applying the first-order optimality condition yields (see Appendix 5.B for the calculation of the gradient)

$$\nabla g(\theta) = b + A\theta = 0_D, \tag{5.36}$$

which returns the only stationary point and global minimum

$$\theta^* = -A^{-1}b, \tag{5.37}$$

generalizing (5.20). Note that matrix A is invertible because all eigenvalues are assumed to be positive.

Example 5.13

Consider the **ridge regression** problem studied in Chapter 4 (cf. (4.41)), namely

$$\min_{\theta \in \Theta} \left\{ \frac{1}{N} \|t_{\mathcal{D}} - X_{\mathcal{D}}\theta\|^2 + \frac{\lambda}{N} \|\theta\|^2 \right\} \tag{5.38}$$

for hyperparameter $\lambda > 0$. We now show that the cost function in (5.38) of ridge regression is a strictly convex quadratic function, and hence the solution can be obtained using formula (5.37). To this end, let us multiply the objective function by $N/2$ to rewrite the problem (5.38) as $\min_{\theta \in \Theta} g(\theta)$, with cost function

$$g(\theta) = \underbrace{\frac{1}{2}\|t_{\mathcal{D}}\|^2}_{=c} + \underbrace{\left(-X_{\mathcal{D}}^T t_{\mathcal{D}} \right)}_{=b}^T \theta + \frac{1}{2}\theta^T \underbrace{\left(X_{\mathcal{D}}^T X_{\mathcal{D}} + \lambda I_D \right)}_{=A} \theta. \tag{5.39}$$

As highlighted, this is a quadratic function of the general form (5.35). Therefore, the Hessian is constant and given as $\nabla^2 g(\theta) = X_{\mathcal{D}}^T X_{\mathcal{D}} + \lambda I$. It can be proved that its eigenvalues are larger than or equal to $\lambda > 0$, and hence the function is strictly convex. It follows from (5.37) that the stationary point

$$\theta^* = -A^{-1}b = (\lambda I + X_{\mathcal{D}}^T X_{\mathcal{D}})^{-1} X_{\mathcal{D}}^T t_{\mathcal{D}} \tag{5.40}$$

is the only global minimum. This coincides with the LS solution (4.42) given in Chapter 4.

Some convex functions. Examples of convex functions in the domain $\theta \in \mathbb{R}^D$ include:

- $g(\theta) = b^T\theta + c$, for any $b \in \mathbb{R}^D$ and $c \in \mathbb{R}$;
- $g(\theta) = \frac{1}{2}\theta^T A\theta + b^T\theta + c$, with $A \succeq 0$;
- $g(\theta) = \|\theta\|^2$;
- $g(\theta) = \log\left(\sum_{i=1}^{D} \exp(a_i\theta_i) \right)$, for any vector $a = [a_1, \ldots, a_D]^T \in \mathbb{R}^D$ – this is known as the **log-sum-exp function**; and
- any linear combination of convex functions with non-negative coefficients.

5.8 Gradient Descent

Apart from special cases such as strictly convex cost functions, global, or even local, minima cannot be computed directly by applying simple formulas. Instead, iterative algorithms based on local optimization steps are deployed with the aim of approaching a local minimum or, at least, a stationary point. Local optimization algorithms produce a sequence $\theta^{(1)}, \theta^{(2)}, \theta^{(3)}, \ldots$ of **iterates** for the model parameter vector in $\mathbb{R}^D$. The updates are local in the sense that, in order to produce iterate $\theta^{(i+1)}$, the algorithms explore only a neighborhood of iterate $\theta^{(i)}$.

5.8.1 Local Search-Based Optimization and Gradient Descent

The exploration of the neighborhood of the current iterate $\theta^{(i)}$ depends on the type of information that is available about the local behavior of the function around $\theta^{(i)}$. Accordingly, we can distinguish the following main types of local search optimizers:

- **Zeroth-order optimization.** The optimizer only has access to the value of the function $g(\theta^{(i)})$ at $\theta^{(i)}$ (and possibly at other points in the neighborhood of $\theta^{(i)}$).
- **First-order optimization.** In addition to $g(\theta^{(i)})$, the optimizer also has access to the gradient $\nabla g(\theta^{(i)})$.
- **Second-order optimization.** In addition to $g(\theta^{(i)})$ and $\nabla g(\theta^{(i)})$, the optimizer also has access to the Hessian $\nabla^2 g(\theta^{(i)})$ or to some approximation thereof.

The available information about the local behavior of the cost function around the current iterate $\theta^{(i)}$ allows the optimizer to construct an approximation of the cost function in a neighborhood of $\theta^{(i)}$. For example, a second-order optimizer could use the second-order Taylor approximation (5.27). The most common local optimization algorithm is first-order optimization via gradient descent and variations thereof.

At each iteration i, **gradient descent (GD)** selects the next iterate $\theta^{(i+1)}$ by following the direction of the negative gradient $-\nabla g(\theta^{(i)})$ at the current iterate $\theta^{(i)}$. This, we recall, is the direction of steepest local descent of the cost function around $\theta^{(i)}$. The aim is to ensure that the sequence of iterates $\theta^{(1)}, \theta^{(2)}, \theta^{(3)}, \ldots$ converges to a stationary point (at which the gradient is zero), in the hope that the latter is also a local minimum. Key design questions pertain to the size of each update step – how far along the negative gradient direction should the next iterate move? – and the corresponding convergence guarantees towards a stationary point.

5.8.2 Local Strictly Convex Quadratic Approximant

As we have discussed in the preceding section, strictly convex quadratic functions (5.35) (with a positive definite Hessian $A \succ 0$) are particularly convenient to minimize, since the (only) global minimum can be computed in closed form as in (5.37). GD constructs and minimizes a strictly convex quadratic approximation of the cost function around the current iterate $\theta^{(i)}$.

Local Strictly Convex Quadratic Approximant. At each iteration i, GD approximates the cost function $g(\theta)$ around the current iterate $\theta^{(i)}$ with the quadratic function

$$\tilde{g}_\gamma(\theta|\theta^{(i)}) = \underbrace{g(\theta^{(i)}) + \nabla g(\theta^{(i)})^T(\theta - \theta^{(i)})}_{\text{first-order Taylor approximation}} + \underbrace{\frac{1}{2\gamma}||\theta - \theta^{(i)}||^2}_{\text{quadratic proximity penalty}} \qquad (5.41)$$

Figure 5.7 A function $g(\theta)$ along with the first-order Taylor approximation and two local strictly convex quadratic approximants $\tilde{g}_\gamma(\theta|\theta^{(i)})$ with different values of the inverse curvature γ at the current iterate $\theta^{(i)} = -0.5$.

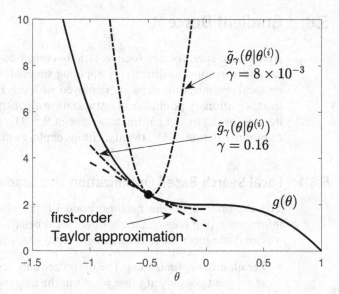

for some fixed constant $\gamma > 0$. This local approximant adds to the first-order Taylor approximation (5.22) around the current iterate $\theta^{(i)}$ a strictly convex quadratic term that penalizes a vector θ in accordance to its squared Euclidean distance to the current iterate $\theta^{(i)}$. An illustration of the first-order Taylor approximation and of the strictly convex approximant (5.41) can be found in Fig. 5.7. (Note that approximating the cost function with the first-order Taylor approximation (5.22) alone would not be useful since, at any non-stationary point, this linear function would be unbounded below.)

It can be easily checked that the Hessian of the approximant (5.41) is given by

$$\nabla \tilde{g}_\gamma(\theta|\theta^{(i)}) = \frac{1}{\gamma} I_D \succ 0. \tag{5.42}$$

Therefore, all eigenvalues of the Hessian are positive and equal to $1/\gamma > 0$. The proximity penalty term hence ensures that the function is quadratic and strictly convex. Therefore, GD can obtain the next iterate by minimizing the approximant (5.41) via (5.37), as we will detail in the next subsection.

Since all eigenvalues of the Hessian are equal to $1/\gamma$, the constant γ plays the role of **inverse curvature** of the quadratic approximation (5.41). As illustrated by the examples in Fig. 5.7, a smaller γ thus implies a larger curvature of the approximant.

Properties of the Local Strictly Convex Quadratic Approximant. The local strictly convex approximant (5.41) has the following properties:

- **Tightness at the current iterate**. The approximant equals the cost function at the current iterate, i.e.,

$$\tilde{g}_\gamma(\theta^{(i)}|\theta^{(i)}) = g(\theta^{(i)}). \tag{5.43}$$

- **Gradient matching at the current iterate**. The gradient of the approximant equals the gradient of the cost function at the current iterate, i.e.,

$$\nabla_\theta \tilde{g}_\gamma(\theta|\theta^{(i)})|_{\theta=\theta^{(i)}} = \nabla g(\theta^{(i)}). \tag{5.44}$$

- **Convexity with uniform curvature**. The second derivative is equal in all directions to $1/\gamma > 0$, and hence a smaller γ implies an approximant with a larger curvature.

5.8.3   Gradient Descent As Successive Convex Approximation Minimization

At each iteration i, GD minimizes the local strictly convex quadratic approximant $\tilde{g}_\gamma(\theta|\theta^{(i)})$. Accordingly, the iterate $\theta^{(i+1)}$ is obtained by solving the problem

$$\theta^{(i+1)} = \arg\min_\theta \tilde{g}_\gamma(\theta|\theta^{(i)}). \tag{5.45}$$

Since the cost function is strictly convex, the unique global optimum is obtained by using (5.37), yielding the next iterate

$$\theta^{(i+1)} = \theta^{(i)} - \gamma \nabla g(\theta^{(i)}). \tag{5.46}$$

This update rule is applied by GD until a stopping criterion is satisfied, as summarized in the following algorithm. Note that, since γ determines the size of the update step, it is known as **learning rate**.

Algorithm 5.1: Gradient descent (GD)

---

initialize $\theta^{(1)}$ and $i = 1$
while *stopping criterion not satisfied* **do**
    obtain next iterate as

$$\theta^{(i+1)} = \theta^{(i)} - \gamma^{(i)} \nabla g(\theta^{(i)})$$

    set $i \leftarrow i + 1$
end
return $\theta^{(i)}$

In the table, we have allowed for the learning rate to possibly change along the iterations i via the notation $\gamma^{(i)}$. As we will see, this turns out to be practically and theoretically useful. Also, as per the table, the algorithm continues until a **stopping criterion** is satisfied. There are several possible choices for the stopping criterion. For instance, one may fix a priori the number of iterations; or stop when the norm of the gradient $||\nabla g(\theta^{(i)})||$ is sufficiently small, which may be taken as an indication that the current iterate is close to a stationary point.

Example 5.14

In order to start building an intuition about the role of the learning rate, consider the two examples in Fig. 5.8. In both cases, the initial point is $\theta^{(1)} = -1$, but the learning rates are different. In the top figure, a larger (constant) learning rate γ is chosen, while a smaller learning rate is selected for the bottom figure. With a larger γ, the approximant has a smaller curvature and this yields longer update steps, which may cause the iterates to overshoot and possibly diverge. In contrast, with a smaller γ, the approximant has a larger curvature, which corresponds to shorter update steps. A small learning rate may cause the iterate to proceed too slowly. In the next section, we will develop theoretical arguments to optimize the choice of the learning rate for GD.

Figure 5.8 An illustration of the impact of the choice of the learning rate γ. The arrows point in the direction of increasing iteration index $i = 1, 2, \ldots .$

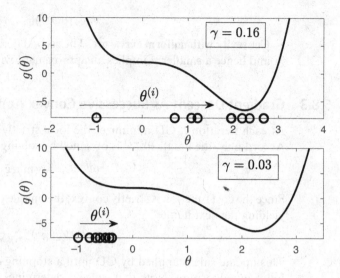

5.8.4   Negative Gradient As a "Force" Field

The negative gradient $-\gamma \nabla g(\theta)$ defines a "force" field in the space $\mathbb{R}^D$ of model parameter vectors θ that points in the direction of steepest local descent of the cost function $g(\theta)$. This is the direction along which the first-order Taylor approximation of the cost function at the given point θ decreases the most when taking a small step from θ. Importantly, the direction of steepest local descent generally does not coincide with the direction towards a stationary point, and hence also towards a local or global minimum. Instead, it corresponds to a "greedy" move that maximizes the current decrease in the cost function without accounting for the end goal of obtaining a (local) minimum.

The degree to which the gradient direction approximates a direction towards a stationary point can be (partly) quantified via the curvature of the function around the current iterate. In particular, the alignment of the gradient direction with the direction to a stationary point is affected by whether the function has a uniform curvature, like a symmetric bowl, or a heterogeneous curvature, like a canyon, along different directions away from the given point.

Example 5.15

To elaborate on this point, compare the negative gradient fields in Fig. 5.4. For the function on the left, which presents an equal curvature in all directions, the negative gradient vectors point in the direction of the minimum; for the one on the right, the negative gradient tends to point in the direction of maximal local curvature of the function. Note that GD does not have access to second-order information, and hence it is generally unable to distinguish between these two situations. This will be a central topic in the next section.

5.9   Convergence of Gradient Descent

The final example in the preceding section suggests that the curvature of the cost function $g(\theta)$, as encoded by the eigenvalues of the Hessian, plays an important role in assessing the performance of GD. In fact, as illustrated in Fig. 5.4, the relationship between the negative gradient and the direction towards a local minimum or a stationary point depends on the curvature of the function. In this section, we elaborate on this important point by analyzing the convergence properties of GD. This study will also yield important insights into the choice of the learning rate.

5.9.1   Why Can GD Converge to a Stationary Point?

To start, let us build some intuition on the change in the cost function caused by a single iteration of GD. This will reveal a key mechanism that enables GD to converge, while also highlighting the importance of the choice of the learning rate γ.

To this end, let us first recall that, by the tightness property (5.43), the convex approximant $\tilde{g}_\gamma(\theta|\theta^{(i)})$ equals the cost function $g(\theta)$ at the current iterate $\theta^{(i)}$. We can hence choose the learning rate γ, i.e., the inverse curvature of the approximant $\tilde{g}_\gamma(\theta|\theta^{(i)})$, to be small enough so that $\tilde{g}_\gamma(\theta|\theta^{(i)})$ is a **global upper bound** on the cost function, satisfying the inequality

$$\tilde{g}_\gamma(\theta|\theta^{(i)}) \geq g(\theta) \tag{5.47}$$

for all $\theta \in \mathbb{R}^D$. This property is illustrated in Fig. 5.7, where we have two local approximants: The first, with a relatively large value of γ ($\gamma = 0.16$), is clearly not a global – nor a local – upper bound; while the second, with a smaller γ ($\gamma = 8 \times 10^{-3}$), is a global upper bound.

With this choice of the learning rate, we will now argue that the value of the cost function $g(\theta^{(i+1)})$ for the next iterate cannot be larger than at the current iterate – and hence we have the **per-iteration descent condition**

$$g(\theta^{(i+1)}) \leq g(\theta^{(i)}). \tag{5.48}$$

The descent condition (5.48) is a direct consequence of two observations depicted in Fig. 5.9: (*i*) By the global upper bound property (5.47), the value $g(\theta^{(i+1)})$ of the cost function must necessarily be no larger than the corresponding value $\tilde{g}_\gamma(\theta^{(i+1)}|\theta^{(i)})$ of the approximant; and (*ii*) since $\theta^{(i+1)}$ is selected as the minimum of the approximant $\tilde{g}_\gamma(\theta|\theta^{(i)})$, the inequality $\tilde{g}_\gamma(\theta^{(i+1)}|\theta^{(i)}) \leq \tilde{g}_\gamma(\theta^{(i)}|\theta^{(i)}) = g(\theta^{(i)})$ holds.

Overall, by (5.48), we can conclude that, if the learning rate γ is properly chosen, the cost function is guaranteed to be non-increasing along the iterations i. We will formalize this conclusion in the rest of this section.

5.9.2   Smoothness and Global Upper Bound Property

How small should the learning rate γ be so that the approximant is a global upper bound on the cost function as in (5.47) so that we can ensure the per-iteration descent property (5.48)? Intuitively, the more curved a function $g(\theta)$ is, the smaller γ must be in order for this condition to be satisfied.

Figure 5.9 GD yields non-increasing values for the cost function if the local strictly convex quadratic approximants are sufficiently curved as to satisfy the global upper bound property (5.47). The solid gray curve is the cost function; dashed functions are the approximants at successive GD iterations; stars represent the optimal values of the approximants; and circles represent the GD iterates.

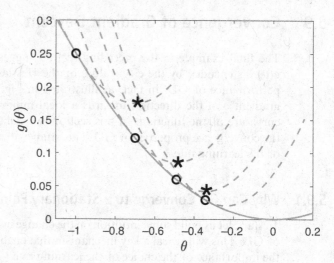

L-Smooth Function. To answer this question, let us define a function $g(\theta)$ as **L-smooth**, for $L > 0$, if the maximum eigenvalue of the Hessian $\nabla^2 g(\theta)$ is no larger than L for all $\theta \in \mathbb{R}^D$. That is, a smooth function has an overall curvature limited by L. As a special case, for $D = 1$, an L-smooth function satisfies the inequality

$$\frac{\mathrm{d}^2 g(\theta)}{\mathrm{d}\theta^2} \leq L \tag{5.49}$$

for all $\theta \in \mathbb{R}$. Smoothness is a global property of the function, as it depends on the Hessian at all values of θ. (One can also talk about local smoothness around a specific point by restricting the Hessian to such point, but we will not need this notion here.)

In terms of terminology, it is also worth noting that a function is smoother the *smaller L* is. So, an L-smooth function is also L'-smooth for $L' > L$; but the reverse is not true. Examples of L-smooth single-variable functions with different values of L can be found in Fig. 5.10. By definition, for instance, the 0.1-smooth function is also 1-smooth or 10-smooth, but the 1-smooth function is not 0.1-smooth.

Example 5.16

For the function $g(\theta_1, \theta_2) = 3\theta_1^2 + 2\theta_2^2 - \theta_1\theta_2$, the Hessian is

$$\nabla^2 g(\theta) = \begin{bmatrix} 6 & -1 \\ -1 & 2 \end{bmatrix} \tag{5.50}$$

and the eigenvalues are 3.6 and 6.4. The function is hence smooth with constant $L = 6.4$. Note that what matters in the definition of smoothness is the direction with the largest curvature, and hence the largest eigenvalue.

Figure 5.10 Examples of L-smooth single-variable functions with different values of the smoothness constant L.

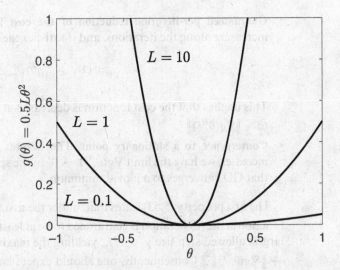

Global Upper Bound Property of the Local Approximant for Smooth Functions. This is the key result of this subsection: For an L-smooth function $g(\theta)$, if the learning rate satisfies the inequality

$$\gamma \leq \frac{1}{L}, \tag{5.51}$$

the local approximant $\tilde{g}_\gamma(\theta|\theta^{(i)})$ satisfies the **global upper bound property** (5.47). Importantly, condition (5.51) indicates that a more curved function – with a larger L – calls for a smaller learning rate γ in order for condition (5.47) to be satisfied.

Proof: Given its importance, it is useful to go over the proof of this key property. To this end, we consider the simpler case of single-variable functions, i.e., $D = 1$, but the same steps apply more generally to any $D \geq 1$. By the first-order Taylor expansion (5.8) and the inequality (5.49), we have the first inequality in

$$g(\theta) \leq g(\theta^{(i)}) + \frac{dg(\theta^{(i)})}{d\theta}(\theta - \theta^{(i)}) + \frac{L}{2}(\theta - \theta^{(i)})^2$$

$$\underset{L \leq 1/\gamma}{\leq} g(\theta^{(i)}) + \frac{dg(\theta^{(i)})}{d\theta}(\theta - \theta^{(i)}) + \frac{1}{2\gamma}(\theta - \theta^{(i)})^2 = \tilde{g}_\gamma(\theta|\theta^{(i)}), \tag{5.52}$$

while the second inequality follows from the assumption (5.51) on the learning rate. This concludes the proof. $\square$

5.9.3 Convergence of Gradient Descent

The main result of this section builds on the global upper bound property (5.47) of the local approximant optimized by GD, and is stated as follows. For an L-smooth function, if the learning rate satisfies the inequality (5.51), we have the following properties of GD:

- **Guaranteed per-iteration reduction of the cost function** $g(\theta)$. The cost function is non-increasing along the iterations, and it satisfies the inequality

$$g(\theta^{(i+1)}) \leq g(\theta^{(i)}) - \frac{\gamma}{2} \left\| \nabla g(\theta^{(i)}) \right\|^2. \tag{5.53}$$

This implies that the cost function is decreased at each iteration by at least an amount equal to $\frac{\gamma}{2} \left\| \nabla g(\theta^{(i)}) \right\|^2$.

- **Convergence to a stationary point**. The iterates converge to a stationary point, i.e., as i increases, we have the limit $\nabla g(\theta^{(i)}) \to 0$. In the special case of convex functions, this implies that GD converges to a global minimum.

The first property (5.53) states that, under the assumption (5.51), the improvement in the cost function at the ith iteration is guaranteed to be at least $\frac{\gamma}{2} \left\| \nabla g(\theta^{(i)}) \right\|^2$, which is maximized for the largest allowed step size $\gamma = 1/L$, yielding the **maximum guaranteed per-iteration improvement** $\frac{1}{2L} \left\| \nabla g(\theta^{(i)}) \right\|^2$. Consequently, one should expect larger improvements in the cost function at iterates $\theta^{(i)}$ at which the gradient has a larger norm $||\nabla g(\theta^{(i)})||^2$. The inequality (5.53) also indicates that, under the condition (5.51), GD eventually stops when the gradient equals the all-zero vector, that is, at a stationary point, as formalized by the second property above.

Example 5.17

Two examples of GD iterates for the functions in Fig. 5.4 can be found in Fig. 5.11 with $\gamma = 0.2/L$. Note that the smoothness constants are given as $L = 4$ for the first function (left figure) and $L = 7$ for the second function (right figure). With the selected learning rate $\gamma = 0.2/L$, which satisfies the condition (5.51), the iterates converge to the only stationary point of the two functions, which is also the only global minimum (the functions are strictly convex and quadratic). Moreover, it is observed that the per-iteration improvement is more pronounced at the earlier iterations, for which the norm of the gradient is larger, while slower progress is made at the later iterates.

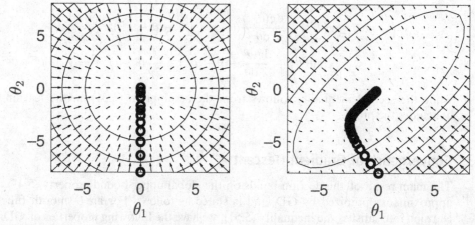

Figure 5.11 Contour lines, negative gradient vectors, and GD iterates starting at $\theta^{(1)} = [0, -8]^T$ for functions $g(\theta) = 2\theta_1^2 + 2\theta_2^2$ (left) and $g(\theta) = 2\theta_1^2 + 2\theta_2^2 - 3\theta_1\theta_2$ (right).

Proof: It is useful to understand how the two convergence properties stated above can be formally derived. Let us first prove that the cost function is non-increasing over the iterations with the following inequalities:

$$g(\theta^{(i+1)}) \leq \tilde{g}_\gamma(\theta^{(i+1)}|\theta^{(i)}) \text{ (by the global upper bound property (5.47))}$$
$$\leq \tilde{g}_\gamma(\theta^{(i)}|\theta^{(i)}) \text{ (since } \theta^{(i+1)} \text{ minimizes } \tilde{g}_\gamma(\theta|\theta^{(i)}) \text{ over } \theta)$$
$$= g(\theta^{(i)}) \text{ (by the tightness property (5.43))}, \tag{5.54}$$

where we have indicated the rationale for each step beside it. We can now do a more refined analysis to evaluate by *how much* the cost function is reduced at each step under the assumption (5.51). To simplify the notation, let us again consider $D = 1$. The result (5.53) follows from these steps:

$$g(\theta^{(i+1)}) \leq \tilde{g}_\gamma(\theta^{(i+1)}|\theta^{(i)}) \text{ (by the upper bound property (5.47))}$$
$$= g(\theta^{(i)}) + \frac{dg(\theta^{(i)})}{d\theta}(\theta^{(i+1)} - \theta^{(i)}) + \frac{1}{2\gamma}(\theta^{(i+1)} - \theta^{(i)})^2$$
$$= g(\theta^{(i)}) + \frac{dg(\theta^{(i)})}{d\theta}\left(-\gamma\frac{dg(\theta^{(i)})}{d\theta}\right) + \frac{1}{2\gamma}\left(-\gamma\frac{dg(\theta^{(i)})}{d\theta}\right)^2$$
$$= g(\theta^{(i)}) - \frac{\gamma}{2}\left(\frac{dg(\theta^{(i)})}{d\theta}\right)^2, \tag{5.55}$$

where in the third equality we have used the GD update (5.46).

Finally, by (5.53), the cost function decreases at each iteration except when $\nabla g(\theta) = 0_D$. Therefore, the algorithm must converge to a stationary point. □

5.9.4 Iteration Complexity

Apart from knowing that the iterates converge, one may also wish to quantify *how many iterations* are needed for GD to converge. This is known as the **iteration complexity** of GD. To elaborate, consider an arbitrary initialization $\theta^{(1)}$, and assume that the goal is to approach a stationary point within some small error $\epsilon > 0$. This requirement can be formulated by imposing that the iterations reach a point $\theta^{(i)}$ that satisfies the inequality $||\nabla g(\theta^{(i)})|| \leq \epsilon$. Let us refer to a point satisfying this inequality as an ϵ-**stationary point**. (A stationary point is obtained with $\epsilon = 0$.) To simplify the discussion, we will assume assume that the cost function has at least one global minimum θ^*.

Setting $\gamma = 1/L$, from (5.53) each iteration decreases the cost function by at least $\frac{1}{2L}\left\|\nabla g(\theta^{(i)})\right\|^2$. Since the algorithm stops as soon as the gradient norm is no larger than ϵ, the improvement in the cost function is no smaller than $1/(2L)\epsilon^2$ in the iterations preceding the attainment of an ϵ-stationary point. Therefore, the number of iterations to reach such a stationary point cannot be larger than $(g(\theta^{(1)}) - g(\theta^*))/(\epsilon^2/(2L))$, since the function cannot be reduced below the optimal value $g(\theta^*)$. It follows that the number of iterations cannot be larger than a constant multiple of $2L/\epsilon^2$.

The arguments in the preceding paragraph suggest that the number of iterations needed to obtain an ϵ-stationary point is of the order $\mathcal{O}(1/\epsilon^2)$ when viewed as a function of the error ϵ. (The notation $\mathcal{O}(f(\epsilon))$ indicates a quantity that equals $f(\epsilon)$ apart from multiplicative and additive constants that do not depend on ϵ). Furthermore, it is proportional to the smoothness constant L, with a larger curvature – a larger L – requiring more iterations.

This conclusion relies on the assumption that the cost function is L-smooth. As seen, this amounts to imposing an upper bound on the maximum eigenvalue of the Hessian. As detailed in Appendix 5.C, a more refined analysis of the iteration complexity can be carried out by making further assumptions about the cost function, typically in the form of additional constraints on the spread of the eigenvalues of the Hessian. In particular, under a lower bound condition on the minimum eigenvalue of the Hessian, one can improve the iteration complexity order to $\mathcal{O}(\log(1/\epsilon))$ – an exponential savings in terms of the number of iterations.

The role of the spread of the eigenvalues in determining the convergence properties is further highlighted in the next subsection.

5.9.5 Choosing the Learning Rate and the Initialization

In this subsection, we build on the discussion so far in this section to provide some insights on the choice of learning rate and of the initialization for GD.

What If the Learning Rate is Too Large? As we have seen in the preceding subsection, the condition (5.51) guarantees a monotonically decreasing cost function towards a stationary point. Condition (5.51) is sufficient, but not necessary in general: GD may still converge to a stationary point when $\gamma > 1/L$, but progress may not always be guaranteed at each iteration as in (5.53). Furthermore, the threshold value of the learning rate ensuring convergence depends on the eigenvalues of the Hessian of the cost function.

To elaborate on these important aspects, consider the case of the single-variable strictly convex quadratic function

$$g(\theta) = c + b\theta + \frac{L}{2}\theta^2 \tag{5.56}$$

with $L > 0$, for which the global minimum (and only stationary point) is $\theta^* = -b/L$ (cf. (5.37)). As shown in Appendix 5.C, the difference between the current iterate $\theta^{(i+1)}$ and the global minimum θ^* for this function evolves as

$$\theta^{(i+1)} - \theta^* = (1 - \gamma \cdot L)^i(\theta^{(1)} - \theta^*) \tag{5.57}$$

as a function of the difference $(\theta^{(1)} - \theta^*)$ between initialization and optimal value. Therefore, if $\gamma < 2/L$, we have $|1 - \gamma \cdot L| < 1$, which implies that the difference $\theta^{(i+1)} - \theta^*$ tends to zero as i increases. Note that this requirement on γ is less strict than the condition $\gamma \leq 1/L$ derived above to guarantee monotonic convergence for general functions.

If the learning rate satisfies the condition $\gamma > 2/L$, we have the inequality $(1 - \gamma \cdot L) < -1$. With this excessively large learning rate, the term $(1 - \gamma \cdot L)^i$ in (5.57) oscillates between positive and negative numbers of increasing magnitude, and hence the iterates diverge.

In the more general case of multi-variable strictly convex quadratic functions (5.35), as detailed in Appendix 5.C, one can similarly study convergence by considering separately the evolution of the iterates along the directions defined by the eigenvectors of the Hessian A. If $\gamma > 2/L$, the learning rate satisfies the inequality $\gamma > 2/\lambda_i$ for at least one eigenvalue λ_i of the

Figure 5.12 Illustration of the effect of the choice of a large learning rate. For this example, corresponding to the strictly convex quadratic function in Fig. 5.5, the learning rate is larger than $2/\lambda_1$, where $\lambda_1 = L = 7$ is the largest eigenvalue of the Hessian, while it is smaller than $2/\lambda_2$ with $\lambda_2 = 1$. Accordingly, the iterates oscillate and diverge in the direction of the first eigenvector. For clarity, in the figure, arrows are drawn to connect the first iterates, while the pattern of oscillation continues in the same way for the following iterations (which are not explicitly connected).

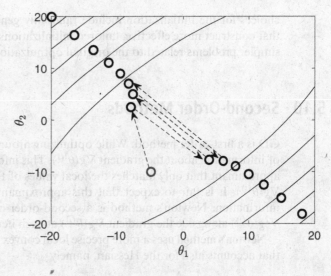

Hessian matrix. Along all the directions defined by the corresponding eigenvectors, iterates will oscillate and eventually diverge.

Example 5.18

An example is shown in Fig. 5.12 for the strictly convex quadratic function illustrated in Fig. 5.5, which has smoothness constant $L = 7$. The two eigenvalues of the Hessian are $\lambda_1 = L = 7$ and $\lambda_2 = 1$. With $\gamma = 0.3$, the learning rate is larger than $2/\lambda_1 = 2/L$, but smaller than $2/\lambda_2$. As shown in Fig. 5.5, this causes the iterates to oscillate and diverge in the direction of the first eigenvector.

Choosing the Learning Rate. The discussion so far suggests that one should choose a learning rate γ inversely proportional to the smoothness constant L. In practice, however, the smoothness constant L – a global property of the cost function – is hard to evaluate. A useful algorithm based on this theoretical result is known as **Armijo's rule**, which operates as follows: (*i*) start each iteration with a large learning rate γ; then (*ii*) decrease γ until the relaxed per-iteration improvement condition

$$g(\theta^{(i+1)}) \leq g(\theta^{(i)}) - \frac{\alpha\gamma}{2} \left\| \nabla g(\theta^{(i)}) \right\|^2 \tag{5.58}$$

is satisfied for some small $0 < \alpha < 1$. Note that a small α will make it easier to satisfy this inequality, and that this condition amounts to a relaxed version of the theoretical per-iteration improvement (5.53).

Choosing the Initialization. While the theoretical guarantees introduced in this section apply for an arbitrary initialization $\theta^{(1)}$, the choice of the initialization $\theta^{(1)}$ is an important aspect of the implementation of GD. This is because GD is a local optimization algorithm, which may end up at a suboptimal stationary point if initialized too far from close-to-optimal solutions. Typical choices for the initialization include randomly generated vectors $\theta^{(1)}$, but there are methods that construct more effective, tailored initializations. This can be done, for instance, by solving simpler problems related to the original optimization of interest.

5.10 Second-Order Methods

GD is a first-order method: While optimizing around the current iterate $\theta^{(i)}$, it only makes use of information about the gradient $\nabla g(\theta^{(i)})$. This information is translated into a local quadratic approximant that only matches the local values of the cost function $g(\theta^{(i)})$ and of the gradient $\nabla g(\theta^{(i)})$. It is fair to expect that this approximant can be improved by using second-order information. Newton's method is a second-order optimization scheme that uses the Hessian $\nabla^2 g(\theta^{(i)})$ alongside the gradient $\nabla g(\theta^{(i)})$ at each iteration i.

Newton's method uses a more precise local convex approximant around the current iterate $\theta^{(i)}$ that accounts also for the Hessian, namely

$$\tilde{g}_{\gamma}^{\text{Newton}}(\theta|\theta^{(i)}) = g(\theta^{(i)}) + \nabla g(\theta^{(i)})(\theta - \theta^{(i)}) + \underbrace{\frac{1}{2\gamma}(\theta - \theta^{(i)})^T \nabla^2 g(\theta^{(i)})(\theta - \theta^{(i)})}_{\text{proximity penalty}}. \quad (5.59)$$

This approximant provides a closer local quadratic approximation by retaining second-order information about the cost function as in the Taylor approximation (5.27). If the Hessian $\nabla^2 g(\theta^{(i)})$ is positive definite, the approximant is a strictly convex quadratic function. Minimizing it via (5.37) yields the update

$$\theta^{(i+1)} = \theta^{(i)} - \gamma (\nabla^2 g(\theta^{(i)}))^{-1} \nabla g(\theta^{(i)}). \quad (5.60)$$

The multiplication of the gradient by the inverse of the Hessian scales down the gradient in directions in which the function is more curved. To see this, consider a cost function with a diagonal Hessian, for which an eigenvalue represents the curvature of the cost function along the corresponding coordinate of vector θ. The multiplication $\nabla^2 g(\theta^{(i)})^{-1} \nabla g(\theta^{(i)})$ divides each entry of the gradient by the inverse of the corresponding eigenvalue of the Hessian. Therefore, coordinates with a larger curvature are scaled down more significantly than coordinates along which the cost function is less curved. By scaling down the gradient in different directions of the optimization space, Newton's method helps reduce the oscillations discussed in the preceding section (see Fig. 5.12).

The operation of Newton's method accordingly be thought of in one of two ways:

- as applying **different learning rates** along different directions in the model parameter space, with reduced learning rates in directions of larger curvature of the cost function; or
- as **equalizing the curvatures** in different directions of the parameter space by pre-processing the gradient prior to an update.

The multiplication by the inverse of the Hessian in Newton's method is an instance of the broader class of **pre-conditioning** techniques. Pre-conditioning amounts to linear operations on the gradient vector applied prior to an update within an iterative optimization scheme.

Computing the inverse of the Hessian is in practice too demanding in terms of computational complexity when D is large, and there exist various simplifications of Newton's method that avoid this step, including the Broyden–Fletcher–Goldfarb–Shanno (BFGS) algorithm.

5.11 Stochastic Gradient Descent

GD applies to any cost function $g(\theta)$. However, training problems of the form (5.2) aim to minimize the finite-sum cost function

$$g(\theta) = \frac{1}{N} \sum_{n=1}^{N} g_n(\theta), \tag{5.61}$$

where the data set size N is typically very large. In this case, computing the gradient

$$\nabla g(\theta) = \frac{1}{N} \sum_{n=1}^{N} \nabla g_n(\theta) \tag{5.62}$$

may be prohibitively complex and time-consuming. Stochastic GD (SGD) addresses this computational shortcoming of GD for finite-sum optimization problems, and is the subject of this section. Furthermore, as we will see, apart from reducing the complexity of the optimization algorithm, SGD also improves the generalization performance of the trained model.

5.11.1 Stochastic Gradient Updates

Unlike GD, **stochastic GD (SGD)** uses at each iteration only a subset of the training data $\mathcal{D}$. This subset of data points, which is generally different for each iteration, is denoted as $\mathcal{S}^{(i)} \subseteq \{1, \ldots, N\}$ and is referred to as a **mini-batch**. As we will see, the size of the mini-batch, S, is an important hyperparameter that defines the performance of SGD. The operation of SGD is detailed in the following algorithm.

Algorithm 5.2: Stochastic gradient descent (SGD)

---

initialize $\theta^{(1)}$ and $i = 1$

while *stopping criterion not satisfied* **do**

    pick a *mini-batch* $\mathcal{S}^{(i)} \subseteq \{1, \ldots, N\}$ of S samples from the data set $\mathcal{D}$

    obtain next iterate as

$$\theta^{(i+1)} \leftarrow \theta^{(i)} - \frac{\gamma^{(i)}}{S} \sum_{n \in \mathcal{S}^{(i)}} \nabla g_n(\theta^{(i)}) \tag{5.63}$$

    set $i \leftarrow i + 1$

end

return $\theta^{(i)}$

As we will detail, the **stochastic gradient**

$$\frac{1}{S} \sum_{n \in \mathcal{S}^{(i)}} \nabla g_n(\theta^{(i)}) \tag{5.64}$$

used by SGD is an estimate of the true gradient (5.62), for which the average is extended to the entire training set. The computational efficiency of SGD stems from the computation of S, instead of N, gradients $\nabla g_n(\theta^{(i)})$. The main question, of course, is the extent to which using the estimate (5.64) in lieu of (5.62) affects the performance of the trained model $\theta^{(i)}$ produced by the algorithm.

5.11.2 Choosing the Mini-batch

Before we address this question, let us discuss how to choose the mini-batch $\mathcal{S}^{(i)} \subseteq \{1, \ldots, N\}$ of examples at each iteration i. There are several ways to do this:

- **Random uniform mini-batching.** The most convenient way to choose the mini-batch in terms of analysis is to assume that each of the S samples in $\mathcal{S}^{(i)}$ is selected from the N samples in set $\mathcal{D}$ uniformly at random and independently of the other samples. Note that this sampling *with replacement* may cause the same sample to appear multiple times in the mini-batch, although this is extremely unlikely when N is large. Under this random uniform mini-batch assumption, we can think of the average gradient (5.64) used by the SGD update as a **stochastic estimate** of the true gradient. The estimate is stochastic because of the randomness that arises from the selection of the mini-batch of samples. Note that, even if $S = N$, this stochastic estimate does not coincide with the true gradient. The properties of SGD with random uniform mini-batches will be studied in the next section, and the approach will be adopted throughout the book.

- **Permutation-based mini-batching.** Other approaches, which do not replace the selected sample when drawing a mini-batch, are more commonly used in practice. A standard solution is to divide the training processes into **epochs**, such that the SGD iterations within an epoch run over the entire data set exactly once. To this end, at the beginning of each epoch, the N samples in the training set $\mathcal{D}$ are randomly permuted; and the mini-batches are then constructed by taking disjoint subsets of S consecutive samples in the order dictated by the permutation. Note that, with this approach, SGD coincides with GD for $S = N$.

5.11.3 Iterations vs. Gradient Computations

SGD was introduced in this section as a way to reduce the **per-iteration** computational complexity of GD. As we discuss in this subsection, this per-iteration advantage translates into benefits in terms of the **wall-clock time** needed to make progress towards a solution of the minimization problem (5.2).

Let us fix the time needed to compute the gradient of any one of the functions $g_n(\theta)$ in (5.2). This time, denoted as ΔT, is assumed here to be constant across all data points n, and is due in practice to the need to retrieve the corresponding data point and to run a differentiation routine. For a certain amount of overall time T available for training, one is then only allowed to compute $T/\Delta T$ gradients. We are then interested in how much reduction in the cost function can be attained with a total amount $T/\Delta T$ of gradient calculations.

GD requires computing gradients for all N examples in the training set before it can make any progress away from the initialization. In contrast, SGD can start updating the model parameter vector as soon as one mini-batch of S data points is processed, and progress is potentially made upon the calculation of the gradient on each mini-batch. So, while SGD has a slower convergence in terms of number of iterations – i.e., along index i (see Appendix 5.C for additional information) – it can converge more quickly when measured in terms of wall-clock time. In other words, the number of gradient computations required to obtain some level of precision can be significantly lower for SGD than for GD.

Example 5.19

For a logistic regression problem (see the next chapter), Fig. 5.13 shows three realizations of the evolution of the cost function $g(\theta^{(i)})$ at the SGD iterates, alongside the corresponding cost function

Figure 5.13 Cost function (training loss) as a function of number of per-sample gradient computations for GD and SGD on a logistic regression model (see Chapter 6) with $N = 80$ and $S = 4$.

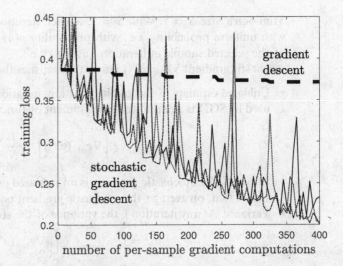

values attained by GD iterates. Since the data set contains $N = 80$ data points, GD needs 80 gradient computations to make one update to the model parameters. In contrast, with a mini-batch size $S = 4$, SGD can make one update for every four gradient computations. Importantly, the evolution of the SGD iterates is random owing to the randomized choice of the mini-batch. The effect of this stochastic behavior will be investigated in the next section.

5.12   Convergence of Stochastic Gradient Descent

Since the direction taken by SGD at each iteration is random, one should expect the iterates to form noisy trajectories in the model parameter space that do not rest at a stationary point. In fact, even when a stationary point is reached and the true gradient is zero, the estimate (5.64) of the gradient is generally non-zero. Therefore, on the one hand, the convergence properties of SGD should depend on the variance of the stochastic gradient (5.64). On the other hand, in order to ensure convergence, one should guarantee that the SGD update $-\frac{\gamma^{(i)}}{S^{(i)}} \sum_{n \in \mathcal{S}^{(i)}} \nabla g_n(\theta^{(i)})$ in (5.63) vanishes as i approaches a stationary point. Since the gradient estimate itself does not go to zero, a reasonable strategy is to decrease the learning rate $\gamma^{(i)}$ along the iterations i. This section elaborates on these important aspects by generalizing the theory developed in Sec. 5.9 for GD.

5.12.1   Properties of the Stochastic Gradient

To understand the convergence of SGD, we first study some basic properties of the distribution of the stochastic gradient (5.64) used by SGD. We start with the case in which the mini-batch is of size $S = 1$, and then we generalize the conclusions to any size $S \geq 1$. Throughout, as mentioned, we assume random uniform mini-batches.

Mini-batch Size $S = 1$. With $S = 1$, at each iteration i, SGD picks one example n at random with uniform probability, i.e., with probability $p(n) = 1/N$. Accordingly, we write the index of the selected sample at iteration i as the rv $n^{(i)} \sim p(n) = 1/N$. Under this assumption, the stochastic gradient $\nabla g_{n^{(i)}}(\theta^{(i)})$ is a rv that satisfies the following properties:

- **Unbiased estimate of the gradient.** At any iteration i, the average of the stochastic gradient used by SGD is equal to the true gradient of function $g(\theta)$, i.e.,

$$E_{n^{(i)} \sim p(n)=\frac{1}{N}} \left[\nabla g_{n^{(i)}}(\theta^{(i)}) \right] = \frac{1}{N} \sum_{n=1}^{N} \nabla g_n(\theta^{(i)}) = \nabla g(\theta^{(i)}). \qquad (5.65)$$

In words, the stochastic gradient is an **unbiased** estimate of the true gradient. This demonstrates that, on average, the stochastic gradient points in the same direction as GD.

- **Variance.** At any iteration i, the variance of the stochastic gradient is given as

$$
\begin{aligned}
v^{(i)} &= E_{n^{(i)} \sim p(n)=\frac{1}{N}} \left[||\nabla g_{n^{(i)}}(\theta^{(i)}) - \nabla g(\theta^{(i)})||^2 \right] \\
&= E_{n^{(i)} \sim p(n)=\frac{1}{N}} \left[||\nabla g_{n^{(i)}}(\theta^{(i)})||^2 \right] - ||\nabla g(\theta^{(i)})||^2. \qquad (5.66)
\end{aligned}
$$

The variance thus depends on how much the individual gradients $\nabla g_n(\theta^{(i)})$ differ from one another across the sample index n: When all gradients are the same – i.e., when $\nabla g_n(\theta^{(i)}) = \nabla g(\theta^{(i)})$ – the variance is zero, i.e., $v^{(i)} = 0$; and it is otherwise positive.

Overall, we can conclude that the stochastic gradient used by SGD points *on average* in the same direction as the gradient, with a variance that depends on the "disagreement" of the gradients across different samples.

Mini-batch Size $S \geq 1$. Generalizing to any mini-batch size S, the properties of the stochastic gradient are stated as:

- **Unbiased estimate of the gradient.** The average of the stochastic gradient is equal to the gradient.

- **Variance.** The variance of the stochastic gradient is given as

$$v^{(i)} = \frac{1}{S^{(i)}} v^{(i)}_{S^{(i)}=1}, \qquad (5.67)$$

where $v^{(i)}_{S^{(i)}=1}$ denotes the variance (5.66) for a mini-batch of size $S^{(i)} = 1$.

It follows that a larger mini-batch size increases computational complexity – since we need to compute multiple gradients – but, in turn, it decreases the variance by a factor equal to $S^{(i)}$.

Example 5.20

Consider the three data points x_1, x_2, and x_3 represented as crosses in Fig. 5.14, as well as the cost function

$$g(\theta) = \frac{1}{3}((\theta - x_1)^2 + (\theta - x_2)^2 + (\theta - x_3)^2), \qquad (5.68)$$

with $g_n(\theta) = (\theta - x_n)^2$. The (only) global minimum of this cost function is shown as a gray star. The figure displays the (deterministic) evolution of the iterate $\theta^{(i)}$ under GD, along with one random

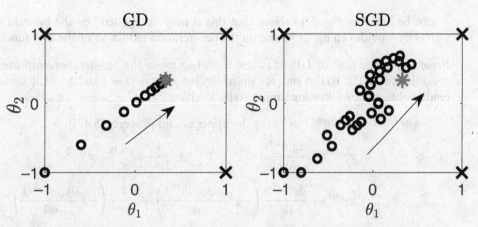

Figure 5.14 Evolution of 30 iterates (circles) for GD and SGD assuming cost function $g(\theta) = \frac{1}{3}((\theta - x_1)^2 + (\theta - x_2)^2 + (\theta - x_3)^2)$, with $g_n(\theta) = (\theta - x_n)^2$ and samples x_1, x_2, and x_3 represented as crosses (constant learning rate $\gamma = 0.1$). The arrows point in the direction if increasing number of iterations $i = 1, 2, \ldots$.

realization of the evolution under SGD with $S = 1$, both with fixed learning rate $\gamma^{(i)} = 0.1$. The "noisiness" of the trajectory of SGD is due to the non-zero variance of the stochastic gradient, which is caused by the differences among the gradients $\nabla g_n(\theta)$ for $n = 1, 2, 3$. In fact, each negative gradient $-\nabla g_n(\theta) = 2(x_n - \theta)$ points in the direction away from θ towards the corresponding data point x_n.

5.12.2 Convergence Properties of SGD

Let us denote as $\mathrm{E}_{\mathcal{S}^{(i)}}[\cdot]$ the expectation with respect to the selection of the mini-batch $\mathcal{S}^{(i)}$, which follows the random uniform mini-batch assumption. Given the properties of the stochastic gradient reviewed in the preceding subsection, we will now prove the following key convergence result for SGD.

Guaranteed Average Per-iteration Reduction of the Cost Function. If $\gamma^{(i)} \leq 1/L$, given the current iterate $\theta^{(i)}$, the next SGD iterate $\theta^{(i+1)}$ satisfies the inequality

$$\mathrm{E}_{\mathcal{S}^{(i)}}[g(\theta^{(i+1)})] \leq \underbrace{g(\theta^{(i)}) - \frac{\gamma^{(i)}}{2}||\nabla g(\theta^{(i)})||^2}_{\text{improvement for GD}} + \frac{\gamma^{(i)}}{2} v^{(i)}, \tag{5.69}$$

where $v^{(i)}$ indicates the variance of the stochastic gradient (given by (5.66) for $S = 1$ and (5.67) for any $S \geq 1$). This result has two important consequences:

- By comparison with (5.53), the average per-iteration reduction of the cost function of SGD guaranteed by (5.69) is smaller than for GD by a factor $\gamma^{(i)} v^{(i)}/2$. This term grows with the variance $v^{(i)}$ of the stochastic gradient.
- The bound (5.69) suggests that, even when we reach a stationary point $\theta^{(i)}$, where $\nabla g(\theta^{(i)}) = 0$, SGD does not stop. Furthermore, the "size" of the region over which SGD wanders around a stationary point is proportional to the term $\gamma^{(i)} v^{(i)}/2$. An illustration of this phenomenon

can be found in Fig. 5.14. Note that this is only "suggested" by the bound (5.69), which merely provides an upper bound on the per-iteration reduction of the cost function.

Proof: As in the case of GD, it is also useful to prove the per-iteration improvement (5.69) condition for SGD. To this end, for simplicity, we assume $D = 1$ and $S = 1$. Extensions can be readily obtained by following similar steps. We have

$$g(\theta^{(i+1)}) \le \tilde{g}_{\gamma^{(i)}}(\theta^{(i+1)}|\theta^{(i)}) \text{ (by the upper bound property (5.47))}$$

$$= g(\theta^{(i)}) + \frac{dg(\theta^{(i)})}{d\theta}(\theta^{(i+1)} - \theta^{(i)}) + \frac{1}{2\gamma^{(i)}}(\theta^{(i+1)} - \theta^{(i)})^2$$

$$= g(\theta^{(i)}) + \frac{dg(\theta^{(i)})}{d\theta}\left(-\gamma^{(i)}\frac{dg_{n^{(i)}}(\theta^{(i)})}{d\theta}\right) + \frac{1}{2\gamma^{(i)}}\left(-\gamma\frac{dg_{n^{(i)}}(\theta^{(i)})}{d\theta}\right)^2, \qquad (5.70)$$

where in the third equality we have used the SGD update (5.63) with $S = 1$. Let us now take an average over the selection of the sample $n^{(i)}$:

$$\mathrm{E}_{n^{(i)} \sim p(n) = \frac{1}{N}}[g(\theta^{(i)}) - g(\theta^{(i+1)})]$$

$$\le \gamma^{(i)}\frac{dg(\theta^{(i)})}{d\theta}\left(\mathrm{E}_{n^{(i)} \sim p(n) = \frac{1}{N}}\left[\frac{dg_{n^{(i)}}(\theta^{(i)})}{d\theta}\right]\right) - \frac{\gamma^{(i)}}{2}\mathrm{E}_{n^{(i)} \sim p(n) = \frac{1}{N}}\left[\left(\frac{dg_{n^{(i)}}(\theta^{(i)})}{d\theta}\right)^2\right]$$

$$= \underbrace{\frac{\gamma^{(i)}}{2}\left(\frac{dg(\theta^{(i)})}{d\theta}\right)^2}_{\text{improvement for GD}} - \frac{\gamma^{(i)}}{2}v^{(i)}, \qquad (5.71)$$

where in the last equality we have used the unbiasedness of the stochastic gradient and the expression for the variance (5.66). This concludes the proof.   □

5.12.3 Choosing the Learning Rate for SGD

As suggested by the bound (5.69), in order to make SGD stop at a stationary point, one needs to ensure that the product $\gamma^{(i)}v^{(i)}$ of learning rate and variance of the stochastic gradient vanishes as the number of iterations i increases. How can this be accomplished? One convenient way is to decrease the step size $\gamma^{(i)}$ with the iteration index i so that we have the limit $\gamma^{(i)}v^{(i)} \to 0$ as $i \to \infty$. However, clearly, the step size should not decrease too quickly lest the iterates stop too early, before reaching a stationary point.

Under the assumption that the variance $v^{(i)}$ is bounded, it can be proved that convergence to a stationary point can be guaranteed under the **Munro–Robbins** conditions

$$\sum_{i=1}^{\infty}\gamma^{(i)} = \infty \quad \text{and} \quad \sum_{i=1}^{\infty}(\gamma^{(i)})^2 < \infty. \qquad (5.72)$$

The first condition ensures that SGD can make sufficient progress towards a stationary point, while the second ensures a vanishing term $\gamma^{(i)}v^{(i)}$. Learning rate schedules of the form

$$\gamma^{(i)} = \frac{1}{i^{\alpha}} \qquad (5.73)$$

with $0.5 < \alpha \leq 1$ satisfy this condition. Accordingly, a common practical choice is the polynomial learning rate

$$\gamma^{(i)} = \frac{\gamma^{(0)}}{(1 + \beta i)^\alpha}. \qquad (5.74)$$

for some $\gamma^{(0)} > 0$, $\beta > 0$, and $0.5 < \alpha \leq 1$.

A conceptually different approach to ensure the limit $\gamma^{(i)} v^{(i)} \to 0$ as $i \to \infty$ would be to use an iteration-dependent mini-batch size $S^{(i)}$. In particular, if we increased the mini-batch size $S^{(i)}$ with i, the variance of the stochastic gradient would vanish by (5.67), ensuring the desired limit.

The choices of the learning rate discussed so far are non-adaptive in the sense that they do not depend on the evolution of the cost function across the iterations. In practice, **adaptive learning rate schedules** are often successfully adopted. A simple heuristic, for instance, is to divide the current learning rate $\gamma^{(i)}$ by 2 when the algorithm is stalled. More sophisticated methods attempt to account for the curvature of the cost function as estimated along the trajectory of iterates. These techniques include momentum, RMSProp and Adam, and are covered in Appendix 5.D.

5.13 Minimizing Population Loss vs. Minimizing Training Loss

It is important to remember that the ultimate goal of learning is to minimize the population loss

$$L_p(\theta) = \mathrm{E}_{(\mathrm{x,t}) \sim p(x,t)}[f(\mathrm{x,t}|\theta)], \qquad (5.75)$$

where $f(\mathrm{x,t}|\theta)$ represents the loss function as in (5.1), and that the regularized training loss is just a proxy for this criterion. This observation has the following important implications.

SGD as a Regularizer. First, an accurate optimization of the training loss may not necessarily improve the population loss. In fact, it may even be counterproductive: An extremely accurate optimization of the training loss may cause overfitting, and impair generalization (see Sec. 4.4). In this regard, SGD may effectively act as an added layer of regularization, preventing the training algorithm from choosing a model parameter θ that is too dependent on the training data set.

A popular – but as yet unproven and not universally accepted – explanation of this phenomenon is that **"narrow" (local) minima** of the training loss do not correspond to model parameters θ that generalize well, i.e., that yield a low population loss; while **"wide" (local) minima** of the training loss generalize better. The idea is that narrow minima are likely to be artifacts of the specific sampling of the training points, while wide minima are more robust to "sampling noise" and may correspond to actual minima of the cost function. SGD tends to choose wide local minima, as narrow minima are not stable points for SGD. In fact, at a narrow minimum θ^*, a small perturbation of θ away from θ^* would cause the training loss to increase significantly, allowing SGD to escape the minimum. In contrast, a wide minimum θ^* keeps the value of the training loss low in a larger neighborhood around θ^*, making it more difficult for SGD to move away from it.

Learning Rate Selection via Validation. A second important consequence of the fact that the end goal is minimizing the population loss is that, rather than choosing the hyperparameters of the training algorithm to minimize the training loss, one should use validation. Following the discussion in Sec. 4.4, this would entail running SGD with different hyperparameters, such as learning rate schedule and mini-batch size; estimating the corresponding population losses via validation; and choosing the hyperparameters that minimize the validation loss.

Interestingly, learning rate schedules that minimize the population loss may be different from those suggested by the theory developed in the preceding section, which assumes training loss as the objective. Recent evidence, at the time of writing, suggests for instance that fixed large learning rates may be preferable to decaying learning rates in terms of performance on a test set.

5.14   Symbolic and Numerical Differentiation

Implementing GD and SGD requires computing the gradients of the cost functions $g_n(\theta)$. There are three main ways to compute gradients, namely symbolic calculation, numerical differentiation, and automatic differentiation. In the rest of this chapter, we will review these methods in turn. A schematic comparison can be found in Table 5.1. We start this section by reviewing the baseline schemes of symbolic and numerical differentiation. Throughout this discussion, to simplify the notation, we will write $\nabla g(\theta)$ for the gradient to be computed, although one should more correctly refer to the gradient $\nabla g_n(\theta)$ for the given nth training example.

5.14.1   Symbolic Differentiation

Symbolic differentiation corresponds to the standard approach of computing the derivatives "by hand", obtaining analytical expressions for the gradient vector $\nabla g(\theta)$. With reference to the taxonomy in Table 5.1, symbolic differentiation leverages the analytical structure of the function $g(\theta)$ via the chain rule of differentiation – i.e., it is **structure-aware**. Furthermore, it provides an **exact** expression for the gradient $\nabla g(\theta)$ that is valid for every value of θ – i.e., it outputs a **function** of θ. This is the approach that we have used in the examples seen so far in this chapter.

5.14.2   Numerical Differentiation

Numerical differentiation obtains an approximation of the value of the gradient by leveraging the definition of the derivative as the ratio of the amount of change in the function's output over the amount of change in the input. Unlike symbolic differentiation, numerical differentiation treats the function $g(\theta)$ as an input–output black box – i.e., it is **structure-agnostic**. Moreover, it is a **single-value** method that produces an **approximation** of the gradient $\nabla g(\theta)$ only for one vector θ.

Table 5.1 How to compute gradients

approach	exact/ approximate	structure	output
symbolic	exact	structure-aware	function
numerical	approximate	structure-agnostic	single value
automatic	exact	structure-aware	single value

For $D = 1$, numerical differentiation uses the approximation

$$\frac{\mathrm{d}g(\theta)}{\mathrm{d}\theta} \simeq \frac{g(\theta + \epsilon) - g(\theta - \epsilon)}{2\epsilon} \tag{5.76}$$

for some small $\epsilon > 0$. Using the Taylor expansion for $g(\theta)$, one can easily conclude that the error of this approximation scales as $\mathcal{O}(\epsilon^2)$. Therefore, as the derivative gets smaller, i.e., as one approaches a stationary point, it is necessary to choose increasingly small values of ϵ to ensure that the approximation error does not become larger than the derivative that is being estimated. This creates problems in terms of numerical stability of the operation.

For any $D \geq 1$, we similarly have the approximation of each partial derivative

$$\frac{\partial g(\theta)}{\partial \theta_i} = [\nabla g(\theta)]_i \simeq \frac{g(\theta + \epsilon e_i) - g(\theta - \epsilon e_i)}{2\epsilon} \tag{5.77}$$

for some small $\epsilon > 0$, where e_i is a one-hot vector with all zeros except for a 1 in position i. Note that we need $2D$ evaluations of the function $g(\theta)$ in order to obtain an estimate of the entire gradient. Therefore, the approach is only viable for sufficiently small values of D.

As a final note, in some applications, one is interested only in computing the directional derivative $\nabla g(\theta)^T u$ for some unitary-norm vector u. This can be directly estimated without explicitly evaluating the entire gradient vector as

$$\nabla g(\theta)^T u \simeq \frac{g(\theta + \epsilon u) - g(\theta - \epsilon u)}{2\epsilon}, \tag{5.78}$$

which requires only two evaluations of the function $g(\theta)$. Note that this expression generalizes (5.77), for which we have $\nabla g(\theta)^T e_i = \partial g(\theta)/\partial \theta_i$.

5.15   Automatic Differentiation

As discussed in the preceding section, symbolic differentiation uses the structure of the function $g(\theta)$ to compute an exact analytical expression for the gradient $\nabla g(\theta)$, which is valid for all values of θ. Numerical differentiation treats the function $g(\theta)$ as a black box to compute an approximation of $\nabla g(\theta)$ for only one vector θ. In contrast, automatic differentiation uses the **structure** of the function $g(\theta)$ to compute the **exact** gradient for **one** value of θ (see Table 5.1).

Like symbolic differentiation, automatic differentiation is based on the chain rule of differentiation. However, it does not compute a symbolic expression for the gradient, but only the numerical value $\nabla g(\theta)$ for a given vector θ. This section presents automatic differentiation, focusing specifically on backpropagation, or **backprop** – a key algorithm for the training of neural networks. Applications will be discussed in the next chapter.

5.15.1 Computational Graph

Like symbolic differentiation, automatic differentiation views the function $g(\theta)$ as a composition of **elementary functions** $\{f_i\}$, whose **local derivatives** (with respect to each function's inputs) are easy to compute. Unless stated otherwise, all elementary functions are assumed to output a single scalar. The composition of elementary functions is represented in a **computational graph**, in which each elementary function represents a node of the graph.

We will also refer to an elementary function as a **computational block** in the graph, or **block** for short when no confusion may arise. The local derivatives of each elementary function

Figure 5.15 Computational graph for function (5.79).

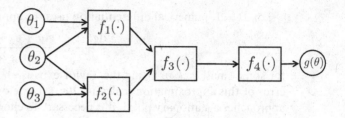

are assumed to be known as part of the definition of the computational block. This implies that, for any given input value, evaluating the local derivatives does not add to the order of complexity of the calculation of the elementary function implemented by the computational block.

The computational graph is a **directed acyclic graph** composed of **input nodes** $\{\theta_i\}_{i=1}^{D}$, denoted as circles; of **computational blocks**, denoted as rectangles, each implementing an elementary function $f_i(\cdot)$ along with its partial derivatives; and of an **output node**, reporting the value of function $g(\theta)$, also denoted as a circle. Each directed edge (arrow) connecting any two nodes of the graph carries a scalar quantity.

Example 5.21

Consider the function

$$g(\theta) = f_4(f_3(f_1(\theta_1, \theta_2), f_2(\theta_2, \theta_3))), \tag{5.79}$$

with $D = 3$, where $f_1(\cdot), f_2(\cdot), f_3(\cdot)$, and $f_4(\cdot)$ are elementary functions with the indicated numbers of scalar inputs. The computational graph for this function is shown in Fig. 5.15.

5.15.2 Introducing Backprop

To introduce backprop, we start by studying a special class of cost functions. To describe it, we fix elementary functions $f_1(\theta), f_2(\theta), \ldots, f_K(\theta)$, with $\theta \in \mathbb{R}^D$, and elementary function $f_{K+1}(f_1, \ldots, f_K)$, and we denote as f_k the output of each elementary function $f_k(\cdot)$. All elementary functions in this example produce a scalar output. In this subsection, we focus on the class of cost functions that can be represented as the composition

$$g(\theta) = f_{K+1}(f_1(\theta), f_2(\theta), \ldots, f_K(\theta)). \tag{5.80}$$

The class of functions given by (5.80) represents a prototypical type of parallel computing architectures, such as **map-reduce**, whereby inputs are processed in parallel by functions $f_k(\cdot)$ for $k = 1, \ldots, K$, and then the outputs of these functions are combined by function $f_{K+1}(\cdot)$. The computational graph corresponding to function (5.80) is shown in Fig. 5.16.

Let us try to compute the partial derivative $\partial g(\theta)/\partial \theta_i$ of the output $g(\theta)$ with respect to the ith entry θ_i of the model parameter vector for cost functions of the form (5.80). The partial derivative quantifies the rate of change of the output $g(\theta)$ over a change in the input θ_i. Intuitively, looking at Fig. 5.16, changes in θ_i affect the output $g(\theta)$ along K different paths:

Figure 5.16 Computational graph for function (5.80).

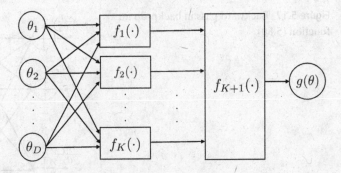

a change in θ_i modifies the output f_1 of the first function, which in turn affects f_{K+1}; but it also affects $f_2, f_3, \ldots$, and f_K, and, through each of them, the output f_{K+1}.

Chain Rule of Differentiation. The chain rule of differentiation formalizes this intuition by stipulating that the partial derivative $\partial g(\theta)/\partial \theta_i$ of the output $g(\theta)$ equals the sum of K contributions, each corresponding to one of the influence paths just described:

$$
\frac{\partial g(\theta)}{\partial \theta_i} = \underbrace{\frac{\partial f_{K+1}(f_1, \ldots, f_K)}{\partial f_1}}_{\text{local derivative for function } f_{K+1}(\cdot)} \cdot \underbrace{\frac{\partial f_1(\theta)}{\partial \theta_i}}_{\text{local derivative for function } f_1(\cdot)}
$$
$$
+ \cdots + \underbrace{\frac{\partial f_{K+1}(f_1, \ldots, f_K)}{\partial f_K}}_{\text{local derivative for function } f_{K+1}(\cdot)} \cdot \underbrace{\frac{\partial f_K(\theta)}{\partial \theta_i}}_{\text{local derivative for function } f_K(\cdot)} \tag{5.81}
$$

For short, we can write this equality as

$$
\frac{\partial g(\theta)}{\partial \theta_i} = \sum_{k=1}^{K} \underbrace{\frac{\partial f_{K+1}(f_1, \ldots, f_K)}{\partial f_k}}_{\text{local derivative for function } f_{K+1}(\cdot)} \cdot \underbrace{\frac{\partial f_k(\theta)}{\partial \theta_i}}_{\text{local derivative for function } f_k(\cdot)} \tag{5.82}
$$

In words, the chain rule of differentiation states that **"derivatives multiply"** along each influence path and that the contributions of all paths sum.

Forward Mode of Automatic Differentiation. There is a simple way to implement the calculation (5.82) automatically, which is known as the **forward mode of automatic differentiation**:

1. Each computational block $f_k(\cdot)$ computes the local partial derivative $\partial f_k(\theta)/\theta_i$ (which is part of its definition and hence easy to evaluate).
2. This "message" is passed to computational block $f_{K+1}(\cdot)$, which multiplies it by the local partial derivative $\partial f_{K+1}(f_1, \ldots, f_K)/\partial f_k$ (again this partial derivative is easy to evaluate by assumption).
3. The computational block $f_{K+1}(\cdot)$ sums up all product terms $\partial f_{K+1}(f_1, \ldots, f_K)/\partial f_k \cdot \partial f_k(\theta)/\partial \theta_i$ to obtain the final result $\partial g(\theta)/\partial \theta_i$.

Importantly, in order to compute the full gradient $\nabla g(\theta)$, this procedure should be repeated for *every* entry θ_i. This makes the forward mode of automatic differentiation inefficient for large

Figure 5.17 Backward pass in backprop for function (5.80).

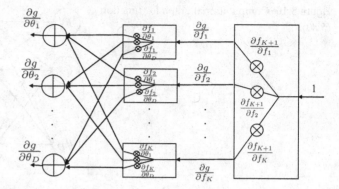

parameter vectors θ, i.e., for a large D. Finally, we note that it is possible to use the forward mode of automatic differentiation to compute any directional derivative (5.24) in one pass, with the partial derivative with respect to θ_i being a special case.

Reverse Mode of Automatic Differentiation. The reverse mode of automatic differentiation, also known as **backpropagation** or **backprop**, addresses the highlighted shortcoming of the forward mode. It has the key advantage that it can compute all the partial derivatives $\partial g(\theta)/\partial \theta_i$ for $i = 1, \ldots, D$ *simultaneously* via parallel processing at the computational blocks. As we detail next, backprop consists of a forward pass followed by a backward pass. The forward pass corresponds to the standard calculation of the function $g(\theta)$ through the computational graph, while the backward pass is illustrated in Fig. 5.17 for functions of the form (5.80), and is detailed in Algorithm 5.3.

Algorithm 5.3: Backprop to compute all partial derivatives (5.82) for the computational graph in Fig. 5.16

Forward pass:

1. Given an input vector θ, by following the direction of the edges in the computational graph, each kth computational block $f_k(\cdot)$ evaluates its output f_k for $k = 1, \ldots, K + 1$.
2. Each computational block $f_k(\cdot)$ also evaluates the local derivatives $\partial f_k(\theta)/\partial \theta_i$ with respect to its inputs $\theta_1, \ldots, \theta_D$ with $k = 1, \ldots, K$.
3. Computational block $f_{K+1}(\cdot)$ evaluates all the local partial derivatives $\partial f_{K+1}(f_1, \ldots, f_K)/\partial f_k$ with respect to all its inputs f_k, with $k = 1, 2, \ldots, K$.

Backward pass:

1. Block $f_{K+1}(\cdot)$ passes each message $\partial f_{K+1}(f_1, \ldots, f_K)/\partial f_k$ back to the respective block $f_k(\cdot)$.
2. Each block $f_k(\cdot)$ multiplies the message $\partial f_{K+1}(f_1, \ldots, f_K)/\partial f_k$ by the corresponding local derivative $\partial f_k(\theta)/\partial \theta_i$.
3. Each block $f_k(\cdot)$ sends the computed message $\partial f_{K+1}(f_1, \ldots, f_K)/\partial f_k \cdot \partial f_k(\theta)/\partial \theta_i$ for $i = 1, \ldots, D$ back to variable node θ_i.
4. Finally, each variable node θ_i sums up all its received upstream messages to obtain (5.82).

As described in the algorithm, backprop requires two successive passes over the computational graph to ensure that each variable node θ_i obtains the exact value of the its partial

Figure 5.18 Computational graph for function $g(\theta) = \theta_1^2 + 2\theta_2^2 - \theta_3^2$.

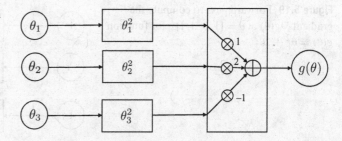

derivative (5.82) at the given input vector θ. Therefore, unlike the forward mode, the complexity of backprop does not scale with D. Instead, backprop is approximately *twice as complex* as merely evaluating the output of the function $g(\theta)$. In this regard, we emphasize again that backprop requires each computational block to have access to its local partial derivatives for its given input.

Example 5.22

In this example, we apply backprop to the function $g(\theta) = \theta_1^2 + 2\theta_2^2 - \theta_3^2$. We specifically wish to compute the gradient $\nabla g(\theta)$ at $\theta = [1, -1, 1]^T$. To start, we choose the computational blocks $f_k(\theta) = \theta_k^2$, for $k = 1, 2, 3$ ($K = 3$) and $f_4(f_1, f_2, f_3) = f_1 + 2f_2 - f_3$, which yields the computational graph in Fig. 5.18. Note that this is a specific instance of the computational graph in Fig. 5.16. The corresponding local partial derivatives are

$$\frac{\partial f_k(\theta)}{\partial \theta_j} = 2\theta_k \mathbb{1}(j = k) \tag{5.83}$$

for $k = 1, 2, 3$, and

$$\frac{\partial f_4(f_1, f_2, f_3)}{\partial f_j} = \begin{cases} 1 & \text{for } j = 1 \\ 2 & \text{for } j = 2 \\ -1 & \text{for } j = 3. \end{cases} \tag{5.84}$$

We emphasize that the selection of the elementary functions is not unique. For instance, one may choose $K = 2$ and combine two of the functions $f_k(\theta)$.

To apply backprop, we first carry out the forward pass, which is illustrated in Fig. 5.19. In the backward pass, we specialize the procedure shown in Fig. 5.17 to obtain, from Fig. 5.20, the gradient $\nabla g([1, -1, 1]^T) = [2, -4, -2]^T$.

5.15.3 General Form of Backprop

In the preceding subsection, we have focused on the special class of functions (5.80). We are now ready to extend backprop to any computational graph. Recall that a computational graph consists of a directed acyclic graph connecting inputs θ, computational blocks, and output $g(\theta)$. The constituent computational blocks implement elementary functions whose local partial derivatives can be easily evaluated alongside the output of the functions for any given input. Backprop on a general computational graph can be formally derived via a recursive application

Figure 5.19 Forward pass to compute the gradient $\nabla g(\theta)$ at $\theta = [1, -1, 1]^T$ for function $g(\theta) = \theta_1^2 + 2\theta_2^2 - \theta_3^2$.

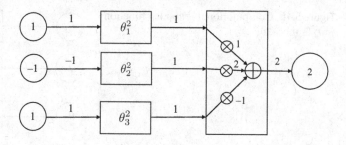

Figure 5.20 Backward pass to compute the gradient $\nabla g(\theta)$ at $\theta = [1, -1, 1]^T$ for function $g(\theta) = \theta_1^2 + 2\theta_2^2 - \theta_3^2$.

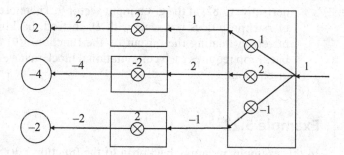

of the chain rule of differentiation. This yields a two-pass procedure, with a forward pass followed by a backward pass.

A High-Level View of Backprop. As detailed in Algorithm 5.4, in the forward pass, each computational block evaluates its outputs and its local partial derivatives. (The table considers for generality computational blocks that may have multiple scalar outputs – we will return to this point later.) In the backward pass, each computational block is first replaced by a **"backward" version** of the block. The backward version of a computational block multiplies the received **"upstream" messages** via the local partial derivatives to produce the "downstream" message to be sent back to the previous nodes in the computational graph. An example is shown in Fig. 5.17, which reports the backward versions of the computational blocks in the computational graph depicted in Fig. 5.16. The common terminology "upstream" refers to the fact that the input of the backward version of a block is closer to the "source" $g(\theta)$ than the "downstream" output of the block in the backward pass. Finally, by summing all upstream messages, each input node θ_i obtains the corresponding partial derivative $\partial g(\theta)/\partial\theta_i$.

Backward Version of Computational Blocks. In order to implement backprop, as detailed in Algorithm 5.4, we need to evaluate the backward version of each computational block. As we detail next, the backward version of any computational block implements a linear transformation of the upstream messages into the downstream messages that depends on the local partial derivatives evaluated at the block during the forward pass. The backward versions of the computational blocks are obtained as follows.

- **Computational blocks with scalar outputs.** Consider first computational blocks implementing an elementary function $f(x_1, \ldots, x_M)$ with M scalar inputs (i.e., an $M \times 1$ vector x) and a single, scalar, output. Forward and backward versions for any such block are illustrated in

---

Algorithm 5.4: Backprop

Forward pass:

1. Given an input vector θ, following the directions of the arrows in the computational graph, each computational block evaluates its output.
2. Each computational block also evaluates the local partial derivatives of each of its scalar outputs with respect to each of of its scalar inputs.

Backward pass:

1. Transform each block into its backward version (to be detailed below).
2. Apply the resulting backward computational graph using 1 as the input of the final computational block to produce the partial derivative $\partial g(\theta)/\partial \theta_i$ at each input node θ_i for $i = 1, \dots, D$.

Figure 5.21 Forward and backward versions of computational blocks $f(x)$ with a vector input x and a scalar output.

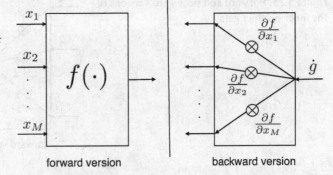

forward version   backward version

Fig. 5.21. As seen in the figure, the backward version has one input ($\dot{g}$ in the figure) and M outputs, with arrows having the opposite direction as compared to the forward block. Importantly, while the (forward) function $f(x_1, \dots, x_M)$ may be non-linear, the backward version of the computational block is always **linear**. The backward input $\dot{g}$ is also known as upstream message, and the M backward outputs are also referred to as downstream messages. In order to obtain the downstream message for the arrow corresponding to (forward) input x_m, the backward block simply multiplies the upstream message $\dot{g}$ by the corresponding local partial derivative $\partial f(x_1, \dots, x_M)/\partial x_m$.

- **Computational blocks with vector outputs.** Let us now turn to computational blocks with multiple outputs, or equivalently with a vector output. We specifically concentrate solely on functions with a single scalar input and multiple outputs, i.e., of the form $f(x) = [f_1(x), \dots, f_L(x)]^T$, where x is a scalar input, and each function $f_l(x)$ outputs a scalar. As seen in Fig. 5.22, while each function $f_l(\cdot)$ is generally non-linear, the backward version of the block is, again, always **linear**: Each lth upstream input $\dot{g}_l$ is multiplied by the local derivative $\mathrm{d}f_l(x)/\mathrm{d}x$ and the L resulting products are summed to produce the downstream output of the backward block.

Figure 5.22 Forward and backward versions of computational blocks $f(x) = [f_1(x), \ldots, f_L(x)]^T$ with a scalar input x and an $L \times 1$ vector output.

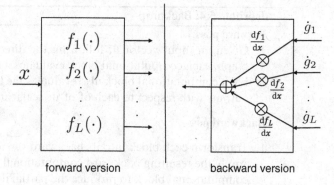

forward version   backward version

Figure 5.23 Forward and backward versions of the inner product gate.

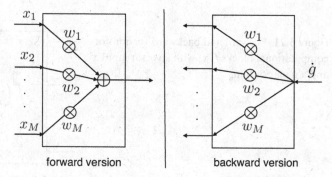

forward version   backward version

Example 5.23

As an example, consider the **inner product gate**. The inner product gate corresponds to the function $f(x_1, \ldots, x_M) = \sum_{m=1}^{M} w_m x_m$ with vector input and scalar output that is defined by some weight vector $w = [w_1, \ldots, w_M]^T$. The local partial derivatives are given as $\partial f(x)/\partial x_m = w_m$. Therefore, using the general rule shown in Fig. 5.21, the backward version of the inner product gate is shown in Fig. 5.23.

Example 5.24

Consider now the **copier gate** that repeats the input to multiple outputs. This is an example of a computational block with a scalar input and a vector output. Since we have $f_l(x) = x$ for all $l = 1, \ldots, L$, the local derivatives are given as $df_l(x)/dx = 1$. Therefore, specializing the general correspondence illustrated in Fig. 5.22, as shown in Fig 5.24, the backward version of the copier gate is given by a sum gate, i.e., by an inner product gate with weights $w = 1_L$.

Figure 5.24 Forward and backward versions of the copier gate.

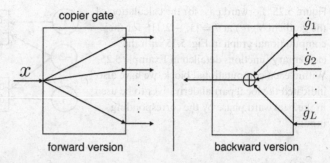

5.15.4 Backprop Computes *All* the Partial Derivatives of the Output $g(\theta)$

Using the chain rule recursively, one can show that, for each computational block, the upstream messages in the backward pass equal the partial derivative of the function $g(\theta)$ with respect to the corresponding (forward) output of the block:

- For every scalar-output functional block $f(\cdot)$, the upstream message $\dot{g}$ in Fig. 5.21 is the partial derivative of function $g(\theta)$ with respect to the output of the function, i.e.,

$$\dot{g} = \frac{\partial g(\theta)}{\partial f}. \tag{5.85}$$

- Similarly, for every vector-output block as in Fig. 5.22, we have

$$\dot{g}_l = \frac{\partial g(\theta)}{\partial f_l} \tag{5.86}$$

for all output functions $f_l(\cdot)$ with $l = 1, 2, \ldots, L$.

Therefore, backprop computes in one backward pass the partial derivatives of the function $g(\theta)$ with respect to *all* intermediate outputs produced by the constituent elementary functions. To see that this implies the **correctness** of backprop in computing the partial derivatives with respect to all inputs θ_i with $i = 1, \ldots, D$, one can add a copier gate for each θ_i that takes as input θ_i and copies it to all the connected computational blocks in the computational graph. By (5.86) and (5.82), the sum of all upstream messages received by a variable node θ_i is indeed the partial derivative $\partial g(\theta)/\partial \theta_i$.

The **complexity** of backprop is approximately *twice* that of computing the value of the function through the forward pass. In fact, the backward pass is "easier" than the forward pass, since it only applies linear functions – composed of sums and multiplications. This contrasts with the forward pass, which generally also includes non-linear computational blocks.

While we have presented the intermediate derivatives (5.85) and (5.86) as auxiliary results needed to obtain the desired partial derivatives $\partial g(\theta)/\partial \theta_i$, they may, in fact, be of independent interest for some applications. A notable example is **"explainable AI"**, in which one wishes to "interpret" the output of a model. A simple (and quite limited) way to do this is to carry out a sensitivity analysis, whereby one gauges which intermediate calculations have the largest impact on the output of the function $g(\theta)$. Modifying (by a small amount) intermediate outputs with a larger partial derivative causes larger changes in the output, yielding evidence regarding the relevance of such intermediate results.

Figure 5.25 Forward pass for the calculation of the gradient $\nabla g(\theta)$ at $\theta = [1, -2, 1]^T$ for the computational graph in Fig. 5.15 with the elementary functions detailed in Example 5.25. Within each computational block, we have also indicated the local partial derivatives to be used in the backward phase by the corresponding input.

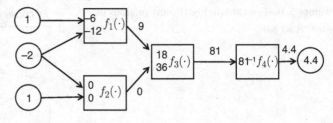

Figure 5.26 Backward pass for the calculation of the gradient $\nabla g(\theta)$ at $\theta = [1, -2, 1]^T$ for the computational graph in Fig. 5.15 with the elementary functions detailed in Example 5.25.

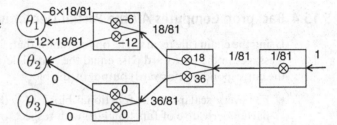

Example 5.25

Consider the computational graph in Fig. 5.15 with functions f_k given as

$$f_1(x_1, x_2) = f_2(x_1, x_2) = f_3(x_1, x_2) = (x_1 + 2x_2)^2, \tag{5.87}$$

with x_1 corresponding to the topmost input; and $f_4(x_1) = \log(x_1)$. We wish to compute the gradient $\nabla g(\theta)$ for $\theta = [1, -2, 1]^T$. The forward pass is shown in Fig. 5.25. Note that the figure also shows the local partial derivatives $\partial f_k(x)/\partial x_1 = 2(x_1 + 2x_2)$ and $\partial f_k(x)/\partial x_2 = 4(x_1 + 2x_2)$ for $k = 1, 2, 3$; as well as $\partial f_4(x)/\partial x_1 = 1/x_1$. These are indicated in the graph close to the corresponding input. The backward pass is depicted in Fig. 5.26. From the results of the backward pass, we finally obtain the gradient as

$$\nabla g(\theta) = \begin{bmatrix} -6 \times \frac{18}{81} \\ -12 \times \frac{18}{81} + 0 \\ 0 \end{bmatrix} = \begin{bmatrix} -\frac{4}{3} \\ -\frac{8}{3} \\ 0 \end{bmatrix} \tag{5.88}$$

for the given input θ.

As discussed, from the backward pass, we can also draw conclusions about the sensitivity of the output value $g(\theta)$ with respect to all intermediate calculations. For instance, modifying the output of function $f_1(\cdot)$ by a small amount has a larger effect on $g(\theta)$ than modifying the output of function $f_3(\cdot)$ by the same amount, since the corresponding partial derivative $\partial g(\theta)/\partial f_1 = 18/81$ is larger than $\partial g(\theta)/\partial f_3 = 1/81$.

5.16 Summary

- Machine learning problems are typically formulated as the minimization of a finite-sum criterion, i.e., as $\min_\theta g(\theta) = \frac{1}{N} \sum_{n=1}^{N} g_n(\theta)$.

- Optimization ideally aims to obtain a global minimum of the cost function $g(\theta)$, but for the types of problems encountered in machine learning one typically accepts local minima, or even stationary points, as solutions. A point θ is stationary if the gradient $\nabla g(\theta)$ evaluated at θ equals the all-zero vector.

- In the special case of convex cost functions, the three solution concepts – local minimum, global minimum, and stationary point – coincide, since all stationary points are local and global minima.

- A notable example in which the global minimum can be computed in closed form is the regularized ERM problem obtained with linear models, ℓ_2 loss, and (possibly) ℓ_2 regularization. This yields an LS problem, which is characterized by a strictly convex quadratic cost function.

- More commonly, in machine learning, minimization is most often tackled via SGD. SGD carries out updates via a stochastic gradient that uses a mini-batch $\mathcal{S}$ of S samples selected from the data set at each iteration.

- With a random uniform choice of the mini-batch, the stochastic gradient is an unbiased estimate of the true gradient, and its variance decreases with the size S of the mini-batch.

- Convergence of SGD can be guaranteed by selecting learning rates $\gamma^{(i)}$ that decrease with the iteration index i. This ensures a decreasing variance of the stochastic update.

- The choice of the learning rate for GD and SGD should depend on the curvature properties of the function in order to avoid divergence and to mitigate oscillations in the parameter space.

- The hyperparameters defining SGD – initialization and learning rate schedule – can be considered as part of the inductive bias, and hence they can be optimized via validation.

- Gradients can be computed numerically by using various approaches, with the most useful for machine learning being automatic differentiation via backprop.

- Backprop relies on a description of the cost function in terms of a computational graph consisting of elementary computational blocks with easy-to-compute local partial derivatives.

- Backprop computes the full gradient to numerical precision for a specific value of the input, with a computational load that is approximately twice the complexity of evaluating the cost function. This is due to the need to carry out forward and backward passes of comparable complexity.

- The backward pass carries out linear – copy, sum, and multiply – operations for any choice of the computational blocks in the computational graph.

5.17 Recommended Resources

This chapter only scratches the surface of the field of optimization, and in-depth presentations can be found in [1] for convex optimization and [2, 3] for more general problems. Useful discussions in the context of machine learning problems can be found in [4, 5, 6]. For automatic differentiation and backprop, the reader is referred to [8]. A summary of variance-reduction techniques for SGD is provided by [9] (see Appendix 5.C).

Problems

5.1 Given the function $g(\theta) = 2\theta^4 - 5\theta^3 + 1$, address the following points. (a*) Plot the function. (b) What are the stationary points? (c) What are the local minima? (d) What are the global minima? (e) Is the function convex? (Use more than one definition of convexity to provide different reasons for your answer.)

5.2 For the function $g(\theta) = 2\theta^4 - 5\theta^3 + 1$ used in Problem 5.1, (a) compute and (b*) plot the first-order Taylor approximation at $\theta^0 = 0$ and $\theta^0 = 3$.

5.3 Consider the quadratic function $g(\theta) = 1 - 2\theta + 4\theta^2$. (a) Is it convex? (b) What are the locally and globally optimal points? (c) Repeat for function $g(\theta) = 1 - 2\theta - 4\theta^2$.

5.4 Consider the function

$$g(\theta_1, \theta_2) = 2\theta_1^2 + 8\theta_2^2 - 4\theta_1\theta_2 - 3\theta_1 + 2\theta_2.$$

(a*) Display the function using a 3D plot. (b) Compute the gradient. (c) What are the stationary points? (d) Compute the Hessian. (e*) Evaluate the eigenvalues and eigenvectors of the Hessian. Comment on the curvature of the function. (f) What are the local minima? (g) What are the global minima? (h) Is the function convex?

5.5 Consider again the function in Problem 5.4. (a) Prove that the function can be written as $g(\theta) = c + b^T\theta + \frac{1}{2}\theta^T A\theta$ for some (A, b, c). (b) Based on this result, what is the gradient and what is the Hessian?

5.6 Consider the function

$$g(\theta_1, \theta_2) = 0.1\theta_1^4 + 2\theta_1^2 + 8\theta_2^2 - 4\theta_1\theta_2 - 3\theta_1 + 2\theta_2.$$

(a*) Plot the function in 3D. (b*) Compute and plot the negative gradient as a vector field. (c) Comment on the impact of the curvature of the function on the direction of the gradient.

5.7 Consider the function $g(\theta_1, \theta_2) = \theta_1 \exp(-3\theta_1 - 2\theta_2)$. (a) Compute the gradient vector. (b*) Display the function using a 3D plot and a contour plot. (c*) Superimposed on the contour figure, plot also the negative gradient field.

5.8 Prove that the only global minimum of the local quadratic approximant $\tilde{g}_\gamma(\theta|\theta^{(i)})$ optimized by GD is given by $\theta^* = \theta^{(i)} - \gamma \nabla g(\theta^{(i)})$.

5.9 Consider the quadratic function $g(\theta_1, \theta_2) = 2\theta_1^2 + 2\theta_2^2 + 3$. (a*) Compute the gradient, and then plot the negative gradient field. (b*) Apply GD with learning rate $\gamma = 0.05$ and initial point $\theta^{(1)} = [-8, -8]^T$ for 50 iterations. (c*) Show the evolution of the iterates $\theta^{(i)}$ in the same figure. (d*) Print out the iterates $\theta^{(i)}$ obtained from the same initialization for $\gamma = 1$ across five iterations. Comment on your results.

5.10 Consider the quadratic function $g(\theta_1, \theta_2)$ in Problem 5.9. (a*) Apply GD with learning rate $\gamma = 0.05$ and initial point $\theta^{(1)} = [-8, -8]^T$ for 50 iterations and plot the evolution of the cost function $g(\theta^{(i)})$ across the iterations. (b*) Plot the evolution of the scaled norm of the gradient $\gamma/2||\nabla g(\theta^{(i)})||^2$. (c) Derive the smoothness constant L. (d*) How does the choice of the learning rate affect the speed of convergence? To validate your answer, plot the evolution of the cost function $g(\theta^{(i)})$ across the iterations for $\gamma = 0.05$ and $\gamma = 0.2$.

5.11 Consider the quadratic function $g(\theta_1, \theta_2) = 2\theta_1^2 + 2\theta_2^2 - 3\theta_1\theta_2 + 3$. (a*) Compute the gradient, and then plot the negative gradient field. (b*) Apply GD with learning rate $\gamma = 0.05$ and initial point $\theta^{(1)} = [0, -8]$ for 50 iterations. Show the evolution of the iterates $\theta^{(i)}$ in the same figure. (c*) What happens if the initialization is $\theta^{(1)} = [-8, -8]^T$? Why?

5.12 Prove that the variance of the stochastic gradient with a mini-batch size of $S^{(i)}$ is equal to the variance with a mini-batch size of 1 divided by $S^{(i)}$, i.e., $v^{(i)} = \frac{1}{S^{(i)}} v^{(i)}_{S^{(i)}=1}$.

5.13 Consider functions $g_1(\theta) = 3\theta^4$, $g_2(\theta) = -\theta^4$, $g_3(\theta) = -2\theta^3$, and $g_4(\theta) = -3\theta^3 + 1$, as well as the global cost function $g(\theta) = \frac{1}{4} \sum_{n=1}^{4} g_n(\theta)$. (a) Compute the expectation of the stochastic gradient $\nabla g_n(\theta)$ obtained by picking the index of the selected sample at iteration as $n \sim p(n) = 1/4$, and check that it equals the true gradient $\nabla g(\theta)$. (b) Compute the variance of the stochastic gradient $\nabla g_n(\theta)$. (c*) Plot the variance in the interval $\theta \in [-1, 1]$ and comment on your result. (d) For $\theta^{(i)} = 2$, what would be the variance of the stochastic gradient if the functions were selected as $g_1(\theta) = \theta^4, g_2(\theta) = \theta^4$, $g_3(\theta) = -2.5\theta^3 + 0.5, g_4(\theta) = -2.5\theta^3 + 0.5$? Why is it smaller/larger than with the original functions? (e) Consider the stochastic gradient $\frac{1}{2} \sum_{n \in S} \nabla g_n(\theta)$ with mini-batch S of size 2 obtained by picking the two indices of the selected samples at iteration i independently and uniformly in the set $\{1, 2, 3, 4\}$ (with replacement). Compute the expectation of the stochastic gradient and check that it equals the true gradient $\nabla g(\theta)$. (f) Prove that the variance of the stochastic gradient is equal to half the value obtained with a mini-batch size equal to 1.

5.14 For the functions $g_1(\theta) = (\theta - 1)^2$ and $g_2(\theta) = (\theta + 1)^2$: (a) Evaluate the local and global minima of the global cost function $g(\theta) = \frac{1}{2} \sum_{n=1}^{2} g_n(\theta)$. (b*) Implement SGD with mini-batch size of 1, initialization $\theta^{(1)} = -1$, and learning rate $\gamma = 0.1$. Show the evolution of the iterates superimposed on the global cost function for 30 iterations. (c*) Then, implement SGD with mini-batch size of 1, initialization $\theta^{(1)} = -1$, and learning rate $\gamma = 0.01$. Show the evolution of the iterates, superimposed on the global cost function, for 100 iterations. Comment on your results.

5.15 For the functions $g_1(\theta) = 3\theta^4$, $g_2(\theta) = -\theta^4$, $g_3(\theta) = -2\theta^3$, and $g_4(\theta) = -3\theta^3 + 1$: (a*) Implement SGD with mini-batch size of 1, initialization $\theta^{(1)} = -1$, and learning rate $\gamma = 0.01$. Show the evolution of the value of the loss function $g(\theta^{(i)})$ for 10 independent runs of the algorithm. (b*) In the same figure, show also the average of $g(\theta^{(i)})$ obtained from 10 independent runs of SGD.

5.16 For the functions $g_1(\theta) = (\theta - 1)^2$ and $g_2(\theta) = (\theta + 1)^2$: (a*) Implement SGD with mini-batch size of 1, initialization $\theta^{(1)} = -1$, and learning rate $\gamma^{(i)} = 0.3/(1 + 0.5i)$. (b*) Show the evolution of the iterates, superimposed on the global cost function, for 100 iterations. Comment on your results.

5.17 (a*) Write a function G=numdiff(f,x,epsilon) that takes as input the name of a function f of a vector of D variables x and outputs the gradient G evaluated using numerical differentiation with precision epsilon. (b*) For the function $g(\theta_1, \theta_2) = \theta_1 \exp(-3\theta_1 - 2\theta_2)$, use the function designed in point (a*) to calculate the gradient at $\theta = [1, 2]^T$. Verify that the function has produced the correct output using analytical differentiation.

5.18 Consider the function

$$g(\theta_1, \theta_2) = \frac{1}{1 + \exp(-\theta_1 - 2\theta_2)} = \sigma(\theta_1 + 2\theta_2).$$

(a) Detail the computational graph that takes θ_1 and θ_2 as inputs and produces the output $g(\theta_1, \theta_2)$, by breaking the computation down in terms of the following computational blocks: inner product gate and sigmoid gate. (b) Using the derived computational graph, detail the forward and backward passes for the computation of the gradient $\nabla g(1, -1)$ using backprop. Compare the result with the gradient computed via analytical differentiation.

5.19 For each of the following computational blocks, derive the corresponding backward-mode block. (a) Sum gate $f(\theta) = \sum_i \theta_i$: show that in the backward pass this acts as a "gradient copier gate". (b) Inner product gate $f(\theta) = \sum_i w_i \theta_i$: show that in the backward pass this acts as a "gradient copy-and-multiply gate". (c) Multiplication gate $f(\theta_1, \theta_2) = \theta_1 \theta_2$: show that in the backward pass this acts as a "swap multiplier" gate. (d) Max gate $f(\theta_1, \theta_2) = \max(\theta_1, \theta_2)$: show that in the backward pass this acts as a "gradient router" gate.

5.20 Consider the function

$$g(\theta_1, \theta_2, \theta_3) = \sigma(2\sigma(\theta_1 + 2\theta_2) - \sigma(-2\theta_2 - \theta_3)).$$

(a) Detail the computational graph that takes θ_1, θ_2, and θ_3 as inputs and produces the output $g(\theta_1, \theta_2, \theta_3)$ by breaking the computation down in terms of inner product gates and sigmoid gates. (b) Using the derived computational graph, detail the forward and backward pass for the computation of the gradient $\nabla g(-1, 1, -1)$ using backprop. Compare the result with the gradient computed via numerical differentiation using the function G = numdiff(f,x,epsilon) designed in Problem 5.17.

5.21 Consider the function

$$x = \sigma(\theta)$$
$$f_1(x) = x^2$$
$$f_2(x) = 2x$$
$$g(\theta) = f_1(x) + f_2(x).$$

(a) Detail the computational graph that takes θ as input and produces the output $g(\theta)$ by breaking the computation down in terms of the indicated elementary functions. (b) Using the derived computational graph, detail the forward and backward passes for the computation of the partial derivatives $\partial g / \partial \theta$ and $\partial g / \partial x$ for $\theta = -1$. (c) Compare the result with the gradient computed via numerical differentiation using the function G=numdiff(f,x,epsilon) designed in Problem 5.17.

Appendices

Appendix 5.A: Alternative Definitions of Convexity for Multi-variable Functions

Generalizing the discussion in Sec. 5.6 on single-variable convex functions, we have the following alternative definitions for convex multi-variable functions.

- A multi-variable function $g(\theta)$ is convex if its first derivative is non-decreasing along any direction, i.e., if the inequality

$$(\nabla g(\theta_1) - \nabla g(\theta_2))^T (\theta_1 - \theta_2) \geq 0 \tag{5.89}$$

holds for any θ_1 and θ_2 in $\mathbb{R}^D$.
- A multi-variable function $g(\theta)$ is convex if it is never below the first-order Taylor approximation (5.22) evaluated at any point $\theta^0 \in \mathbb{R}^D$.
- A multi-variable function $g(\theta)$ is convex if the value $g(\alpha \theta^1 + (1 - \alpha)\theta^2)$ on the segment connecting two points θ^1 and θ^2 in $\mathbb{R}^D$, for $\alpha \in [0, 1]$, is never above the corresponding value $\alpha g(\theta^1) + (1 - \alpha)g(\theta^2)$ on the segment drawn between points $g(\theta^1)$ and $g(\theta^2)$, as expressed by inequality (5.17).

Appendix 5.B: Some Useful Gradients

For vectors and matrices of suitable dimensions, the following are useful formulas for the computation of gradients:

$$\nabla_\theta \left(b^T \theta \right) = b \tag{5.90}$$

$$\nabla_\theta \left(\theta^T A \theta \right) = 2A\theta \text{ for a symmetric matrix } A \tag{5.91}$$

$$\nabla_\Theta \log \det(\Theta) = \Theta^{-1} \text{ for a positive definite matrix } \Theta \tag{5.92}$$

$$\nabla_\Theta \mathrm{tr}(A\Theta) = A. \tag{5.93}$$

Note that the last two gradients are formatted in the form of matrices: The gradient $\nabla_A f(A)$, where A is a matrix, is defined in a manner analogous to (5.21) in that each entry $[\nabla_A f(A)]_{i,j}$ is given by the partial derivative $[\nabla_A f(A)]_{i,j} = \partial f(A)/\partial [A]_{i,j}$.

Appendix 5.C: Convergence Speed of Gradient-Based Methods

In this appendix, we first summarize known results on the convergence speed of GD and SGD under standard assumptions. Then, we provide some basic technical details behind these results by analyzing the convergence of GD for a strictly convex quadratic cost function. Finally, we briefly discuss how to bridge the performance gap in terms of convergence speed between GD and SGD.

Iteration and Communication Complexity of GD and SGD. In order to investigate the speed of convergence of gradient-based methods, let us make here the following assumptions, in addition the blanket assumption of differentiability made throughout the chapter:

- **Global and local smoothness**. The function $g(\theta)$ is L-smooth and each function $g_n(\theta)$ is L_n-smooth. Note that we have the inequalities $L_{\max}/n \leq L \leq L_{\max}$, where $L_{\max} = \max\{L_1, \ldots, L_n\}$.
- **Strong convexity**. The function $g(\theta)$ is μ-strongly convex, i.e., the function $g(\theta) - \frac{\mu}{2}\|\theta\|^2$ is convex, for some $\mu > 0$; and each function $g_n(\theta)$ is convex. Note that we have the inequality $\mu \leq L$.
- **Existence of a single global minimum**. Strong convexity implies strict convexity, and hence function $g(\theta)$ has at most one single global minimum θ^*. We assume that this minimum exists.

The **iteration complexity** of an iterative algorithm is defined in this appendix as the number of iterations i needed to ensure that the average optimality gap satisfies

$$\mathrm{E}[g(\theta^{(i+1)})] - g(\theta^*) \leq \epsilon \tag{5.94}$$

for some $\epsilon > 0$, where the average is taken over the randomness of the algorithm, namely the mini-batch size selection in SGD. The iteration complexity will turn out to depend on the quantity $\log_{10}(1/\epsilon)$, which can be thought of as the **number of digits of accuracy** with which the algorithm is required to approximate the optimal value $g(\theta^*)$ by (5.94). For instance, if $\epsilon = 10^{-3}$, the number of digits of accuracy equals $\log_{10}(1/\epsilon) = 3$.

The main results in terms of iteration complexity orders under the stated assumptions are summarized in Table 5.2. In the table, the **per-iteration computational complexity** refers to the number of gradients to be computed per iteration; $\kappa = L/\mu \geq 1$ is the **condition**

Table 5.2 Iteration complexity and per-iteration computational complexity of GD and SGD for smooth, strictly convex cost functions

algorithm	iteration complexity	per-iteration computational complexity
GD (fixed step size)	$\mathcal{O}(\kappa \log(1/\epsilon))$	N
SGD (decreasing step size)	$\mathcal{O}(\kappa_{\max}/\epsilon)$	1

number of the cost function $g(\theta)$; and we have also defined $\kappa_{\max} = L_{\max}/\mu \geq \kappa$. Based on the table, under the given assumptions, GD has **linear iteration complexity** in the sense that the iteration complexity is linear in the number of digits of accuracy. In contrast, SGD (with step sizes decreasing according to the Munro–Robbins conditions (5.72)) has a **sublinear iteration complexity**, as the number of iteration grows exponentially with the number of digits of accuracy. However, the per-iteration computational complexity of GD is N times that of SGD.

Iteration Complexity of GD for a Strictly Convex Quadratic Cost Function. We now wish to derive the iteration complexity order $\mathcal{O}(\kappa \log(1/\epsilon))$ of GD in Table 5.2. This will highlight the role of the condition number κ of the function $g(\theta)$. We will specifically analyze the convergence of GD for the problem of minimizing the strictly convex quadratic function (5.35) with $A \succ 0$. Denoting the minimum eigenvalue of matrix A as μ and the maximum eigenvalue as L, the function $g(\theta)$ is L-smooth and μ-strongly convex, satisfying the assumptions underlying the results in Table 5.2. Furthermore, given the positive definiteness of A, we have the inequalities $0 < \mu \leq L$. More general strongly convex functions can be approximated using the quadratic function (5.35) around the (unique) global optimal point.

Given that the gradient is given as in (5.36), the GD update with a fixed learning rate γ can be written as

$$\begin{aligned} \theta^{(i+1)} &= \theta^{(i)} - \gamma \nabla g(\theta^{(i)}) \\ &= \theta^{(i)} - \gamma(A\theta^{(i)} + b) \\ &= (I_D - \gamma A)\theta^{(i)} - \gamma b. \end{aligned} \tag{5.95}$$

Using this expression, we can evaluate the current error $(\theta^{(i+1)} - \theta^*)$ between the updated iterate and the optimal solution $\theta^* = -A^{-1}b$ in (5.37) as a function of the previous error $(\theta^{(i)} - \theta^*)$ by summing and subtracting θ^*. This yields the recursion

$$\begin{aligned} \theta^{(i+1)} - \theta^* &= (I_D - \gamma A)\theta^{(i)} - \gamma b - (-A^{-1}b) \\ &= (I_D - \gamma A)\theta^{(i)} - (I_D - \gamma A)(-A^{-1}b) \\ &= (I_D - \gamma A)(\theta^{(i)} - \theta^*). \end{aligned} \tag{5.96}$$

This relationship can be applied recursively i times to relate the error after i iterations, $\theta^{(i+1)} - \theta^*$, to the initial error, $\theta^{(1)} - \theta^*$, as

$$\theta^{(i+1)} - \theta^* = (I_D - \gamma A)^i (\theta^{(1)} - \theta^*) \tag{5.97}$$

for $i = 1, 2, \ldots$, where $(I_D - \gamma A)^i$ represents the ith power of matrix $(I_D - \gamma A)$ (i.e., the i-fold multiplication of this matrix by itself). We can now study the evolution of the norm $\|\theta^{(i+1)} - \theta^*\|$ of the error after i iterations as a function of the initial error $\|\theta^{(1)} - \theta^*\|$.

To this end, let us write the eigenvector decomposition of matrix A as $A = U\Lambda U^T$, and note that we have the equality

$$(I_D - \gamma A)^i = (I_D - \gamma U\Lambda U^T)^i$$
$$= U(I_D - \gamma \Lambda)^i U^T \qquad (5.98)$$

by the orthogonality of U, which implies the equalities $U^T U = UU^T = I_D$. From (5.97), we now have

$$||\theta^{(i+1)} - \theta^*|| = ||(I_D - \gamma A)^i(\theta^{(1)} - \theta^*)||$$
$$= ||(I_D - \gamma \Lambda)^i(\theta^{(1)} - \theta^*)||$$
$$\leq \lambda_{\max}(\gamma)^i ||\theta^{(1)} - \theta^*||, \qquad (5.99)$$

where the second equality follows again from the orthogonality of U, and the inequality is obtained by defining $\lambda_{\max}(\gamma)$ as the maximum among the absolute values of the diagonal element of matrix $(I_D - \gamma \Lambda)$.

As indicated by the notation, the quantity $\lambda_{\max}(\gamma)$ depends on the learning rate γ, and it can be directly computed as

$$\lambda_{\max}(\gamma) = \max\{1 - \gamma\mu, \gamma L - 1\}. \qquad (5.100)$$

This is because all eigenvalues of A are in the interval $[\mu, L]$. Since, by (5.99), a smaller $\lambda_{\max}(\gamma)$ implies a faster convergence, we can optimize this function over γ, by imposing that the two terms in the maximum in (5.100) are equal, to obtain the optimal learning rate $\gamma^* = 2/(L+\mu)$ and $\lambda_{\max}(\gamma^*) = (L-\mu)/(L+\mu)$. Plugging this result back into (5.99) yields the inequality

$$||\theta^{(i+1)} - \theta^*|| \leq \left(\frac{L-\mu}{L+\mu}\right)^i ||\theta^{(1)} - \theta^*||$$
$$= \left(\frac{\kappa - 1}{\kappa + 1}\right)^i ||\theta^{(1)} - \theta^*||. \qquad (5.101)$$

In order to obtain the iteration complexity, we impose the condition (5.94) as $||\theta^{(i+1)} - \theta^*||^2 \leq 2\epsilon/L$, or equivalently $||\theta^{(i+1)} - \theta^*|| \leq (2\epsilon/L)^{1/2} = \epsilon'$. This inequality implies (5.94) by the smoothness of the cost function (this is left as an exercise), and it can be stated as

$$\left(\frac{\kappa - 1}{\kappa + 1}\right)^i ||\theta^{(1)} - \theta^*|| \leq \epsilon', \qquad (5.102)$$

which in turn implies the inequality

$$i \geq \frac{\log\left(\frac{||\theta^{(1)} - \theta^*||}{\epsilon'}\right)}{\log\left(\frac{\kappa+1}{\kappa-1}\right)}. \qquad (5.103)$$

Noting that $1/(\log(\kappa+1) - \log(\kappa-1)) < \kappa$, this condition is also satisfied if

$$i \geq \kappa \log\left(\frac{||\theta^{(1)} - \theta^*||}{\epsilon'}\right), \qquad (5.104)$$

which recovers the desired $\mathcal{O}(\kappa \log(1/\epsilon))$ order.

Recovering the Iteration Complexity of GD via SGD. As discussed in this appendix, and summarized in Table 5.2, the iteration complexity of SGD is sublinear, while that of GD is linear in the number of digits of accuracy $\log_{10}(1/\epsilon)$. This indicates that the required number of iterations grows linearly with the number of digits of accuracy for GD, while the increase is exponential for SGD. It turns out that it is possible to recover the same convergence speed, in terms of number of iterations, as GD, via suitable modifications of SGD. This can be done in one of two ways, both aimed at improving the accuracy of the stochastic gradient as an estimate of the true gradient of the cost function. The first approach requires additional computations, while the price to pay for the second is memory.

As a notable example of the first type of scheme, **stochastic variance reduced gradient (SVRG)** periodically computes the *full* gradient to act as a control variate with the aim of reducing the variance of the stochastic gradient. As key representatives of the second class, **stochastic average gradient (SAG)** and **SAGA** keep in memory the last computed gradient for all data points. These outdated gradients are used alongside the new gradients for the current mini-batch to evaluate the model parameter update.

For both types of schemes, the stochastic gradient (5.64) is substituted with the following alternative estimate of the full gradient $1/N \sum_{n=1}^{N} \nabla g_n(\theta^{(i)})$:

$$\frac{1}{N} \sum_{n=1}^{N} \hat{\nabla}_n^{(i)} + \frac{1}{S} \sum_{n \in \mathcal{S}^{(i)}} \left(\nabla g_n(\theta^{(i)}) - \hat{\nabla}_n^{(i)} \right). \tag{5.105}$$

In (5.105), the vectors $\hat{\nabla}_n^{(i)}$, with $n = 1, \ldots, N$, are of the same size as the gradients $\nabla g_n(\theta^{(i)})$, and play the role of control variates aimed at reducing the variance of the gradient estimate. Clearly, if we have the equality $\hat{\nabla}_n^{(i)} = \nabla g_n(\theta^{(i)})$, the estimate (5.105) is exact. Hence, intuitively, if the control variate vector $\hat{\nabla}_n^{(i)}$ is a good estimate of the gradient $\nabla g_n(\theta^{(i)})$, then the variance of (5.105) is reduced as compared to (5.64). In SVRG, the control variate vector $\hat{\nabla}_n^{(i)}$ equals the gradient $\nabla g_n(\theta^{(j)})$ evaluated at a periodically chosen iterate $\theta^{(j)}$ with $j \leq i$; while in SAGA it equals the gradient $\nabla g_n(\theta^{(j)})$ evaluated at the last iterate $\theta^{(j)}$ for which data point n was included in the mini-batch $\mathcal{S}^{(j)}$. For further details, see Recommended Resources, Sec. 5.17.

Appendix 5.D: Momentum, RMSProp, and Adam

SGD inherits from GD the limitation of relying solely on first-order information. This makes it impossible to gauge how the average stochastic gradient direction is related to the set of directions towards stationary points. As we have seen, this can lead to an oscillatory behavior of the iterates and to longer convergence times (see, e.g., Fig. 5.12). Furthermore, the variance of the stochastic gradient due to mini-batching may cause the iterates' trajectory in the model parameter space to be excessively noisy. The first issue can be addressed by Newton's method (Sec. 5.5), but this requires the computation of the Hessian, which is infeasible for sufficiently large models. The second can be mitigated by properly optimizing the learning rate or by using the more sophisticated techniques mentioned at the end of Appendix 5.C. Momentum, RMSProp, and Adam are SGD-based schemes that can be interpreted as ways to address both issues via tractable heuristics.

Momentum. Momentum tries to reduce abrupt direction changes and oscillations in the iterates' trajectory by keeping memory of the direction of the previous update and by controlling deviations of the current update direction from it. Note that, in contrast, standard GD and SGD implementations follow update directions that may not be consistent with previous iterations.

Denoting the stochastic gradient as

$$\Delta^{(i)} = \frac{1}{S^{(i)}} \sum_{n \in \mathcal{S}^{(i)}} \nabla g_n(\theta^{(i)}), \tag{5.106}$$

the momentum update can be expressed as

$$u^{(i+1)} \leftarrow \alpha u^{(i)} + (1 - \alpha)\gamma^{(i)}\Delta^{(i)} \tag{5.107}$$

$$\theta^{(i+1)} \leftarrow \theta^{(i)} - u^{(i+1)}, \tag{5.108}$$

where $\alpha > 0$ is a hyperparameter (e.g., $\alpha = 0.9$). The hyperparameter α determines the contribution of the previous update direction $u^{(i)}$ on the current update direction $u^{(i+1)}$, which depends also on the current stochastic gradient $\Delta^{(i)}$. Momentum mitigates oscillations by averaging previous gradients through the momentum vector $u^{(i)}$.

RMSProp. RMSProp applies a different learning rate to each entry of the model parameter vector θ. This can help account for the possible different curvatures of the objective function along each coordinate of the parameter vector. As such, it can be interpreted as a form of pre-conditioning (see Sec. 5.10). The curvature for each coordinate is approximated by evaluating a moving average of the square of the gradients in each direction. Specifically, RMSProp applies the updates

$$[g^{(i+1)}]_d \leftarrow \beta[g^{(i)}]_d + (1 - \beta)([\Delta^{(i)}]_d)^2 \tag{5.109}$$

$$[\theta^{(i+1)}]_d \leftarrow [\theta^{(i)}]_d - \frac{\gamma}{\sqrt{[g^{(i+1)}]_d + \epsilon}}[\Delta^{(i)}]_d, \tag{5.110}$$

for all entries $d = 1, \ldots, D$ of the model parameter vector θ, given some fixed hyperparameters β, γ, and ϵ. Typical choices for the hyperparameters are $\beta = 0.9$, $\gamma = 10^{-3}$, and $\epsilon = 10^{-6}$.

Adam. Adaptive moment estimation (Adam) combines momentum with the per-coordinate adaptive learning rate of RMSProp. Accordingly, the updates are given as

$$u^{(i+1)} \leftarrow \alpha u^{(i)} + (1 - \alpha)\Delta^{(i)} \tag{5.111}$$

$$[g^{(i+1)}]_d \leftarrow \beta[g^{(i)}]_d + (1 - \beta)([\Delta^{(i)}]_d)^2 \tag{5.112}$$

$$[\theta^{(i+1)}]_d \leftarrow [\theta^{(i)}]_d - \frac{\gamma}{\sqrt{[g^{(i+1)}]_d + \epsilon}}[u^{(i+1)}]_d \tag{5.113}$$

for all entries $d = 1, \ldots, D$. Typical choices for the hyperparameters are $\alpha = 0.9$ and $\beta = 0.99$. It is worth noting that, while extensively used in practice, Adam has limited convergence guarantees.

Appendix 5.E: Jensen's Inequality

Given a convex function $f(\cdot)$ and an rv $x \sim p(x)$ defined on the domain of the function, **Jensen's inequality** is stated as

$$E_{x \sim p(x)}[f(x)] \geq f(E_{x \sim p(x)}[x]). \tag{5.114}$$

In words, the average of the function is no smaller than the function of the average. This is a direct consequence of inequality (5.17). To see the connection between (5.17) and (5.114), consider an rv that takes value θ^1 with probability α and value θ^2 with probability $1 - \alpha$, and apply Jensen's inequality (5.114) to it.

Table 5.3 Gradient vectors and Jacobian matrices

	scalar θ	vector θ ($D \times 1$)
scalar $f(\theta)$	scalar $\frac{df}{d\theta}$	gradient vector $\nabla f(\theta)$ ($D \times 1$)
vector $f(\theta)$ ($L \times 1$)	vector $\frac{df}{d\theta}$ ($L \times 1$)	Jacobian matrix $\nabla f(\theta)$ ($D \times L$)

Appendix 5.F: Jacobian Matrix

For vector-valued functions of multiple variables, the counterpart of the gradient is the Jacobian matrix. This appendix provides a short introduction.

Definition. The Jacobian matrix of an $L \times 1$ function $f(\theta) = [f_1(\theta), \ldots, f_L(\theta)]^T$ with $\theta \in \mathbb{R}^D$ is defined as the $D \times L$ matrix

$$\nabla f(\theta) = \begin{bmatrix} \frac{\partial f_1(\theta)}{\partial \theta_1} & \frac{\partial f_2(\theta)}{\partial \theta_1} & \cdots & \frac{\partial f_L(\theta)}{\partial \theta_1} \\ \frac{\partial f_1(\theta)}{\partial \theta_2} & \frac{\partial f_2(\theta)}{\partial \theta_2} & \cdots & \frac{\partial f_L(\theta)}{\partial \theta_2} \\ \vdots & & & \vdots \\ \frac{\partial f_1(\theta)}{\partial \theta_D} & \frac{\partial f_2(\theta)}{\partial \theta_D} & \cdots & \frac{\partial f_L(\theta)}{\partial \theta_D} \end{bmatrix} = [\nabla f_1(\theta), \nabla f_2(\theta), \ldots, \nabla f_L(\theta)]. \quad (5.115)$$

As such, the Jacobian collects the gradients of the component functions $f_i(\theta)$ as columns. Therefore, for $L = 1$, the Jacobian is equivalent to the gradient of the scalar function $f(\theta) = f_1(\theta)$. Note that the Jacobian is often also defined as the transpose of (5.115). The relationship between gradients and Jacobians is summarized in Table 5.3.

Example 5.26

For $f(\theta) = A\theta + b$, with $L \times D$ matrix A and $L \times 1$ vector b, the Jacobian matrix is the $D \times L$ matrix $\nabla f(\theta) = A^T$.

Chain Rule of Differentiation. Consider the composition $f_2(f_1(\theta))$, with $f_2 : \mathbb{R}^L \to \mathbb{R}$ (i.e., f_2 takes as input an L-dimensional vector and outputs a scalar) and $f_1 : \mathbb{R}^D \to \mathbb{R}^L$. This composition is a function from $\mathbb{R}^D$ to $\mathbb{R}$. Its $D \times 1$ gradient can be computed as

$$\nabla f_2(f_1(\theta)) = \underbrace{(\nabla f_1(\theta))}_{\text{Jacobian } (D \times L)} \times \underbrace{\nabla_{f_1} f_2}_{\text{gradient } (L \times 1)} . \quad (5.116)$$

Accordingly, in the case of multi-valued and/or multivariate functions, one needs to multiply the relevant gradients and Jacobians (from right to left) in an "inverse order" as compared to the composition (from inner to outer) and be careful with the dimensionality of the terms being multiplied.

Example 5.27

If $f_2(f_1(\theta)) = f_2(A\theta + b)$, we have

$$\nabla f_2(A\theta + b) = A^T \nabla_{f_1} f_2(f_1)|_{f_1 = Ax + b}. \quad (5.117)$$

Bibliography

[1] S. Boyd, S. P. Boyd, and L. Vandenberghe, *Convex Optimization*. Cambridge University Press, 2004.

[2] D. P. Bertsekas, *Nonlinear Programming*. Athena Scientific, 2016.

[3] G. Scutari and Y. Sun, "Parallel and distributed successive convex approximation methods for big-data optimization," in *Multi-agent Optimization*. Springer, 2018, pp. 141–308.

[4] J. Watt, R. Borhani, and A. Katsaggelos, *Machine Learning Refined: Foundations, Algorithms, and Applications*. Cambridge University Press, 2020.

[5] I. Goodfellow, Y. Bengio, and A. Courville, *Deep Learning*. The MIT Press, 2016, www.deeplearningbook.org.

[6] R. M. Gower, M. Schmidt, F. Bach, and P. Richtárik, "Variance-reduced methods for machine learning," *Proceedings of the IEEE*, vol. 108, no. 11, pp. 1968–1983, 2020.

[7] Z. Li and S. Arora, "An exponential learning rate schedule for deep learning," in *Eighth International Conference on Learning Representations*. ICLR, 2020.

[8] A. Griewank and A. Walther, *Evaluating Derivatives: Principles and Techniques of Algorithmic Differentiation*. SIAM, 2008.

[9] A. Defazio, F. Bach, and S. Lacoste-Julien, "SAGA: A fast incremental gradient method with support for non-strongly convex composite objectives," arXiv preprint arXiv:1407.0202, 2014.

6 Supervised Learning: Beyond Least Squares

6.1 Overview

In this chapter, we use the optimization tools presented in Chapter 5 to develop supervised learning algorithms that move beyond the simple settings studied in Chapter 4 for which the training problem could be solved exactly, typically by addressing an LS problem. We will focus specifically on binary and multi-class classification, with a brief discussion at the end of the chapter about the (direct) extension to regression problems. Following Chapter 4, the presentation will mostly concentrate on parametric model classes, but we will also touch upon mixture models and non-parametric methods.

Learning Objectives and Organization of the Chapter. After reading this chapter, the reader should be able to:

- train and apply linear models for binary classification with hard predictors (Sec. 6.2);
- train and apply linear models for binary classification via logistic regression and kernel methods (Sec. 6.3);
- train and apply multi-layer neural networks for binary classification (Sec. 6.4);
- train and apply generative models for binary classification (Sec. 6.5);
- extend training and inference from binary to multi-class classification for linear models (Sec. 6.6), neural networks (Sec. 6.7), and generative models (Sec. 6.8);
- understand mixture models (Sec. 6.9);
- interpret more advanced neural network architectures beyond feedforward multi-layer models (Sec. 6.10);
- train and apply K-nearest neighbors techniques (Sec. 6.11); and
- elaborate on the application of this chapter's techniques to regression (Sec. 6.12).

6.2 Discriminative Linear Models for Binary Classification: Hard Predictors

In this section, we consider the problem of training **linear hard predictors** for binary classification. In binary classification, the input x is an arbitrary vector in $\mathbb{R}^D$ and the target variable $t \in \{0, 1\}$ is binary. Linear models output predictions that depend on the **decision variable**

$$\theta^T u(x) = \sum_{d=1}^{D'} \theta_d u_d(x), \tag{6.1}$$

where $u(x) = [u_1(x), \ldots, u_{D'}(x)]^T$ is a $D' \times 1$ vector of D' scalar features $u_d(x)$, with $d = 1, \ldots, D'$, and $\theta = [\theta_1, \ldots, \theta_{D'}]^T$ is the $D' \times 1$ vector of model parameters. Note that, as compared to Chapter 4, we will denote as D' the number of features in order to leave the notation D for the size of the input vector x. By (6.1), in linear models, each feature $u_d(x)$ contributes to the decision variable with a weight defined by a distinct model parameter θ_d. Importantly, the (scalar) features $u_d(x)$ can be arbitrary functions of x – not necessarily linear. What *is* linear is the dependence of the decision variable (6.1) on the model parameters θ. We will also see that linear classifiers correspond to linear decision surfaces in the feature space (see Fig. 6.2 for a preview).

6.2.1 Model Class

The model class $\mathcal{H} = \{\hat{t}(x|\theta) : \theta \in \mathbb{R}^{D'}\}$ of linear hard predictors includes predictors $\hat{t}(x|\theta)$ of the form

$$\hat{t}(x|\theta) = \text{step}(\theta^T u(x)) = \begin{cases} 1 & \text{if } \theta^T u(x) > 0 \\ 0 & \text{if } \theta^T u(x) < 0, \end{cases} \tag{6.2}$$

where we have defined the step function $\text{step}(z) = 1$ if $x > 0$ and $\text{step}(z) = 0$ if $x < 0$. Accordingly, if the decision variable is positive, class 1 – labeled as $t = 1$ – is selected; and, if it is negative, class 0 – labeled as $t = 0$ – is chosen. The decision is undefined when we have $\theta^T u(x) = 0$, as the classifier is unable to make a decision for either class. In practice, in the unlikely circumstance that this condition occurs, the classifier can make an arbitrary decision, or indicate that the decision is undefined.

Based on (6.2), linear models can be represented by the block diagram in Fig. 6.1. In it, there are two layers. The first layer takes x as input and produces as output the fixed feature vector $u(x)$. The second layer carries out classification by computing the decision variable (6.1) and evaluating the hard prediction (6.2). While the operation of the first layer is fixed, the weights θ of the second layer can be optimized through training.

Geometry of Linear Classifiers. Geometrically, the decision rule (6.2) determines two disjoint regions in the feature space $\mathbb{R}^{D'}$, one corresponding to vectors $u(x)$ that are assigned the prediction $\hat{t}(x|\theta) = 1$, and the other corresponding to vectors $u(x)$ that are assigned $\hat{t}(x|\theta) = 0$. The two regions are partitioned by the **decision hyperplane**, which consists of all vectors u in the feature space $\mathbb{R}^{D'}$ that satisfy the equation

$$\theta^T u = 0. \tag{6.3}$$

Figure 6.1 Computational graph of a linear hard predictor for binary classification. While the feature-extraction functions $u_d(\cdot)$ with $d = 1, \ldots, D'$ in the first layer are fixed as part of the inductive bias, the model parameter vector θ in the second layer can be trained.

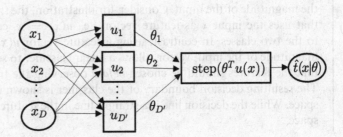

Figure 6.2 The decision hyperplane – a line in a two-dimensional feature space ($D' = 2$) – of a linear classifier is orthogonal to the model parameter vector θ. The figure also shows the geometric classification margin $|\theta^T u(x)|/\|\theta\|$.

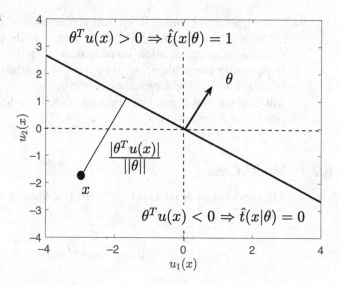

Note that feature vectors on the decision hyperplane correspond to undefined predictions. The decision hyperplane is a line for $D' = 2$, a plane for $D' = 3$, and so on, and it is defined by the orthogonal direction determined by vector θ. This is illustrated in Fig. 6.2.

It is important to emphasize that the surface separating the two decision regions is indeed a hyperplane in the feature space, but not necessarily in the original input space $\mathbb{R}^D$. This is because the feature vector $u(x)$ may be a non-linear function of x.

On the Choice of Features. The choice of the features in vector $u(x)$ should be generally based on prior knowledge about the problem – a task known as **feature engineering** – and/or on validation. Features may even be selected at random by applying stochastic linear transformations to the input – an approach known as **"extreme machine learning"**.

In some cases, useful features vectors have larger dimensionality than the original data, i.e., $D' > D$. In other settings, especially when D is very large, we may prefer the feature dimension D' to be smaller than D in order to avoid overfitting.

Example 6.1

Assume that we have prior information that inputs in different classes can be distinguished based on the magnitude of the input. Consider, for illustration, the training set in Fig. 6.3. A linear classifier that uses the input x as feature vector, i.e., $u(x) = x$, cannot separate the samples belonging to the two classes. In contrast, using a feature vector $u(x) = [x_1^2, x_2^2]^T$ which is sensitive to the magnitude of the input vector, allows a linear classifier to separate them perfectly. (Note, however, that the features should be chosen *before* observing the training data or using validation data.) The resulting decision boundary of the classifier is shown in Fig. 6.3 as a dashed line in the input space. While the decision line is a straight line in the feature space, it is curved in the original input space.

Figure 6.3 Data for a binary classification problem with samples in either class represented as circles or crosses. The vector of features $u(x) = [x_1^2, x_2^2]^T$ allows the training of an effective binary linear classifier, whose decision line is dashed. Note that the decision line is indeed a straight line in the feature space, but it is curved in the original input space, as shown in the figure.

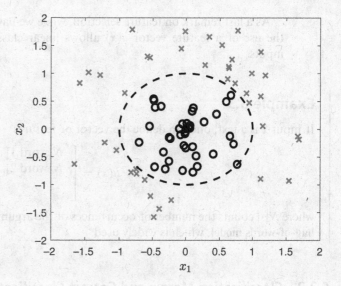

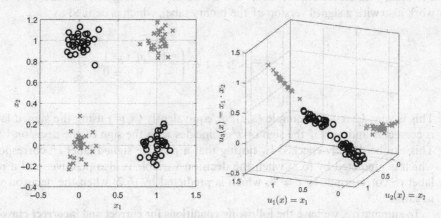

Figure 6.4 Data for the XOR binary classification problem with samples in either class represented as circles or crosses. The three-dimensional vector of features $u(x) = [x_1, x_2, x_1 x_2]^T$ allows the training of an effective binary linear classifier (right), while this is not the case in the original two-dimensional space (left).

Example 6.2

An example in which it is useful to choose a feature vector larger than the input vector, i.e., $D' > D$, is the **XOR (exclusive OR)** problem illustrated in Fig. 6.4. The figure shows data points x_n and corresponding labels t_n both in the input space with $D = 2$ and in the space of features $u(x) = [x_1, x_2, x_1 x_2]^T$ with $D' = 3$. In this three-dimensional feature space, there exists a plane that perfectly separates the two classes, while a line that partitions the two classes does not exist in the original two-dimensional space.

As a last remark on feature selection, while we have presented input x as a numerical vector, the use of a feature vector $u(x)$ allows linear classifiers to be applied also to non-numeric inputs.

Example 6.3

If input x is a text, one can define the vector of features

$$u(x) = \begin{bmatrix} N[\text{word } 1] \\ N[\text{word } 2] \\ \vdots \end{bmatrix}, \tag{6.4}$$

where $N[\cdot]$ counts the number of occurrences of the argument word in text x. This is known as the **bag-of-words model**, which is widely used.

6.2.2   Classification Margin and Correct Classification Condition

For binary classification, the label t is a Bernoulli rv, taking values 0 and 1. It is often useful to work also with a **signed** version of the binary label, which is defined as

$$t^{\pm} = 2t - 1 = \begin{cases} 1 & \text{if } t = 1 \\ -1 & \text{if } t = 0. \end{cases}$$

This is why: Given an example (x, t), or equivalently $(x, t^{\pm})$ using the signed label, a decision is correct if and only if the sign of $t^{\pm}$ coincides with the sign of the decision variable $\theta^T u(x)$. This can be directly checked by noting that a positive signed label $t^{\pm}$ corresponds to $t = 1$, which is predicted by (6.2) when the decision variable is also positive; and a negative signed label $t^{\pm}$ corresponds to $t = 0$, which is predicted by (6.2) when the decision variable is also negative.

To summarize, we have the following **conditions for correct and incorrect classification**:

$$\text{correct classification: } t^{\pm} \cdot (\theta^T u(x)) > 0$$
$$\text{incorrect classification: } t^{\pm} \cdot (\theta^T u(x)) < 0. \tag{6.5}$$

Given its role in (6.5), the quantity $t^{\pm} \cdot (\theta^T u(x))$ is known as the **classification margin**. By (6.5), the sign of the classification margin $t^{\pm} \cdot (\theta^T u(x))$ indicates whether the decision is correct or not for an example (x, t). Furthermore, a more positive classification margin $t^{\pm}(\theta^T u(x))$ indicates a more "confident" correct decision; a more negative classification margin indicates a more "confident" incorrect decision.

Geometrically, when divided by the norm θ, the absolute value $|\theta^T u(x)|$ of the classification margin $t^{\pm} \cdot (\theta^T u(x))$ defines the (Euclidean) distance of the feature vector $u(x)$ from the decision hyperplane in $\mathbb{R}^{D'}$. This is also known as the **geometric classification margin**, and is illustrated in Fig. 6.2 for a point x.

6.2.3 Surrogate Loss Function

Detection Error Loss and Classification Margin. As discussed in Chapter 4, the standard choice for the loss function in classification (and detection) is the detection-error loss, which, by (6.5), can be expressed in terms of the classification margin as

$$\ell(t, \hat{t}(x|\theta)) = \mathbb{1}(t \neq \hat{t}(x|\theta))$$
$$= \begin{cases} 0 & \text{if correct prediction } (t^{\pm} \cdot (\theta^T u(x)) > 0) \\ 1 & \text{if incorrect prediction } (t^{\pm} \cdot (\theta^T u(x)) < 0). \end{cases} \tag{6.6}$$

So, the detection-error loss is 0 if the prediction is correct, and it equals 1 if prediction is incorrect. The value of the loss for the case $\theta^T u(x) = 0$ is irrelevant in practice, and it can be set, e.g., to 0.5, which corresponds to the average loss of a random uniform decision. Note that we can also write the detection-error loss (6.6) more succinctly as

$$\ell(t, \hat{t}(x|\theta)) = \text{step}(-t^{\pm} \cdot (\theta^T u(x))), \tag{6.7}$$

with the step function defined as in (6.2).

Detection-Error Loss and (S)GD. One could attempt to tackle the regularized ERM problem (see Chapter 4) by adopting the detection-error loss (6.7). Training with this loss function, however, is problematic. In fact, as seen in the last chapter, GD or SGD updates the model parameters θ in the direction of the negative gradient of the loss function. But, as illustrated in Fig. 6.5, the detection-error loss function is flat almost everywhere: All correctly classified examples have loss equal to 0, and all incorrectly classified ones have loss equal to 1. As a result, the gradient of the detection-error loss is

$$\nabla_\theta \ell(t_n, \hat{t}(x_n|\theta)) = 0_{D'} \tag{6.8}$$

for almost all vectors θ. Therefore, GD and SGD would apply zero updates and be stuck at the initialization.

Figure 6.5 Surrogate loss functions.

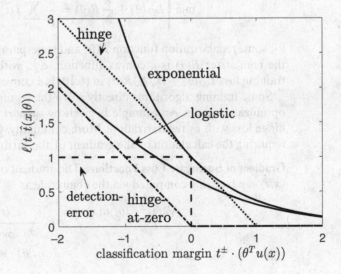

Table 6.1 Detection-error loss and examples of surrogate losses for binary classification $(y = t^{\pm} \cdot (\theta^T u(x)))$

surrogate loss	$f(y)$
exponential	$\exp(-y)$
hinge	$\max(0, 1 - y)$
hinge-at-zero	$\max(0, -y)$
logistic	$\log(1 + \exp(-y))$

Surrogate Loss Function. In order to apply GD or SGD, we therefore need to replace the detection-error loss with an alternative, or **surrogate**, loss function that is similar enough to the detection-error loss while also providing useful gradient updates. To this end, note from Fig. 6.5 that the (flipped) step function $\text{step}(-y)$ in the detection-loss function (6.7) is non-increasing in y and has a zero derivative $\text{dstep}(-y)/\text{d}y$ for almost every value of y (except $y = 0$). We can obtain a **surrogate loss function** by replacing the (flipped) step function $\text{step}(-y)$ with a continuously differentiable, convex, non-increasing function $f(\cdot)$, yielding

$$\ell(t, \hat{t}(x|\theta)) = f(t^{\pm}(\theta^T u(x))). \tag{6.9}$$

As illustrated in Fig. 6.5 and summarized in Table 6.1, examples of surrogate loss functions include the exponential loss, the hinge loss, the hinge-at-zero loss, and the logistic loss. All these convex surrogate losses yield a non-zero gradient at least for misclassified examples, i.e., at least for $y < 0$.

6.2.4 Training Binary Classifiers via SGD

Using a surrogate loss function $f(\cdot)$, the regularized ERM problem for the model class of linear hard predictors (6.2) can be written as

$$\min_{\theta} \left\{ L_{\mathcal{D}}(\theta) + \frac{\lambda}{N} R(\theta) = \frac{1}{N} \sum_{n=1}^{N} f(t_n^{\pm} \cdot (\theta^T u(x_n))) + \frac{\lambda}{N} R(\theta) \right\} \tag{6.10}$$

for some regularization function $R(\theta)$ and hyperparameter $\lambda \geq 0$. In the common case in which the regularizer $R(\theta)$ is a convex function, e.g., with ℓ_1 or ℓ_2 regularization, the regularized training loss $L_{\mathcal{D}}(\theta) + (\lambda/N)R(\theta)$ in (6.10) is a convex function of the model parameter θ.

Some training algorithms directly solve the regularized ERM problem (6.10) using convex optimization tools. An example is given by **support vector machines (SVMs)** that assume the hinge loss with ℓ_2 regularization. More commonly, GD, SGD, and variants thereof are used, requiring the calculation of the gradient of the surrogate loss function.

Gradient of Surrogate Loss Functions. The gradient of the loss function for any training sample (x, t) can be easily computed via the chain rule as

$$\nabla_{\theta} f(t^{\pm} \cdot (\theta^T u(x))) = \underbrace{t^{\pm} \cdot f'(t^{\pm} \cdot (\theta^T u(x)))}_{\delta(x, t|\theta), \text{ signed classification error}} \cdot u(x)$$

$$= \delta(x, t|\theta) \cdot u(x), \tag{6.11}$$

where we have defined the derivate of the surrogate function as $f'(y) = df(y)/dy$. As indicated in (6.11) and explained later in the next subsection, it is useful to think of the term $\delta(x, t|\theta) = t^{\pm} \cdot f'(t^{\pm} \cdot (\theta^T u(x)))$ as a **signed** measure of the classification error obtained by the model vector θ.

Delta Rule. A GD or SGD scheme that uses the gradient (6.11) is said to follow a **delta rule** to update the model parameter vector θ. The name follows from the form of the update rule. In fact, assuming an SGD mini-batch of size $S = 1$ with selected sample $(x_{n(i)}, t_{n(i)})$ at iteration i, the update rule is of the form

$$\theta^{(i+1)} = \theta^{(i)} - \gamma^{(i)} \delta(x_{n(i)}, t_{n(i)}|\theta^{(i)}) \cdot u(x_{n(i)}). \tag{6.12}$$

In the update (6.12), each dth weight $[\theta^{(i+1)}]_d = \theta_d^{(i+1)}$, for $d = 1, \ldots, D$, is updated by a term that depends on the product of the corresponding dth input feature $u_d(x_{n(i)})$ and the common scalar error ("delta") signal $\delta(x_{n(i)}, t_{n(i)}|\theta^{(i)})$.

Neurons and Synapses. The delta rule (6.12) is often described by drawing an analogy with biological systems. According to this perspective, one views the block computing the classification decision (6.2) as a single **neuron** processing D' sensory inputs – the features – through D' **synapses**. Each dth synapse processes input $u_d(x)$ by multiplying it by the **synaptic weight** θ_d. All processed inputs are then combined to produce the decision variable (6.1) and the prediction (6.2).

With this in mind, the delta rule (6.12) can be understood as updating the weight of each dth synapse θ_d as a function of two terms:

- a **pre-synaptic** term, the input feature $u_d(x_{n(i)})$; and
- the **post-synaptic** error term $\delta(x_{n(i)}, t_{n(i)}|\theta^{(i)})$.

The "delta" error term $\delta(x_{n(i)}, t_{n(i)}|\theta^{(i)})$ provides feedback about the effectiveness of the current weights $\theta^{(i)}$ on the basis of the observation of sample $(x_{n(i)}, t_{n(i)})$ to all synapses. Interestingly, the update (6.12) can be applied **locally** at each synapse, since each dth synapse only needs to be informed about its current weight $\theta_d^{(i)}$, its input feature $u_d(x_{n(i)})$ (pre-synaptic term), and the error signal $\delta(x_{n(i)}, t_{n(i)}|\theta^{(i)})$ (post-synaptic error term). Note that the error signal is shared among all synapses.

6.2.5 Perceptron Algorithm

As an example of SGD-based training of linear models via the delta rule (6.12), we now consider the case of the hinge-at-zero loss, which yields the **perceptron algorithm**. We will later also specialize the delta rule to the logistic loss. The perceptron algorithm is an early training scheme, introduced in 1958, that applies SGD on the ERM problem (6.10) with the hinge-at-zero loss. (The original implementation of the perceptron required a five-ton computer the size of a room, and was hailed by its inventor, Frank Rosenblatt, as the "the first machine which is capable of having an original idea", with *The New Yorker* echoing this view: "Indeed, it strikes us as the first serious rival to the human brain ever devised.") For the hinge-at-zero surrogate $f(y) = \max(0, -y)$, we have the derivative

$$f'(y) = \begin{cases} 0 & \text{if } y > 0 \\ -1 & \text{if } y < 0. \end{cases} \tag{6.13}$$

Note that $f'(y)$ is not defined at $y = 0$, and we can arbitrarily set the derivative to any number between 0 and 1 at this point (technically, this corresponds to using a subgradient). We hence have the **signed classification error**

$$\delta(x, t | \theta) = \begin{cases} 0 & \text{if prediction with model parameter vector } \theta \text{ is correct} \\ -1 & \text{if prediction with model parameter vector } \theta \text{ is incorrect and } t = 1 \\ 1 & \text{if prediction with model parameter vector } \theta \text{ is incorrect and } t = 0, \end{cases}$$

(6.14)

where the correct and incorrect classification conditions are as in (6.5). This can be expressed in a compact way as

$$\delta(x, t | \theta) = \text{step}(\theta^T u(x)) - t,$$

(6.15)

as can be easily checked, where we recall that $\text{step}(\theta^T u(x))$ represents the hard decision (6.2) made with model parameter θ.

By (6.11), the gradient of the loss is given as

$$\nabla f(t^{\pm} \cdot (\theta^T u(x))) = \delta(x, t | \theta) \cdot u(x) = (\text{step}(\theta^T u(x)) - t) \cdot u(x).$$

(6.16)

Using this formula for the gradient yields the SGD scheme summarized in Algorithm 6.1, assuming a mini-batch size 1.

Algorithm 6.1: The perceptron algorithm

---

initialize $\theta^{(1)}$ and $i = 1$
while *stopping criterion not satisfied* **do**
> pick a random sample $(x_{n^{(i)}}, t_{n^{(i)}})$ from the data set uniformly at random
> apply the current predictor $\hat{t}(x_{n^{(i)}} | \theta^{(i)})$ in (6.2) to obtain prediction $\hat{t}_{n^{(i)}}$
> obtain the next iterate as
>
> $$\theta^{(i+1)} \leftarrow \theta^{(i)} - \gamma^{(i)} (\hat{t}_{n^{(i)}} - t_{n^{(i)}}) \cdot u(x_{n^{(i)}})$$ (6.17)
>
> set $i \leftarrow i + 1$
end
return $\theta^{(i)}$

Properties of the Perceptron Algorithm. The perceptron algorithm has the following useful property: After each iteration i, the classification margin of the selected sample $(x_{n^{(i)}}, t_{n^{(i)}})$ is unchanged if the sample is correctly predicted by the current model $\theta^{(i)}$, and is decreased if the sample is misclassified under $\theta^{(i)}$. In fact, in the latter case, the update (6.17) can be expressed as $\theta^{(i+1)} = \theta^{(i)} + \gamma^{(i)} t_{n^{(i)}}^{\pm} \cdot u(x_{n^{(i)}})$, and we have

$$\underbrace{t_{n^{(i)}}^{\pm} \cdot (\theta^{(i+1)})^T u(x_{n^{(i)}})}_{\text{classification margin at iteration } i+1} = t_{n^{(i)}}^{\pm} \cdot (\theta^{(i)} + \gamma^{(i)} t_{n^{(i)}}^{\pm} \cdot u(x_{n^{(i)}}))^T u(x_{n^{(i)}})$$

$$= \underbrace{t_{n^{(i)}}^{\pm} \cdot (\theta^{(i)})^T u(x_{n^{(i)}})}_{\text{classification margin at iteration } i} + \underbrace{\gamma^{(i)} ||u(x_{n^{(i)}})||^2}_{>0},$$

(6.18)

and hence the classification margin at iteration $i + 1$ is larger than that at iteration i as long as the feature vector is non-zero, i.e., as long as $||u(x_{n^{(i)}})|| > 0$.

Example 6.4

We are given training data

$$\mathcal{D} = \{(x_n, t_n)_{n=1}^4\} = \{(2,1), (-2,0), (-0.5,0), (5,1)\}, \tag{6.19}$$

and we consider a single feature $u(x) = x$. The initialization is $\theta^{(1)} = -1$. We would like to run the perceptron algorithm by selecting the examples in the order shown in (6.19) with constant learning rate $\gamma^{(i)} = 0.1$.

At iteration $i = 1$, we use sample $(2,1)$. We have the decision variable $(\theta^{(1)} \cdot 2) = -2$, which implies the prediction $\hat{t}(2|-1) = 0 \neq t_1 = 1$. Since the decision is incorrect and $t_1^{\pm} = 1$, the update is

$$\theta^{(2)} \leftarrow \theta^{(1)} + \gamma^{(1)} x_1 = -1 + 0.1 \cdot 2 = -0.8. \tag{6.20}$$

Note that the classification margin for the selected sample has increased from $1 \cdot ((-1) \cdot 2) = -2$ to $1 \cdot ((-0.8) \cdot 2) = -1.6$.

At iteration $i = 2$, we use sample $(-2,0)$. The decision variable is $(-0.8) \cdot (-2) = 1.6$, which implies the prediction $\hat{t}(-2|-0.8) = 1 \neq t_2 = 0$. Since the decision is again wrong, and we have $t_1^{\pm} = -1$, the update is

$$\theta^{(3)} \leftarrow \theta^{(2)} - \gamma^{(2)} x_2 = -0.8 - 0.1 \cdot (-2) = -0.6. \tag{6.21}$$

The classification margin for the second sample has increased from $-1 \cdot ((-0.8) \cdot (-2)) = -1.6$ to $-1 \cdot ((-0.6) \cdot (-2)) = -1.2$. We leave the next two iterations as an exercise for the reader.

6.3 Discriminative Linear Models for Binary Classification: Soft Predictors and Logistic Regression

As introduced in Sec. 3.3, probabilistic models provide a soft prediction for the label t, from which a hard decision can be obtained from any loss function (cf. (3.16)). A soft prediction is meant to capture the aleatoric uncertainty that exists about the target variable t when observing the input x = x (see Sec. 4.5). Like the linear models (6.2) for hard prediction, probabilistic linear models produce a prediction that depends on the input x only through a linear decision variable $\theta^T u(x)$ (where linearity is with respect to the model parameters θ). In this section, we first introduce probabilistic binary linear classifiers, and then focus on a specific class of such classifiers, namely logistic regression models.

6.3.1 Model Class

The model class of probabilistic binary linear classifiers

$$\mathcal{H} = \{p(t|x,\theta) : \theta \in \mathbb{R}^{D'}\} \tag{6.22}$$

contains soft predictors of the form

$$p(t|x,\theta) = \exp(-f(\ \underbrace{t^{\pm} \cdot (\theta^T u(x))}_{\text{classification margin}}\ )) \tag{6.23}$$

for a surrogate loss function $f(\cdot)$. As discussed in the preceding section, the surrogate loss function must be convex and non-decreasing. Here, in order for (6.23) to be a probability, we also add the assumption that it should satisfy the normalization condition $p(t = 0|x,\theta) + p(t = 1|x,\theta) = 1$ for all values of x and θ. In practice, this equality can be guaranteed by adding a suitable normalizing constant to the loss function.

According to (6.23), the score assigned to target value t for input x depends on the corresponding value of the surrogate loss $f(t^\pm \cdot (\theta^T u(x)))$ that would be incurred if t were the correct value of the label: The larger the loss, the smaller the corresponding score. In fact, with this choice, the log-loss is exactly equal to the surrogate loss:

$$-\log p(t|x,\theta) = f(t^\pm(\theta^T u(x))). \tag{6.24}$$

Once again, we see the close connection between loss functions for hard and soft predictors, which was first discussed in Sec. 3.8.

6.3.2   Training Probabilistic Models via SGD

In light of (6.24), the MAP learning problem for probabilistic linear models can be written as in (6.10), for some prior $p(\theta)$ and corresponding regularization function $R(\theta) = -\log p(\theta)$. Recall that the maximum likelihood (ML) learning problem corresponds to the special case with $\lambda = 0$. Therefore, the MAP and ML problems can be tackled via SGD by using the delta rule introduced in the preceding section. In the next subsection, we will detail the resulting training algorithm for the special case of logistic regression.

6.3.3   Logistic Regression: Model Class

Model Class. As an important special case of probabilistic linear predictors, we now study **logistic regression**, which corresponds to the choice of the **logistic loss**

$$f(y) = -\log \sigma(y) = \log(1 + \exp(-y)) \tag{6.25}$$

in (6.23). As a matter of terminology, it should be emphasized that logistic *regression* performs classification despite what the name may seem to imply. With this choice for the loss, the soft predictors (6.23) in class $\mathcal{H}$ are given as

$$p(t|x,\theta) = \frac{1}{1 + \exp(-t^\pm \cdot \theta^T u(x))} = \sigma(t^\pm \cdot (\theta^T u(x))), \tag{6.26}$$

where we have used the definition of the **sigmoid function** $\sigma(x) = 1/(1 + \exp(-x))$. More extensively, we can write the soft predictor (6.26) as

$$p(t = 1|x,\theta) = \sigma(\theta^T u(x)) \text{ and}$$
$$p(t = 0|x,\theta) = \sigma(-\theta^T u(x)) = 1 - \sigma(\theta^T u(x)), \tag{6.27}$$

where we have used the useful relationship

$$\sigma(-x) = 1 - \sigma(x). \tag{6.28}$$

By (6.27), an equivalent way to think about the logistic-regression model class is that it assumes the Bernoulli conditional distribution

$$(t|x = x) \sim \text{Bern}(\sigma(\theta^T u(x))). \tag{6.29}$$

Figure 6.6 A linear soft predictor for binary classification modeled via logistic regression.

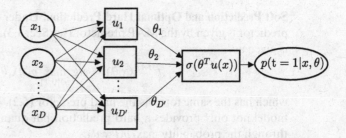

Figure 6.7 Probabilistic model $p(t = 1|x, \theta) = \sigma(\theta^T u(x))$ assumed by logistic regression, alongside the MAP hard predictor (6.31) (gray line), which corresponds to the hard predictor (6.2).

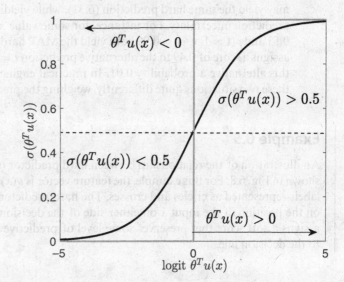

Overall, logistic regression operates as illustrated in Fig. 6.6. Comparing this block diagram with that of the linear hard predictor in Fig. 6.1, we observe that linear regression replaces the "hard" step non-linearity with the "soft" sigmoid function to produce its prediction $p(t = 1|x, \theta) = \sigma(\theta^T u(x))$.

The soft predictor $p(t = 1|x, \theta) = \sigma(\theta^T u(x))$ in (6.26) is shown in Fig. 6.7 as a function of the decision variable $\theta^T u(x)$. As seen in the figure, when the decision variable $\theta^T u(x)$ is large in absolute value, say when $|\theta^T u(x)| > 5$, the prediction is essentially deterministic and the predictor is very confident in its output, resembling the hard predictor (6.2) in Fig. 6.1. The hard decision of the latter is also shown in the figure as a gray line.

Logit, or Log-Odds. In logistic regression, the decision variable $\theta^T u(x)$ is also referred to as **logit** or **log-odds**. This is because, as can be easily proved, we have the equality

$$\theta^T u(x) = \log\left(\frac{p(t = 1|x, \theta)}{p(t = 0|x, \theta)}\right), \tag{6.30}$$

where the right-hand side is the logarithm of the odds (2.4) for binary label t given input x. The log-odds can be interpreted the LLR for the problem of detecting the event t = 1 against t = 0 when the observation is x = x.

Soft Prediction and Optimal Hard Prediction. Under the detection-error loss, the optimal hard predictor is given by the MAP predictor (see Sec. 3.3). For logistic regression, from Fig. 6.7, this is given as

$$\hat{t}^{MAP}(x|\theta) = \arg\max_{t\in\{0,1\}} p(t|x,\theta) = \text{step}(\theta^T u(x)), \tag{6.31}$$

which has the same form as the hard predictor (6.2). In this regard, recall that the probabilistic model not only provides a hard prediction but quantifies the confidence of the hard decision through the probability $\max_t p(t|x,\theta)$.

Let us elaborate further on this point. An important observation is that two soft predictors may yield the same hard prediction (6.31), while yielding substantially different estimates of the prediction uncertainty. For instance, for some value x, the two soft predictions $p(t = 1|x,\theta) = 0.51$ and $p(t = 1|x,\theta) = 0.99$ both yield the MAP hard prediction $\hat{t}^{MAP}(x|\theta) = 1$, but the former assigns a score of 0.49 to the alternative prediction $t = 0$, while the latter, more confidently, gives this alternative a probability 0.01. In practical engineering applications, one may wish to treat these two situations quite differently, weighing the predictive uncertainty in the decision process.

Example 6.5

An illustration of the relationship between hard predictor (6.31) and soft predictor $p(t = 1|x,\theta)$ is shown in Fig. 6.8. For this example, the feature vector is $u(x) = x$, and the training set is shown with labels represented as circles and crosses. The hard predictor outputs a definite decision depending on the position of the input x on either side of the decision line (dashed); while the soft predictor assigns a soft score that preserves some level of predictive uncertainty for points x that are close to the decision line.

6.3.4   Logistic Regression: Training via SGD

We now consider the problem of ML training of logistic regression via SGD. The extension to MAP learning is straightforward.

Gradient of the Log-Loss. For logistic regression, by (6.24) and (6.25), the log-loss is given by the logistic loss

$$f(t^{\pm}\cdot(\theta^T u(x))) = -\log\sigma(t^{\pm}\cdot(\theta^T u(x)))$$
$$= \log(1 + \exp(-t^{\pm}\cdot(\theta^T u(x)))). \tag{6.32}$$

From the general formula (6.11), we obtain the gradient

$$\nabla_\theta(-\log\sigma(t^{\pm}(\theta^T u(x)))) = \underbrace{(\sigma(\theta^T u(x)) - t)}_{\delta(x,t|\theta),\ \text{mean error}}\cdot\underbrace{u(x)}_{\text{feature vector}}. \tag{6.33}$$

A derivation of this formula can be found in Appendix 6.A.

The **signed error** $\delta(x,t|\theta) = \sigma(\theta^T u(x)) - t$ in the gradient (6.33) is given by the difference between the mean of the rv $t \sim p(t|x,\theta)$, which is distributed according to the model, and the actual label t, i.e.,

$$\delta(x,t|\theta) = \mathbb{E}_{t\sim p(t|x,\theta)}[t] - t = \sigma(\theta^T u(x)) - t. \tag{6.34}$$

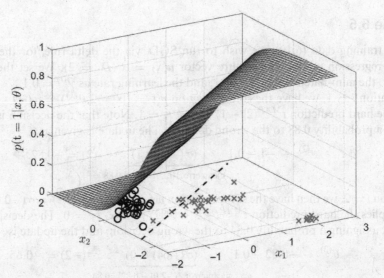

Figure 6.8 Soft predictor $p(t = 1|x, \theta)$ for logistic regression trained using the displayed training set (with crosses representing label $t = 1$ and circles label $t = 0$). The decision line defining the hard, MAP, predictor (6.31) is shown as a dashed line.

It is hence also referred to as the **mean error**. The mean error replaces the step function in the signed classification error (6.15) for the perceptron with the "soft" non-linearity given by the sigmoid. As a result, it behaves as follows:

- We have $\delta(x, t|\theta) = 0$ if the prediction is certain and correct, i.e., if $t = 1$ and $\theta^T u(x) \to \infty$, or if $t = 0$ and $\theta^T u(x) \to -\infty$; and $\delta(x, t|\theta) = -t^{\pm}$ if the prediction is certain but incorrect.
- Otherwise – and this is the novel element with respect to the perceptron – we have $0 < \delta(x, t|\theta) < 1$ if $t = 0$; and $-1 < \delta(x, t) < 0$ if $t = 1$, providing a "soft" measure of the signed error.

SGD-Based Training. The SGD-based training algorithm for logistic regression is summarized in Algorithm 6.2.

Algorithm 6.2: SGD-based ML training for logistic regression

initialize $\theta^{(1)}$ and $i = 1$
while *stopping criterion not satisfied* **do**
  pick a mini-batch $\mathcal{S}^{(i)} \subseteq \{1, \ldots, N\}$ of $S^{(i)}$ samples from the data set uniformly at random
  obtain next iterate via SGD as

$$\theta^{(i+1)} \leftarrow \theta^{(i)} - \gamma^{(i)} \frac{1}{S^{(i)}} \sum_{n \in \mathcal{S}^{(i)}} \underbrace{\left(\sigma\left(\left(\theta^{(i)} \right)^T u(x_n) \right) - t_n \right)}_{\text{mean error } \delta(x_n, t_n|\theta^{(i)})} \cdot \underbrace{u(x_n)}_{\text{feature vector}}$$

  set $i \leftarrow i + 1$
end
return $\theta^{(i)}$

Example 6.6

With the training data (6.19), we wish to run SGD, via the delta rule, for the ML training of a logistic regression model with feature vector $u(x) = x$ ($D' = 2$). We set the initialization as $\theta^{(1)} = -1$, the mini-batch size as $S^{(i)} = 1$, and the learning rate as $\gamma^{(i)} = 0.1$.

At iteration $i = 1$, we have the soft prediction $p(t = 1|x = 2, \theta^{(1)}) = \sigma(-1 \cdot 2) = 0.12$, which implies the hard prediction $\hat{t}^{MAP}(2|-1) = 0 \neq t_1 = 1$. Note that the decision is quite confident, assigning a probability 0.88 to the wrong decision. The update is given as

$$\theta^{(2)} \leftarrow -1 - 0.1 \underbrace{(\sigma(-2) - 1)}_{\text{mean error } \delta(2, 1|-1) = -0.88} \cdot (2) = -0.82. \tag{6.35}$$

At iteration $i = 2$, we then have the soft prediction $p(t = 1|x = -2, \theta^{(2)}) = \sigma(-0.82 \cdot (-2)) = 0.84$, which implies the hard prediction $\hat{t}^{MAP}(-2|-0.82) = 1 \neq t_2 = 0$. The decision is again quite confident, assigning a probability 0.84 to the wrong decision; and the update is

$$\theta^{(3)} \leftarrow -0.82 - 0.1 \underbrace{(\sigma(1.64) - 0)}_{\text{mean error } \delta(-2, 0|-0.82) = 0.84} \cdot (-2) = -0.65. \tag{6.36}$$

6.3.5   Kernelized Binary Classifiers

The complexity of training and deploying linear models scales with the number of features, D'. Therefore, when D' is large, they may become too complex to implement. As introduced in Sec. 4.11 for regression, it is possible to implement linear models in an efficient manner even in the presence of large-dimensional features, via "kernelization". Kernel methods are non-parametric, and they are based on a measure of correlation, defined by a kernel function, rather than on the explicit specification and computation of a feature vector $u(x)$. In this subsection, we briefly elaborate on kernel methods for binary classification.

Consider either hard linear models (6.2) trained via regularized ERM (6.10), or probabilistic linear models (6.23) trained via MAP learning, through the delta rule. As discussed, both settings yield the same mathematical problem and iterative algorithm. We choose a regularizer of the form $R(\theta) = r(||\theta||)$, where $r(\cdot)$ is an increasing function of its input. The most classical example is the ℓ_2 regularization $R(\theta) = ||\theta||^2$, which corresponds to choosing $r(a) = a^2$. For MAP learning, this amounts to choosing a prior distribution of the form $p(\theta) \propto \exp(-r(||\theta||))$, i.e., a Gaussian prior.

By (6.10), the function to be minimized is hence of the form

$$L_{\mathcal{D}}(\theta) + \frac{\lambda}{N} R(\theta) = \frac{1}{N} \sum_{n=1}^{N} f(t^{\pm} \theta^T u(x_n)) + \frac{\lambda}{N} r(||\theta||), \tag{6.37}$$

where $f(\cdot)$ represents a convex surrogate loss. The key observation is that any minimizing vector θ for cost function (6.37) must have the form $\theta = \sum_{n=1}^{N} \vartheta_n u(x_n)$ for some coefficients α_n. This is because any additional contribution orthogonal to the feature vectors $\{u(x_n)\}_{n=1}^{N}$ would not change the decision variables $\theta^T u(x_n)$ while increasing the norm $||\theta||$.

It follows that an optimal soft or hard linear predictor depends on the decision variable

$$\theta^T u(x) = \sum_{n=1}^{N} \vartheta_n u(x_n)^T u(x) = \sum_{n=1}^{N} \vartheta_n \cdot \kappa(x, x_n), \tag{6.38}$$

where we have defined the kernel function as in (4.68), i.e.,

$$\kappa(x, x_n) = u(x)^T u(x_n). \tag{6.39}$$

Through (6.38), the optimal predictor can be expressed in terms of the correlations $\kappa(x, x_n)$ between input x and each data point x_n, without requiring the specification of a feature vector $u(\cdot)$. As part of their inductive bias, kernel methods choose a kernel function and assume predictors with decision variables of the form (6.38).

Sparse Kernel Methods. As discussed in Sec. 4.11, sparse kernel methods include in the sum (6.38) only a limited number of data points. An important example is given by SVMs, which assume the hinge loss and ℓ_2 regularization. It can be proved that the solution of the MAP problem has the form (6.38), with sum limited to the **"support vectors"** $u(x_n)$ at the minimal distance from the decision hyperplane.

6.3.6   Generalized Linear Models

Logistic regression is part of a larger family of probabilistic models that are known as **generalized linear models (GLMs)**. This family includes also the polynomial regression model studied in Chapter 4, softmax regression to be studied later in this chapter, and many more. We will return to GLM in Chapter 9.

6.4   Discriminative Non-linear Models for Binary Classification: Multi-layer Neural Networks

We now turn to extensions of the linear predictive models studied in the preceding two sections, whereby the predictions depend on a *non-linear* function of the model parameters θ rather than on the logit $\theta^T u(x)$, which is a linear function of θ. The most popular class of non-linear models consists of **feedforward multi-layer neural networks**, which are the subject of this section. Accordingly, throughout this section, unless stated otherwise, we will use the term "neural networks" to refer to feedforward multi-layer neural networks (see Sec. 6.10 for a discussion of more general neural architectures).

As illustrated in Fig. 6.9, the model class $\mathcal{H}$ of neural networks contains predictors that make decisions based on a vector of features $u(x|\theta)$ that is *not* predetermined as part of the inductive bias as in linear models. Instead, the vector of features $u(x|\theta)$ depends on model parameter vector θ through the multi-layer computational graph in Fig. 6.9, and it can be

Figure 6.9 Computational graph describing a multi-layer feedforward neural network.

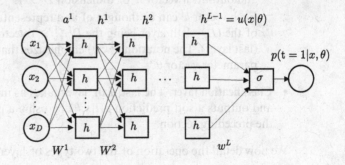

optimized during training alongside the classification hyperplane in the feature vector space. This enables optimization based on data not only for the classification hyperplane, but also for the mapping between input x and feature vector $u(x)$. Neural networks are hence particularly useful for problems in which it is hard to establish a priori good features, such as for computer vision.

(Historically, multi-layer neural networks have been known since the time of the perceptron, but were initially deemed to be too hard to train. In their famous book on perceptrons, Minsky and Papert wrote:

> We consider it to be an important research problem to elucidate (or reject) our intuitive judgement that the extension [to multi-layer neural networks] is sterile. Perhaps some powerful convergence theorem will be discovered, or some profound reason for the failure to produce an interesting "learning theorem" for the multilayered machine will be found.

Things changed with the (re)discovery of backprop for the implementation of SGD-based learning in the mid 1980s, and more recently with the development of dedicated open-source software libraries.)

A note on terminology: In some studies, the input x is referred to as the (input) feature vector, while the vector $u(x|\theta)$ produced by the neural network is referred to as "representation" of the input. This terminology reflects the fact that one could indeed add a first, fixed computational block that extracts a "feature" vector x based on the actual input data fed to the network. This preliminary feature extraction step is good practice if one has information about a useful data representation, and is necessary if the input data are non-numerical, such as text. Here, we will use the term "input" for x, and will refer to $u(x)$ as the "feature vector" for consistency with the terminology for linear models in the previous sections of this chapter.

6.4.1 Multi-layer Feedforward Neural Network Model Class

The model class of feedforward neural networks includes models whose computational graph is illustrated in Fig. 6.9 and described next. When the number of layers, L, is large, the model class may also be referred to as implementing **deep learning**, although we will not be using this term explicitly in this chapter. As seen in Fig. 6.9, a multi-layer feedforward neural network – or "neural network" for short – consists of $L - 1$ feature-extraction layers and one final classification layer, which are composed as follows:

- $L - 1$ **feature-extraction layers**.
  - The first layer takes input x and produces a vector h^1 of dimension $D^1 \times 1$.
  - Each successive lth layer takes input h^{l-1}, of dimension $D^{l-1} \times 1$, from the previous layer and outputs a vector h^l of dimension $D^l \times 1$.
  - Each vector h^l can be thought of as a representation of the input x, with the output h^{L-1} of the $(L-1)$th layer being the $D^{L-1} \times 1$ vector of features $u(x|\theta)$ used by the classifier (last layer). The notation $u(x|\theta)$ emphasizes that the feature vector depends on the model parameter vector θ.

- **Classification layer**. The last, Lth, layer takes as input the vector of features $h^{L-1} = u(x|\theta)$, and outputs a soft prediction $p(t|x,\theta)$ by using a probabilistic linear classifier, as studied in the preceding section.

We now detail the operation of the two types of layers.

Feature-Extraction Layers. Each feature-extraction layer $l = 1, 2, \ldots, L - 1$ takes as input a $D^{l-1} \times 1$ vector h^{l-1} from the previous layer and applies the cascade of two transformations:

- the **matrix–vector multiplication**

$$a^l = W^l h^{l-1}, \tag{6.40}$$

with a $D^l \times D^{l-1}$ matrix of trainable weights W^l, which produces the $D^l \times 1$ vector of **pre-activations** a^l; and

- the application of the **element-wise non-linear transformation**

$$h^l = \begin{bmatrix} h^l_1 \\ \vdots \\ h^l_{D^l} \end{bmatrix} = \begin{bmatrix} h(a^l_1) \\ \vdots \\ h(a^l_{D^l}) \end{bmatrix} \tag{6.41}$$

for some **activation function** $h(\cdot)$, which produces the vector of **post-activations** h^l.

One can think of the feature-extraction layers as producing more and more informative representations $h^1, h^2, \ldots, h^{L-1} = u(x|\theta)$ of the input vector x. "Informativeness" here refers to the ultimate goal of the network to classify the data, i.e., to infer the target variable t. Note that the feature vector $u(x|\theta)$ only depends on matrices $W^1, \ldots, W^{L-1}$, while the model parameter vector θ also includes the last layer's weight vector w^L. To emphasize this dependence, we may also write $u(x|W^{1:L-1})$, with $W^{1:L-1}$ being a shortcut notation for the set of matrices $W^1, \ldots, W^{L-1}$. Incidentally, this hierarchy of representations is in line with dominant theories of the operation of the neocortex in processing sensory inputs.

Activation Function. The activation function $h(\cdot)$ is a single-variable function that takes a scalar as the input – an entry in the pre-activation vector for a layer – and outputs a scalar – an entry in the post-activation vector. There are various choices for the activation function, such as those detailed in Table 6.2 and illustrated in Fig. 6.10.

Among other considerations around the choice of the activation function, a particularly important aspect is the usefulness of the resulting gradients for learning via SGD. Early implementations of neural networks used a sigmoid activation function. This choice, however, yields vanishing derivatives for pre-activations a whose absolute value is sufficiently large (i.e., larger than 5 or smaller than -5). In contrast, a ReLU yields non-vanishing derivatives for all positive pre-activations a, while giving a zero derivative for negative pre-activations. This helps explain the preference for ReLU as compared to the sigmoid activation function. The leaky ReLU provides non-vanishing gradients for both positive and negative pre-activations as long as one chooses $\alpha > 0$.

Table 6.2 Examples of activation functions

activation function	$h(a)$
sigmoid	$\sigma(a)$
hyperbolic tangent	$\tanh(a)$
rectified linear unit (ReLU)	$\max(0, a)$
leaky ReLU	$\max(\alpha a, a)$ with $\alpha \in [0, 1]$

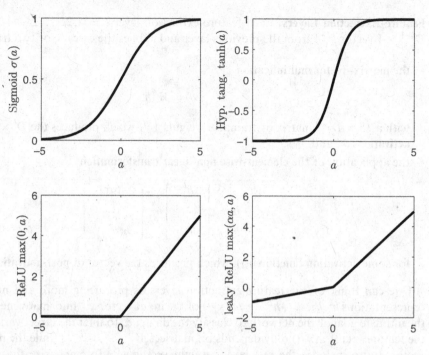

Figure 6.10 Examples of activation functions (the leaky ReLU is shown for $\alpha = 0.2$).

Classification Layer. The classification layer – the last, Lth, layer – takes the vector of features $u(x|\theta) = h^{L-1}$ and applies a probabilistic linear classifier. Using the logistic regression classifier, this yields

$$p(\mathsf{t} = 1|x, \theta) = \sigma(a^L), \tag{6.42}$$

where we have defined the **logit**

$$a^L = (w^L)^T h^{L-1} \tag{6.43}$$

for a $D^{L-1} \times 1$ trainable vector w^L.

Summarizing, and Counting the Number of Parameters. Table 6.3 provides a summary of the operation of a neural network and of the involved parameters. Accordingly, the model parameter vector θ is defined as

$$\theta = \{W^1, W^2, \ldots, W^{L-1}, w^L\}, \tag{6.44}$$

and we have $\sum_{l=1}^{L-1} (D^{l-1} \times D^l) + D^{L-1}$ parameters to train, with $D^0 = D$. The number L of layers and the sizes D^l of the intermediate representations, with $l = 1, 2, \ldots, L-1$, are hyperparameters to be set using domain knowledge or validation.

Relationship to Linear Models. If the activation function $h(\cdot)$ is linear – i.e., $h(a) = a$ – then the overall operation of a neural network is equivalent to that of the linear predictors studied in the preceding section. This is because, with this choice of activation function, the relationship between decision variable and model parameter vector becomes linear. The distinctive new element introduced by neural network models is hence the presence of non-linear activation functions.

Table 6.3 Summary of operation of a (fully connected) feedforward multi-layer neural network

layer	input	pre-activation	post-activation
1	$x\,(D \times 1)$	$a^1 = W^1 x\,(D^1 \times 1)$	$h^1 = h(a^1)\,(D^1 \times 1)$
2	$h^1\,(D^1 \times 1)$	$a^2 = W^2 h^1\,(D^2 \times 1)$	$h^2 = h(a^2)\,(D^2 \times 1)$
$\vdots$		$\vdots$	$\vdots$
l	$h^{l-1}\,(D^{l-1} \times 1)$	$a^l = W^l h^{l-1}\,(D^l \times 1)$	$h^l = h(a^l)\,(D^l \times 1)$
$\vdots$		$\vdots$	$\vdots$
$L-1$	$h^{L-2}\,(D^{L-2} \times 1)$	$a^{L-1} = W^{L-1} h^{L-2}\,(D^{L-1} \times 1)$	$h^{L-1} = h(a^{L-1})\,(D^{L-1} \times 1)$
L	$u(x|\theta) = h^{L-1}\,(D^{L-1} \times 1)$	$a^L = (w^L)^T u(x|\theta)\,(1 \times 1)$	$p(\mathrm{t} = 1|x,\theta) = \sigma(a^L)$

A linear model implementing logistic regression can be thought of as a special case of a neural network with one layer ($L = 1$) – namely the classification layer – taking as input the vector of features $u(x)$. Equivalently, it can be thought of as a two-layer neural network ($L = 2$), in which the first layer has fixed weights, and hence the vector of features $u(x|\theta)$ is fixed. Note that this second interpretation only holds if the feature vector $u(x|\theta)$ can indeed be computed as the composition of a matrix–vector multiplication and an element-wise non-linearity.

Connectivity. The multi-layer feedforward neural network model presented in this subsection assumes **full connectivity** between two successive layers. This is in the sense that each output of a layer depends, in general, on all of the layer's inputs. In some problems, it is useful to allow for **partial connectivity** by including edges in the computational graph of Fig. 6.9 only between specific subsets of units in successive layers. These subsets may be selected as a result of model selection via validation – a process also known as **pruning**.

6.4.2 Neurons, Synapses, and Computational Graph

The name "neural network" highlights a useful interpretation of the model in Fig. 6.9 as a network of individual computational blocks referred to as **neurons**. The terminology reflects the biological inspiration for the model, which captures the locality and parallelism of common models of brain operation in neuroscience. In the computational graph in Fig. 6.9, for the feature-extraction layers, a neuron corresponds to the computational block that evaluates an entry of the post-activation vector; while the only neuron in the classification layer outputs the final soft prediction. Links between neurons in successive layers are referred to as **synapses**.

Accordingly, for each feature-extraction layer $l = 1, \ldots, L - 1$, we have D^l neurons, while for the classification layer we have a single neuron. Each neuron in the feature-extraction layers takes as input the output vector h^{l-1} from the previous layer – or the input x for the first layer $l = 1$ – and outputs one of the elements of the post-activation vector h^l. The single neuron in the classification layer takes as input the feature vector $h^{L-1} = u(x|\theta)$ and outputs the soft prediction $p(\mathrm{t} = 1|x,\theta)$.

Importantly, all neurons in the same layer can operate in parallel, without the need to exchange information. However, the operation of the network is sequential: Neurons in layer l need to wait for the outputs produced by the neurons in layer $l - 1$ to commence computation.

Figure 6.11 Computational graph for neuron j in a feature-extraction layer $l = 1, \ldots, L-1$.

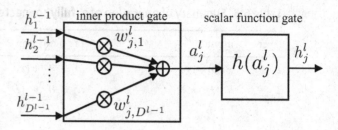

Feature-Extraction Layers. Let us first elaborate on the operation of the neurons in the feature-extraction layers $l = 1, \ldots, L-1$. To this end, denote as $w_j^l = [w_{j,1}^l, \ldots, w_{j,D^{l-1}}^l]^T$ the $D^{l-1} \times 1$ vector containing the jth row of the weight matrix W^l. Each neuron $j = 1, \ldots, D^l$ in layer l takes as input the $D^{l-1} \times 1$ vector h^{l-1} from the previous layer, and outputs

$$h_j^l = h \left(\underbrace{(w_j^l)^T h^{l-1}}_{a_j^l} \right). \tag{6.45}$$

Therefore, as illustrated in Fig. 6.11, each neuron j in feature-extraction layer $l = 1, 2, \ldots, L-1$ consists of the composition of an inner product gate, computing the activation a_j^l, and a scalar function gate, implementing the activation function. We refer to Sec. 5.15 for additional discussion on computational blocks, or gates, in the context of backprop.

Classification Layer. The neuron in the classification layer, i.e., in the Lth layer, takes as input the feature vector $h^{L-1} = u(x|\theta)$, computes the logit

$$a^L = (w^L)^T h^{L-1}, \tag{6.46}$$

and produces the probability

$$p(\mathrm{t} = 1|x, \theta) = \sigma \left(a^L \right). \tag{6.47}$$

Copier Gates. To complete the list of computational gates that are needed to describe basic (feedforward) neural network architectures, we need to introduce copier gates. A copier gate simply reproduces its scalar input to its multiple scalar outputs. This gate is needed since the output of a neuron in layers $l = 1, 2, \ldots, L-2$ is broadcast to all neurons in the next layer $l+1$ (or to a subset of such neurons in the presence of partial connectivity). This is illustrated in Fig. 6.12, which shows a neural network with $L = 3$ layers, $D^1 = D^2 = 2$, and a two-dimensional input ($D = 2$). In this case, the computational graph that produces the soft predictor $p(\mathrm{t} = 1|x, \theta)$ requires the use of a copier gate in order to transfer the outputs of the neurons in layer $l = 1$ to the inputs of the neurons in layer $l = 2$.

Other Gates. More complex network structures, some of which are discussed in Sec. 6.10, require additional gates. For instance, several state-of-the-art architectures include blocks that evaluate the product of the outputs of neurons. This operation – also known as **gating** – enables the implementation of **attention** mechanisms.

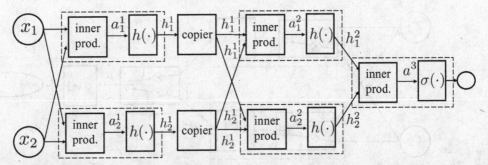

Figure 6.12 Computational graph for a three-layer neural network, illustrating the role of the copier gate.

Introducing "Biases". It is common to add "constant neurons" that output 1 in each layer. This allows the computation of pre-activation functions of the form

$$a^l = W^l h^{l-1} + b^l, \qquad (6.48)$$

where we have defined the $D^l \times 1$ "bias" vector $b^l = [b_1^l, \ldots, b_{D^l}^l]^T$. The parameter b_j^l represents the weight assigned by each jth neuron to the "constant neuron" in the previous layer $l - 1$. We will not consider biases explicitly in what follows, although the analysis carries over directly to this case.

On Biological Plausibility. While the model class of neural networks is indeed inspired by the locality and parallelism of the operation of biological brains, it should be emphasized that the analogy is rather limited. Biological neurons operate over time in a complex dynamic fashion, exchanging action potentials, or spikes. In contrast, neurons in a neural network operate in a static fashion, exchanging real numbers. In this regard, the use of a ReLU non-linearity may be considered as a step in the direction of accounting for some biological constraints. In fact, with a ReLU activation, the output of each neuron is forced to be non-negative, and hence it may be seen as a rough approximation of the (non-negative) *rate* at which a biological neuron counterpart produces spikes in a certain interval of time.

Example 6.7

Consider a neural network with $L = 2$ layers and a two-dimensional input ($D = 2$), having weight matrix

$$W^1 = \begin{bmatrix} 1 & -1 \\ -2 & 1 \end{bmatrix} \qquad (6.49)$$

and classification vector

$$w^2 = \begin{bmatrix} -1 \\ 1 \end{bmatrix}. \qquad (6.50)$$

The computational graph that produces the soft predictor $p(t = 1|x, \theta)$ is shown in Fig. 6.13.

Given this network, assuming ReLU activation functions $h(\cdot)$ for the feature-extraction layers, let us carry out inference for the input vector $x = [1, -1]^T$ by evaluating soft and hard predictors. To this end, running the network in the forward direction yields the results in Fig. 6.14. Accordingly, the network predicts $\hat{t}^{MAP}(x|\theta) = 0$ with confidence level $p(t = 0|x, \theta) = 0.88$.

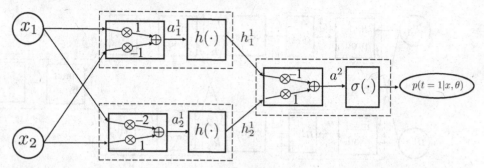

Figure 6.13 Computational graph producing the prediction $p(t = 1|x, \theta)$ for the two-layer neural network with weight matrix (6.49) and weight vector (6.50).

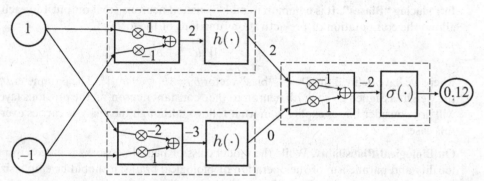

Figure 6.14 Inference on the two-layer neural network with weight matrix (6.49), weight vector (6.50), and ReLU activation functions, given the input vector $x = [1, -1]^T$.

Batch Normalization. We finally remark that in some implementations of neural network models, the prediction $p(t|x, \theta)$ depends not only on the current input x but also on the inputs for a whole batch of data points. Specifically, with **batch normalization**, the batch of data is used to normalize the inputs to each layer so that the empirical mean over the batch is 0 and the empirical variance over the batch is 1. See Recommended Resources, Sec. 6.14, for further discussion.

6.4.3   Training Neural Networks via Backprop: Problem Definition and Computational Graph

The MAP learning problem for neural networks can be stated as

$$\min_{\theta} \left\{ L_{\mathcal{D}}(\theta) + \frac{\lambda}{N} R(\theta) = \frac{1}{N} \sum_{n=1}^{N} \ell_n(\theta) + \frac{\lambda}{N} R(\theta) \right\}, \tag{6.51}$$

where we have written the log-loss for the nth data point (x_n, t_n) in the more compact notation

$$\ell_n(\theta) = -\log \sigma(t_n^{\pm} \cdot \underbrace{(w^L)^T u(x_n|\theta)}_{\text{logit } a^L}). \tag{6.52}$$

Recall that the feature vector $u(x|\theta)$ depends on the weight matrices $W^1, W^2, \ldots, W^{L-1}$ for the feature-extraction layers, and that the vector θ of model parameters $\theta = \{W^1, W^2, \ldots, W^{L-1}, w^L\}$ includes also the weights w^L of the classification layer. As mentioned, we can use the notation $u(x|W^{1:L})$ to make this dependence more explicit.

To implement MAP learning via SGD, we have to compute the gradient

$$\nabla_\theta \ell(\theta) = \nabla_\theta(-\log \sigma(t^{\pm} \cdot ((w^L)^T u(x|W^{1:L-1})))) \tag{6.53}$$

of the log-loss for any sample (x, t) with respect to all the model parameters in θ. As in (6.53), for the rest of this section we will drop the sample index n to lighten the notation. The gradient $\nabla_\theta \ell(\theta)$ includes:

- all the partial derivatives $\partial \ell(\theta)/\partial w^l_{j,d}$ for neurons $j = 1, \ldots, D^l$ in the feature-extraction layers $l = 1, 2, \ldots, L - 1$ (setting $D^0 = D$) with $d = 1, \ldots, D^{l-1}$; as well as
- all partial derivatives $\partial \ell(\theta)/\partial w^L_d$ for $d = 1, \ldots, D^{L-1}$ for the last layer.

To compute the gradient (6.53) for a given θ, an efficient solution is to use the reverse mode of automatic differentiation, also known as **backprop**. This was described in detail in Sec. 5.15 for general computational graphs, and we will specialize it in the rest of this section as a means to evaluate the gradient (6.53) for neural networks.

To enable the use of backprop, we first need to define the **computational graph** that obtains the loss function $\ell(\theta)$. It is easy to see that this is the same computational graph that produces the soft prediction $p(t = 1|x, \theta)$, with the only difference that the sigmoid scalar gate in the last layer is replaced by a scalar gate that computes the loss function $\ell(\theta) = -\log \sigma(t^{\pm} a^L)$.

Example 6.8

For the neural network in Example 6.7, as we have seen, the computational graph that produces the predictive distribution $p(t = 1|x, \theta)$ is shown in Fig. 6.13. The corresponding computational graph that computes the loss function $\ell(\theta)$ on a training example (x, t) is shown in Fig. 6.15. Note that the computational graph in Fig. 6.13, which is used for inference, depends only on the input x. In contrast, the computational graph in Fig. 6.15, which is to be used for training, depends also on the target variable t.

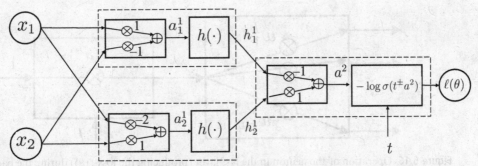

Figure 6.15 Computational graph producing the loss function $\ell(\theta)$ on a training example (x, t) for the two-layer neural network with weight matrix (6.49) and weight vector (6.50).

6.4.4   Training Neural Networks via Backprop: Forward Pass and Backward Pass

Evaluating the gradient (6.53) via backprop requires us to carry out forward and backward passes. The forward pass evaluates the loss function $\ell(\theta)$ by following the computational graph described in the preceding subsection. Then, in the backward pass, we replace each computational block in the computational graph by its corresponding backward version. Since a neural network is composed of inner product, scalar function, and copier gates, we only need the backward versions of these gates, which were all detailed in Sec. 5.15.

Accordingly, in the backward pass, the neuron in the last layer, which computes $\ell(\theta)$, is replaced by the backward block illustrated in Fig. 6.16. In fact, the backward version of the inner product gate is as shown in Fig. 5.23, and the local derivative for the scalar gate producing $\ell(\theta) = -\log\sigma(t^{\pm}a^L)$ is given as

$$\frac{d}{da^L}(-\log\sigma(t^{\pm}a^L)) = \sigma(a^L) - t = \delta^L. \tag{6.54}$$

In (6.54), we have defined δ^L as the **error** computed at the last layer. Note that the error $\delta^L = \sigma(a^L)$ is a scalar quantity and appears also in the delta rule (6.33) for logistic regression.

In a similar way, each internal neuron in the feature-extraction layers applies the backward versions of the scalar (activation) gate and of the inner product gate, as seen in Fig. 6.17. Useful examples of the derivatives of the (scalar) activation function are summarized in Table 6.4.

Finally, the backward version of the copier block can be seen in Fig. 5.24. Note that the backward pass uses the same weights of the forward pass.

With reference to Figs. 6.16 and 6.17, having completed backprop, by the properties of automatic differentiation (Sec. 5.15), we obtain the partial derivatives

$$\delta_j^l = \frac{\partial\ell(\theta)}{\partial a_j^l} \tag{6.55}$$

Table 6.4   Derivatives of some activation functions

activation function	$h'(a)$
sigmoid	$\sigma(a)(1-\sigma(a)) = \sigma(a)\sigma(-a)$
rectified linear unit (ReLU)	$h'(a) = \mathbb{1}(a>0)$
leaky ReLU	$\mathbb{1}(a>0) + \alpha\mathbb{1}(a<0)$

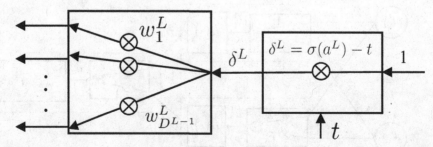

Figure 6.16 Operation of the neuron in the last layer (producing the loss $\ell(\theta)$) during the backward pass of backprop. The term δ^L is the error, also known as mean error using the terminology introduced for logistic regression (cf. (6.33)).

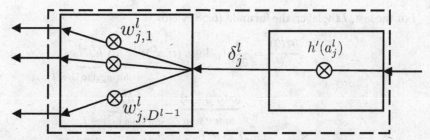

Figure 6.17 Operation of each neuron j in layer $l = 1, \ldots, L-1$ in the feature-extraction layers during the backward pass of backprop. The term δ_j^l is known as backpropagated error.

with respect to the pre-activation of each neuron j in layer $l = 1, \ldots, L-1$, as well the derivative

$$\delta^L = \frac{\partial \ell(\theta)}{\partial a^L} \tag{6.56}$$

for layer L. By analogy with logistic regression, the term δ^L is referred to as the **error** (or mean error, see (6.33)), while the terms δ_j^l are known as **backpropagated errors**, each corresponding to a neuron j in layer $l = 1, \ldots, L-1$.

6.4.5 Training Neural Networks via Backprop: Combining Forward and Backward Passes

The remaining question is: How do we obtain the desired partial derivatives with respect to the weight vector θ from (6.55) and (6.56) in order to compute the gradient (6.53)? This can be done using the chain rule, as we discuss in this subsection.

Consider any neuron j in layer l, from either feature-extraction or classification layers. Let us momentarily drop the subscript j and superscript l in order to highlight the generality of the rule. Accordingly, we write the pre-activation of any neuron as

$$a = (w^T)h^-, \tag{6.57}$$

where h^- is the vector of inputs from the previous layer and w is the weight vector for the given neuron. Let us also write δ for the error or the backpropagated error, depending on whether we consider the last layer or intermediate layers, respectively

The chain rule says that the partial derivative with respect to (wrt) the dth entry of vector w is given as

$$\underbrace{\frac{\partial \ell(\theta)}{\partial w_d}}_{\text{loss wrt weight}} = \underbrace{\frac{\partial \ell(\theta)}{\partial a}}_{\text{loss wrt pre-activation}} \cdot \underbrace{\frac{\partial a}{\partial w_d}}_{\text{pre-activation wrt weight}}$$

$$= \delta \cdot h_d^-, \tag{6.58}$$

where h_d^- is the dth element of vector h^- in (6.57). It follows that the partial derivative with respect to a weight is the product of two terms: the (backpropagated) error δ from the backward pass for the given neuron and the input h_d^- from the forward pass corresponding to the given weight w_d. We now specialize this general formula to the last layer and to the classification layers.

For the last, Lth, layer, the formula (6.58) yields

$$\frac{\partial \ell(\theta)}{\partial w_d^L} = \frac{\partial}{\partial a^L}(-\log \sigma(t^\pm a^L)) \cdot \underbrace{h_d^{L-1}}_{d\text{th input to layer } L}$$

$$= \underbrace{(\sigma(a^L) - t)}_{\delta^L, \text{error at layer } L} \cdot \underbrace{h_d^{L-1}}_{d\text{th input to layer } L}$$

$$= \delta^L \cdot h_d^{L-1}. \tag{6.59}$$

For layers $l = 1, 2, \ldots, L-1$, using the same formula, the partial derivative for the dth weight of neuron j yields

$$\frac{\partial \ell(\theta)}{\partial w_{j,d}^l} = \delta_j^l \cdot h_d^{l-1}. \tag{6.60}$$

In short, for every jth neuron, we have

$$\text{derivative wrt } (j, d)\text{th weight} = j\text{th backprop. error} \cdot d\text{th input}, \tag{6.61}$$

where the backpropagated error is obtained from the backward pass and the dth input from the forward pass.

6.4.6 · Training Neural Networks via Backprop: A Summary

To sum up, backprop computes the gradient $\nabla \ell(\theta)$ on an example (x, t) as follows.

1. **Forward pass**. Given (x, t), compute $a^1, h^1, a^2, h^2, \ldots, a^L$ from first to last layer on the computational graph.
2. **Backward pass**. Using the backward counterpart of the computational graph, from the last to the first layer, compute the (scalar) error $\delta^L = (\sigma(a^L) - t)$ and the backpropagated error vectors $\delta^{L-1}, \ldots, \delta^1$ with $\delta^l = [\delta_1^l, \ldots, \delta_{D^l}^l]^T$, and evaluate the partial derivatives

$$\frac{\partial \ell(\theta)}{\partial w_d^L} = \delta^L \cdot h_d^{L-1} \text{ and} \tag{6.62}$$

$$\frac{\partial \ell(\theta)}{\partial w_{j,d}^l} = \delta_j^l \cdot h_d^{l-1}. \tag{6.63}$$

Backprop can be equivalently formulated in **vector form** as follows. Recall that $\odot$ is the element-wise product, and consider the application of function $h(\cdot)$ element-wise.

1. **Forward pass**. Given (x, t), compute $a^1, h^1, a^2, h^2, \ldots, a^L$ from first to last layer on the computational graph.
2. **Backward pass**. From the last to the first layer, compute backpropagated errors

$$\delta^L = (\sigma(a^L) - t) \tag{6.64}$$

$$\delta^{L-1} = (h'(a^{L-1}) \odot w^L)\delta^L \tag{6.65}$$

$$\delta^l = h'(a^l) \odot ((W^{l+1})^T \delta^{l+1}) \text{ for } l = L-2, \ldots, 1, \tag{6.66}$$

where we set $h^0 = x$ and $h'(\cdot)$ represents the element-wise derivative. Then, evaluate the gradients

$$\nabla_{w^L}\ell(\theta) = \delta^L \cdot h^{L-1} \tag{6.67}$$

$$\nabla_{W^l}\ell(\theta) = \delta^l \cdot (h^{l-1})^T \text{ for } l = L-1, L-2, \dots, 1. \tag{6.68}$$

Example 6.9

For the neural network with computational graph in Fig. 6.15, we now evaluate the gradient of the loss at the indicated values of the model parameters for input $x = [1, -1]^T$ and label $t = 0$. First, the forward pass is illustrated in Fig. 6.18. The inputs to each layer are shown in bold font to emphasize that they will be needed to compute the gradient via (6.59) and (6.60).

Then, we perform the backward pass as shown in Fig. 6.19. Note that we have the error $\delta^2 = \sigma(a^2) - t = 0.12 - 0 = 0.12$, and that we have used the fact that ReLUs have local derivatives equal to 1 if the input (in the forward phase) is positive and 0 otherwise (see Table 6.4). The error and backpropagated errors are shown in bold font. Parts of the graph that are not relevant for the calculation of the gradient are not shown.

Finally, we combine results from the two passes to compute the partial derivatives by multiplying inputs and backpropagated errors as in (6.59) and (6.60) to obtain

$$\nabla_{W^1}\ell(\theta) = \begin{bmatrix} \delta_1^1 \cdot x_1 & \delta_1^1 \cdot x_2 \\ \delta_2^1 \cdot x_1 & \delta_2^1 \cdot x_2 \end{bmatrix} = \begin{bmatrix} -0.12 \cdot 1 & -0.12 \cdot (-1) \\ 0 \cdot 1 & 0 \cdot (-1) \end{bmatrix}$$

$$= \begin{bmatrix} -0.12 & 0.12 \\ 0 & 0 \end{bmatrix} \tag{6.69}$$

and

$$\nabla_{w^2}\ell(\theta) = \begin{bmatrix} \delta^2 \cdot h_1^1 \\ \delta^2 \cdot h_2^1 \end{bmatrix} = \begin{bmatrix} 0.12 \cdot 2 \\ 0.12 \cdot 0 \end{bmatrix}$$

$$= \begin{bmatrix} 0.24 \\ 0 \end{bmatrix}. \tag{6.70}$$

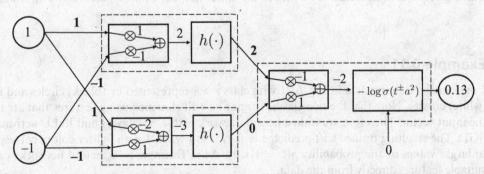

Figure 6.18 For the computational graph in Fig. 6.15, the forward pass is shown for an example with input $x = [1, -1]^T$ and label $t = 0$. Inputs to each neuron are shown in bold to highlight their role in the computation of the partial derivatives in (6.59) and (6.60).

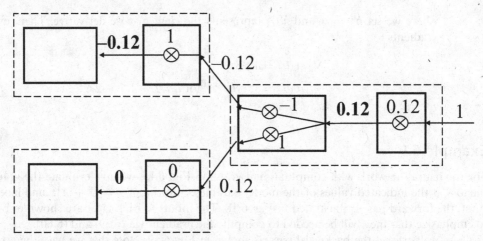

Figure 6.19 For the computational graph in Fig. 6.15, the backward pass is shown for an example with input $x = [1, -1]^T$ and label $t = 0$. Error and backpropagated errors are shown in bold to highlight their role in the computation of the partial derivatives in (6.59) and (6.60). Parts of the graph that are not relevant for the calculation of the gradient are not shown.

Example 6.10

Consider again the XOR problem illustrated in Fig. 6.4. As shown in the bottom part of the figure, with hand-crafted features of size 3, one can effectively separate the data from the two classes. Can a neural network learn useful features directly from training data? To address this question, we train a two-layer network ($L = 2$) with D^1 ReLU neurons in the hidden layer. Recall that D^1 represents the size of the vector of learned features $u(x|\theta)$ to be used by the second, classification, layer. For a data set of size $N = 100$, equally divided between the two classes, we implement SGD with batch size 50 and a decaying learning rate. Fig. 6.20 shows three realizations of the training loss vs. the number of SGD iterations starting from a random realization.

It is observed that, with the choice $D^1 = 3$, the network can indeed be trained to effectively separate the two classes. This is not the case for $D^1 = 2$, although, for a limited number of iterations, the smaller number of parameters to be trained can yield some advantage. With $D^1 = 5$, training requires more iterations owing to the larger number of parameters to be optimized.

Example 6.11

Consider the training data in Fig. 6.21 with class $t = 1$ represented by (black) circles and $t = 0$ by (white) crosses. Note that the two classes cannot be well classified using features that are linear in the input vector. Therefore, we train a neural network with $L = 2$ layers and ReLU activations via SGD. The resulting trained soft predictor is shown in Fig. 6.21 with lighter colors corresponding to larger values of the probability $p(t = 1|x, \theta)$. As in Example 6.10, neural networks can learn suitable features directly from the data.

Figure 6.20 Evolution of the training error across SGD iterations for a neural network with two layers ($L = 2$) and D^1 hidden neurons, given training data for the XOR problem in Fig. 6.4.

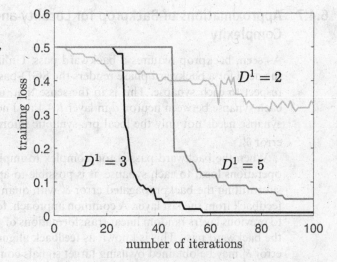

Figure 6.21 Data set for binary classification, with class $t = 0$ represented by (black) circles and $t = 1$ by (white) crosses, along with the soft prediction $p(\mathrm{t} = 1|x, \theta)$ obtained from a trained neural network with $L = 2$ layers.

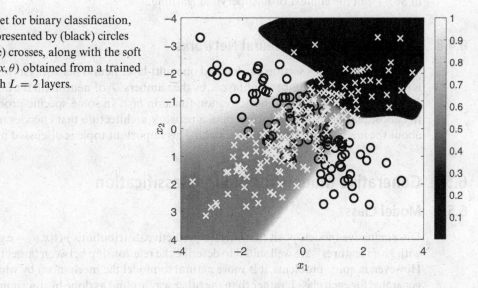

Backprop Computes the Partial Derivatives with Respect to All Variables. As discussed in Sec. 5.15, after the backward pass, one can directly obtain the partial derivatives of the loss $\ell(\theta)$ not only with respect to θ, but also with respect to the input x, the pre-activations, and the outputs of each neuron. Some of these gradients may be of separate interest for some applications. For instance, the gradient $\nabla_x \ell(\theta)$ with respect to the input x is useful for **adversarial learning**. In adversarial learning, at inference time, an adversary may try to modify the input x so as to maximize the loss of the classifier – a task that can be accomplished by leveraging the gradient $\nabla_x \ell(\theta)$. More discussion on adversarial learning can be found in Chapter 15.

6.4.7 Approximations of Backprop for Locality and Reduced Computational Complexity

As seen, backprop requires a backward pass. Unlike the delta rule (6.12) for linear models, the need for a backward phase renders the SGD-based rule for neural networks **non-local** with respect to each synapse. This is in the sense that, in order to apply the update rule (6.60) at each synapse between neuron d in layer $l-1$ and neuron j in layer l, for $l = 2, \ldots, L = 1$, the synapse needs not only the local pre-synaptic information h_d^{l-1}, but also the backpropagated error δ_j^l.

When the backward pass is too complex to implement, and/or one wishes to keep update operations local to each synapse, it is possible to approximate the partial derivative (6.60) by substituting the backpropagated error δ_j^l with quantities that use local information or limited feedback from the last layer. A common approach, for instance, is for the last layer to feed back to previous layers random linear transformations of the error signal $\sigma(a^L) - t$ without requiring the backward pass. This is known as **feedback alignment**. Alternatively, approximations of the error δ_j^l may be obtained by using target signals computed using only information at the given layer l, which are referred to as **local errors**. More discussion on local learning rules can be found in Sec. 7.7 in the context of unsupervised learning.

6.4.8 Beyond Feedforward Neural Networks

The discussion in this section has focused on multi-layer neural networks in which the structure is determined by the number L of layers, by the numbers D^l of neurons in the intermediate layers $l = 1, 2, \ldots, L-1$, and by the activation function $h(\cdot)$. In some specific problems, it is useful to choose, as part of the inductive bias, a network architecture that encodes more information about the mechanism generating the data. This important topic is discussed in Sec. 6.10.

6.5 Generative Models for Binary Classification

6.5.1 Model Class

Discriminative models posit that certain predictive distributions $p(t|x, \theta)$ – e.g., a linear model with given features – are well suited to describe the relationship between target t and input x = x. However, in some problems, it is more natural to model the mechanism by which the data x are generated for each class t, rather than the other way around as done by discriminative models. As an example, consider again the data set in Fig. 6.21: While modeling the mapping between input x and output t is not straightforward, the input data x within each class, i.e., separately for each value of t, appear to be generated according to a distinct multivariate Gaussian distribution.

To address scenarios in which modeling the generation of x from t is more natural than assuming a predictive distribution, generative models assume a model class $\mathcal{H}$ of *joint* distributions $p(x, t|\theta)$ of input vector x and target t. For binary classification, the joint distribution can be written as

$$p(x, t|\theta) = p(t|\theta) \cdot p(x|t, \theta) = \text{Bern}(t|\pi) \cdot p(x|\phi_t), \tag{6.71}$$

where the distribution of the binary label t is Bernoulli with an unknown parameter π, and the **per-class data distribution** $p(x|t, \theta) = p(x|\phi_t)$ depends on a **per-class model parameter vector** ϕ_t. The model parameters are hence given as $\theta = \{\pi, \phi_0, \phi_1\}$. The inductive bias posited by generative models thus amounts to the selection of the per-class distributions $p(x|\phi_t)$ for $t = 0, 1$.

As mentioned, while discriminative models describe the distribution of the output given the input, generative models conversely describe the distribution of the input given the output.

Example 6.12

For the example in Fig. 6.21, it seems reasonable to assume class-dependent distributions of the form

$$p(x|\phi_t) = \mathcal{N}(x|\mu_t, \Sigma_t), \tag{6.72}$$

where the class-dependent model parameters are $\phi_t = \{\mu_t, \Sigma_t\}$ for $t = 0, 1$. Note that, as seen in the preceding subsection, we could apply complex discriminative models such as neural networks, but it may be simpler, and more natural, to adopt a generative model for examples such as this one.

When to Use Generative Models. Generative models assume an inductive bias that is stronger than that of discriminative models, in the sense that they also model the distribution of the input x through the marginal $p(x|\theta)$ of the joint distribution $p(x, t|\theta)$. As discussed, this can be useful if the problem is such that modeling the mechanism generating x from t is more efficient than the other way around. Specifically, if the assumed generative model reflects the (unknown) population distribution of the data, this approach can bring benefits in terms of capacity of the model, reducing bias. Conversely, if the assumed generative model is a poor approximation of the population distribution, the resulting bias can hinder the overall population loss attained by the trained predictor. Furthermore, when data are more naturally modeled by a generative process, using a generative model may help to reduce the number of parameters, potentially mitigating overfitting.

Another advantage of generative models is that they can make use of *unlabeled* data in addition to the labeled data necessary for supervised learning. In fact, using samples that include only input x_n, one can optimize the model parameters that affect the marginal $p(x|\theta)$, e.g., with the goal of minimizing the marginal log-loss $-\log p(x_n|\theta)$. This is an instance of **semi-supervised learning**, which combines unsupervised learning – i.e., learning from unlabeled data – with supervised learning. We will discuss how to deal with training problems of this type, in which the target variables are hidden, in the next chapter.

6.5.2 Ancestral Data Sampling with a Trained Generative Model

Here is another important distinction between discriminative and generative models: A trained generative model $p(x, t|\theta)$ can be used not only to extract a soft predictor $p(t|x, \theta)$, but also to generate "new" data points (x, t) from the model. This is not possible with discriminative models, which only allow us to generate an output prediction t for a given input x = x. In fact, the joint distribution (6.71) defines the following mechanism for data generation, which is known as **ancestral data sampling**:

• Sample a binary class

$$t \sim \text{Bern}(t|\pi) \tag{6.73}$$

for the given probability $\pi \in [0, 1]$.

- Then, for the sampled class t $= t$, draw an input x from a class-dependent parametric distribution

$$x \sim p(x|\phi_t) \tag{6.74}$$

for the given parameters ϕ_0 and ϕ_1.

6.5.3   Inference with a Trained Generative Model

Given a trained generative model $p(x, t|\theta)$, one can obtain the optimal soft predictor by computing the conditional distribution

$$
\begin{aligned}
p(\mathrm{t} = 1|x, \theta) &= \frac{p(x, \mathrm{t} = 1|\theta)}{p(x|\theta)} \\
&= \frac{p(x, \mathrm{t} = 1|\theta)}{p(x, \mathrm{t} = 1|\theta) + p(x, \mathrm{t} = 0|\theta)} \\
&= \frac{\pi \cdot p(x|\phi_1)}{\pi \cdot p(x|\phi_1) + (1 - \pi) \cdot p(x|\phi_0)}.
\end{aligned} \tag{6.75}
$$

Furthermore, under the detection-error loss, the optimal hard predictor yields the MAP prediction

$$\hat{t}^{MAP}(x|\theta) = \begin{cases} 1 & \text{if } p(\mathrm{t} = 1|x, \theta) \geq 0.5 \\ 0 & \text{otherwise.} \end{cases} \tag{6.76}$$

6.5.4   Training Generative Models

By (6.71), the log-loss for an example (x, t) is given as

$$-\log p(x, t|\theta) = -\log \mathrm{Bern}(t|\pi) - \log p(x|\phi_t). \tag{6.77}$$

The log-loss is hence the sum of two separate contributions, one dependent only on π and one only on ϕ_t. This allows us to optimize over parameters π, ϕ_0, and ϕ_1 separately, as long as the regularization function also factorizes in the same way. This is the case for separable regularization functions of the form

$$R(\pi, \phi_0, \phi_1) = R(\pi) + R(\phi_0) + R(\phi_1), \tag{6.78}$$

where we have reused the same symbol for generally different regularization functions.

Accordingly, by (6.77), the training log-loss can be written as

$$
\begin{aligned}
L_{\mathcal{D}}(\theta) &= \frac{1}{N} \sum_{n=1}^{N} (-\log p(x_n, t_n|\theta)) \\
&= \underbrace{\frac{1}{N} \sum_{n=1}^{N} (-\log \mathrm{Bern}(t_n|\pi))}_{L_{\mathcal{D}}(\pi)} + \underbrace{\frac{1}{N} \sum_{\substack{n=1: \\ t_n=0}}^{N} (-\log p(x_n|\phi_0))}_{L_{\mathcal{D}}(\phi_0)} + \underbrace{\frac{1}{N} \sum_{\substack{n=1: \\ t_n=1}}^{N} (-\log p(x_n|\phi_1))}_{L_{\mathcal{D}}(\phi_1)},
\end{aligned} \tag{6.79}
$$

which decomposes into three cost functions that can be separately optimized over π, η_0, and η_1. Importantly, the cost function $L_{\mathcal{D}}(\pi)$ depends on the entire data set, while the cost function $L_{\mathcal{D}}(\phi_t)$ for $t = 0, 1$ only accounts for the number $N[t]$ of training examples (x_n, t_n) with $t_n = t$.

This property, known as **data fragmentation**, can create problems, particularly if there is an imbalance in the data set in terms of the number of data points available for each class.

Example 6.13

With the training set

$$\mathcal{D} = \left\{ \begin{array}{l} (2.7, 1), (-3.1, 0), (1.5, 1), (6.1, 1), (2.8, 1), \\ (-0.2, 0), (0.1, 0), (3.2, 1), (1.1, 1), (-0.1, 0) \end{array} \right\}, \tag{6.80}$$

we have $N = 10$, $N[0] = 4$, and $N[1] = 6$. Therefore, parameter ϕ_0 is trained using only $N[0] = 4$ examples from the data set, and parameter ϕ_1 will use the remaining $N[1] = 6$ examples.

Based on the discussion so far, assuming a separable regularization function (6.78), the training of generative models requires the separate solution of the three problems

$$\min_{\pi} L_{\mathcal{D}}(\pi) + \frac{\lambda}{N} R(\pi),$$

$$\min_{\phi_0} L_{\mathcal{D}}(\phi_0) + \frac{\lambda}{N} R(\phi_0), \text{ and}$$

$$\min_{\phi_1} L_{\mathcal{D}}(\phi_1) + \frac{\lambda}{N} R(\phi_1). \tag{6.81}$$

ML Learning. Let us now elaborate on ML learning, which sets $\lambda = 0$ in problems (6.81). In this case, the first problem in (6.81) has the simple closed-form solution

$$\pi^{ML} = \frac{N[1]}{N}. \tag{6.82}$$

This can be easily checked by solving the first-order stationary condition, since the log-loss for Bernoulli rvs can be proved to be a convex function (see Appendix 6.B). The ML solution (6.82) amounts to an empirical average of the observed labels $\{t_n\}_{n=1}^{N}$.

As discussed in Chapter 4, ML can suffer from overfitting when the amount of data N is limited. For instance, when N is small, it may happen by chance that no observation $t_n = 1$ is available. In this case, ML returns the estimate $\pi^{ML} = 0$, and the model predicts that it is impossible to ever observe $t_n = 1$. This is an instance of the **black-swan problem**: The very existence of a black swan seems impossible if you have never seen one.

While the ML solution for π is given by (6.82), the ML solutions for the per-class model parameters ϕ_t, with $t = 0, 1$ depend on the choice of the class-dependent distribution $p(x|\phi_t)$, and will be discussed later for specific models.

MAP Learning. The outlined overfitting problems of ML learning can be mitigated by implementing MAP learning, i.e., by letting $\lambda > 0$ in problems (6.81). For the first problem in (6.81), as we will detail in Sec. 10.3, it is in fact possible to choose a prior distribution $p(\pi)$, and hence a regularization function $R(\pi)$, that effectively amounts to adding $a > 1$ and $b > 1$ prior "pseudo-observations" to the subsets of actual observations $t_n = 1$ and $t_n = 0$, respectively. (The prior $p(\pi)$ is a beta distribution.) With this choice of prior, the MAP estimate is given as

$$\pi^{MAP} = \frac{N[1] + a - 1}{N + a + b - 2}. \tag{6.83}$$

By allowing a priori for the possibility of observing $t_n = 1$ through the assumption of a pseudo-observations equal to $t = 1$, MAP learning accounts for the epistemic uncertainty in the value of π caused by the availability of limited data. This point will be further elaborated on in Chapter 12 in the context of Bayesian learning.

6.5.5   Quadratic Discriminant Analysis

Quadratic discriminant analysis (QDA) chooses Gaussian conditional distributions, so that we have the model

$$t \sim \text{Bern}(\pi)$$
$$(x|t = t, \phi_t) \sim p(x|\phi_t) = \mathcal{N}(x|\mu_t, \Sigma_t) \tag{6.84}$$

with class-dependent mean vector μ_t and covariance matrix Σ_t, i.e., with per-class model parameters $\phi_t = \{\mu_t, \Sigma_t\}$. With this choice, the ML problems in (6.81), with $\lambda = 0$, for ϕ_0 and ϕ_1 can be easily solved in closed form, yielding

$$\mu_t^{ML} = \frac{1}{N[t]} \sum_{\substack{n=1:\\ t_n=t}}^{N} x_n \text{ and} \tag{6.85}$$

$$\Sigma_t^{ML} = \frac{1}{N[t]} \sum_{\substack{n=1:\\ t_n=t}}^{N} (x_n - \mu_t^{ML})(x_n - \mu_t^{ML})^T. \tag{6.86}$$

In a manner similar to (6.82), these ML solutions amount to empirical averages of the relevant observations labeled as $t = 0$ or $t = 1$. They can be obtained as in (6.82) by applying first-order optimality conditions (see Appendix 6.B).

Example 6.14

We are given data set $\mathcal{D} = \{(x_n, t_n)_{n=1}^4\} = \{(2.1, 1), (-1.5, 0), (-0.3, 0), (5.2, 1)\}$. ML training for a QDA model with class-dependent distributions $\mathcal{N}(\mu_t, \sigma_t^2)$ for $t = 0, 1$ yields

$$\pi^{ML} = \frac{N[1]}{N} = \frac{2}{4} = 0.5 \tag{6.87}$$

$$\mu_0^{ML} = \frac{1}{N[0]} \sum_{\substack{n=1:\\ t_n=0}}^{N} x_n = \frac{1}{2}(-1.5 - 0.3) = -0.9 \tag{6.88}$$

$$\mu_1^{ML} = \frac{1}{N[1]} \sum_{\substack{n=1:\\ t_n=1}}^{N} x_n = \frac{1}{2}(2.1 + 5.2) = 3.65 \tag{6.89}$$

$$(\sigma_0^2)^{ML} = \frac{1}{N[0]} \sum_{\substack{n=1:\\ t_n=0}}^{N} (x_n - \mu_0^{ML})^2 = \frac{1}{2}((-1.5 + 0.9)^2 + (-0.3 + 0.9)^2) = 0.36 \tag{6.90}$$

Figure 6.22 Data set for binary classification, with class $t = 1$ represented by (black) circles and $t = 0$ by (white) crosses, along with the soft prediction $p(t = 1|x, \theta)$ obtained from QDA.

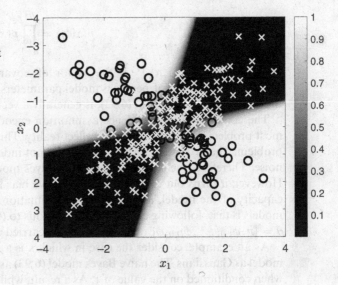

$$(\sigma_1^2)^{ML} = \frac{1}{N[1]} \sum_{\substack{n=1: \\ t_n=1}}^{N} (x_n - \mu_1^{ML})^2 = \frac{1}{2}((2.1 - 3.65)^2 + (5.2 - 3.65)^2) = 2.4. \quad (6.91)$$

Using this trained model to classify a test input $x = 0.1$, we obtain the soft predictor (6.75) as

$$p(t = 1|x, \theta^{ML}) = \frac{\pi^{ML} \mathcal{N}(x|\mu_1^{ML}, (\sigma_1^2)^{ML})}{\pi^{ML} \mathcal{N}(x|\mu_1^{ML}, (\sigma_1^2)^{ML}) + (1 - \pi^{ML}) \mathcal{N}(x|\mu_0^{ML}, (\sigma_0^2)^{ML})}$$

$$= \frac{0.5 \mathcal{N}(x|3.65, 2.4)}{0.5 \mathcal{N}(x|3.65, 2.4) + 0.5 \mathcal{N}(x| - 0.9, 0.36)} = 0.1 < 0.5, \quad (6.92)$$

and, therefore, the optimal hard prediction (under the detection-error loss) is $\hat{t}^{MAP}(x|\theta^{ML}) = 0$.

Example 6.15

For the data set in Fig. 6.21, Fig. 6.22 shows the trained soft predictor $p(t = 1|x, \theta^{ML})$ using QDA. The natural match between the assumed model and the data is reflected in the clear and symmetric demarcation between the two classes produced by the soft predictor.

6.5.6 Naive Bayes Classification

Naive Bayes refers to a general class of generative models whereby the input x is assumed to have independent entries when conditioned on the class index t. Mathematically, **naive Bayes** classifiers for binary classification are based on generative models $p(x, t|\theta)$ with class-dependent parametric distributions

$$p(x|\phi_t) = \prod_{d=1}^{D} p(x_d|\phi_{d,t}), \tag{6.93}$$

where $x = [x_1, \ldots, x_D]^T$ is the vector of covariates, and $\phi_t = [\phi_{1,t}^T, \ldots, \phi_{D,t}^T]^T$ is the corresponding vector of per-class model parameters. Note that each term $\phi_{d,t}$, which describes the corresponding factor in (6.93), is generally a vector of parameters.

The conditional independence assumption encoded by the model (6.93) is strong, and, in most problems, it clearly does not reflect reality. Think, for instance, of an image classification problem: The pixel values for the image are not independent, unless the images represent pure noise. Therefore, the choice of the naive Bayes model generally causes a non-negligible bias. However, it turns out to be a useful inductive bias for many problems as a way to reduce the capacity of the model, and, with it, the estimation error. Another advantage of naive Bayes models is that, following calculations analogous to (6.79), the optimization over all parameters $\theta = \{\pi, \phi_{1,0}, \ldots, \phi_{D,0}, \phi_{1,1}, \ldots, \phi_{D,1}\}$ can be carried out separately for ML training.

As an example, consider the case in which x is a $D \times 1$ vector of continuous rvs, which we model as Gaussians. The naive Bayes model (6.93) assumes that these variables are independent when conditioned on the value of t. As a result, while QDA trains two $D \times 1$ mean vectors and two $D \times D$ covariance matrices, naive Bayes optimizes over two $D \times 1$ mean vectors and two $D \times 1$ vectors of variances, one for each class.

Example 6.16

The **USPS (United States Postal Service)** data set contains 16×16 grayscale images of handwritten digits. Examples of images in this data set can be found in Fig. 6.23. We digitize all images so that each pixel equals 0 (black) or 1 (white). Here, we take only images of handwritten digits 2 and 9, and we consider the problem of classifying these images as a "2" or a "9". To this end, we train a naive Bayes classifier. As mentioned, naive Bayes assumes that each pixel x_d – a Bernoulli rv with probability $\phi_{d,t}$ under class $t \in \{0, 1\}$ – is independent of all other pixels. The two vectors of estimated probabilities ϕ_t^{ML} for $t = 0, 1$ via ML are shown in Fig. 6.24. The figure suggests that, while the naive Bayes model (6.93) is simplistic, the trained model is able to capture differences between the two per-class input data distributions that are useful for classification.

Figure 6.23 Examples of images in the USPS data set of handwritten digits.

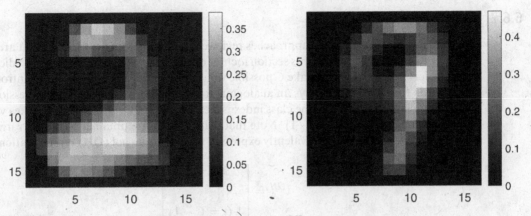

Figure 6.24 Per-class model parameters (probabilities) $\phi_{d,t}^{ML}$ estimated via ML for the classes of handwritten digits 2 and 9 from the USPS data set obtained via naive Bayes training.

6.6 Discriminative Linear Models for Multi-class Classification: Softmax Regression

This section extends linear classifiers from binary classification to multi-class classification. In multi-class classification, the target variable t can take C possible values, rather than only two as in binary classification. As such, each data point x belongs to one of C classes, with the index of the class being specified by a categorical label $t \in \{0, 1, \ldots, C-1\}$. In this section, we will first briefly describe how multiple binary classifiers can be repurposed to serve as multi-class classifiers, and then we focus on the most common approach – softmax regression – that generalizes logistic regression to the case of $C > 2$ classes.

6.6.1 Using Binary Classifiers for Multi-class Classification

A simple approach to tackle the multi-class classification problem is to train several binary classifiers and suitably combine their outputs. There are two main ways to do this:

- **One-vs-all**. Train C binary classifiers to distinguish one class against all others.

- **All-pairs**. Train $C \cdot (C-1)$ binary classifiers to distinguish inputs from each pair of classes.

Both approaches may be computationally inefficient since they require training multiple classifiers. Furthermore, they require some heuristics to combine the outputs of the trained binary classifiers. For instance, for the one-vs-all approach, one could select the class that has the largest score under the trained soft predictor against all other classes.

There are also alternative methods such as techniques based on distributed output representations and **error-correction codes**. With these solutions, each class is represented by a bit string, with each bit being predicted by a distinct binary classifier; and error-correcting codes are used to account for mistakes made by the individual binary predictors in order to produce the final multi-class decision.

6.6.2   Softmax Function

Given the shortcomings of the approaches reviewed in the preceding subsection that are based on binary classifiers, the rest of this section focuses on models that directly output a prediction as a categorical variable t that can take C possible values, one for each class. We start by introducing the softmax function, which plays an analogous role to the sigmoid for logistic regression.

In multi-class classification, the class index is a categorical rv $t \sim \text{Cat}(t|\pi)$ that takes value in a discrete finite set $\{0, 1, \dots, C-1\}$. Note that, with $C = 2$, we obtain a Bernoulli variable. A categorical rv t can also be equivalently expressed using the **one-hot (OH) representation**

$$t^{OH} = \begin{bmatrix} \mathbb{1}(t=0) \\ \vdots \\ \mathbb{1}(t=C-1) \end{bmatrix}, \tag{6.94}$$

so that, when $t = t \in \{0, 1, \dots, C-1\}$, all entries in the one-hot vector t^{OH} are equal to zero except for a 1 in the position corresponding to value t (see Sec. 2.2).

Binary classification models produce a real number a – the logit – that is converted into a probability via a sigmoid function (see (6.27), where $a = \theta^T u(x)$). For multi-class classification, a similar role is played by the softmax function. Specifically, given a real vector $a = [a_0, \dots, a_{C-1}]^T \in \mathbb{R}^C$, the **softmax function** outputs the probability vector

$$\text{softmax}(a) = \text{softmax}(a_0, \dots, a_{C-1})$$

$$= \begin{bmatrix} \frac{\exp(a_0)}{\sum_{k=0}^{C-1} \exp(a_k)} \\ \frac{\exp(a_1)}{\sum_{k=0}^{C-1} \exp(a_k)} \\ \vdots \\ \frac{\exp(a_{C-1})}{\sum_{k=0}^{C-1} \exp(a_k)} \end{bmatrix} \propto \begin{bmatrix} \exp(a_0) \\ \exp(a_1) \\ \vdots \\ \exp(a_{C-1}) \end{bmatrix}. \tag{6.95}$$

The "proportional to" $\propto$ notation here hides the common normalizing term $Z = \sum_{k=0}^{C-1} \exp(a_k)$, which is implicit in the constraint that the probabilities sum to 1, i.e.,

$$\frac{1}{Z} \sum_{k=0}^{C-1} \exp(a_k) = 1. \tag{6.96}$$

Through (6.95), the softmax function converts an arbitrary real vector a into a probability vector in such a way that larger entries in a correspond to (exponentially) larger probabilities.

The softmax function is indeed a *"soft" max*, since $\lim_{\alpha \to \infty} \text{softmax}(\alpha \cdot a_0, \dots, \alpha \cdot a_{C-1})$ equals a one-hot vector in which the entry equal to 1 is in the position corresponding to the class corresponding to the maximum value of a_c, i.e., $\arg\max_c a_c$ (assuming that there is only one maximal value).

As a remark on notation, we will also use the definition

$$\text{softmax}_c(a_0, \dots, a_{C-1}) = \frac{\exp(a_c)}{\sum_{k=0}^{C-1} \exp(a_k)} \tag{6.97}$$

for the cth component of vector output by the softmax function (6.95).

Figure 6.25 Computational graph for softmax regression.

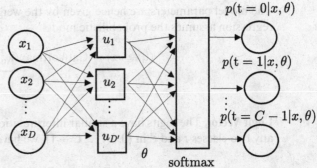

6.6.3   Softmax Regression: Model Class

Softmax regression assumes probabilistic models that implement the computational graph shown in Fig. 6.25. Accordingly, softmax regression first computes a fixed $D' \times 1$ feature vector $u(x)$, as for logistic regression. Then, it evaluates the vector of C decision variables, known as **logits**:

$$\begin{bmatrix} w_0^T u(x) \\ w_1^T u(x) \\ \vdots \\ w_{C-1}^T u(x) \end{bmatrix} = Wu(x), \tag{6.98}$$

where w_c for $c = 0, 1, \ldots, C-1$ is a $D' \times 1$ vector, and we have defined the $C \times D'$ **weight matrix**

$$W = \begin{bmatrix} w_0^T \\ w_1^T \\ \vdots \\ w_{C-1}^T \end{bmatrix}. \tag{6.99}$$

Note that we need C logits for softmax regression, while logistic regression uses the single logit $\theta^T u(x)$.

The logits are converted into a vector of probabilities through the softmax function as

$$\begin{bmatrix} p(t = 0|x, \theta) \\ p(t = 1|x, \theta) \\ \vdots \\ p(t = C-1|x, \theta) \end{bmatrix} = \text{softmax}(w_0^T u(x), \ldots, w_{C-1}^T u(x)) \tag{6.100}$$

$$\propto \begin{bmatrix} \exp(w_0^T u(x)) \\ \exp(w_1^T u(x)) \\ \vdots \\ \exp(w_{C-1}^T u(x)) \end{bmatrix}. \tag{6.101}$$

The model parameters are hence given by the weight matrix, i.e., $\theta = \{W\}$. All in all, softmax regression assumes the probabilistic model

$$(t|x = x) \sim \text{Cat}(\text{softmax}(\underbrace{w_0^T u(x), \ldots, w_{C-1}^T u(x)}_{\text{logits, } Wu(x)})). \tag{6.102}$$

Logit Vector. The logits have a similar interpretation as for logistic regression. In particular, for any two classes t and t' in $\{0, 1, \ldots, C-1\}$, we have

$$w_t^T u(x) - w_{t'}^T u(x) = \log\left(\frac{p(t|x,\theta)}{p(t'|x,\theta)}\right), \tag{6.103}$$

as can be directly checked. So, the difference between logits gives the corresponding **LLR** for the problem of detecting the event $\mathsf{t} = t$ against $\mathsf{t} = t'$ when the observation is $\mathsf{x} = x$.

Optimal Hard Prediction. Under the detection-error loss, the optimal hard predictor for logistic regression is given by the MAP predictor

$$\hat{t}^{MAP}(x|\theta) = \arg\max_{t\in\{0,1,\ldots,C-1\}} p(t|x,\theta) = \arg\max_t w_t^T u(x). \tag{6.104}$$

That is, the hard predictor selects the class with the largest logit.

6.6.4 Training Softmax Regression

The **log-loss** for softmax regression is given as

$$-\log p(t|x,\theta) = -\log \text{softmax}_t(Wu(x))$$

$$= -w_t^T u(x) + \log\left(\sum_{k=0}^{C-1} \exp(w_k^T u(x))\right). \tag{6.105}$$

In a manner similar to logistic regression, the log-loss is a **convex** function of the model parameter matrix W, since it is given by a linear function summed to a log-sum-exp function (see Sec. 5.7.6). The training log-loss is given as

$$L_{\mathcal{D}}(\theta) = \frac{1}{N}\sum_{n=1}^{N}\left(-w_{t_n}^T u(x_n) + \log\left(\sum_{k=0}^{C-1} \exp(w_k^T u(x_n))\right)\right), \tag{6.106}$$

and is also convex in W.

For a sample (x, t), the **gradient of the log-loss** with respect to each vector w_c is computed as

$$\nabla_{w_c}\left(-w_t^T u(x) + \log\left(\sum_{k=0}^{C-1} \exp(w_k^T u(x))\right)\right) = \underbrace{(\text{softmax}_c(Wu(x)) - \mathbb{1}(c = t))}_{\delta_c(x,t|\theta),\text{ mean error for }c\text{th class}} \times \underbrace{u(x)}_{\text{feature vector}}. \tag{6.107}$$

Therefore, the gradient is the product of the **per-class mean error** $\delta_c(x, t|\theta)$ and the feature vector.

The mean errors can be collected in the vector

$$\delta(x, t|\theta) = \begin{bmatrix} \delta_0(x, t|\theta) \\ \delta_1(x, t|\theta) \\ \vdots \\ \delta_{C-1}(x, t|\theta) \end{bmatrix} = \text{softmax}(Wu(x)) - t^{OH}. \tag{6.108}$$

Hence, in a manner similar to logistic regression, the mean error vector is equal to the difference

$$\delta(x, t|\theta) = \text{E}_{\text{t}\sim p(t|x,\theta)}[\text{t}^{OH}] - t^{OH} \tag{6.109}$$

between the average one-hot representation of t under the model and the one-hot representation of the actual label t (cf. (6.34)).

SGD-based training can be derived based on the gradient (6.107). In particular, the update for each vector w_c, with $c = 0, 1, \ldots, C - 1$, at any iteration i is given as

$$w_c^{(i+1)} \leftarrow w_c^{(i)} - \gamma^{(i)} \frac{1}{S^{(i)}} \sum_{n \in \mathcal{S}^{(i)}} \underbrace{\delta_c(x_n, t_n|\theta^{(i)})}_{\text{per-class mean error}} \cdot \underbrace{u(x_n)}_{\text{feature vector}}, \tag{6.110}$$

where we recall that the current model parameters $\theta^{(i)}$ coincide with the weight matrix $W^{(i)}$. This update can be equivalently stated in matrix form as

$$W^{(i+1)} \leftarrow W^{(i)} - \gamma^{(i)} \frac{1}{S^{(i)}} \sum_{n \in \mathcal{S}^{(i)}} \underbrace{\delta(x_n, t_n|\theta^{(i)})}_{\text{mean error vector}} \underbrace{u(x_n)^T}_{\text{feature vector}}. \tag{6.111}$$

6.7 Discriminative Non-linear Models for Multi-class Classification: Multi-layer Neural Networks

Having extended linear models from binary to multi-class classification, in this section we similarly generalize multi-layer neural networks.

6.7.1 Model Class

A neural network model for multi-class classification operates just like one for binary classification except for the last layer, in which we replace the logistic regression block with a softmax regression block. Specifically, as illustrated in Fig. 6.26, the first $L - 1$ layers extract a feature vector $u(x|\theta)$ that depends on weight matrices $W^1, \ldots, W^{L-1}$ as in the architecture for binary classification in Fig. 6.9. The last, Lth, layer computes the probabilities

$$\begin{bmatrix} p(\text{t} = 0|x, \theta) \\ p(\text{t} = 1|x, \theta) \\ \vdots \\ p(\text{t} = C - 1|x, \theta) \end{bmatrix} = \text{softmax}((w_1^L)^T u(x|\theta), \ldots, (w_{C-1}^L)^T u(x|\theta))$$

$$= \text{softmax}(W^L u(x|\theta)), \tag{6.112}$$

Figure 6.26 Computational graph for a feedforward multi-layer neural network implementing multi-class classification.

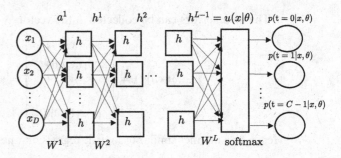

Figure 6.27 Computational graph for the last layer of a feedforward multi-layer neural network implementing multi-class classification.

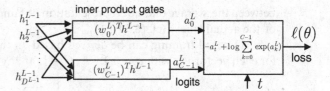

based on the logit vector

$$a^L = W^L u(x|\theta) \tag{6.113}$$

corresponding to the pre-activations of the last layer. In (6.112), we have collected all vectors $\{w_c^L\}$ by rows in matrix W^L. Hence the model parameter includes all the weight matrices as $\theta = \{W^1, \ldots, W^L\}$.

6.7.2   Training Neural Networks

In order to apply backprop, we only need to specify the different operation of the last layer of the computational graph, since the rest of the network is defined in the same way as for binary classification. To this end, in the forward pass, for each example (x, t), the last layer computes the log-loss (cf. (6.105))

$$\ell(\theta) = -(w_t^L)^T u(x|\theta) + \log\left(\sum_{k=0}^{C-1} \exp((w_k^L)^T u(x|\theta))\right)$$

$$= -a_t^L + \log\left(\sum_{k=0}^{C-1} \exp(a_k^L)\right), \tag{6.114}$$

where we have used the vector of logits $a^L = [a_0^L, \ldots, a_{C-1}^L]^T$ defined in (6.113). The operation of the last layer in the forward pass is shown in Fig. 6.27.

In the backward pass, all the computational blocks in the computational graph in Fig. 6.27 are replaced by their corresponding backward versions. For the last layer, whose computational graph is illustrated in Fig. 6.27, this results in the operation depicted in Fig. 6.28 in the backward pass. To derive it, one needs the backward version of the inner product gates in Fig. 5.23; and for the last, scalar, gate, the gradient

Figure 6.28 Backward version of the block in Fig. 6.27.

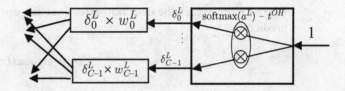

$$\nabla_{a^L}\ell(\theta) = \text{softmax}(a^L) - t^{OH} = \delta^L \,, \tag{6.115}$$

where $\delta^L = [\delta_0, \ldots, \delta_{C-1}]^T$ represents the **error vector** computed at the last layer. SGD-based training can then be applied following the same steps as for binary classification.

6.8   Generative Models for Multi-class Classification

As we have seen in Sec. 6.5, generative models assume a model class $\mathcal{H}$ of joint distributions $p(x, t|\theta)$ of input vector x and target t. The joint distribution for multi-class classification is of the form

$$p(x, t|\theta) = \text{Cat}(t|\pi) \cdot p(x|\phi_t) \tag{6.116}$$

for some probability vector $\pi = [\pi_0, \ldots, \pi_{C-1}]^T$ with $\sum_{c=0}^{C-1} \pi_c = 1$ and per-class model parameters $\phi_0, \phi_1, \ldots, \phi_{C-1}$. Accordingly, the vector of model parameters is given as $\theta = \{\pi, \phi_0, \phi_1, \ldots, \phi_{C-1}\}$. This defines the following **ancestral sampling** mechanism underlying data generation:

1. Sample a categorical rv for the class index

$$\text{t} \sim \text{Cat}(t|\pi); \tag{6.117}$$

2. then, for the sampled class $\text{t} = t \in \{0, 1, \ldots, C-1\}$, draw an input x from a class-dependent parametric distribution

$$\text{x} \sim p(x|\phi_t). \tag{6.118}$$

6.8.1   Inference with a Trained Generative Model

Given a trained generative model $p(x, t|\theta)$, we can obtain a soft predictor by computing the conditional distribution

$$
\begin{aligned}
p(\text{t} = c|x, \theta) &= \frac{p(x, \text{t} = c|\theta)}{p(x|\theta)} \\
&= \frac{p(x, \text{t} = c|\theta)}{\sum_{k=0}^{C-1} p(x, \text{t} = k|\theta)} \\
&= \frac{\pi_c \cdot p(x|\phi_c)}{\sum_{k=0}^{C-1} \pi_k \cdot p(\text{x} = k|\phi_k)}.
\end{aligned}
\tag{6.119}
$$

Furthermore, under the detection-error loss, the optimal hard predictor is given by the MAP predictor

$$\hat{t}^{MAP}(x|\theta) = \arg\max_t\{\pi_t \cdot p(x|\phi_t)\}. \tag{6.120}$$

6.8.2   Training Generative Models

The **log-loss** for an example (x, t) is given as

$$-\log p(x, t|\theta) = -\log\mathrm{Cat}(t|\pi) - \log p(x|\phi_t). \tag{6.121}$$

As in the case of binary classification, as long as the regularization function factorizes, it is possible to optimize separately over parameters π, $\phi_0, \phi_1, \ldots, \phi_{C-1}$. To see this, note that the **training log-loss**

$$\begin{aligned}
L_{\mathcal{D}}(\theta) &= \frac{1}{N}\sum_{n=1}^{N}(-\log p(x_n, t_n|\theta)) \\
&= \underbrace{\frac{1}{N}\sum_{n=1}^{N}(-\log\mathrm{Cat}(t_n|\pi))}_{=L_{\mathcal{D}}(\pi)} + \sum_{c=0}^{C-1}\underbrace{\frac{1}{N}\sum_{\substack{n=1: \\ t_n=c}}^{N}(-\log p(x_n|\phi_c))}_{=L_{\mathcal{D}}(\phi_c)}
\end{aligned} \tag{6.122}$$

decomposes into $C + 1$ separate functions.

Based on (6.122), training generative models requires the solution of $C + 1$ optimization problems. Furthermore, this decomposition indicates that the problem of **data fragmentation** for multi-class classification is more pronounced than for binary classification, since the loss function $L_{\mathcal{D}}(\phi_t)$ only accounts for the number $N[t]$ of training examples (x_n, t_n) with $t_n = t$, and t can take C possible values.

ML and MAP training can be done by following the same steps as for binary classification. In particular, the ML solution for the parameters of the categorical distribution is given as

$$\pi_c^{ML} = \frac{N[c]}{N} \tag{6.123}$$

for $c = 0, 1, \ldots, C - 1$, as can be easily checked by applying the first-order stationary condition.

6.9   Mixture Models

So far in this chapter, we have discussed parametric models of the form $p(t|x, \theta)$ (or $p(x, t|\theta)$ for generative models), for which training aims to identify a single model parameter vector θ to be used for inference. Reliance on a single parameter vector θ is, however, not always desirable. In this chapter, we briefly discuss **mixture models**, which rely instead on the "opinions" of multiple models.

Example 6.17

As an example, consider a class $\mathcal{H}$ of fact checkers that provide the predictive distribution

$$p(t = \text{true/ false}|x = \text{statement}, \theta = \text{fact-checking system}) \qquad (6.124)$$

that a certain input statement x is true or false. Note that θ here is a discrete model parameter that identifies a fact-checking algorithm. Should we use a single fact checker for all statements, or instead account for the "opinions" of all fact checkers in a weighted fashion? Intuitively, the latter option is more appealing. While some fact-checkers may be more accurate for, say, historical events, others may be more accurate for scientific queries. Even considering different fact-checkers targeting the same field of knowledge, the presence of disagreements in the predictions of different algorithms may be a sign of an existing epistemic uncertainty that should be accounted for in order to avoid overconfident predictions.

Model Class. Mixtures of soft predictors are of the form

$$q(t|x) = \sum_{k=1}^{K} \pi_k p(t|x, \theta_k). \qquad (6.125)$$

Accordingly, a mixture model "mixes" the soft predictions produced with K values of the model parameters $\{(\theta_k)_{k=1}^{K}\}$ with weights given by a probability vector $\pi = [\pi_1, \ldots, \pi_K]$, with $\sum_k \pi_k = 1$. One can think of the probability π_k as capturing the degree of reliability of a certain predictor $p(t|x, \theta_k)$. The model parameters to be trained thus include the probability vector π as well as the model parameters θ_k for $k = 1, \ldots, K$. Note that it is also possible to mix hard predictors, or to mix intermediate quantities that determine the final prediction. The reader is referred to the recommended resources for details.

As illustrated by the fact-checking example, at a high level there are two main reasons why one may want to use mixture models:

- **Epistemic uncertainty.** Given a limited availability of data, there may be several models $p(t|x, \theta_k)$ that have similar performance on the training set, while disagreeing outside it. A mixture (6.125) can properly account for the resulting state of epistemic uncertainty by producing a score that is the (weighted) average of all disagreeing predictors.

- **Specialization.** Each component model $p(t|x, \theta_k)$ may be specialized to some different parts of the input and/or output space. In this case, a mixture (6.125) can provide a better "coverage" of the domain of the problem.

As we will discuss in Chapter 12 in the context of Bayesian learning, in gauging the benefits of mixture models, it is useful to distinguish between **well-specified** and **misspecified** model classes $\mathcal{H}$. In the former case, there exists a ground-truth value of the model parameter vector θ_0 such that $p(t|x, \theta_0) = p(t|x)$, where $p(t|x) = p(x, t)/p(x)$ is the population-optimal unconstrained soft predictor; while, in the latter case, the conditional population distribution $p(t|x)$ cannot be realized by any of the soft predictors in the model class $\mathcal{H}$. Epistemic uncertainty is an issue both for well-specified and for misspecified model classes; while specialization can be potentially useful for misspecified models in which there is no single model that can represent the population distribution (irrespective of the amount of data available).

Training Mixture Models. In principle, the training of mixture models can be done via SGD, but this is generally more complex than for the conventional models studied so far. To see why, note that, even if the log-loss $-\log p(t|x, \theta)$ is convex in θ, that is generally no longer the case for the log-loss $-\log(\sum_{k=1}^{K} \pi_k p(t|x, \theta_k))$ of a mixture model. An effective way to train mixture models is to view them as including a **latent, i.e., unobserved, categorical variables** taking C values according to the probability vector π. Techniques to train models with latent variables will be discussed in the next chapter.

There are also various simplified ways to construct mixture models. For example, one could train K models with independent random initializations and random seeds for the selection of mini-batches in SGD, obtaining K model parameters θ_k for $k = 1, \ldots, K$. Then, the corresponding predictive distributions would be averaged to obtain the mixture predictor (6.125). When used on deep neural networks, this approach is known as **deep ensembling**. Alternatively, one could train K models with **resampled data sets**, whereby each data point is selected at random with replacement from the original data set, and then average the predictive distribution in the same way. This approach is known as **bagging** or **aggregated bootstrap**. Yet another approach is to sequentially add models $p(t|x, \theta_k)$ to the mixture – a method known as **boosting**. Boosting trains the next model to be added to the mixture by accounting for the errors made by the current mixed model (6.125).

Example 6.18

For binary classification, suppose that the class $\mathcal{H}$ includes four discriminative models of the form

$$p(t = 1|x, \theta = 1) = \mathbb{1}(x \geq -1)$$
$$p(t = 1|x, \theta = 2) = \mathbb{1}(x \leq 1)$$
$$p(t = 1|x, \theta = 3) = \mathbb{1}(x \leq -1)$$
$$p(t = 1|x, \theta = 4) = \mathbb{1}(x \geq 1). \tag{6.126}$$

The training data set is $\mathcal{D} = \{(-0.2, 1), (-0.6, 1), (0.3, 1), (0.8, 1)\}$. Note that we only have data for the interval $[-0.6, 0.8]$ and that all labels are equal to 1. ML training would choose either $\theta = 1$ or $\theta = 2$ since they both correctly classify all the examples in data set $\mathcal{D}$ (while $\theta = 3$ and $\theta = 4$ incorrectly classify all the examples). Choosing either value is, however, not well justified since both models work equally well on the available data – there is epistemic uncertainty regarding the value of the model parameter. A mixture model could instead set $\pi_1 = \pi_2 = 0.5$, properly accounting for the equal reliability that the two models have on the available data. This leads to very different predictions from the ML solution. With, say, $\theta^{ML} = 1$, we have the soft predictor $p(t = 1|x, \theta^{ML}) = 1$ for all $x \geq -1$. In contrast, with the mixture predictor, we have the soft predictor $0.5p(t = 1|x, \theta = 1) + 0.5p(t = 1|x, \theta = 2)$, which equals 0.5 for all $|x| > 1$ – yielding a more reasonable representation of the epistemic uncertainty.

6.10 Beyond Feedforward Multi-layer Neural Networks

The neural network models studied in Secs. 6.4 and 6.7 do not leverage any specific property of the relationship between input and output that is being learned. The only underlying assumption made in the definition of the inductive bias is that the input–output relationship can be well represented by a neural network of sufficient capacity. Since neural networks are **"universal approximators"**, the assumption is generally valid. However, encoding additional prior information on the structure of the input–output relationship in the architecture of the neural network model can help reduce the size of the model parameter vector to be trained, thereby decreasing the estimation error. If the prior information is accurate, and hence the structure reflects well the actual relationship between input and output, the reduction in estimation error comes at no cost in terms of bias (see Sec. 4.5). The net result is that the number of samples required to achieve a certain level of accuracy can be significantly reduced.

In many problems, the input–output mapping can be reasonably assumed, as part of the inductive bias, to have invariance or equivariance properties that can be conveniently encoded by the model architecture. This section provides a brief introduction to this important topic.

6.10.1 Invariance

Consider a general mapping $t = f(x)$ between input x and output t. This may correspond, e.g., to a predictor or to a feature extractor. In this section, we do not explicitly indicate the dependence of this function on model parameters in order to simplify the notation, but function $f(\cdot)$ is to be interpreted as a machine learning model to be trained. A function $t = f(x)$ is **invariant** to some transformation $T(\cdot)$ if we have the equality

$$f(T(x)) = f(x), \tag{6.127}$$

that is, if applying the transformation to the input does not change the output. Notable instances of invariance include permutation invariance and translation invariance.

Permutation Invariance. In permutation invariance, permuting the order of the entries of the input vector x leaves the output unchanged. The transformation $T(\cdot)$ is hence any permutation of the entries of the input vector. As an example, consider statistics across a population of individuals: In this case, the entries of vector x may correspond to features of different individuals, and the order in which individuals are arranged in the input vector does not affect the output statistics.

A basic class of functions that preserve permutation invariance is of the form

$$f(x) = a(m(x_1), m(x_2), \ldots, m(x_D)), \tag{6.128}$$

where $m(\cdot)$ is a (trainable) "mapping" function, which is applied separately to all entries of the input vector $x = [x_1, \ldots, x_D]^T$, and $a(\cdot)$ is an "aggregation" function whose output does not depend on the order of its inputs. Common examples of aggregation, also known as **pooling**, functions $a(\cdot)$ are mean, sum, and max. Importantly, the mapping functions are shared among all inputs, which is implemented by **"tying"** the trainable parameters to be the same across all mapping functions. (When computing the partial derivative with respect to a tied parameter, the respective partial derivatives for each instance of the parameter must be summed.)

Translational Invariance. Translational invariance applies to tasks such as object recognition in image processing or keyword spotting in speech processing. In either case, the output of the model – indicating whether a certain object or keyword is present in the input signal – does not depend on *where* in the input signal the object or keyword is found. Translation invariance can be accounted for in a similar way as permutation invariance. To this end, vector x is partitioned into subvectors; the same trainable function is applied to each subvector; and the resulting outputs are aggregated using permutation-invariant pooling functions such as mean, sum, and max.

A notable class of neural networks that encodes translation invariance is given by **convolutional neural networks (CNNs)**. Along with translational invariance, CNNs assume **locality** and **compositionality** as part of the inductive bias. Locality refers to the fact that the mapping functions apply to specific limited neighborhoods of entries of vector x. For instance, for image inputs, the mapping functions $m(\cdot)$ in (6.128) take as input two-dimensional tiles of the image. The mapping function applied by CNNs to each subvector is an inner product with a trainable "filter" vector, effectively implementing a convolution operation on the input. The reuse of the same filter across the input image ensures that the complexity of the model, in terms of number of parameters, is decoupled from the dimension of the input. Compositionality refers to the fact that, in signals such as images, features of interest appear at different scales. This is leveraged in CNNs by alternating convolutional and subsampling layers.

6.10.2 Equivariance

A function $t = f(x)$ between a D-dimensional input and a D-dimensional output is **equivariant** with respect to some transformation $T(\cdot)$ if we have the equality

$$f(T(x)) = T(f(x)), \tag{6.129}$$

that is, if applying the transformation to the input causes the output to undergo the same transformation. A notable instance is **permutation equivariance**, whereby permuting the order of entries of the input vector x produces outputs that are permuted in the same way. This class of functions is relevant, for instance, when learning recommendations for individuals in a given population: The entries of vector x correspond to features of different individuals, and permuting the order in which individuals are arranged in the input vector should permute the corresponding recommendations in the same way.

A basic class of functions that preserve permutation equivariance is of the form

$$f(x) = \begin{bmatrix} c(x_1, a(x_{-1})) \\ c(x_2, a(x_{-2})) \\ \vdots \\ c(x_D, a(x_{-D})) \end{bmatrix}, \tag{6.130}$$

where $a(\cdot)$ is a permutation-invariant pooling function, such as mean, max, or sum, and $c(\cdot)$ is a (trainable) combining function. In this expression, x_{-d} refers to a $(D-1) \times 1$ vector that contains all entries of x except the dth. Importantly, the aggregation and combining functions are shared among all outputs. As discussed, this is implemented by tying the trainable parameters to be the same across all such functions.

Intuitively, function (6.130) implements a **message-passing** architecture in which the output $c(x_d, a(x_{-d}))$ for each "individual" d is obtained by combining its input x_d with an aggregate $a(x_{-d})$ of the inputs x_{-d} of all other individuals. Functions of this form can be stacked in **multi-layer architectures** that preserve permutation equivariance. The resulting multi-layer model effectively implements multiple rounds of message passing.

Models in this class encompass **deep sets**, which treat the input vector as a set, i.e., they disregard the order of the input entries; **graph neural networks (GNNs)**, which specify aggregation functions that limit message passing among individuals connected on a given graph; and **recurrent neural networks (RNNs)**, which are applied to sequential data. RNNs typically also include **gating** mechanisms that block or enable various signal paths through the network.

6.11 *K*-Nearest Neighbors Classification

Parametric models, of the form $p(t|x, \theta)$ (or $p(x, t|\theta)$), are specified by a finite-dimensional vector θ. In contrast, non-parametric techniques make lighter assumptions about the data generation mechanism, such as smoothness of the predictive function. An example of non-parametric methods is given by the kernel regression approach derived in Sec. 6.3.5 from parametric linear regression models. The predictions made by non-parametric models, such as kernel-based regression, generally depend on the entire training data set $\mathcal{D}$ (or a subset thereof). In this section, we briefly review another important class of non-parametric techniques, namely nearest neighbors methods for classification.

K-Nearest Neighbors Classifier. For a test point x, define as $\mathcal{D}_K(x)$ the set of K training points (x_n, t_n) in the training set $\mathcal{D}$ that have the closest inputs x_n to x in terms of Euclidean distance. The *K*-nearest neighbors classifier (*K*-NN) is defined as

$$\hat{t}_K(x) = \begin{cases} 1 & \text{if most samples } (x_n, t_n) \in \mathcal{D}_K(x) \text{ have } t_n = 1 \\ 0 & \text{if most samples } (x_n, t_n) \in \mathcal{D}_K(x) \text{ have } t_n = 0. \end{cases} \tag{6.131}$$

In (6.131), ties can be broken arbitrarily. For example, given a data set $\mathcal{D} = \{(-0.6, 0), (2.1, 0), (7.1, 1), (3.8, 1)\}$, for the text point $x = 6$ we have $\mathcal{D}_1(6) = \{(7.1, 1)\}$, and the 1-NN predictor is $\hat{t}_1(6) = 1$; and, since $\mathcal{D}_2(6) = \{(3.8, 1), (7.1, 1)\}$, we have the 2-NN predictor $\hat{t}_2(6) = 1$ given that both samples in $\mathcal{D}_2(6)$ have label $t = 1$.

Bias and Estimation Error. The choice of the hyperparameter K affects the generalization performance of K-NN in a way that can be interpreted in terms of the tension between bias and estimation error. Specifically, a small K may lead to overfitting, which is manifested in non-smooth prediction functions that rely on a limited set of data points for each input, while a large K may lead to underfitting, causing the function to be excessively smooth.

Curse of Dimensionality. Unlike parametric methods, non-parametric methods like K-NN suffer from the curse of dimensionality: As the dimension D of the input x increases, it takes an exponentially larger number of training samples to provide a sufficient "cover" for the space around any possible test point. More discussion on this aspect can be found in Sec. 7.3.

6.12 Applications to Regression

This chapter has focused on classification, but the main tools we have developed apply also to regression problems. A short discussion on this point is provided in this final section.

Linear Models. As in the case of classification, linear models for regression produce (continuous-valued) predictions that depend on the input only through a vector $Wu(x)$, with matrix W to be trained. Furthermore, the loss function is a convex function of W. An example is given by the polynomial regression model based on ridge regression studied in Chapter 4. More general models can be obtained within the generalized linear model (GLM) framework, which is reviewed in Chapter 9. SGD-based learning methods can accordingly be developed that have the same general delta-rule structure derived for classification.

Neural Network Models. Neural network models for regression define the feature vector $u(x|\theta)$ as the output of a sequence of layers with trainable weights, as for classification. Only the last layer is modified to produce a continuous, rather than a discrete, prediction as a function of the product $Wu(x|\theta)$. Therefore, the same training techniques – namely, SGD via backprop – apply directly to regression problems.

Generative Models. Generative models for regression assume a pdf, rather than a pmf, for the distribution $p(t|\vartheta)$ of the target variable t given model parameters ϑ. Furthermore, the distribution $p(x|t,\phi)$ depends on a parameter vector θ and directly on the target variable t. For instance, it may be selected as $p(x|t,\theta) = \mathcal{N}(x|Wt,I)$, where matrix W is part of the model parameters θ. Following the same arguments reviewed in this chapter for classification, ML training of the parameters ϑ and ϕ can be carried out separately (as long as the regularizer also factorizes across the two vectors of parameters).

Non-parametric Models. Finally, K-NN methods can be extended to regression problems by assuming that the predictor $\hat{t}_K(x)$ is a function only of the target variables t and the input x within the set $\mathcal{D}_K(x)$ of the K nearest neighbors within the training set.

6.13 Summary

- Discriminative linear models for binary classification produce hard predictors of the form $\hat{t}(x|\theta) = \text{step}(\theta^T u(x))$ and soft predictors of the form $p(t = 1|x,\theta) = \sigma(\theta^T u(x))$ (for logistic regression), where $u(x)$ is a fixed vector of features. The decision variable $\theta^T u(x)$, or logit, is a linear function of the model parameters θ.

- Training is typically done via SGD, and the gradient for both hard and soft predictors has the general form $\nabla_\theta \ell = \delta(x,t) \cdot u(x)$ for any training sample (x,t), where $\delta(x,t)$ measures the signed error between prediction and true label t. The resulting update is known as a delta rule.

- Representative algorithms for hard and soft predictors include the perceptron and logistic regression, respectively.

- Neural network models include $L - 1$ feature-extraction layers of neurons that produce a feature vector $u(x|\theta)$ as the input to the last, classification, layer.

- Calculation of the (stochastic) gradient for neural networks is typically done via backprop, whose complexity is about double that of evaluating the loss (or of producing a prediction).

- Discriminative models for classification are extended to multi-class classification by substituting the sigmoid function with a softmax function. The softmax function maps an arbitrary vector to a probability vector of the same dimension.

- The structure of the (stochastic) gradient for multi-class classification is a natural extension of that for binary classification. In particular, for linear models (i.e., for softmax regression), the gradient is given by the (outer) product of the mean error vector and of the feature vector, and the gradient for neural network models is obtained via backprop.

- Generative models for both binary and multi-class classification may offer a stronger inductive bias when it is more natural to model the class-conditioned data distribution. Training of these models can be done separately for each parameter (unless the regularization functions are coupled).

- Mixture models may be useful when there are multiple models that fit the data almost equally well, a hallmark of the presence of epistemic uncertainty, and/or when the model is misspecified.

- Non-parametric models such as K-NN are asymptotically consistent because of their low bias, but suffer from the curse of dimensionality.

6.14   Recommended Resources

The material in this chapter overlaps with standard textbooks on machine learning such as [1, 2]. For more details on probabilistic models, the reader is referred to [3, 4, 5]. The book by Minsky and Papert mentioned in the text is [6]. A useful reference for programming deep neural network models via automatic differentiation libraries is [7]. The capacity of local learning algorithms for neural networks is elaborated on in [8]. For readers interested in connections to neuroscience, I recommend the textbook [9] and the recent books [10, 11, 12] aimed at a general audience. A helpful review on invariance and equivariance properties of neural networks is given by [13]. Kernel methods are extensively covered in [14]. Important early work on the information-theoretic analysis of non-parametric models can be found in [15]. Mixture models are detailed in the book [16].

Problems

6.1   Consider the classifier

$$\hat{t}(x|\theta) = \text{step}(x_1 + 3x_2 + 1) = \begin{cases} 1 & \text{if } x_1 + 3x_2 + 1 > 0 \\ 0 & \text{if } x_1 + 3x_2 + 1 < 0. \end{cases}$$

(a) Identify the vector of features $u(x)$, the vector of model parameters θ, and the decision variable. (b) In the (x_1, x_2) plane, plot (by hand) the decision line, and indicate the two decision regions for $\hat{t}(x|\theta) = 0$ and $\hat{t}(x|\theta) = 1$. (c) Then, for the point $x = [x_1 = 1, x_2 = 6]^T$, evaluate the classifier's decision $\hat{t}(x|\theta)$. (d) For the pair $(x = [x_1 = 1, x_2 = 6]^T, 1)$, evaluate the classification margin. (e) For the pair $(x = [x_1 = 1.1, x_2 = 0]^T, 0)$, evaluate the classification margin.

6.2   Consider the classifier

$$\hat{t}(x|\theta) = \text{step}(x_1^2 + 4x_2^2 - 1) = \begin{cases} 1 & \text{if } x_1^2 + 4x_2^2 - 1 > 0 \\ 0 & \text{if } x_1^2 + 4x_2^2 - 1 < 0. \end{cases}$$

(a) Identify the vector of features $u(x)$, the vector of model parameters θ, and the decision variable. (b) In the (x_1, x_2) plane, plot (by hand) the decision line, and indicate the two decision regions for $\hat{t}(x|\theta) = 0$ and $\hat{t}(x|\theta) = 1$. (c) Then, for the point $x = [x_1 = 1, x_2 = 2]^T$, evaluate the classifier's decision $\hat{t}(x|\theta)$. (d) Evaluate the classification margin for the pair $(x = [x_1 = 1, x_2 = 2]^T, 0)$.

6.3   Load the binary iris data set (`irisdatasetbinary`) from the book's website, which consists of 80 training pairs (x, t) and 40 test pairs, with x being two-dimensional and $t \in \{0, 1\}$. (a*) Visualize the examples in the training set on the (x_1, x_2) plane. (b*) Implement the perceptron algorithm with feature vector $u(x) = x$ by selecting at each step a random example from the training set. (c*) At each iteration, draw the decision line of the perceptron in the same figure as the training samples. Run the algorithm for a fixed rate $\gamma^{(i)} = 0.1$ for 100 iterations starting from the initial point $\theta^{(1)} = [-1, -1]^T$. Describe the operation of the algorithm at each iteration. (d*) Plot the evolution of the training hinge-at-zero loss and of the validation hinge-at-zero loss as the number of iterations increases. (e*) Plot the evolution of the training detection-error loss and of the validation detection-error loss as the number of iterations increases for the same setting.

6.4 Derive an SGD-based algorithm that minimizes the training hinge loss, instead of using the hinge-at-zero loss as the perceptron algorithm. Compare the resulting algorithm to the perceptron by following similar steps to the previous problem.

6.5 Consider a one-dimensional linear classifier with feature $u(x) = x$ and weight $\theta = -0.1$. (a*) Plot the logit $\theta^T u(x)$ as a function of x in one figure, as well as the soft predictor $p(t = 1|x, \theta)$ provided by logistic regression in another figure. You can consider the range of values $x \in [-50, 50]$. (b*) Repeat for $\theta = -1$, and comment on the difference between the two probabilistic models obtained with $\theta = -0.1$ and $\theta = -1$. (c) Given the two logistic regression models, indicate the decision regions of the optimal hard predictor under the detection error loss.

6.6 (a) Prove the equality

$$\theta^T u(x) = \log \left(\frac{p(t = 1|x, \theta)}{p(t = 0|x, \theta)} \right),$$

where the right-hand side is the logit, or log-odds. (b) Then, prove that the log-loss for logistic regression can be written as

$$f(t^\pm \cdot (\theta^T u(x))) = \log(1 + \exp(-t^\pm \cdot (\theta^T u(x)))).$$

(c) Finally, prove that the gradient is given as

$$\nabla_\theta (-\log \sigma (t^\pm \cdot (\theta^T u(x)))) = (\sigma (\theta^T u(x)) - t) \cdot u(x).$$

6.7 Load the binary iris data set (`irisdatasetbinary`), also considered in Problem 6.3. (a*) Visualize the examples in the training set on the (x_1, x_2) plane. (b*) Implement logistic regression assuming the feature vector $u(x) = x$ by selecting at each step a random example from the training set. (c*) At each iteration, draw the decision line in the same figure as the training samples. Run the algorithm for a fixed rate $\gamma^{(i)} = 0.1$ for 100 iterations starting from the initial point $\theta^{(1)} = [-1, -1]^T$. (d*) Repeat with the larger batch size $S = 10$. Comment on the comparison between the two implementations. (e*) Produce a 3D plot that shows the score assigned by the trained soft predictor $p(t = 1|x, \theta)$ on the portion of the plane occupied by the training data.

6.8 Modify the update rule applied by logistic regression to include an ℓ_2 regularizer.

6.9 Consider a neural network with $L = 2$ layers and a two-dimensional input ($D = 2$), having weight matrix

$$W^1 = \begin{bmatrix} 1 & -2 \\ 2 & -1 \end{bmatrix}$$

and vector

$$w^2 = \begin{bmatrix} 1 \\ -1 \end{bmatrix}.$$

Assume that the element-wise non-linearity is a ReLU. Compute the soft predictor $p(t = 1|x, \theta)$ for $x = [1, 1]^T$.

6.10 Consider a neural network with $L = 3$ layers and a two-dimensional input ($D = 2$), having weight matrices

$$W^1 = \begin{bmatrix} 1 & -2 \\ 2 & -1 \end{bmatrix} \text{ and}$$

$$W^2 = \begin{bmatrix} 0 & 1 \\ 1 & 1 \end{bmatrix},$$

and vector

$$w^3 = \begin{bmatrix} 1 \\ -1 \end{bmatrix}.$$

Assume that the element-wise non-linearity is a ReLU. (a) Compute the soft predictor $p(t = 1|x, \theta)$ for $x = [1, -0.5]^T$. (b) Compute the optimal hard predictor under the detection-error loss. (c*) Implement the neural network and confirm your results. (d) Evaluate the gradient of the log-loss for the example $(x = [1, -0.5]^T, t = 1)$ with respect to the model parameters $\theta = \{W_1, W_2, w_3\}$ at the given values by using backpropagation. (e) Using these results, evaluate the gradient of the log-loss with respect to the input x. (f) Use the numerical differentiation routine developed in the last chapter (see Problem 5.17) to validate your calculations. (g) Using the gradient with respect to the model parameters, apply one step of SGD with learning rate 0.1.

6.11 Assume that you have N i.i.d. Bernoulli observations $t_n \sim \text{Bern}(\theta)$ with $\theta \in [0, 1]$ for $n = 1, \ldots, N$. Find the ML solution for θ.

6.12 Assume that you have N i.i.d. Gaussian observations $x_n \sim \mathcal{N}(\theta, 1)$ for $n = 1, \ldots, N$. Find the ML solution for parameter θ.

6.13 We are given a trained generative model

$$t \sim \text{Bern}(0.5)$$
$$(x|t = t, \phi_t) \sim \text{Bern}(\phi_t)$$

with $\phi_0 = 0.1$ and $\phi_1 = 0.9$. (a) What is the optimal soft predictor for the target variable t when $x = 1$? (b) What is the optimal hard predictor under the detection-error loss for the target variable t when $x = 1$? (c*) Generate 10 samples (x, t) from the model.

6.14 (a) Consider a generative model

$$t \sim \text{Bern}(\pi)$$
$$(x|t = t) \sim \text{Bern}(\phi_t),$$

and obtain the general expression for the ML trained parameters $\{\pi, \phi_0, \phi_1\}$ as a function of N examples in a training set $\mathcal{D} = \{(x_n, t_n)\}_{n=1}^N$. (b) Given the training set $\mathcal{D} = \{(x_n, t_n)\} = \{(1, 1), (0, 0), (1, 0), (1, 1), (0, 0), (1, 0), (0, 0)\}$, find the ML parameters $\{\pi^{ML}, \phi_0^{ML}, \phi_1^{ML}\}$. (c) For the trained generative model, find the optimal soft prediction for $x = 1$.

6.15 Load the data set twobumps training data set from the book's website. This consists of 50 examples per class in the training set. (a*) Plot the examples using different markers for the two classes. (b*) Train a QDA model. Display the soft predictor using contour lines in the graph that contains the examples. (c*) Show a 3D version of the graph.

6.16 Consider the neural network in Problem 6.9, but substitute the last logistic regression layer with a Gaussian layer, i.e., set $p(t|x, W) = \mathcal{N}(t|a^L, \beta^{-1})$ with $a^L = (w^L)^T h^{L-1}$ and a fixed precision $\beta = 1$. Using the weights in the example, derive the gradient over the weights for the data point $(x = [1, -1]^T, t = 2.3)$.

6.17 Given a data set $\mathcal{D} = \{(-1.7, 0), (-7.1, 1), (0.8, 0), (8.1, 1)\}$, what are the K-NN predictors for $K = 1$ and $K = 2$ when $x = -1$?

Appendices

Appendix 6.A: Derivation of the Logistic Loss Gradient Formula (6.33)

For logistic regression, the gradient of the log-loss with respect to vector θ can be computed directly as

$$
\begin{aligned}
\nabla(-\log \sigma(t^{\pm}(\theta^T u(x)))) &= -\frac{\nabla \sigma(t^{\pm}(\theta^T u(x)))}{\sigma(t^{\pm}(\theta^T u(x)))} \\
&= -\frac{\sigma(t^{\pm}(\theta^T u(x)))(1 - \sigma(t^{\pm}(\theta^T u(x))))}{\sigma(t^{\pm}(\theta^T u(x)))} \cdot t^{\pm} u(x) \\
&= (\sigma(t^{\pm}(\theta^T u(x))) - 1) \cdot t^{\pm} u(x).
\end{aligned} \tag{6.132}
$$

For $t = 0$, we get

$$
\begin{aligned}
\nabla(-\log \sigma(t^{\pm}(\theta^T u(x)))) &= (1 - \sigma(-(\theta^T u(x))))u(x) \\
&= (\sigma(\theta^T u(x)) - 0)u(x),
\end{aligned} \tag{6.133}
$$

and for $t = 1$, we obtain

$$
\nabla(-\log \sigma(t^{\pm}(\theta^T u(x)))) = (\sigma(\theta^T u(x)) - 1)u(x). \tag{6.134}
$$

Therefore, the desired unified formula (6.33) follows.

Appendix 6.B: ML Learning for the Bernoulli Distribution

The training log-loss for the parameter π is given as

$$
\begin{aligned}
L_{\mathcal{D}}(\pi) &= \frac{1}{N} \sum_{n=1}^{N} (-t_n \log(\pi) - (1 - t_n) \log(1 - \pi)) \\
&= \frac{N[1]}{N}(-\log(\pi)) + \frac{N[0]}{N}(-\log(1 - \pi)).
\end{aligned} \tag{6.135}
$$

It can be seen that the function is convex in π since $-\log(x)$ is a convex function for $x > 0$. Therefore, imposing the first-order optimality condition is necessary and sufficient to identify globally optimal solutions. (Since this is a constrained problem, one should also check that the solution of the first-order optimality condition meets the constraint.) Since $N[0] = N - N[1]$, we have

$$
\frac{dL_{\mathcal{D}}(\pi)}{d\pi} = -\frac{1}{\pi}\frac{N[1]}{N} + \frac{1}{1 - \pi}\left(1 - \frac{N[1]}{N}\right) = 0. \tag{6.136}
$$

This equation is satisfied only for $\pi = N[1]/N$, which is thus the only global optimal solution.

Bibliography

[1] C. M. Bishop, *Pattern Recognition and Machine Learning*. Springer, 2006.

[2] T. Hastie, R. Tibshirani, and J. Friedman, *The Elements of Statistical Learning*. Springer, 2001.

[3] K. P. Murphy, *Machine Learning: A Probabilistic Perspective*. The MIT Press, 2012.

[4] S. Theodoridis, *Machine Learning: A Bayesian and Optimization Perspective*. Academic Press, 2015.

[5] D. Koller and N. Friedman, *Probabilistic Graphical Models: Principles and Techniques*. The MIT Press, 2009.

[6] M. Minsky and S. Papert, *Perceptrons: An Introduction to Computational Geometry*. The MIT Press, 1969.

[7] A. Zhang, Z. C. Lipton, M. Li, and A. J. Smola, *Dive into Deep Learning*. https://d2l.ai, 2020.

[8] P. Baldi, *Deep Learning in Science*. Cambridge University Press, 2021.

[9] P. Dayan and L. F. Abbott, *Theoretical Neuroscience: Computational and Mathematical Modeling of Neural Systems*. The MIT Press, 2001.

[10] M. Humphries, *The Spike: An Epic Journey through the Brain in 2.1 Seconds*. Princeton University Press, 2021.

[11] J. Hawkins, *A Thousand Brains*. Basic Books, 2021.

[12] G. Lindsay, *Models of the Mind*. Bloomsbury Sigma, 2021.

[13] M. M. Bronstein, J. Bruna, Y. LeCun, A. Szlam, and P. Vandergheynst, "Geometric deep learning: Going beyond Euclidean data," *IEEE Signal Processing Magazine*, vol. 34, no. 4, pp. 18–42, 2017.

[14] B. Schölkopf and A. Smola, *Learning with Kernels: Support Vector Machines, Regularization, Optimization, and Beyond*. The MIT Press, 2002.

[15] T. Cover and P. Hart, "Nearest neighbor pattern classification," *IEEE Transactions on Information Theory*, vol. 13, no. 1, pp. 21–27, 1967.

[16] Z.-H. Zhou, *Ensemble Methods: Foundations and Algorithms*. CRC Press, 2012.

7 Unsupervised Learning

7.1 Overview

The previous chapter, as well as Chapter 4, have focused on supervised learning problems, which assume the availability of a labeled training set $\mathcal{D}$. A labeled data set consists of examples in the form of pairs (x, t) of input x and desired output t. Unsupervised learning tasks operate over **unlabeled data sets**, which provide no "supervision" as to what the desired output t should be for an input x. The general goal of unsupervised learning is to discover properties of the unknown population distribution $p(x)$ underlying the generation of data x automatically, without the use of a supervising signal.

This discovery often relies on the detection of similarity patterns in the data, such as

- discovering "similar" data points for **clustering**;
- generating data "similar" to data points for **inpainting**;
- detecting inputs "dissimilar" to the data observed so far for **anomaly detection**; or
- representing "similar" inputs with "similar" lower-dimensional vectors for **compression**.

While training criteria for supervised learning have to do with adherence of the model to the input–output relationships observed in the data (and to regularizing penalties), unsupervised learning problems are more diverse, and their specific formulation depends on the given task of interest. Clustering and anomaly detection, for instance, clearly entail different design problems.

The scope of unsupervised learning is much broader than for supervised learning, and many fundamental questions about theory and practice remain open. The French computer scientist Yann LeCun went as far as saying "Some of us see unsupervised learning as the key towards machines with common sense."

At the time of writing, a striking recent example of unsupervised learning involves data collected by a camera mounted on the head of a child: Processing the video streams without any supervision, the system aims to infer the existence of different visual classes, such as "cars" and "chairs". This task can be accomplished via identification of similar objects encountered multiple times in the video stream without any supervision. (It may be debatable whether or not a child actually learns in this fashion, but this is a topic for a different book.)

Learning Objectives and Organization of the Chapter. By the end of this chapter, the reader should be able to:

- identify unsupervised learning tasks (Sec. 7.2);
- apply parametric and non-parametric density estimators, most notably contrastive density learning (CDL) and kernel density estimation (KDE) (Sec. 7.3);
- understand the different types of latent-variable models (Sec. 7.4);
- understand and implement autoencoders in the form of principal component analysis (PCA), sparse dictionary learning, and neural autoencoders (Sec. 7.5);

- apply discriminative models in the form of contrastive representation learning (CRL) (Sec. 7.6);
- interpret and train undirected generative models based on restricted Boltzmann machines (RBMs) and more general energy-based models (Sec. 7.7);
- understand directed generative models (Sec. 7.8);
- implement the K-means clustering algorithm (Sec. 7.9); and
- train directed generative models via the expectation maximization (EM) algorithm (Sec. 7.10).

7.2 Unsupervised Learning Tasks

Given an unlabeled training set $\mathcal{D} = \{x_1, \ldots, x_N\}$ consisting of samples generated as

$$x_n \underset{\text{i.i.d.}}{\sim} p(x), \, n = 1, \ldots, N \tag{7.1}$$

for some unknown population distribution $p(x)$, unsupervised learning pursues one of several possible objectives, among which are the following:

- **Density estimation**. Estimate the density $p(x)$ for applications such as anomaly detection or compression (see Sec. 7.3).
- **Clustering**. Partition all points in $\mathcal{D}$ into groups of similar objects (see Sec. 7.9).
- **Dimensionality reduction, representation, and feature extraction**. Represent each data point x_n in a more convenient space – possibly of lower dimensionality – in order to highlight independent explanatory factors, to ease visualization and interpretation, or for use as features in successive tasks (see Sec. 7.4).
- **Generation of new samples**. Produce samples approximately distributed according to distribution $p(x)$, e.g., for inpainting, to produce artificial scenes for use in games or films, or for language generation (see Sec. 7.8).

In the rest of this chapter, we overview methods to tackle these tasks.

7.3 Density Estimation

In density estimation, given a training set $\mathcal{D} = \{x_1, \ldots, x_N\} = \{(x_n)_{n=1}^N\}$ generated as in (7.1), the goal is to obtain an estimate $p_{\mathcal{D}}^{est}(x)$ of the population distribution $p(x)$. The term "density" reflects the focus on continuous rvs x and probability density functions $p(x)$. The estimate $p_{\mathcal{D}}^{est}(x)$ can be useful for several applications, including:

- **Anomaly detection**. Values x for which the estimated distribution $p_{\mathcal{D}}^{est}(x)$ is low (relative to other values of x), or, equivalently, for which the information-theoretic surprise $-\log p_{\mathcal{D}}^{est}(x)$ is large (see Sec. 3.6.1), can be treated as **outliers**, triggering an alarm or a response by an end user.
- **Compression**. As discussed in Sec. 3.6.1, compression can be carried out by assigning fewer bits to values x that occur frequently, i.e., for which the probability $p_{\mathcal{D}}^{est}(x)$ is large; and more bits to infrequent values x, which have a small probability $p_{\mathcal{D}}^{est}(x)$. This approach is common

to many compression schemes, from the simplest, such as Morse code, to the more complex, such as JPEG or MPEG.

In this section, two different classes of methods for density estimation will be considered, namely non-parametric and parametric estimators.

- **Non-parametric density estimation**. Non-parametric density estimators assume inductive biases that impose minimal conditions, such as smoothness, on the distribution to be learned. As an important representative, kernel density estimation (KDE) will be presented.
- **Parametric density estimation**. Parametric methods assume that the distribution can be described by a parametric probabilistic model. A state-of-the-art representative in this class is contrastive density learning (CDL).

We will also briefly discuss the related, and important, problem of **density ratio learning**, in which one is interested in estimating a ratio between two distributions, rather than a single distribution.

7.3.1 Histogram-Based Density Estimation

Discrete rvs. While, as discussed, density estimation focuses on continuous rvs, we start by studying the simpler case of a categorical rv $x \in \{0, 1, \ldots, K-1\}$. A standard non-parametric estimator of the pmf is the **histogram**

$$p_{\mathcal{D}}^{hist}(x) = \frac{|\{n : x_n = x\}|}{N} = \frac{N[x]}{N}, \tag{7.2}$$

where $N[x]$ is the number of data points in $\mathcal{D}$ that are equal to x. This estimate corresponds to the empirical distribution $p_{\mathcal{D}}(x)$ of the data, i.e., $p_{\mathcal{D}}^{hist}(x) = p_{\mathcal{D}}(x)$. As we have seen in Sec. 6.8, the histogram also corresponds to the ML estimate of the parameters of the categorical variable. In this sense, we can think of the histogram as a parametric method when applied to discrete rvs.

More generally, for a categorical $D \times 1$ random vector $x = [x_1, \ldots, x_D]^T$ with $x_d \in \{0, 1, \ldots, K-1\}$ for $d = 1, \ldots, D$, the histogram is defined as in (7.2) by counting the number $N[x]$ of times a given configuration $x \in \{0, 1, \ldots, K-1\}^D$, out of the K^D possible configurations for vector x, is observed.

Continuous rvs. For a continuous variable $x \in \mathbb{R}$, the histogram is obtained by first quantizing the range of values of the variable with a quantization step Δ, yielding the quantized value $\mathcal{Q}(x)$; and then using the quantized, categorical, rv $\mathcal{Q}(x)$ to evaluate the histogram in a manner similar to (7.2).

To elaborate, consider first a rv $x \in [0, 1]$, and fix a quantization step $\Delta = 1/K$ for some integer $K > 1$. The **uniform quantizer** can be written as

$$\mathcal{Q}(x) = \left\lfloor \frac{x}{\Delta} \right\rfloor \Delta, \tag{7.3}$$

where $\lfloor \cdot \rfloor$ corresponds to the floor operation that returns the largest integer that is smaller than its argument. This assigns value 0 to all inputs x in the first **quantization bin** $[0, \Delta)$; value Δ to all inputs x in the second quantization bin $[\Delta, 2\Delta)$; and so on. Note that the K **quantization levels** are $\{0, \Delta, \ldots, (K-1)\Delta\}$.

Generalizing, consider now a rv x $\in [a, b]$ for some real numbers a and b with $a < b$, and fix the quantization step $\Delta = (b - a)/K$ for some integer $K > 1$. The **uniform quantizer** is given as

$$\mathcal{Q}(x) = \left\lfloor \frac{x - a}{\Delta} \right\rfloor \Delta + a. \tag{7.4}$$

This assigns value a to $x \in [a, a + \Delta)$ in the first quantization bin; value $a + \Delta$ to $x \in [a + \Delta, a + 2\Delta)$ to the second quantization bin; and so on. Therefore, the quantization levels are $\{a, a + \Delta, \ldots, a + (K - 1)\Delta\}$. Note that, if the original input x is not bounded, one can truncate or "squash" it in order to ensure the condition x $\in [a, b]$ with sufficiently large probability.

Finally, the **histogram estimator** for continuous rvs is defined as

$$p_{\mathcal{D}}^{hist}(x) = \frac{|\{n : \ \mathcal{Q}(x_n) = \mathcal{Q}(x)\}|}{N\Delta}. \tag{7.5}$$

The term $|\{n : \ \mathcal{Q}(x_n) = \mathcal{Q}(x)\}|/N$ is akin to the histogram (7.2) for discrete rvs, and measures the fraction of data points that fall in the same quantization bin as x. The division in (7.5) by the quantization step Δ is needed because $p_{\mathcal{D}}^{hist}(x)$ is the estimate of a probability density, and thus the product $p_{\mathcal{D}}^{hist}(x)\Delta$ must approximate the probability of x falling in its corresponding quantization bin of width Δ.

For a $D \times 1$ continuous rv x, one can apply quantization with step size Δ separately to each of the entries, and then evaluate the histogram as for discrete random vectors. Note that in this case one should divide in (7.5) by Δ^D, since the product $p_{\mathcal{D}}^{hist}(x)\Delta^D$ should approximate a probability.

Finally, it is possible to use **non-uniform** quantization strategies, such as the **Lloyd–Max** method, which coincides with K-means clustering, to be discussed in Sec. 7.9.

Bias vs Estimation Error. The generalization performance of density estimation refers to the quality of the estimate of the distribution on the entire domain space of the input x, and hence also outside the training set. In this regard, for continuous rvs, the histogram assumes as inductive bias that the distribution can be approximated by a *piecewise constant* function with a certain step size Δ, which can be thought of as a **hyperparameter** for the histogram method. This approximation may not be valid, particularly if the step size Δ is large enough, causing a potentially significant bias. Reducing the step size Δ decreases the bias, but also the number of data points per quantization bin, causing the variance of the estimate $p_{\mathcal{D}}^{hist}(x)$ to grow. So, the choice of the quantization step Δ must strike a balance between decreasing the bias, by reducing Δ, and decreasing the estimation error, by increasing Δ.

Example 7.1

As illustrated in Fig. 7.1, consider a mixture-of-Gaussians pdf as the population distribution $p(x)$ (dashed line). From this distribution, $N = 40$ samples are drawn (circles). The top figure shows that the histogram estimator, here with $\Delta = 0.3$, exhibits a significant bias due to the assumption that the distribution is piecewise constant.

7.3.2 Kernel Density Estimation

For many problems, as in Example 7.1, it is more natural to assume that the distribution is *smooth*, rather than piecewise constant as assumed by histogram methods. In such cases, a better

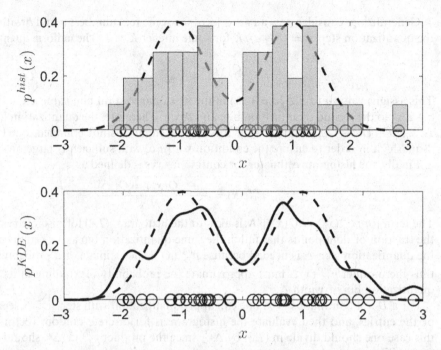

Figure 7.1 For a mixture-of-Gaussians population distribution $p(x)$ (dashed lines), from which $N = 40$ samples are drawn (circles). top: histogram estimator with $\Delta = 0.3$; bottom: KDE with a Gaussian kernel and $h = 0.04$.

density estimator can be obtained via an alternative non-parametric method known as **kernel density estimation (KDE)**. As the underlying inductive bias, KDE only assumes the smoothness of the distribution, with the degree of smoothness controlled by a **bandwidth hyperparameter** $h > 0$.

Kernel Function. KDE relies on the selection of a **kernel function** $\kappa_h(x)$, which is determined by hyperparameter h and dictates the inductive bias assumed by KDE. Despite the terminology, the kernel function $\kappa_h(x)$ used by KDE is conceptually different from the positive semi-definite kernels $\kappa(x, x')$ that are central to kernel methods for regression (Sec. 4.11) and classification (Sec. 6.3.5). The kernel function $\kappa_h(x)$ depends on a single input vector x, instead of two; it is non-negative; and it has the shape of a "smooth bump". An example is the Gaussian kernel $\kappa_h(x) = \mathcal{N}(x|0, h)$. To distinguish this type of kernel function from positive semi-definite kernels $\kappa(x, x')$, we will henceforward refer to them as **non-negative kernels** when confusion may arise.

KDE. Given a (non-negative) kernel function $\kappa_h(x)$ and a training set $\mathcal{D}$, the KDE of the probability density is given as

$$p_{\mathcal{D}}^{KDE}(x) = \frac{1}{N} \sum_{n=1}^{N} \kappa_h(x - x_n). \tag{7.6}$$

KDE thus estimates the population distribution $p(x)$ via a combination of shifted kernel functions, each centered at a data point x_n. A larger bandwidth h implies a smoother, i.e., less curved, distribution, while a smaller h allows for more general distributions that can vary more quickly. As in the case of other non-parametric techniques studied in earlier chapters, such as kernel methods (Secs. 4.11 and 6.3.5) and K-NN predictors (Sec. 6.11), the output (7.6) of KDE depends on the entire training set $\mathcal{D}$.

Example 7.2

For the mixture-of-Gaussians pdf, the bottom part of Fig. 7.1 shows the KDE obtained with a Gaussian kernel and $h = 0.04$ based on the $N = 40$ training points in the figure. KDE clearly provides a closer estimate of the density than does the histogram (top figure).

Bias vs. Estimation Error. In a manner similar to the choice of the step size Δ for the histogram, the selection of the bandwidth hyperparameter h is subject to the bias-estimation error trade-off. Choosing a small value h, and hence a narrow kernel, amounts to making weak assumptions about the curvature of the function, causing little bias. Conversely, a larger value of h, and hence a broad kernel, constrains the estimated distribution (7.6) to have a small curvature. This entails a bias if the population distribution exhibits larger variations than allowed by (7.6). On the flip side, a small h requires more data in order to reduce the estimation error, since each shifted kernel $\kappa_h(x - x_n)$ covers a smaller portion of the input space. In contrast, a broader kernel – with a larger h – covers a more extensive portion of the input space, reducing the data requirements and the estimation error.

7.3.3 Contrastive Density Learning

Having introduced non-parametric density estimators, we now turn to an important representative of *parametric* methods, namely **contrastive density learning (CDL)**. CDL converts the unsupervised learning problem of density estimation into a classification task – a supervised learning problem. This idea – transforming an unsupervised learning problem into a supervised learning one to facilitate training – is common to many unsupervised learning techniques to be discussed in this chapter. We will again focus on continuous rvs x.

CDL estimates the density $p(x)$ by training a classifier to distinguish samples drawn from $p(x)$ from samples generated from a "noise" reference distribution $r(x)$. To this end, one starts by generating N_r noise samples $\mathcal{D}_r = \{x_{N+1}, \ldots, x_{N+N_r}\}$ in an i.i.d. manner, with $x_n \sim r(x)$ for the chosen noise distribution $r(x)$. Based on these samples and on the original data set $\mathcal{D}$, a training set for binary classification is constructed as

$$\tilde{\mathcal{D}} = \{(x_1, 1), \ldots, (x_N, 1), (x_{N+1}, 0), \ldots, (x_{N+N_r}, 0)\}, \tag{7.7}$$

which includes label $t_n = 1$ if x_n is a true data point – i.e., for $n = 1, \ldots, N$ – and $t_n = 0$ if x_n is a noise data point – i.e., for $n = N+1, \ldots, N+N_r$. Note that, unlike supervised learning, the target variable t is not part of the observed data, but it is rather artificially constructed from data for the purpose of defining an auxiliary classification problem. Using the augmented, labeled, data set $\tilde{\mathcal{D}}$, CDL trains a discriminative classifier $p(t|x, \theta)$. This can be done, for example, by using logistic regression or a neural network (see the previous chapter). As shown next, the trained classifier $p(t|x, \theta)$ can be leveraged to obtain an estimate of the desired distribution $p(x)$.

By construction of the data set and of the underlying population distribution $p(x, t) = p(x)p(t|x)$, we have $p(x|t = 1) = p(x)$, since data samples are labeled as $t = 1$; we also have $p(x|t = 0) = r(x)$, since noise samples are labeled as $t = 0$. Furthermore, if the model class is large enough and the classifier is well trained, the discriminative distribution $p(t|x, \theta)$ approximates the true posterior $p(t|x) = p(x, t)/p(x)$, that is, we have $p(t|x, \theta) \simeq p(t|x)$. In this case, we can write the approximation

$$p(t = 1|x, \theta) \simeq p(t = 1|x)$$

$$= \frac{p(t = 1)p(x|t = 1)}{p(t = 0)p(x|t = 0) + p(t = 1)p(x|t = 1)}$$

$$= \frac{p(t = 1)p(x)}{p(t = 0)r(x) + p(t = 1)p(x)}. \tag{7.8}$$

Define the ratio $\delta = p(t = 1)/p(t = 0)$. From (7.8), we then obtain the approximation

$$p(t = 1|x, \theta) \simeq \frac{\delta p(x)}{r(x) + \delta p(x)}, \tag{7.9}$$

as well as

$$p(t = 0|x, \theta) = 1 - p(t = 1|x, \theta) \simeq \frac{r(x)}{r(x) + \delta p(x)}. \tag{7.10}$$

These approximations, in turn, imply that the odds $p(t = 1|x, \theta)/p(t = 0|x, \theta)$ satisfies the approximate equality

$$\frac{p(t = 1|x, \theta)}{p(t = 0|x, \theta)} \simeq \delta \frac{p(x)}{r(x)}. \tag{7.11}$$

Solving for $p(x)$ finally gives the CDL estimate

$$p_{\mathcal{D}}^{CDL}(x) \propto r(x) \frac{p(t = 1|x, \theta)}{p(t = 0|x, \theta)}. \tag{7.12}$$

Intuitively, in the CDL estimate (7.12), the classifier "pushes up" the noise distribution $r(x)$ at values of x for which the classifier is more confident that the label is $t = 1$, i.e., that x is a true data point.

The right-hand side in (7.12) is not guaranteed to be normalized, as reflected by the proportionality sign "$\propto$". An additional normalization step is therefore needed to obtain the CDL estimate as

$$p_{\mathcal{D}}^{CDL}(x) = \frac{1}{Z} r(x) \frac{p(t = 1|x, \theta)}{p(t = 0|x, \theta)}, \tag{7.13}$$

where the normalizing constant is given as $Z = \int r(x) \frac{p(t=1|x,\theta)}{p(t=0|x,\theta)} dx$, with the integral carried out on the domain of rv x. As we will see later in this section, normalization is not an issue when CDL is used to estimate ratios of densities.

Based on the derivation here, for the CDL estimate $p_{\mathcal{D}}^{CDL}(x)$ to be accurate, we need the following conditions to hold:

- The classifier should provide a close approximation of the true posterior distribution: This is necessary in order to ensure that the approximation $p(t|x, \theta) \simeq p(t|x)$ is close. Note that, since we have assumed $p(t = 1)/p(t = 0) = \delta$, this condition also requires N_r to be approximately equal to N/δ.
- The noise distribution $r(x)$ should span the entire input space of interest: This is necessary since, otherwise, the probability $p(t = 0|x) = r(x)/(r(x) + p(x))$ would equal zero for regions of the input space in which we have $r(x) = 0$ and $p(x) > 0$, causing the odds $p(t = 1|x, \theta)/p(t = 0|x, \theta)$ to go to infinity. In principle, any such choice for $r(x)$ yields a

correct estimate under the assumption that the equality $p(t|x,\theta) = p(t|x)$ holds. In practice, it is common to choose a uniform distribution in support of $p(x)$.

Example 7.3

Fig. 7.2 shows samples from a cross-shaped data set $\mathcal{D}$ (black markers), along with the contour lines of a CDL-based density estimate. To obtain it, noise samples (in gray) were generated uniformly in the square region shown in the figure, and a binary classifier was implemented via a three-layer neural network with eight neurons in both hidden layers and ReLU activation functions.

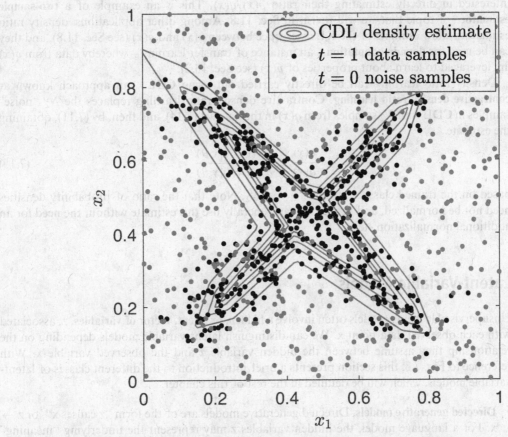

Figure 7.2 Cross-shaped data set (black markers) and the contour lines of a CDL-based density estimate obtained using the shown noise samples (in gray) and a three-layer neural network classifier.

7.3.4   Other Density Estimation Methods

Other non-parametric density estimators include **nearest neighbors methods**, which are related to the predictors studied in Sec. 6.11. Among parametric methods, one of the most popular in recent years is based on latent variable models, whereby the estimated density is obtained by transforming a base density, e.g., Gaussian, of some "hidden" variables. The transformation is determined by trainable parameters. The approach is known as the method of **normalizing flows**, and is an example of latent-variable techniques to be discussed in the next section.

7.3.5   Contrastive Density Ratio Learning

In a number of problems, we are given a training set $\mathcal{D}$ containing data from *two* different distributions $p(x)$ and $q(x)$:

$$
\begin{aligned}
x_n &\underset{\text{i.i.d.}}{\sim} p(x),\ n = 1, \ldots, N_1 \\
x_n &\underset{\text{i.i.d.}}{\sim} q(x),\ n = N_1 + 1, \ldots, N_1 + N_2.
\end{aligned}
\tag{7.14}
$$

Using this data set, instead of estimating the densities $p(x)$ and $q(x)$ separately, we may be interested in directly estimating their ratio $p(x)/q(x)$. This is an example of a **two-sample estimator** – a topic that we will revisit in Sec. 11.8. Among other applications, density ratio estimators allow estimate of the KL divergence between $p(x)$ and $q(x)$ (see Sec. 11.8); and they can be used for **domain adaptation** – an instance of transfer learning – whereby data from $q(x)$ are leveraged to learn about properties of $p(x)$ (see Sec. 13.2).

Density ratio learning can be directly carried out using CDL – an approach known as **contrastive density ratio learning**. Contrastive density ratio learning replaces the N_2 "noise" samples in CDL with N_2 samples from $q(x)$ in the data set (7.14), and then, by (7.11), obtaining the estimate

$$
\frac{p(x)}{q(x)} \simeq \frac{1}{\delta} \frac{p(\mathsf{t} = 1|x, \theta)}{p(\mathsf{t} = 0|x, \theta)}
\tag{7.15}
$$

based on the trained classifier, where $N_1 = \delta N_2$. Note that the ratio of probability densities need not be normalized, and hence one can generally use this estimate without the need for an additional normalization step.

7.4   Latent-Variable Models

Unsupervised learning models often involve a **hidden, or latent, vector of variables,** z, associated with each observed data point x. We can distinguish latent-variable models depending on the relationship they assume between the hidden variable z and the observed variable x. With reference to Fig. 7.3, this section presents a brief introduction to the different classes of latent-variable models, which will be detailed in the rest of this chapter.

- **Directed generative models.** Directed generative models are of the form "z causes x", or z → x. For a language model, the hidden variables z may represent the underlying "meaning" of a text x that "causes" text x to be generated. In a similar manner, for speech generation, the observed audio signal x may be modeled as being produced by the underlying latent

Figure 7.3 Illustration of latent-variable models for unsupervised learning: (a) directed generative models; (b) undirected generative models; (c) discriminative models; and (d) autoencoders.

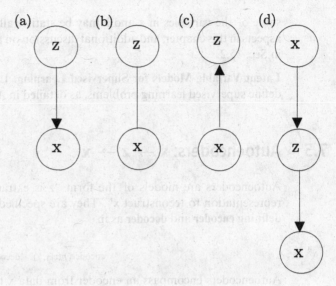

semantic information z. For more examples, consider digital images x that can be modeled as measurements of the continuous-valued, latent, visual inputs z; or the observed spiking signals x that are produced by the retina in response to hidden continuous-valued sensory inputs z.

- **Undirected generative models.** Undirected models are of the form "z is correlated with x". For instance, for a movie recommendation system, the hidden variables z may represent a user's profile – e.g., "likes foreign films", "dislikes war films" – that are correlated with a vector x of movie ratings.

- **Discriminative models.** Discriminative models are of the form "z is extracted from x". As an example, in the cocktail party problem (see Sec. 3.4), from mixed audio signals x one wishes to extract independent audio components z representing individual voices.

- **Autoencoders.** Autoencoders are of the form "z is extracted from x in order to provide a useful representation to reconstruct x". For instance, in compression, one encodes an image x into a smaller vector z, so that x can be decoded from it, typically within some (tolerable) distortion.

Semantics of the Latent Variables. Importantly, for all models, the latent variables z are **"anonymous"** in the sense that their semantics are not specified a priori, but are instead extracted from data in an unsupervised fashion. For instance, in the preceding movie recommendation example, while the observable vector x consists of handcrafted features, namely the ratings of different movies, the meaning of the entries of latent vector z is not specified as part of the inductive bias. The training process, if successful, identifies latent variables that are semantically relevant for the task of interest.

Probabilistic Graphical Models. All the mentioned latent-variable models can encode additional structure via probabilistic graphical models. In particular, both x and z can have an internal structure that can be enforced by the inductive bias. As an example, the latent structure of a language may be modeled as a Markov chain to encode the sequential nature of text. Furthermore, the links between x and z can also have a specific structure. For instance, only some of the variables in x and z may be statistically dependent. We will study some of these aspects in this chapter, and additional discussion on probabilistic graphical models can be found in Sec. 15.2.

Latent Variable Models for Supervised Learning. Latent variable models can also be used to define supervised learning problems, as detailed in Appendix 7.F.

7.5   Autoencoders: $x \to z \to x$

Autoencoders are models of the form "z is extracted from x in order to provide a useful representation to reconstruct x". They are specified by two distinct conditional distributions, defining **encoder** and **decoder** as in

$$x \underset{\text{encoder: } p(z|x,\theta)}{\to} z \underset{\text{decoder:} p(x|z,\theta)}{\to} x. \tag{7.16}$$

Autoencoders encompass an encoder from data x to hidden variables z, as well as a decoder from hidden variables z back to data x. Both encoder and decoder depend on the vector of model parameters θ. They may have distinct model parameters, or may more generally share some – or all – of their parameters.

This section will specifically focus on deterministic models for autoencoders, which are defined as

$$x \underset{\text{encoder: } F(\cdot|\theta)}{\to} z = F(x|\theta) \underset{\text{decoder: } G(\cdot|\theta)}{\to} \hat{x} = G(z|\theta), \tag{7.17}$$

where $\hat{x}$ is the reconstruction of the input through the cascade of encoder and decoder. By (7.17), the model class of autoencoders includes models consisting of

$$\text{encoder } z = F(x|\theta) \tag{7.18}$$
$$\text{and decoder } \hat{x} = G(z|\theta), \tag{7.19}$$

with both functions being generally dependent on the model parameters θ. For probabilistic autoencoders, the reader is referred to Sec. 10.13.4.

7.5.1   Training Autoencoders

Autoencoders turn the original unsupervised learning problem of extracting the latent variable z from observation x into a supervised learning problem by selecting as target vector the input x itself. Training autoencoders can hence be described as a form of **self-supervised learning**. Self-supervised learning more generally describes machine learning problems in which the target variables are computed as a function of the input x, and no separate supervisory signal is available in the data. (Self-supervised learning does not necessarily require the reconstruction of the input as the target. For instance, one could augment the data set with various transformations of the training data, and use the identity of the transformation as the supervisory signal.)

When training autoencoders, one wishes to determine model parameters θ such that the reconstruction produced by the decoder is close to the input of the encoder. In formulas, the goal is to ensure the approximate equality $\hat{x} = G(F(x|\theta)|\theta) \simeq x$, where $G(F(x|\theta)|\theta)$ represents the cascade of encoder and decoder applied to input x. Intuitively, if this approximation holds, the latent variables z must capture the essential features of x needed to reconstruct it. Formally, the optimization can be stated as the ERM problem

$$\min_{\theta} \frac{1}{N} \sum_{n=1}^{N} \ell(x_n, \underbrace{G(F(x_n|\theta)|\theta))}_{\hat{x}}, \tag{7.20}$$

where $\ell(x, \hat{x})$ is a loss function. A typical choice is the ℓ_2 loss, which yields $\ell(x, \hat{x}) = ||x - \hat{x}||^2$.

If no constraints are imposed on encoder and decoder, the ERM problem (7.20) is clearly solved by choosing encoder and decoder as identity operators, i.e., $F(x|\theta) = x$ and $G(z|\theta) = z$. This yields a zero loss for all examples, i.e., $\ell(x_n, G(F(x_n|\theta)|\theta)) = \ell(x_n, x_n) = 0$, but it provides a trivial solution that fails to provide a meaningful, informative representation z of a data point x – we simply have $z = x$. In order to address this issue, one needs to impose constraints on encoder and decoder, and, through them, on the representation z. These constraints place a **bottleneck** between the input x and its reconstruction that forces the model to ensure that vector z is informative about x.

The rest of this section will review three standard examples of autoencoders, namely principal component analysis (PCA), sparse dictionary learning, and neural autoencoders.

7.5.2 Principal Component Analysis

Inductive Bias: Decoder. PCA assumes that each observation $x \in \mathbb{R}^D$ can be well approximated as a linear combination of $M < D$ feature vectors $w_m \in \mathbb{R}^D$, known as **principal components**. Mathematically, PCA assumes the **linear decoder**

$$\text{decoder: } \hat{x} = G(z|\theta) = \sum_{m=1}^{M} z_m w_m = Wz, \tag{7.21}$$

which reconstructs x through a weighted sum of component vectors w_m, each multiplied by a scalar weight $z_m \in \mathbb{R}$. In (7.21), we have defined the **latent vector** $z = [z_1, z_2, \ldots, z_M]^T \in \mathbb{R}^M$, containing the weights of all components $\{w_m\}_{m=1}^{M}$; as well as the $D \times M$ **dictionary matrix** $W = [w_1, \ldots, w_M]$ that collects all components as columns. The "dictionary" of columns of matrix W represents the model parameters, i.e., we have $\theta = \{W\}$. In PCA, the dictionary is said to be **undercomplete**, since we have the inequality $M < D$, and hence not all vectors $x \in \mathbb{R}^D$ can be produced through the decoder (7.21).

PCA also assumes that the principal components are orthogonal and normalized, i.e., that the equalities

$$w_{m'}^T w_m = 0 \text{ for } m' \neq m, \text{ and} \tag{7.22}$$

$$w_m^T w_m = ||w_m||^2 = 1 \tag{7.23}$$

hold. Since the columns of W are orthogonal and unitary, the dictionary matrix W is orthonormal, and hence it satisfies the equality $W^T W = I_M$.

Overall, the inductive bias of PCA posits that the data distribution is well supported by a **linear subspace** of dimension $M < D$ in the input space $\mathbb{R}^D$ of dimension D. A linear subspace, or subspace for short, of dimension M is a subset of the input space $\mathbb{R}^D$ obtained by taking all possible linear combinations of M orthonormal vectors. The subspace is said to be **generated** by the M orthonormal vectors. Geometrically, when $D = 2$, a subspace of dimension $M = 1$ corresponds to a line passing through the origin of the plane $\mathbb{R}^2$, which is generated by a vector of unitary norm on the line. When $D = 3$, a linear subspace with dimension $M = 1$ is a line passing through the origin in $\mathbb{R}^3$, and a subspace with dimension $M = 2$ is a plane passing through the origin (generated by any two orthonormal vectors on the plane); and so on for higher dimensions D.

The inductive bias assumed by PCA, which views data points x as lying in a subspace, is generally well suited only for data distributions that have **zero mean**, so that their distribution is centered at the origin of the input space. Accordingly, before applying PCA, one should first subtract the empirical average from the data. This is done by pre-processing each data point x_n by computing the difference $x_n - \hat{\mu}$ between it and the empirical mean $\hat{\mu} = \frac{1}{N} \sum_{n=1}^{N} x_n$. Note that, after PCA reconstruction is carried out, one can always add back the mean vector $\hat{\mu}$.

Inductive Bias: Encoder. We have seen that PCA applies the linear decoder (7.21) for a given dictionary matrix W, with the latter being subject to optimization during training. We still need to specify the encoding function $z = F(x|\theta)$ that extracts the vector z of coefficients to be used in the linear combination (7.21). PCA assumes that the **encoder** is also **linear**, and given as

$$\text{encoder: } z = F(x|\theta) = W^T x = \begin{bmatrix} w_1^T x \\ w_2^T x \\ \vdots \\ w_M^T x \end{bmatrix}. \tag{7.24}$$

The contribution z_m for each component is hence obtained by computing the inner product $w_m^T x$. Geometrically, one can interpret the vector $z_m w_m$ in the decoder (7.21) as the orthogonal projection of x onto the component w_m (see Fig. 7.4). The parameters of encoder (7.24) and decoder (7.21) are **tied**, as they both coincide with the dictionary matrix W.

In this discussion, we have introduced the linear encoder (7.24) as part of the inductive bias assumed by PCA. One may wonder: Given the linear decoder (7.21), could another encoder be preferable to (7.24)? To answer, let us assume the ℓ_2 loss $\ell(x, Wz) = ||x - Wz||^2$ between input x and reconstruction (7.21). An optimal encoder should obtain the M-dimensional vector z that minimizes the loss $\ell(x, Wz)$. It can be easily seen that this loss function is strictly convex and quadratic in z, which implies, by the results in Sec. 5.7.6, that the solution is given exactly

Figure 7.4 Two-dimensional data set with (scaled) principal component w_1 represented as a vector (arrow). Also shown is the reconstruction $z_1 w_1$ for a given input x (solid black).

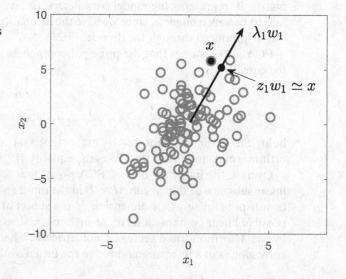

by the encoder (7.24). So, under the ℓ_2 loss, the linear encoder (7.24) is indeed optimal for the decoder (7.21).

Training PCA. Given the linear encoder (7.24) and linear decoder (7.21), under the ℓ_2 loss, the ERM problem for PCA can be written as

$$\min_W \frac{1}{N} \sum_{n=1}^N ||x_n - W \underbrace{W^T x_n}_{z_n}||^2, \tag{7.25}$$

where the $D \times M$ matrix W is constrained to be orthonormal, that is, to have orthogonal columns of unitary norm. Optimization (7.25) corresponds to a (fictitious) supervised learning problem in which the target is given by the input x_n and the predictor is given by the cascade $WW^T x_n$ of encoder and encoder. Note that this predictor is non-linear in the model parameter matrix W. This formulation reflects the self-supervised nature of PCA.

Due to the given non-linearity of the predictor, the optimization (7.25) is a **non-convex** problem, in the sense that the objective is a non-convex function of matrix W. Defining the $D \times D$ **empirical covariance matrix**

$$\hat{\Sigma}_\mathcal{D} = \frac{1}{N} \sum_{n=1}^N x_n x_n^T, \tag{7.26}$$

the solution of the ERM problem (7.25) turns out to be conceptually simple. The optimal matrix W is given by collecting in its columns the M eigenvectors of matrix $\hat{\Sigma}_\mathcal{D}$ corresponding to the M largest eigenvalues. (See Problem 7.4 for hints on how to solve the ERM problem (7.25).) For reference, we summarize PCA in the following algorithm.

Algorithm 7.1: Principal component analysis (PCA)

compute the empirical data covariance matrix $\hat{\Sigma}_\mathcal{D}$
compute the eigenvalue decomposition

$$\hat{\Sigma}_\mathcal{D} = U \Lambda U^T,$$

where Λ is a $D \times D$ diagonal matrix with diagonal elements $\{\lambda_1, \ldots, \lambda_D\}$ being the eigenvalues and U is a $D \times D$ matrix collecting by columns the eigenvectors of $\hat{\Sigma}_\mathcal{D}$
choose the columns of matrix W as the M eigenvectors corresponding to the M largest eigenvalues of $\hat{\Sigma}_\mathcal{D}$
return W

Example 7.4

For the (unlabeled) data set in Fig. 7.4, the figure shows the principal component vector w_1, scaled by the corresponding eigenvalue λ_1, as an arrow. Note that in this problem we have $D = 2$ and $M = 1$. For a given input data point x (solid black), the figure also illustrates the reconstruction $z_1 w_1$ obtained by applying the PCA-based autoencoder. This reconstruction is obtained by first applying the encoder (7.24), which computes the inner product z_1 between x and w_1, and then applying the decoder (7.21), which evaluates the reconstructed input $z_1 w_1$. Note that vector $z_1 w_1$

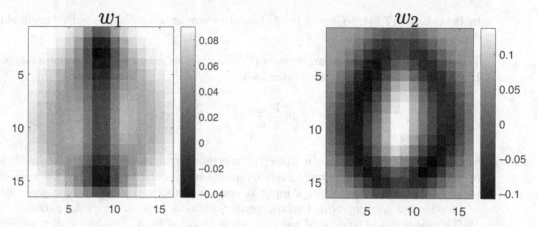

Figure 7.5 First two principal components ($M = 2$) for digits in the USPS data set corresponding to images of handwritten digits 0 and 1.

is the orthogonal projection of x onto the first component w_1, and that z_1 is the length of vector $z_1 w_1$ since vector w_1 has unitary norm.

Example 7.5

Consider the USPS data set, also encountered in Sec. 6.5.6, consisting of 16×16 grayscale images of handwritten digits. Examples of images in this data set can be found in Fig. 6.23. A 16×16 image can be represented by a 16×16 matrix or, consistently with the format assumed by PCA, as a 256×1 vector obtained by stacking the columns of the matrix one on top of the other. Considering a subset of the USPS data set consisting of only digits 0 and 1, the first two principal components ($M = 2$) are shown in Fig. 7.5, where, from darker to brighter, the pixels' shading represents increasing numbers, starting from negative values. Note that both components resemble a mix of both digits 0 and 1, with the first component w_1 having larger values at pixels that tend to be large for the digit-0 images and w_2 having larger values at pixels that tend to be large for digit-1 images. As discussed, the interpretation of the semantics of latent vector z can only be done *post hoc*, that is, upon completion of the unsupervised learning process.

The entries of the latent vector $z_n = [z_{n,1}, z_{n,2}]^T$ provide the contribution of each of these components in reconstructing an input x_n. To elaborate further on this point, Fig. 7.6 displays the latent vectors z_n obtained from all images x_n in the training data set by using different markers for images belonging to different classes. It is emphasized that PCA does not have access to the labels, which are used here exclusively to aid interpretation of the output of PCA. The figure reveals that images belonging to different classes have clearly distinct contributions along the principal components. In line with the discussion in the preceding paragraph, digit-0 images tend to have a larger component z_1, and digit-1 images a larger component z_2.

On Applications of Autoencoders. The preceding example illustrates some of the different possible applications of autoencoders:

Figure 7.6 Latent vectors z_n obtained by PCA with principal components shown in Fig. 7.5 for digit-0 and digit-1 images in the USPS data set.

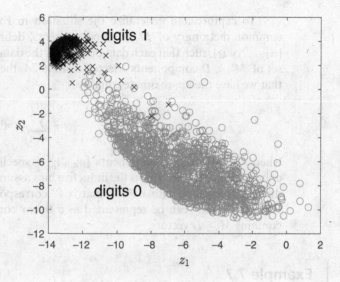

- They can be used to compress an input x into a smaller vector z.

- They can provide a representation z of input x that may be easier to visualize and interpret.

- They can be used to extract a reduced-dimension feature vector z for a downstream task such as classification. For instance, for Example 7.5, the classifier would operate in an $M = 2$-dimensional space instead of the original $D = 256$ dimensions.

7.5.3 Sparse Dictionary Learning

PCA assumes that all data points can be represented by using a *common* set of $M < D$ principal components $\{w_m\}_{m=1}^{M}$. Geometrically, PCA hence assumes that the data lie in a common subspace of dimension $M < D$, which is generated by the orthonormal columns of matrix W. (More precisely, the underlying population distribution is assumed to be supported on this subspace.) This may, however, not be the case.

Example 7.6

For the two-dimensional ($D = 2$) data set in Fig. 7.7, PCA fails since there is no single direction w_1 along which one can obtain faithful reconstructions for all data points. Geometrically, there is no line passing through the origin (a subspace of dimension $M = 1$) along which most of the points in the data set are aligned. However, as shown in the figure, we can identify three *non-orthogonal* vectors w_m, with $m = 1, 2, 3$, such that each point in the training set can be well described by a scaled version of *one* of them. In other words, while there is no single line suitable for all points, there are several lines, here $M = 3$, such that each data point can be well represented by one of them. Geometrically, the data set is not arranged along a single line, but it approximately lies on a *union of lines*.

To capture and generalize the situation in Example 7.6, let us assume that there is a common dictionary of $M \geq D$ components, defined by a $D \times M$ dictionary matrix $W = [w_1, \ldots, w_M]$, such that each data point x_n in the data set is well described by a generally *different* set of $M' < D$ components selected from it. Mathematically, for each data point x_n, we posit that we have the approximate equality

$$x_n \simeq \sum_{m=1}^{M'} z_{n,m} w_{n,m}, \tag{7.27}$$

where the $M' < M$ components $\{w_{n,m}\}_{m=1}^{M'}$ specific to data point x_n are selected from the column of matrix W. This is the inductive bias assumed by sparse dictionary learning. In sparse dictionary learning, unlike PCA, matrix W corresponds to an **overcomplete** dictionary, since all vectors $x \in \mathbb{R}^D$ can be represented as a linear combination of at most D vectors, while W contains $M \geq D$ vectors.

Example 7.7

For the data set in Fig. 7.7, an overcomplete dictionary W can be constructed as $W = [w_1, w_2, w_3]$ by including the three vectors w_1, w_2, and w_3 shown in the figure as columns. Note that we have $M = 3$. This choice is effective in the sense that each data point can be well represented via the linear combination (7.27) by choosing a single component ($M' = 1$) in the dictionary W.

Inductive Bias. In sparse dictionary learning, one therefore assumes as part of the inductive bias that there exists a suitable overcomplete $D \times M$ dictionary matrix W, with $M > D$, such that each data point x can be well reconstructed as a linear combination (7.27) of $M' < D$ columns of W. Since we can have $M > D$, the columns of the overcomplete dictionary cannot in general be orthogonal (see, e.g., Fig. 7.7), and one only requires that they be normalized, i.e., that they

Figure 7.7 For the displayed data set, PCA with $M = 1$ fails since there is no single vector w_1 along which all data points can be well represented. However, for each data point, there is a vector taken from the dictionary $W = [w_1, w_2, w_3]$ that is well suited to represent the data point. This makes the underlying data distribution amenable to the use of sparse dictionary learning with $M = 3$ and $M' = 1$.

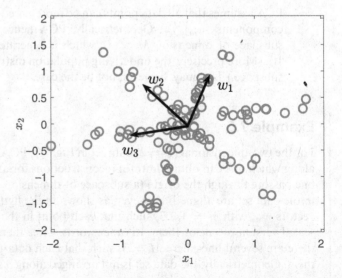

satisfy the equality

$$w_m^T w_m = ||w_m||^2 = 1. \tag{7.28}$$

Mathematically, from (7.27), sparse dictionary learning assumes the **linear decoder**

$$\hat{x} = Wz, \text{ with } ||z||_0 \leq M', \tag{7.29}$$

where $||z||_0$ is the ℓ_0 (pseudo-)norm that counts the number of non-zero elements in vector z. Accordingly, the key difference with respect to the linear decoder assumed by PCA is that the latent vector z is **sparse** – only (at most) $M' < M$ entries of z are allowed to be non-zero for each x.

As seen in the example in Fig. 7.7, geometrically, sparse dictionary learning assumes that data lie in a **union of subspaces**, with each subspace spanned by a subset of $M' < D$ columns of the dictionary matrix W. In contrast, PCA assumes the data to be supported on a single subspace of dimension $M < D$. As with PCA, this assumption is more likely to be well justified if the data have zero mean, and hence if the pre-processing step of subtracting the empirical mean has been carried out before sparse dictionary learning is applied.

Optimal Encoder. Given the decoder (7.29), the optimal encoder should obtain the M-dimensional vector z that minimizes the loss $\ell(x, Wz)$ under the sparsity constraint $||z||_0 \leq M'$. Assuming the ℓ_2 loss, this amounts to the problem

$$\text{encoder: } z_n = \arg \min_{z:||z||_0 \leq M'} ||x_n - Wz||^2 \tag{7.30}$$

for each data point x_n. Unlike PCA, this problem cannot be solved in closed form, and the encoder is generally **non-linear**.

Training Sparse Dictionary Learning. Under the ℓ_2 loss, the resulting ERM problem is given as the joint optimization

$$\min_{W, \{z_n\}_{n=1}^N} \sum_{n=1}^N ||x_n - Wz_n||^2 \tag{7.31}$$

over the model parameters θ given by the $D \times M$ dictionary W, whose columns are constrained to be normalized as in (7.28), and the $M \times 1$ latent vectors z_n, which are constrained to be sparse as in problem (7.30).

The ERM problem (7.31) is typically solved by using **alternate optimization**: At each iteration, in the first step, for a fixed set of latent vectors $\{z_n\}_{n=1}^N$, one optimizes over W; while, in the second step, for a fixed matrix W, one optimizes separately over each latent vector z_n for $n = 1, \ldots, N$. The two steps are repeated iteratively until a convergence criterion is satisfied. Each step can be carried out in different ways.

1. For the **first step**, if one removes the normalization constraint (7.28), the optimization over W is a standard LS problem (with strictly convex quadratic cost function), which has the following closed-form solution. Denoting as Z the $N \times M$ matrix that includes by rows the vectors $\{z_n\}_{n=1}^N$ and as X_D the standard $N \times D$ data matrix, the optimal solution is given as $\tilde{W} = X_D^T Z(Z^T Z)^{-1}$, assuming N to be larger than M and matrix $Z^T Z$ to be invertible. Having calculated this matrix, one can then normalize each column to satisfy the constraint (7.28), which yields a suboptimal solution W to the original problem (7.31) of minimizing over W.

2. The **second step** requires solving the separate problems (7.30) for each $n = 1, \ldots, N$. Various efficient approximate solutions exist, such as orthogonal matching pursuit (OMP), which is detailed next.

Orthogonal Matching Pursuit. OMP is carried out separately for each n, and hence we can drop the subscript n in order to focus on a general data point x and corresponding latent vector z. The algorithm greedily selects one column of the dictionary after another until M' columns have been selected, as detailed in the following algorithm.

Algorithm 7.2: Orthogonal matching pursuit (OMP)

initialize the residual as $r^{(0)} = x$, and the set of selected columns of W as $\mathcal{M}^{(0)} = \emptyset$
  (empty set)
for $m' = 1, \ldots, M'$ **do**

  find index m of the column w_m in the set $\{1, \ldots, M\} \setminus \mathcal{M}^{(m'-1)}$ of unselected columns
    of matrix W that solves the problem $\max_m |w_m^T r^{(m'-1)}|$ (ties can be resolved
    arbitrarily)
  add the optimized index m to the set of selected columns by updating
    $\mathcal{M}^{(m')} = \mathcal{M}^{(m'-1)} \cup \{m\}$
  define as V the matrix containing the columns of matrix W indexed by set $\mathcal{M}^{(m')}$, and
    minimize the approximation error $\min_{\tilde{z}} ||x - V\tilde{z}||$, obtaining $\tilde{z} = (V^T V)^{-1} V^T x$
    update the residual as $r^{(m')} = r^{(m'-1)} - V\tilde{z}$

end
return $M \times 1$ vector z containing the elements of vector $\tilde{z}$ in the positions indexed by set
  $\mathcal{M}^{(M')}$ and zero elsewhere

7.5.4   Neural Autoencoders

Neural autoencoders model both encoder and decoder with multi-layer feedforward networks. The **encoder** takes x as input and produces output z as the result of L feature-extraction layers indexed as $l = 1, 2, \ldots, L$. The Lth layer contains M neurons, and the M-dimensional vector collectively produced by these neurons corresponds to the latent vector z. The dimension M is selected to be smaller than the dimension of the input, i.e., $M < D$, so that the encoder reduces the dimensionality of the input.

Then, the **decoder** takes vector z as input, and outputs an estimate of x through another series of layers. A common modeling choice is to use the same number of layers, L, for encoder and decoder, and to tie the weights of the two networks. Accordingly, the weights for layer l in the encoder are reused in layer $L - l + 1$ of the decoder. This generalizes the operation of PCA, which can be interpreted as a neural autoencoder with linear models in lieu of neural networks as encoders and decoders. End-to-end training of neural encoder and decoder can be carried out using backprop, following the same steps discussed in Chapter 6.

7.6   Discriminative Models: x → z

Discriminative models are of the form "z is extracted from x" in the sense that they directly model the encoder

$$\text{x} \underset{\text{encoder } p(z|x,\theta)}{\longrightarrow} \text{z} \tag{7.32}$$

that obtains the latent rv z from the observation x. As such, discriminative models do not provide a model for the marginal distribution of x. Instead, they only cater to the extraction of the hidden variables from data.

As a notable representative, this section discusses contrastive representation learning (CRL). In a manner similar to CDL (see Sec. 7.3), CRL converts the unsupervised learning problem at hand into a supervised classification problem.

7.6.1 Contrastive Representation Learning

Contrastive representation learning (CRL) trains a deterministic encoder $z = F(x|\theta)$ for some parametric function $F(\cdot|\theta)$. This corresponds, as with the autoencoders studied in the preceding section, to a special case of the probabilistic model (7.32). In a manner similar to CDL (see Sec. 7.3), CRL sets up an auxiliary supervised problem by extracting label information from the data set $\mathcal{D} = \{x_n\}_{n=1}^N$.

Auxiliary Positive and Negative Examples. This is done as follows. Given each example x_n in the training set, CRL generates C auxiliary examples, namely

- $C - 1$ **negative examples** $x_{k,n}^-$, with $k = 1, \ldots, C - 1$; and
- one **positive example** x_n^+.

The positive and negative examples are typically obtained as functions of the samples in the original training set $\mathcal{D}$, and, in the standard implementation of CRL, they have the same dimensionality as x_n. The key requirement is that the positive example x_n^+ should be "similar" to the original data point x_n, while the negative examples $x_{k,n}^-$ should be "dissimilar" to it. For instance, the positive example x_n^+ may be derived through a transformation of x_n, such as a rotation, cropping, or noise addition, while negative examples $x_{k,n}^-$ may be selected by picking other examples from the training set $\mathcal{D}$ at random.

The principle underlying CRL is to optimize the encoder $F(\cdot|\theta)$ so that the representation $z_n = F(x_n|\theta)$ of a data point x_n is:

- "similar" to the representation $z_n^+ = F(x_n^+|\theta)$ of the positive example; and
- "dissimilar" to the representations $z_{k,n}^- = F(x_{k,n}^-|\theta)$ for all the negative examples indexed by $k = 1, \ldots, C - 1$.

The specific problem formulation depends on how positive and negative examples are generated and on the measure of similarity between representations.

As mentioned, a transformation is typically applied to input x_n to produce the positive example x_n^+. The choice of this transformation can enforce desired **invariance** properties for the trained encoder $F(\cdot|\theta)$ (see Sec. 6.10). In fact, the encoder should ideally produce "similar" representations upon the application of the transformation. An example is shown in Fig. 7.8, in which the positive example is obtained by mimicking a different point of view on the same object of interest. This type of transformation encodes the desired property that the encoder produce similar representations for all images of objects in the same class, irrespective of the viewpoint. As we detail next, a typical similarity metric is the inner product between representations, i.e., the cosine similarity (see Sec. 2.5).

original   positive   negative

Figure 7.8 Examples of positive and negative examples used in contrastive representation learning (CRL).

Training for CRL. Based on the auxiliary samples described above, CRL constructs a (fictitious) multi-class supervised learning problem as follows. Consider a **multi-class classifier** (see Sec. 6.6) that uses as logits the inner products

$$a_n^+ = F(x_n|\theta)^T F(x_n^+|\theta) \text{ and}$$
$$\{a_{k,n}^- = F(x_n|\theta)^T F(x_{k,n}^-|\theta)\}_{k=1}^{C-1}. \tag{7.33}$$

The first logit, a_n^+, measures the similarity between the representation of x_n and that of the positive example via the corresponding inner product; while the other logits, $\{a_{k,n}^-\}_{k=1}^{C-1}$, quantify the similarity with the representations of the negative examples. Based on the logits, the classifier chooses a class, which corresponds either to the positive example or to one of the negative examples. Intuitively, if the classifier selects with high probability the class corresponding to the first logit, the encoder will satisfy the requirements listed above in terms of the similarity of the representation to the positive example and the dissimilarity between the representation of x_n and that of the negative examples. In fact, in this case, the inner product with the positive example is larger than the inner products for the negative examples.

Denote as $t = +$ the label corresponding to the positive example. In line with the discussion in the preceding paragraph, in order to optimize the encoder parameter vector θ, CRL assumes an auxiliary data set given by examples of the form $(x_n, +)$ for $n = 1, \ldots, N$. That is, the positive label is assigned to all examples in order to supervise the trained encoder to yield larger inner products (7.33) for the positive example.

Mathematically, from Sec. 6.6, the training log-loss for the described multi-class classifier is given as (cf. (6.106))

$$L_{\mathcal{D}}(\theta) = -\frac{1}{N} \sum_{n=1}^{N} \log \left(\frac{\exp(a_n^+)}{\exp(a_n^+) + \sum_{k=1}^{C-1} \exp(a_{k,n}^-)} \right)$$
$$= \frac{1}{N} \sum_{n=1}^{N} \left(-a_n^+ + \log \left(\exp(a_n^+) + \sum_{k=1}^{C-1} \exp(a_{k,n}^-) \right) \right). \tag{7.34}$$

Writing $a_n = [a_n^+, a_{1,n}^-, \ldots, a_{C-1,n}^-]^T$ and introducing the $C \times 1$ one-hot vector $t^{OH} = [1, 0, 0, \ldots, 0]$ of the label, the gradient for SGD is then given as (cf. (6.107))

$$\nabla_\theta \left(-a_n^+ + \log \left(\exp(a_n^+) + \sum_{k=1}^{C-1} \exp(a_{k,n}^-) \right) \right) = \nabla_\theta a_n \cdot \left(\text{softmax}(a_n) - t^{OH} \right). \tag{7.35}$$

The Jacobian $\nabla_\theta a_n$ (see Appendix 5.F) depends on the choice of the encoder $F(\cdot|\theta)$. For instance, if the latter is selected as a neural network, backprop can be used to compute $\nabla_\theta a_n$. With (7.35), one can directly set up an SGD-based training algorithm for CRL.

7.7 Undirected Generative Models: x ↔ z

Model Class. Undirected generative models are of the form "z is correlated with x", accounting for the statistical dependence between x and z via the joint distribution $p(x, z|\theta)$, i.e.,

$$(x, z) \sim \underbrace{p(x, z|\theta)}_{\text{joint distribution}}. \tag{7.36}$$

The parametric joint distribution $p(x, z|\theta)$ is meant to capture the affinity, or compatibility, of any given configuration (x, z) for observation $x = x$ and latent variables $z = z$. This is in the sense that a larger distribution $p(x, z|\theta)$ indicates a more pronounced compatibility of the joint configuration (x, z).

Example 7.8

Consider a **recommendation system** for a video streaming platform that includes D movies. Assume that users give binary "like/dislike" ratings to movies they have watched. For each user, we can then define a rv x that collects binary movie ratings for the user, i.e.,

$$x = \begin{bmatrix} x_1 = \text{movie 1 - like (1)/ dislike (0)} \\ x_2 = \text{movie 2 - like (1)/ dislike (0)} \\ \vdots \\ x_D = \text{movie } D \text{ - like (1)/ dislike (0)} \end{bmatrix}. \tag{7.37}$$

Assume for simplicity that, for the subset of users whose data are to be used for training, we have ratings for all D movies, and hence vector x is observed.

During testing, any user will have watched only a subset of the movies, and hence only a subset of the entries in rv x are specified. In order to enable prediction of unavailable ratings, we introduce a latent rv z of size $M < D$ as a binary vector that describes the profile of the user. The idea is that this smaller vector provides a summary of the preferences of the user that can be used to predict her preferences outside the subset of watched movies.

Importantly, the vector z is hidden. The list of features making up the profile of a user in z is hence not constructed by hand via the selection of specific genres and movie categories, but is instead inferred from training data by observing the ratings of a subset of users. This is possible since users with similar rating vectors can reasonably be "clustered" together into groups, with each group being characterized by a profile vector of dimension M smaller than the number, D, of movies.

The model is defined by a joint distribution $p(x, z|\theta)$ relating rating vector x and user profile vector z. The goal of training is to ensure that this distribution is large when the ratings x are consistent with the user profile z.

Once training is completed, one can try to "reverse engineer", in a *post-hoc* fashion, the profile vectors z. One may, for instance, discover that each entry of the vector corresponds to a different genre, e.g., we may be able to interpret the entries of the latent vector as

$$z = \begin{bmatrix} z_1 = \text{comedies - like (1)/ dislike (0)} \\ z_2 = \text{thrillers - like (1)/ dislike (0)} \\ \vdots \\ z_M = \text{horror - like (1)/ dislike (0)} \end{bmatrix}. \tag{7.38}$$

During testing, based on a limited number of ratings from a new user, the latent vector for the user can be inferred from the trained joint distribution $p(x, z|\theta)$, making it possible to predict the unavailable ratings.

Challenges in Training Undirected Generative Models. Computing the log-loss $-\log p(x|\theta)$ for an observation x requires marginalizing over the latent variables z as in

$$-\log p(x|\theta) = -\log\left(\sum_z p(x, z|\theta)\right), \tag{7.39}$$

where the sum over all possible values of z is replaced by an integral for continuous latent variables. The need to marginalize the latent variables causes complications in the solution of the ML training problem. These will be addressed in the rest of this section by focusing first on a specific class of models – restricted Boltzmann machines (RBMs) – and then generalizing the discussion to energy-based models.

7.7.1   Restricted Boltzmann Machines: Model Class

Model Class. Restricted Boltzmann machines (RBMs) assume binary latent variables $z \in \{0, 1\}^M$ and discrete or continuous observed variables x. As in the recommendation system (Example 7.8), here we consider binary observations $x \in \{0, 1\}^D$. As part of the inductive bias, RBMs adopt joint distributions of the form

$$p(x, z|\theta) \propto \exp(a^T x + b^T z + x^T W z)$$

$$= \exp\left(\sum_{i=1}^D a_i x_i + \sum_{j=1}^M b_j z_j + \sum_{i=1}^D \sum_{j=1}^M w_{ij} x_i z_j\right) \tag{7.40}$$

for $D \times 1$ vector a, $M \times 1$ vector b, and $D \times M$ matrix W. The model parameters are given as $\theta = \{a, b, W\}$. (Note that, when no latent rvs are present, RBMs reduce to **Hopfield networks**, which are widely used as a model of associated memory.)

The impact of the model parameters θ on the value of the joint distribution (7.40) of a configuration (x, z) can be understood as follows:

- The term $a_i x_i$ indicates that configurations in which rv x_i equals 1 ($x_i = 1$) tend to yield a larger value $p(x, z|\theta)$ if a_i is large and positive, while if a_i is large and negative, configurations with $x_i = 0$ tend to have a larger distribution $p(x, z|\theta)$.

- The term $b_j z_j$ indicates that rv z_j is more likely to equal 1 if b_j is large and positive, while it will tend to equal 0 if b_j is large and negative.

- Finally, the term $w_{ij} x_i z_j$ indicates that *both* rvs x_i and z_j will tend to equal 1 if w_{ij} is large and positive, while at least one of them will tend to equal 0 if w_{ij} is large and negative.

Most importantly, from this discussion, we can think of model parameter w_{ij} as a measure of **compatibility** between visible variable x_i and latent variable z_j. Therefore, RBMs model the compatibility between visible "units" x and hidden "units" z via the $D \times M$ matrix W.

Figure 7.9 A probabilistic graphical model (Markov network) describing the RBM joint distribution (7.40) with $D = 3$ and $M = 4$. Shaded units correspond to observations, while empty units represent latent variables.

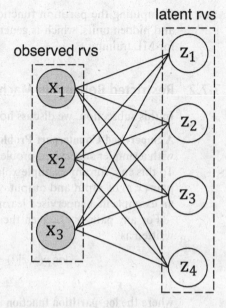

Example 7.9

For the movie recommendation example, a large w_{ij} implies that movie i is well described by feature j: If the user "likes" feature j, i.e., if $z_j = 1$, they are also likely to have a positive opinion about movie i, i.e., $x_i = 1$.

The relationship defined by the joint distribution (7.40) can be described by means of the **probabilistic graphical model** in Fig. 7.9. This bipartite graph indicates that the joint distribution contains only terms that include pairs of observed and latent variables, x_i and z_j, apart from terms involving either x_i or z_j separately. No larger subsets of variables appear in individual terms of the joint distribution. More discussion on probabilistic graphical models can be found in Sec. 15.2.

Partition Function. In order to ensure normalization, the full form of the joint distribution (7.40) is given as

$$p(x, z | \theta) = \frac{1}{Z(\theta)} \exp(a^T x + b^T z + x^T W z), \tag{7.41}$$

where the **partition function** $Z(\theta)$ is

$$Z(\theta) = \sum_{x \in \{0,1\}^D} \sum_{z \in \{0,1\}^M} \exp(a^T x + b^T z + x^T W z). \tag{7.42}$$

Computing the partition function requires summing over all possible 2^{M+D} values of visible and hidden units, which is generally impractical. We will see next how to handle this problem for ML training.

7.7.2 Restricted Boltzmann Machines: Training

In this subsection, we discuss how to train RBMs.

A Supervised Counterpart Problem. In order to study the problem of training RBMs, we begin with a simpler supervised problem in which the latent variable z is observed, rather than latent. To this end, for any example x, let us fix a value of the latent vector z. With a fixed z, we have a pair (x,z) of input and output, which is referred to as **complete data** and can be thought of as an example in a supervised learning problem.

For any pair (x,z), given the model (7.41), the **supervised log-loss** for RBMs can hence be written as

$$-\log p(x,z|\theta) = -a^T x - b^T z - x^T W z + \underbrace{A(\theta)}_{\text{log-partition function}}, \tag{7.43}$$

where the **log-partition function** is defined as

$$A(\theta) = \log Z(\theta)$$

$$= \log\left(\sum_{x\in\{0,1\}^D}\sum_{z\in\{0,1\}^M} \exp(a^T x + b^T z + x^T W z)\right). \tag{7.44}$$

Since the log-partition function is a log-sum-exp function, it is convex in the model parameters θ (see Sec. 5.7.6). It follows that the supervised log-loss (7.43) for RBM is also a **convex** function of the model parameter θ. In this regard, we note that RBMs are a special case of distributions from the **exponential family**, which will be covered in Chapter 9.

By (7.43), if the latent vector z were actually observed, the ML training problem would amount to the supervised learning of a generative model, which could be addressed in a manner similar to the previous chapter.

Training RBMs. However, in RBMs, the latent vector is not observed. The ML problem for RBMs is accordingly defined as

$$\min_{\theta}\left(-\frac{1}{N}\sum_{n=1}^N \log p(x_n|\theta)\right), \tag{7.45}$$

where the log-loss is given as

$$-\log p(x|\theta) = -\log\left(\sum_{z\in\{0,1\}^M} p(x,z|\theta)\right). \tag{7.46}$$

Unlike the supervised log-loss $-\log p(x,z|\theta)$ in (7.43), the "unsupervised" log-loss $-\log p(x|\theta)$ is not convex in θ. In this section, we will study a training algorithm based on SGD.

Fisher's Identity. To use SGD, we need to compute the gradient $\nabla_\theta(-\log p(x|\theta))$ of the log-loss. For any generative model $p(x,z|\theta)$, not limited to RBMs, we show in Appendix 7.A that we have the important equality, known as **Fisher's identity**,

$$\nabla_\theta(-\log p(x|\theta)) = \mathrm{E}_{z \sim p(z|x,\theta)}\left[-\nabla_\theta \log p(x,z|\theta)\right].$$ (7.47)

Fisher's identity says that the gradient of the (unsupervised) log-loss is given by the *average of the gradient of the supervised log-loss* $-\log p(x,z|\theta)$ over the posterior $p(z|x,\theta)$ of the latent variables. This formula has an intuitive and elegant explanation: The right-hand side of (7.47) replaces the actual, unknown, value of the latent rv z with an average over the *optimal soft predictor* $p(z|x,\theta)$ of rv z given the observation x = x. We will see in the rest of this chapter that the idea of using predictions in lieu of the true, unknown, values of the latent variables is common to many training algorithms for latent-variable models.

Estimating the Gradient (7.47). We now discuss how to evaluate the gradient (7.47) for RBMs. To this end, we need to compute the posterior $p(z|x,\theta)$, the gradient $\nabla_\theta(-\log p(x,z|\theta))$ of the supervised log-loss, and finally the average in (7.47). We describe how to carry out each step in turn in the rest of this subsection.

The optimal soft predictor $p(z|x,\theta)$ for the latent vector z is obtained by noting from the joint distribution (7.40) that the entries of the latent vector z are independent when conditioned on x = x, i.e.,

$$p(z|x,\theta) = \prod_{j=1}^{M} p(z_j|x,\theta),$$ (7.48)

with

$$p(z_j = 1|x,\theta) = \sigma(w_j^T x + b_j),$$ (7.49)

where w_j is the jth column of matrix W and $\sigma(x)$ is the sigmoid function. This can be checked by direct computation of the posterior distribution (7.48).

Example 7.10

In the movie recommendation example, the $D \times 1$ vector w_j has large positive entries for movies that are compatible for feature j. Therefore, the probability (7.49) indicates that a user tends to like (latent) feature j, i.e., $z_j = 1$, if it is present in many of the compatible films they like.

Next, we compute the gradient $\nabla_\theta(-\log p(x,z|\theta))$ of the supervised log-loss for the complete data (x,z). The gradient includes the partial derivatives with respect to a_i, b_j, and w_{ij} for $i = 1, \ldots, D$ and $j = 1, \ldots, M$. Denoting as $p(x_i|\theta)$, $p(z_j|\theta)$, and $p(x_i, z_j|\theta)$ the corresponding marginals of the joint distribution (7.41), we have the desired partial derivatives

$$\frac{\partial}{\partial a_i}(-\log p(x,z|\theta)) = \mathrm{E}_{x_i \sim p(x_i|\theta)}[x_i] - x_i$$ (7.50)

$$\frac{\partial}{\partial b_j}(-\log p(x,z|\theta)) = \mathrm{E}_{z_j \sim p(z_j|\theta)}[z_j] - z_j$$ (7.51)

$$\frac{\partial}{\partial w_{ij}}(-\log p(x,z|\theta)) = \mathrm{E}_{(x_i,z_j) \sim p(x_i,z_j|\theta)}[x_i z_j] - x_i z_j.$$ (7.52)

The partial derivatives have the form of **mean error signals** between average statistics under the model and actual observations in the complete data (x,z). The mean error signals are *local* to each unit and edge in the probabilistic graphical model represented in Fig. 7.9. This is in the sense that the partial derivative (7.52) with respect to the weight w_{ij} associated with the edge between observed unit i and latent unit j depends solely on rvs x_i and z_j; while the partial derivatives with respect to a_i and b_j in (7.50) and (7.51) depend respectively on rvs x_i and z_j.

Lastly, we need to average the partial derivatives computed in (7.50)–(7.52), and hence the gradient $\nabla_\theta(-\log p(x, z|\theta))$, over the posterior $p(z|x, \theta)$ in (7.48) to obtain the gradient (7.47) by Fisher's identity. This yields

$$\mathrm{E}_{z \sim p(z|x, \theta)}\left[\frac{\partial}{\partial a_i}(-\log p(x, z|\theta))\right] = \mathrm{E}_{x_i \sim p(x_i|\theta)}[x_i] - x_i \tag{7.53}$$

$$\mathrm{E}_{z \sim p(z|x, \theta)}\left[\frac{\partial}{\partial b_j}(-\log p(x, z|\theta))\right] = \mathrm{E}_{z_j \sim p(z_j|\theta)}[z_j] - \sigma(w_j^T x + b_j) \tag{7.54}$$

$$\mathrm{E}_{z \sim p(z|x, \theta)}\left[\frac{\partial}{\partial w_{ij}}(-\log p(x, z|\theta))\right] = \mathrm{E}_{x_i, z_j \sim p(x_i, z_j|\theta)}[x_i z_j] - x_i \sigma(w_j^T x + b_j). \tag{7.55}$$

In order to evaluate these expressions, one needs to compute the averages in (7.53)–(7.55) with respect to the relevant marginals of the joint distribution $p(x, z|\theta)$. This calculation is tractable only if M and D are sufficiently small, since the complexity of evaluating the log-partition function (7.44), and hence the joint distribution $p(x, z|\theta)$, scales exponentially with $M + D$.

Contrastive Divergence. When this computational complexity is excessive, a standard approach is to apply a method known as **contrastive divergence**. Contrastive divergence starts with the derivatives in (7.50)–(7.52), and approximates the expectations of the first terms (with respect to $p(x, z|\theta)$) and of the second terms (with respect to $p(z|x, \theta)$) in two separate phases that involve only **local** operations.

In the **"positive" phase**, the observed units are "clamped" (i.e., set to be equal) to the data vector x, and each latent unit j generates a sample z_j from the conditional distribution $p(z_j|x, \theta)$ in (7.49). This sample is used to estimate the expectation of the second terms in (7.50)–(7.52) with respect to $p(z|x, \theta)$. For instance, for the last partial derivative, we have the estimate $\mathrm{E}_{z_j \sim p(z_j|x, \theta)}[x_i z_j] \simeq x_i z_j$, where x_i is the ith entry of observation x, and z_j is the produced sample.

In the **"negative" phase**, contrastive divergence generates a **"negative" sample** (x, z), approximately distributed from the joint distribution $p(x, z|\theta)$ by alternating sampling between latent and observed units from the respective marginal distributions. These sampling operations can again be implemented locally at each unit. The resulting sample is then used to approximate the expectations in the first terms of the partial derivatives (7.50)–(7.52). We refer to Appendix 7.B for further details.

Based on this description, an implementation of SGD based on contrastive divergence is **doubly stochastic**, as the model parameter update relies on randomly selected examples from the data set, as well as on randomly drawn "negative" samples.

Contrastive divergence is an example of **contrastive Hebbian learning rules**. Hebbian learning stipulates that units that are simultaneously activated upon the presentation of an input should strengthen their connection ("Neurons that fire together, wire together."). The second term in (7.52) implements Hebbian learning, as it yields an increase in weight w_{ij} when x_i and z_j are both equal to 1 in the positive sample (which is produced with the observed units clamped to the data). Conversely, the first term in (7.52) leads to a reduction of the weight w_{ij} when x_i and z_j are both equal to 1 in the negative samples, implementing "contrastive" Hebbian learning.

7.7.3   Energy-Based Models: Model Class

RBMs are special cases of energy-based models. **Energy-based models** define the joint distribution $p(x, z|\theta)$ of observed and latent variables as

$$p(x,z|\theta) \propto \exp(-\mathcal{E}(x,z|\theta)), \tag{7.56}$$

where the **energy function** $\mathcal{E}(x,z|\theta)$ is a measure of *incompatibility* between configurations of rvs x and z. In the case of RBMs, by (7.40), the energy function is given as

$$\mathcal{E}(x,z|\theta) = -a^T x - b^T z - x^T W z. \tag{7.57}$$

In more general energy-based models than RBM, the energy function is not necessarily linear in the model parameters. As in RBMs, the energy is typically written as a sum of separate factors, each involving a subset of variables in vectors x and z. This factorization can be represented via the graphical formalism of probabilistic graphical models, as in Fig. 7.9 for RBMs (see Sec. 15.2).

Generalizing RBMs, energy-based models encode information about the plausibility of different configurations of subsets of rvs using the associated energy value: A large energy entails an implausible configuration, while a small energy identifies likely configurations. For example, to describe the fact that a subset of rvs tends to be equal with high probability, configurations in which this condition is *not* satisfied should have high energy. (The association of high-probability states with low-energy states comes from thermodynamics, in which equilibrium of a closed system is reached when the internal energy is minimal.)

Energy-based models can be thought of as **unnormalized probabilistic models**. This is in the sense that the energy function $\mathcal{E}(x,z|\theta)$ determines the unnormalized distribution $\exp(-\mathcal{E}(x,z|\theta))$, and the corresponding normalized distribution is obtained as

$$p(x,z|\theta) = \frac{\exp(-\mathcal{E}(x,z|\theta))}{\sum_{x,z} \exp(-\mathcal{E}(x,z|\theta))}, \tag{7.58}$$

where the sum is replaced by an integral for continuous rvs. In energy-based models, no constraint is imposed on the tractability of the normalizing constant, i.e., of the denominator in (7.58). In fact, as we will see next, training of energy-based models does not rely on the availability of the normalizing constant. This makes energy-based models more flexible and expressive than conventional (normalized) probabilistic models, while posing new challenges for training.

7.7.4 Energy-Based Models: Training

There are three main ways to train energy-based models: ML, score matching, and noise contrastive estimation. We briefly review each in turn.

Maximum Likelihood. Using Fisher's identity (7.47), the gradient of the log-loss with respect to the model parameters for energy-based models can be computed as

$$
\begin{aligned}
\nabla_\theta(-\log p(x|\theta)) &= E_{z \sim p(z|x,\theta)}\left[-\nabla_\theta \log p(x,z|\theta)\right] \\
&= E_{(x,z) \sim p(x,z|\theta)}[\nabla_\theta \mathcal{E}(x,z|\theta)] - E_{z \sim p(z|x,\theta)}[\nabla_\theta \mathcal{E}(x,z|\theta)]. \tag{7.59}
\end{aligned}
$$

The gradient (7.59) generalizes the expression (7.53) of the gradient for RBM. Furthermore, in the standard case in which the energy decomposes into a sum of terms involving different subsets of rvs, as for RBMs, the gradient of the energy can be evaluated using local computations.

By definition, the gradient $\nabla_\theta \mathcal{E}(x,z|\theta)$ of the energy points in a direction of the model parameter space along which the energy of the given configuration (x,z) is increased. Therefore, moving the model parameter vector θ along this direction decreases the probability of the configuration (x,z). With this in mind, we can distinguish two contributions in the gradient (7.59):

- a (so-called) **"negative"** component $\mathrm{E}_{x,z \sim p(x,z|\theta)}[\nabla_\theta \mathcal{E}(x,z|\theta)]$, which points in a direction that, on average, decreases the probability of configurations (x,z) drawn from the model distribution $p(x,z|\theta)$; and

- a (so-called) **"positive"** component $\mathrm{E}_{z \sim p(z|x,\theta)}[\nabla_\theta \mathcal{E}(x,z|\theta)]$, which points in the direction that increases the probability of the observation x.

So, the "positive" component increases the probability of the actual observation x, while the "negative" component decreases the probability, loosely speaking, of all other configurations. Note that the "positive/negative" terminology may be confusing given the signs of the respective terms in (7.59).

In order to approximate the expectations in (7.59), generalizing the discussion for RBMs in the preceding subsection, we can use empirical estimates obtained by drawing "negative" examples approximately distributed as $p(x,z|\theta)$. (This phase is taken by some authors to model the working of the brain while dreaming!) In practice, samples from $p(x,z|\theta)$ can be drawn using **Markov chain Monte Carlo (MCMC)** methods, of which contrastive divergence is a special case. These are reviewed in Chapter 12 in the context of Bayesian learning.

As a notable and practical example, we describe here an instance of stochastic gradient MCMC (SG-MCMC) methods, namely **stochastic gradient Langevin dynamics (SGLD)**. SGLD can be understood without prior exposure to Bayesian learning or MCMC, and has the advantage of requiring minor modifications to SGD. Specifically, SGLD generates "negative" samples by applying the updates

$$(x^{(i+1)}, z^{(i+1)}) \leftarrow (x^{(i)}, z^{(i)}) + \gamma^{(i)}\left(\nabla_{x,z} \log p(x^{(i)}, z^{(i)}|\theta)\right) + \mathrm{v}^{(i)}, \qquad (7.60)$$

where the added noise $\mathrm{v}^{(i)} \sim \mathcal{N}(0, 2\gamma^{(i)} I_{M+D})$ is independent across the iterations $i = 1, 2, \ldots$ and the initialization $(x^{(1)}, z^{(1)})$ is arbitrary. The notation (7.60) indicates that the vectors $x^{(i)}$ and $z^{(i)}$ are updated based on the corresponding gradients $\nabla_x \log p(x^{(i)}, z^{(i)}|\theta)$ and $\nabla_z \log p(x^{(i)}, z^{(i)}|\theta)$, respectively, with the addition of Gaussian noise, whose variance is twice as large as the learning rate $\gamma^{(i)} > 0$. It is worth emphasizing that the model parameter θ is fixed in (7.60) and that the gradient updates are carried out in the space of observation and latent rvs (x,z). As is common with MCMC methods, the first few generated samples are discarded, and later samples are kept to estimate the expectations in (7.59).

Importantly, the gradient $\nabla_{x,z} \log p(x,z|\theta)$ in (7.60) does not depend on the normalizing constant of the joint distribution (7.59), and it equals the negative gradient of the energy, i.e.,

$$\nabla_{x,z} \log p(x,z|\theta) = -\nabla_{x,z} \mathcal{E}(x,z|\theta). \qquad (7.61)$$

To see this, note that the normalizing term in (7.58) does not depend on the pair (x,z) at which the probability is evaluated. The gradient $\nabla_{x,z} \log p(x,z|\theta)$ is also known as **score vector** in this context.

Score Matching. Score matching is an alternative to ML training that leverages the identity (7.61), as well as the following observation. Two energy functions $\mathcal{E}(x,z|\theta)$ and $\mathcal{E}'(x,z|\theta) = \mathcal{E}(x,z|\theta) + c(\theta)$ that differ by an additive constant $c(\theta)$ yield the same probabilistic model upon normalization. For this to be true, as indicated by the notation, the additive constant $c(\theta)$ should not depend on (x,z), but it may depend on θ. Therefore, if two energy functions $\mathcal{E}(x,z|\theta)$ and $\mathcal{E}'(x,z|\theta)$ have the same score vectors, i.e., if

$$\nabla_{x,z} \mathcal{E}(x,z|\theta) = \nabla_{x,z} \mathcal{E}'(x,z|\theta) \qquad (7.62)$$

for all pairs (x, z), then the two probabilistic models are equivalent. Score matching is based on the principle of minimizing the norm of the difference between the score vector under the model and the score vector of the population distribution. Manipulations of this criterion yield a loss function that can be evaluated based only on data samples. For details, see Recommended Resources, Sec. 7.12.

Noise Contrastive Estimation. Noise contrastive estimation (NCE) can be derived by following the general principle of CDL. To this end, one needs to consider two binary classifiers – the first to distinguish between noise samples and data, and the second between noise and model samples. The learning criterion being minimized by NCE is the average KL divergence between the predictive distributions of the two classifiers. For details, see Recommended Resources, Sec. 7.12.

7.8   Directed Generative Models: z → x

Model Class. Directed generative models are of the form "z causes x", and they are specified by two parametric distributions, namely

$$\text{latent prior distribution}: p(z|\theta) \text{ and} \tag{7.63}$$

$$\text{decoder}: p(x|z, \theta), \tag{7.64}$$

both of which generally depend on the model parameter vector θ. Accordingly, directed generative models view the process of generating the data through the ancestral sampling mechanism

$$z \sim \underbrace{p(z|\theta)}_{\text{latent prior distribution}} \xrightarrow[\text{decoder: } p(x|z,\theta)]{} x, \tag{7.65}$$

whereby a latent rv z is generated from its prior and "pushed through" the random transformation defined by the decoder $p(x|z, \theta)$ to yield the observation x.

Intuitively, and loosely speaking, in order to ensure that rvs x drawn from the population distribution $p(x)$ can be generated via the model (7.65), the latent vector z should capture the main factors of variation that "explain" the observation x. For instance, for an image x, the latent vector may be interpreted, *post hoc*, as describing aspects such as position and shape of objects.

Example 7.11

Consider the problem of document clustering, in which the observed rv x is a numerical **embedding** of a document. An embedding may be obtained, for instance, via bag-of-words features (6.4), i.e., via the frequency of occurrence of selected words. The corresponding latent vector z is meant to capture meaningful categories that describe the generation of a document, such as topic and language. Being latent, the vector z is not observed, and the semantic properties of its entries are not predetermined but instead inferred from data. The latent prior $p(z|\theta)$ represents the distribution of document "categories" in the given data set; and the decoder $p(x|z, \theta)$ generates a document embedding x given its "topic" z.

In many applications involving continuous-valued observations x, an underlying assumption of directed generative models is that the population distribution $p(x)$ is approximately supported on a **manifold**, i.e., on a differentiable surface, that is included in the original space $\mathbb{R}^D$ of the input x. The manifold can be conveniently represented by a latent variable z that lives in a smaller-dimensional space, from which the larger-dimensional vector x can be reconstructed. This is known as the **manifold assumption**.

Training Directed Generative Models. In order to compute the distribution of an observation $\mathrm{x} = x$ under model (7.65), one needs to marginalize over the hidden variables as in

$$p(x|\theta) = \sum_z p(z|\theta)p(x|z,\theta) = \mathrm{E}_{\mathrm{z}\sim p(z|\theta)}[p(x|\mathrm{z},\theta)], \tag{7.66}$$

where the sum is to be substituted with an integral for continuous hidden variables. Therefore, the log-loss for directed generative models is given as

$$-\log p(x|\theta) = -\log(\mathrm{E}_{\mathrm{z}\sim p(z|\theta)}[p(x|\mathrm{z},\theta)]), \tag{7.67}$$

which is akin to the form for mixture models encountered in Sec. 6.9.

To train directed generative models, we would like to address the ML problem

$$\min_\theta \left\{ L_{\mathcal{D}}(\theta) = \frac{1}{N}\sum_{n=1}^N \underbrace{-\log\left(\mathrm{E}_{\mathrm{z}_n\sim p(z_n|\theta)}[p(x_n|\mathrm{z}_n,\theta)]\right)}_{-\log p(x_n|\theta)} \right\} \tag{7.68}$$

of minimizing the training log-loss. While we could use SGD as in the preceding section, in the next two sections we develop an approach specialized to the log-loss (7.67) of directed generative models that, when computationally feasible, is more effective: the EM algorithm. To this end, we will start in the next section with K-means clustering, which can be thought of as a specific, simplified instantiation of the EM algorithm.

7.9   Training Directed Generative Models: K-Means Clustering

In a clustering problem, given a data set $\mathcal{D} = \{(x_n)_{n=1}^N\}$, one would like to assign every vector $x_n \in \mathbb{R}^D$ to one of K clusters. For each data point x_n, the cluster index $z_n = k \in \{1,\ldots,K\}$ indicates that data point x_n is assigned to the kth cluster. The one-hot notation

$$z_n^{OH} = [z_{1,n},\ldots,z_{K,n}]^T \tag{7.69}$$

will also be used, where we set $z_{k,n} = 1$ if x_n is assigned to cluster k, and $z_{k,n} = 0$ otherwise. Unlike classification, in which examples of pairs (x_n, t_n) of inputs x_n and desired output t_n are given, here the data set contains no indication as to the mapping between input x_n and latent variable z_n. Instead, clustering should be based directly on the "similarity" between data points x_n. Specifically, K-means aims to assign to the same cluster z_n data points x_n that are close to each other in terms of Euclidean distance.

Before proceeding, it may be useful to note that, as formalized above, clustering does not necessarily aim at generalization. That is, in some applications of clustering, one is solely interested in the quality of the partition of the training data, and not in the quality of a separate test data set. However, as will be clarified at the end of this section, the K-means clustering algorithm can also be thought of as an approach for training a probabilistic model that may be used to generalize the clustering decisions to any point in the domain space.

7.9.1 Model Class

The model assumed by K-means is defined by K **prototype representative vectors** $\mu_k \in \mathbb{R}^D$, one for each cluster $k = 1, \ldots, K$, which determine the model parameters $\theta = \{(\mu_k)_{k=1}^K\}$. Note that there are $K \cdot D$ scalar parameters in vector θ. The prototype vectors are also referred to as **cluster centers**.

7.9.2 Supervised and Unsupervised Losses

In order to optimize the parameter models $\theta = \{(\mu_k)_{k=1}^K\}$, one should introduce a loss function $\ell(x_n | \theta)$ to describe how well the current cluster prototypes "fit" the training point x_n. To this end, let us first consider the simpler case in which the corresponding assignment variable z_n is fixed and given. This simpler setting can be interpreted as the supervised counterpart of the problem under study in the sense that, in it, the latent variables are observed.

K-means clustering adopts the squared Euclidean distance to measure similarity between data points. Accordingly, for a given cluster allocation $z_n = k$, the supervised loss for data point x_n is defined as the squared Euclidean distance between a data point x_n and the assigned cluster prototype vector μ_k, i.e.,

$$\ell_s(x_n, z_n = k | \theta) = ||x_n - \mu_k||^2. \tag{7.70}$$

This "supervised" loss can be equivalently expressed using the one-hot cluster index (7.69) as

$$\ell_s(x_n, z_n | \theta) = \sum_{k=1}^K z_{k,n} ||x_n - \mu_k||^2. \tag{7.71}$$

As mentioned, the loss (7.71) can be thought of as a supervised counterpart of the actual clustering loss, since it is computed for a fixed value of the latent variable z_n, which acts as a known label. How to define the actual loss $\ell(x_n | \theta)$? The idea is to *estimate* the latent variable z_n based on x_n and the model parameters θ. Specifically, K-means uses a hard prediction of z_n: Each point x_n is assigned to the cluster $z_n = k$ that minimizes the supervised loss $\ell_s(x_n, z_n | \theta)$, i.e.,

$$k = \arg \min_{j \in \{1, \ldots, K\}} ||x_n - \mu_j||^2. \tag{7.72}$$

With this hard predictor of z_n, the actual, "unsupervised", loss obtained for data point x_n is defined as

$$\ell(x_n | \theta) = \min_{k \in \{1, \ldots, K\}} ||x_n - \mu_k||^2$$
$$= \min_{z_n \in \{1, \ldots, K\}} \ell_s(x_n, z_n | \theta). \tag{7.73}$$

7.9.3   ERM Problem

K-means clustering aims to find model parameters $\theta = \{(\mu_k)_{k=1}^{K}\}$ that solve the ERM problem

$$\min_{\theta=\{(\mu_k)_{k=1}^{K}\}} \left\{ L_{\mathcal{D}}(\theta) = \frac{1}{N} \sum_{n=1}^{N} \ell(x_n|\theta) \right\}, \qquad (7.74)$$

which averages the unsupervised loss $\ell(x_n|\theta)$ in (7.73) over the training set. Note that, when distance metrics other than the Euclidean distance are used, one obtains the more general **K-medoids algorithm**, which is also applicable to discrete or structured data.

7.9.4 K-Means Clustering

The derivation of K-means clustering is based on an insight that is common to many latent-variable models: When the latent variables are fixed, the ERM training problem (7.74) becomes supervised – since the latent variables play the role of labels – and it can be solved with standard methods studied in the previous chapters.

Specifically, in the case of K-means, for fixed cluster assignments $\{(z_n)_{n=1}^{N}\}$ the problem of minimizing the training supervised loss in (7.71) is an LS program, which yields the optimal solution

$$\mu_k = \frac{\sum_{n=1}^{N} z_{k,n} x_n}{\sum_{n=1}^{N} z_{k,n}} = \frac{\sum_{\substack{n=1: \\ z_n=k}}^{N} x_n}{N[k]} \text{ for } k = 1, \ldots, K, \qquad (7.75)$$

where $N[k]$ is the number of data points assigned to cluster k. The optimal cluster representative μ_k for each cluster k is hence the mean of the data points assigned to cluster k. Note that this coincides with the ML solution for Gaussian distributions derived in Sec. 6.5.

The K-means algorithm iterates between optimizing cluster assignments (latent variables) and prototypes (model parameters), and is detailed in Algorithm 7.3.

Example 7.12

An example with $K = 2$ and $D = 2$ is illustrated in Fig. 7.10. The top row shows an unlabeled training data set and a specific choice for the $K = 2$ initial prototype vectors $\{(\mu_k^{(1)})_{k=1}^{2}\}$, with black and gray shading used to distinguish the two clusters. In the first iteration of K-means, as seen in the second row, the E step updates the latent variables z_n by assigning each data point x_n to the closest prototype – shading is used to indicate the assignment to a specific cluster. In the M step, the cluster centers are then moved to the "barycenters" of the corresponding clusters, that is, to the empirical average of the points assigned to a cluster. The process is continued for two more iterations in the last rows.

7.9.5   Interpreting K-Means

K-means clustering alternates between inference and supervised learning. In the E step, the model parameters $\theta = \{(\mu_k)_{k=1}^{K}\}$ are fixed, and the algorithm infers the latent variables $\{(z_n)_{n=1}^{N}\}$. In other words, in the E step, a cluster index is "imputed" to each training point.

Algorithm 7.3: K-means clustering

initialize cluster representatives $\{(\mu_k^{(1)})_{k=1}^K\}$ and set $i = 1$

while *stopping criterion not satisfied* **do**

Expectation step, or E step (aka *imputation*): Given fixed vectors $\{(\mu_k^{(i)})_{k=1}^K\}$, for every data point x_n, optimally assign the cluster index as

$$z_{k,n}^{(i+1)} = \begin{cases} 1 & \text{for } k = \arg\min_j ||x_n - \mu_j^{(i)}||^2 \\ 0 & \text{otherwise} \end{cases} \tag{7.76}$$

with ties broken arbitrarily

Maximization step, or M step: Given fixed cluster assignments $\{(z_n^{(i+1)})_{n=1}^N\}$, optimize the cluster representatives $\{(\mu_k)_{k=1}^K\}$ as

$$\mu_k^{(i+1)} = \frac{\sum_{n=1}^N z_{k,n}^{(i+1)} x_n}{\sum_{n=1}^N z_{k,n}^{(i+1)}} \tag{7.77}$$

set $i \leftarrow i + 1$

end

return $\theta^{(i)}$

Then, in the M step, the latent variables $\{(z_n)_{n=1}^N\}$ are fixed, and the algorithm trains the model parameters θ based on the complete data defined by the actual observations $\{(x_n)_{n=1}^N\}$ along with the inferred variables $\{(z_n)_{n=1}^N\}$.

7.9.6 K-Means As Alternate Optimization and Convergence

Both E step and M step in K-means can be interpreted as minimizing the common cost function

$$\frac{1}{N} \sum_{n=1}^N \ell_s(x_n, z_n | \theta), \tag{7.78}$$

which is the empirical average of the supervised training loss across the training set. In fact, the E step minimizes (7.78) over the cluster assignment variables $\{(z_n)_{n=1}^N\}$ for fixed cluster prototypes; and the M step over cluster prototypes $\theta = \{(\mu_k)_{k=1}^K\}$ for fixed latent variables.

In other words, K-means clustering applies alternate optimization to the **bi-level optimization problem** (7.74), which can be equivalently restated as

$$\min_\theta \frac{1}{N} \sum_{n=1}^N \left\{ \min_{z_n \in \{1,\ldots,K\}} \ell_s(x_n, z_n | \theta) \right\}. \tag{7.79}$$

In (7.79), the outer optimization is over the parameter model θ, while the inner optimizations are applied separately for each cluster assignment variable z_n for $n = 1, \ldots, N$. **Alternate optimization** refers to iterative schemes whereby each iteration optimizes first over a subset of

Figure 7.10 K-means algorithm. Top row: unlabeled data set $\mathcal{D}$ and initialized cluster prototypes $\mu_1^{(1)}$ and $\mu_2^{(1)}$; second row: updated cluster assignment variables z_n after the E step at iteration $i = 1$, with shading referring to the respective cluster prototypes; and third and fourth rows: as for the second row but for iterations $i = 2$ and $i = 3$.

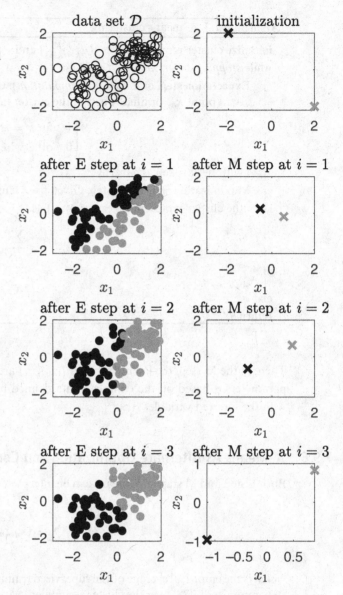

variables while fixing the rest of the variables; and then optimizes over the rest of the variables while fixing the first set. We have encountered an example of alternate optimization in Sec. 7.5.3 when discussing sparse dictionary learning. Following alternate optimization, K-means clustering carries out the inner minimizations during the E step for fixed θ; while the M step addresses the outer minimization for fixed cluster assignments.

As a result of the use of alternate optimization, the value of the cost function (7.78) cannot be increased by either the E or the M step, and hence the value of the cost function (7.78) is non-increasing across the iterations of K-means clustering. Given that the loss function is non-negative, this ensures convergence, albeit typically to a suboptimal point with respect to the minimum of the ERM problem (7.74).

7.9.7   Connection with Directed Generative Models

To conclude this section and prepare for the next, let us now connect K-means clustering with the ML problem (7.68) for the directed generative models introduced in the preceding section. This link hinges on the fact that K-means clustering can be interpreted as assuming a directed generative model in which the latent variable is categorical, taking K possible values, and the per-class model distribution $p(x|z = k, \theta)$ is Gaussian with mean μ_k and a fixed variance. This can be checked by comparing the cost function (7.78) with the supervised log-loss obtained with this choice of distribution. More precisely, as detailed in the next section, the ERM problem (7.74) tackled by K-means can be thought of as an approximation of the ML optimization (7.68) under this model.

With regard to this connection, it is useful to view the number K of clusters as a **hyper-parameter** that is part of the inductive bias. As such, it can be selected using domain knowledge or validation. When the goal is solely to effectively partition the training data set, one may also rely on criteria based on training data such as the "homogeneity" of the resulting clusters.

7.10   Training Directed Generative Models: Expectation Maximization

The **EM algorithm** provides an efficient way to tackle the ML learning problem (7.68) for directed generative models. Like GD, it is based on the iterative optimization of convex approximants of the cost function, here the training log-loss. Unlike GD, the convex approximants are tailored to the latent-variable structure of the log-loss. Therefore, they generally provide tighter bounds on the log-loss, allowing updates to make more progress as compared to GD.

As we will see, constructing the convex approximant (E step) requires solving the inference problem of predicting the latent variables $\{z_n\}$ for the current iterate of model parameter vector θ; while optimizing the approximant (M step) amounts to updating the model parameters θ for the given fixed predictor of the latent rvs $\{z_n\}$. Therefore, EM exhibits the same general structure of the K-means algorithm presented in the preceding section, with the key caveat that EM relies on a soft predictor $q(z|x)$, rather than on a hard predictor, of the latent variables. Intuitively, a soft predictor is generally preferable, and anyway more flexible, since it can account for uncertainty in the optimal selection of the latent variables, especially in the early stages of the training process.

7.10.1   Supervised vs. Unsupervised Training Loss

As we have done for K-means in the preceding section, it is useful to start by contrasting the unsupervised ML problem (7.68) with its supervised counterpart, in which the latent variable z_n is observed for each sample x_n. The resulting supervised ML problem for the given generative model $p(x, z|\theta)$ is given as

$$\min_{\theta} \left\{ \frac{1}{N} \sum_{n=1}^{N} \underbrace{(-\log p(x_n, z_n|\theta))}_{\text{supervised log-loss}} \right\}. \tag{7.80}$$

Example 7.13

As a simple example of the fact that optimizing the unsupervised training log-loss in (7.68) is more complex than optimizing its supervised counterpart (7.80), consider the mixture-of-Gaussians model defined as

$$z \sim p(z) = \text{Bern}(z|0.5)$$
$$(x|z = 0) \sim p(x|z = 0) = \mathcal{N}(x|2, 1)$$
$$(x|z = 1) \sim p(x|z = 1) = \mathcal{N}(x|\theta, 1), \tag{7.81}$$

where the model parameter θ determines the mean of the distribution of x given $z = 1$. Assuming that we also observe the latent z for each observation x, the corresponding supervised log-loss for a pair (x, z) is given by the function

$$-\log p(x, z|\theta) = -z \log \mathcal{N}(x|\theta, 1) - (1 - z) \log \mathcal{N}(x|2, 1) - \log(0.5)$$
$$= \frac{z}{2}(\theta - x)^2 + \text{constant}, \tag{7.82}$$

where we have omitted additive constants independent of θ. In contrast, the unsupervised training log-loss is given as

$$-\log p(x|\theta) = -\log(0.5\mathcal{N}(x|2, 1) + 0.5\mathcal{N}(x|\theta, 1)). \tag{7.83}$$

Figure 7.11 shows the unsupervised training log-loss (7.68) and its supervised counterpart (7.80) for $N = 100$ samples that are randomly generated from the mixture-of-Gaussians model (7.81) with true value of the model parameter $\theta = 2$. Even for this simple example, the unsupervised training loss is seen to be more "complex" to optimize than its supervised counterpart. In fact, for this specific problem, the supervised log-loss is convex, while it is non-convex for unsupervised learning. What this example shows is that, even in the most benign case in which the log-loss is

Figure 7.11 Unsupervised training log-loss (7.68) and its supervised counterpart (7.80) for the mixture-of-Gaussians model (7.81). The underlying $N = 100$ samples are randomly generated from the mixture-of-Gaussians model with true value of the model parameter $\theta = 2$.

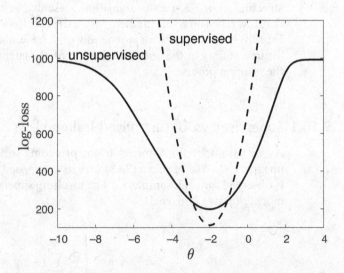

convex when the latents $\{z_n\}$ are observed, the unsupervised loss is non-convex, making the training problem more involved. This illustrates some of the additional challenges faced by unsupervised learning.

7.10.2   Free Energy and Optimal Soft Prediction

As mentioned, central to the development of EM is the use of a soft predictor $q(z|x)$ to stand in for the latent variables. Let us fix a parameter θ and the corresponding joint distribution $p(x, z|\theta)$. For a given joint distribution $p(x, z|\theta)$, the optimal soft predictor is given by the posterior of the latent rv given the observation, i.e., $q^*(z|x) = p(z|x, \theta)$. We recall from Sec. 3.9 that this distribution can be obtained as the minimizer of the **free energy** between $q(z|x)$ and $p(x, z|\theta)$, which is defined as (see (3.95))

$$\mathrm{F}(q(z|x)||p(x,z|\theta)) = \underbrace{\mathrm{E}_{z \sim q(z|x)}\left[-\log p(x,z|\theta)\right]}_{\text{average supervised log-loss}} - \underbrace{\mathrm{E}_{z \sim q(z|x)}\left[-\log q(z|x)\right]}_{\mathrm{H}(q(z|x))}. \tag{7.84}$$

The first term in the free energy (7.84) represents the average supervised log-loss, with the average taken with respect to the soft predictor $q(z|x)$, while the second term penalizes soft predictors that are too confident, i.e., whose entropy is too low (see Sec. 3.9).

To elaborate on the free energy (7.84), consider the case in which the latent rv z is discrete and the soft predictor $q(z|x)$ assigns a score of 1 to a single value z', i.e., we have $q(z|x) = \mathbb{1}(z - z')$. Then, the average supervised log-loss in (7.84) is given as $-\log p(x, z'|\theta)$, which is the log-loss that we would have incurred had the latent variable been observed to equal z'. A soft predictor of the form $q(z|x) = \mathbb{1}(z - z')$ has zero entropy $\mathrm{H}(q(z|x))$, and hence it does not generally minimize the free energy $\mathrm{F}(q(z|x)||p(x, z|\theta))$ because of the second term in (7.84). With a more general soft predictor $q(z|x)$, the average supervised log-loss in (7.84) evaluates an expectation of the supervised log-loss, with each value $-\log p(x, z|\theta)$ weighted by the score $q(z|x)$.

7.10.3   Free Energy As a Bound on the Log-Loss

In the preceding subsection, we have seen that the free energy can be used as a criterion to optimize the soft predictor of the latent rv given the observation, and that, by its definition (7.84), it is directly related to the supervised log-loss. In this subsection, we will take the key step of demonstrating that the free energy is also strongly linked to the "unsupervised" log-loss $-\log p(x|\theta)$ whose empirical average we wish to minimize in the ML problem (7.68). Specifically, we will show that the free energy provides a global upper bound on the training log-loss $-\log p(x|\theta)$. This bounding property will enable the use of the free energy as a local convex approximant within an iterative procedure akin to the successive convex approximation scheme applied by GD (see Sec. 5.8).

Free Energy As an Upper Bound on the Unsupervised Log-Loss. The discussion in Sec. 3.9 implies that the free energy can also be written as

$$\underbrace{-\log p(x|\theta)}_{\text{unsupervised log-loss}} = \underbrace{\mathrm{F}(q(z|x)||p(x,z|\theta))}_{\text{free energy}} - \mathrm{KL}(q(z|x)||p(z|x,\theta)). \tag{7.85}$$

In fact, in Sec. 3.9, the free energy was introduced by subtracting $\log p(x|\theta)$ from $\text{KL}(q(z|x)||p(z|x,\theta))$ (having made the necessary changes in the notation). By the equality (7.85), given that the KL divergence is always non-negative (by Gibbs' inequality), the free energy is an upper bound on the unsupervised log-loss, i.e.,

$$\underbrace{-\log p(x|\theta)}_{\text{unsupervised log-loss}} \leq \underbrace{\text{F}(q(z|x)||p(x,z|\theta))}_{\text{free energy}}. \tag{7.86}$$

Note that this is true for *any* soft predictor $q(z|x)$. Furthermore, with the optimal soft predictor $q^*(z|x) = p(z|x,\theta)$, since the KL divergence term in (7.85) is zero we have the equality

$$-\log p(x|\theta) = \text{F}(p(z|x,\theta)||p(x,z|\theta)). \tag{7.87}$$

A picture of the relationship between the log-loss $-\log p(x|\theta)$ and the free energy can be found in Fig. 7.12. The figure illustrates the facts that the free energy is an upper bound on the log-loss, as per (7.86), and that it is tight at a given value of θ when the soft predictor $q(z|x)$ is selected to equal the optimal predictor $p(z|x,\theta)$, by (7.87).

Alternative proofs of the inequality (7.86) can be found in Appendix 7.C. An estimation-based perspective on (7.86), alongside a **multi-sample** version of this inequality, can be found in Appendix 7.D.

Evidence Lower Bound (ELBO) = Negative Free Energy. The negative of the free energy is also known as **evidence lower bound (ELBO)**. This because the inequality (7.86) can be equivalently interpreted as the following bound on the log-likelihood of the "evidence", i.e., of the observation x:

$$\underbrace{\log p(x|\theta)}_{\text{log-probability of the "evidence" } x} \geq \underbrace{-\text{F}(q(z|x),p(x,z|\theta))}_{\text{ELBO}}. \tag{7.88}$$

This terminology is quite common in machine learning.

Figure 7.12 Illustration of the relationship between free energy and unsupervised log-loss. The figure serves as a visual aid to emphasize that the free energy is an upper bound on the unsupervised log-loss, as per (7.86), and that it is tight at a given value of θ when the soft predictor $q(z|x)$ is selected to equal the optimal predictor $p(z|x,\theta)$, by (7.87). (Note that the unsupervised log-loss is guaranteed to be non-negative only for discrete rv x, but the upper bounding and tightness conditions apply also for continuous rvs.)

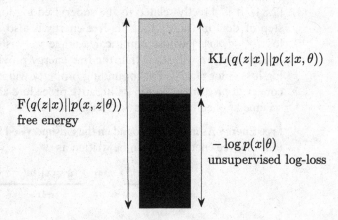

$\text{KL}(q(z|x)||p(z|x,\theta))$

$\text{F}(q(z|x)||p(x,z|\theta))$
free energy

$-\log p(x|\theta)$
unsupervised log-loss

7.10.4 Introducing EM As the Successive Optimization of Local Approximants

The EM algorithm is an iterative procedure for the minimization of the training log-loss $L_{\mathcal{D}}(\theta)$ in the ML problem (7.68) that produces a sequence of iterates $\theta^{(1)}, \theta^{(2)}, \ldots$ by implementing a successive convex approximation strategy. Around the current iterate $\theta^{(i)}$, EM obtains a convex approximant of the training log-loss with the following properties: (a) it is a global upper bound on the training log-loss; and (b) it is tight at the current iterate. Following Sec. 5.8, an iterative optimization procedure that minimizes an approximant with these properties at each iteration yields non-increasing values of the cost function (see Fig. 5.9 for an illustration). In this subsection, we show how to construct such an approximant for the training log-loss using the free energy functional by leveraging the properties (7.86) and (7.87).

Let us start by translating the two desired conditions (a) and (b) for an approximant of the training log-loss $L_{\mathcal{D}}(\theta)$ around the current iterate $\theta^{(i)}$ into mathematical constraints. The approximant $\tilde{L}_{\mathcal{D}}(\theta|\theta^{(i)})$ should satisfy (a) the **global upper bound** condition

$$\tilde{L}_{\mathcal{D}}(\theta|\theta^{(i)}) \geq L_{\mathcal{D}}(\theta), \tag{7.89}$$

for all $\theta \in \mathbb{R}^D$, and (b) the **local tightness** condition

$$\tilde{L}_{\mathcal{D}}(\theta^{(i)}|\theta^{(i)}) = L_{\mathcal{D}}(\theta^{(i)}). \tag{7.90}$$

Using (7.86) and (7.87), both conditions can be satisfied by using the free energy as the local approximant by defining

$$\tilde{L}_{\mathcal{D}}(\theta|\theta^{(i)}) = \frac{1}{N} \sum_{n=1}^{N} \mathrm{F}(p(z_n|x_n, \theta^{(i)})||p(x_n, z_n|\theta)). \tag{7.91}$$

Therefore, to obtain an approximant satisfying conditions (a) and (b), given the current iterate $\theta^{(i)}$, we can first solve the inference problem of computing the optimal soft predictor $p(z|x, \theta^{(i)})$ and then plug this predictor into the free energy to obtain (7.91).

7.10.5 EM Algorithm

EM Algorithm: First Version. We are now ready to present a first version of the EM algorithm. EM produces a sequence $\theta^{(1)}, \theta^{(2)}, \ldots$ of iterates, with each next iterate $\theta^{(i+1)}$ obtained by minimizing the approximant $\tilde{L}_{\mathcal{D}}(\theta|\theta^{(i)})$ in (7.91). In this regard, the approximant (7.91) can be written as

$$\tilde{L}_{\mathcal{D}}(\theta|\theta^{(i)}) = \underbrace{\frac{1}{N} \sum_{n=1}^{N} \mathrm{E}_{z_n \sim p(z_n|x_n, \theta^{(i)})}[-\log p(x_n, z_n|\theta)]}_{\mathrm{Q}(\theta|\theta^{(i)})} + \text{constant}, \tag{7.92}$$

where the constant term does not depend on θ. Therefore, minimizing the approximant over θ is equivalent to minimizing the function $\mathrm{Q}(\theta|\theta^{(i)})$ defined in (7.92). Overall, the EM algorithm operates as detailed in Algorithm 7.4 by iteratively optimizing the function $\mathrm{Q}(\theta|\theta^{(i)})$.

EM Algorithm: Final Version. Hidden in the formulation of the EM algorithm given in Algorithm 7.4 is the fact that the EM algorithm carries out two distinct steps at each iteration. The first step is the computation of the optimal soft predictor $p(z_n|x_n, \theta^{(i)})$ for every data point x_n with $n = 1, \ldots, N$. This is known as the **expectation, or E, step**. The second step is the

---

Algorithm 7.4: EM algorithm: first version

---

initialize $\theta^{(1)}$ and $i = 1$
while *stopping criterion not satisfied* **do**
> obtain next iterate as

$$\theta^{(i+1)} = \arg \min_\theta \left\{ Q(\theta|\theta^{(i)}) = \frac{1}{N} \sum_{n=1}^{N} E_{z_n \sim p(z|x,\theta^{(i)})}[-\log p(x_n, z_n|\theta)] \right\}$$

> set $i \leftarrow i + 1$

end
return $\theta^{(i)}$

---

optimization of the resulting cost function $Q(\theta|\theta^{(i)})$, which is known as the **maximization, or M, step**. Expanding on the preceding table, we can therefore provide a more detailed description of the EM algorithm as in Algorithm 7.5.

Algorithm 7.5: EM algorithm

---

initialize $\theta^{(1)}$ and $i = 1$
while *stopping criterion not satisfied* **do**
> **Expectation step, or E step** (aka *soft imputation*): Given fixed model parameter $\theta^{(i)}$, for every data point x_n, obtain the optimal soft predictor of z_n, namely $p(z_n|x_n, \theta^{(i)})$
> **Maximization step, or M step**: Given fixed posteriors $\{p(z_n|x_n, \theta^{(i)})\}_{n=1}^{N}$, obtain the next iterate as

$$\theta^{(i+1)} = \arg \min_\theta \left\{ Q(\theta|\theta^{(i)}) = \frac{1}{N} \sum_{n=1}^{N} E_{z_n \sim p(z|x,\theta^{(i)})}[-\log p(x_n, z_n|\theta)] \right\}$$

> set $i \leftarrow i + 1$

end
return $\theta^{(i)}$

EM Algorithm as Alternate Optimization. The derivation of the EM algorithm presented above views it as carrying out the iterative optimization of local approximants that serve as locally tight global upper bounds. This type of optimization approach – also common to GD – is known as **majorization minimization**. We now review a different, but equivalent, perspective on EM that interprets it as an **alternate optimization** procedure. This is akin to the interpretation given in the preceding section for K-means clustering.

To frame EM as alternate minimization, consider the **bi-level optimization problem**

$$\min_\theta \frac{1}{N} \sum_{n=1}^{N} \left\{ \min_{q_n(z_n|x_n)} F(q_n(z_n|x_n) || p(x_n, z_n|\theta)) \right\}. \tag{7.93}$$

In it, the outer minimization is carried out over the model parameter vector θ, and the inner optimizations are over the soft predictors $q_n(z_n|x_n)$, one for each data point $n = 1, \ldots, N$. Note that the separate minimization of the soft predictors for every n in (7.93) comes with no loss

Figure 7.13 Diagram representing the distributions involved in the EM algorithm, which includes the joint distribution $p(x, z|\theta)$ and the soft predictors $q_n(z_n|x_n)$.

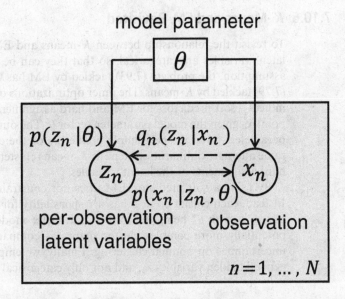

of optimality, since each soft predictor $q_n(z_n|x_n)$ affects only the term corresponding to the nth data point.

We can now address problem (7.93) via alternate optimization by first optimizing over the soft predictors $q_n(z_n|x_n)$ for all $n = 1, \ldots, N$ given a fixed model parameter vector θ; and then over the model parameter θ given fixed soft predictors $q_n(z_n|x_n)$ for all $n = 1, \ldots, N$. An illustration of all distributions involved in this process is in Fig. 7.13. As we discuss next, the first optimization coincides with the E step, while the second is equivalent to the M step.

Given the current model parameter iterate $\theta^{(i)}$, the first optimization over the soft predictors amounts to the separate minimizations

$$\min_{q_n(z_n|x_n)} \text{F}(q_n(z_n|x_n)||p(x_n, z_n|\theta^{(i)})) \tag{7.94}$$

for each example $n = 1, \ldots, N$, which yield as solutions the posteriors $p(z_n|x_n, \theta^{(i)})$. This recovers the E step. For fixed soft predictors $q_n^{(i)}(z_n|x_n) = p(z_n|x_n, \theta^{(i)})$, as we have seen, the optimization in (7.93) over the model parameter vector θ is equivalent to the supervised problem

$$\min_{\theta} \text{Q}(\theta|\theta^{(i)}), \tag{7.95}$$

in the sense that it returns the same optimal solution. This amounts to the M step.

All in all, we can interpret EM as applying alternate optimization with respect to model parameters θ on the one hand, and soft predictors $\{q_n(z_n|x_n)\}$ on the other hand, to the **training free energy**

$$\frac{1}{N} \sum_{n=1}^{N} \text{F}(q_n(z_n|x_n)||p(x_n, z_n|\theta)). \tag{7.96}$$

7.10.6 K-Means and EM, Revisited

To revisit the relationship between K-means and EM, consider the special case in which the latent variables are categorical, so that they can be interpreted as cluster indices. Under this assumption, the problem (7.93) tackled by EM has the same bi-level structure as the problem (7.79) tackled by K-means. The inner optimizations obtain the optimal predictors of the cluster indices – soft predictors for EM and hard assignments for K-means – separately for each data point x_n given the model parameter vector θ. The outer optimization minimizes over the model parameters given the per-sample predictors of the cluster indices. Therefore, EM follows the same alternate optimization steps as K-means clustering, but with "soft" imputations in lieu of hard assignments of the latent variables.

By using a soft predictor, EM does not constrain each point x_n to lie in a single cluster. Instead, each cluster $z_n = k$ has a **responsibility** for point x_n given by the probability $p(z_n = k|x_n, \theta^*)$, with θ^* being the trained parameter model. This flexibility allows EM to discover potentially more complex cluster shapes as compared to K-means, and to capture residual uncertainties on optimal clustering. Finally, we emphasize again that EM can accommodate general hidden variables z_n, and not only categorical variables like K-means clustering.

7.10.7 EM for Mixture of Gaussians

Consider the mixture-of-Gaussians model defined as

$$z \sim p(z|\theta) = \text{Cat}(z|\pi)$$
$$(x|z = k) \sim p(x|z = k, \theta) = \mathcal{N}(x|\mu_k, \Sigma_k) \text{ for } k = 0, 1, \ldots, K - 1, \tag{7.97}$$

with model parameters $\theta = \{\pi, \mu_0, \ldots, \mu_{K-1}, \Sigma_0, \ldots, \Sigma_{K-1}\}$. The latent-variable model (7.97) can be considered as the unsupervised version of the QDA model for the fully observed, or supervised, case studied in Sec. 6.5.5. This is in the sense that, when the variable z is observable, the model reduces to QDA. We now develop the EM algorithm for this model by detailing E and M steps.

E Step. In the E step, given the current model parameters $\theta^{(i)}$, we compute the posterior of the latent variable for each data point n as

$$p(z_{k,n} = 1|x_n, \theta^{(i)}) = \frac{p(x_n, z_{k,n} = 1|\theta^{(i)})}{p(x_n|\theta^{(i)})} = \frac{\pi_k^{(i)} \mathcal{N}(x_n|\mu_k^{(i)}, \Sigma_k^{(i)})}{\sum_{j=0}^{K-1} \pi_j^{(i)} \mathcal{N}(x_n|\mu_j^{(i)}, \Sigma_j^{(i)})} = \bar{z}_{k,n}^{(i)}. \tag{7.98}$$

The probability $\bar{z}_{k,n}^{(i)}$, which is equal to the average $\bar{z}_{k,n}^{(i)} = \text{E}_{z_n \sim p(z_n|x_n, \theta^{(i)})}[z_{k,n}]$, represents the **responsibility** of cluster k in explaining data point x_n.

M Step. In the M step, given the soft predictors, or responsibilities, in (7.98), we solve the supervised problem (7.95). To this end, we need the supervised log-loss

$$-\log p(x_n, z_n|\theta) = \sum_{k=0}^{K-1} z_{k,n}(-\log \pi_k - \log \mathcal{N}(x_n|\mu_k, \Sigma_k)), \tag{7.99}$$

and its average

$$E_{z_n \sim p(z_n|x_n, \theta^{(i)})}\left[\sum_{k=0}^{K-1} z_{k,n}(-\log \pi_k - \log \mathcal{N}(x_n|\mu_k, \Sigma_k))\right]$$

$$= \sum_{k=0}^{K-1} \bar{z}_{k,n}^{(i)}(-\log \pi_k - \log \mathcal{N}(x_n|\mu_k, \Sigma_k)). \tag{7.100}$$

Using (7.100) in the definition of function $Q(\theta|\theta^{(i)})$ in (7.92), the problem to be solved in the M step is the same as addressed for QDA in Sec. 6.5.5, with the caveat that the soft imputation variable $\bar{z}_{k,n}^{(i)}$ is used in lieu of $z_{k,n}$, i.e.,

$$\min_\theta \sum_{n=1}^{N} \sum_{k=0}^{K-1} \bar{z}_{k,n}^{(i)}(-\log \pi_k - \log \mathcal{N}(x_n|\mu_k, \Sigma_k)). \tag{7.101}$$

By imposing the first-order optimality condition as in Sec. 6.5.5, we finally obtain the next iterate $\theta^{(i+1)} = \{\pi_k^{(i+1)}, \mu_k^{(i+1)}, \Sigma_k^{(i+1)}\}_{k=0}^{K-1}$ with

$$\pi_k^{(i+1)} = \frac{\sum_{n=1}^{N} \bar{z}_{k,n}^{(i)}}{N} = \frac{\bar{N}[k]}{N} \tag{7.102}$$

$$\mu_k^{(i+1)} = \frac{1}{\bar{N}[k]} \sum_{n=1}^{N} \bar{z}_{k,n}^{(i)} x_n \tag{7.103}$$

$$\Sigma_k^{(i+1)} = \frac{1}{\bar{N}[k]} \sum_{n=1}^{N} \bar{z}_{k,n}^{(i)} (x_n - \mu_k^{(i+1)})(x_n - \mu_k^{(i+1)})^T, \tag{7.104}$$

where $\bar{N}[k] = \sum_{n=1}^{N} \bar{z}_{k,n}^{(i)}$.

From this derivation, setting $\Sigma_k = \epsilon I$ as a known parameter and letting $\epsilon \to 0$ can be seen to recover the K-means algorithm.

Example 7.14

Consider the data set in the upper left panel of Fig. 7.14. At a glance, we have two clusters, which correspond to the two directions over which data are supported. K-means clustering with $K = 2$ would not successfully identify the two clusters, given its reliance on the assumption of well-separated clusters in terms of Euclidean distance. We train a mixture-of-Gaussians model with $K = 2$ via EM as detailed in this subsection. The right-hand column shows the evolution of the model by displaying the contour lines for the component Gaussian distributions along the iteration index i. The left-hand column illustrates the evolution of the responsibilities $\bar{z}_{k,n}^{(i)}$ for the two classes, with darker shading corresponding to a larger probability of belonging to the cluster defined by the mean vector in black. It is observed that EM can separate the two classes into two clusters with shapes defined by the corresponding covariance matrices.

Figure 7.14 EM algorithm used to train a mixture-of-Gaussians model. Top row: data set $\mathcal{D}$ and initialized model represented by the contour lines for the component Gaussian distributions; second row: updated responsibilities $\bar{z}_{k,n}^{(i)}$ after the E step at iteration $i = 1$, with shading referring to the respective clusters in the right-hand column; and third and fourth rows: as for the second row but for iterations $i = 4$ and $i = 8$.

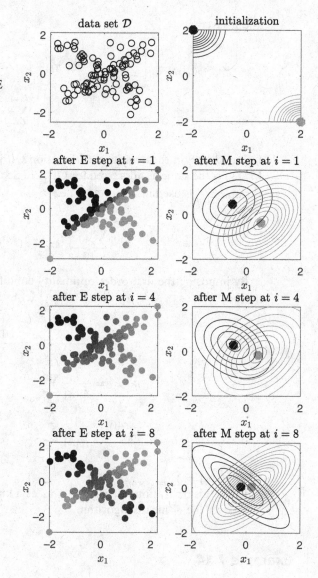

7.10.8 Probabilistic PCA

As another example of an application of EM, **probabilistic PCA (PPCA)** describes the data in terms of a latent vector comprising $M < D$ features via the directed generative model

$$z_n \sim \mathcal{N}(0, I_M)$$
$$(x_n | z_n = z) \sim \mathcal{N}(Wz + \mu, \sigma^2 I_D), \tag{7.105}$$

with model parameters $\theta = \{W, \mu, \sigma\}$. Equivalently, PPCA assumes the stochastic decoder

$$x_n = W z_n + \mu + q_n = \sum_{m=1}^{M} w_m z_{n,m} + \mu + q_n, \qquad (7.106)$$

with $z_{n,m} = [z_n]_m$, independent additive noise $q_n \sim \mathcal{N}(0, \sigma^2 I_D)$, and dictionary matrix $W = [w_1 w_2 \cdots w_M]$. As compared to the deterministic decoder (7.21) of PCA, the stochastic decoder (7.106) explicitly accounts for the approximation error in the reconstruction of x_n from z_n via the noise term q_n. The development of the EM algorithms follows from the same general steps detailed above for the mixture-of-Gaussians model.

7.10.9 Convergence of EM

By conditions (7.89) and (7.90), at each iteration EM minimizes an upper bound on the training log-loss $L_{\mathcal{D}}(\theta)$ that is tight at the current iterate. By the same arguments as presented in Sec. 5.8, these properties guarantee that the value $L_{\mathcal{D}}(\theta^{(i)})$ of the training log-loss cannot increase from one iteration to the next, which, in turn, implies that EM converges. The last open question is whether the iterates converge to a stationary point of the training log-loss $L_{\mathcal{D}}(\theta)$: When the EM algorithm stops, does it stop at a stationary point of $L_{\mathcal{D}}(\theta)$?

To address this question, we first observe that the local approximant $\tilde{L}_{\mathcal{D}}(\theta|\theta^{(i)})$ not only matches the value of the training log-loss $L_{\mathcal{D}}(\theta)$ at the current iterate as per (7.89), but it also reproduces the gradient of the training log-loss as in

$$\nabla_\theta \tilde{L}_{\mathcal{D}}(\theta|\theta^{(i)})|_{\theta=\theta^{(i)}} = \nabla_\theta L_{\mathcal{D}}(\theta)|_{\theta=\theta^{(i)}}. \qquad (7.107)$$

This follows directly from Fisher's identity (7.47) as detailed in Appendix 7.E.

Consider now a **fixed point** of the EM algorithm, i.e., an iterate $\theta^{(i)}$ at which EM stops, in the sense that the next iterate equals the current one, $\theta^{(i+1)} = \theta^{(i)}$. Since the next iterate minimizes the approximant $\tilde{L}_{\mathcal{D}}(\theta|\theta^{(i)})$, it must be a stationary point of the approximant, i.e.,

$$\nabla_\theta \tilde{L}_{\mathcal{D}}(\theta|\theta^{(i)})|_{\theta=\theta^{(i+1)}} = 0_D. \qquad (7.108)$$

Therefore, given that $\theta^{(i)}$ is a fixed point of EM, we have, by (7.107),

$$\nabla_\theta \tilde{L}_{\mathcal{D}}(\theta|\theta^{(i)})|_{\theta=\theta^{(i+1)}} = \nabla_\theta \tilde{L}_{\mathcal{D}}(\theta|\theta^{(i)})|_{\theta=\theta^{(i)}} = \nabla_\theta L_{\mathcal{D}}(\theta)|_{\theta=\theta^{(i)}} = 0_D. \qquad (7.109)$$

This confirms that a fixed point of the EM algorithm is a stationary point of the training log-loss.

7.10.10 Scaling EM

The EM algorithm is generally impractical to implement for large-scale problems, with the main computational challenges being:

- **E step**. Computing the posterior $p(z|x, \theta^{(i)})$ is intractable when the latent vector z is defined over a large domain.
- **M step**. Averaging over the latent vectors $z_n \sim p(z_n|x_n, \theta^{(i)})$ and carrying out exact optimization over θ in the M step are computationally infeasible for large-dimensional problems.

Common solutions to these problems include:

- **E step**. Parametrize the soft predictor as a **variational distribution** $q(z|x, \varphi)$ dependent on a vector of parameters φ, and minimize over vector φ via GD to obtain an approximation of the optimal soft predictor.
- **M step**. Approximate the expectation over the latent vectors via a Monte Carlo average by drawing samples from the soft predictor, and use GD to optimize over θ.

All these solutions are adopted by **variational EM (VEM)**, which is detailed in Chapter 10.

7.10.11 Beyond EM

EM tackles the ML problem. ML learning may, however, not always provide the best training framework, and alternative approaches to train directed generative models exist that adopt different learning criteria.

A first problem with ML is that it tends to provide inclusive and "blurry" estimates of the population distribution, since it strongly penalizes models that assign low probability to some data points (see Sec. 11.2 for further discussion). An example of this phenomenon is shown in Fig. 7.15, which displays the true underlying population distribution $p(x)$, along with a mixture-of-Gaussians distribution, with $K = 2$, trained via EM. It is observed that the trained distribution covers the support of the population distribution well, but fails to capture its fine-grained properties in terms of its modes.

Another limitation of ML is that it requires an explicit likelihood function $p(x|z, \theta)$ in order to enable optimization. This may be problematic in applications such as image generation, in which powerful models inject random noise z as an input to a large neural network to generate an image x. For these models, it is generally not possible to give an explicit form for the likelihood $p(x|z, \theta)$. Solutions to this problem will be covered in Chapter 11.

Figure 7.15 Underlying population distribution $p(x)$, along with the a mixture-of-Gaussians distribution, with $K = 2$, trained via EM.

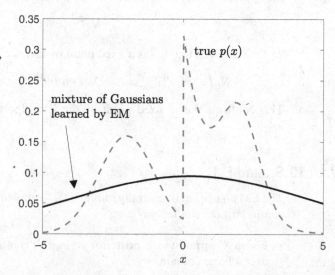

7.10.12 EM As Behavioral Principle for Biological Systems

In this subsection, we draw an intriguing connection between EM and a prominent "unified brain theory" that has the ambition of explaining the relationship among perception, action, and learning. The **free energy principle** postulates that living systems optimize jointly internal models of the "environment", predictors of sensory stimuli, and action policies so as to minimize their information-theoretic surprise. The rationale for optimizing this objective relates to the idea that living beings have states they expect to find themselves in under normal conditions – say in the water for a fish, or within a certain range of temperatures – and such states should be "unsurprising" under the system's model of the environment. The system should actively ensure that it remains in such states of low information-theoretic surprise.

The free energy principle subsumes two conceptual frameworks: *perceptual inference* and *active inference*. Perceptual inference refers to the "Bayesian brain" idea that the brain produces probabilistic predictions of the states of the environment that are most likely to have caused the observation of the sensory data. These predictions minimize the information-theoretic surprise associated with perception. (They are also, according to the theory, how living beings perceive the world.) Active inference posits that living organisms act on their environment so as to change the sensory input in a manner that minimizes information-theoretic surprise. As we discuss next, perceptual inference can be formulated mathematically as optimizing the same criterion as the EM algorithm.

An internal *generative model* of the environment is defined by a joint distribution $p(x, z|\theta)$ relating the unobservable true state of the world z to the sensory observation x. The joint distribution $p(x, z|\theta)$, which can be learned via parameter θ, should account for: (*i*) the system's experience, or prior knowledge, about the states z that are most desirable for the system, which should be given a larger distribution value $p(z|\theta)$; and (*ii*) the "causal" relationship $p(x|z, \theta)$ between state of the environment z and sensory stimuli x.

The prediction of the state of the environment z given the sensory observation x is described by a stochastic mapping $q(z|x)$. This is known as "recognition density" in the literature on the free energy principle.

The information-theoretic surprise of an observation x is given by its log-loss $-\log p(x|\theta)$, which is, however, hard to compute. By (7.86) the free energy $F(q(z|x)||p(x, z|\theta))$ is an upper bound on the surprise, which has the important advantage of depending directly on the generative model $p(x, z|\theta)$ and on the recognition density $q(z|x)$. For a given set of observations $\{(x_n)_{n=1}^{N}\}$, perceptual inference stipulates that the parameters θ of the generative model and the inferential mapping $q(z|x)$ are jointly optimized by addressing the problem of minimizing the training free energy. As anticipated, this problem coincides with the minimization (7.93) tackled by the EM algorithm, showing the direct connection between EM and the free energy principle. A summary of the correspondence between quantities relevant for perceptual inference and variables used in the EM algorithm can be found in Table 7.1.

Active inference posits that the system can also act on the environment with the aim of modifying the distribution of the sensory input x so as to further decrease the free energy. In order to account for the causal effect of actions on the environment, the generative model is extended to account for the dynamics of the environment over time. See Recommended Resources, Sec. 7.12, for details.

Table 7.1 Correspondence between quantities introduced by the free energy principle and variables used by EM

neuroscience	machine learning model
sensory observations	x
state of the environment	z
belief about desirable states of the environment	$p(z\|\theta)$
belief about how environmental states map to sensory data	$p(x\|z,\theta)$
inference of environmental states given sensory observations	$q(z\|x)$
surprise	$-\log p(x\|\theta)$
bound on the surprise – free energy	$F(q(z\|x)\|\|p(x,z\|\theta))$

7.10.13 Directed Generative Models for Density Estimation

As discussed in this section, directed generative models can be used for clustering and data generation. In this final subsection, we briefly show that they can also be used to carry out density estimation. To this end, consider the model

$$z \sim p(z)$$
$$x = G(z|\theta) \tag{7.110}$$

for some *invertible* decoding function $G(\cdot|\theta)$. We denote the inverse of function $G(\cdot|\theta)$ as $G^{-1}(\cdot|\theta)$. By the change-of-variable formula, we can write the distribution of x as

$$p(x) = p(z = G^{-1}(x|\theta)) \cdot \det(\nabla_\theta G^{-1}(x|\theta)). \tag{7.111}$$

Therefore, if the determinant of the Jacobian $\det(\nabla_\theta G^{-1}(x|\theta))$ is easy to compute, one can obtain the distribution of x from the trained model $G(\cdot|\theta)$. This approach is known as **normalizing flows** (since the distribution of z "flows" through $G(\cdot|\theta)$), and is a subject of active research at the time of writing.

7.11   Summary

- Unsupervised learning operates over unlabeled data sets to extract useful information about the underlying, unknown, data distribution. Specific tasks include density estimation, feature extraction, and data generation.
- Density estimation can be performed via non-parametric methods, such as KDE, or parametric techniques, such as CDL.
- Most model classes used for unsupervised learning include per-data point latent variables. The main classes of models are illustrated in Fig. 7.3, and they include autoencoders, discriminative models, undirected generative models, and directed generative models.
- Many unsupervised learning methods for latent-variable models use auxiliary supervised learning tasks as subroutines to tackle the original unsupervised learning problem:
  - CDL for density estimation trains a binary classifier to distinguish true samples from noise.
  - Training algorithms for autoencoders use self-supervised learning whereby the input is used as target.

- CRL for discriminative models solves a softmax regression problem with artificially defined "positive" and "negative" examples, and the label of the positive example is used as target.
- The gradient used for RBMs – an example of undirected generative models – depends on the gradient of the "supervised" log-loss computed from data, and on the optimal soft predictors of the latent variables.
- The M step of the EM algorithm for directed generative models addresses a supervised learning problem with "soft imputation" based on optimized soft predictors of the latent variables.

- A common aspect of unsupervised learning methods based on latent-variable models is the reliance on the optimal hard or soft prediction of the latent variables for fixed model parameters. Most notably, the EM algorithm carries out optimal soft prediction of the latent variables at each iteration in the E step.

7.12 Recommended Resources

The study mentioned in the introduction involving a child-mounted camera is reported in [1]. Classical topics such as histogram, KDE, K-means clustering, and the EM algorithm can be found in the standard texts [2, 3]. A more in-depth discussion on latent-variable models is available in [4, 5]. For an overview of CRL, please see [6], and for energy-based models a useful reference is [7]. Score matching and NCE are discussed in [8, 9]. Informative references on the ELBO and extensions thereof include [10]. The free energy is presented as a behavioral principle in [11], and useful summaries can be found in [12, 13]. Among unsupervised learning methods that do not follow the models covered in this chapter, an important example is t-SNE, introduced in [14], which carries out dimensionality reduction by matching "neighbor selection" probabilities in the original and reduced-dimension space. Properties of the log-loss, including the data processing inequality, are detailed in [15].

Problems

7.1 Load the data set xshape from the book's website. It consists of $N = 500$ two-dimensional points $\{(x_n)_{n=1}^{N}\}$ ($D = 2$). (a*) Plot the points on the (x_1, x_2) plane. Note that no labels are provided. (b*) Divide the region of the plane with $x_1 \in [0, 1]$ and $x_2 \in [0, 1]$ into squares of size 0.1×0.1. Evaluate the histogram by computing the fraction of sample in each square. Illustrate the histogram with a 3D plot on the same figure containing the samples. (c*) Consider a kernel $\kappa_h(x) = \mathcal{N}(x|0, hI)$ with bandwidth $h = 0.01$ and draw a 3D plot of the resulting KDE. (d*) Repeat point (c*) with $h = 0.001$ and comment on your results.

7.2 For the `horizontal` data set, which you can find on the book's website, we consider the use of CDL. (a*) To this end, let us first generate $N = 100$ noise samples in the region of the plane $x_1 \in [-40, 40]$ and $x_2 \in [-40, 40]$. (b*) With these samples, use logistic regression to train a classifier with feature vector $u(x) = [1, x_1^2, x_2^2]^T$, and plot the resulting density estimate produced by CDL. (c*) What happens if the noise samples are selected to be uniform in the region of the plane with $x_1 \in [-20, 20]$ and $x_2 \in [-20, 20]$? Explain your result.

7.3 We are given a data set of $N = 1000$ data points for a random variable x that can take values in the interval [0,1]. We choose a uniform noise distribution $r(x)$ in the same interval, and generate 1000 noise samples from it. Given these 2000 samples, we train a soft classifier. The trained classifier assigns to the data label t $= 1$ a probability 0.2 in the interval [0,0.5] and a probability 0.6 in the interval [0.5,1]. What is the estimated distribution for $x = 0.6$ produced by CDL?

7.4 Prove that the ERM problem for PCA can be written as

$$\max_{W} \sum_{n=1}^{N} \text{tr}\left(W^T \hat{\Sigma}_{\mathcal{D}} W\right),$$

where $\hat{\Sigma}_{\mathcal{D}}$ is the empirical covariance matrix (7.26). [Hints: we have $\text{tr}(AB) = \text{tr}(BA)$ for matrices that commute; we have $W^T W = I_M$ since the columns of matrix W are orthonormal; and the trace is a linear operator.] Argue that the optimal solution is given by the matrix of M principal eigenvectors of matrix $\hat{\Sigma}_{\mathcal{D}}$.

7.5 Load the data set `usps29` from the book's website. It contains $N = 1750$ examples of grayscale 16×16 images of handwritten digits 2 and 9. (a*) Plot a few images in grayscale. (b*) Compute the first two principal components. (c*) Show the first two principal components as images. (d*) Reconstruct the first image in the training set using the first two principal components, and compare original and reconstructed images. (e*) In a plane with axes representing the contributions of the first two components, represent the entire data set by distinguishing with images of 2s (the first 929 images) and 9s (the remaining images) different markers. (f*) Reconstruct the first image using the first two principal components and compare original and reconstructed images.

7.6 Can the columns of a dictionary matrix W be orthogonal for an overcomplete dictionary? Why or why not?

7.7 Load the data set `xshape` from the book's website. It consists of $N = 500$ two-dimensional points $\{(x_n)_{n=1}^{N}\}_{n=1}^{N}$ ($D = 2$). (a*) Remove the empirical mean from the training set. (b*) Plot the resulting points on the (x_1, x_2) plane. Note that no labels are provided. (c*) Compute the first principal component on the resulting training set. (d*) Draw the first principal component in the figure containing the training points as a vector. (e*) Reconstruct the 100th training point using the first principal component. Why does PCA fail to provide an accurate reconstruction of this training point? (f*) Implement sparse dictionary learning with $M = 2$ and $M' = 1$. Draw the columns of matrix W in a plot containing the data points.

7.8 (a) Prove that, in RBMs, the latent units are conditionally independent given the visible units. (b) Prove also that the visible units are conditionally independent given the latent units. (c) Write the corresponding conditional distributions.

7.9 From the book's website, load the data set `3dclustering`, which includes $N = 150$ points in a three-dimensional space ($D = 3$). (a*) Plot the data set. (b*) Implement K-means clustering with $K = 2$ clusters. For initialization, pick K prototypes randomly in the data set. For each iteration, produce a figure that shows the points in the two classes with different shading or colors. In the same figure, also indicate the current cluster centers.

7.10 For the `3dclustering` dataset on the book's website, we wish to implement the EM algorithm to train a mixture-of-Gaussians model with $K = 2$. (a*) To this end, pick K points randomly in the data set for the class means μ_k, with $k = 1,2$; initialize the covariance matrices as $\Sigma_k = 10 \cdot I_3$ and the a priori probability of each class as $\pi_k = 1/K$. (b*) At each iteration, show the current cluster means and represent with shading or colors the current a posteriori probabilities, i.e., responsibilities, $p(z_n = k|x_n, \theta)$. (c*) Repeat the experiment to check the impact of the random initialization of cluster centers.

7.11 Consider the data set obtained in Problem 4.14 that includes images of digits 0 and 1 from the data set `USPSdataset9296`. (a*) Using these data, train the generative directed naive Bayes model defined as

$$z \sim \text{Bern}(\pi)$$
$$(x|z = z) \sim \prod_{ij} \text{Bern}(p_{ij|z}),$$

where the product is taken over the pixels of the image, and $p_{ij|z}$ is the probability corresponding to pixel (i, j) when $z = z$. (b*) Use the trained model to assign a value z_n to each data point x_n by using a MAP predictor. (c*) Do you see a correspondence between the values of z_n and the true labels t_n? Note that the model was not given the labels during training.

7.12 Consider the following generative directed model:

$$z = \text{Bern}(0.5) + 1,$$

where $\text{Bern}(0.5)$ represents a rv with the indicated pmf, and

$$(x|z = k) \sim \mathcal{N}(k\theta, 0.5),$$

where the model parameter is θ. Note that z takes values 1 and 2 with equal probability. We observe a sample x = 2.75. We wish to develop the EM algorithm to tackle the problem of ML estimation of the parameter θ given this single observation ($N = 1$). (a) Write the likelihood function $p(x = 2.75|\theta)$ of θ for the given observation. (b*) Plot the log-loss $-\log p(x = 2.75|\theta)$ for θ in the interval $[-5, 5]$ and check that it is not convex. (c) In the E step, we need to compute $p(z|x = 2.75, \theta^{(i)})$. Write the probability $p(z = 2|x = 2.75, \theta^{(i)})$ as a function of $\theta^{(i)}$ (note that $p(z = 1|x = 2.75, \theta^{(i)}) = 1 - p(z = 2|x = 2.75, \theta^{(i)})$). (d) In the M step, we instead need to solve the problem of minimizing

$$E_{z \sim p(z|x=2.75, \theta^{(i)})}[-\log p(x = 2.75, z|\theta)]$$

over θ. Formulate and solve the problem to obtain $\theta^{(i+1)}$ as a function of $\theta^{(i)}$.

7.13 Consider again images 0 and 1 from `USPSdataset9296` (see Problem 7.11). (a) Assuming a linear encoder $F(x|\theta) = Ax$ with matrix A of dimensions 2×256, compute the gradient of the loss function for CRL with respect to A. (b*) Implement gradient descent and plot the two-dimensional vector of representations for both images representing 0 and 1. (c*) Compare your results from point (b*) with the same plot obtained with PCA.

Appendices

Appendix 7.A: Proof of Fisher's Identity

In this appendix, we prove **Fisher's identity** (7.47). This formula states that the gradient of the unsupervised log-loss with respect to θ equals the average of the gradient of the supervised log-loss, where the average is taken with respect to the optimal soft predictor $p(z|x, \theta)$ computed at the given value θ of interest. The following derivation applies for discrete rvs z but the formula applies more generally:

$$\nabla_\theta(-\log p(x|\theta)) = \nabla_\theta\left(-\log\left(\sum_z p(x,z|\theta)\right)\right)$$

$$= -\sum_z \frac{\nabla_\theta p(x,z|\theta)}{\sum_{z'} p(x,z'|\theta)}$$

$$= -\sum_z \frac{\nabla_\theta p(x,z|\theta)}{p(x|\theta)} \cdot \frac{p(x,z|\theta)}{p(x,z|\theta)}$$

$$= -\sum_z \nabla_\theta \log p(x,z|\theta) \frac{p(x,z|\theta)}{p(x|\theta)}$$

$$= \sum_z p(z|x,\theta)(-\nabla_\theta \log p(x,z|\theta)), \tag{7.112}$$

where we have used the elementary identity $\nabla_\theta \log p(x,z|\theta) = \nabla_\theta p(x,z|\theta)/p(x,z|\theta)$.

Appendix 7.B: Contrastive Divergence

If one had access to "negative" samples (x,z) drawn from the joint distribution $p(x,z|\theta)$, the expectations for the first terms in the partial derivatives (7.50)–(7.52) for RBM could be estimated via empirical averages. However, generating such samples is generally of intractable complexity. In practice, one can resort to generating samples *approximately* drawn from distribution $p(x,z|\theta)$. This can be done via MCMC methods such as SGLD, presented in Sec. 7.7.3. A particularly effective MCMC method for RBMs is contrastive divergence.

In the "negative" phase, **contrastive divergence** applies an MCMC method known as Gibbs sampling. **Gibbs sampling** produces a sequence of samples (x,z), here described by their respective realizations, by alternately drawing sample x for a fixed z, and sample z for a fixed x. Specifically, at each iteration, given the current sample z, Gibbs sampling draws a new sample x of the observations from the conditional distribution $p(x|z,\theta)$; and then, given the updated x, it draws a new sample z from the conditional distribution $p(z|x,\theta)$. Sampling from both conditional distributions is tractable in RBMs, since the entries of rv x are conditionally independent given $z = z$; and the entries of rv z are conditional independent given x $= x$ (cf. (7.48)).

The most typical, and simplest, implementation of contrastive divergence applies a single iteration of Gibbs sampling starting from the observation x. Specifically, given a data point x_n from the training set $\mathcal{D}$, contrastive divergence produces a negative sample (x,z) via the following steps:

1. Sample latent rv z $\sim \prod_{j=1}^M \text{Bern}(z_j|\sigma(w_j^T x_n + b_j))$, where z_j is the jth entry of vector z and w_j is the jth column of matrix W (each rv z_j is hence sampled independently of all other latent rvs).
2. For the sampled latent rv z $= z$, sample x $\sim \prod_{i=1}^D \text{Bern}(x_i|\sigma(\tilde{w}_i^T z + a_i))$, where x_i is the ith entry of vector x and $\tilde{w}_i$ is the ith row of matrix W (each rv x_i is hence sampled independently of all other observed rvs).

The obtained realized sample (x,z) is then used to estimate the expectations in the first terms of (7.50)–(7.52). For instance, for the last partial derivative, we have the estimate $E_{(x_i,z_j)\sim p(x_i,z_j|\theta)}[x_i z_j] \simeq x_i z_j$, where x_i and z_j are the relevant entries of the sample (x,z).

Appendix 7.C: Alternative Proofs of the Free Energy (or ELBO) Inequality (7.86)

This appendix collects two alternative derivations of the key inequality (7.86).

- **From the chain rule of probability.** The chain rule of probability implies the equality $p(x|\theta) = p(x,z|\theta)/p(z|x,\theta)$. Using this identity, the log-loss can be written as

$$-\log p(x|\theta) = -\log\left(\frac{p(x,z|\theta)}{p(z|x,\theta)}\right)$$

$$= -E_{z \sim q(z|x)}\left[\log\left(\frac{p(x,z|\theta)}{p(z|x,\theta)}\right)\right], \qquad (7.113)$$

where the first equality holds for any value z, which in turn implies that the last equality applies for any distribution $q(z|x)$. Multiplying and dividing by $q(z|x)$ and rearranging the terms yields (7.85), concluding the proof.

- **From the data processing inequality.** Define $\pi(x) = \delta(x - x')$ – with $\delta(\cdot)$ being a Kronecker or Dirac delta function, respectively for discrete or continuous variables – as a distribution concentrated at a point $x' \in \mathbb{R}^D$. We have

$$-\log p(x'|\theta) = \text{KL}\left(\pi(x)||p(x|\theta)\right)$$

$$\leq \text{KL}\left(\pi(x)q(z|x)||p(x,z|\theta)\right)$$

$$= E_{z \sim q(z|x')}\left[\log\left(\frac{p(x',z|\theta)}{q(z|x')}\right)\right] = \text{F}(q(z|x')||p(x',z|\theta)). \qquad (7.114)$$

The inequality in the second line is an instance of the data processing inequality (see Recommended Resources, Sec. 7.12).

Appendix 7.D: An Estimation-Based Perspective on the Free Energy, and Multi-sample Generalizations

In this appendix, we view the free energy via the problem of estimating the marginal of the observation for directed generative models, i.e., $p(x|\theta) = E_{z \sim p(z|\theta)}\left[p(x|z,\theta)\right]$. We will see that this perspective yields a generalization of the free energy – or equivalently of its negative, the ELBO – that is based on a more accurate estimate of the marginal $p(x|\theta)$.

Free Energy as an Average Estimate of the Log-Loss. Suppose that we wish to evaluate the marginal $p(x|\theta)$, but computing the expectation over the latent variables $p(z|\theta)$ is intractable. A simple estimate can be obtained as follows: Draw one sample $z \sim p(z|\theta)$ for the latent rv, and then output the estimate

$$p(x|\theta) \simeq p(x|z,\theta). \qquad (7.115)$$

One could, of course, improve this estimate by drawing multiple i.i.d. samples $z \sim p(z|\theta)$ and evaluating the corresponding empirical average of the conditional distribution $p(x|z,\theta)$, but we will limit ourselves to the one-sample estimator for the time being. Multi-sample estimators will be studied later in this appendix.

The one-sample estimator (7.115) is unbiased, in the sense that its average equals the marginal $p(x|\theta)$ – this is straightforward to verify. But it can have high variance. To see this, assume that there is a set of values z that "explain well" data x, that is, for which the conditional distribution $p(x|z,\theta)$ is large, while for the rest of the values z, $p(x|z,\theta)$ is small. By drawing a single value z

from the prior, which is independent x, we may get either a large value or a small value for the rv $p(x|z,\theta)$ in (7.115), causing the variance to be large.

Apart from drawing multiple samples, we can improve the one-sample estimate (7.115) by using **importance weighting**: Instead of sampling the latent variable from the prior $z \sim p(z|\theta)$, which is independent of x, we sample it from an auxiliary distribution $q(z|x)$. The conditional distribution $q(z|x)$ is arbitrary and, crucially, it can be chosen to be a function of x. Using the sample $z \sim q(z|x)$, we then estimate the marginal of the observation as

$$p(x|\theta) \simeq \frac{p(x,z|\theta)}{q(z|x)}. \tag{7.116}$$

It can be easily checked that this estimate is also unbiased, and that its variance can be smaller than for the original estimator. Notably, if we set $q(z|x) = p(z|x,\theta)$, the estimate is exact.

Taking the average of the negative logarithm of the importance-weighted one-sample estimator (7.116) yields the free energy, i.e., we have

$$F(q(z|x)||p(x,z|\theta)) = E_{z \sim q(z|x)}\left[-\log\left(\frac{p(x,z|\theta)}{q(z|x)}\right)\right]. \tag{7.117}$$

Accordingly, the "sample free energy" term $-\log(p(x,z|\theta)/q(z|x))$ can be thought of as a one-sample estimate of the log-loss $-\log p(x|\theta)$. Unlike the one-sample estimates (7.115) and (7.116), this estimate is biased since we have the equality (7.85).

A Generalization of the Free Energy. We have seen that the free energy can be interpreted as the expectation of a biased estimator of the log-loss $-\log p(x|\theta)$, which is in turn obtained from an unbiased estimator of the marginal $p(x|\theta)$. We can therefore generalize the definition of free energy by considering any unbiased estimator of $p(x|\theta)$, as we detail next.

To this end, consider an arbitrary unbiased estimator $t(x,z|\theta)$ of $p(x|\theta)$, where $z \sim q(z|x)$ for some conditional distribution $q(z|x)$, i.e., we have $E_{z \sim q(z|x)}[t(x,z|\theta)] = p(x|\theta)$. For the free energy, as discussed, we set $t(x,z|\theta) = p(x,z|\theta)/q(z|x)$. The term $\log t(x,z|\theta)$ is a biased estimate of the log-loss $-\log p(x|\theta)$ since we have

$$E_{z \sim q(z|x)}[-\log t(x,z|\theta)] = -\log p(x|\theta) + E_{z \sim q(z|x)}\left[-\log\left(\frac{t(x,z|\theta)}{p(x|\theta)}\right)\right]. \tag{7.118}$$

Furthermore, by Jensen's inequality, we have the bound

$$E_{z \sim q(z|x)}[-\log t(x,z|\theta)] \geq -\log p(x|\theta), \tag{7.119}$$

which shows that, like the free energy, the left-hand side in (7.118) is an upper bound on the log-loss $-\log p(x|\theta)$.

In conclusion, given an arbitrary unbiased estimator $t(x,z|\theta)$ of the marginal $p(x|\theta)$, with $z \sim q(z|x)$, the expectation $E_{z \sim q(z|x)}[-\log t(x,z|\theta)]$ on the right-hand side of (7.118) can be thought of as a generalization of the free energy. This is in the sense that, by (7.119), it provides an upper bound on the log-loss. Equivalently, the negative of the left-hand side of (7.118) is a generalization of the ELBO, in the sense that it provides a lower bound on the log-probability of the evidence, $\log p(x|\theta)$.

Multi-sample Free Energy and Multi-sample ELBO. Given any unbiased estimator $t(x,z|\theta)$ of $p(x|\theta)$ with latent variable z drawn as $z \sim q(z|x)$, we can obtain an unbiased estimator with a lower variance by drawing $S > 1$ i.i.d. samples $z_s \sim q(z|x)$ with $s = 1, \ldots, S$, and then using the empirical average $1/S \sum_{s=1}^{S} t(x,z_s|\theta)$. In fact, the variance of this multi-sample estimate is

S times smaller than the variance of the original estimator with $S = 1$. Following the same reasoning for the case $S = 1$ yields the upper bound on the log-loss

$$-\log p(x|\theta) \leq -\mathrm{E}_{\{z_s\}_{s=1}^S \underset{\text{i.i.d.}}{\sim} q(z|x)}\left[\log\left(\frac{1}{S}\sum_{s=1}^S t(x, z_s|\theta)\right)\right].\tag{7.120}$$

It can be checked that, by Jensen's inequality, the right-hand side of (7.120) provides a tighter bound on the log-loss $-\log p(x|\theta)$ as S increases. Furthermore, as $S \to \infty$, this expression tends to the log-loss $-\log p(x|\theta)$ by the law of large numbers.

As an example, we can consider the multi-sample generalization of the free energy by setting $t(x, z|\theta) = \log(p(x, z|\theta)/q(z|x))$, which yields

$$\mathrm{F}_S(q(z|x)||p(x, z|\theta)) = -\mathrm{E}_{\{z_s\}_{s=1}^S \underset{\text{i.i.d.}}{\sim} q(z|x)}\left[\log\left(\frac{1}{S}\sum_{s=1}^S \frac{p(x, z_s|\theta)}{q(z_s|x)}\right)\right].\tag{7.121}$$

The negative of the multi-sample free energy $\mathrm{F}_S(q(z|x)||p(x, z|\theta))$ is known as **multi-sample ELBO**. Based on the discussion in the preceding paragraph, the multi-sample free energy provides a tighter upper bound on the log-loss, compared to the conventional free energy, and it is asymptotically, as $S \to \infty$, equal to the log-loss. Being a tighter bound on the log-loss for any choice of $q(z|x)$, the multi-sample free energy may provide a more effective criterion for the optimization of the model parameter θ when allowing for a suboptimal soft predictor $q(z|x)$ (see Sec. 7.10.10).

A further generalization of the multi-sample free energy/ ELBO can be obtained in terms of **Rényi's divergence**, and is given as

$$\frac{1}{1-\alpha}\log\left(\mathrm{E}_{\{z_s\}_{s=1}^S \underset{\text{i.i.d.}}{\sim} q(z|x)}\left[\left(\frac{1}{S}\sum_{s=1}^S \frac{p(x, z_s|\theta)}{q(z_s|x)}\right)^{1-\alpha}\right]\right),\tag{7.122}$$

for any $\alpha \in [0, 1)$. This bound equals the multi-sample free energy $\mathrm{F}_S(q(z|x)||p(x, z|\theta))$ in (7.121) for $\alpha = 0$.

Appendix 7.E: Proof of the Equality (7.107)

In this appendix, we prove the equality (7.107), which indicates a match between the gradient of the approximant optimized by EM and the gradient of the training log-loss at the current iterate. First note that, by (7.92), we have the equality $\nabla_\theta \tilde{L}(\theta|\theta^{(i)}) = \nabla_\theta Q(\theta|\theta^{(i)})$. Furthermore, we have

$$\nabla_\theta Q(\theta|\theta^{(i)}) = \frac{1}{N}\sum_{n=1}^N \mathrm{E}_{z_n \sim p(z_n|x_n, \theta^{(i)})}[-\nabla_\theta \log p(x_n, z_n|\theta)].\tag{7.123}$$

Evaluating the gradient for $\theta = \theta^{(i)}$, we finally get

$$\begin{aligned}\nabla_\theta Q(\theta|\theta^{(i)})|_{\theta=\theta^{(i)}} &= \frac{1}{N}\sum_{n=1}^N \mathrm{E}_{z_n \sim p(z_n|x_n, \theta^{(i)})}[-\nabla_\theta \log p(x_n, z_n|\theta^{(i)})]\\ &= \frac{1}{N}\sum_{n=1}^N (-\nabla_\theta \log p(x_n|\theta^{(i)}))\\ &= \nabla_\theta L_\mathcal{D}(\theta^{(i)}),\end{aligned}\tag{7.124}$$

where we have used Fisher's identity (7.47) in the second equality.

Appendix 7.F: Latent-Variable Models for Supervised Learning

In this appendix, we briefly discuss two ways of defining supervised learning models based on latent variables, namely mixture discriminative models and RBM-based discriminative models.

Mixture Discriminative Models. From Sec. 6.9, recall that a mixture discriminative model for supervised learning is of the form

$$q(t|x) = \sum_{c=0}^{C-1} \pi_c p(t|x, \theta_c), \tag{7.125}$$

which consists of a linear combination of C models, each with weight given by probability π_c, with $\sum_{c=0}^{C-1} \pi_c = 1$ and $\pi_c \geq 0$. The model can be equivalently defined by introducing a latent variable $z \sim \text{Cat}(\pi)$ with $\pi = [\pi_0, \ldots, \pi_{C-1}]^T$ as

$$q(t|x) = \mathrm{E}_{z \sim \text{Cat}(\pi)}[p(t|x, z, \theta)], \tag{7.126}$$

where we have written, with a slight abuse of notation, $p(t|x, z, \theta) = p(t|x, \theta_z)$.

For the mixture model (7.126), the log-loss is given as

$$-\log q(t|x) = -\log\left(\mathrm{E}_{z \sim p(z|\pi)}[p(t|x, z, \theta)]\right). \tag{7.127}$$

Apart from the presence of the conditioning on x, the form of the log-loss is the same as that considered in this chapter for unsupervised learning. Therefore, a learning algorithm based on EM can be directly derived by following the steps described in Sec. 7.10.

RBM-Based Discriminative Models. RBM-based discriminative models consist of RBM models $(\tilde{x}, z) \sim p(\tilde{x}, z|\theta)$, where the observable vector $\tilde{x}$ is split into two parts as $\tilde{x} = [x, t]$. In it, the vector x represents the input, while t is the target variable. Supervised learning can be implemented by adopting the techniques introduced in Sec. 7.7. For inference, one fixes the input $x = x$ and predicts t using optimal, or approximately optimal, soft prediction.

Bibliography

[1] A. E. Orhan, V. V. Gupta, and B. M. Lake, "Self-supervised learning through the eyes of a child," arXiv preprint arXiv:2007.16189, 2020.

[2] C. M. Bishop, *Pattern Recognition and Machine Learning*. Springer, 2006.

[3] T. Hastie, R. Tibshirani, and J. Friedman, *The Elements of Statistical Learning*. Springer, 2001.

[4] K. P. Murphy, *Machine Learning: A Probabilistic Perspective*. The MIT Press, 2012.

[5] D. Koller and N. Friedman, *Probabilistic Graphical Models: Principles and Techniques*. The MIT Press, 2009.

[6] P. H. Le-Khac, G. Healy, and A. F. Smeaton, "Contrastive representation learning: A

framework and review," *IEEE Access*, vol. 8, pp. 193907-193934, 2020.

[7] Y. LeCun, S. Chopra, R. Hadsell, M. Ranzato, and F. Huang, "A tutorial on energy-based learning," in *Predicting Structured Data*. The MIT Press, 2009.

[8] A. Hyvärinen, "Some extensions of score matching," *Computational Statistics & Data Analysis*, vol. 51, no. 5, pp. 2499–2512, 2007.

[9] M. Gutmann and A. Hyvärinen, "Noise-contrastive estimation: A new estimation principle for unnormalized statistical models," in *Proceedings of the 13th International Conference on Artificial Intelligence and Statistics*. JMLR Workshop and Conference Proceedings, 2010, pp. 297–304.

[10] A. Mnih and D. Rezende, "Variational inference for Monte Carlo objectives," in *The 33rd International Conference on Machine Learning*. ICML, 2016, pp. 2188–2196.

[11] K. Friston, "The free-energy principle: A unified brain theory?" *Nature Reviews Neuroscience*, vol. 11, no. 2, pp. 127–138, 2010.

[12] C. L. Buckley, C. S. Kim, S. McGregor, and A. K. Seth, "The free energy principle for action and perception: A mathematical review," *Journal of Mathematical Psychology*, vol. 81, pp. 55–79, 2017.

[13] G. Lindsay, *Models of the Mind*. Bloomsbury Sigma, 2021.

[14] L. Van der Maaten and G. Hinton, "Visualizing data using t-SNE," *Journal of Machine Learning Research*, vol. 9, no. 11, 2008.

[15] Y. Polyanskiy and Y. Wu, *Lecture Notes on Information Theory*. http://people.lids.mit.edu/yp/homepage/data/itlectures_v5.pdf, May 15, 2019.

Part III

Advanced Tools and Algorithms

8 Statistical Learning Theory

8.1 Overview

As introduced in Chapter 4, setting up a learning problem requires the selection of an inductive bias, which consists of a model class $\mathcal{H}$ and a training algorithm. By the no-free-lunch theorem, this first step is essential in order to make **generalization** possible. A trained model generalizes if it performs well *outside* the training set, on average with respect to the unknown population distribution. This chapter introduces theoretical tools that aim at understanding when a given combination of population distribution, inductive bias, and size of training set can yield successful generalization.

Throughout this chapter, we will focus on model classes $\mathcal{H}$ of hard predictors $\hat{t}(x)$ of the target variable t given an observation of the input x $= x$. The learner is given a training set $\mathcal{D} = \{(x_n, t_n)_{n=1}^N\}$, consisting of N data points generated i.i.d. from the unknown population distribution $p(x, t)$. It is important to emphasize that, in the theory to be developed, we will view the training set $\mathcal{D}$ as a set of rvs drawn i.i.d. from $p(x, t)$ rather than a fixed data set. The performance of a trained predictor $\hat{t}(\cdot) \in \mathcal{H}$ is measured by the **population loss**

$$L_p(\hat{t}(\cdot)) = \mathrm{E}_{(x, t) \sim p(x, t)}[\ell(t, \hat{t}(x))], \tag{8.1}$$

which evaluates the performance "outside the training set", that is, on a randomly generated test sample $(x, t) \sim p(x, t)$. Training algorithms tackle the minimization of criteria that depend on the **training loss**

$$L_{\mathcal{D}}(\hat{t}(\cdot)) = \frac{1}{N} \sum_{n=1}^N \ell(t_n, \hat{t}(x_n)), \tag{8.2}$$

which is an empirical estimate of the population loss. Most notably, ERM refers to the problem of minimizing the training loss $L_{\mathcal{D}}(\hat{t}(\cdot))$. The key question that we will address in this chapter is: under which conditions do solutions of the ERM problem yield predictors that have a small population loss?

As discussed in Sec. 4.5, the performance of a trained predictor in terms of population loss can be decomposed into an irreducible **bias**, which is solely related to the choice of the model class $\mathcal{H}$, and an **estimation error**. The estimation error is the focus of statistical learning theory. The estimation error measures the difference between the population losses obtained by the trained predictor and by the optimal within-class predictor, which serves as the ideal baseline. As such, the estimation error depends on the population distribution, on the inductive bias, and on the size of the training set.

Ideally, one would indeed like to capture at a theoretical level the dependence of the estimation error on the choice of the inductive bias – model class and training algorithm – on the one hand, and on the population distribution and number of training points N, on the other. Furthermore, since the population distribution is not known, in practice a more useful theory would rely not on a specific choice of the population distribution, but on a class of population distributions that is known to include the ground-truth distribution underlying the generation of the data.

As will be discussed in this chapter, a key role in this analysis is played by the **generalization error**, which measures the difference between the training and population losses for a trained predictor. Intuitively, this metric indicates how much one can rely on the training loss as a proxy learning criterion for the population loss. If the generalization error is small, a small training loss implies that the population loss is also small. This in turn indicates that the trained predictor generalizes well outside the training set.

The field of statistical learning theory is very broad, and there are textbooks entirely devoted to it (see Recommended Resources, Sec. 8.8). The goal of this chapter is to introduce the basics of the theory by focusing on the most common approach, which revolves around the **capacity** of a model class $\mathcal{H}$. The framework relates the capacity of a model class to the generalization error, and, through it, to the estimation error of ERM. Its reliance on the model capacity limits the usefulness of this approach when it comes to explaining the generalization capability of large models such as deep neural networks (see Sec. 4.6). However, it is a useful first step.

The framework – known as **PAC learning** – is agnostic to the distribution of the data (i.e., it applies to worst-case population distributions) and to the choice of the training algorithm (i.e., it assumes the ideal solution of the ERM problem). This chapter also includes a brief introduction to alternative and more advanced statistical learning frameworks that attempt to capture the impact of the training algorithm and, possibly, of the specific population distribution.

Learning Objectives and Organization of the Chapter. By the end of this chapter, the reader should be able to:

- understand the key definitions of bias and estimation error, as well as their roles in defining the generalization performance of ERM (Sec. 8.2);
- formulate the PAC learning framework and use the concept of sample complexity (Sec. 8.3);
- apply the PAC learning framework to finite model classes (Sec. 8.4);
- apply the PAC learning framework and the concept of Vapnik–Chervonenkis (VC) dimension to more general model classes (Sec. 8.5); and
- navigate the landscapes of the more advanced analytical frameworks of PAC Bayes learning and information theory for generalization (Sec. 8.6).

8.2   Benchmarks and Decomposition of the Optimality Error

In this section, we first recall the decomposition of the optimality error in terms of bias and estimation error, and then review ERM.

8.2.1   Bias vs. Estimation Error Decomposition of the Optimality Error

In this subsection, we recall the bias vs. estimation error decomposition introduced in Sec. 4.5.

Benchmark Predictors. How well does a learning algorithm perform? As introduced in Sec. 4.5, we have two key benchmark predictors:

- **Population-optimal unconstrained predictor**. The population-optimal unconstrained predictor $\hat{t}^*(\cdot)$ minimizes the population loss without any constraint on the model class, i.e.,

$$\hat{t}^*(\cdot) = \arg\min_{\hat{t}(\cdot)} L_p(\hat{t}(\cdot)). \tag{8.3}$$

- **Population-optimal within-class predictor**. The population-optimal within-class predictor $\hat{t}^*_{\mathcal{H}}(\cdot)$ minimizes the population loss within a given model class $\mathcal{H}$, i.e.,

$$\hat{t}^*_{\mathcal{H}}(\cdot) = \arg\min_{\hat{t}(\cdot)\in\mathcal{H}} L_p(\hat{t}(\cdot)). \tag{8.4}$$

For simplicity, we will assume in this chapter that the two reference predictors $\hat{t}^*(\cdot)$ and $\hat{t}^*_{\mathcal{H}}(\cdot)$ are unique. By (8.3) and (8.4), the unconstrained predictor $\hat{t}^*(\cdot)$ can be instantiated in a richer domain than the within-class predictor $\hat{t}^*_{\mathcal{H}}(\cdot)$, whose domain is limited to the model class $\mathcal{H}$. Hence, the unconstrained minimum population loss $L_p(\hat{t}^*(\cdot))$ cannot be larger than the minimum within-class population loss $L_p(\hat{t}^*_{\mathcal{H}}(\cdot))$:

$$L_p(\hat{t}^*(\cdot)) \leq L_p(\hat{t}^*_{\mathcal{H}}(\cdot)). \tag{8.5}$$

Furthermore, we have the equality

$$L_p(\hat{t}^*(\cdot)) = L_p(\hat{t}^*_{\mathcal{H}}(\cdot)) \tag{8.6}$$

if and only if the model class $\mathcal{H}$ is large enough so that the population-optimal unconstrained predictor is in it, i.e., $\hat{t}^*(\cdot) = \hat{t}^*_{\mathcal{H}}(\cdot) \in \mathcal{H}$.

Example 8.1

Consider the problem of binary classification with a binary input, i.e., we have $x, t \in \{0, 1\}$. The population distribution $p(x, t)$ is such that the input variable is distributed as $x \sim \text{Bern}(p)$, with $0 < p < 0.5$, and the target equals the input, i.e., $t = x$. Recall that the population distribution is unknown to the learner. Under the detection-error loss $\ell(t, \hat{t}) = \mathbb{1}(\hat{t} \neq t)$, the population-optimal predictor is clearly $\hat{t}^*(x) = x$, yielding the minimum unconstrained population loss $L_p(\hat{t}^*(\cdot)) = 0$.

Let us fix now the "constant-predictor" model class

$$\mathcal{H} = \{\hat{t}(x) = \hat{t} \in \{0, 1\} \text{ for all } x \in \{0, 1\}\}, \tag{8.7}$$

which contains predictors that output the same binary value $\hat{t}$ irrespective of the input. By comparing the population losses obtained by the choices $\hat{t} \in \{0, 1\}$ corresponding to two predictors in $\mathcal{H}$, the population-optimal within-class predictor can be seen to be $\hat{t}^*_{\mathcal{H}}(x) = 0$. This yields the minimum within-class population loss

$$L_p(\hat{t}^*_{\mathcal{H}}(\cdot)) = (1 - p) \cdot 0 + p \cdot 1 = p > L_p(\hat{t}^*(\cdot)) = 0, \tag{8.8}$$

in line with the general inequality (8.5). Note that the inequality is strict because the population-optimal unconstrained predictor $\hat{t}^*(\cdot)$ does not belong to the model class $\mathcal{H}$.

Decomposition of the Optimality Error. From now on, we will often write predictors $\hat{t}(\cdot)$ as $\hat{t}$ in order to simplify the notation. We have seen in Sec. 4.5 that, for any predictor $\hat{t}$, we can decompose the optimality error $L_p(\hat{t}) - L_p(\hat{t}^*)$ as

$$\underbrace{L_p(\hat{t}) - L_p(\hat{t}^*)}_{\text{optimality error}} = \underbrace{(L_p(\hat{t}_{\mathcal{H}}^*) - L_p(\hat{t}^*))}_{\text{bias}} + \underbrace{(L_p(\hat{t}) - L_p(\hat{t}_{\mathcal{H}}^*))}_{\text{estimation error}}. \tag{8.9}$$

The **bias** $L_p(\hat{t}_{\mathcal{H}}^*) - L_p(\hat{t}^*)$, also known as **approximation error**, depends on the population distribution $p(x,t)$ and on the choice of the model class $\mathcal{H}$ (see, e.g., Example 8.1). In contrast, the **estimation error** $L_p(\hat{t}) - L_p(\hat{t}_{\mathcal{H}}^*)$ depends on the population distribution $p(x,t)$, the model class $\mathcal{H}$, the training algorithm producing the predictor $\hat{t}$ from the training data, and, through it, the training data $\mathcal{D}$ itself.

8.2.2 Training Algorithms and ERM

Given the training data set

$$\mathcal{D} = \{(x_n, t_n)_{n=1}^N\} \underset{\text{i.i.d.}}{\sim} p(x,t), \tag{8.10}$$

a training algorithm returns a predictor $\hat{t}_{\mathcal{D}} \in \mathcal{H}$. The trained model $\hat{t}_{\mathcal{D}}$ is stochastic, owing to randomness of the data set $\mathcal{D}$. In some statistical learning theory frameworks, the mapping between data $\mathcal{D}$ and predictor $\hat{t}_{\mathcal{D}}$ is also randomized, as we will discuss in Sec. 8.6. ERM minimizes the training loss $L_{\mathcal{D}}(\hat{t})$ as

$$\hat{t}_{\mathcal{D}}^{ERM} = \arg\min_{\hat{t} \in \mathcal{H}} L_{\mathcal{D}}(\hat{t}) \tag{8.11}$$

with training loss

$$L_{\mathcal{D}}(\hat{t}) = \frac{1}{N} \sum_{n=1}^N \ell(t_n, \hat{t}(x_n)). \tag{8.12}$$

Since ERM generally does not obtain the population-optimal within-class predictor, and in any case cannot improve over it in terms of population loss, we have the inequalities

$$L_p(\hat{t}_{\mathcal{D}}^{ERM}) \geq L_p(\hat{t}_{\mathcal{H}}^*) \geq L_p(\hat{t}^*). \tag{8.13}$$

Example 8.2

Consider again the problem of binary classification with a binary input, i.e., $x, t \in \{0,1\}$, as in Example 8.1. The population distribution is such that the input variable is still distributed as $x \sim \text{Bern}(p)$, with $0 < p < 0.5$; but now the target variable equals the input, $t = x$, with probability $0.5 < q \leq 1$, while it is different from it, $t \neq x$, with probability $1 - q$. Since it is more likely for the target to equal the input, under the detection-error loss, the population-optimal unconstrained predictor is $\hat{t}^*(x) = x$ as in Example 8.1, with corresponding unconstrained minimum population loss $L_p(\hat{t}^*) = 1 - q$.

Proceeding as in Example 8.1, for the constant-predictor model class $\mathcal{H}$ in (8.7), the population-optimal within-class predictor is $\hat{t}_{\mathcal{H}}(x) = 0$, with minimum within-class population loss

$$\begin{aligned} L_p(\hat{t}_{\mathcal{H}}^*) &= (1-p)\Pr[t \neq 0 | x = 0] + p\Pr[t \neq 0 | x = 1] \\ &= (1-p)(1-q) + pq > 1 - q = L_p(\hat{t}^*), \end{aligned} \tag{8.14}$$

which is again consistent with the general inequality (8.5).

Let us now consider training. Given training data set $\mathcal{D} = \{(1,0),(0,1),(0,1)\}$, with $N = 3$, ERM trains a model in class $\mathcal{H}$ by minimizing the training loss

$$L_{\mathcal{D}}(\hat{t}) = \frac{1}{3}(\mathbb{1}(\hat{t} \neq 0) + 2 \cdot \mathbb{1}(\hat{t} \neq 1)). \tag{8.15}$$

The ERM predictor is hence $\hat{t}_{\mathcal{D}}^{ERM} = 1$, which gives the training loss $L_{\mathcal{D}}(\hat{t}_{\mathcal{D}}^{ERM}) = 1/3$. The resulting population loss of the ERM predictor is

$$\begin{aligned} L_p(\hat{t}_{\mathcal{D}}^{ERM}) &= (1-p)\mathrm{Pr}[t \neq 1|x = 0] + p\mathrm{Pr}[t \neq 1|x = 1] \\ &= (1-p)q + p(1-q). \end{aligned} \tag{8.16}$$

To compare population-optimal and ERM predictors, let us set as an example $p = 0.3$ and $q = 0.7$. With this choice, the minimum unconstrained population loss is $L_p(\hat{t}^*) = 0.3$; the minimum within-class population loss is $L_p(\hat{t}_{\mathcal{H}}^*) = 0.42$; and the ERM population loss is $L_p(\hat{t}_{\mathcal{D}}^{ERM}) = 0.58$. These losses satisfy, strictly, all the inequalities (8.13). We also have the decomposition of the optimality error (8.9) as in

$$\underbrace{L_p(\hat{t}_{\mathcal{D}}^{ERM})}_{0.58} - \underbrace{L_p(\hat{t}^*)}_{0.3} = \underbrace{(L_p(\hat{t}_{\mathcal{H}}^*) - L_p(\hat{t}^*))}_{\text{bias}=0.12} + \underbrace{(L_p(\hat{t}_{\mathcal{D}}^{ERM}) - L_p(\hat{t}_{\mathcal{H}}^*))}_{\text{estimation error}=0.16}. \tag{8.17}$$

8.3   Probably Approximately Correct Learning

The bias $L_p(\hat{t}_{\mathcal{H}}^*) - L_p(\hat{t}^*)$ in the decomposition (8.9) of the optimality error is generally difficult to quantify since it depends on the population-optimal unconstrained predictor $\hat{t}^*$. Furthermore, this term does not capture the effectiveness of the training algorithm for a given training set size, but only the suitability of the model class in the regime of arbitrarily large data sets. As mentioned, statistical learning theory focuses on the other term in the optimality error decomposition (8.9), namely the estimation error $L_p(\hat{t}_{\mathcal{D}}) - L_p(\hat{t}_{\mathcal{H}}^*)$.

If the estimation error is small, the training algorithm obtains close to the best possible within-class population loss. This in turn implies that the learner has enough data available to identify a close-to-optimal predictor within the model class. Quantifying the amount of data required to ensure a small estimation error is a key theoretical concern of statistical learning theory. As we discuss next, since the estimation error is random, being a function of the training set $\mathcal{D}$, one needs to first specify what is meant by "small" in order to make progress.

8.3.1   PAC Condition

Because of the randomness of the data set $\mathcal{D}$, a learning algorithm $\hat{t}_{\mathcal{D}}$ can only minimize the population loss $L_p(\hat{t}_{\mathcal{D}})$, and hence also the estimation error, with the following qualifications:

- A learning algorithm minimizes the estimation error **approximately**, achieving a population loss $L_p(\hat{t}_{\mathcal{D}})$ that is within some approximation level $\epsilon > 0$ of the minimum unconstrained population loss $L_p(\hat{t}_{\mathcal{H}}^*)$ – i.e., the estimation error satisfies the inequality $L_p(\hat{t}_{\mathcal{D}}) - L_p(\hat{t}_{\mathcal{H}}^*) \leq \epsilon$.
- It does so **probably**, with a probability at least $1 - \delta$ for some $\delta \in (0,1)$, in the sense that we have the inequality

$$L_p(\hat{t}_{\mathcal{D}}) - L_p(\hat{t}_{\mathcal{H}}^*) \leq \epsilon \text{ with probability at least } 1 - \delta. \tag{8.18}$$

Figure 8.1 Illustration of the PAC condition (8.18). The population loss $L_p(\hat{t})$ is shown as a function of the choice of the predictor $\hat{t}$, along with the population-optimal within-class predictor $\hat{t}_{\mathcal{H}}$ and a high-probability interval (with probability $1 - \delta$) of values taken by the trained predictor $\hat{t}_{\mathcal{D}}$. For the PAC condition to hold, the high-probability interval of values taken by the population loss $L_p(\hat{t}_{\mathcal{D}})$ should be within the approximation level ϵ of the within-class minimum population loss $L_p(\hat{t}_{\mathcal{H}}^*)$.

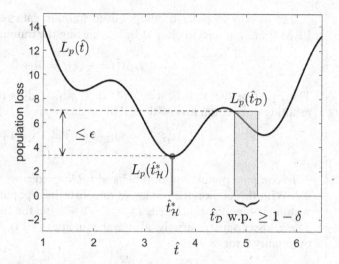

The probabilistic inequality (8.18) is referred to as the **probably approximately correct (PAC)** condition. The PAC condition is illustrated in Fig. 8.1, which shows the population loss $L_p(\hat{t})$ as a function of the choice of the predictor $\hat{t}$, represented as a scalar for simplicity of visualization. The trained predictor $\hat{t}_{\mathcal{D}}$ is generally different from the population-optimal within-class predictor $\hat{t}_{\mathcal{H}}^*$, and is random because of its dependence on $\mathcal{D}$. The figure illustrates an interval of values taken by $\hat{t}_{\mathcal{D}}$ with probability $1 - \delta$. For the PAC condition (8.18) to hold, the corresponding values of the population loss $L_p(\hat{t}_{\mathcal{D}})$, given $\hat{t}_{\mathcal{D}}$ within this high-probability interval, should be within the approximation level ϵ of the within-class minimum population loss $L_p(\hat{t}_{\mathcal{H}}^*)$.

8.3.2 PAC Training Algorithm

We now formalize the definition of a training algorithm satisfying the PAC condition (8.18) for an estimation error ϵ and confidence level $1 - \delta$. To this end, we fix a model class $\mathcal{H}$, a loss function $\ell(t, \hat{t})$, and a *family* of population distributions $p(x, t)$. The consideration of a family of population distributions reflects the fact that the true population distribution is unknown.

When operating on data sets $\mathcal{D}$ of N examples drawn from one of the population distributions in the family, a training algorithm A produces a predictor $\hat{t}_{\mathcal{D}}^A(\cdot)$ in $\mathcal{H}$. The training algorithm A is said to be (N, ϵ, δ) **PAC** if it satisfies the PAC condition (8.18) for accuracy parameter $\epsilon > 0$ and confidence parameter $\delta \in (0, 1)$, that is, if we have the inequality

$$\Pr_{\mathcal{D} \underset{\text{i.i.d.}}{\sim} p(x,t)}[L_p(\hat{t}_{\mathcal{D}}^A) \leq L_p(\hat{t}_{\mathcal{H}}^*) + \epsilon] \geq 1 - \delta \tag{8.19}$$

for *any* population distribution $p(x, t)$ in the family.

The PAC requirement (8.19) is a worst-case constraint in the sense that it imposes the inequality

$$\min_{p(x,t)} \left\{ \Pr_{\mathcal{D} \underset{\text{i.i.d.}}{\sim} p(x,t)}[L_p(\hat{t}_{\mathcal{D}}^A) \leq L_p(\hat{t}_{\mathcal{H}}^*) + \epsilon] \right\} \geq 1 - \delta, \tag{8.20}$$

where the minimum is taken over all possible population distributions in the set of interest.

Figure 8.2 Illustration of the sample complexity $N_{\mathcal{H}}^A(\epsilon, \delta)$ of a training algorithm A. When $N \geq N_{\mathcal{H}}^A(\epsilon, \delta)$, the $(1 - \delta)$-probability interval of values for the population loss of the trained predictor $\hat{t}_{\mathcal{D}}^A$ (across all population distributions in the family) is within ϵ of the population loss of the population-optimal within-class predictor $\hat{t}_{\mathcal{H}}^*$.

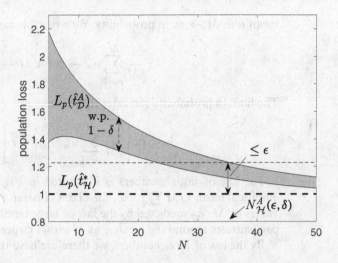

8.3.3 Sample Complexity

For a training algorithm A, the amount of data needed to achieve a certain pair of accuracy–confidence parameters (ϵ, δ) in the PAC condition (8.19) is known as **sample complexity**. More precisely, for model class $\mathcal{H}$, loss function $\ell(t, \hat{t})$, and family of population distributions $p(x, t)$, a training algorithm A has sample complexity $N_{\mathcal{H}}^A(\epsilon, \delta)$ if, for the given parameters $\epsilon > 0$ and $\delta \in (0, 1)$, it is (N, ϵ, δ) PAC for all $N \geq N_{\mathcal{H}}^A(\epsilon, \delta)$.

According to this definition, the sample complexity $N_{\mathcal{H}}^A(\epsilon, \delta)$ is not uniquely specified: If $N_{\mathcal{H}}^A(\epsilon, \delta)$ satisfies the condition that the training algorithm is (N, ϵ, δ) PAC for all $N \geq N_{\mathcal{H}}^A(\epsilon, \delta)$, clearly all values of N larger than $N_{\mathcal{H}}^A(\epsilon, \delta)$ meet this condition too. One could then define the sample complexity as the minimum value of N for which this condition is satisfied, but we will not pursue this more formal approach here.

The definition of sample complexity of a training algorithm A is illustrated in Fig. 8.2. The figure shows the population loss of the population-optimal within-class predictor $\hat{t}_{\mathcal{H}}^*$, as well as a high-probability interval of values for the population loss of the trained predictor $\hat{t}_{\mathcal{D}}^A$ for a training algorithm A, as a function of the data set size N. To comply with the definition of PAC training algorithm, this interval should include the largest such interval across all possible population distributions in the given family. The training algorithm is (N, ϵ, δ) PAC if the upper bound of the interval is within ϵ of the minimum within-class population loss $L_p(\hat{t}_{\mathcal{H}}^*)$ when the data set has at least N data points. Accordingly, the algorithm A has a sample complexity $N_{\mathcal{H}}^A(\epsilon, \delta)$ if, for any $N \geq N_{\mathcal{H}}^A(\epsilon, \delta)$, the upper bound on the interval is within ϵ of $L_p(\hat{t}_{\mathcal{H}}^*)$.

8.3.4 Is ERM PAC?

As we discuss in this subsection, intuitively, for a sufficiently large N, ERM should be PAC by the law of large numbers. To understand this intuition, let us start by reviewing the law of large numbers.

The (weak) **law of large numbers** says that, for i.i.d. rvs $u_1, u_2, \ldots, u_M \sim p(u)$ with some mean $E_{u_i \sim p(u)}[u_i] = \mu$ and finite variance, the empirical average $1/M \sum_{m=1}^M u_m$ tends to the ensemble mean μ as $M \to \infty$ in probability. We write this asymptotic condition as the limit

$$\frac{1}{M} \sum_{m=1}^{M} u_m \to \mu \text{ for } M \to \infty \text{ in probability.} \tag{8.21}$$

The limit in probability indicates that for any $\epsilon > 0$ we have the conventional limit

$$\Pr\left[\left| \frac{1}{M} \sum_{m=1}^{M} u_m - \mu \right| > \epsilon \right] \to 0 \text{ for } M \to \infty. \tag{8.22}$$

The law of large numbers is illustrated in Fig. 8.3, which shows the distribution of the empirical mean $1/M \sum_{m=1}^{M} u_m$ for i.i.d. Gaussian rvs with mean μ and variance equal to 1 for values of M. As predicted by the law of large numbers, the distribution of the empirical mean concentrates around the mean μ as M grows larger.

By the law of large numbers, we therefore have the limit

$$L_{\mathcal{D}}(\hat{t}) = \frac{1}{N} \sum_{n=1}^{N} \ell(t_n, \hat{t}(x_n)) \to L_p(\hat{t}) \text{ for } N \to \infty \text{ in probability,} \tag{8.23}$$

separately for each predictor $\hat{t}$. Hence, intuitively, the ERM solution $\hat{t}_{\mathcal{D}}^{ERM}$ – which minimizes the training loss $L_{\mathcal{D}}(\hat{t})$ – should also minimize the population loss $L_p(\hat{t})$ with high probability as the data set size N increases. In other words, the ERM solution $\hat{t}_{\mathcal{D}}^{ERM}$ should tend to the population-optimal unconstrained solution $\hat{t}_{\mathcal{H}}^*$ as N grows large. But how large do we need N to be – i.e., what can we say about the sample complexity of ERM?

Example 8.3

Consider the model class of threshold functions

$$\mathcal{H} = \left\{ \hat{t}(x|\theta) = \begin{cases} 0, & \text{if } x < \theta \\ 1, & \text{if } x \geq \theta \end{cases} = \mathbb{1}(x \geq \theta) \right\}, \tag{8.24}$$

Figure 8.3 Illustration of the law of large numbers.

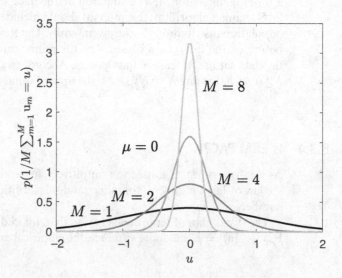

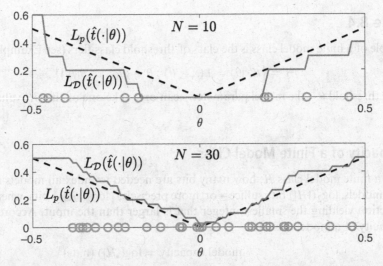

Figure 8.4 Training and population losses as a function of the model parameter θ for the class of threshold predictors considered in Example 8.3.

where input x and model parameter θ are real numbers ($D = 1$). Assume that the population distribution $p(x, t)$ is given as

$$p(x, t) = p(x)\mathbb{1}(t = \hat{t}(x|\theta^* = 0)) \tag{8.25}$$

for some arbitrary marginal distribution $p(x)$. Note that the conditional distribution $p(t|x) = \mathbb{1}(t = \hat{t}(x|\theta^* = 0))$ is included in $\mathcal{H}$, or, more precisely, the deterministic mapping defining $p(t|x)$ is in $\mathcal{H}$. In such cases, we say that the class of population distributions (8.25) is **realizable** for the model class $\mathcal{H}$. Under this assumption, for the detection-error loss, the population-optimal unconstrained predictor and within-class predictor coincide, and they are given by $\hat{t}^*(\cdot) = \hat{t}^*_{\mathcal{H}}(\cdot) = \hat{t}(\cdot|\theta^* = 0)$. It follows that the contribution of the bias to the optimality error is zero.

Let us now consider training. To elaborate, we assume that the marginal $p(x)$ is uniform in the interval $[-0.5, 0.5]$, and we generate N training points, shown as circles in Fig. 8.4 for $N = 10$ and $N = 30$. The resulting training loss is displayed as a gray line as a function of θ. The population loss is also shown in Fig. 8.4 for reference, as a dashed line. When N is large enough, the training loss becomes a good approximation of the population loss. As a result, the parameter θ^{ERM} selected by ERM, which minimizes the training loss, tends to the optimal threshold $\theta^* = 0$.

8.4   PAC Learning for Finite Model Classes

To address the sample complexity of ERM, we will first consider model classes that include a finite number of models. We write the cardinality of the model class $\mathcal{H}$, i.e., the number of models in $\mathcal{H}$, as $|\mathcal{H}|$. For finite model classes, individual models in the model class are often referred to in the literature as **hypotheses**, and the model class as the **hypothesis class**.

Example 8.4

An example of a finite model class is the class of threshold classifiers (see Example 8.3) of the form

$$\mathcal{H} = \left\{ \hat{t}(x|\theta) = \mathbb{1}(x \geq \theta) : \theta \in \{\theta_1, \ldots, \theta_{|\mathcal{H}|}\} \right\}, \tag{8.26}$$

where the threshold θ – the model parameter – can only take one of the $|\mathcal{H}|$ values $\{\theta_1, \ldots, \theta_{|\mathcal{H}|}\}$.

8.4.1 Capacity of a Finite Model Class

For a finite model class $\mathcal{H}$, how many bits are needed to index all models in it? Since there are $|\mathcal{H}|$ models, $\log_2(|\mathcal{H}|)$ bits suffice – or, more precisely, $\lceil \log_2(|\mathcal{H}|) \rceil$ bits, where $\lceil \cdot \rceil$ is the "ceiling" function yielding the smallest integer that is larger than the input. Accordingly, we define the capacity of a model class as

$$\begin{aligned} \text{model capacity} &= \log(|\mathcal{H}|) \text{ (nats)} \\ &= \log_2(|\mathcal{H}|) \text{ (bits)}, \end{aligned} \tag{8.27}$$

which is the number of nats or bits required to select any individual model within the model class.

Intuitively, a larger model capacity entails a larger sample complexity, since it is more challenging to pin down the optimal model within a larger class. This intuition is made precise, and elaborated on, in the rest of this section.

8.4.2 Sample Complexity of ERM for Finite Model Classes

The main result in this section is a characterization of an upper bound on the estimation error for ERM when applied to finite model classes, which in turn yields a measure of sample complexity for ERM. We will first state the results, and then provide a proof.

ERM is PAC for Finite Model Classes. For the family of all possible population distributions, under the detection-error loss – or any other loss bounded in the interval $[0, 1]$ – and given any finite model class $\mathcal{H}$, ERM is (N, ϵ, δ) PAC with estimation error

$$\epsilon = \sqrt{\frac{2\log(|\mathcal{H}|) + 2\log(2/\delta)}{N}} \tag{8.28}$$

for any confidence parameter $\delta \in (0, 1)$. Equivalently, ERM satisfies the estimation error bound

$$L_p(\hat{t}_{\mathcal{D}}^{ERM}) - L_p(\hat{t}_{\mathcal{H}}^*) \leq \sqrt{\frac{2\log(|\mathcal{H}|) + 2\log(2/\delta)}{N}} \tag{8.29}$$

with probability no smaller than $1 - \delta$.

Sample Complexity of ERM for Finite Model Classes. The important result above can also be restated in terms of sample complexity, by solving the equality (8.28) in terms of N. For the family of all possible population distributions, under the detection-error loss – or any other loss bounded in the interval $[0, 1]$ – and given any finite model class $\mathcal{H}$, ERM has sample complexity

$$N_{\mathcal{H}}^{ERM}(\epsilon, \delta) = \left\lceil \frac{2\log(|\mathcal{H}|) + 2\log(2/\delta)}{\epsilon^2} \right\rceil. \tag{8.30}$$

The key observation is that, according to (8.30), the sample complexity of ERM is proportional to the model capacity $\log(|\mathcal{H}|)$, as well as to the number of confidence decimal digits $\log(1/\delta)$ (see Appendix 5.C), while being inversely proportional to the square of the accuracy parameter ϵ.

8.4.3 Proof of the ERM Sample Complexity (8.30) and Generalization Error

The proof of the key result (8.30) is based on the analysis of the **generalization error** of ERM. Given a training algorithm A, such as ERM, the generalization error

$$\text{generalization error} = L_p(\hat{t}_{\mathcal{D}}^A) - L_{\mathcal{D}}(\hat{t}_{\mathcal{D}}^A) \tag{8.31}$$

is the difference between population and training losses obtained by the trained predictor $\hat{t}_{\mathcal{D}}^A$. As anticipated in Sec. 8.1, the generalization error quantifies the extent to which the training loss can be trusted as a proxy of the population loss for training. Therefore, intuitively, it plays a central role in the analysis of the performance of ERM, which minimizes the training loss.

To analyze the generalization error of ERM, the proof of the result (8.30) uses two main ingredients: the law of large numbers – in a stronger form known as Hoeffding's inequality – and the union bound. The **union bound** is a standard result in probability, which states that the probability of a union of events is no larger than the sum of the probabilities of the individual events. In the rest of this subsection, we first review Hoeffding's inequality; then we apply this inequality and the union bound to study the generalization error of ERM; and we finally return to the problem of proving the sample complexity formula (8.30).

Hoeffding's Inequality. For i.i.d. rvs $u_1, u_2, \ldots, u_M \sim p(u)$ such that $\text{E}[u_i] = \mu$ and $\Pr[a \le u_i \le b] = 1$ for some real numbers $a \le b$, we have the inequality

$$\Pr\left[\left|\frac{1}{M}\sum_{m=1}^{M} u_m - \mu\right| > \epsilon\right] \le 2\exp\left(-\frac{2M\epsilon^2}{(b-a)^2}\right). \tag{8.32}$$

Hoeffding's inequality (8.32) bounds the probability that the empirical average rv $1/M \sum_{m=1}^{M} u_m$ deviates from its mean μ by more than $\epsilon > 0$. It shows that this probability decreases exponentially fast with M. Therefore, Hoeffding's inequality implies the (weak) **law of large numbers** limit (8.22), while also quantifying *how quickly* the limit (8.22) is realized as a function of M.

Analysis of the Generalization Error. While one may be justified in thinking otherwise, the training loss $L_{\mathcal{D}}(\hat{t}^{ERM}) = 1/N \sum_{n=1}^{N} \ell(t_n, \hat{t}_{\mathcal{D}}^{ERM}(x_n))$ obtained by the ERM algorithm is not in a form that enables the direct use of Hoeffding's inequality. In fact, the component rvs $\ell(t_n, \hat{t}_{\mathcal{D}}^{ERM}(x_n))$ for $n = 1, \ldots, N$ are not independent, since they are all functions of the same training data set $\mathcal{D}$ through the ERM predictor $\hat{t}_{\mathcal{D}}^{ERM}(\cdot)$. We will now see how to get around this problem to obtain a probabilistic bound on the absolute value of the generalization error.

To proceed, we would like to bound the probability that the absolute value of the generalization error of ERM is larger than some threshold $\xi > 0$, i.e.,

$$\Pr\left[|L_p(\hat{t}_{\mathcal{D}}^{ERM}) - L_{\mathcal{D}}(\hat{t}_{\mathcal{D}}^{ERM})| > \xi\right]. \tag{8.33}$$

This quantity can be upper bounded by the probability that there exists ($\exists$) at least one predictor in $\mathcal{H}$ that satisfies the inequality, i.e.,

$$\Pr\left[|L_p(\hat{t}_{\mathcal{D}}^{ERM}) - L_{\mathcal{D}}(\hat{t}_{\mathcal{D}}^{ERM})| > \xi\right] \leq \Pr\left[\exists \hat{t} \in \mathcal{H} : |L_p(\hat{t}) - L_{\mathcal{D}}(\hat{t})| > \xi\right]$$

$$= \Pr\left[\bigcup_{\hat{t} \in \mathcal{H}} \left\{|L_p(\hat{t}) - L_{\mathcal{D}}(\hat{t})| > \xi\right\}\right], \qquad (8.34)$$

where the last equality follows from the interpretation of a union of events as a logical OR. Inequality (8.34) follows since, if there is no predictor in $\mathcal{H}$ that satisfies the condition $|L_p(\hat{t}_{\mathcal{D}}^{ERM}) - L_{\mathcal{D}}(\hat{t}_{\mathcal{D}}^{ERM})| > \xi$, ERM clearly cannot find one.

Now, we can use the union bound first and then Hoeffding's inequality to obtain the inequalities

$$\Pr\left[\bigcup_{\hat{t} \in \mathcal{H}} \left\{|L_p(\hat{t}) - L_{\mathcal{D}}(\hat{t})| > \xi\right\}\right] \leq \sum_{\hat{t} \in \mathcal{H}} \Pr\left[|L_p(\hat{t}) - L_{\mathcal{D}}(\hat{t})| > \xi\right]$$

$$\leq 2|\mathcal{H}|\exp\left(-2N\xi^2\right). \qquad (8.35)$$

Importantly, for the second step in (8.35), we are allowed to apply Hoeffding's inequality since the predictor $\hat{t} \in \mathcal{H}$ is fixed, and hence independent of the training data set $\mathcal{D}$. As a result, the component rvs $\ell(t_n, \hat{t}(x_n))$ with $n = 1, \ldots, N$ in the empirical average $L_{\mathcal{D}}(\hat{t})$ are i.i.d. and bounded by assumption in the interval $[a = 0, b = 1]$.

Overall, via (8.34) and (8.35), we have proved the **generalization error bound**

$$\Pr\left[|L_p(\hat{t}_{\mathcal{D}}^{ERM}) - L_{\mathcal{D}}(\hat{t}_{\mathcal{D}}^{ERM})| > \xi\right] \leq 2|\mathcal{H}|\exp\left(-2N\xi^2\right), \qquad (8.36)$$

which applies to *any* population distribution.

Concluding the Proof. The generalization error bound (8.36) states that the absolute value of the generalization error can be upper bounded by $\xi > 0$ with probability $\delta = 2|\mathcal{H}|\exp(-2N\xi^2)$. We now show that this implies that the estimation error is upper bounded by 2ξ with the same probability. In fact, by the generalization error bound (8.36), with probability at least $1 - \delta$, we have the inequalities

$$L_p(\hat{t}_{\mathcal{D}}^{ERM}) \overset{\substack{\text{generalization error bound}}}{\leq} L_{\mathcal{D}}(\hat{t}_{\mathcal{D}}^{ERM}) + \xi$$

$$\overset{\substack{\text{ERM definition}}}{\leq} L_{\mathcal{D}}(\hat{t}_{\mathcal{H}}^*) + \xi$$

$$\overset{\substack{\text{generalization error bound}}}{\leq} L_p(\hat{t}_{\mathcal{H}}^*) + 2\xi, \qquad (8.37)$$

where the second equality follows from the definition of ERM. In fact, ERM minimizes the training loss, and hence any other predictor, including $\hat{t}_{\mathcal{H}}^*$, cannot obtain a smaller training loss. We now only need to choose $\xi = \epsilon/2$ to complete the proof of (8.30).

8.5   PAC Learning for General Model Classes: VC Dimension

The main conclusion drawn in the preceding section – that ERM has a sample complexity that scales with the capacity of the model class – hinges on the assumption of a *finite* model class.

Can ERM yield successful learning (in the PAC sense) even for infinite model classes such as linear classifiers or neural networks? If so, what is the "correct" measure of capacity for such model classes?

8.5.1   Quantizing a Model Class

Let us first consider a naive approach to this question. Assume that we have a parametric class $\mathcal{H}$ of predictors $\hat{t}(\cdot|\theta)$ with model parameter vector $\theta \in \mathbb{R}^D$. Note that this class contains a continuum of models. We can reduce this infinite model class to a finite one by quantizing each element of the model parameter vector θ with 2^b quantization levels (see Sec. 7.3), requiring b bits per element. The "quantized" model class $\mathcal{H}_b$ is finite, including 2^{bD} models, i.e., $|\mathcal{H}_b| = 2^{bD}$.

Having converted the original infinite model class $\mathcal{H}$ to the quantized, finite class $\mathcal{H}_b$, we can now apply to the latter the theory developed in the preceding section. In particular, the capacity (8.27) of the finite class is $\log_2(|\mathcal{H}|) = Db$ (bits) or $\log(|\mathcal{H}|) = bD\log(2)$ (nats). Therefore, by (8.30), the ERM sample complexity is

$$N_{\mathcal{H}}^{ERM}(\epsilon,\delta) = \left\lceil \frac{2bD\log(2) + 2\log(2/\delta)}{\epsilon^2} \right\rceil, \tag{8.38}$$

which scales proportionally to the number of parameters D and to the bit resolution b.

The sample complexity (8.38) seems to imply that learning a continuous model class requires an infinite number of samples, since a continuous model is described exactly only if we use an infinite number of bits, i.e., if $b \to \infty$. This conclusion is fortunately not true in general. To see why, the theory must to be extended to introduce a more refined notion of capacity of a model class – the **Vapnik–Chervonenkis (VC)** dimension. We will specifically conclude in the rest of this section that, in a nutshell, the sample complexity result (8.30) proved for finite model classes still holds for general model classes by substituting the model capacity $\log(|\mathcal{H}|)$ with the VC dimension. Throughout this section, we will focus on binary classification.

8.5.2   Capacity of a Model Class, Revisited

Consider again a finite class of models: What does it mean that its capacity is $\log_2(|\mathcal{H}|)$ bits as in (8.27)? Information theory stipulates that an information source has a capacity of b bits if it can produce *any* binary vector of b bits (possibly after a remapping of the output alphabet to bits). For example, a source has capacity $b = 2$ bits if it can produce four messages, e.g., $\{00, 01, 10, 11\}$, four hand gestures, four words, etc.

With this definition in mind, fix a data set of N inputs $\mathcal{X} = \{x_1, \ldots, x_N\}$ and a model $\hat{t}(\cdot) \in \mathcal{H}$. Note that, since we focus on binary classification, the output of the model is binary, i.e., $\hat{t}(\cdot) \in \{0, 1\}$. The model $\hat{t}(\cdot)$ produces a set of N binary predictions $\{\hat{t}(x_1), \ldots, \hat{t}(x_N)\}$ on $\mathcal{X}$. We can now think of the model class as a source of capacity N bits if, for *some* data set $\mathcal{X} = \{x_1, \ldots, x_N\}$, it can produce *all* possible 2^N binary vectors $\{\hat{t}(x_1), \ldots, \hat{t}(x_N)\}$ of predictions by running over all models $\hat{t}(\cdot) \in \mathcal{H}$.

Let us apply this definition to see how it relates to the capacity (8.27) of a finite model class. For a finite model class, there are only $|\mathcal{H}|$ models to choose from, and hence, based on the definition in the preceding paragraph, the maximum capacity is $N = \log_2(|\mathcal{H}|)$ bits. Take, for instance, the case $|\mathcal{H}| = 2$, where we only have two models in the class $\mathcal{H} = \{\hat{t}_1(\cdot), \hat{t}_2(\cdot)\}$. Unless the two models are equivalent, we can produce both one-bit messages if $\hat{t}_1(x) = 0$ and $\hat{t}_2(x) = 1$ (or vice versa) for some input x. But we can never produce all messages of two bits since one can only choose between two models, and hence only two distinct configurations of the predictions are possible for any input x. Therefore, by the given definition, the capacity is $N = \log_2(|\mathcal{H}|) = 1$ bit. The same arguments apply to any $N = \log_2(|\mathcal{H}|)$.

It is important to emphasize that the capacity of a model class is a purely combinatorial concept: It counts the *number* of possible configurations of the output, irrespective of whether they are *semantically* relevant for some application. This follows from Shannon's oft-quoted paradigm-shifting insight into communicating systems: "Frequently the messages have meaning; that is, they refer to or are correlated according to some system with certain physical or conceptual entities. These semantic aspects of communication are irrelevant to the engineering problem. The significant aspect is that the actual message is one selected from a set of possible messages."

How does the definition of model capacity given in this subsection apply to infinite model classes? We start with an example.

Example 8.5

Consider the set of all linear binary classifiers on the plane, with decision line passing through the origin (see Sec. 6.2), i.e.,

$$\mathcal{H} = \left\{ \hat{t}(x|\theta) = \text{step}(\theta^T x) = \begin{cases} 0, & \text{if } \theta^T x < 0 \\ 1, & \text{if } \theta^T x > 0 \end{cases} : \theta \in \mathbb{R}^2 \right\}, \tag{8.39}$$

which contains a continuum of models indexed by vector $\theta \in \mathbb{R}^2$. With this model class, we can obtain all messages of size $N = 1$: As illustrated in Fig. 8.5, we can label any, and hence also some, point x_1 as either 0 or 1. For $N = 2$, as seen in Fig. 8.6, we can label *some* pair of points x_1 and x_2 with any pairs of binary labels. However, it can be easily seen that there is no data set of three data points for which we can obtain all eight messages of $N = 3$ bits. Therefore, the capacity of this model is 2 bits. Note that two is also the number of model parameters in vector θ for this model class.

8.5.3 Definition of VC Dimension

In this subsection, we formalize the concept of model capacity outlined in this subsection through the VC dimension. To introduce the VC dimension, we first need the following definition. A model class $\mathcal{H}$ is said to **shatter** a set of inputs $\mathcal{X} = \{x_1, \ldots, x_N\}$ if, no matter how the corresponding binary labels $\{t_1, \ldots, t_N\}$ are selected, there exists a model $\hat{t} \in \mathcal{H}$ that ensures $\hat{t}(x_n) = t_n$ for all $n = 1, \ldots, N$. This definition is in line with the concept of capacity described in the preceding subsection: The set of inputs $\mathcal{X} = \{x_1, \ldots, x_N\}$ is shattered by $\mathcal{H}$ if the models in $\mathcal{H}$ can produce all possible 2^N messages of N bits when applied to $\mathcal{X}$.

Figure 8.5 Binary linear classifiers (with decision line passing through the origin) can produce all messages of 1 bit.

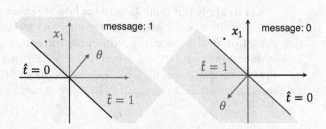

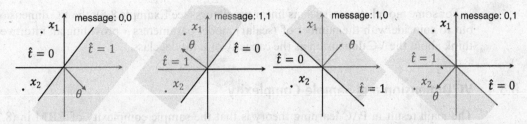

Figure 8.6 Binary linear classifiers (with decision line passing through the origin) can produce messages of 2 bits.

The **VC dimension** VCdim($\mathcal{H}$) (measured in bits) of the model $\mathcal{H}$ is the size of the largest set $\mathcal{X}$ that can be shattered by $\mathcal{H}$. The VC dimension VCdim($\mathcal{H}$) hence coincides with the capacity of the model introduced in the preceding subsection. In particular, for finite classes, we have the inequality VCdim($\mathcal{H}$) $\leq$ log$_2$(|$\mathcal{H}$|), since |$\mathcal{H}$| hypotheses can create at most |$\mathcal{H}$| different label configurations.

Therefore, to prove that a model has VCdim($\mathcal{H}$) $= N$, we need to carry out the following two steps:

1. demonstrate the existence of a set $\mathcal{X}$ with |$\mathcal{X}$| $= N$ that is shattered by $\mathcal{H}$; and
2. prove that no set $\mathcal{X}$ of dimension $N + 1$ exists that is shattered by $\mathcal{H}$.

Example 8.6

The finite threshold function model class

$$\mathcal{H} = \left\{ \hat{t}(x|\theta) = \mathbb{1}(x \geq \theta) : \theta \in \{\theta_1, \ldots, \theta_{|\mathcal{H}|}\} \right\}, \tag{8.40}$$

with $x \in \mathbb{R}$, has VCdim($\mathcal{H}$)$= 1$. To see this, in *step 1* we observe that any set $\mathcal{X}$ of one sample ($N = 1$) can be shattered – and hence there is clearly at least one such set; and in *step 2*, we note that there are no sets of $N = 2$ points that can be shattered. In fact, for any set $\mathcal{X} = \{x_1, x_2\}$ of two points with $x_1 \leq x_2$, the label assignment $(t_1, t_2) = (1, 0)$ cannot be realized by any choice of the threshold θ.

Example 8.7

The model

$$\mathcal{H} = \{ \hat{t}(x|a, b) = \mathbb{1}(a \leq x \leq b) : a \leq b \}, \tag{8.41}$$

which assigns the label $t = 1$ within an interval $[a, b]$ and the label $t = 0$ outside it, has VCdim($\mathcal{H}$) $= 2$. In fact, for *step 1* we note that any set of $N = 2$ points can be shattered – and hence there also exists one such set; and for *step 2* we observe that there are no sets $\mathcal{X}$ of $N = 3$ points that can be shattered. In fact, for any set $\mathcal{X} = \{x_1, x_2, x_3\}$ of three points with $x_1 \leq x_2 \leq x_3$, the label assignment $(t_1, t_2, t_3) = (1, 0, 1)$ cannot be realized.

For some model classes, such as linear predictors (see Example 8.5), the VC dimension turns out to coincide with the number of (scalar) model parameters – providing an intuitive way to think about the VC dimension as the capacity of a model class.

8.5.4 VC Dimension and Sample Complexity

The main result in PAC learning theory is that the sample complexity of ERM in (8.30) still holds, apart from different multiplicative constants, for all model classes $\mathcal{H}$ with finite VC dimension, as long as we replace the model capacity $\log(|\mathcal{H}|)$ with $\text{VCdim}(\mathcal{H})$. But there is more: Can we find a training algorithm that has a smaller sample complexity? It turns out that the answer to this question is, essentially, no. In fact, for a model $\mathcal{H}$ with finite $\text{VCdim}(\mathcal{H}) < \infty$, the sample complexity of any training algorithm A is lower bounded as

$$N_{\mathcal{H}}^A(\epsilon, \delta) \geq C \frac{\text{VCdim}(\mathcal{H}) + \log(1/\delta)}{\epsilon^2} \tag{8.42}$$

for some multiplicative constant $C > 0$. This result demonstrates that, if PAC learning is possible for a given model $\mathcal{H}$, ERM obtains close-to-optimal sample complexity. We emphasize that this result is proved and valid only within the worst-case framework of PAC learning.

The results summarized in this subsection require more sophisticated tools to be proved, and we refer to Recommended Resources, Sec. 8.8, for details.

8.6   PAC Bayes and Information-Theoretic Bounds

In the previous sections, we have introduced the PAC learning framework. As we have seen, this statistical learning approach connects generalization to the capacity of a model class, and is agnostic to population distribution and training algorithm. In this section, we briefly discuss alternative methodologies that move away from a characterization of generalization that is based purely on capacity considerations to account also for training algorithm and/or population distribution. These frameworks revolve around the analysis of the generalization error (8.31), which, as discussed in Secs. 8.1 and 8.4, relates the performance of a training algorithm on the training set to the corresponding performance outside it, i.e., to the population loss.

8.6.1 To Recap: Generalization Error in PAC Learning

As a brief recap from Sec. 8.4, given *any* training algorithm A and *any* population distribution, the generalization error (8.31) satisfies the bound (8.36), i.e.,

$$\Pr\left[|L_p(\hat{t}_{\mathcal{D}}^A) - L_{\mathcal{D}}(\hat{t}_{\mathcal{D}}^A)| > \xi\right] \leq 2|\mathcal{H}| \exp\left(-2N\xi^2\right) = \exp\left(-2N\xi^2 + \log(2|\mathcal{H}|)\right) \tag{8.43}$$

for any $\xi > 0$, if the model class $\mathcal{H}$ is finite. Note that condition (8.43) was proved in Sec. 8.4 with reference to ERM, but the same proof steps apply, with no change, to any training algorithm A. The generalization error provided by PAC learning theory is hence agnostic to the training algorithm and to the population distribution, depending only on the model class capacity.

By (8.43), the generalization error can be upper bounded as

$$L_p(\hat{t}_{\mathcal{D}}^A) - L_{\mathcal{D}}(\hat{t}_{\mathcal{D}}^A) \leq \sqrt{\frac{\log(|\mathcal{H}|) + \log(2/\delta)}{2N}} \tag{8.44}$$

with probability at least $1 - \delta$. It follows that the number of examples N needed to achieve a certain generalization error with some probability must scale with the capacity $\log(|\mathcal{H}|)$ of the model class – and more generally with its VC dimension, as concluded in Sec. 8.5.

Owing to its sole reliance on the model capacity, the PAC learning approach is of limited use when it comes to explaining the generalization properties of specific training algorithms, such as SGD, when applied to particular types of data, such as images. In fact, as summarized in Sec. 4.6, in many such cases, excellent generalization properties have been empirically demonstrated even with significantly overparametrized model classes. We now briefly review two alternative statistical learning frameworks that may prove useful for mitigating some of the mentioned limitations of PAC learning; the first is known as PAC Bayes and the second is based on information theory.

8.6.2   Stochastic Training Algorithms

Both PAC Bayes and information-theoretic frameworks assume **probabilistic training algorithms**. Formally, given a training set $\mathcal{D}$, a probabilistic training algorithm outputs a model $\hat{t} \sim q(\hat{t}|\mathcal{D})$, randomly drawn from the model class $\mathcal{H}$, by sampling from the data-dependent distribution $q(\hat{t}|\mathcal{D})$. As an example, the distribution $q(\hat{t}|\mathcal{D})$ may describe the output of SGD, which, owing to the stochastic choice of mini-batches and possibly of the initialization, is random. Clearly, a standard deterministic training algorithm, producing (with probability 1) a predictor $\hat{t}_{\mathcal{D}}$, is a special instance of the broader class of probabilistic algorithms. We will see in Chapter 12 that stochastic training algorithms play a central role within the framework of Bayesian learning, wherein the distribution $q(\hat{t}|\mathcal{D})$ represents the posterior of the model given the training data. Here, more generally, the distribution $q(\hat{t}|\mathcal{D})$ describes an arbitrary stochastic training algorithm.

Under the random selection of the model defined by the training algorithm $q(\hat{t}|\mathcal{D})$, we can define the **average population loss**

$$L_p(q(\hat{t}|\mathcal{D})) = \mathrm{E}_{\hat{t} \sim q(\hat{t}|\mathcal{D})} \mathrm{E}_{(x,t) \sim p(x,t)}[\ell(t, \hat{t}(x))] \tag{8.45}$$

as the performance criterion of interest. This quantity represents the population loss accrued *on average* when the model is generated via the stochastic algorithm as $\hat{t} \sim q(\hat{t}|\mathcal{D})$. (Note that one may be interested in controlling not only the mean but also other probabilistic properties of the rv $\mathrm{E}_{(x,t) \sim p(x,t)}[\ell(t, \hat{t}(x))]$, where the randomness is due to the choice of the predictor $\hat{t}$, such as some quantiles.)

Taking the average population loss (8.45) as the learning objective, the **average training loss** can be correspondingly defined as

$$L_{\mathcal{D}}(q(\hat{t}|\mathcal{D})) = \mathrm{E}_{\hat{t} \sim q(\hat{t}|\mathcal{D})} \left[\frac{1}{N} \sum_{n=1}^{N} \ell(t_n, \hat{t}(x_n)) \right]. \tag{8.46}$$

This quantity measures the training loss obtained from data $\mathcal{D}$ *on average* over the choice of the predictor $\hat{t}$.

8.6.3   PAC Bayes Generalization Bounds

PAC Bayes and information-theoretic methods aim to bound the

$$\text{average generalization error} = L_p(q(\hat{t}|\mathcal{D})) - L_{\mathcal{D}}(q(\hat{t}|\mathcal{D})), \tag{8.47}$$

where the average is taken, as per (8.45) and (8.46), with respect to the stochastic training algorithm producing $\hat{t} \sim q(\hat{t}|\mathcal{D})$. As such, both frameworks provide generalization error bounds that are **training algorithm-dependent**, through the distribution $q(\hat{t}|\mathcal{D})$, and not agnostic to it as is PAC learning. Specifically, PAC Bayes produces bounds on the average generalization error that hold *with high probability* with respect to the choice of the data $\mathcal{D}$ for *any* population distribution (like PAC learning theory). In contrast, information-theoretic bounds concern the *expectation* of the average generalization with respect to a *given* population distribution. So, PAC Bayes bounds are agnostic to the population distribution, while the results of information-theoretic approaches depend on the population distribution. A summary of the three approaches can be found in Table 8.1.

For illustration, we state here, without proof, a standard **PAC Bayes bound**. Let $q(\hat{t})$ be a prior distribution on the predictors in the model class $\mathcal{H}$, and $q(\hat{t}|\mathcal{D})$ a stochastic training algorithm in the same space. The **prior distribution** $q(\hat{t})$ must be selected before observing the data $\mathcal{D}$, and is arbitrary. In contrast, as discussed, the distribution $q(\hat{t}|\mathcal{D})$ represents the training algorithm that maps the training data $\mathcal{D}$ on a probability distribution over the models in class $\mathcal{H}$. Then, with probability no smaller than $1 - \delta$, for any $\delta \in (0, 1)$, the generalization error satisfies the inequality

$$L_p(q(\hat{t}|\mathcal{D})) - L_{\mathcal{D}}(q(\hat{t}|\mathcal{D})) \leq \sqrt{\frac{\mathrm{KL}(q(\hat{t}|\mathcal{D})||q(\hat{t})) + \log(\frac{2\sqrt{N}}{\delta})}{2N}} \tag{8.48}$$

for *all* training algorithms $q(\hat{t}|\mathcal{D})$ and *any* population distributions $p(x, t)$.

The PAC Bayesian bound (8.48) has an interesting interpretation in terms of **stability**, or **sensitivity**, of the given training algorithm $q(\hat{t}|\mathcal{D})$. In fact, with a proper choice of the prior $q(\hat{t})$, a large KL term $\mathrm{KL}(q(\hat{t}|\mathcal{D})||q(\hat{t}))$ tends to indicate that the distribution $q(\hat{t}|\mathcal{D})$ produced by the training algorithm is strongly dependent on the data $\mathcal{D}$, as it deviates more significantly from the prior $q(\hat{t})$. A learning algorithm whose output models are more sensitive to the training data is more prone to overfitting, which in turn can cause the generalization error to be large. So, the PAC Bayes bound (8.48) captures the notion that a training algorithm that is more sensitive to the training data requires a larger N in order to avoid overfitting.

Another interesting property of the PAC Bayes bound is that condition (8.48) is guaranteed to hold with probability no smaller than $1 - \delta$ **uniformly** across *all* training algorithms. This justifies the use of the right-hand side of (8.48) as an optimization criterion for the distribution $q(\hat{t}|\mathcal{D})$. Using this optimization criterion for training yields the **information risk minimization** formulation of the training problem.

Finally, let us relate the PAC Bayesian bound (8.48) and the PAC bound (8.44). To this end, consider a finite model class and select the prior $q(\hat{t})$ as uniform over a finite model class $\mathcal{H}$, i.e.,

Table 8.1 Dependence of statistical learning frameworks on model class, training algorithm, and population distribution

framework	model class	training algorithm	population distribution
PAC	dependent	agnostic	agnostic
PAC Bayes	dependent	dependent	agnostic
information theory	dependent	dependent	dependent

$q(\hat{t}) = 1/|\mathcal{H}|$. With this choice, the KL divergence in (8.48) can be bounded as

$$
\begin{aligned}
\mathrm{KL}(q(\hat{t}|\mathcal{D})||q(\hat{t})) &= \mathrm{E}_{\hat{t}\sim q(\hat{t}|\mathcal{D})}[\log q(\hat{t}|\mathcal{D})] + \log(|\mathcal{H}|) \\
&= -\mathrm{H}(q(\hat{t}|\mathcal{D})) + \log(|\mathcal{H}|) \leq \log(|\mathcal{H}|),
\end{aligned} \tag{8.49}
$$

where the last inequality follows from the non-negativity of the entropy. Hence, the PAC Bayes bound (8.48) yields the inequality

$$
L_p(q(\hat{t}|\mathcal{D})) - L_{\mathcal{D}}(q(\hat{t}|\mathcal{D})) \leq \sqrt{\frac{\log(|\mathcal{H}|) + \log(\frac{2\sqrt{N}}{\delta})}{2N}}, \tag{8.50}
$$

which mirrors the PAC bound (8.44) in terms of its dependence on the model capacity $\log(|\mathcal{H}|)$.

8.6.4 Information-Theoretic Generalization Bounds

Information-theoretic generalization bounds concern the expectation of the generalization error with respect to an arbitrary, fixed population distribution $p(x,t)$. Under suitable assumptions, the main result within this framework shows that the average generalization error can be bounded as

$$
\mathrm{E}_{\mathcal{D}\sim p(\mathcal{D})}[L_p(q(\hat{t}|\mathcal{D})) - L_{\mathcal{D}}(q(\hat{t}|\mathcal{D}))] \leq \sqrt{\frac{2\sigma^2}{N}\mathrm{I}(\mathcal{D};\hat{t})}, \tag{8.51}
$$

where σ^2 is a parameter that depends on the loss function and on the population distribution. Note that the average on the left-hand side of (8.51) is taken with respect to the training set $\mathcal{D} \sim p(\mathcal{D})$, whose data points are drawn i.i.d. from the population distribution $p(x,t)$. Furthermore, the mutual information $\mathrm{I}(\mathcal{D};\hat{t})$ is computed with respect to the joint distribution $p(\mathcal{D},\hat{t}) = p(\mathcal{D})q(\hat{t}|\mathcal{D})$.

The bound (8.51) depends on the mutual information $\mathrm{I}(\mathcal{D};\hat{t})$ between training set and trained model. In a manner similar to the PAC Bayesian bound (8.48), this term can be interpreted as a measure of the **sensitivity** of the trained model $\hat{t}$ to the training set $\mathcal{D}$. The bound hence reflects again the increasing data requirements of more sensitive training algorithms. Another way to think about the mutual information $\mathrm{I}(\mathcal{D};\hat{t})$ is as the amount of information about the training set $\mathcal{D}$ that is stored in the predictor $\hat{t}$.

8.7 Summary

- Statistical learning theory deals with the following key question: Given an inductive bias, how many training examples (N) are needed to learn to a given level of generalization accuracy?
- The generalization accuracy is ideally measured by the estimation error, i.e., by the difference between the population loss of a given training algorithm and the population loss of the population-optimal within-class predictor.
- A key metric that contributes to the estimation error is the generalization error, i.e., the difference between the training and population losses for a given algorithm.
- The analysis of estimation error and generalization error may focus on a specific combination of training algorithm and family of population distributions, or it may be agnostic to either or both algorithm and population distribution.

- The PAC learning framework provides bounds on the generalization error that are agnostic to the specific training algorithm and to the family of population distributions. The bounds are probabilistic, and are specified by accuracy and confidence parameters.
- PAC learning bounds depend on the capacity of the model, which is defined as the logarithm of the number of models for finite model classes and as the VC dimension for general model classes.
- PAC learning also provides bounds on the estimation error for ERM, which depend on the capacity of the model.
- Alternative theoretical frameworks, such as PAC Bayes learning, produce algorithm-dependent bounds that are agnostic to the population distribution; while others, such as information-theoretic frameworks, depend also on the population distribution.
- A thread that is common to many statistical learning frameworks is the reliance on a measure of training algorithm stability, which quantifies the sensitivity of the algorithm to the training data.

8.8   Recommended Resources

In-depth presentations on statistical learning theory can be found in [1, 2, 3], while detailed discussions on the capacity of neural network models are offered by [4]. Useful reviews on PAC Bayes learning include [5], and information-theoretic bounds are covered in [6].

Problems

8.1   We have the class of population distributions $t \sim p(t) = \mathcal{N}(t|\mu, 1)$, where μ is a real number. (There is no input x in this problem.) The model class $\mathcal{H}$ is given by the constant predictors $\hat{t} \in \mathbb{R}$, and the ℓ_2 loss is adopted. (a) Obtain the expression for the ERM predictor $\hat{t}_{\mathcal{D}}^{ERM}$ for any training data set $\mathcal{D} = \{(t_n)_{n=1}^N\}$. (b) Write the expression for the estimation error as a function of μ and $\hat{t}_{\mathcal{D}}^{ERM}$. Note that the estimation error is an rv because of the dependence on the training data. (c*) Draw a random realization of the data set $\mathcal{D}$ with $N = 20$ data points generated from the pdf $\mathcal{N}(t|\mu, 1)$ with ground-truth value $\mu = 1$. Calculate the random value of the estimation error. (d*) Generate 1000 independent data sets $\mathcal{D}$ following the procedure in point (c*). Using these data, estimate the value ϵ such that the probability that the estimation error is less than ϵ is larger than $1 - \delta$ with $\delta = 0.1$. [Hint: You have generated 1000 data sets, and so you can estimate 1000 independent realizations of the estimation error. You can therefore take the $1000 \cdot \delta$th largest value as an estimate of ϵ.] (e*) What happens if you increase N?

8.2   We study a binary classification problem with detection-error loss and model class given by the set of symmetric intervals

$$\mathcal{H} = \{\hat{t}(x|\theta) = \mathbb{1}(|x| \leq \theta)\}.$$

(a) Assuming that the population distribution of the data is given by $p(x, t) = p(x)\mathbb{1}(t = \hat{t}(x|1))$ with $p(x)$ being uniform in the interval $[-2, 2]$, calculate the population loss $L_p(\theta)$ for $\theta \geq 0$. (b) We observe a data set of $N = 4$ data points with $x_1 = -1.6$, $x_2 = -0.4$,

$x_3 = 0.6$, and $x_4 = 1.4$, with labels assigned according to the population distribution. Compute the resulting training loss $L_{\mathcal{D}}(\theta)$ for $\theta \geq 0$. (c) Plot by hand and compare the population and training losses. (d) Describe the corresponding set of ERM solutions. (e) Compute the VC dimension of the model class.

8.3 Consider the model class of binary classifiers defined on a discrete and finite set $\mathcal{X}$

$$\mathcal{H} = \{\hat{t}(x|\theta) = \mathbb{1}(x = \theta) \text{ for all } \theta \in \mathcal{X}\}$$

under the detection-error loss. Obtain the sample complexity $N_{\mathcal{H}}^{ERM}(\epsilon, \delta)$ of ERM using PAC learning theory.

8.4 Calculate the VC dimension of the model class $\mathcal{H} = \{\hat{t}(x|\theta) = \mathbb{1}(||x|| \leq \theta) \text{ for } \theta \geq 0\}$ with $x \in \mathbb{R}^2$.

8.5 Consider the model class of binary linear classifiers with a feature vector $u(x) = [x^T, 1]^T$, with $x \in \mathbb{R}^2$ so that the model parameter vector is three-dimensional, i.e., $D = 3$ (see Chapter 6). (a) Evaluate the capacity of a quantized version of this model class where each model parameter is quantized with b bits. (b) Evaluate the VC dimension and comment on its dependence on D, and comment on the comparison with point (a).

8.6 For the population distribution $t \sim p(t) = \mathcal{N}(t|\mu, 1)$, where μ is a real number, and model class $\mathcal{H}$ given by the constant predictors $\hat{t} \in \mathbb{R}$ (see Problem 8.1), assume the probabilistic training algorithm $q(\hat{t}|\mathcal{D})$ that outputs $\hat{t} = 1/N \sum_{n=1}^{N} t_n + z_n$, where $z_n \sim \mathcal{N}(0, \beta^{-1})$ is independent of the training data $\mathcal{D} = \{(t_n)_{n=1}^{N}\}$. (a) For prior $q(\hat{t}) = \mathcal{N}(\hat{t}|0, 1)$, compute the KL divergence that appears in the PAC Bayes bound, and comment on the role of the noise variance β^{-1} for generalization. (b) Compute the mutual information appearing in the information-theoretic analysis of generalization, and comment again on the role of β^{-1}.

Bibliography

[1] S. Shalev-Shwartz and S. Ben-David, *Understanding Machine Learning: From Theory to Algorithms*. Cambridge University Press, 2014.

[2] M. Mohri, A. Rostamizadeh, and A. Talwalkar, *Foundations of Machine Learning*. The MIT Press, 2018.

[3] V. Vapnik, *The Nature of Statistical Learning Theory*. Springer Science+Business Media, 2013.

[4] P. Baldi, *Deep Learning in Science*. Cambridge University Press, 2021.

[5] B. Guedj, "A primer on PAC-Bayesian learning," arXiv preprint arXiv:1901.05353, 2019.

[6] M. Raginsky, A. Rakhlin, and A. Xu, "Information-theoretic stability and generalization," in *Information-Theoretic Methods in Data Science*. Cambridge University Press, 2021, pp. 302–329.

9 Exponential Family of Distributions

9.1 Overview

The previous chapters have adopted a limited range of probabilistic models, namely Bernoulli and categorical distributions for discrete rvs and Gaussian distributions for continuous rvs. While these are common modeling choices, they clearly do not represent many important situations of interest for machine learning applications. For instance, discrete data may a priori take arbitrarily large values, making categorical models unsuitable. Continuous data may need to satisfy certain constraints, such as non-negativity, rendering Gaussian models far from ideal.

Bernoulli, categorical, and Gaussian distributions share several desirable common features:

- The **gradient of the log-loss** with respect to the model parameters can be expressed in terms of a **mean error** that measures the difference between mean under the model (for the model parameter vector at which the gradient is computed) and observation (see Sec. 6.3.3).
- **Maximum likelihood (ML) learning** of the distribution's parameters can be carried out by evaluating empirical averages of functions of the observations (see Sec. 6.5).
- **Information-theoretic metrics** such as (differential) entropy and KL divergence can be explicitly evaluated as a function of the model parameters (see Sec. 3.5).

In this chapter, a general family of distributions – the **exponential family** – will be introduced whose members satisfy the same desirable properties, and include, among many others, Bernoulli, categorical, and Gaussian distributions. For example, the exponential family of distributions also encompasses Poisson and geometric distributions, whose support is discrete and includes all integers, as well as exponential and gamma distributions, whose support is continuous and includes only non-negative values. (A distribution that does not belong to the family of exponential distributions is given by a uniform pdf in an interval dependent on model parameters.)

Thanks to the mentioned properties in terms of log-loss, ML learning, and information-theoretic measures, the methods studied in the previous chapters can be extended to a much larger class of problems by replacing Bernoulli, categorical, and Gaussian distributions with a probabilistic model in the exponential family. Some of these extensions will be explored in the following chapters.

Learning Objectives and Organization of the Chapter. By the end of this chapter, the reader should be able to:

- recognize distributions in the exponential family and identify key quantities such as sufficient statistics and log-partition function (Sec. 9.2);
- understand the information-theoretic optimality of exponential-family distributions from the viewpoint of the maximum-entropy principle (Sec. 9.3);
- derive the gradient of the log-loss for distributions in the exponential family (Sec. 9.4);

- use the gradient of the log-loss to obtain ML learning algorithms for distributions in the exponential family (Sec. 9.5);
- compute and use information-theoretic metrics for the exponential family (Sec. 9.6);
- understand the definition of the Fisher information matrix (FIM) for parametric models, as well as the use of natural GD as an alternative to GD (Sec. 9.7);
- compute and use the FIM for distributions in the exponential family (Sec. 9.7); and
- understand and use conditional distribution models obtained from exponential-family distributions, also known as generalized linear models (GLMs) (Sec. 9.8).

9.2 Definitions and Examples

To start, a remark about notation: this chapter will describe parametric probability distributions that are specified by one of two types of parameters, namely **natural parameters** – denoted as η – and **mean parameters** – denoted as μ. Accordingly, the notation θ used thus far for the model parameter vector will be specialized to either η or μ depending on the type of parameters under consideration.

With this aspect clarified, we begin by recalling from the previous chapters that Bernoulli, categorical, and Gaussian distributions have the useful property that the **log-loss** $-\log p(x|\eta)$ is a **convex** function of the model parameters η. (The reason for the use of the notation η will be made clear next.) For example, for a Gaussian rv x $\sim \mathcal{N}(\eta, 1)$, the log-loss is given as

$$-\log p(x|\eta) = \frac{1}{2}(x - \eta)^2 + \text{constant independent of } \eta, \qquad (9.1)$$

which is a (strictly) convex (quadratic) function of the parameter η (see Sec. 5.6). As we will see next, all distributions in the exponential family share this key property.

9.2.1 Definition of the Exponential Family

The exponential family contains pmfs and pdfs whose **log-loss** can be written as

$$-\log p(x|\eta) = -\underbrace{\eta^T s(x)}_{\sum_{k=1}^{K} \eta_k s_k(x)} - \underbrace{M(x)}_{\text{log-base measure}} + \underbrace{A(\eta)}_{\text{log-partition function}}, \qquad (9.2)$$

where we have defined

- the $K \times 1$ vector of **sufficient statistics**

$$s(x) = \begin{bmatrix} s_1(x) \\ \vdots \\ s_K(x) \end{bmatrix}, \qquad (9.3)$$

- the **log-base measure** function $M(x)$,
- the $K \times 1$ **natural parameter vector**

$$\eta = \begin{bmatrix} \eta_1 \\ \vdots \\ \eta_K \end{bmatrix} \text{ and} \qquad (9.4)$$

- the **log-partition function** $A(\eta)$.

A probabilistic model $p(x|\eta)$ in the exponential family is hence *specified* by the choice of the vector of sufficient statistics $s(x)$ and by the log-base measure function $M(x)$, and is *parametrized* by the natural parameter vector η. The log-partition function, as we will see below, is a function of η, $s(x)$, and $M(x)$, and it serves the purpose of normalizing the distribution. The terminology "sufficient statistics" indicates that vector $s(x)$, alongside function $M(x)$, are *sufficient* to specify a probabilistic model in the exponential family.

To reiterate this important point, by (9.2) the distribution $p(x|\eta)$ depends on x only through vector $s(x)$ and function $M(x)$. One can hence think of $s(x)$ and $M(x)$ as collectively defining a vector of features that determines the distribution of rv x $\sim p(x|\eta)$.

As should be clear from (9.2), there is some ambiguity in the definition of sufficient statistics. For instance, if vector $s(x)$ is a sufficient statistics vector, then so would any scalar multiple of it – say $c \cdot s(x)$ for $c \in \mathbb{R}$ – at the cost of redefining natural parameters accordingly as η/c. In the following, we will adopt the most common choices for the sufficient statistics of standard distributions.

Log-Partition Function. The log-partition function $A(\eta)$ ensures that the distribution $p(x|\eta)$ is normalized, and is determined by the functions $s(x)$ and $M(x)$. To elaborate, by (9.2) we can write an exponential-family probabilistic model specified by sufficient statistics $s(x)$ and log-base measure function $M(x)$ as

$$p(x|\eta) = \exp\left(\eta^T s(x) + M(x) - A(\eta)\right)$$
$$= \frac{1}{\exp(A(\eta))} \exp\left(\eta^T s(x) + M(x)\right). \tag{9.5}$$

The function $\exp(A(\eta))$ is known as a **partition function** – whence the name log-partition function for $A(\eta)$. In order to guarantee the normalization condition $\int p(x|\eta)\mathrm{d}x = 1$ for continuous rvs and $\sum_x p(x|\eta) = 1$ for discrete rvs, we need to choose the log-partition function as

$$A(\eta) = \log\left(\int \exp\left(\eta^T s(x) + M(x)\right) \mathrm{d}x\right) \tag{9.6}$$

for continuous rvs and

$$A(\eta) = \log\left(\sum_x \exp\left(\eta^T s(x) + M(x)\right)\right) \tag{9.7}$$

for discrete rvs.

Domain of the Natural Parameters. The vector of natural parameters η can take any value that ensures that the distribution can be normalized, i.e., that the log-partition function is finite, $A(\eta) < \infty$. This feasible set – or **domain** of the natural parameters – can be shown to be a convex set. A convex set is such that it contains all segments between any two points in the set. For the class of distributions in the **regular** exponential family, the domain is assumed to be open, that is, not to include its boundary. We will implicitly restrict our attention to regular exponential-family distributions throughout this chapter.

Convexity of the Log-Loss. The log-partition function (9.6) or (9.7) has a "log-sum-exp" form, and is hence convex in η (see Sec. 5.7). Therefore, the log-loss (9.2) is the sum of a linear function in η, namely $-\eta^T s(x)$, and a convex function in η, namely $A(\eta)$. By the properties of convex functions reviewed in Sec. 5.7, this implies that the log-loss is also convex in η.

Recap and More Compact Representation. To summarize, the exponential family contains pmfs and pdfs that are specified by sufficient statistics $s(x)$ and log-base measure function $M(x)$, and take the exponential form

$$p(x|\eta) \propto \exp\left(\underbrace{\eta^T s(x)}_{\text{linear function of } \eta} + \underbrace{M(x)}_{\text{general function of } x}\right), \tag{9.8}$$

where the "proportional to" sign $\propto$ makes the normalizing multiplicative constant $\exp(-A(\eta))$ implicit. When the sufficient statistics $s(x)$ and log-based measure function $M(x)$ are clear from the context, we will write a distribution in the exponential family as

$$p(x|\eta) = \text{ExpFam}(x|\eta). \tag{9.9}$$

For the purpose of analytical calculations, it is often convenient to write the distribution (9.8) in terms of the **augmented sufficient statistics** vector

$$\tilde{s}(x) = \left[\begin{array}{c} s(x) \\ M(x) \end{array}\right], \tag{9.10}$$

which collects both sufficient statistics and log-base measure function. This corresponds to the feature vector mentioned earlier in this subsection that fully specifies the model. Defining the **augmented natural parameter vector**

$$\tilde{\eta} = \left[\begin{array}{c} \eta \\ 1 \end{array}\right], \tag{9.11}$$

this yields the compact representation

$$-\log p(x|\eta) = -\underbrace{\tilde{\eta}^T \tilde{s}(x)}_{\sum_{k=1}^K \tilde{\eta}_k \tilde{s}_k(x)} + \underbrace{A(\eta)}_{\text{convex function of } \eta} \tag{9.12}$$

or equivalently

$$p(x|\eta) \propto \exp\left(\tilde{\eta}^T \tilde{s}(x)\right). \tag{9.13}$$

9.2.2 Examples

In this section, we provide some examples of distributions in the exponential family.

Example 9.1

Gaussian distribution with fixed variance. For the Gaussian distribution $\mathcal{N}(\mu, \beta^{-1})$ with fixed precision β, the log-loss can be written as

$$-\log \mathcal{N}(x|\mu, \beta^{-1}) = -\underbrace{\beta\mu}_{\eta}\underbrace{x}_{s(x)} - \left(\underbrace{-\frac{\beta}{2}x^2 - \frac{1}{2}\log(2\pi\beta^{-1})}_{M(x)}\right) + \left(\underbrace{\frac{\beta\mu^2}{2}}_{A(\eta)}\right). \tag{9.14}$$

Accordingly, the sufficient statistic is one-dimensional ($K = 1$) and given as $s(x) = x$. Furthermore, the **log-partition function** can be expressed in terms of the **natural parameter** $\eta = \beta\mu$ as

$$A(\eta) = \frac{\beta\mu^2}{2} = \frac{1}{2}\frac{\eta^2}{\beta}, \tag{9.15}$$

which is indeed a (strictly) convex function of η for all $\eta \in \mathbb{R}$.

Based on the previous paragraph, the Gaussian distribution $\mathcal{N}(x|\mu, \beta^{-1})$ with a fixed variance β^{-1} can be parametrized by the natural parameter $\eta = \beta\mu$. It can, of course, also be described by the **mean parameter**

$$\mu = \mathrm{E}_{x \sim \mathcal{N}(\mu, \beta^{-1})}[s(x)] = \mathrm{E}_{x \sim \mathcal{N}(\mu, \beta^{-1})}[x], \tag{9.16}$$

which turns out to be the mean of the sufficient statistic. The natural parameter η and the mean parameter μ are in a *one-to-one correspondence*, as one can be recovered from the other through the equality $\beta\mu = \eta$ (where β is fixed).

Example 9.2

Bernoulli distribution. The Bernoulli distribution can be written as

$$\mathrm{Bern}(x|\mu) = \mu^x (1 - \mu)^{1-x}, \tag{9.17}$$

where $x \in \{0, 1\}$ and we have the **mean parameter**

$$\mu = \mathrm{E}_{x \sim \mathrm{Bern}(x|\mu)}[x] = \Pr[x = 1]. \tag{9.18}$$

Therefore, the log-loss is

$$-\log \mathrm{Bern}(x|\mu) = \underbrace{-\log\left(\frac{\mu}{1-\mu}\right)}_{\eta} \underbrace{x}_{s(x)} + \underbrace{(-\log(1-\mu))}_{A(\eta)}, \tag{9.19}$$

where $M(x) = 0$ and the sufficient statistic is again one-dimensional ($K = 1$) and given as $s(x) = x$. It follows that the **natural parameter** is the **logit** or **log-odds** (see Sec. 6.3)

$$\eta = \log\left(\frac{\mathrm{Bern}(1|\mu)}{\mathrm{Bern}(0|\mu)}\right) = \log\left(\frac{\mu}{1-\mu}\right). \tag{9.20}$$

The mean parameter (9.18) is, as in Example 9.1, the expectation of the sufficient statistic, and is in a *one-to-one correspondence* with the natural parameter η: Inverting the equality (9.20), we obtain

$$\mu = \sigma(\eta) = \frac{1}{1 + e^{-\eta}}. \tag{9.21}$$

Therefore, by (9.19), the **log-partition function** can be expressed in terms of the natural parameter η as

$$A(\eta) = \log(1 + e^\eta), \tag{9.22}$$

which is a (strictly) convex function of $\eta \in \mathbb{R}$.

Example 9.3

General Gaussian distribution. For a Gaussian distribution $\mathcal{N}(v, \beta^{-1})$ with parameters (v, β), we have the log-loss

$$-\log \mathcal{N}(x|v, \beta^{-1}) = -\left(\underbrace{\beta v}_{\eta_1} \underbrace{x}_{s_1(x)} + \underbrace{\left(-\frac{\beta}{2}\right)}_{\eta_2} \underbrace{x^2}_{s_2(x)} + \underbrace{\left(\frac{v^2 \beta}{2} + \frac{1}{2} \log(2\pi \beta^{-1})\right)}_{A(\eta)} \right), \tag{9.23}$$

where $M(x) = 0$. For this example, we have a two-dimensional vector of **sufficient statistics**, i.e., $K = 2$, namely $s(x) = [x, x^2]^T$, and, correspondingly, a two-dimensional **natural parameter vector**

$$\eta = \begin{bmatrix} \eta_1 = \beta v \\ \eta_2 = -\frac{\beta}{2} \end{bmatrix}. \tag{9.24}$$

Following the previous examples, the two-dimensional vector of **mean parameters** is defined as the vector of averages of the sufficient statistics under the model, i.e.,

$$\mu = \begin{bmatrix} \mathrm{E}_{\mathrm{x} \sim \mathcal{N}(v, \beta^{-1})}[s_1(\mathrm{x})] \\ \mathrm{E}_{\mathrm{x} \sim \mathcal{N}(v, \beta^{-1})}[s_2(\mathrm{x})] \end{bmatrix} = \begin{bmatrix} \mathrm{E}_{\mathrm{x} \sim \mathcal{N}(v, \beta^{-1})}[\mathrm{x}] \\ \mathrm{E}_{\mathrm{x} \sim \mathcal{N}(v, \beta^{-1})}[\mathrm{x}^2] \end{bmatrix} = \begin{bmatrix} v \\ v^2 + \beta^{-1} \end{bmatrix}. \tag{9.25}$$

This vector is in a *one-to-one correspondence* with the vector of natural parameters η in (9.24). Furthermore, we can write the **log-partition function** as

$$\begin{aligned} A(\eta) &= \frac{v^2 \beta}{2} + \frac{1}{2} \log(2\pi \beta^{-1}) \\ &= -\frac{\eta_1^2}{4\eta_2} + \frac{1}{2} \log\left(-\frac{\pi}{\eta_2}\right), \end{aligned} \tag{9.26}$$

which is strictly convex in the domain $\eta_1, \eta_2 \in \mathbb{R} \times \mathbb{R}^-$, with $\mathbb{R}^-$ denoting the (open) set of strictly negative numbers.

Generalizing these examples, among many others, the exponential family also includes the following distributions:

- **pmfs**: binomial, negative binomial, geometric, Poisson; and
- **pdfs**: lognormal, gamma, inverse gamma, chi-squared, exponential, beta, Dirichlet, Pareto, Laplace.

Example 9.4

Poisson distribution. As an example of a distribution that we have not considered before in this book, let us introduce the Poisson distribution, which is used extensively in fields as diverse as neuroscience and communication networks. The Poisson distribution is supported over the set of integers, and is defined as

$$\mathrm{Poiss}(x|\lambda) = \frac{\lambda^x \exp(-\lambda)}{x!}, \tag{9.27}$$

where $x \in \{0, 1, 2, \dots\}$. The log-loss is

$$- \log \text{Poiss}(x|\lambda) = - \underbrace{\log(\lambda)}_{\eta} \underbrace{x}_{s(x)} - \underbrace{(- \log(x!))}_{M(x)} + \underbrace{\lambda}_{A(\eta)}. \tag{9.28}$$

The **mean parameter** is given by the mean of the sufficient statistic, $\mu = \text{E}_{\text{x}\sim\text{Poiss}(x|\lambda)}[\text{x}] = \lambda$, and is in a *one-to-one correspondence* with the natural parameter $\eta = \log(\lambda)$. Therefore, the **log-partition function** can be expressed in terms of the natural parameter η as

$$A(\eta) = \exp(\eta), \tag{9.29}$$

which is a (strictly) convex function of $\eta \in \mathbb{R}$.

Example 9.5

Joint Bernoulli–Gaussian distribution. For a more complex example, consider the rv $\text{x} = [\text{x}_1, \text{x}_2]^T$ with $\text{x}_1 \sim \text{Bern}(x|\mu)$ and $(\text{x}_2|\text{x}_1 = x_1) \sim \mathcal{N}(\nu_{x_1}, \beta^{-1})$, where the precision β is fixed and the model parameters are (μ, ν_0, ν_1). Note that, when both rvs x_1 and x_2 are observed, this joint distribution may be used as a generative model for binary classification (see Sec. 6.5). Following the same steps as in the preceding examples, one can see that this joint distribution is in the exponential family, with sufficient statistics given by $s(x) = [x_1, x_2(1 - x_1), x_2 x_1]^T$ and natural parameter vector $\eta = \left[\log(\mu/(1 - \mu)), \beta\nu_0, \beta\nu_1\right]^T$. The marginal distribution of x_1 under this joint distribution is a mixture of Gaussians, which is not in the exponential family. This example shows that the exponential family is not "closed" with respect to the operation of marginalization; that is, the marginal distribution of a distribution in the exponential family is not necessarily also in the exponential family.

A reference list of some basic distributions in the exponential family can be found in Table 9.1. With reference to the table, when the sufficient statistics $s(x)$ are defined by a square matrix, as is the case with the multivariate Gaussian distribution $\mathcal{N}(\nu, \Theta^{-1})$ with mean vector ν and precision matrix Θ, the corresponding natural parameters η are also expressed in the form a matrix of the same dimension and the inner product is written as the trace $\text{tr}(\eta^T s(x))$.

9.2.3 Natural and Mean Parameters

As illustrated by the examples in the preceding section, distributions in the exponential family can be specified by a vector η of natural parameters or by a vector μ of **mean parameters**, which is defined as

$$\mu = \text{E}_{\text{x}\sim p(x|\eta)}[s(\text{x})], \tag{9.30}$$

Table 9.1 Sufficient statistics $s(x)$, natural parameters η, and mean parameters μ for some distributions in the exponential family

distribution	$s(x)$	η	μ
$\text{Bern}(\mu)$	x	$\log\left(\frac{p}{1-p}\right)$ (logit)	$\sigma(\eta)$
$\text{Cat}(\mu)$	x^{OH} (one-hot vector)	$\eta_k = \log(a\mu_k)$ for $a > 0$ (logits)	$\mu = \text{softmax}(\eta)$
$\mathcal{N}(\nu, \Theta^{-1})$, fixed Θ	x	$\Theta\nu$	ν
$\mathcal{N}(\nu, \Theta^{-1})$	x and xx^T	$\Theta\nu$ and $-\frac{1}{2}\Theta$	ν and $\Theta^{-1} + \nu\nu^T$

i.e., as the vector of averages of the sufficient statistics $s(x)$. Given the outlined **duality** between natural and mean parameters, we can write a distribution in the exponential family as a function of the natural parameters η – which we denote as $p(x|\eta) = \text{ExpFam}(x|\eta)$ – or as a function of the mean parameters μ – which we indicate as $p(x|\mu) = \text{ExpFam}(x|\mu)$. We have used, and will use, both formulations, hence overloading the notations $p(x|\cdot)$ and $\text{ExpFam}(x|\cdot)$.

Minimal Exponential Family. While the vector of mean parameters, μ, is uniquely defined for a given distribution in the exponential family, there may be multiple natural parameters η yielding the same distribution. A class of distributions $p(x|\eta) = \text{ExpFam}(x|\eta)$ in the exponential family is said to be **minimal** if no two natural parameter vectors yield the same distribution.

A minimal class of distributions is equivalently characterized by the following condition: There is no natural parameter vector η_0 in the domain for which the inner product $\eta_0^T s(x)$ is constant. In fact, if such a vector existed, we could add it to any other natural parameter η without changing the distribution. Furthermore, clearly, the same property would apply to any vector $b \cdot \eta_0$ for any scalar $b \in \mathbb{R}$. Therefore, for non-minimal classes, there is an infinitely large set of natural parameters $\eta + b \cdot \eta_0$ all producing the same distribution. Technically, this set defines an affine subspace.

For minimal classes of distributions in the exponential family, there is a *one-to-one correspondence* between natural and mean parameters: Each natural parameter vector yields a different distribution $p(x|\eta)$; and each mean parameter vector μ is associated with a single natural parameter vector η. Note that the reverse condition – that a single mean parameter vector is associated with any natural parameter vector – is always true.

Under a minimal distribution class, the natural parameters are **identifiable** in the sense that there is a single natural parameter that is consistent with a given distribution in the class. In contrast, for non-minimal classes, the natural parameters can only be identified as any vector lying in the mentioned affine subspace.

We will see in Sec. 9.4 that the minimality of a model is equivalent to the **strict convexity of the log-partition function**, i.e., to the condition $\nabla^2 A(\eta) \succ 0$ for all vectors η in the domain (see Sec. 5.7.6 for the definition of strict convexity).

Finally, for minimal models in the exponential family, there exists a general explicit expression for the log-loss as a function of the mean parameters, as detailed in Appendix 9.B.

All examples in the previous sections were instances of minimal classes of exponential-family distributions. But not all classes of distributions in the exponential family are minimal. The following is a common example.

Example 9.6

Categorical (or multinoulli) distribution. Consider the categorical distribution of a rv x that can take C values $\{0, 1, \ldots, C - 1\}$ with respective probabilities

$$\mu_k = \Pr[\text{x} = k] \text{ for } k = 0, \ldots, C - 1. \tag{9.31}$$

Defining the $C \times 1$ vector of probabilities $\mu = [\mu_0, \ldots, \mu_{C-1}]^T$, the categorical pmf can be written as

$$\text{Cat}(x|\mu) = \prod_{k=0}^{C-1} \mu_k^{\mathbb{1}(x=k)} = \frac{1}{a} \prod_{k=0}^{C-1} (a\mu_k)^{\mathbb{1}(x=k)}, \tag{9.32}$$

for *any* choice of the constant $a > 0$. Accordingly, the **log-loss** is

$$-\log(\text{Cat}(x|\mu)) = -\sum_{k=0}^{C-1} \underbrace{\log(a\mu_k)}_{\eta_k} \underbrace{\mathbb{1}(x = k)}_{s_k(x)} + \underbrace{\log(a)}_{A(\eta)=\log\left(\sum_{k=0}^{C-1} e^{\eta_k}\right)}, \tag{9.33}$$

with $M(x) = 0$. Therefore, we have the $C \times 1$ **sufficient statistics vector** $(K = C)$

$$s(x) = \begin{bmatrix} \mathbb{1}(x = 0) \\ \vdots \\ \mathbb{1}(x = C - 1) \end{bmatrix}, \tag{9.34}$$

which corresponds to the $C \times 1$ **one-hot vector** x^{OH}. The **natural parameter** vector is $\eta = [\eta_0, \dots, \eta_{C-1}]^T$, and the **mean parameter vector** μ collects the probabilities $\mu_k = \mathrm{E}_{x \sim \text{Cat}(x|\mu)}[s_k(x)] = \Pr[x = k]$ for $k = 0, 1, \dots, C - 1$.

The parametrization (9.32) is not minimal, since we can always add a vector of the form $[\log(a), \dots, \log(a)]^T$ to the natural parameter vector η without changing the distribution. This is because, as mentioned, the constant $a > 0$ is arbitrary in the natural parameters $\eta_k = \log(a\mu_k) = \log(\mu_k) + \log(a)$ of the pmf. As a result, there are infinitely many natural parameter vectors yielding the same categorical distribution $\text{Cat}(x|\mu)$, i.e.,

$$\eta = \begin{bmatrix} \log(\mu_0) \\ \vdots \\ \log(\mu_{C-1}) \end{bmatrix} + b \cdot 1_C, \tag{9.35}$$

where constant b is arbitrary, and we recall that 1_C represents the $C \times 1$ vector containing all ones.

Even for a non-minimal class of distributions, there is a single mean parameter vector μ for each natural parameter vector η. By (9.35), in the case of a categorical distribution, the relationship between the two vectors is given by the **softmax function**

$$\mu = \text{softmax}(\eta) = \begin{bmatrix} \frac{e^{\eta_0}}{\sum_{k=0}^{C-1} e^{\eta_k}} \\ \vdots \\ \frac{e^{\eta_{C-1}}}{\sum_{k=0}^{C-1} e^{\eta_k}} \end{bmatrix}.$$

Recalling the discussion in Sec. 6.6, this relationship reveals that the natural parameter vector corresponds to the **logits** associated with the probability vector μ. Adding a constant vector to the logit vector does not change the resulting categorical distribution.

An example with $C = 3$ is given by rv $x \sim \text{Cat}(x|\mu = [0.1, 0.8, 0.1]^T)$, which has mean parameters $\mu_0 = \Pr[x = 0] = 0.1$, $\mu_1 = \Pr[x = 1] = 0.8$, $\mu_2 = \Pr[x = 2] = 0.1$; natural parameters (logits) $\eta_0 = \eta_2 = \log(a \cdot 0.1) = -2.30$ and $\eta_1 = \log(a \cdot 0.8) = -0.22$, where we set $a = 1$; and we have the equality

$$\text{softmax}\left(\begin{bmatrix} -2.30 \\ -0.22 \\ -2.30 \end{bmatrix}\right) = \frac{1}{\sum_{k=0}^{2} e^{\eta_k}} \begin{bmatrix} e^{\eta_0} \\ e^{\eta_1} \\ e^{\eta_2} \end{bmatrix} = \begin{bmatrix} 0.1 \\ 0.8 \\ 0.1 \end{bmatrix} = \mu. \tag{9.36}$$

9.3 Exponential-Family Distributions As Maximum-Entropy Models

How can one justify the use of distributions from the exponential family apart from their analytical tractability? In this section, we will see that models in the exponential family can be derived as solutions of a **maximum-entropy problem**, whereby a distribution is sought that preserves the maximum "uncertainty" while satisfying certain constraints on its moments.

To elaborate, suppose that the only information available about a random quantity x is given by the expectations of some functions – i.e., statistics – of x. As an example, we may wish to model the waiting time of a new customer at a certain office based on knowledge of the average waiting time measured across a large number of customers. Mathematically, the goal is to define a model $p(x)$ for the distribution of rv x based on the observation of a collection of expectations

$$E_{x \sim p(x)}[s_k(x)] = \mu_k \tag{9.37}$$

of given functions $s_k(x)$ for $k = 1, \ldots, K$. Hence, the problem is one of **density estimation**, where, unlike the setting in Sec. 7.3, we are not given samples from x, but instead the expectations (9.37).

A well-established principle for addressing this problem is to choose the distribution $p(x)$ that is least predictable, or "more random", under the given average constraints (9.37). Recall from Sec. 3.6 that the entropy $H(p(x))$ is a measure of unpredictability of a random variable $x \sim p(x)$, in that it measures the minimum average prediction log-loss when all that is known is its distribution $p(x)$. Taking the entropy as a measure of the randomness of the distribution $p(x)$, the outlined problem can be formulated as the maximization

$$\max_{p(x)} H(p(x)) \text{ s.t. } E_{x \sim p(x)}[s_k(x)] = \mu_k \text{ for } k = 1, \ldots, K, \tag{9.38}$$

where "s.t." stands for "subject to", indicating the constraints that the solution of the optimization problem must meet. The problem may also contain additional constraints limiting the support of the distribution. For instance, to enforce the non-negativity of rv x, one can include the constraint $E_{x \sim p(x)}[\mathbb{1}(x \geq 0)] = 1$.

It can be proved that the exponential-family distribution $p(x|\eta)$ in (9.5) solves this problem, where each natural parameter η_k is the optimal **Lagrange multiplier** associated with the kth constraint in (9.38) and the log-base measure $M(x)$ can account for the mentioned additional support-defining constraints. Informally, the Lagrange multiplier reflects the impact of each constraint on the optimal solution. See Recommended Resources, Sec. 9.10, for details and for an exact definition of Lagrange multipliers.

All in all, the optimality of the exponential family in terms of the maximum-entropy problem (9.38) offers a theoretical justification for its adoption as a model: The exponential family "makes the fewest assumptions", in the sense of retaining the maximum uncertainty, while being consistent with the available information about the expectations of some functions of the quantity of interest.

The maximum-entropy viewpoint also offers a perspective on the fact that a distribution in the exponential family is specified by its mean parameters. In fact, the mean parameters determine the constraints in the maximum-entropy problem (9.38), which in turn specify the solution (9.5). The corresponding connection between natural parameters and mean parameters through the Lagrange multipliers provides an interesting mathematical link between these two quantities.

9.4   Gradient of the Log-Loss

We have seen in Sec. 9.2 that distributions in the exponential family have the useful property that their log-loss is convex in the natural parameters. This section demonstrates another key property of exponential-family distributions, namely the general link that exists between gradient of the log-loss and mean error measures. For the special case of Bernoulli and categorical distributions, this property played a central role in Chapter 6 in deriving ML learning algorithms based on GD for logistic and softmax regression, as well as for neural networks. We will see here that the same relationship applies more generally to all distributions in the exponential family. We will also use this relationship to further discuss the connection between mean and natural parameters.

9.4.1   Gradient and Mean Error

The key result of this section is that the partial derivative of the log-loss $-\log p(x|\eta)$ with respect to each natural parameter η_k, for a fixed value x, is given by

$$\frac{\partial(-\log p(x|\eta))}{\partial \eta_k} = \underbrace{\mu_k - s_k(x)}_{\text{mean error for } s_k(x)} . \tag{9.39}$$

Equivalently, in vector form, the gradient with respect to the natural parameters is

$$\nabla_\eta(-\log p(x|\eta)) = \underbrace{\mu - s(x)}_{\text{mean error for } s(x)} . \tag{9.40}$$

That is, the **gradient of the log-loss** equals the **mean error** $\mu - s(x)$ between the mean vector μ, which contains the expectations of the sufficient statistics, and the actual value $s(x)$ of the sufficient statistics.

Example 9.7

For the Gaussian distribution $\mathcal{N}(x|v,I)$, by Table 9.1, both natural and mean parameter vectors are equal to v, i.e., $\eta = \mu = v$, and the sufficient statistics vector is $s(x) = x$. Accordingly, using the general formula (9.40), the gradient of the log-loss is given as

$$\nabla_v(-\log \mathcal{N}(x|v,I)) = v - x. \tag{9.41}$$

Following the discussion in Chapter 5, the negative gradient of the log-loss is a pointer towards the direction $x - v$ in natural parameter space (or equivalently mean parameter space in this example) that locally minimizes the log-loss $-\log \mathcal{N}(x|v,I)$. An example is illustrated in Fig. 9.1 for $x = [0.5, 0.5]^T$. Note that the negative gradient points towards the direction of the data point x.

9.4.2   Proof of the Relationship between Gradient and Mean Error

Given the importance of the formula (9.40) in deriving ML training solutions, understanding its proof is useful and instructive. This is the focus of this subsection.

By using the expression for the log-loss (9.2) for exponential-family distributions, we compute the partial derivative with respect to each natural parameter η_k as

Figure 9.1 Negative gradient of the log-loss, $\nabla_\nu \log \mathcal{N}(x|\nu, I) = x - \nu$, for $x = [0.5, 0.5]^T$.

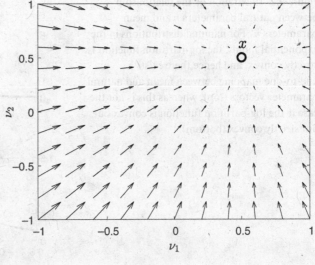

$$\frac{\partial(-\log p(x|\eta))}{\partial \eta_k} = -s_k(x) + \frac{\partial A(\eta)}{\partial \eta_k}. \tag{9.42}$$

Moreover, the partial derivative of the log-partition function can be obtained from its definition (9.6) or (9.7) as (see Appendix 9.C)

$$\frac{\partial A(\eta)}{\partial \eta_k} = \mathrm{E}_{x \sim p(x|\eta)}\left[s_k(x)\right] = \mu_k \tag{9.43}$$

or, in vector form,

$$\nabla_\eta A(\eta) = \mathrm{E}_{x \sim p(x|\eta)}\left[s(x)\right] = \mu. \tag{9.44}$$

The identity (9.44) states that the gradient of the log-partition function $A(\eta)$ equals the mean parameter vector μ associated with the natural parameter η. This essential relationship will be further explored in the next subsection. Plugging (9.43) into (9.42) concludes the proof of the gradient formula (9.40).

9.4.3 From Natural Parameters to Mean Parameters and, Possibly, Back

By (9.44), given a natural parameter vector η, the (unique) corresponding mean parameter is given by the gradient $\nabla_\eta A(\eta)$ of the log-partition function evaluated at η. So, the mean parameter μ_k measures the rate of change of the log-partition function $A(\eta)$ as one varies the corresponding natural parameter η_k, i.e., $\mu_k = \partial A(\eta)/\partial \eta_k$. Under what conditions can we invert this mapping, obtaining a single natural parameter vector η for a given mean parameter vector μ? We recall from Sec. 9.2 that the invertibility of the map between mean and natural parameters is the defining property of **minimal parametrizations** for exponential-family distributions.

Fig. 9.2 illustrates the relationship between natural parameter and mean parameter. By (9.44), as seen, the mean parameter μ corresponding to a natural parameter η is the slope of the log-partition function $A(\eta)$ at η. Therefore, the mean parameter associated with a natural parameter is unique. The reverse is true as long as the log-partition function $A(\eta)$ has a strictly positive curvature – i.e., if the Hessian is positive definite,

Figure 9.2 Illustration of the relationship between natural parameters η and mean parameters μ. For minimal distributions in the exponential family, the log-partition function is strictly convex, and hence there exists a one-to-one mapping between mean and natural parameter vectors (top), whereas this is not the case if the log-partition function is convex but not strictly convex (bottom).

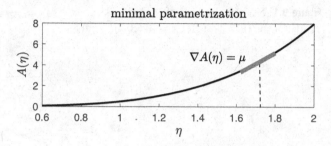

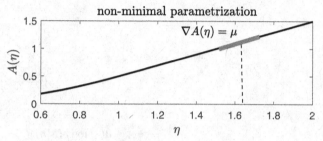

$$\nabla^2_\eta A(\eta) \succ 0 \text{ for all values of } \eta \text{ in the domain,} \qquad (9.45)$$

or equivalently if the log-partition function is strictly convex. In fact, this implies that the function $A(\eta)$ does not contain linear segments along which the derivative remains constant (see Chapter 5).

Since the strict convexity of the log-partition function implies a one-to-one mapping between mean and natural parameters, the condition (9.45) provides an equivalent definition of minimal parametrizations and a useful means to verify if a class of distributions $p(x|\eta)$ is minimal.

9.5   ML Learning

This section studies the problem of ML learning for probabilistic models in the exponential family. As a direct consequence of the gradient formula (9.40), we will see that ML learning for exponential-family distributions can be formulated as imposing a **mean matching** condition between mean parameters and corresponding empirical averages of the sufficient statistics. For simple models, such as Bernoulli and categorical pmfs, as well as for Gaussian pdfs, the mean matching condition can be solved exactly as we have seen in Secs. 6.5 and 6.8. For more complex models, such as RBM, iterative approximate solutions are necessary, as discussed in Sec. 7.7.

9.5.1   Gradient of the Training Log-Loss

By the definition of the log-loss (9.2), given training data $\mathcal{D} = \{(x_n)_{n=1}^N\}$, the training log-loss is given as

$$L_{\mathcal{D}}(\eta) = -\frac{1}{N}\sum_{n=1}^{N}\log p(x_n|\eta)$$

$$= -\eta^T\underbrace{\left(\frac{1}{N}\sum_{n=1}^{N}s(x_n)\right)}_{s(\mathcal{D})} - \frac{1}{N}\sum_{n=1}^{N}M(x_n) + A(\eta). \qquad (9.46)$$

Therefore, the training log-loss, viewed as a function of the model parameter vector η, depends on the training set $\mathcal{D}$ only through the **empirical average of the sufficient statistics**

$$s(\mathcal{D}) = \frac{1}{N}\sum_{n=1}^{N}s(x_n). \qquad (9.47)$$

Note, in fact, that the term $N^{-1}\sum_{n=1}^{N}M(x_n)$ does not depend on the parameter vector η.

Furthermore, using the formula (9.40) for the gradient of the log-loss, the gradient of the training log-loss can be directly computed as

$$\nabla_{\eta}L_{\mathcal{D}}(\eta) = \frac{1}{N}\sum_{n=1}^{N}\nabla_{\eta}(-\log p(x_n|\eta))$$

$$= \frac{1}{N}\sum_{n=1}^{N}\underbrace{(\mu - s(x_n))}_{\text{mean error for } s(x_n)}$$

$$= \underbrace{\mu - s(\mathcal{D})}_{\text{mean error for } s(\mathcal{D})}, \qquad (9.48)$$

where it is recalled that $\mu = \nabla_{\eta}A(\eta)$ is the mean parameter associated with the natural parameter vector η. The gradient of the log-loss hence equals the **mean error** obtained as the difference between the ensemble average under the model, i.e., μ, and the empirical average $s(\mathcal{D}) = N^{-1}\sum_{n=1}^{N}s(x_n)$ of the observations for the sufficient statistics in the data set $\mathcal{D}$. As the training log-loss, its gradient also depends on the data only through the empirical average of the sufficient statistics, $s(\mathcal{D})$.

9.5.2 Mean Matching Condition

Since the log-loss $-\log p(x|\eta)$ is convex in η, the stationarity condition $\nabla_{\eta}L_{\mathcal{D}}(\eta) = 0$ is necessary and sufficient for global optimality (see Chapter 5). This is, strictly speaking, only true in case η is unconstrained, i.e., when the domain is $\mathbb{R}^K$, and we will make this assumption throughout this subsection. By (9.48), the stationarity condition is equivalent to imposing that the mean error is zero. This, in turn, means that a natural parameter η^{ML} is the ML solution if its corresponding mean parameter matches the empirical average of the sufficient statistics, i.e., if we have the **mean matching condition**

$$\mu^{ML} = s(\mathcal{D}). \qquad (9.49)$$

The condition (9.49) states that, at an ML solution η^{ML}, the mean of the sufficient statistics vector under the model distribution $p(x|\eta^{ML})$ matches the corresponding mean under the empirical distribution $p_{\mathcal{D}}(x)$, that is,

$$\underbrace{E_{x \sim p(x | \eta^{ML})}[s(x)]}_{\mu^{ML}} = \underbrace{E_{x \sim p_{\mathcal{D}}(x)}[s(x)]}_{s(\mathcal{D})}. \tag{9.50}$$

Having obtained the ML estimate of the mean parameters from (9.50), for minimal distributions the ML estimate η^{ML} is unique, while for non-minimal families there will be multiple equivalent solutions η^{ML} for the ML problem (see Sec. 9.4).

Example 9.8

Let us revisit the problem, first studied in Sec. 6.5, of ML training for a Bernoulli distribution $\mathrm{Bern}(\mu)$. From Table 9.1, the mean parameter of the $\mathrm{Bern}(\mu)$ pmf is μ, and the natural parameter is the logit. Using the mean matching condition (9.49), the ML estimate of the mean parameter is

$$\mu^{ML} = \frac{1}{N} \sum_{n=1}^{N} s(x_n) = \frac{1}{N} \sum_{n=1}^{N} x_n = \frac{N[1]}{N}, \tag{9.51}$$

which equals the fraction of observations equal to 1 ($N[1]$ is the number of observations equal to 1). Since this is a minimal distribution, the ML estimate is unique, and it is given by the logit $\eta^{ML} = \log(\mu^{ML}/(1 - \mu^{ML}))$.

Example 9.9

As another example, let us reconsider the categorical distribution $\mathrm{Cat}(\mu)$, for which ML learning was investigated in Sec. 6.8. From Table 9.1, the mean parameter vector is the vector of probabilities, μ, and the natural parameter vector is given by the logits. Using mean matching, the ML estimate is given as

$$\mu^{ML} = \frac{1}{N} \begin{bmatrix} N[0] \\ \vdots \\ N[C-1] \end{bmatrix}, \tag{9.52}$$

which corresponds to the empirical distribution $p_{\mathcal{D}}(x)$ of the data set $\mathcal{D}$ ($N[k]$ is the number of observations equal to k). In fact, we have $\mu_k^{ML} = p_{\mathcal{D}}(k)$ for $k = 0, \ldots, C-1$, where $\mu^{ML} = [\mu_0^{ML}, \ldots, \mu_{C-1}^{ML}]^T$.

Since the parametrization is not minimal, the ML solution for the natural parameter vector is not unique. In fact, there are infinitely many ML solutions for the natural parameters, namely any logit vector

$$\eta^{ML} = \begin{bmatrix} \log(\mu_0^{ML}) \\ \vdots \\ \log(\mu_{C-1}^{ML}) \end{bmatrix} + b \cdot 1_C \tag{9.53}$$

for some constant b.

9.5.3 Gradient Descent-Based ML Learning

Apart from simple examples, the mean matching condition (9.50) that characterizes ML learning for exponential-family distributions is generally not tractable. This is because the mean parameter vector μ is often defined implicitly in terms of the parameters of the distribution, and hence the equality (9.50) cannot be directly solved to yield the ML solution. In such cases, a typical approach is to apply GD to minimize the training loss. GD iteratively updates the natural parameters as

$$\eta^{(i+1)} = \eta^{(i)} - \gamma \nabla_\eta L_\mathcal{D}(\eta^{(i)}) = \eta^{(i)} - \gamma(\mu^{(i)} - s(\mathcal{D})) \tag{9.54}$$

from one iteration i to the next $i + 1$, where $\gamma > 0$ is a learning rate, and we have used the formula (9.48) for the gradient of the training log-loss. In this context, we can think of GD as a way to approximately satisfy the mean matching condition (9.50).

Implementing the GD Rule (9.54). Implementing the GD update (9.54) comes with two computational challenges. First, by (9.47), evaluating the empirical mean $s(\mathcal{D})$ requires accessing the entire data set. This problem can be easily addressed by approximating the expectation via an empirical average $s(\mathcal{S}^{(i)})$ over a mini-batch $\mathcal{S}^{(i)} \subset \mathcal{D}$ of samples from the training set. This amounts to the standard SGD approach studied in Chapter 5.

The second issue is that the mean parameter $\mu^{(i)}$ corresponding to the current iterate $\eta^{(i)}$ may be difficult to evaluate exactly (see Example 9.10). For such cases, assume that we can draw a sample $x^{(i)} \sim p(x|\eta^{(i)})$ from the current model $p(x|\eta^{(i)})$. Then, by its definition (9.30), we can estimate the mean parameter as

$$\hat{\mu}^{(i)} = s(x^{(i)}). \tag{9.55}$$

Note that it is straightforward to generalize the estimate by averaging over multiple samples from the model $p(x|\eta^{(i)})$.

With these two approximations, the update (9.54) can be approximated as

$$\eta^{(i+1)} = \eta^{(i)} - \gamma(\hat{\mu}^{(i)} - s(\mathcal{S}^{(i)})). \tag{9.56}$$

This update is **doubly stochastic** since it is based on a mini-batch $\mathcal{S}^{(i)}$ of samples drawn from the training set, as well as on the sample $x^{(i)} \sim p(x|\eta^{(i)})$ from the model used to produce the estimate $\hat{\mu}^{(i)}$. An example of this approach was described in Sec. 7.7 for the problem of training RBMs. We revisit this example next.

Example 9.10

The log-loss of the **Boltzmann distribution** is defined as

$$-\log p(x|\eta) = -x^T W x - a^T x + A(\eta)$$

$$= -\sum_{i=1}^{D} \sum_{j=1}^{D} w_{ij} x_i x_j - \sum_{i=1}^{D} a_i x_i + A(\eta), \tag{9.57}$$

where $\eta = (a, W)$ are the natural parameters, with $[W]_{i,j} = w_{ij}$ and $[a]_i = a_i$, and the sufficient statistics $s(x)$ include vector x, as well as the matrix of products $x_i x_j$ for $i,j \in \{1, \ldots, D\}$. A special case of this distribution is the RBM model studied in Sec. 7.7. For the Boltzmann distribution, the mean parameters are given by the expectations $\mathrm{E}_{\mathrm{x}\sim p(x|\eta)}[\mathrm{x}_i]$ and $\mathrm{E}_{\mathrm{x}\sim p(x|\eta)}[\mathrm{x}_i\mathrm{x}_j]$, which are difficult to compute exactly. As in the contrastive divergence method for RBMs, the gradient update (9.54) can be approximated using the doubly stochastic estimate in (9.56), whereby samples from the model are (approximately) drawn via Monte Carlo techniques.

9.6   Information-Theoretic Metrics

Distributions in the exponential family have yet another useful property – namely, that information-theoretic metrics can be efficiently evaluated as a function of natural and/or mean parameters. This is especially important in the many learning methods, to be covered in this part of the book, that rely on information-theoretic metrics such as entropy and KL divergence.

9.6.1   Entropy

Let us first consider the entropy, with the convention discussed in Chapter 3 of also using the term "entropy" for the differential entropy of continuous rvs. Using the log-loss (9.2), the **entropy** of an exponential-family distribution can be computed as

$$\begin{aligned}
\mathrm{H}(p(x|\eta)) &= \mathrm{E}_{\mathrm{x}\sim p(x|\eta)}[-\log p(\mathrm{x}|\eta)] \\
&= -\eta^T\mu - \mathrm{E}_{\mathrm{x}\sim p(x|\eta)}[M(\mathrm{x})] + A(\eta),
\end{aligned} \tag{9.58}$$

where, as usual, μ is the mean parameter vector corresponding to the natural parameter vector η. This relation shows that entropy and log-partition function are strongly related to one another. Technically, as detailed in Appendix 9.D, one can say that negative entropy and log-partition function are **dual** to one another. The general expression (9.58) can be evaluated explicitly as a function of μ or η for many common distributions in the exponential family. Examples are given in Table 9.2.

9.6.2   KL Divergence

In a manner similar to the entropy, using the log-loss (9.2), the general form of the **KL divergence** between distributions in the same class within the exponential family is

$$\begin{aligned}
\mathrm{KL}(\mathrm{ExpFam}(x|\eta_1)||\mathrm{ExpFam}(x|\eta_2)) &= \mathrm{E}_{\mathrm{x}\sim p(x|\eta_1)}\left[\log\left(\frac{p(\mathrm{x}|\eta_1)}{p(\mathrm{x}|\eta_2)}\right)\right] \\
&= A(\eta_2) - A(\eta_1) - (\eta_2 - \eta_1)^T\mu_1,
\end{aligned} \tag{9.59}$$

Table 9.2 Entropy for some distributions in the exponential family

distribution	$\mathrm{H}(\mathrm{ExpFam}(x|\eta))$
$\mathrm{Bern}(\mu)$	$-\mu\log\mu - (1-\mu)\log(1-\mu)$
$\mathrm{Cat}(\mu)$	$-\sum_{k=0}^{C-1}\mu_k\log(\mu_k)$
$\mathcal{N}(\nu,\Theta^{-1})$	$\frac{1}{2}\log\det((2\pi e)\Theta^{-1})$

Table 9.3 KL divergence for some distributions in the exponential family

$p(x)$	$q(x)$	$\text{KL}(p \| q)$
$\text{Bern}(\mu)$	$\text{Bern}(\bar{\mu})$	$\mu \log\left(\frac{\mu}{\bar{\mu}}\right) + (1-\mu)\log\left(\frac{1-\mu}{1-\bar{\mu}}\right)$
$\text{Cat}(\mu)$	$\text{Cat}(\bar{\mu})$	$\sum_{k=0}^{C-1} \mu_k \log\left(\frac{\mu_k}{\bar{\mu}_k}\right)$
$\mathcal{N}(\nu, \Sigma)$	$\mathcal{N}(\bar{\nu}, \bar{\Sigma})$	$\frac{1}{2}\left[\text{tr}\left(\bar{\Sigma}^{-1}\Sigma\right) + \log\left(\frac{\det(\bar{\Sigma})}{\det(\Sigma)}\right) + (\bar{\nu} - \nu)^T \bar{\Sigma}^{-1}(\bar{\nu} - \nu) - D \right]$

where μ_1 is the mean parameter vector corresponding to η_1. This formula shows that the KL divergence is also strongly related to the partition function. As we detail in Appendix 9.D, the right-hand side of (9.59) can be interpreted as a measure of distance between natural parameter vectors η_1 and η_2 – namely the **Bregman divergence** – that is defined by the partition function. The formula (9.59) can be computed explicitly for many common distributions in the exponential family, and some examples can be found in Table 9.3.

9.6.3 Cross Entropy

Using the derived expressions of KL divergence and entropy, one can also directly compute the **cross entropy** between distributions in the same class within the exponential family as

$$\text{H}(\text{ExpFam}(x|\eta_1) \| \text{ExpFam}(x|\eta_2)) = \text{KL}(\text{ExpFam}(x|\eta_1) \| \text{ExpFam}(x|\eta_2)) + \text{H}(p(x|\eta_1)).$$
(9.60)

9.7 Fisher Information Matrix

The **Fisher information matrix (FIM)** is a quantity of central importance in **estimation theory** and statistics with deep connections to information-theoretic metrics. As we will see in this section, for general probabilistic models including arbitrary parametric distributions $p(x|\theta)$ – not necessarily in the exponential family – the FIM evaluated at vector θ allows one to approximate the KL divergence $\text{KL}(p(x|\theta) \| p(x|\theta'))$ with a quadratic function of the difference $\theta - \theta'$ when the latter is sufficiently small (in terms of the norm of $\theta - \theta'$). In this way, the FIM translates a distance between distributions, the KL divergence, into a distance between parameter vectors θ and θ'.

The FIM plays an important role in machine learning for the definition of local optimization strategies that go beyond GD by accounting for the geometry of the model distribution space, as well as for approximate Bayesian inference (see Sec. 10.4). The former application of the FIM yields the **natural GD** method, which is also described in this section.

Well-Specified Model Class. We will introduce the FIM by studying the case in which the model class is **well specified**. A well-specified model class is such that the population distribution can be expressed as $p(x) = p(x|\theta)$ for some ground-truth vector θ. Well-specified model classes are commonly assumed in statistics, and are considered here to justify some of the derivations. However, the key results, such as the quadratic approximation provided by the KL divergence, do not require the model class to be well specified, as we will emphasize.

In the rest of this section, we first provide a general definition of the FIM for any probabilistic model, not necessarily in the exponential family. Then, we specialize the discussion to exponential-family distributions, and finally we briefly introduce natural GD.

9.7.1   Definition of the Fisher Information Matrix

To introduce the FIM, consider a general probabilistic model $p(x|\theta)$, not necessarily in the exponential family, and assume that the model is well specified. We will also implicitly assume that all the necessary technical conditions are satisfied; see Recommended Resources, Sec. 9.10, for details. These conditions are typically referred to as **regularity** assumptions, and they include the assumption that the support of the distribution does not depend on θ. We will see in this subsection that the FIM can be introduced as a measure of the curvature of the population log-loss, evaluated at the ground-truth model parameter vector θ, for well-specified models.

Under the assumption that the population distribution $p(x)$ equals $p(x|\theta)$ for some ground-truth value θ, i.e., that the model class is well specified, the cross entropy

$$\mathrm{H}(p(x|\theta)||p(x|\theta')) = \mathrm{E}_{\mathrm{x}\sim p(x|\theta)}[-\log p(x|\theta')] \tag{9.61}$$

is the population log-loss $L_p(\theta')$ for model parameter θ'. We also recall that the cross entropy is related to the KL divergence through

$$\mathrm{KL}(p(x|\theta)||p(x|\theta')) = \mathrm{H}(p(x|\theta)||p(x|\theta')) - \mathrm{H}(p(x|\theta)). \tag{9.62}$$

Since the population distribution is included in the model class, the population-optimal unconstrained predictor and the population-optimal within-class predictors coincide, and they are given by $p(x|\theta')$ for the value of model parameter θ' that minimizes the cross entropy (9.61) or, equivalently, the KL divergence (9.61). The minimizer is clearly given by $\theta' = \theta$, since this solution yields a zero KL divergence. To simplify the discussion, we assume in this subsection that $\theta' = \theta$ is the only minimizer of the population log-loss. As we will see next, the FIM measures the curvature of the cross entropy, or KL divergence, at the optimal, ground-truth, model parameter vector $\theta' = \theta$.

We have just seen that the choice $\theta' = \theta$ minimizes the cross entropy (9.61) and the KL divergence (9.61). From Chapter 5, this simple observation has two useful consequences, with the second offering a way to introduce the FIM:

- The **first-order necessary optimality condition** requires that $\theta' = \theta$ be a stationary point of the cross entropy and of the KL divergence when interpreted as functions of θ' for a fixed θ. Mathematically, this stationarity condition amounts to the equality

$$\nabla_{\theta'}\mathrm{H}(p(x|\theta)||p(x|\theta'))|_{\theta'=\theta} = \nabla_{\theta'}\mathrm{KL}(p(x|\theta)||p(x|\theta'))|_{\theta'=\theta}$$

$$= \mathrm{E}_{\mathrm{x}\sim p(x|\theta)}\left[\underbrace{-\nabla_\theta \log p(x|\theta)}_{\text{gradient of the log-loss}}\right] = 0. \tag{9.63}$$

  This demonstrates that the gradient of the log-loss $\nabla_\theta \log p(x|\theta)$ has zero mean when averaged over $p(x|\theta)$.
- The **second-order necessary optimality condition** $\nabla^2_{\theta'}\mathrm{H}(p(x|\theta)||p(x|\theta'))|_{\theta'=\theta} \succeq 0$ requires the curvature of the cross entropy and of the KL divergence to be non-negative at $\theta' = \theta$. Mathematically, we have the condition

$$\nabla^2_{\theta'} H(p(x|\theta)||p(x|\theta'))|_{\theta'=\theta} = \nabla^2_{\theta'} KL(p(x|\theta)||p(x|\theta'))|_{\theta'=\theta}$$

$$= \underbrace{E_{x\sim p(x|\theta)}[-\nabla^2_\theta \log p(x|\theta)]}_{\text{FIM}(\theta)} \succeq 0. \qquad (9.64)$$

Therefore, the **Fisher information matrix (FIM)**

$$\text{FIM}(\theta) = E_{x\sim p(x|\theta)}[-\nabla^2_\theta \log p(x|\theta)] \qquad (9.65)$$

measures the (non-negative) curvature of the population log-loss at the optimal point $\theta' = \theta$.

We have just introduced the FIM (9.65) as a measure of the curvature of the population log-loss for well-specified models at the ground-truth model parameter vector. While one can take (9.65) merely as a mathematical definition of the FIM, this viewpoint offers a useful interpretation of the significance of the FIM in inference and learning problems. A larger curvature implies that the population log-loss is more concentrated around the ground-truth value θ. This, in turn, can be interpreted as indicating that it is "easier" to estimate the parameter value θ based on data generated from the distribution $p(x|\theta)$, since the population log-loss strongly penalizes deviations from θ. In broader terms, the FIM can be thought of as quantifying the amount of "information" that data generated from the model $p(x|\theta)$ provides about the value of the model parameter θ.

When the FIM is positive definite, i.e., when $\text{FIM}(\theta) \succ 0$, the **inverse of the FIM**, $\text{FIM}(\theta)^{-1}$, can conversely be interpreted as a measure of the difficulty of estimating the parameter value θ based on data generated from the distribution $p(x|\theta)$. This is further discussed in Appendix 9.E, which examines the **asymptotic performance of the ML estimator** of parameter θ for well-specified models. It is shown that the asymptotic **squared error** of the ML estimator is the trace of the inverse of the FIM. In the non-asymptotic regime, the role of the FIM as a lower bound on the achievable squared estimation error is formalized by the **Cramer–Rao bound**, for which the reader is referred to Recommended Resources, Sec. 9.10.

Example 9.11

Given a Bern(μ) rv, the FIM for the mean parameter μ can be computed as

$$\text{FIM}(\mu) = \frac{1}{\mu(1-\mu)} \qquad (9.66)$$

for $\mu \in (0,1)$. The FIM is hence maximal for $\mu \to 0$ and $\mu \to 1$, tending to infinity under both limits. The population log-loss is hence maximally concentrated around the ground-truth values $\mu = 0$ and $\mu = 1$, respectively. As a result, based on data generated from the pmf Bern(μ), values of μ closer to 0 or 1 are "easier" to estimate in terms of squared loss. In contrast, the FIM takes the smallest value for $\mu = 0.5$, at which point the population log-loss has the smallest curvature. This is the most "difficult" value to estimate.

Example 9.12

Given a $\mathcal{N}(\mu, \beta^{-1})$ rv with a fixed precision β, the FIM for the mean parameter μ is $\text{FIM}(\mu) = \beta$: When data are generated as $x \sim \mathcal{N}(\mu, \beta^{-1})$, all values of the mean μ are equally "difficult" to estimate (in terms of squared loss), as the curvature of the log-loss at the ground-truth mean μ is uniform across all values μ. Furthermore, the amount of "information" we have about μ, based on data generated from the pdf $\mathcal{N}(\mu, \beta^{-1})$, increases with the precision β.

9.7.2 FIM and Quadratic Approximation of the KL Divergence

By using the optimality conditions (9.63) and (9.64), as well as the fact that the KL divergence is zero when $\theta = \theta'$, we can obtain the useful **second-order Taylor approximation of the KL divergence**,

$$\text{KL}(p(x|\theta)||p(x|\theta')) \simeq \frac{1}{2}(\theta - \theta')^T \text{FIM}(\theta)(\theta - \theta'), \tag{9.67}$$

around any point θ. Accordingly, the matrix $\text{FIM}(\theta)$ describes the curvature of the KL divergence around θ.

The result (9.67) is in line with the discussion in the preceding subsection regarding the problem of estimating the ground-truth parameter θ. By (9.67), when the FIM is large, the KL divergence increases quickly as we move the model parameter θ' away from θ in the parameter space, making it "easier" to distinguish distribution $p(x|\theta)$ from $p(x|\theta')$.

It is emphasized that, while the derivation in this section has assumed well-specified models, the approximation (9.67) holds in full generality (under suitable assumptions). For machine learning applications, the relationship (9.67) provides a useful and intuitive way to think about the FIM as providing a local quadratic approximation for the KL divergence that connects a distance between distributions with a distance in the model parameter space.

9.7.3 FIM As the Covariance of the Gradient of the Log-Loss

Another important general property of the FIM (9.65) is that it can also be written as the **covariance of the gradient of the log-loss**, i.e., we have

$$\begin{aligned}
\text{FIM}(\theta) &= \text{E}_{\text{x} \sim p(x|\theta)}[-\nabla_\theta^2 \log p(\text{x}|\theta)] \\
&= \text{E}_{\text{x} \sim p(x|\theta)}[(\nabla_\theta \log p(\text{x}|\theta))(\nabla_\theta \log p(\text{x}|\theta))^T].
\end{aligned} \tag{9.68}$$

Note that the expectation in (9.68) is the covariance matrix since the mean of the gradient of the log-loss is zero by (9.63). Therefore, we can take the covariance of the gradient of the log-loss in (9.68) as an equivalent definition of the FIM. Intuitively, a "larger" gradient on average over the distribution $p(x|\theta)$ provides a more "informative" learning signal that enables a more accurate estimate of the parameter θ based on samples $\text{x} \sim p(x|\theta)$. A proof of this equality and further discussion on the FIM for general probabilistic models can be found in Appendix 9.E.

9.7.4 Natural Gradient Descent

As we have seen in Sec. 5.10, second-order methods can be more effective than first-order techniques such as GD and SGD in that they account for the curvature of the cost function in the model parameter space. When applied to the training log-loss $L_\mathcal{D}(\theta) = N^{-1} \sum_{n=1}^N (-\log p(x_n|\theta))$, second-order methods ideally rely on the Hessian

$$\nabla_\theta^2 L_\mathcal{D}(\theta) = \frac{1}{N} \sum_{n=1}^N (-\nabla_\theta^2 \log p(x_n|\theta)). \tag{9.69}$$

The Hessian may be hard to compute, especially for a large data set $\mathcal{D}$. We will now see that, by using the FIM, one can obtain an approximation of the Hessian that does not depend on $\mathcal{D}$. The resulting iterative optimization scheme is known as **natural GD**.

To derive natural GD, let us make the working assumption that the data points x_n are i.i.d. rvs drawn from the model distribution $p(x|\theta)$. Note that, in practice, this is clearly not true – the assumption is only made for the sake of justifying a specific approximation of the Hessian. Under this assumption, when the number of data points N increases, by the law of large numbers, the Hessian $\nabla_\theta^2 L_\mathcal{D}(\theta)$ in (9.69) tends with high probability to the expectation of the rv $-\nabla_\theta^2 \log p(\mathrm{x}|\theta)$, with $\mathrm{x} \sim p(x|\theta)$. This is exactly the FIM (9.65).

Based on this asymptotic analysis, one may be justified in replacing the Hessian (9.69) with its asymptotic counterpart given by $\mathrm{FIM}(\theta)$. Importantly, the FIM does not depend on the training set $\mathcal{D}$, but only on the probabilistic model $p(x|\theta)$. Therefore, the FIM may be easier to obtain than the exact Hessian, particularly for exponential-family distributions for which the FIM has the special properties to be discussed in the next subsection. The resulting second-order method is known as natural GD.

By (9.67), natural GD can be understood as adapting to the geometry implied by the KL divergence in the space of distributions. This is unlike standard GD, which implicitly assumes the standard Euclidean geometry. Details and more discussion on natural GD can be found in Appendix 9.F, which also introduces the related method of **mirror descent**.

9.7.5 FIM for the Exponential Family

Having introduced the FIM in the context of general probabilistic models, in this subsection we specialize the use of the FIM to exponential-family distributions.

FIM for the Natural Parameters. The FIM for exponential-family distributions can be evaluated with respect to the natural or mean parameters. Using the general definition (9.65) along with the expression (9.2) for the log-loss, the **FIM for the natural parameter vector** η can be directly computed as

$$\mathrm{FIM}(\eta) = \mathrm{E}_{\mathrm{x} \sim p(x|\eta)}[-\nabla_\eta^2 \log p(\mathrm{x}|\eta)] = \nabla_\eta^2 A(\eta), \tag{9.70}$$

that is, the FIM is the **Hessian of the log-partition function**. So, the more curved the log-partition function is as a function of the natural parameters η, the "easier" it is to estimate η in terms of squared error based on samples $\mathrm{x} \sim p(x|\eta)$. Looking at Fig. 9.2, this becomes quite intuitive: If the log-partition function is locally linear, i.e., if it has a zero curvature (Fig. 9.2 (bottom)), the natural parameters are not even *identifiable* in the sense that there are multiple natural parameter vectors yielding the same distribution.

Furthermore, evaluating the Hessian of the log-partition function using the definition (9.6) or (9.7), we obtain that the FIM can also be written as the **covariance of the sufficient statistic vector** $s(\mathrm{x})$, i.e.,

$$\mathrm{FIM}(\eta) = \mathrm{E}_{\mathrm{x} \sim p(x|\eta)}\left[(s(\mathrm{x}) - \mu)(s(\mathrm{x}) - \mu)^T\right], \tag{9.71}$$

where $\mu = \nabla_\eta A(\eta)$ is the parameter vector corresponding to natural parameters η. Proofs are in Appendix 9.E.

Some implications of these important results are as follows. First, by (9.70), a distribution in the exponential family has a **minimal** parametrization if the FIM is positive definite, $\mathrm{FIM}(\eta) \succ 0$, for all η in the domain (and hence the log-partition function is strictly convex; see Sec. 9.4). Second, applying the equality (9.70) to the second-order approximation (9.67) yields

$$\mathrm{KL}(p(x|\eta)||p(x|\eta')) \simeq \frac{1}{2}(\eta - \eta')^T \nabla_\eta^2 A(\eta)(\eta - \eta'). \tag{9.72}$$

This shows that the curvature of the log-partition function coincides with that of the KL divergence $\mathrm{KL}(p(x|\eta)||p(x|\eta'))$ when evaluated at $\eta' = \eta$ in the space of natural parameters.

Table 9.4 Fisher information matrix (FIM) for some distributions in the exponential family

distribution	$\text{FIM}(\mu)$
$\text{Bern}(\mu)$	$\frac{1}{\mu(1-\mu)}$
$\mathcal{N}(\mu, \Theta^{-1})$, fixed Θ	Θ^{-1}
$\mathcal{N}(\mu, \Theta^{-1})$	$\begin{bmatrix} \Theta^{-1} & 0 \\ 0 & \frac{1}{2}\Theta^{-2} \end{bmatrix}$

FIM for the Mean Parameters. The characterizations (9.70) and (9.71) apply to the FIM for the natural parameters. For **minimal** exponential-family distributions, it turns out that there is a simple link between FIM for natural and mean parameters, namely

$$\text{FIM}(\mu) = \text{FIM}(\eta)^{-1}, \tag{9.73}$$

where η is the unique natural parameter vector associated with μ. In words, the FIM for the mean parameters is the inverse of that for natural parameters. This relationship further strengthens the perspective that the two sets of parameters are **dual** to one another.

The general formula (9.70) or (9.73) can be computed explicitly for many common distributions in the exponential family, as in the examples in Table 9.4.

Example 9.13

The $\text{Bern}(\mu)$ pmf is minimal, and thus from (9.66) in Example 9.11 and (9.73), we get

$$\text{FIM}(\eta) = \mu(1 - \mu) = \sigma(\eta)\sigma(-\eta). \tag{9.74}$$

The most "difficult" values of the logit η to estimate, in terms of squared error, are hence obtained in the limits $\eta \to -\infty$ or $\eta \to \infty$, in a manner dual to the mean parameter μ.

Example 9.14

The $\mathcal{N}(\mu, \beta^{-1})$ pdf is also minimal, and thus, from Example 9.12 and (9.73), the FIM with respect to the natural parameter $\eta = \beta\mu$ is $\text{FIM}(\eta) = \beta^{-1}$. Increasing the precision β hence causes the achievable squared error for the estimate $\eta = \beta\mu$ to increase, while reducing the squared error for the estimation of the mean parameter μ.

9.8   Generalized Linear Models

As seen in this chapter, the exponential family provides a flexible class of distributions to model pmfs and pdfs. In many problems in machine learning, however, one also needs to model *conditional* distributions. A useful extension of the exponential family to conditional distributions is given by **generalized linear models (GLMs)**, which are the subject of this section. As we will discuss, several GLMs were encountered in previous chapters, namely linear predictors) with a Gaussian likelihood as studied in Chapter 4, as well as logistic and softmax regression models in Chapter 6.

9.8.1 Definitions

Following the notation adopted in this chapter, let us define as $\text{ExpFam}(t|\mu)$ any distribution in the exponential family with mean parameter vector μ for a rv t. A GLM defines a model class of conditional distributions for a target rv t given the observation x $= x$ of an input rv x that have the form

$$p(t|x, W) = \text{ExpFam}(t|\mu = g(Wu(x))), \tag{9.75}$$

where $u(x)$ is a vector of **features** of the input x, W is a matrix defining the **model parameters**, and $g(\cdot)$ is an *invertible* function, known as the **response function**. The inverse $g^{-1}(\cdot)$ of the response function is referred to as the **link function**.

Intuitively, GLMs generalize deterministic linear models of the form $t = g(Wu(x))$ – which compose a linear transformation with a non-linearity – by "adding noise" around the mean $\mu = g(Wu(x))$, with the noise being drawn from a distribution in the exponential family.

Denoting as $\text{ExpFam}(t|\eta)$ any distribution in the exponential family with natural parameter vector η, a GLM in the **canonical form** is written as

$$p(t|x, W) = \text{ExpFam}(t|\eta = Wu(x)), \tag{9.76}$$

so that the natural parameter vector is the linear function $\eta = Wu(x)$ of the model parameters W. Canonical models are obtained from the more general form (9.75) of GLMs by setting the response function as $g(\eta) = \nabla_\eta A(\eta)$. In fact, in this case, using the mapping (9.44), the natural parameter vector corresponding to the mean parameter $\mu = g(Wu(x)) = \nabla_\eta A(Wu(x))$ in (9.75) is given by $\eta = Wu(x)$.

In the rest of this section, we will first show that logistic and softmax regression are special cases of GLMs (see also Problem 9.8); then, to illustrate the flexibility of GLMs, we will briefly introduce probit regression and GLMs for generative modeling in unsupervised learning.

9.8.2 Logistic Regression As a Generalized Linear Model

As discussed in Chapter 6, logistic regression assumes that the label t is conditionally distributed as

$$(t|x = x, \theta) \sim \text{Bern}(\sigma(\theta^T u(x))), \tag{9.77}$$

so that we have the predictive distribution $p(t = 1|x, \theta) = \sigma(\theta^T u(x))$ for model parameter θ. Therefore, using Table 9.1, logistic regression is a GLM in canonical form (9.76) with exponential-family distribution given by the Bernoulli distribution and natural parameter given by the logit $\eta = \theta^T u(x)$. Note that the matrix W in (9.75) reduces to vector θ for this GLM, since the mean and natural parameters of the underlying exponential-family distribution are scalar quantities.

The gradient (6.33) with respect to the model parameter θ derived in Chapter 6 can be directly obtained from the general formula (9.40) and the chain rule of differentiation as

$$\nabla_\theta(-\log \text{Bern}(\sigma(\theta^T u(x)))) = \frac{d}{d\eta}(-\log \text{Bern}(\eta)|_{\eta=\theta^T u(x)}) \cdot \nabla_\theta \eta$$

$$= \underbrace{(\sigma(\theta^T u(x)) - t)}_{\delta(x,t|\theta), \text{ mean error}} \cdot \underbrace{u(x)}_{\text{feature vector}} . \qquad (9.78)$$

9.8.3   Softmax Regression As a Generalized Linear Model

As also seen in Chapter 6, softmax regression assumes that the label t is distributed as

$$(t|x = x, W) \sim \text{Cat}(t|\eta = Wu(x)) \qquad (9.79)$$

for model parameter matrix W. Therefore, the conditional distribution $p(t|x, W)$ is a GLM in canonical form (9.76), with exponential-family distribution given by the categorical distribution and natural parameters given by the logit vector $\eta = Wu(x)$. As in the case of logistic regression, it can be readily checked that the gradient (6.107) with respect to W derived in Chapter 6 can be obtained from the expression (9.40) via the chain rule.

9.8.4   Probit Regression

As a variant of logistic regression, probit regression replaces the sigmoid response function, which yields a GLM in canonical form, with the cumulative distribution function $\Phi(\cdot)$ of a standard Gaussian distribution $\mathcal{N}(0, 1)$, i.e., $\Phi(x) = \text{Pr}_{x \sim \mathcal{N}(0,1)}[x \leq x]$. The resulting GLM $(t|x = x) \sim p(t = 1|x, \theta) = \Phi(\theta^T u(x))$ sets $\mu = \Phi(\theta^T u(x))$ in the definition (9.75), and is not in canonical form.

The prediction t produced by probit regression can be sampled by first adding Gaussian noise $\mathcal{N}(0, 1)$ to the decision variable $\theta^T u(x)$ and then applying a hard threshold function step$(\cdot)$, i.e., we have t = step$(\theta^T u(x) + z)$ with $z \sim \mathcal{N}(0, 1)$.

9.8.5   Generalized Linear Models for Unsupervised Learning

GLMs can also be used to define the decoder of directed generative models for unsupervised learning (see Sec. 7.8), yielding **exponential-family PCA models**. In these models, the input x in (9.76) plays the role of the latent rv and is given a prior distribution. An example of exponential-family PCA is given by the PPCA model described in Sec. 7.10.

9.9   Summary

- The exponential family contains parametric pmfs and pdfs that have convex log-loss functions, as well as gradients of the log-loss and information-theoretic measures with general analytical expressions that depend on the model parameters.
- A distribution in the exponential family can be expressed in terms of natural parameters η or mean parameters μ; accordingly, we can write $p(x|\eta) = \text{ExpFam}(x|\eta)$ or $p(x|\mu) = \text{ExpFam}(x|\mu)$, where $\text{ExpFam}(\cdot|\cdot)$ denotes any distribution in the exponential family.

- For every natural parameter vector η, there is a unique mean vector $\mu = \mathrm{E}_{x \sim \mathrm{ExpFam}(x|\eta)}[s(x)]$. For a mean vector μ, there may be more than one natural parameters η unless the distribution class is minimal (and hence the natural parameter vector η is identifiable).
- For all distributions in the exponential family, the training log-loss and hence also the ML estimates of the model parameters depend only on the empirical average of the sufficient statistics $s(\mathcal{D}) = \frac{1}{N} \sum_{n=1}^{N} s(x_n)$.
- The gradient of the training log-loss is given by the mean error $\nabla_\eta L_{\mathcal{D}}(\eta) = \mu - s(\mathcal{D})$. Therefore, ML is obtained via mean matching, or, when mean matching is not feasible, via gradient descent.
- Information-theoretic measures and FIM can be computed as a function of natural and mean parameters.
- Extensions of distributions in the exponential family to conditional distributions yield GLMs.

9.10 Recommended Resources

A standard reference on the exponential family is the monograph [1], while GLMs are presented in the book [2]. Information-theoretic aspects of the exponential family are covered in [3, 4]. Introductory books on estimation theory include [5, 6]. A comprehensive reference on natural GD is [7], and other useful readings on the subject include [8, 9]. Some discussion on the FIM for mismatched models can be found in [10].

Problems

9.1 Consider the exponential pdf defined as

$$p(x) = \lambda \exp(-\lambda x) \tag{9.80}$$

for $x \geq 0$ and $p(x) = 0$ otherwise. The model parameter λ is strictly positive. (a) Show that this distribution belongs to the exponential family by identifying sufficient statistic, natural parameters, log-base measure function, and log-partition function. To resolve the sign ambiguity, choose the natural parameter to be negative. [Hint: We can write $p(x) = \lambda \exp(-\lambda x) \mathbb{1}(x > 0)$.] (b) Check that the log-partition function is convex and obtain the corresponding domain. (c) Is the parametrization minimal? (d) Express the distribution in the compact form that uses the augmented sufficient statistics.

9.2 Formulate the maximum-entropy problem that recovers the exponential distribution (9.80) as its solution.

9.3 For the exponential pdf (9.80), (a) obtain the mean parameter by differentiating the log-partition function. (b*) Plot log-loss and its derivative as a function of the natural parameter. (c*) Plot the log-partition function vs. the natural parameter, and indicate in the figure the mean parameter, corresponding to some selected values of the natural parameter.

9.4 For the exponential pdf (9.80), (a*) generate a random data set of $N = 100$ observations, and (b*) evaluate the resulting ML estimate of the natural parameter.

9.5 For the exponential pdf (9.80), (a) compute the entropy as a function of parameter λ by using the general formula in terms of natural and mean parameters. (b) Compute the

KL divergence between two exponential pdfs with different parameters λ and λ' by using the general formula in terms of natural and mean parameters. (c) Compute the FIM for the exponential pdf with respect to the natural parameters in four different ways: as the average of the Hessian of the log-loss; as the covariance of the gradient of the log-loss; as the Hessian of the log-partition function; and as the covariance of the sufficient statistics.

9.6 Consider the "continuous Bernoulli" pdf

$$p(x) \propto \lambda^x (1-\lambda)^{1-x}$$

for $x \in [0,1]$ and $p(x) = 0$ otherwise. The model parameter λ is constrained in the open interval $\lambda \in (0,1)$. Identify sufficient statistic, natural parameters, and log-base measure function.

9.7 Propose an alternative, minimal parametrization of the categorical distribution by representing the log-loss in terms of $C-1$ natural parameters.

9.8 Consider the regression problem studied in Chapter 4, which is characterized by the conditional distribution $p(t|x,\theta) = \mathcal{N}(t|\theta^T u(x), 1)$ for a given feature vector $u(x)$. (a) Is this a GLM? (b) If so, identify the response function. (c) Is it in canonical form? (d) Compute the gradient $\nabla_\theta(-\log p(t|x,\theta))$ of the log-loss $-\log p(t|x,\theta)$.

9.9 Consider a regression problem in which the goal is to train a model relating a non-negative quantity t, representing, e.g., the time of the next update on a social network, and a vector of potentially correlated quantities $x \in \mathbb{R}^D$. To this end, we focus on a probabilistic discriminative model given by

$$p(t|x,\theta) = \lambda(x,\theta)\exp(-\lambda(x,\theta)t)$$

for $t \geq 0$ and $p(t|x,\theta) = 0$ otherwise, where $\lambda(x,\theta) = \sigma(\theta^T x)$ with $\theta \in \mathbb{R}^D$. (a) Is this a GLM? (b) If so, identify the response function. (c) Is it in canonical form? (d) Compute the gradient $\nabla_\theta(-\log p(t|x,\theta))$ of the log-loss $-\log p(t|x,\theta)$.

9.10 Prove that the rv $(t|x = x,\theta) \sim p(t = 1|x,\theta) = \Phi(\theta^T u(x))$ produced by probit regression, where $\Phi(\cdot)$ is the cumulative distribution function of a standard Gaussian pdf $\mathcal{N}(0,1)$, can be sampled by first adding Gaussian noise $\mathcal{N}(0,1)$ to the decision variable $\theta^T u(x)$ and then applying a hard threshold function step$(\cdot)$.

9.11 (a) Show that the distribution

$$p(t|x,\theta) = \frac{\exp(t\theta^T x)\exp(-\exp(\theta^T x))}{t!}$$

with $t \in \{0,1,2,\dots\}$ defines a GLM in canonical form by identifying the underlying exponential-family distribution. (b) Obtain the response function.

Appendices

Appendix 9.A: Duality and Exponential Family

This appendix develops useful properties of the exponential family in terms of **convex duality**, which will be leveraged in some of the following appendices.

Convex (Fenchel) Dual of a Convex Function. We will first need the definition of the convex dual, or Fenchel dual, of a convex function $f(x)$ with vector input $x \in \mathbb{R}^K$ and scalar output in $\mathbb{R}$. This definition hinges on the following key property of convex functions: A convex function can be identified by specifying the intercepts of its tangents for *all* possible values of the slope of

Figure 9.3 Illustration of the convex dual function $f^*(y)$ of a convex function $f(x)$ with $K = 1$. The convex dual function represents the negative intercept of the (unique) tangent line to the function that has slope (more generally, gradient) y.

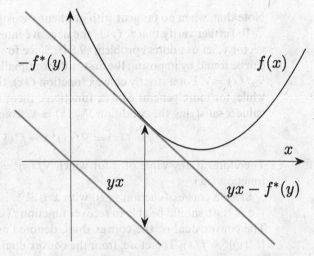

the tangents. So, an equivalent **dual** representation of a convex function can be given in terms of all its intercepts.

To unpack this statement, consider a convex function $f(x)$ with input in $\mathbb{R}$ ($K = 1$), and fix some value $y \in \mathbb{R}$ for the derivative. Note that there may or may not be a point $x' \in \mathbb{R}$ such that the derivative of function $f(x)$ equals y, i.e., such that $df(x')/dx = y$. Suppose that there is at least one such value of x' (there may be more than one). Then, as illustrated in Fig. 9.3, for a convex function $f(x)$, there exists a *unique* line

$$\tilde{f}(x|y) = yx + b, \tag{9.81}$$

with slope equal to the derivative y, that is **tangent** to the function $f(x)$. The single tangent line $\tilde{f}(x|y)$ coincides with the original function $f(x)$ for all values x' that share the derivative y (i.e., such that $df(x')/dx = y$).

Since the tangent function $\tilde{f}(x|y)$ is unique – assuming that it exists for the given slope y – the corresponding **intercept** b is a function of the slope y. Following the notation for convex duality, we will write the negative of this function as $f^*(y)$, i.e., $f^*(y) = -b$. As we will formalize below, the function $f^*(y)$ defines the **convex dual** of function $f(x)$.

How can we mathematically define the intercept of a tangent line with slope y? To this end, consider the line $\tilde{g}(x|y) = yx$, which has the same slope y of the tangent line $\tilde{f}(x|y)$ but passes through the origin. As illustrated in Fig. 9.3, the intercept b describes by how much the line $\tilde{g}(x|y) = yx$ should be "pushed up" to meet the function $f(x)$. Mathematically, we can translate this observation into the problem

$$b = \min_{x \in \mathbb{R}} \{f(x) - yx\} = -f^*(y). \tag{9.82}$$

In the previous paragraphs, we have argued that, for any convex function $f(x)$ from $\mathbb{R}$ to $\mathbb{R}$, given a derivative value $y \in \mathbb{R}$, the tangent line

$$\tilde{f}(x|y) = yx - f^*(y) \tag{9.83}$$

is unique, if it exists, and the intercept is given by the function (9.82) of y. The negative intercept function $f^*(y)$ is known as the convex dual of function $f(x)$. Generalizing (9.82) to any $K \geq 1$, the **convex dual of function** $f(x)$ is mathematically defined as

$$f^*(y) = \max_{x \in \mathbb{R}^K} \left\{ y^T x - f(x) \right\}. \tag{9.84}$$

Note that, when no tangent with gradient y exists, by (9.84) the dual function takes value $+\infty$.

To further verify that $f^*(y)$ is the negative intercept for the tangent line $\tilde{f}(x|y)$ with gradient vector y, let us address problem (9.84). Since function $f(x)$ is convex, global optimal solutions can be found by imposing the first-order optimality condition $\nabla_x(x^T y - f(\hat{x})) = 0$, which yields $\nabla_x f(x) = y$. For a strictly convex function $f(x)$, there is at most one solution for this condition, while, for more general convex functions, there may be multiple such values of x. Given any value x satisfying the condition $\nabla_x f(x) = y$, from (9.84) we have

$$f(x) = \nabla f(x)^T x - f^*(\nabla f(x)) = \tilde{f}(x|\nabla f(x)). \tag{9.85}$$

Therefore, at any value of x for which $\nabla f(x) = y$, the tangent line $\tilde{f}(x|y)$ in (9.83) meets the function $f(x)$.

Since a convex function $f(x)$, with $x \in \mathbb{R}^K$, is fully specified by its convex dual $f^*(y)$, with $y \in \mathbb{R}^K$, we should be able to recover function $f(x)$ from function $f^*(y)$. This is indeed possible: The convex dual of the convex dual, denoted as $\{f^*(y)\}^*$, is the function itself, i.e., we have $\{f^*(y)\}^* = f(x)$. Therefore, from the convex dual, one can recover the original function.

Duality and the Exponential Family. Let us now see how convex duality applies to exponential-family distributions. The starting point is that the log-partition function $A(\eta)$ is convex, and, by (9.44), its gradient $\nabla A(\eta)$ is given by the mean parameter μ, i.e., $\nabla A(\eta) = \mu$. An illustration can be found in Fig. 9.2. Therefore, from (9.84) the convex dual of the log-partition function can be expressed as a function of the mean parameters μ as

$$A^*(\mu) = \max_{\eta} \left\{ \mu^T \eta - A(\eta) \right\}. \tag{9.86}$$

Under the assumption of **minimality**, the log-partition function is strictly convex, and the unique minimizer in (9.86) is the single natural parameter vector η corresponding to the mean parameter μ, i.e., $\nabla A(\eta) = \mu$. Therefore, given a pair of mean and natural parameters (μ, η) related trough the mapping (9.44), the equality

$$A^*(\mu) = \mu^T \eta - A(\eta) \tag{9.87}$$

holds. A useful way to recall this relationship is in the form

$$A(\eta) + A^*(\mu) = \mu^T \eta. \tag{9.88}$$

The relationship (9.88) can be used to prove that, for minimal distributions, we have the inverse mapping between natural and mean parameters given by

$$\nabla_\mu A^*(\mu) = \eta. \tag{9.89}$$

Together with (9.44), the equality (9.89) describes the one-to-one correspondence between natural and mean parameters for minimal parametrizations.

Bregman Divergence. In the following appendices, we will also need another concept from convex analysis, namely the Bregman divergence between two vectors x and y in $\mathbb{R}^K$. Given a convex function $f(\cdot)$ from $\mathbb{R}^K$ to $\mathbb{R}$, the **Bregman divergence** is defined as

$$B_f(x,y) = f(x) - f(y) - (x - y)^T \nabla f(y). \tag{9.90}$$

So, the Bregman divergence measures the deviation of the convex function $f(x)$ from its linear approximation computed at y. Using the convexity of $f(x)$, by the equivalent definition (5.16), this implies that the Bregman divergence is non-negative, $B_f(x,y) \geq 0$, and that we have $B_f(x,x) = 0$. It can also be seen that $B_f(x,y)$ is convex in x, but generally not in y.

Examples of Bregman divergences include the quadratic Euclidean distance $B_f(x,y) = ||x-y||^2$, which is obtained with the convex function $f(x) = ||x||^2$; and the KL divergence $B_f(x,y) = \text{KL}(x||y)$, which applies to probability vectors $x = [x_1, \ldots, x_K]^T$ and $y = [y_1, \ldots, y_K]^K$ and is obtained with the convex function $f(x) = \sum_{k=1}^{K} x_k \log(x_k)$.

Appendix 9.B: Distribution and Log-Loss As a Function of the Mean Parameters

One of the most useful properties of the Gaussian distribution when it comes to machine learning applications is its strong association with the Euclidean distance. Specifically, the log-loss for a Gaussian distribution $\mathcal{N}(\mu, 1)$ can be expressed as

$$-\log p(x|\mu) = \frac{1}{2}||x - \mu||^2 + \text{constant}, \tag{9.91}$$

where the constant does not depend on μ. The log-loss is thus a function of the squared Euclidean distance $||x - \mu||^2$ defined in the space of the mean parameter μ. We will see in this appendix that this association between distributions in the exponential family and distance measures in the mean parameter space applies more generally to all **minimal** distributions.

Specifically, under the assumption of minimality, we will prove that an exponential-family distribution $\text{ExpFam}(x|\mu)$ with mean parameter μ and log-partition function $A(\eta)$ can be expressed as

$$\text{ExpFam}(x|\mu) = \exp(-B_{A^*}(s(x), \mu) + G(x)), \tag{9.92}$$

and their log-loss as

$$\log \text{ExpFam}(x|\mu) = B_{A^*}(s(x), \mu) + \text{constant}, \tag{9.93}$$

for a suitable function $G(x)$ independent of the model parameters. Hence all minimal distributions in the exponential family have associated with them a measure of distance in the mean parameter space that is given by the Bregman divergence $B_{A^*}(s(x), \mu)$ defined by the convex dual $A^*(\mu)$ of the log-partition function. Generalizing the Gaussian case (9.91), the distance is between the sufficient statistic vector $s(x)$ and the mean parameter μ.

To provide some examples, as summarized in Table 9.5, the Gaussian distribution is associated with the (weighted) squared Euclidean distance, while Bernoulli and categorical variables are associated with the KL divergence. In the table, we have denoted

$$\text{KL}(x||\mu) = x \log\left(\frac{x}{\mu}\right) + (1 - x) \log\left(\frac{1-x}{1-\mu}\right) \tag{9.94}$$

for the Bernoulli distribution; and

$$\text{KL}(x||\mu) = \sum_{k=0}^{C-1} x_k \log\left(\frac{x_k}{\mu_k}\right) \tag{9.95}$$

for the categorical distribution. We have also used the notation

$$||x||_\Theta^2 = x^T \Theta x. \tag{9.96}$$

Table 9.5 Distribution in the exponential family, log-partition function, convex dual of the log-partition function, and Bregman divergence

distribution	$A(\eta)$	$A^*(\mu)$	$B_{A^*}(s(x), \mu)$
$\text{Bern}(\mu)$	$\log(1 + e^\eta)$	$\mu \log \mu + (1 - \mu) \log(1 - \mu)$	$\text{KL}(x \| \mu)$
$\text{Cat}(\mu)$	$\log\left(\sum_{k=0}^{C-1} e^{\eta_k}\right)$	$\sum_{k=0}^{C-1} \mu_k \log \mu_k$	$\text{KL}(x \| \mu)$
$\mathcal{N}(v, \Theta^{-1})$ with fixed Θ^{-1}	$\frac{1}{2}\eta^T \Theta^{-1} \eta$	$\frac{1}{2}\mu^T \Theta \mu$	$\frac{1}{2}\|s(x) - \mu\|_\Theta^2$

Proof of (9.92). Defining as η the natural parameter vector corresponding to the mean parameter vector μ, the equality (9.92) follows from these steps:

$$
\begin{aligned}
\text{ExpFam}(x|\eta) &= \exp\left(\eta^T s(x) + M(x) - A(\eta)\right) \\
&= \exp\left(\eta^T s(x) - (\eta^T \mu - A^*(\mu)) + M(x)\right) \\
&= \exp\left(A^*(\mu) + \eta^T (s(x) - \mu) + M(x)\right) \\
&= \exp\left(-\underbrace{(A^*(s(x)) - A^*(\mu) - (s(x) - \mu)^T \nabla A^*(\mu))}_{B_{A^*}(s(x), \mu)} + \underbrace{(A^*(s(x)) + M(x))}_{G(x)}\right),
\end{aligned}
$$
(9.97)

where in the second equality we have used (9.88), and in the last equality we have added and subtracted $A^*(s(x))$ and used the definition (9.90) of Bregman divergence.

Appendix 9.C: Gradient of the Log-Partition Function

In this appendix, we validate the relationship between mean parameters and gradient of the log-partition function in (9.44). Focusing first on discrete rvs, this is done by using the definition of log-partition function (9.7) through the direct calculation

$$
\begin{aligned}
\frac{\partial A(\eta)}{\partial \eta_k} &= \frac{\sum_x s_k(x) \exp\left(\eta^T s(x) + M(x)\right)}{\sum_{x'} \exp\left(\eta^T s(x') + M(x')\right)} \\
&= \mathbb{E}_{\mathbf{x} \sim p(x|\eta)}[s_k(\mathbf{x})] = \mu_k.
\end{aligned}
$$
(9.98)

The same derivation applies for continuous rvs by replacing (9.7) with (9.6).

Appendix 9.D: Entropy and KL Divergence for the Exponential Family

Using the background developed in Appendix 9.A, this appendix describes important duality properties of entropy (9.58) and KL divergence (9.59) for **minimal** exponential-family distributions.

Entropy. The negative entropy for a distribution in the exponential family and the log-partition function are convex dual to one another in the following sense. By (9.58), if $M(x) = 0$ for all x in the domain, we have the equality

$$A(\eta) - \mathrm{H}(\mathrm{ExpFam}(x|\eta)) = \eta^T \mu. \tag{9.99}$$

Under the assumption of a minimal parametrization, comparing this equality with (9.88) reveals that, under the mentioned conditions, the convex dual of the partition function is the negative entropy, i.e.,

$$A^*(\mu) = -\mathrm{H}(p(x|\eta)). \tag{9.100}$$

KL Divergence. The KL divergence (9.59) between $p(x|\eta_1) = \mathrm{ExpFam}(x|\eta_1)$ and $p(x|\eta_2) = \mathrm{ExpFam}(x|\eta_2)$ from the same exponential-family distribution can be related to the Bregman divergence. Specifically, by the definition of Bregman divergence (9.90), we can rewrite (9.59) as

$$\mathrm{KL}(\mathrm{ExpFam}(x|\eta_1)||\mathrm{ExpFam}(x|\eta_2)) = B_A(\eta_2, \eta_1)$$
$$= A(\eta_2) - A(\eta_1) - (\eta_2 - \eta_1)^T \mu_1. \tag{9.101}$$

Therefore, the KL divergence is a Bregman divergence in the natural parameter space. Furthermore, assuming a minimal parametrization and denoting as μ_1 and μ_2 the mean parameters corresponding to natural parameters η_1 and η_2, respectively, by the duality relationship (9.88) we can also express the KL divergence in terms of Bregman divergence in the mean parameter space as

$$\mathrm{KL}(\mathrm{ExpFam}(x|\mu_1)||\mathrm{ExpFam}(x|\mu_2)) = B_{A^*}(\mu_1, \mu_2)$$
$$= A^*(\mu_1) - A^*(\mu_2) - (\mu_1 - \mu_2)^T \eta_2. \tag{9.102}$$

Appendix 9.E: Results on the Fisher Information Matrix

In this appendix, results mentioned in the main text about the FIM are proved, and some additional discussion on the role of the FIM in estimation is presented. First, we demonstrate the validity of (9.68) for general probabilistic models; then, we prove (9.70) and (9.71) for the exponential family; and, finally, we briefly elaborate on the FIM as the asymptotic covariance of an ML estimator for well-specified models.

FIM As Covariance of the Gradient of the Log-Loss. First, we prove the equality (9.68), which is valid for general parametric distributions $p(x|\theta)$ (satisfying suitable technical conditions; see Sec. 9.7). To this end, first note that the Hessian $\nabla_\theta^2 \log p(x|\theta)$ can be directly computed using rules of differentiation as

$$\nabla_\theta^2 \log p(x|\theta) = \nabla_\theta(\nabla_\theta \log p(x|\theta))$$
$$= \nabla_\theta \left(\frac{\nabla_\theta p(x|\theta)}{p(x|\theta)} \right)$$
$$= -\left(\frac{\nabla_\theta p(x|\theta)}{p(x|\theta)} \right) \left(\frac{\nabla_\theta p(x|\theta)}{p(x|\theta)} \right)^T + \frac{\nabla_\theta^2 p(x|\theta)}{p(x|\theta)}$$
$$= -(\nabla_\theta \log p(x|\theta))(\nabla_\theta \log p(x|\theta))^T + \frac{\nabla_\theta^2 p(x|\theta)}{p(x|\theta)}. \tag{9.103}$$

Now, taking the expectation over $p(x|\theta)$, the first term recovers the desired result, while the second equals an all-zero vector. To see this, consider for simplicity of notation a discrete rv x, and observe that we have

$$E_{x \sim p(x|\theta)} \left[\frac{\nabla_\theta^2 p(x|\theta)}{p(x|\theta)} \right] = \sum_x \nabla_\theta^2 p(x|\theta)$$

$$= \nabla_\theta^2 \underbrace{\sum_x p(x|\theta)}_{1} = 0. \tag{9.104}$$

This result also applies to continuous rvs by replacing sums with integrals.

FIM As Hessian of the Log-Partition Function and As Covariance of the Sufficient Statistics for the Exponential Family. We now concentrate on proving the equalities (9.70) and (9.71). To this end, using the expression for the gradient of the log-loss (9.40), we have

$$\begin{aligned}
\text{FIM}(\eta) &= E_{x \sim p(x|\eta)}[-\nabla_\eta^2 \log p(x|\eta)] \\
&= E_{x \sim p(x|\eta)}[-\nabla_\eta (\nabla_\eta \log p(x|\eta))] \\
&= E_{x \sim p(x|\eta)}[-\nabla_\eta s(x) + \nabla_\eta \mu] \\
&= \nabla_\eta \mu \\
&= \nabla_\eta (\nabla_\eta A(\eta)) \\
&= \nabla_\eta^2 A(\eta), \tag{9.105}
\end{aligned}$$

where we have also used the mapping (9.44) between mean and natural parameters. Furthermore, again using (9.40), we have

$$\begin{aligned}
\text{FIM}(\eta) &= E_{x \sim p(x|\eta)}[(\nabla_\eta \log p(x|\eta))(\nabla_\eta \log p(x|\eta))^T] \\
&= E_{x \sim p(x|\mu)}[(s(x) - \mu)(s(x) - \mu)^T], \tag{9.106}
\end{aligned}$$

concluding the proof.

FIM As Asymptotic Covariance of the ML Estimator. It was explained in Sec. 9.7 that the matrix FIM(θ) measures how "easy" it is to estimate θ based on data from the model $p(x|\theta)$ in terms of the quadratic error. We now provide a more formal statement of this property by studying the asymptotic performance of the ML estimator of parameter vector θ under a well-specified model. In statistics – and sometimes also in machine learning (see Chapter 8) – it is often assumed that the model class

$$\mathcal{H} = \{p(x,t|\theta) : \theta \in \Theta\} \tag{9.107}$$

includes the (unknown) population distribution $p(x,t)$ for some ground-truth (unknown) value of the parameter vector θ_0, i.e., $p(x,t|\theta_0) = p(x,t)$. As mentioned, when this condition holds, we say that the model class is **well specified**, or that the population distribution is realizable by the model class. In this case, one can ask how well the true parameter θ_0 can be estimated by the ML estimate θ^{ML}.

The key result of interest here is that, under suitable technical assumptions (see Recommended Resources, Sec. 9.10), as the number of data points goes to infinity, i.e., as $N \to \infty$, the following properties hold for the ML estimator θ^{ML}:

- ML provides a **consistent** estimate, that is, we have the limit $\theta^{ML} \to \theta_0$ with high probability.
- ML is **asymptotically Gaussian** with covariance matrix given by the inverse of the FIM, that is, the distribution of the rv $\sqrt{N}(\theta^{ML} - \theta_0)$ tends to the pdf $\mathcal{N}(0, \text{FIM}(\theta_0)^{-1})$.

This implies that the inverse of the FIM quantifies the asymptotic quadratic error of the ML estimator θ^{ML}. Specifically, in the limit as $N \rightarrow \infty$, the quadratic error $\mathrm{E}[||\theta^{ML} - \theta_0||^2]$ tends to the trace $\mathrm{tr}(\mathrm{FIM}(\theta_0)^{-1})$.

Appendix 9.F: Natural Gradient Descent and Mirror Descent

In this appendix, natural GD and mirror descent are briefly introduced. Unless stated otherwise, the discussion applies to any probabilistic model, and not solely to the exponential family. We start by deriving a general relationship between the gradient vectors with respect to the natural parameters and the gradient vectors with respect to the mean parameter through the FIM. Then, we present natural GD for general distributions, followed by the specialization of natural GD to exponential-family models. Finally, mirror descent is described.

Relating Gradients over Natural and Mean Parameters. By (9.44) and (9.70), the FIM is the **Jacobian** (see Appendix 5.F) of the mean parameters with respect to natural parameters, i.e.,

$$\mathrm{FIM}(\eta) = \nabla_\eta^2 A(\eta) = \nabla_\eta \mu. \tag{9.108}$$

This relationship is useful for relating gradients with respect to mean and natural parameters.

To elaborate on this point, consider a function $g(\eta)$ of the natural parameters. A typical example is a function of the form $g(\eta) = \mathrm{E}_{x \sim p(x|\eta)}[f(x)]$ for some function $f(\cdot)$. The gradients $\nabla_\eta f(\eta)$ and $\nabla_\mu f(\eta)$ with respect to natural and mean parameters can be related via the chain rule of differentiation as

$$\nabla_\eta f(\eta) = \nabla_\eta \mu \cdot \nabla_\mu f(\eta) = \mathrm{FIM}(\eta) \cdot \nabla_\mu f(\eta). \tag{9.109}$$

Therefore, multiplying by the FIM allows one to convert the gradient with respect to μ into the gradient with respect to η. Furthermore, if the class of distributions is minimal, the FIM is invertible and one can also write

$$\nabla_\mu f(\eta) = \mathrm{FIM}(\eta)^{-1} \cdot \nabla_\eta f(\eta) = \mathrm{FIM}(\mu) \cdot \nabla_\eta f(\eta), \tag{9.110}$$

where we have used (9.73).

Natural Gradient Descent. In GD, as detailed in Chapter 5, at each step one minimizes the following strictly convex quadratic approximant of the cost function:

$$\tilde{g}_\gamma(\theta|\theta^{(i)}) = \underbrace{g(\theta^{(i)}) + \nabla g(\theta^{(i)})^T(\theta - \theta^{(i)})}_{\text{first-order Taylor approximation}} + \underbrace{\frac{1}{2\gamma}||\theta - \theta^{(i)}||^2}_{\text{proximity penalty}}. \tag{9.111}$$

Therefore, the distance between θ and the previous iterate is measured by the squared Euclidean distance $||\theta - \theta^{(i)}||^2$. An issue with this approach is that, for probabilistic models, the Euclidean distance may not reflect actual changes in the distribution. Therefore, the choice of the learning rate must be conservative in order to avoid unstable updates.

As an example, consider the optimization over the mean μ of a Gaussian distribution $\mathcal{N}(\mu, \sigma^2)$ with fixed variance σ^2. For a given squared Euclidean distance $||\mu - \mu^{(i)}||^2$, the two distributions $\mathcal{N}(\mu, \sigma^2)$ and $\mathcal{N}(\mu^{(i)}, \sigma^2)$ may be more or less distinguishable depending on the value of σ^2: If σ^2 is large compared to $||\mu - \mu^{(i)}||^2$, the two distributions are very similar, while the opposite is true when σ^2 is smaller.

We would therefore prefer to use proximity penalties that more accurately reflect the actual distance in the distribution space. A natural idea is to replace the squared Euclidean distance

$||\theta - \theta^{(i)}||^2$ with a more relevant measure of the distance between the two distributions $p(x|\theta)$ and $p(x|\theta^{(i)})$, such as the KL divergence $\text{KL}(p(x|\theta^{(i)})||p(x|\theta))$ or $\text{KL}(p(x|\theta)||p(x|\theta^{(i)}))$.

To this end, natural GD approximates the KL divergence via the second-order Taylor expansion (9.67), i.e.,

$$\text{KL}(p(x|\theta^{(i)})||p(x|\theta)) \simeq \frac{1}{2}(\theta - \theta^{(i)})^T \text{FIM}(\theta^{(i)})(\theta - \theta^{(i)})$$

$$= \frac{1}{2}||\theta - \theta^{(i)}||^2_{\text{FIM}(\theta^{(i)})}, \tag{9.112}$$

where we have again used the definition (9.96). The approximation (9.112) holds as long as the norm of the update $\Delta\theta^{(i)} = \theta - \theta^{(i)}$ is sufficiently small. Note that, if the FIM is replaced by an identity matrix, we recover the standard squared Euclidean distance used by GD.

Overall, assuming a non-singular FIM, natural GD minimizes at each iteration the strictly convex quadratic approximant of the cost function

$$\tilde{g}_\gamma(\theta|\theta^{(i)}) = \underbrace{g(\theta^{(i)}) + \nabla g(\theta^{(i)})^T(\theta - \theta^{(i)})}_{\text{first-order Taylor approximation}} + \underbrace{\frac{1}{2\gamma}||\theta - \theta^{(i)}||^2_{\text{FIM}(\theta^{(i)})}}_{\text{proximity penalty}}, \tag{9.113}$$

yielding the update

$$\theta^{(i+1)} = \theta^{(i)} - \gamma \text{FIM}(\theta^{(i)})^{-1} \nabla g(\theta^{(i)}), \tag{9.114}$$

where the inverse of the FIM pre-conditions the gradient.

Natural GD is hence generally more complex than GD because it requires us to compute and invert the FIM. On the flip side, by operating using a metric that is tailored to the space of distributions, it generally allows the use of larger step sizes, potentially speeding up convergence. Note that natural GD can be generalized from the space of distributions with the KL divergence metric to other geometries via the notion of Riemannian manifold (see Recommended Resources, Sec. 9.10).

Natural GD for the Exponential Family. The natural-gradient update (9.114) simplifies for **minimal** exponential-family distributions. To see this, consider natural GD with respect to the natural parameters, whose update (9.114) is given as

$$\eta^{(i+1)} = \eta^{(i)} - \gamma \text{FIM}(\eta^{(i)})^{-1} \nabla_\eta g(\eta^{(i)}). \tag{9.115}$$

By (9.110), one obtains the equivalent simplified update

$$\eta^{(i+1)} = \eta^{(i)} - \gamma \nabla_\mu g(\eta^{(i)}). \tag{9.116}$$

So, natural GD on the *natural* parameters follows the direction of the gradient with respect to the *mean* parameters.

Mirror Descent. Following the same philosophy as natural GD, at each iteration **mirror descent** minimizes the convex approximant

$$\tilde{g}_\gamma(\theta|\theta^{(i)}) = \underbrace{g(\theta^{(i)}) + \nabla g(\theta^{(i)})^T(\theta - \theta^{(i)})}_{\text{first-order Taylor approximation}} + \underbrace{\frac{1}{\gamma}\text{KL}(p(x|\theta)||p(x|\theta^{(i)}))}_{\text{proximity penalty}}. \tag{9.117}$$

Note that this applies to any distribution parametrized by a vector θ, and not only to the exponential family. Furthermore, observe the reversal of the order of the distributions within the KL divergence as compared to natural GD.

To elaborate on the minimization of the approximant (9.117), consider the important case of categorical distributions, which yields the approximant

$$\tilde{g}_\gamma(\mu|\mu^{(i)}) = \underbrace{g(\mu^{(i)}) + \nabla g(\mu^{(i)})^T (\mu - \mu^{(i)})}_{\text{first-order Taylor approximation}} + \underbrace{\frac{1}{\gamma} \text{KL}(\text{Cat}(x|\mu)||\text{Cat}(x|\mu^{(i)}))}_{\text{proximity penalty}}. \tag{9.118}$$

Minimizing this expression is an instance of free energy minimization, which can be seen to yield the **exponentiated gradient** update

$$\mu_k^{(i+1)} = \frac{\mu_k^{(i)} \exp\left(-\gamma [\nabla g(\mu^{(i)})]_k\right)}{\sum_{c=0}^{K-1} \mu_c^{(i)} \exp\left(-\gamma [\nabla g(\mu^{(i)})]_c\right)}. \tag{9.119}$$

We refer to Sec. 11.3 for further discussion on free energy minimization.

The terminology "mirror GD" reflects the fact that updates such as (9.119) can be thought of as being applied through a "mirror" defined by a specific invertible transformation. In particular, for (9.119), the transformation is given by the logarithm.

Mirror descent can be generalized to account for different geometrical constraints on the space of model parameters by replacing the KL divergence in the approximant (9.117) with any other Bregman divergence $B_f(\theta, \theta^{(i)})$. The invertible function defining the mirror transformation is then given by the gradient of convex function $f(\cdot)$, so that the updates take the form $\nabla f(\theta^{(i+1)}) = \nabla f(\theta^{(i)}) - \gamma \nabla g(\theta^i)$.

Bibliography

[1] M. J. Wainwright and M. I. Jordan, *Graphical Models, Exponential Families, and Variational Inference*. Now Publishers, 2008.

[2] A. J. Dobson and A. G. Barnett, *An Introduction to Generalized Linear Models*. CRC Press, 2018.

[3] I. Csiszár and P. C. Shields, *Information Theory and Statistics: A Tutorial*. Now Publishers, 2004.

[4] T. M. Cover and J. A. Thomas, *Elements of Information Theory*. John Wiley & Sons, 2006.

[5] S. M. Kay, *Fundamentals of Statistical Signal Processing*. Prentice Hall PTR, 1993.

[6] H. V. Poor, *An Introduction to Signal Detection and Estimation*. Springer Science+Business Media, 2013.

[7] S.-I. Amari, "Natural gradient works efficiently in learning," *Neural Computation*, vol. 10, no. 2, pp. 251–276, 1998.

[8] H. Salimbeni, S. Eleftheriadis, and J. Hensman, "Natural gradients in practice: Non-conjugate variational inference in Gaussian process models," in *International Conference on Artificial Intelligence and Statistics*, Proceedings of Machine Learning Research, vol. 84. PMLR 2018, pp. 689–697.

[9] J. Martens, "New insights and perspectives on the natural gradient method," arXiv preprint arXiv:1412.1193, 2014.

[10] S. Fortunati, F. Gini, M. S. Greco, and C. D. Richmond, "Performance bounds for parameter estimation under misspecified models: Fundamental findings and applications," *IEEE Signal Processing Magazine*, vol. 34, no. 6, pp. 142–157, 2017.

10 Variational Inference and Variational Expectation Maximization

10.1 Overview

This chapter focuses on three key problems that underlie the formulation of many machine learning methods for inference and learning, namely variational inference (VI), amortized VI, and variational expectation maximization (VEM). We have already encountered these problems in simplified forms in previous chapters, and they will be essential in developing the more advanced techniques to be covered in the rest of the book. Notably, VI and amortized VI underpin optimal Bayesian inference, which was used, e.g., in Chapter 6 to design optimal predictors for generative models; and VEM generalizes the EM algorithm that was introduced in Chapter 7 for training directed generative latent-variable models. Among the new training algorithms enabled by these formulations, this chapter will review variational autoencoders (VAEs).

Learning Objectives and Organization of the Chapter. By the end of this chapter, the reader should be able to:

- understand and recognize VI, amortized VI, and VEM problems, along with their applications (Sec. 10.2);
- address exact Bayesian inference – which corresponds to the ideal solution of the VI problem – when computationally possible, such as in the important case of conjugate exponential-family models (Sec. 10.3);
- apply the Laplace approximation as a first step towards tractable Bayesian inference (Sec. 10.4);
- understand the basic taxonomy of VI methods (Sec. 10.5);
- understand and derive mean-field VI methods to address the VI problem under the assumption that the soft predictor factorizes into a product of individual unconstrained distributions (Sec. 10.6);
- apply parametric VI methods (Sec. 10.7) via black-box VI (Sec. 10.8), reparametrization-based VI (Sec. 10.9), or a combination of both (Sec. 10.10);
- understand particle-based VI techniques, and apply Stein variational gradient descent (SVGD) as a specific instance of these solutions (Sec. 10.11);
- understand how to "amortize" the problem of VI across all values x of the input via amortized VI (Sec. 10.12); and
- address the VEM problem, including variational autoencoders (VAEs) as a special case (Sec. 10.13).

10.2   Variational Inference, Amortized Variational Inference, and Variational EM

In this section, we introduce the problems of VI, amortized VI, and VEM, and discuss some of their connections with the methods studied so far in the book.

10.2.1 Variational Inference

VI formalizes the problem of designing a soft predictor $q(\cdot|x)$ within some set of distributions $\mathcal{Q}_x = \{q(\cdot|x)\}$ for a fixed value x of the input. Specifically, given a joint distribution $p(x, z)$, a fixed value x of the input, and a set of distributions $\mathcal{Q}_x = \{q(\cdot|x)\}$, the **VI problem** amounts to the minimization of the **free energy** with respect to the soft predictor $q(\cdot|x) \in \mathcal{Q}_x$, i.e.,

$$\min_{q(\cdot|x)\in\mathcal{Q}_x} \mathrm{F}(q(z|x)||p(x,z)). \tag{10.1}$$

In the context of VI, the conditional distribution $q(z|x)$, which is subject to optimization, is referred to as the **variational posterior distribution**, or **variational posterior** for short. The qualifier "variational" indicates that the distribution $q(z|x)$ is under optimization.

We have seen in Chapter 3 that, if the set $\mathcal{Q}_x$ includes all possible soft predictors, i.e., all conditional distributions $q(\cdot|x)$ for the given fixed x, then the optimal variational posterior $q^*(z|x)$, which minimizes (10.1), is given by the posterior distribution

$$q^*(z|x) = p(z|x) = \frac{p(x,z)}{p(x)}. \tag{10.2}$$

Computing the posterior $p(z|x)$ is infeasible when marginalizing over the rv z to compute the marginal distribution $p(x)$ at the denominator in (10.2) is of prohibitive complexity.

In previous chapters we have, however, assumed that the posterior (10.2) is computable, so the VI problem (10.1) can be solved exactly under the assumption of an unconstrained distribution class $\mathcal{Q}_x$. We have specifically encountered two main applications of optimal soft predictors:

- In Chapter 6 (see Secs. 6.5 and 6.8), we have demonstrated that optimal soft predictors play a key role in the inference (or testing) phase in supervised generative models. Once a generative model $p(x, t|\theta)$ is trained, prediction requires the computation of the posterior $p(t|x, \theta)$ of the target t given the input x $= x$.
- In Chapter 7 (see Sec. 7.8), we have seen that optimal soft prediction is a central subroutine for the problem of training directed generative models in unsupervised learning. In particular, in order to train a model $p(x|\theta) = \mathrm{E}_{z\sim p(z|\theta)}[p(x|z, \theta)]$ with latent variables z, the EM algorithm requires the computation of the posterior $p(z|x, \theta)$ at each iteration (E step).

In this chapter, we study scalable tools that tackle the VI problem (10.1) by restricting the set $\mathcal{Q}_x$ of allowed variational posteriors so as to reduce computational complexity.

10.2.2 Amortized VI

The VI problem (10.1) optimizes the soft predictor $q(\cdot|x)$ for a fixed value x of the input. When the goal is to produce a prediction for *all* values x, this approach is impractical since the VI problem would have to be addressed *separately* for all inputs x of interest. The amortized VI problem provides a solution to this issue by optimizing a free energy functional over conditional distributions $q(\cdot|\cdot)$, which are functions of both input and target variables. As a result, a solution of the amortized VI problem is not a soft predictor $q(\cdot|x)$ applicable only to a fixed value x of the input; rather, it is a function of x that provides a soft prediction $q(\cdot|x)$ for all input values x.

Formally, given a joint distribution $p(x,z)$ and a class $Q = \{q(\cdot|\cdot)\}$ of conditional distributions $q(\cdot|\cdot)$, the **amortized VI problem** minimizes the **average free energy** with respect to the soft predictor within the set Q, i.e.,

$$\min_{q(\cdot|\cdot)\in Q} E_{x\sim p(x)}[F(q(z|x)\,||\,p(x,z))]. \tag{10.3}$$

Unlike in the VI problem (10.1), the free energy in the amortized VI problem (10.3) is averaged over the input marginal $p(x)$. This way, the optimal soft predictor obtained from amortized VI minimizes the free energy on average across all values of the input $x \sim p(x)$. The soft predictor $q(z|x)$, which is under optimization in (10.3), is referred to as the **amortized variational posterior**.

As we discussed for the VI problem, if the set Q includes all possible soft predictors, i.e., all conditional distributions $q(\cdot|\cdot)$, the solution of this problem is given by the posterior distribution (10.2). In this chapter, we will study scalable solutions in which the class Q of amortized variational posteriors is constrained so as to reduce computational complexity.

10.2.3 Variational Expectation Maximization

As recalled in Sec. 10.2.1, optimal soft prediction is a central subroutine for the problem of training generative models $p(x,z|\theta)$ of observation x and latent rv z. Specifically, using the EM algorithm, this problem can be formulated as the joint optimization over model parameters θ and soft predictor $q(\cdot|\cdot)$ of a training free energy metric (cf. (7.93)). The VEM problem provides a broad framework for the definition of training algorithms that, generalizing EM, can be used for the training of models with latent rvs.

Given a generative model $p(x,z|\theta)$, a training data set $\mathcal{D} = \{x_1,\ldots,x_N\}$, and a class Q of **amortized variational posteriors** $q(\cdot|\cdot)$, the **VEM problem** minimizes the **training free energy** with respect to the model parameter θ and the soft predictor within the set Q, i.e.,

$$\min_{\theta}\min_{q(\cdot|\cdot)\in Q} \frac{1}{N}\sum_{n=1}^{N} F(q(z_n|x_n)\,||\,p(x_n,z_n|\theta)). \tag{10.4}$$

Equivalently, the VEM problem can be written in terms of the empirical distribution $p_{\mathcal{D}}(x)$ as

$$\min_{\theta}\min_{q(\cdot|\cdot)\in Q} E_{x\sim p_{\mathcal{D}}(x)}[F(q(z|x)\,||\,p(x,z|\theta))], \tag{10.5}$$

where we recall that the empirical distribution is given as $p_{\mathcal{D}}(x) = 1/N\sum_{n=1}^{N}\delta(x - x_n)$, with $\delta(x)$ representing a Kronecker or Dirac delta function for discrete or continuous rvs, respectively (see Appendix 4.A).

As discussed in the previous subsections, with an unconstrained set Q of variational posteriors, the inner minimization in (10.4) yields the posterior distribution $p(z|x,\theta) = p(x,z|\theta)/p(x|\theta)$. With this choice, by (7.87) we have the identity

$$F(p(z|x,\theta)\,||\,p(x,z|\theta)) = -\log p(x|\theta). \tag{10.6}$$

Therefore, the optimization of the resulting training free energy in (10.4) coincides with the ML problem of minimizing the unsupervised training log-loss $L_{\mathcal{D}}(\theta) = 1/N\sum_{n=1}^{N}(-\log p(x_n|\theta))$. This is indeed the main insight that yields the EM algorithm as a means to address the ML problem.

VI and amortized VI can be obtained as special cases of the VEM problem in which the model parameter θ is fixed and not subject to optimization. In particular, the VI problem (10.1) is a special case of VEM (10.4) with $N = 1$ and a fixed θ. Amortized VI (10.3) is obtained from the VEM formulation (10.5) by replacing the empirical distribution $p_{\mathcal{D}}(x)$ with the input marginal $p(x)$ while fixing the model parameter θ.

This chapter will focus on settings in which the inner optimization in the VEM problem (10.4) cannot be solved exactly for an unconstrained set $\mathcal{Q}$, calling for restrictions on the set $\mathcal{Q}$ of amortized variational posteriors.

10.3 Exact Bayesian Inference

For reference, in this section, we study the ideal situation in which exact Bayesian inference is tractable. In this case, the optimal soft predictor $p(z|x)$ in (10.2) can be computed, and there is no need to address the VI problem (10.1) within a restricted class of distributions. The computation of the posterior $p(z|x)$ is only feasible if the latent vector has a small dimensionality so that the marginal $p(x)$ of the observation in (10.2) can be be evaluated numerically; or if the joint distribution $p(x, z)$ has a special structure that enables an explicit analytical evaluation of the posterior (10.2). This section will focus on the latter case by describing a family of distributions that have the desired property of tractability of the posterior.

Example 10.1

Consider the joint distribution $p(x, z) = p(z)p(x|z)$ defined as

$$z \sim p(z) = \mathcal{N}(z|0, I)$$
$$(x|z = z) \sim p(x|z) = \mathcal{N}(x|\mu(z), I), \tag{10.7}$$

where the mean vector $\mu(z)$ is a function of the latent variable z. The posterior $p(z|x)$ can be computed in closed form if the mean function $\mu(z)$ is linear, i.e., if we have $\mu(z) = Az$ for some matrix A. In fact, we encountered this example in Chapter 2, where we saw that the posterior is also Gaussian. We will refer to this setting as the **Gaussian–Gaussian model**, since both the prior $p(z)$ of the latent rv z and the likelihood $p(x|z)$ are Gaussian. If $\mu(z)$ is a non-linear function of z, e.g., defined by a neural network, the problem of computing $p(z|x)$ for any given value of x is generally of prohibitive complexity, unless the dimension of z is small enough to enable numerical integration, or Monte Carlo estimation, of the marginal $p(x) = E_{z \sim p(z)}[p(x|z)]$.

10.3.1 Conjugate Exponential Family

The Gaussian–Gaussian model is a special case of a general class of structured joint distributions $p(x, z)$ that admit an efficient computation of the posterior: the **conjugate exponential family**. For joint distributions in the conjugate exponential family, the posterior $p(z|x)$ is in the same class of distributions as the prior $p(z)$. For example, in the Gaussian–Gaussian model, the posterior is Gaussian like the prior.

The conjugate exponential family is defined by specific pairs of conditional distribution $p(x|z)$ and prior $p(z)$. The conditional distribution $p(x|z)$ is also referred to as "likelihood" in this context to emphasize that the problem of interest is estimation of the latent rv z via the posterior $p(z|x)$.

Likelihood. For any value of z, the **likelihood** $p(x|z)$ for a conjugate exponential family pair takes the form of the exponential-family distribution (see Chapter 9)

$$p(x|z) = \exp\Big(\eta_l(z)^T s_l(x) + M_l(x) - A_l(z)\Big), \tag{10.8}$$

with $D \times 1$ sufficient statistics vector $s_l(x)$ and log-base measure function $M_l(x)$. The subscript l indicates that these quantities are related to the likelihood $p(x|z)$. The dependence of rv x on rv z is through the $D \times 1$ natural parameter vector $\eta_l(z)$, which is a general function of z. The log-partition function $A_l(\eta(z))$, which depends on the natural parameters $\eta_l(z)$, is written as $A_l(z)$ for simplicity of notation.

Conjugate Prior. How should we select the prior $p(z)$ so as to ensure that the posterior $p(z|x)$ belongs to the same class of distributions as $p(z)$? To address this question, consider a **prior** distribution $p(z)$ belonging to a class of distributions in the exponential family that is generally distinct from the likelihood. Accordingly, we write

$$p(z) = \exp\Big(\eta_p^T s_p(z) + M_p(z) - A_p(\eta_p)\Big), \tag{10.9}$$

where the subscript p identifies natural parameters, sufficient statistics, log-base measure function, and log-partition function for the prior. We assume that the prior $p(z)$ is a standard distribution for which the partition function $A_p(\eta_p)$ is available in closed form, so that evaluating $p(z)$ does not require any numerical integration. As we discussed in Chapter 9, this class of distributions includes many standard choices for both pmfs and pdfs.

Combining likelihood (10.8) and prior (10.9) we obtain the posterior

$$p(z|x) \propto \underbrace{p(z)}_{\text{prior}} \cdot \underbrace{p(x|z)}_{\text{likelihood}}$$

$$\propto \exp\Big(s_l(x)^T \eta_l(z) + \eta_p^T s_p(z) + M_p(z) - A_l(z)\Big), \tag{10.10}$$

where we have highlighted only the terms dependent on z, since x is fixed as the conditioning value. This distribution is generally not easy to normalize, since the sufficient statistics $(\eta_l(z), s_p(z))$ and the log-base measure $M_p(z) - A_l(z)$ do not correspond to one of the standard distributions in the exponential family.

Returning to the question posed above, can we choose the sufficient statistics of the prior so as to ensure that the posterior (10.10) is in the same class of exponential-family distributions as the prior (10.9)? This requires the posterior (10.10) to be characterized by the same sufficient statistics $s_p(z)$ and log-based measure $M_p(z)$ as the prior. It can be easily verified that this condition holds if we choose the sufficient statistics of the prior as the $(D + 1) \times 1$ vector

$$s_p(z) = \begin{bmatrix} \eta_l(z) \\ -A_l(z) \end{bmatrix}. \tag{10.11}$$

In fact, plugging (10.11) into the expression for the posterior (10.10), we have

$$p(z|x) \propto \exp\Big(s_l(x)^T \eta_l(z) + \eta_p^T s_p(z) + M_p(z) - A_l(z)\Big)$$

$$\propto \exp\Big(\eta_{post}(x)^T s_p(z) + M_p(z)\Big), \tag{10.12}$$

where

$$\eta_{post}(x) = \eta_p + \begin{bmatrix} s_l(x) \\ 1 \end{bmatrix}. \tag{10.13}$$

The posterior is hence in the same class of distributions as the prior, which is characterized by the pair $(s_p(z), M_p(z))$ of sufficient statistics and log-base measure function.

Conjugate Exponential-Family Model. The choice of **prior** (10.9) with sufficient statistics (10.11) is said to be **conjugate** with respect to the likelihood $p(x|z)$ in (10.8). The **conjugate exponential family** consists of all joint distributions $p(x, z)$ with likelihood of the form (10.8) and prior distributions of the form (10.9) with (10.11).

The key advantage of conjugate exponential-family models is that computing the posterior is straightforward: All we need to do is to obtain the natural parameters $\eta_{post}(x)$ via (10.13) by adding to the natural parameters η_p of the prior a $D + 1$ vector that consists of the sufficient statistics of the likelihood, $s_l(x)$, and 1 as the last component. It is emphasized that, since the prior is easy to normalize by assumption, so is the posterior.

By the derivations in this subsection, for every likelihood in the exponential family, one can always find a conjugate prior. The Gaussian–Gaussian is one example, and we will now study other examples.

10.3.2 Beta–Bernoulli Model

The beta–Bernoulli model consists of a Bernoulli likelihood with a beta prior, which produces a beta posterior.

Likelihood. The likelihood $p(x|z) = \text{Bern}(x|z)$ is Bernoulli, with latent variable z identifying the mean parameter, i.e., the probability of the event x = 1 conditioned on z = z. Formally, we have $p(\text{x} = 1|z) = z$ and $p(\text{x} = 0|z) = 1 - z$. Referring to Sec. 9.2 for details on the definition of the Bernoulli pmf as an exponential-family distribution and using the notation (10.8), the **likelihood** is therefore written as

$$p(x|z) = \text{Bern}(x|z) = \exp\left(\underbrace{\log\left(\frac{z}{1-z}\right)}_{\eta_l(z)} \underbrace{x}_{s_l(x)} - \underbrace{(-\log(1-z))}_{A_l(z)} \right). \tag{10.14}$$

Conjugate Prior. By (10.11), the conjugate prior has sufficient statistics

$$s_p(z) = \begin{bmatrix} \log\left(\frac{z}{1-z}\right) \\ \log(1-z) \end{bmatrix}, \tag{10.15}$$

and, writing $\eta_p = [\eta_{p,1}, \eta_{p,2}]^T$ for the vector of natural parameters of the prior, the **conjugate prior** (10.9) is given by

$$\begin{aligned} p(z) &\propto \exp\left(\eta_{p,1} \log\left(\frac{z}{1-z}\right) + \eta_{p,2} \log(1-z) \right) \\ &= z^{\eta_{p,1}}(1-z)^{-\eta_{p,1}+\eta_{p,2}} \\ &= z^{a-1}(1-z)^{b-1} \\ &= \text{Beta}(z|a,b), \end{aligned} \tag{10.16}$$

Figure 10.1 Beta distribution Beta$(z|a,b)$ with various choices for the shape parameters a and b.

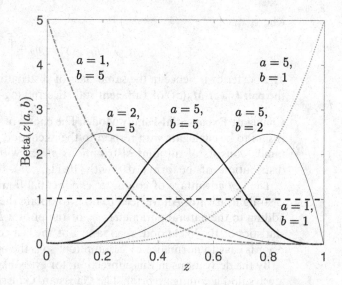

which corresponds to a beta distribution with parameters

$$a = \eta_{p,1} + 1 \text{ and } b = -\eta_{p,1} + \eta_{p,2} + 1. \tag{10.17}$$

The parameters (a,b) – both constrained to be non-negative – define the shape of the beta prior, as illustrated in Fig. 10.1. Note that the support of the beta distribution is the interval $[0,1]$. Therefore, the beta distribution is a pdf of a probability.

The average and the mode, which is the value with the maximum density, of the beta distribution are given as

$$\mathrm{E}_{z\sim\mathrm{Beta}(a,b)}[z] = \frac{a}{a+b}$$

$$\mathrm{mode}_{z\sim\mathrm{Beta}(a,b)}[z] = \frac{a-1}{a+b-2} \text{ when } a > 1, b > 1. \tag{10.18}$$

Therefore, as illustrated in Fig. 10.1, increasing parameter a skews the distribution towards larger values of z, while increasing b skews the distribution towards smaller values of z. Furthermore, the variance of the prior,

$$\mathrm{Var}[z] = \frac{ab}{(a+b)^2(a+b+1)}, \tag{10.19}$$

decreases with the sum $a + b$.

Posterior. Because of conjugacy, the posterior is also a beta distribution and its natural parameters are given by (10.13). Using the relationship (10.17) between the natural parameters and the parameters (a,b) of a beta distribution, we can write the posterior as $p(z|x) = \mathrm{Beta}(z|a_{post}(x), b_{post}(x))$, with

$$a_{post}(x) = \eta_{p,1} + x + 1 = a + x$$
$$b_{post}(x) = -\eta_{p,1} - x + \eta_{p,2} + 1 + 1 = b + (1 - x). \tag{10.20}$$

In summary, the **posterior** for the beta–Bernoulli model is given by the beta distribution

$$p(z|x) = \text{Beta}(z|a + \mathbb{1}(x = 1), b + \mathbb{1}(x = 0)). \tag{10.21}$$

In words, the parameters of the posterior are obtained by simply adding 1 to prior parameter a if $x = 1$, and adding 1 to prior parameter b if $x = 0$. This skews the posterior distribution towards larger values of z if $x = 1$, while it skews it towards smaller values of z if $x = 0$.

Multiple Observations. Suppose that we make two conditionally independent observations x_1 and x_2 from this model, so that the joint distribution is

$$p(x_1, x_2, z) = \text{Beta}(z|a, b) \cdot \text{Bern}(x_1|z) \cdot \text{Bern}(x_2|z). \tag{10.22}$$

We will now see that we can account for each of the observations sequentially, updating the posterior via (10.21) in two successive steps. To prove this, let us write the posterior as

$$p(z|x_1, x_2) \propto (\text{Beta}(z|a, b) \cdot \text{Bern}(x_1|z)) \cdot \text{Bern}(x_2|z)$$
$$\propto \underbrace{\text{Beta}(z|a + \mathbb{1}(x_1 = 1), b + \mathbb{1}(x_1 = 0))}_{p(z|x_1)} \cdot \text{Bern}(x_2|z), \tag{10.23}$$

where we have used (10.21) to account for the first observation x_1. The factorization (10.23) implies that we can treat $p(z|x_1)$ as the prior as we make the second independent observation $x_2 \sim p(x_2|z)$. By applying (10.21) again to account for the second observation, we obtain the overall posterior

$$p(z|x_1, x_2) = \text{Beta}\left(z \,\middle|\, a + \sum_{j=1}^{2} \mathbb{1}(x_j = 1), b + \sum_{j=1}^{2} \mathbb{1}(x_j = 0)\right). \tag{10.24}$$

Therefore, for every new observation, we simply add 1 to either parameter a or b depending on whether the corresponding observation is 1 or 0, respectively.

This procedure is not limited to two observations. Consider the joint distribution of latent rv z and L conditionally independent observations $x_1, \ldots, x_L$ given by

$$p(x_1, \ldots, x_L, z) = \text{Beta}(z|a, b) \prod_{j=1}^{L} \text{Bern}(x_j|z). \tag{10.25}$$

By recursively repeating the approach outlined above for two observations L times, it follows that the posterior $p(z|x_1, \ldots, x_L)$ is

$$p(z|x_1, \ldots, x_L) = \text{Beta}\left(z \,\middle|\, a + \underbrace{\sum_{j=1}^{L} \mathbb{1}(x_j = 1)}_{L[1]}, b + \underbrace{\sum_{j=1}^{L} \mathbb{1}(x_j = 0)}_{L[0]}\right). \tag{10.26}$$

Noting that $L[1]$ and $L[0]$ count the number of observations equal to 1 and 0, respectively, by (10.26), we can then interpret the prior parameters a and b as **"pseudo-counts"**, i.e., as prior observations of 1's and 0's made before making the actual observations $x_1, \ldots, x_L$.

Figure 10.2 Posterior distributions $p(z|x)$, with $x = (x_1, \ldots, x_L)$, for the probability z of a positive review in the two-vendor example (Example 10.2) for two different choices of the beta prior.

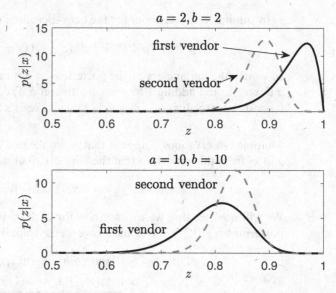

Example 10.2

On a street in the East Village in Manhattan, there are two stores offering similar products at the same price: The first has 30 positive reviews and 0 negative reviews, while the second has 90 positive reviews and 10 negative reviews. Which one to choose?

Let us introduce a latent variable $z \sim \text{Beta}(z|a, b)$ that describes the probability of a vendor receiving a positive review. Conditioned on the latent rv z, we model the available reviews $x_1, \ldots, x_L$ as independent binary observations, with $x_i = 1$ representing a positive review. By (10.26), the posterior $p(z|x_1, \ldots, x_L)$ is hence given as $p(z|x_1, \ldots, x_L) = \text{Beta}(z|a + L[1], b + L[0])$, where $L[1]$ counts the number of positive reviews and $L[0]$ counts the number of negative reviews. Recall that we have $L[1] = 30$ with $L = 30$ for the first seller, and $L[1] = 90$ with $L = 100$ for the second.

The posterior $p(z|x_1, \ldots, x_L)$ describes our current belief on the unknown probability z that the next review will be positive. As illustrated in Fig. 10.2, this depends strongly on the prior. Assume that a priori we have no reason to favor one vendor over the other – an information we encode in the prior distribution by choosing $a = b$. If the prior has a small variance (10.19), i.e., when the number of pseudo-counts $a + b$ is large, the posterior for the second vendor has a larger mode (bottom figure), while the opposite is true when the prior is weaker, i.e., when $a + b$ is smaller (top figure). This should suggest caution in choosing the first vendor based on the available evidence: Unless we have good reason to give much weight to the few reviews available for the first vendor, i.e., to choose a weak prior, the second vendor appears to be preferable.

10.3.3 Table of Conjugate Exponential Family Distributions

As studied in this section, for all distributions in the exponential family, there exists a conjugate prior on the model parameters. Table 10.1 provides some examples, including the Gaussian–Gaussian and the beta–Bernoulli models studied in this section. In the table, we have written

Table 10.1 Examples of conjugate exponential-family models

likelihood $p(x|z) = \prod_{j=1}^{L} p(x_j|z)$	prior $p(z)$	posterior $p(z|x)$
$\text{Bern}(x_j|z)$, with $z = \text{E}[x_j]$	$\text{Beta}(a,b)$	$\text{Beta}\,(a + L[1], b + L[0])$
$\text{Cat}(x_j|z)$ with $z_k = \text{E}[x_{k,j}^{OH}]$	$\text{Dirichlet}(\alpha)$	$\text{Dirichlet}\left(\alpha + \sum_{j=1}^{L} x_j^{OH}\right)$
$\mathcal{N}(x_j|z, \Theta^{-1})$, with fixed Θ	$\mathcal{N}(\mu_0, \Theta_0^{-1})$	$\mathcal{N}\left((\Theta')^{-1}\left(\Theta_0^{-1}\mu_0 + \Theta^{-1}\sum_{j=1}^{L} x_j\right), (\Theta')^{-1}\right)$

$x = (x_1, \ldots, x_L)$ for a collection of conditionally i.i.d. observations; and we have used the one-hot notation $x_j^{OH} = [x_{0,j}^{OH}, \ldots, x_{C-1,j}^{OH}]^T$ with $x_{k,j}^{OH} = \mathbb{1}(x_j = k)$, as well as the notation $\Theta' = \Theta_0 + L\Theta$. The **Dirichlet–categorical** model is detailed in Appendix 10.A.

10.3.4 Conjugate Models outside the Exponential Family

While conjugate exponential-family models include the most common conjugate pairs of likelihood and prior, there are also conjugate models with likelihood not in the exponential family. An example is the uniform likelihood distribution $p(x|z) = \mathcal{U}(x|[0,z])$ with support interval $[0,z]$ dependent on the latent rv z. As discussed in Chapter 9, this distribution is not in the exponential family. However, for this likelihood, the Pareto prior $p(z)$ is conjugate, since the posterior $p(z|x)$ turns out to also be a Pareto distribution.

10.4 Laplace Approximation

When exact Bayesian inference is intractable, a simple approach to obtain an approximation of the posterior distribution $p(z|x)$ for a *fixed value of* x is given by the Laplace approximation. The Laplace approximation is introduced in this section to serve as another benchmark for VI and amortized VI.

The Laplace approximation fits a Gaussian distribution around the maximum a posteriori (MAP) solution

$$z^{MAP} = \arg\max_z p(z|x) = \arg\max_z p(x,z), \tag{10.27}$$

which corresponds to the point z at which the posterior $p(z|x)$ is maximized. Note that the MAP solution is obtained for the given fixed value x. As per (10.27), computing the MAP solution z^{MAP} does not require knowledge of the posterior $p(z|x)$ – which would defeat the purpose of approximating it in the first place – since the maximizers of the posterior also maximize the joint distribution $p(x,z)$. In the following, we assume that the posterior has a single mode, and hence the MAP solution (10.27) is unique. This is the setting in which the Laplace approximation is most useful.

The Laplace approximation produces the Gaussian distribution

$$q(z|x) = \mathcal{N}(z|z^{MAP}, \Theta^{-1}) \tag{10.28}$$

as an approximation of the true posterior $p(z|x)$, where the mean is the MAP solution and the precision matrix Θ is to be optimized. Specifically, the precision matrix is chosen so that the log-loss $-\log q(z|x)$ of the approximation matches the unknown true log-loss $-\log p(z|x)$ in a neighborhood of the MAP solution z^{MAP}. This is done by matching the corresponding second-order Taylor expansions. Importantly, as we will see, this goal is accomplished without requiring knowledge of the posterior distribution $p(z|x)$ but only of the joint distribution $p(x,z)$.

By the first-order necessary optimality condition, the gradients of the log-losses $-\log q(z|x)$ and $-\log p(z|x)$ are equal to all-zero vectors, and hence matching the second-order approximations requires equating the two Hessians. The Hessian of the log-loss $-\log q(z|x)$ can easily be seen to equal the precision matrix Θ, i.e., we have $\nabla_z^2(-\log q(z|x)) = \Theta$ (for any value z). Furthermore, the Hessian of the unknown posterior distribution equals the Hessian of the joint distribution, i.e.,

$$\nabla_z^2(-\log p(z|x)) = \nabla_z^2(-\log p(x,z)). \tag{10.29}$$

This is because the multiplicative normalization term needed to obtain the posterior from the joint distribution does not depend on z.

Overall, the **Laplace approximation** is given by the Gaussian distribution (10.28), with precision matrix selected as

$$\Theta = \nabla_z^2(-\log p(x,z)). \tag{10.30}$$

The Laplace approximation can be quite accurate for posterior distributions that have a single mode.

Example 10.3

Consider the following joint distribution $p(x,z) = p(z)p(x|z)$:

$$z \sim p(z) = \mathrm{Beta}(z|a,b)$$

$$(x|z = z) \sim p(x|z) = \mathrm{Exp}(x|z) = \frac{1}{z}\exp\left(-\frac{x}{z}\right)\mathbb{1}(x \geq 0), \tag{10.31}$$

where the conditional distribution is the exponential pdf $\mathrm{Exp}(x|z)$ with mean parameter z. Note that this model is not conjugate, although both prior and likelihood are distributions in the exponential family. We will refer to the joint distribution (10.31) as the **beta–exponential model**.

By the general formula (10.2), the posterior for the beta–exponential model is given as

$$p(z|x) = \frac{\mathrm{Beta}(z|a,b)\cdot\mathrm{Exp}(x|z)}{\int\mathrm{Beta}(z'|a,b)\cdot\mathrm{Exp}(x|z')\mathrm{d}z'}, \tag{10.32}$$

where the integral in the denominator can be approximated numerically through a one-dimensional integration.

Fig. 10.3 shows the posterior (10.32) along with the Laplace approximation for the beta–exponential model (10.31). The approximation is seen to be more accurate when the posterior distribution is closer to a Gaussian distribution.

10.5   Introducing Variational Inference

As discussed in Sec. 10.2, for a fixed input x the posterior distribution $p(z|x)$ solves the VI problem (10.1) when no constraints are imposed on the set $\mathcal{Q}_x$ of variational posteriors $q(z|x)$. More generally, VI obtains an approximation of the posterior $p(z|x)$ by

Figure 10.3 Laplace approximation (solid line) and true posterior (dashed line) for the beta–exponential model (10.31) with different parameter values a and b for the beta prior.

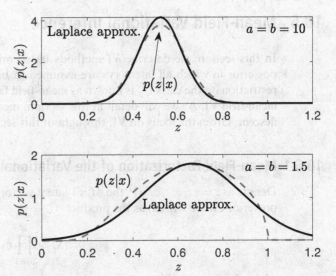

- restricting the **space $\mathcal{Q}_x$ of variational posteriors** $q(z|x)$ over which optimization of the free energy is carried out; and
- applying **local optimization** algorithms to tackle the resulting VI problem (10.1).

These steps contribute in different ways to the accuracy of the approximation of the posterior $p(z|x)$ produced by VI. The choice of the set $\mathcal{Q}_x$ determines how closely, even with an ideal optimizer, one can approximate the posterior $p(z|x)$. If the space $\mathcal{Q}_x$ of variational posteriors is too small, it may not include any close approximant of the true posterior distribution, and hence the output of VI will be, inevitably, inaccurate. The optimization algorithm dictates how close to the minimizer of the free energy within the set $\mathcal{Q}_x$ one can get when addressing the VI problem.

Let us look more closely at the most typical options for both steps. First, the space of variational posteriors $\mathcal{Q}_x$ can be restricted by imposing:

- a **factorization** of the variational posterior $q(z|x)$;
- a **parametrization** of the variational posterior $q(z|x)$; or
- both types of constraints.

Furthermore, standard algorithms used for the minimization of the free energy include:

- **SGD**;
- **coordinate descent**; and
- combinations of both methods.

In the next sections, we will explore several VI solutions that combine one of the restrictions on the set $\mathcal{Q}_x$ with one of the discussed optimization approaches listed above.

10.6   Mean-Field Variational Inference

In this section, we describe VI methods based on a specific factorization of the variational posterior in which all latent rvs are assumed to be independent given the observations. This restriction on the class $\mathcal{Q}_x$ is known as **mean-field factorization**, and the resulting VI strategy as **mean-field VI**. As we will detail in this section, mean-field VI uses optimization via **coordinate descent**. Given the focus on VI, throughout this section, the value of x is assumed to be *fixed*.

10.6.1   Mean-Field Factorization of the Variational Posterior

Denote as $z = [z_1, \ldots, z_M]^T$ the $M \times 1$ latent vector. Mean-field VI assumes that the variational posterior can be written as the product

$$q(z|x) = \prod_{m=1}^{M} q_m(z_m|x), \tag{10.33}$$

where factor $q_m(z_m|x)$ is the marginal variational posterior distribution for latent rv z_m. By (10.33), mean-field VI makes the strong assumption that the latent variables are independent given the observation $\mathrm{x} = x$. This assumption can drastically reduce the complexity of (approximate) Bayesian inference, at the cost of causing an irreducible bias when the true posterior distribution $p(z|x)$ cannot be well approximated by distributions of the fully factorized form in (10.33).

Example 10.4

To see why mean-field VI can reduce complexity, consider the case in which the latent rvs $z_m \in \{0, 1\}$ are binary. The unconstrained variational posterior would have $2^M - 1$ parameters to optimize on, i.e., the probabilities $q(z|x)$ for all possible values of z except one (recall that the value x is fixed). The subtraction of 1 accounts for the constraint that the probabilities must sum to 1. For this case, mean-field VI restricts optimization to the subset $\mathcal{Q}_x$ of distributions of the form

$$q(z|x) = \prod_{m=1}^{M} \mathrm{Bern}(z_m|\mu_m), \tag{10.34}$$

which depends on the M probabilities $\mu = [\mu_1, \ldots, \mu_M]^T \in [0, 1]^M$, yielding an exponential reduction in the complexity, from $2^M - 1$ to M. Note that parameters μ generally depend on x, although this is not explicitly indicated by the notation since x is fixed throughout this section.

10.6.2   Coordinate Descent Optimization

In mean-field VI, the optimization of the free energy in (10.1) is done by means of coordinate descent. For each iteration $i = 1, 2, \ldots$, let us denote the current iterate for the variational posterior as

$$q^{(i-1)}(z|x) = \prod_{m=1}^{M} q_m^{(i-1)}(z_m|x), \tag{10.35}$$

where $q_m^{(i-1)}(z_m|x)$ represents the current iterate for the mth factor $q_m(z_m|x)$ in (10.33). At the current iteration i, we pick one of the factors, that is, one coordinate $m \in \{1, \ldots, M\}$. As we will detail next, only the selected factor $q_m^{(i-1)}(z_m|x)$ is updated to produce a new factor $q_m^{(i)}(z_m|x)$, while all other factors $q_{m'}^{(i-1)}(z_{m'}|x)$ with $m' \neq m$ are preserved by setting $q_{m'}^{(i)}(z_{m'}|x) = q_{m'}^{(i-1)}(z_{m'}|x)$. A typical approach is to successively select all factors one by one across M iterations. With this choice, one set of M iterations, including updates to all M factors, is typically referred to as a **training epoch**.

Recalling that the goal of the VI problem (10.1) is to minimize the free energy, at each iteration, in order to update the selected mth factor, we solve the problem of minimizing the free energy over $q_m(z_m|x)$ when all the other factors are fixed to their current values, i.e.,

$$\min_{q_m(\cdot|x)} F(q(z|x) = q_m(z_m|x) \cdot q_{-m}^{(i-1)}(z_{-m}|x) \| p(x,z)), \tag{10.36}$$

where we have denoted as

$$q_{-m}^{(i-1)}(z_{-m}|x) = \prod_{m' \neq m} q_{m'}^{(i-1)}(z_{m'}|x) \tag{10.37}$$

the product of all factors except the mth. Writing the optimal solution of problem (10.36) as $q_m^*(z_m|x)$, we obtain the next iterate by setting $q_m^{(i)}(z_m|x) = q_m^*(z_m|x)$ and $q_{m'}^{(i)}(z_{m'}|x) = q_{m'}^{(i-1)}(z_{m'}|x)$ for all $m' \neq m$, and we move on to the next iteration.

To fully specify mean-field VI, we finally need to describe how to obtain $q_m^*(z_m|x)$ as the minimizer of the free energy minimization problem (10.36). As we show in Appendix 10.B, this can be computed as

$$q_m^*(z_m|x) \propto \exp\left(E_{z_{-m}}[\log p(x, z_m, z_{-m})]\right), \tag{10.38}$$

where the average is taken with respect to all the latent rvs except the mth, i.e., over the rv $z_{-m} \sim q_{-m}^{(i-1)}(z_{-m}|x)$ with (10.37). Note that in (10.38) we have made explicit the dependence of the joint distribution $p(x, z)$ separately on both z_m and z_{-m}. The name "*mean-field*" refers to the fact that, in the update (10.38), the impact of the other latent variables on the mth is only through an expectation.

By (10.38), computing the optimal factor $q_m^*(z_m|x)$ requires averaging the negative supervised log-loss $\log p(x, z)$ over all other variables z_{-m} and then normalizing. In general, the average over z_{-m} in (10.38) may be computationally problematic, unless rv z_{-m} is low-dimensional or unless it takes a discrete and small number of values. As a first simplification, if the joint distribution $p(x, z)$ is in the exponential family, this expectation amounts to computing the average of the augmented sufficient statistics (see Sec. 9.2). Another simplification is possible when the log-loss $-\log p(x, z)$ can be written as a sum of terms, each dependent on a small subset of variables. This structure can be imposed via probabilistic graphical models, as exemplified by the application discussed in the next subsection.

10.6.3 Mean-Field VI for the Ising Model

Definition. In the **Ising model**, the latent variables $\{(z_m)_{m=1}^M\}$ are bipolar – i.e., $z_m \in \{-1, +1\}$ for $m = 1, \ldots, M$ – and there are M corresponding observations $\{(x_m)_{m=1}^M\}$, also bipolar – i.e., $x_m \in \{-1, +1\}$ for $m = 1, \ldots, M$. Each observation x_m is associated with the corresponding latent variable z_m, and it can be interpreted as a noisy version of z_m. The statistical dependence among the latent variables is described by the graph in Fig. 10.4, which has one node for each

Figure 10.4 The Ising model represented as a probabilistic graphical model (Markov random field).

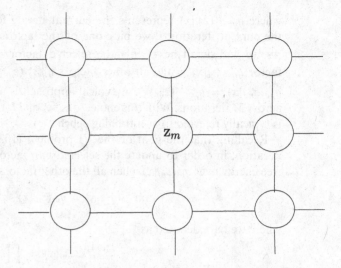

rv z_m and an edge between any two "correlated" rvs. Accordingly, in the Ising model, each rv z_m is modeled as being "correlated" with its four direct neighbors. The graph in Fig. 10.4 is an example of a probabilistic graphical model formalism known as **Markov random fields** (see Chapter 15).

As an application of the Ising model, consider the problem of **image denoising**. In it, the latent variable z_m represents the ground-truth, unknown value of the mth pixel in the image we wish to recover. The image is in black and white, with -1 corresponding to a white pixel and $+1$ to a black pixel. Each pixel is modeled as being correlated with the four immediately adjacent pixels in the four directions up, down, left, and right, as per the graphical model in Fig. 10.4. The observation $x_m \in \{-1, +1\}$ corresponds to a noisy version of the mth pixel z_m, where noise may randomly flip a pixel from white to black or vice versa.

Let us define the Ising model mathematically. To this end, we denote as $\mathcal{E}$ the set of edges in the graph. We recall that an edge (m, m') exists between any two adjacent latent rvs z_m and $z_{m'}$. In the **Ising model**, the joint distribution is given by the **exponential-family pmf**

$$p(x, z) \propto \exp\left(\eta_1 \sum_{(m, m') \in \mathcal{E}} z_m z_{m'} + \eta_2 \sum_{m=1}^{M} z_m x_m \right)$$

$$= \prod_{(m, m') \in \mathcal{E}} \exp(\eta_1 z_m z_{m'}) \cdot \prod_{m=1}^{M} \exp(\eta_2 z_m x_m), \tag{10.39}$$

with natural parameters η_1 and η_2 and sufficient statistics given by the sums $\sum_{(m, m') \in \mathcal{E}} z_m z_{m'}$ and $\sum_{m=1}^{M} z_m x_m$. According to this model, a large natural parameter $\eta_1 > 0$ yields a larger probability when two adjacent rvs z_m and $z_{m'}$, with $(m, m') \in \mathcal{E}$, are equal; and, similarly, a large $\eta_2 > 0$ favours configurations in which $z_m = x_m$, that is, with low "observation noise". Note that parameters η_1 and η_2 are assumed to be fixed, and they are not subject to training in this example.

A useful way to understand the role of the natural parameter η_2 is to observe that, according to this model, we have the conditional distribution $p(x|z) = \prod_{m=1}^{M} p(x_m|z_m)$ with

$$p(\mathsf{x}_m \neq \mathsf{z}_m | \mathsf{z}_m = z_m) = \sigma(-2\eta_2). \tag{10.40}$$

Therefore, the rv x_m is obtained from the corresponding latent rv z_m by flipping z_m with probability $\sigma(-2\eta_2)$, independently of all other coordinates $m' \neq m$.

Mean-Field VI. For the Ising model, it can be seen that the true posterior $p(z|x)$ cannot be written in the product form assumed by mean-field VI. Therefore, mean-field VI can only obtain an approximation of the true posterior.

At each iteration, mean-field VI updates the selected factor using (10.38) as

$$q_m^*(z_m | x) \propto \exp\left(\eta_1 z_m \sum_{m':(m,m')\in\mathcal{E}} \underbrace{\mathbb{E}_{\mathsf{z}_{m'} \sim q_{m'}^{(i-1)}(z_{m'}|x)}[\mathsf{z}_{m'}]}_{\mu_{m'}^{(i-1)}} + \eta_2 x_m z_m \right). \tag{10.41}$$

Note that evaluating (10.41) requires averaging over only the four neighboring variables $z_{m'}$ with $(m, m') \in \mathcal{E}$. Therefore, the complexity of computing (10.41) does not scale with the number of latent variables, M, unlike the general mean-field VI update (10.38).

Since the variables are bipolar, in (10.41) we have the mean

$$\mu_{m'}^{(i-1)} = q_{m'}^{(i-1)}(z_{m'} = 1 | x) - q_{m'}^{(i-1)}(z_{m'} = -1 | x) = 2q_{m'}^{(i-1)}(z_{m'} = 1 | x) - 1. \tag{10.42}$$

Imposing the normalizing condition $q_m^*(z_m = 1 | x) + q_m^*(z_m = -1 | x) = 1$, the normalization constant for (10.41) is given as

$$\exp\left(\eta_1 \sum_{m':(m,m')\in\mathcal{E}} \mu_{m'}^{(i-1)} + \eta_2 x_m \right) + \exp\left(-\eta_1 \sum_{m':(m,m')\in\mathcal{E}} \mu_{m'}^{(i-1)} - \eta_2 x_m \right), \tag{10.43}$$

which finally yields the updated (10.41) as

$$q_m^*(z_m = 1 | x) = \sigma\left(2\left(\eta_1 \sum_{m':(m,m')\in\mathcal{E}} \mu_{m'}^{(i-1)} + \eta_2 x_m \right) \right). \tag{10.44}$$

Example 10.5

Consider a 4×4 binary image z observed as noisy matrix x, where the joint distribution of x and z is given by the Ising model. In this small-dimensional example, it is easy to generate an image x distributed according to the model, as well as to compute the exact posterior $p(z|x)$ by enumeration of all possible images z. The KL divergence $\mathrm{KL}(p(z|x)||q(z|x))$ between the true posterior and the mean-field VI approximation obtained at the end of each epoch – with one epoch encompassing one iteration across all 16 latent variables – is shown in Fig. 10.5 for $\eta_1 = 0.15$ and various values of η_2.

As η_2 increases, the posterior distribution tends to concentrate around the observation x, since x is modeled as an accurate measurement of z. As a result, the mean-field approximation is more faithful to the real posterior, given that a product distribution can capture a pmf concentrated at a single value. Accordingly, with a large η_2, the KL divergence is small. For smaller values of η_2, however, the bias due to the mean-field assumption yields a larger floor on the achievable KL divergence.

Figure 10.5 KL divergence $\mathrm{KL}(p(z|x)||q(z|x))$ between the true posterior $p(z|x)$ and the mean-field VI approximation $q(z|x)$ obtained at the end of each epoch – with one epoch encompassing one iteration across all variables – under the Ising model (Example 10.5).

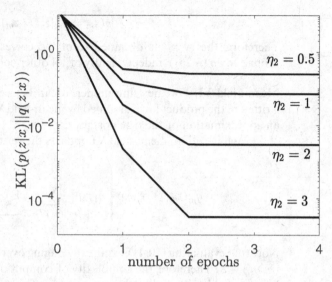

10.6.4 Generalizing Mean-Field VI

Mean-field VI is a **message-passing algorithm** in the sense that, at each iteration, the scheduled node z_m can be thought of as passing its updated variational posterior $q_m^*(z_m|x)$ to the adjacent nodes for further processing. This property of mean-field VI is consequence of the selected factorization, in which each factor includes only one of the variables z_m. Accordingly, mean-field VI can be generalized by considering factorizations over the latent variables in which the scopes of the factors, i.e., the set of latent rvs they depend on, are possibly overlapping. This yields more complex message-passing algorithms, and is a vast area of research (see Recommended Resources, Sec. 10.15).

10.7   Introducing Parametric Variational Inference

As studied in the preceding section, mean-field VI reduces the complexity of solving the VI problem by restricting the set of variational posteriors $\mathcal{Q}_x$ to distributions that factorize across all the latent rvs and by applying coordinate descent. In contrast, in **parametric VI**, the optimization domain $\mathcal{Q}_x$ consists of variational posteriors $q(z|x, \varphi)$ that are parametrized by a vector φ. The **variational parameter vector** φ is typically optimized via GD.

As a typical example of parametric variational posteriors, one could choose the set $\mathcal{Q}_x$ to include multivariate Gaussian distributions defined by a variational parameter vector φ including mean and covariance matrix. With a choice such as this for the variational posteriors, parametric VI can account for statistical dependencies among the latent rvs in the posterior distribution. This is unlike mean-field VI, which assumes the statistical independence of rvs under the variational posterior, while not imposing any constraint on the form of the factors.

VI Problem. With parametric VI, the VI problem (10.1) amounts to the minimization of the free energy with respect to the variational parameters φ, i.e.,

$$\min_{\varphi} \mathrm{F}(q(z|x,\varphi)\|p(x,z)). \tag{10.45}$$

As in the preceding section, given the focus on VI, the value x is to be considered as fixed. Therefore, the parameter φ obtained by addressing the VI problem (10.45) is a function of x. We will discuss later how the optimization can be "amortized" across all values x by tackling the amortized VI problem (10.3).

The optimization (10.45) is implicitly constrained within a given **feasibility domain**. For instance, if the variational parameter vector φ includes probabilities for binary variables as in Example 10.5, all entries should be in the interval $[0,1]$. As another example, in the mentioned case of a Gaussian variational posterior, the covariance matrix must be positive semi-definite.

Gradient Descent. A standard solution for parametric VI is to use **GD**. To implement GD, one needs to compute the **gradient** of the free energy, $\nabla_{\varphi}\mathrm{F}(q(z|x,\varphi)\|p(x,z))$. The complexity of this computation depends on the joint distribution $p(x,z)$ and on the variational posterior family $q(z|x,\varphi)$. In the next section, we review a general-purpose method known as black-box VI, which applies broadly to most common models $p(x,z)$ and variational posterior families $q(z|x,\varphi)$; while the following section describes a potentially more efficient solution that assumes the variational posterior $q(z|x,\varphi)$ to have a specific "reparametrizable" form.

10.8   Black-Box Variational Inference

Black-box VI is a general-purpose parametric VI scheme that relies on the use of a specific estimator for the gradient $\nabla_{\varphi}\mathrm{F}(q(z|x,\varphi)\|p(x,z))$ of the free energy that is known as the REINFORCE gradient (the text in all caps is commonly used to refer to this strategy, and we will follow this convention). In order to facilitate its introduction, in the next subsection we will describe the REINFORCE gradient in the context of a related general problem, namely stochastic optimization. We will then show how to apply the REINFORCE gradient to the VI problem, obtaining black-box VI.

10.8.1   Stochastic Optimization

In order to build the necessary background to tackle the VI problem, we first study the general **stochastic optimization problem**

$$\min_{\varphi} \mathrm{E}_{z\sim q(z|\varphi)}[g(z)] \tag{10.46}$$

for some scalar **cost function** $g(z)$ of a vector z. The term "stochastic" emphasizes that the minimization in (10.46) is with respect to the parameter vector φ of the averaging distribution $q(z|\varphi)$. Note that problem (10.46) does not optimize directly over the argument z of the function $g(z)$, but minimizes function $g(z)$ indirectly through the distribution of the rv $z \sim q(z|\varphi)$. In accordance with the terminology used for the VI problem, we will refer to distribution $q(z|\varphi)$ as the **variational distribution**, and to the vector φ as the **variational parameter vector**. As we will detail later, the VI problem has a form similar to (10.46).

We will be specifically interested in GD-based solutions, which require computing the gradient

$$\nabla_\varphi E_{z\sim q(z|\varphi)}[g(z)]. \tag{10.47}$$

The gradient (10.47) captures how local changes in the variational parameter φ affect the average of function $g(z)$ in the objective of the stochastic optimization problem (10.46).

Example 10.6

Consider the stochastic optimization problem (10.46) with function

$$g(z) = \frac{1}{2}(z_1^2 + 5z_2^2) \tag{10.48}$$

and variational distribution $q(z|\varphi) = \mathcal{N}(z|\varphi, \sigma^2 I_2)$, where $\varphi = [\varphi_1, \varphi_2]^T$ and $z = [z_1, z_2]^T$. Note that the variance σ^2 is assumed to be fixed, and not part of the variational parameter vector. For this example, we can explicitly compute the expectation in (10.46), yielding

$$\min_\varphi \left\{ E_{z\sim q(z|\varphi)}[g(z)] = \frac{1}{2}E_{z\sim q(z|\varphi)}[z_1^2 + 5z_2^2] = \frac{1}{2}(\varphi_1^2 + 5\varphi_2^2 + 6\sigma^2) \right\}. \tag{10.49}$$

From this expression, the gradient can be directly evaluated as

$$\nabla_\varphi E_{z\sim q(z|\varphi)}[g(z)] = \begin{bmatrix} \varphi_1 \\ 5\varphi_2 \end{bmatrix}, \tag{10.50}$$

and the optimal solution is given as $\varphi_1 = \varphi_2 = 0$. So, the optimal averaging distribution is $q(z|\varphi = 0_2) = \mathcal{N}(z|0_2, \sigma^2 I_2)$.

10.8.2 REINFORCE Gradient

In general, unlike the previous example, it is not possible to compute the objective $E_{z\sim q(z|\varphi)}[g(z)]$, as well as its gradient (10.47), as an explicit function of the variational parameter vector φ. As minimal concessions, in order to develop the REINFORCE gradient, we will assume that the following operations are feasible at a reasonable computational cost:

- **evaluating** the function $g(z)$ for any given z;
- **drawing** one or more samples from the variational distribution $q(z|\varphi)$ for any given variational parameter vector φ; and
- **evaluating** the gradient $\nabla_\varphi \log q(z|\varphi)$ of the variational distribution for any given z and φ.

Following standard terminology, in this chapter we will refer to the gradient $\nabla_\varphi \log q(z|\varphi)$ as the **score vector**.

Let us fix some variational parameter vector φ at which we wish to compute the gradient (10.47). As mentioned, we assume that we can draw S i.i.d. samples $z_s \sim q(z|\varphi)$ for $s = 1, \ldots, S$ from the variational distribution $q(z|\varphi)$. As a naive solution, one may consider first estimating the expectation $E_{z\sim q(z|\varphi)}[g(z)]$ of the cost function via the empirical average $1/S \sum_{s=1}^S g(z_s)$. However, this would lead to an impasse, since the empirical average of the cost function does not depend explicitly on φ, and hence the gradient over φ cannot be computed.

The issue at hand is that the gradient (10.47) depends on how much distribution $q(z|\varphi)$ – and, through it, the generated samples – varies with φ. Generating the samples first hides their dependence on the variational parameter φ. The REINFORCE gradient method addresses this problem.

(Before we continue, a note on terminology. As the name suggests, the REINFORCE gradient originates in the reinforcement-learning literature. The connection to reinforcement learning is as follows: In reinforcement learning, the agent takes an action $z \sim q(z|\varphi)$ by following a stochastic policy $q(z|\varphi)$. The goal of the agent is to optimize the policy parameters φ such that the average loss $\mathrm{E}_{z \sim q(z|\varphi)}[g(z)]$ is minimized. This corresponds exactly to the stochastic optimization problem (10.46).)

The derivation of the REINFORCE gradient starts with the following equality:

$$\nabla_\varphi \mathrm{E}_{z \sim q(z|\varphi)}[g(z)] = \mathrm{E}_{z \sim q(z|\varphi)}\left[\underbrace{g(z)}_{\text{cost function}} \cdot \underbrace{\nabla_\varphi \log q(z|\varphi)}_{\text{score vector}} \right]. \tag{10.51}$$

The proof of this equality follows directly from the trivial identity

$$\nabla_\varphi \log q(z|\varphi) = \frac{\nabla_\varphi q(z|\varphi)}{q(z|\varphi)} \tag{10.52}$$

and is detailed in the Appendix 10.C. In words, the formula (10.51) says that the desired gradient (10.47) is the average, over distribution $q(z|\varphi)$, of the score vector $\nabla_\varphi \log q(z|\varphi)$, weighted by the cost function $g(z)$.

To interpret (10.51), recall that the score vector $\nabla_\varphi \log q(z|\varphi)$ points in the direction that maximally increases the probability of value z in the space of parameters φ (see, e.g., Fig. 9.1). In (10.51), the direction corresponding to each value z is weighted by the cost function $g(z)$. Therefore, values z with a larger cost $g(z)$ contribute more significantly to the overall direction of the gradient (10.51). All in all, the REINFORCE gradient, when applied to updating the vector φ, "pushes" the resulting updated distribution $q(z|\varphi)$ towards values z that yield larger values $g(z)$.

Algorithmically, the key advantage of the expression (10.51) for the gradient is that it lends itself to a direct empirical estimate via the available S samples $z_s \sim q(z|\varphi)$ for $s = 1, \ldots, S$. Note that, unlike in the naive solution attempted above, here we estimate the gradient directly, rather than differentiating an estimate of the average cost function. All in all, the REINFORCE gradient is an estimate of the gradient (10.47) obtained as detailed in Algorithm 10.1.

As indicated in the table, the REINFORCE gradient introduces optional constants known as baselines. Each **baseline** c_s is a constant that may not depend on the corresponding sample z_s, but it may be a function of the other samples $z_{s'}$ with $s' \neq s$. As we will see in the next subsection, the baselines can be useful to reduce the variance of the REINFORCE estimator.

10.8.3 Some Properties of the REINFORCE Gradient

As per the requirements stated at the beginning of this section, the REINFORCE gradient estimate (10.54) only requires that function $g(z)$ and score vector $\nabla_\varphi \log q(z|\varphi)$ be computable for any value z. Note in particular that the method does not even require the derivatives of the cost function $g(z)$ – which is, perhaps, surprising in light of the fact that problem (10.46) aims to minimize the average cost function.

Algorithm 10.1: REINFORCE gradient: estimate of the gradient (10.47) for the stochastic optimization problem (10.46)

---

draw S i.i.d. samples $z_s \sim q(z|\varphi)$ for $s = 1, \ldots, S$
estimate the gradient as

$$\nabla_\varphi E_{z \sim q(z|\varphi)}[g(z)] \simeq \hat{\nabla}_\varphi E_{z \sim q(z|\varphi)}[g(z)] \tag{10.53}$$

with

$$\hat{\nabla}_\varphi E_{z \sim q(z|\varphi)}[g(z)] = \frac{1}{S} \sum_{s=1}^{S} (g(z_s) - c_s) \cdot \nabla_\varphi \log q(z_s|\varphi), \tag{10.54}$$

where, for each $s = 1, \ldots, S$, c_s is an a arbitrary constant, known as a baseline, which cannot depend on z_s but may depend on $\{z_{s'}\}_{s' \neq s}$

The REINFORCE estimator (10.54) can easily be proved to be **unbiased**. To see this, take for simplicity the case $S = 1$. Then, for any baseline c, the mean of the REINFORCE estimator equals the true gradient, i.e.,

$$E_{z \sim q(z|\varphi)}[(g(z) - c) \cdot \nabla_\varphi \log q(z|\varphi)] = \nabla_\varphi E_{z \sim q(z|\varphi)}[g(z)], \tag{10.55}$$

where the equality follows directly from (10.51) and from the zero-mean property of the score vector, i.e., $E_{z \sim q(z|\varphi)}[\nabla_\varphi \log q(z|\varphi)] = 0$ (see (9.63)).

As we will discuss later in this section, the **variance** of the REINFORCE gradient tends to be quite large. Intuitively this is because the REINFORCE gradient uses only zeroth-order information about the function $g(z)$, as it only relies on the value of the cost function $g(z)$ at some samples z rather than on the gradient of $g(z)$. That said, a proper choice for the baselines can reduce the variance.

Example 10.7

Continuing Example 10.6, with function $g(z)$ in (10.48) and variational distribution $q(z|\varphi) = \mathcal{N}(z|\varphi, \sigma^2 I_2)$, we can obtain an estimate of the gradient using the REINFORCE gradient with $S = 1$ by:

- drawing one sample $z \sim \mathcal{N}(z|\varphi, \sigma^2 I_2)$; and
- estimating the gradient as

$$\hat{\nabla}_\varphi E_{z \sim \mathcal{N}(z|\varphi, \sigma^2 I_2)}[g(z)] = (g(z) - c) \cdot \nabla_\varphi \log \mathcal{N}(z|\varphi, \sigma^2 I_2)$$
$$= (g(z) - c) \cdot \frac{z - \varphi}{\sigma^2}, \tag{10.56}$$

where we have used the general formula (9.40) for the gradient of the log-loss for an exponential family. So, the REINFORCE gradient points in the direction $z - \varphi$, starting from the variational parameter φ and pointing towards the randomly drawn sample z to an extent that is proportional to the difference $g(z) - c$.

10.8.4 REINFORCE Gradient Algorithm

We have seen in the previous two subsections that the REINFORCE gradient can be used to obtain an unbiased estimate of the gradient (10.47). We can now use this estimate in order to define a GD-based iterative algorithm for the solution of the stochastic optimization problem (10.46). The resulting procedure is known as the **REINFORCE gradient algorithm**, and is described in Algorithm 10.2.

Algorithm 10.2: REINFORCE gradient algorithm (with $S = 1$) for problem (10.46)

initialize $\varphi^{(1)}$ and $i = 1$

while *stopping criterion not satisfied* **do**

 draw one sample $z^{(i)} \sim q(z|\varphi^{(i)})$

 given a learning rate $\gamma^{(i)} > 0$ and a baseline $c^{(i)}$, obtain the next iterate as

$$\varphi^{(i+1)} \leftarrow \varphi^{(i)} + \gamma^{(i)}(c^{(i)} - g(z^{(i)})) \cdot \nabla_\varphi \log q(z^{(i)}|\varphi^{(i)}) \qquad (10.57)$$

 set $i \leftarrow i + 1$

end

return $\varphi^{(i)}$

The REINFORCE gradient algorithm applies a form of SGD in which stochasticity is not caused by the selection of data points from a training set, but instead by the sampling of latent rvs. The approach can be thought of as a **perturbation-based optimization scheme** in the following way. Given the current parameter vector $\varphi^{(i)}$, a random sample $z^{(i)} \sim q(z|\varphi^{(i)})$ is generated to "explore" values of the latent rv z to which the current variational distribution $q(z|\varphi^{(i)})$ assigns a sufficiently large value $q(z|\varphi^{(i)})$. A sample $z^{(i)}$ that yields a small cost $g(z^{(i)})$ – more precisely, a positive value of the difference $(c^{(i)} - g(z^{(i)}))$ – is "reinforced", in the sense that the updated probability distribution $q(z^{(i)}|\varphi^{(i+1)})$ is increased in comparison to $q(z^{(i)}|\varphi^{(i)})$ (as long as the learning rate $\gamma^{(i)}$ is sufficiently small).

Being a perturbation-based scheme, the REINFORCE gradient method generally suffers from the **curse of dimensionality**, since the exploration of the space of the latent variables z requires an exponentially larger number of samples as the dimension of rv z increases. This translates into a higher variance for the REINFORCE gradient as the dimension of the vector z grows larger.

10.8.5 Selecting the Baseline

We finally need to discuss how to select the baseline $c^{(i)}$ in the update (10.57) applied by the REINFORCE gradient algorithm. The parameter φ is updated in (10.57) so as to increase or decrease the variational distribution $q(z|\varphi)$ at the sampled value $z^{(i)}$ depending on the sign of the difference $(c^{(i)} - g(z^{(i)}))$. A positive sign leads to an increase in the distribution $q(z^{(i)}|\varphi)$ at the next iterate $\varphi^{(i+1)}$, while a negative sign decreases it. We would therefore like the difference $(c^{(i)} - g(z^{(i)}))$ to be positive when the cost $g(z^{(i)})$ is small, since the goal in problem (10.46) is to minimize the average value of the function $g(z)$. To quantify *how* small, a useful idea is to take the average $1/L \sum_{j=1}^{L} g(z^{(i-j)})$ of L prior iterates as a benchmark. This suggests setting the baseline as

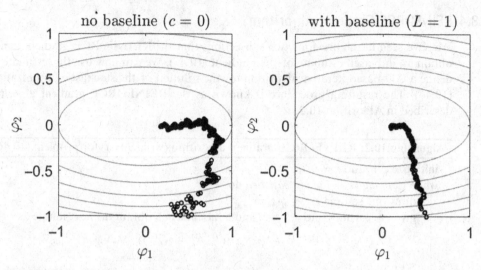

Figure 10.6 Evolution of the iterates for the REINFORCE algorithm with and without baseline.

$$c^{(i)} = \frac{1}{L} \sum_{j=1}^{L} g(z^{(i-j)}). \tag{10.58}$$

This way, the REINFORCE gradient increases the variational distribution $q(z|\varphi)$ at $z^{(i)}$ when the cost $g(z^{(i)})$ is smaller than the average (10.58) of the cost function $g(z)$ obtained at the L previous iterates.

Ultimately, since the REINFORCE gradient is unbiased, irrespective of the choice of baselines (Sec. 10.8.3), the baselines do not affect the mean of the REINFORCE gradient, but they provide a useful handle to control the **variance** of the gradient estimate. This is of critical important in practice, since the variance of the gradient estimate determines the "noisiness" of the trajectories of the REINFORCE gradient updates (10.57). In Appendix 10.D, we derive an optimized formula for the baseline $c^{(i)}$ that targets the minimization of the variance of the REINFORCE gradient.

Example 10.8

Again continuing Example 10.6 and 10.7, Fig. 10.6 shows the evolution of one realization of the (random) iterates produced by the REINFORCE gradient algorithm, both without baseline, i.e., with $c^{(i)} = 0$, and with the baseline (10.58), where $L = 1$. We set $\gamma^{(i)} = 0.002$, $S = 1$, $\sigma = 0.1$, and initialization $\varphi^{(1)} = [0.5, -1]^T$. The iterates are superimposed on the contour line of the cost function in (10.49). Note the reduced "noisiness" of the updates with the baseline.

10.8.6 Black-Box Variational Inference

Let us now return to the VI problem (10.1). In order to use the REINFORCE gradient algorithm, we need to relate it to the stochastic optimization problem (10.46). To this end, we fix, as we have done throughout this section, some value x. The key observation is that the free energy – the cost function in the VI problem (10.1) – can be written as

$$F(q(z|x,\varphi)||p(x,z)) = E_{z \sim q(z|x,\varphi)}\left[\underbrace{\log\left(\frac{q(z|x,\varphi)}{p(x,z)}\right)}_{g(z|\varphi)}\right]. \qquad (10.59)$$

Therefore, the problem of minimizing the free energy is in the form of an optimization over an averaging distribution, which is consistent with the stochastic optimization (10.46). But there is a caveat: The function being averaged depends on the parameter φ too, as highlighted by the notation

$$g(z|\varphi) = \log\left(\frac{q(z|x,\varphi)}{p(x,z)}\right). \qquad (10.60)$$

As shown in Appendix 10.C, despite this difference, the REINFORCE gradient formula (10.55) still applies, yielding the equality

$$\nabla_\varphi F(q(z|x,\varphi)||p(x,z)) = E_{z \sim q(z|x,\varphi)}\left[\left(\log\left(\frac{q(z|x,\varphi)}{p(x,z)}\right) - c\right) \cdot \nabla_\varphi \log q(z|x,\varphi)\right], \qquad (10.61)$$

for any constant c. We can therefore follow the REINFORCE approach to estimate the gradient of the free energy. The direct application of the REINFORCE gradient algorithm yields the procedure for parametric VI in Algorithm 10.3, which is known as **black-box VI**.

Algorithm 10.3: Black-box VI (with $S = 1$)

---

initialize $\varphi^{(1)}$ and $i = 1$
while *stopping criterion not satisfied* **do**

> draw one sample from the current variational distribution as $z^{(i)} \sim q(z|\varphi^{(i)})$
> given a learning rate $\gamma^{(i)} > 0$ and a baseline $c^{(i)}$, obtain the next iterate as
>
> $$\varphi^{(i+1)} \leftarrow \varphi^{(i)} + \gamma^{(i)}\left(c^{(i)} - \log\left(\frac{q(z^{(i)}|x,\varphi^{(i)})}{p(x,z^{(i)})}\right)\right) \cdot \nabla_\varphi \log q(z^{(i)}|x,\varphi^{(i)}) \qquad (10.62)$$
>
> set $i \leftarrow i + 1$

end
return $\varphi^{(i)}$

By construction (see Sec. 10.8.2), black-box VI is applicable to any choice of joint distribution $p(x,z)$ and variational distribution class $q(z|x,\varphi)$ such that the cost function (10.60) and the score vector $\nabla_\varphi \log q(z|x,\varphi)$ can be tractably computed, and samples from the variational posterior $q(z|x,\varphi)$ can be efficiently drawn. Furthermore, by (10.58), the baseline can be set as

$$c^{(i)} = \frac{1}{L}\sum_{j=1}^{L}\log\left(\frac{q(z^{(i-j)}|x,\varphi^{(i-j)})}{p(x,z^{(i-j)})}\right), \qquad (10.63)$$

although this is not the only option (see Appendix 10.D).

Example 10.9

Let us apply black-box VI to the beta–exponential model (10.31) by assuming a beta variational posterior $q(z|x,\varphi) = \text{Beta}(z|\varphi_1,\varphi_2)$ with $\varphi = [\varphi_1,\varphi_2]^T$. Note that the variational posterior implicitly depends on the given fixed value x. Reading off sufficient statistics $s(z)$ and mean parameters μ from relevant tables concerning exponential-family distributions, the general formula (9.40) for the score vector for the exponential family yields

$$\nabla_\varphi \log \text{Beta}(z|\varphi_1,\varphi_2) = s(z) - \mu$$

$$= \underbrace{\begin{bmatrix} \log z \\ \log(1-z) \end{bmatrix}}_{s(z)} - \underbrace{\begin{bmatrix} \psi(\varphi_1) - \psi(\varphi_1 + \varphi_2) \\ \psi(\varphi_2) - \psi(\varphi_1 + \varphi_2) \end{bmatrix}}_{\mu} \qquad (10.64)$$

with digamma function $\psi(\cdot)$. Using this expression, we are ready to implement black-box VI.

In Fig. 10.7, the top row shows the variational posteriors obtained via black-box VI (solid lines), as well as the true posterior distribution (dashed lines), for $x = 1$ and two different choices for the parameters a and b of the prior beta distribution. We have used $S = 30$ samples for the estimate of the gradient, $c^{(i)} = 0$ (no baseline), and 2000 iterations of black-box VI. The iterates for one specific realization are shown in the bottom row in the space of variational parameters φ. We have

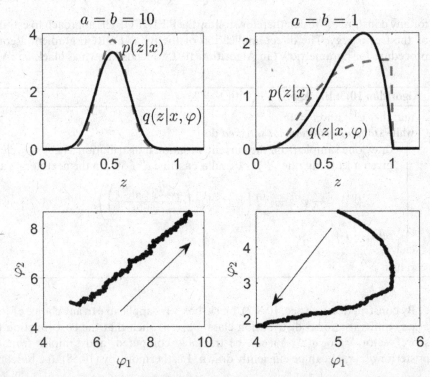

Figure 10.7 Top row: true posterior (dashed line) and variational approximation within the class of beta variational distributions obtained via black-box VI (solid line). Bottom row: corresponding trajectory of the variational parameters φ for one realization (bottom figures) given different values of the parameters a and b of the beta prior distribution. The arrows point in the direction of increasing iterations for the black-box VI algorithm (Algorithm 10.3).

set $\gamma^{(i)} = 0.03$ for $a = b = 1$, and $\gamma^{(i)} = 0.1$ for $a = b = 10$. The quality of the approximation obtained by black-box VI is seen to depend on how well a beta distribution – the selected variational posterior distribution – can describe the true posterior.

Black-box VI is "*black box*" because it does not leverage information about the structure of the variational posterior $q(z|x, \varphi)$ and of the joint distribution $p(x, z)$, except for the mentioned assumptions about computability of their log-ratio (10.60), differentiability of the log-loss $-\log q(z|x, \varphi)$ with respect to the variational parameters φ, and sampling from $q(z|x, \varphi)$. In particular, an interesting aspect of black-box VI is that its updates (10.62) do not depend on the gradient of the log-loss $-\log p(x, z)$ of the joint distribution. While this makes black-box particularly easy to implement on a broad range of problems, it may cause noisy updates with excessive variance, particularly in large-dimensional problems and even with optimized baselines.

10.9 Reparametrization-Based Variational Inference

As discussed in the preceding section, black-box VI is a general-purpose scheme for the approximate solution of the VI problem (10.1) that applies broadly to variational posteriors that can be sampled from and have differentiable log-loss. In this section, we introduce an alternative iterative optimization strategy for the VI problem that can potentially yield less "noisy" trajectories in the variational parameter space, but is only applicable to a special class of variational posteriors. Specifically, **reparametrization-based VI** assumes that the variational posterior is "reparametrizable" in a sense that will be detailed in the next subsection. In a manner similar to black-box VI, we will introduce the method by first focusing on the stochastic optimization problem (10.46), and then returning to the VI problem.

10.9.1 Reparametrizability of a Parametric Variational Posterior

To introduce the idea of reparametrization, let us start again by considering the stochastic optimization problem (10.46), for which we would like to find an effective way to estimate the gradient (10.47). The reparametrization-based gradient estimate is applicable when the variational distribution $q(z|\varphi)$ is reparametrizable.

A variational distribution $q(z|\varphi)$ is **reparametrizable** if there exist (*i*) an **auxiliary distribution** $q(e)$ not dependent on φ, and (*ii*) a function $r(e|\varphi)$ that is **differentiable** in φ for every fixed value of e, such that samples for rv $z \sim q(z|\varphi)$ can be generated via the following two steps:

1. **draw** a sample from the auxiliary distribution $q(e)$ as $e \sim q(e)$; and
2. **compute**

$$z = r(e|\varphi). \tag{10.65}$$

Given the use of an auxiliary source of randomness, reparametrization is also known as **noise outsourcing**.

A typical example, but not the only one, of a reparametrizable variational distribution is the **Gaussian distribution**

$$q(z|\varphi) = \mathcal{N}(z|v, \mathrm{Diag}(\sigma^2)), \tag{10.66}$$

for which the $2M \times 1$ variational parameter vector $\varphi = [\nu^T, \sigma^T]^T$ contains the $M \times 1$ mean vector ν and the $M \times 1$ vector of standard deviations σ. The notation σ^2 represents the $M \times 1$ vector of variances obtained via the element-wise squaring of the standard deviation vector σ. This distribution is reparametrizable, since we can generate a sample $z \sim \mathcal{N}(z|\nu, \text{Diag}(\sigma^2))$ as follows:

1. **draw** a sample from the auxiliary distribution $\mathcal{N}(e|0_M, I_M)$ as

$$e \sim q(e) = \mathcal{N}(e|0_M, I_M); \text{ and} \tag{10.67}$$

2. **compute**

$$z = r(e|\varphi) = \nu + \sigma \odot e = \nu + \text{Diag}(\sigma)e, \tag{10.68}$$

where $\odot$ denotes the element-wise product.

The two steps (10.67) and (10.68) define a valid reparametrization because (*i*) the auxiliary distribution $q(e)$ does not depend on the variational parameter vector φ, and (*ii*) the function $r(e|\varphi)$ is linear, and hence differentiable, in φ.

The Gaussian distribution (10.66) is by far the most common reparametrizable distribution, but there are other examples. One is the **Weibull distribution**, which belongs to the exponential family and is useful for modeling non-negative continuous quantities. Another is given by the **Gumbel–softmax distribution**, which can be used to approximately reparametrize categorical rvs. See Recommended Resources, Sec. 10.15, for further details.

10.9.2 Reparametrization Trick

Assuming that the variational posterior is reparametrizable, the stochastic optimization problem (10.46) can be expressed as the minimization

$$\min_{\varphi} \text{E}_{e \sim q(e)}[g(\underbrace{r(e|\varphi)}_{z})]. \tag{10.69}$$

The key advantage of this formulation is that, unlike the original problem (10.46), the averaging distribution $q(e)$ does not depend on the parameter vector φ. Furthermore, by the chain rule of differentiation, we can directly compute the gradient of the function being averaged in (10.69) as

$$\nabla_{\varphi} g(r(e|\varphi)) = \nabla_{\varphi} r(e|\varphi) \cdot \nabla_z g(z)|_{z=r(e|\varphi)}, \tag{10.70}$$

where $\nabla_{\varphi} r(e|\varphi)$ is the $Q \times M$ Jacobian of function $r(e|\varphi)$ for fixed e, with $Q \times 1$ being the dimension of φ and $M \times 1$ being the dimension of z (see Appendix 5.F for the definition of the Jacobian).

Therefore, we can obtain an unbiased estimate of the gradient by following the procedure described in Algorithm 10.4, which is known as the **reparametrization trick** or **reparametrization-based gradient estimate**.

Unlike the REINFORCE gradient, the reparametrization-based gradient (10.71) leverages the reparametrizable structure of the variational distribution, and depends on the gradient $\nabla_z g(z)$ of the function to be optimized. For these reasons, intuitively, it tends to have a lower variance than the REINFORCE gradient.

Algorithm 10.4: Reparametrization-based gradient: estimate of the gradient (10.47) for problem (10.46)

---

draw S i.i.d. samples from the auxiliary distribution $q(e)$ as $e_s \sim q(e)$ for $s = 1, \ldots, S$

estimate the gradient as $\nabla_\varphi \mathrm{E}_{z \sim q(z|\varphi)}[g(z)] \simeq \hat{\nabla}_\varphi \mathrm{E}_{z \sim q(z|\varphi)}[g(z)]$ with

$$\hat{\nabla}_\varphi \mathrm{E}_{z \sim q(z|\varphi)}[g(z)] = \frac{1}{S} \sum_{s=1}^{S} \nabla_\varphi r(e_s|\varphi) \cdot \nabla_z g(z_s)|_{z_s = r(e_s|\varphi)}. \tag{10.71}$$

10.9.3 Gaussian Reparametrization Trick

Let us now specialize the reparametrization-based gradient in Algorithm 10.4 to the Gaussian variational distribution (10.66). To this end, it is convenient to parametrize the vector of standard deviations as

$$\sigma = \exp(\varrho), \tag{10.72}$$

where the $\exp(\cdot)$ function is applied element-wise to the $M \times 1$ vector ϱ. This is done in order to facilitate optimization, since, unlike σ, vector ϱ is not constrained to be positive. Note, in fact, that GD updates do not guarantee the satisfaction of constraints such as positivity, as they are designed for unconstrained problems. With this choice, the variational parameter vector is given as the $2M \times 1$ vector $\varphi = [\nu^T, \varrho^T]^T$ (and hence we have $Q = 2M$).

In order to evaluate the gradient estimate (10.71), we need to compute the Jacobian of function $r(e|\varphi)$ in (10.68). Given the linearity of function $r(e|\varphi)$, this readily yields the $2M \times M$ matrix

$$\nabla_\varphi r(e|\varphi) = \begin{bmatrix} \nabla_\nu r(e|\varphi) \\ \nabla_\varrho r(e|\varphi) \end{bmatrix}, \tag{10.73}$$

with

$$\begin{aligned} \nabla_\nu r(e|\varphi) &= I_M \\ \nabla_\varrho r(e|\varphi) &= \mathrm{Diag}(\exp(\varrho) \odot e). \end{aligned} \tag{10.74}$$

The resulting reparametrization-based estimate $\hat{\nabla}_\varphi \mathrm{E}_{z \sim \mathcal{N}(z|\nu, \mathrm{Diag}(\exp(2\varrho)))}[g(z)]$ in (10.71) is summarized in the following table, where it is stated for simplicity of notation for $S = 1$ samples, as the extension to any number of samples is straightforward.

10.9.4 Reparametrization-Based Gradient Algorithm

Using the reparametrization-based gradient in Algorithm 10.4, we can now derive a strategy to address the stochastic optimization problem (10.46) via GD. The approach is known as the **reparametrization-based gradient algorithm**, and is detailed in Algorithm 10.5. The algorithm can be readily specialized to the case of a Gaussian variational distribution via Algorithm 10.6.

Algorithm 10.5: Reparametrization-based gradient algorithm (with $S = 1$) for problem (10.46)

---

initialize $\varphi^{(1)}$ and $i = 1$
while *stopping criterion not satisfied* **do**
  draw one sample from the auxiliary distribution $q(e)$ as $e^{(i)} \sim q(e)$
  given a learning rate $\gamma^{(i)} > 0$, obtain the next iterate as

$$\varphi^{(i+1)} \leftarrow \varphi^{(i)} - \gamma^{(i)} \nabla_\varphi r(e^{(i)}|\varphi) \cdot \nabla_z g(z)|_{z=r(e^{(i)}|\varphi)} \tag{10.75}$$

  set $i \leftarrow i + 1$
end
return $\varphi^{(i)}$

Algorithm 10.6: Reparametrization-based gradient with the Gaussian variational distribution $q(z|\varphi) = \mathcal{N}(z|v, \mathrm{Diag}(\exp(2\varrho)))$: estimate of the gradient (10.47) for problem (10.46)

---

draw sample $e \sim \mathcal{N}(e|0_M, I_M)$
compute $z = v + \exp(\varrho) \odot e$
  estimate the gradient as

$$\hat{\nabla}_\varphi E_{z \sim q(z|\varphi)}[g(z)] = \left[\begin{array}{c} \hat{\nabla}_v E_{z \sim \mathcal{N}(z|v, \mathrm{Diag}(\exp(2\rho)))}[g(z)] \\ \hat{\nabla}_\varrho E_{z \sim \mathcal{N}(z|v, \mathrm{Diag}(\exp(2\rho)))}[g(z)] \end{array} \right], \tag{10.76}$$

with

$$\hat{\nabla}_v E_{z \sim \mathcal{N}(z|v, \mathrm{Diag}(\exp(2\rho)))}[g(z)] = \nabla_z g(z) \tag{10.77}$$

$$\hat{\nabla}_\varrho E_{z \sim \mathcal{N}(z|v, \mathrm{Diag}(\exp(2\rho)))}[g(z)] = \mathrm{Diag}(\exp(\varrho) \odot e) \nabla_z g(z)$$

$$= \exp(\varrho) \odot e \odot \nabla_z g(z) \tag{10.78}$$

Example 10.10

Consider again the stochastic optimization problem stated in Example 10.6, and assume a Gaussian variational distribution $q(z|\varphi) = \mathcal{N}(z|v, \mathrm{Diag}(\exp(2\varrho)))$ for the reparametrization-based gradient algorithm (with $\gamma^{(i)} = 0.05$). For reference, we also implement the REINFORCE gradient algorithm (with baseline (10.58), $L = 1$, $\sigma = 1$, and $\gamma^{(i)} = 0.3$). Figure 10.8 shows the evolution of the loss $g(v)$ evaluated at the mean of the distribution $q(z|\varphi)$ for both schemes over 100 iterations, for 10 realizations of the stochastic gradient updates. The reparametrization gradient is seen to have a much reduced variance, as is clear from the faster, and less noisy, convergence displayed in the figure.

10.9.5 Reparametrization-Based Variational Inference

In this subsection, we finally detail the application of the reparametrization-based gradient algorithm to the VI problem (10.1). As we have observed, the free energy can be written as

Figure 10.8 Cost function $g(\nu)$ evaluated at the mean of the distribution $q(z|\varphi)$ for both the REINFORCE gradient algorithm (with baseline, $\sigma = 1$ and $\gamma^{(i)} = 0.3$) and the reparametrization-based gradient algorithm (with $\gamma^{(i)} = 0.05$) over 100 iterations. Ten independent realizations of the algorithms are displayed.

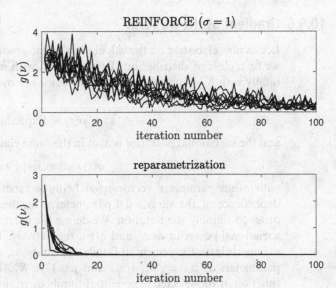

the expectation in (10.59), where, unlike (10.46), the function $g(z|\varphi)$ being averaged depends on the variational parameters φ. Therefore, a direct application of the reparametrization-based gradient algorithm is not possible, and some novel solutions are required. Throughout this subsection, we will assume that the variational posterior $q(z|x, \varphi)$, for the given fixed x is reparametrizable.

The approach we will take to address the dependence of the cost function $g(z|\varphi)$ in (10.60) on the variational parameters φ relies on rewriting the free energy in the original VI problem (10.1) as in (3.99), which we report here for convenience as

$$
\mathrm{F}(q(z|x, \varphi)||p(x, z)) = \underbrace{\mathrm{E}_{z \sim q(z|x, \varphi)}\left[-\log p(x|z)\right]}_{\text{average log-loss of the soft predictor } p(x|z)}
$$
$$
+ \underbrace{\mathrm{KL}(q(z|x, \varphi)||p(z))}_{\text{KL divergence between variational posterior and prior}}. \tag{10.79}
$$

This decomposition is useful because:

- the gradient over φ of the first term can be estimated using the reparametrization trick, since the function being averaged, $\log p(x|z)$, does not depend on the variational parameter φ; and

- if we choose variational posterior $q(z|x, \varphi)$ and prior $p(z)$ to be from the same class in the exponential family, the KL divergence can be computed in closed form (see Sec. 9.6), and so can its gradient over φ.

For instance, as we will show, if we choose a Gaussian prior and a Gaussian variational posterior, we can estimate the gradient of the first term using the reparametrization trick, while the gradient of the second term can be computed in closed form.

10.9.6 Gradient of the KL Term

Let us now elaborate on the calculation of the gradient of the KL term in (10.79). To this end, we fix a class of distributions $\text{ExpFam}(z|\mu)$ with **minimal** parametrization in the exponential family with $K \times 1$ mean vector μ. The prior is a member of this class with a given fixed mean parameter vector μ_p, i.e.,

$$p(z) = \text{ExpFam}(z|\mu_p), \tag{10.80}$$

and the variational posterior is also in the same class,

$$q(z|x,\varphi) = \text{ExpFam}(z|\mu(\varphi)), \tag{10.81}$$

with mean parameter vector $\mu(\varphi)$ being a function of the variational parameter φ. The dependence of the variational parameter φ on the fixed input x is not indicated explicitly in order to simplify the notation. We define the corresponding natural parameters for prior and variational posterior as η_p and $\eta(\varphi)$, respectively. These are unique by the minimality of the assumed class of exponential-family distributions; and they are related to the respective mean parameters as $\mu_p = \nabla_{\eta_p} A(\eta_p)$ and $\mu(\varphi) = \nabla_\eta A(\eta)|_{\eta=\eta(\varphi)}$, where $A(\eta)$ is the log-partition function of the selected exponential-family distribution (see Sec. 9.2.3).

Using the analytical form of the KL divergence studied in Sec. 9.6, the gradient of the KL divergence is obtained as (see Appendix 10.E)

$$\nabla_\varphi \text{KL}(\text{ExpFam}(z|\mu(\varphi))||\text{ExpFam}(z|\mu_p)) = \nabla_\varphi \mu(\varphi) \cdot (\eta(\varphi) - \eta_p), \tag{10.82}$$

where $\nabla_\varphi \mu(\varphi)$ is the $Q \times K$ Jacobian of the $K \times 1$ mean vector function $\mu(\varphi)$ with respect to the $Q \times 1$ variational parameter φ. As special cases of (10.82), if the variational parameters φ coincide with the mean parameters μ of the variational posterior distribution, i.e., if $\mu(\varphi) = \varphi$, then the Jacobian equals the identity matrix, $\nabla_\varphi \mu(\varphi) = I_K$ ($K = Q$); while, if the variational parameters φ coincide with the natural parameters η of the variational posterior distribution, i.e., if $\mu(\varphi) = \nabla_\eta A(\eta)|_{\eta=\varphi}$, the Jacobian $\nabla_\varphi \mu(\varphi)$ equals the Fisher information matrix $\text{FIM}(\eta)$ (see Sec. 9.7).

10.9.7 Reparametrization-Based VI with Gaussian Prior and Variational Posterior

In the previous subsections, we have provided the necessary background to derive reparametrization-based VI algorithms for the case in which prior and variational posterior are selected within the same class of exponential-family distributions, under the assumption that the latter is reparametrizable. In this subsection, we elaborate on the most common special case in which prior and variational posterior are Gaussian, to derive the standard implementation of reparametrization-based VI.

To proceed, we choose as prior the standard Gaussian

$$p(z) = \mathcal{N}(z|0_M, I_M), \tag{10.83}$$

while the variational posterior is given as in (10.66) with (10.72), i.e., we set

$$q(z|x,\varphi) = \mathcal{N}(z|v, \text{Diag}(\exp(2\varrho))) \tag{10.84}$$

with $\varphi = [v^T, \varrho^T]^T$, where x is the given fixed value. To specify the reparametrization-based VI procedure, we need to compute the gradient of the free energy (10.79). As discussed, for the first term we can directly apply the reparametrization-based gradient estimate in Algorithm 10.6, while for the second term we can leverage the general formula (10.82) specialized to Gaussian prior and variational posterior. This is detailed next.

Gradient of the KL between Gaussians. For a Gaussian distribution $\mathcal{N}(\nu, \Sigma)$ with mean vector ν and covariance matrix Σ, from Table 9.1 the mean parameters are given as $\mu = (\nu, \Sigma + \nu\nu^T)$, that is, as the collection of vector ν and matrix $\Sigma + \nu\nu^T$. Using the general formula (10.82) along with the mappings from mean to natural parameters in Table 9.1, we have the gradients

$$\nabla_\nu \mathrm{KL}(\mathcal{N}(\nu, \Sigma) || \mathcal{N}(\nu_p, \Sigma_p)) = \Sigma_p^{-1}(\nu - \nu_p) \tag{10.85}$$

and

$$\nabla_\Sigma \mathrm{KL}(\mathcal{N}(\nu, \Sigma) || \mathcal{N}(\nu_p, \Sigma_p)) = \frac{1}{2}\left(\Sigma_p^{-1} - \Sigma^{-1}\right), \tag{10.86}$$

where we have assumed that the covariance matrices Σ_p and Σ are invertible (which follows from the assumed minimality of the distributions).

Specializing the gradients (10.85) and (10.86) to the choices (10.83) and (10.84), where $\Sigma_p = I_M$ and $\Sigma = \mathrm{Diag}(\sigma^2)$ for a vector of variances σ^2 ($\sigma = \exp(\varrho)$ in (10.84)), we have

$$\nabla_\nu \mathrm{KL}(\mathcal{N}(\nu, \mathrm{Diag}(\sigma^2)) || \mathcal{N}(0_M, I_M)) = \nu \tag{10.87}$$

and

$$\nabla_\sigma \mathrm{KL}(\mathcal{N}(\nu, \mathrm{Diag}(\sigma^2)) || \mathcal{N}(0_M, I_M)) = \sigma - \sigma^{-1}, \tag{10.88}$$

where vector σ^{-1} is obtained by inverting element-wise the entries of vector σ. (Note the change of variables from Σ to σ in (10.88).)

Algorithm 10.7: Reparametrization-based VI with Gaussian prior and Gaussian variational posterior (with $S = 1$)

---

initialize $\varphi^{(1)} = [(\nu^{(1)})^T, (\varrho^{(1)})^T]^T$ and $i = 1$
while *stopping criterion not satisfied* **do**

    draw a sample $e^{(i)} \sim \mathcal{N}(0_M, I_M)$
    compute $z^{(i)} = \nu^{(i)} + \exp(\varrho^{(i)}) \odot e^{(i)}$
    estimate the $2M \times 1$ gradient vector as

$$\hat{\nabla}_\varphi^{(i)} = \begin{bmatrix} \hat{\nabla}_\nu^{(i)} \\ \hat{\nabla}_\varrho^{(i)} \end{bmatrix} \tag{10.89}$$

    with

$$\hat{\nabla}_\nu^{(i)} = \nabla_z(-\log p(x|z^{(i)})) + \nu^{(i)}$$
$$\hat{\nabla}_\varrho^{(i)} = \exp(\varrho^{(i)}) \odot e^{(i)} \odot \nabla_z(-\log p(x|z^{(i)}))$$
$$\qquad + \exp(\varrho^{(i)}) \odot (\exp(\varrho^{(i)}) - \exp(-\varrho^{(i)})) \tag{10.90}$$

    given a learning rate $\gamma^{(i)} > 0$, obtain the next iterate as

$$\varphi^{(i+1)} \leftarrow \varphi^{(i)} - \gamma^{(i)} \hat{\nabla}_\varphi^{(i)} \tag{10.91}$$

    set $i \leftarrow i + 1$
end
return $\varphi^{(i)}$

Reparametrization-Based VI with Gaussian Prior and Variational Posterior. We are now ready to state the overall algorithm that tackles the VI problem via the reparametrization-based gradient estimate under Gaussian prior and variational posterior. This is summarized in Algorithm 10.7. To recap, the gradient of the free energy (10.79) is obtained by *estimating* the gradient of the first term via the reparametrization trick (10.71) with Jacobian matrix (10.74), while the gradient of the KL divergence is calculated *exactly* via (10.87) and (10.88) with $\sigma = \exp(\varrho)$.

Example 10.11

We now consider an example for which the posterior distribution can be computed exactly in order to validate reparametrization-based VI. We adopt the joint distribution given by $z \sim \mathcal{N}(0, 1)$ and $(x|z = z) \sim \mathcal{N}(z, \beta^{-1})$ for a fixed precision $\beta = 0.5$. From Sec. 3.4, we know that the posterior distribution is given as

$$p(z|x) = \mathcal{N}\left(z \left| \frac{\beta x}{1 + \beta}, \frac{1}{1 + \beta}\right.\right). \tag{10.92}$$

We apply reparametrization-based VI via Algorithm 10.7, where we fix $x = 1$. In the top part of Fig. 10.9, we plot the evolution of the variational posterior $q(z|x = 1, \varphi^{(i)})$ for different values of the iteration index i (solid lines), along with the target posterior $p(z|x)$ (dashed line). The hyperparameters in Algorithm 10.7 are set as $\varphi^{(1)} = [-2, 0.1]^T$ and $\gamma^{(i)} = 0.01$. The dashed arrow points in the direction of growing values of the iteration index i, showing an increasingly accurate estimate of the correct prior. In the bottom part of the same figure, this result is corroborated by illustrating the KL divergence (cf. (3.48))

$$KL(q(z|x, \varphi^{(i)})||p(z|x)) = \frac{1}{2}\left[(1 + \beta)(\sigma^{(i)})^2 + (1 + \beta)\left(\frac{\beta x}{1 + \beta} - \nu^{(i)}\right)^2 - 1\right.$$
$$\left. \log\left((1 + \beta)(\sigma^{(i)})^2\right)\right] \tag{10.93}$$

along the iterations i.

10.9.8 Broadening the Applicability of Reparametrization-Based VI

In this section, we have assumed that the gradient of the KL divergence term in (10.79) can be efficiently evaluated. When this is not the case, one can still use reparametrization-based VI, although the resulting gradient may have a larger variance. In fact, it can be shown that, despite the dependence of function $g(z|\varphi)$ in (10.60) on the variational parameter φ, one can obtain an unbiased estimate of the gradient of the free energy by applying the reparametrization trick to $g(z|\varphi)$ while fixing the parameter vector φ in $g(z|\varphi)$ to its current iterate (rather than differentiating over it). Such estimates can also be extended to multi-sample versions of the free energy (see Appendix 7.D). We refer to Recommended Resources, Sec. 10.15, for further details.

Figure 10.9 Top: evolution of the variational posterior $q(z|x, \varphi^{(i)})$ for different values of the iteration index i (solid lines), along with the target posterior $p(z|x)$ (dashed line), with the dashed arrow pointing in the direction of growing values of the iteration index i ($x = 1$); bottom: KL divergence $KL(q(z|x, \varphi^{(i)})||p(z|x))$ in (10.93) as a function of the iteration index i ($x = 1$).

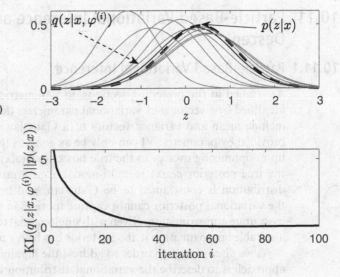

10.10 Combining Factorization and Parametrization for Variational Inference

In the previous sections, we have seen two approaches to formulate and solve the VI problem (10.1). Mean-field VI restricts the space of variational posteriors to fully factorized distributions, and carries out optimization via coordinate descent. In contrast, parametrization-based VI constrains the set of variational posteriors within a parametric family, and optimization is done by SGD. The purpose of this short section is to point out that these two approaches can be combined.

A combination of the two approaches would restrict the space $\mathcal{Q}_x$ of variational posteriors to distributions that factorize as

$$q(z|x, \varphi) = \prod_{m'=1}^{M'} q_{m'}(z_{m'}|x, \varphi), \tag{10.94}$$

where $z_{m'}$ for $m' = 1, \ldots, M'$ are disjoint subsets of the M latent rvs in vector z such that the concatenation of vectors $z_{m'}$ recovers vector z. In this factorization, each distribution $q_m(z_m|x, \varphi)$ is a parametric function of the variational parameters φ. More general factorizations over the latent variables in which the scopes of the factors are possibly overlapping are also possible by means of energy-based models (see Sec. 15.2). Minimization of the free energy can be carried out by SGD, or by coordinate-wise SGD if each factor $q_m(z_m|x, \varphi)$ depends on different subsets of elements of the variational vector φ_m. The derivation of the resulting VI schemes follows from a direct application of the tools derived in the previous two sections.

10.11   Particle-Based Variational Inference and Stein Variational Gradient Descent

10.11.1 Particle-Based Variational Inference

As studied in the preceding sections, in parametric VI, the variational posterior $q(z|x, \varphi)$ is identified by a vector φ of variational parameters (for a fixed x). This vector may, for instance, include mean and variance vectors of a Gaussian variational posterior. The approximation provided by parametric VI can only be as good as the "closest" variational posterior $q(z|x, \varphi)$, upon optimizing over φ, to the true posterior $p(z|x)$. For instance, consider the case in which the true posterior $p(z|x)$ is multi-modal, e.g., a mixture of Gaussians, while the variational distribution is constrained to be Gaussian, and hence to have a single mode. In this case, the variational posterior cannot capture more than one mode of the posterior, resulting in an inaccurate approximation. (Even with single-modal true posteriors, a Gaussian may not provide a suitable approximation if the posterior has heavy tails.)

As we discuss next, in order to address the highlighted bias of parametric VI, an alternative approach is to describe the variational distribution with a number of **particles** in the space of the latent variables through a **kernel density estimator (KDE)** (see Sec. 7.3). In this way, the variational posterior can assume a flexible shape, which is only constrained by the number of particles and by the choice of the kernel function.

KDE. In Sec. 7.3, we have seen that KDE approximates a distribution $q(z)$ with a number S of particles $\{(z_s)_{s=1}^S\}$ as

$$q^{KDE}(z) = \frac{1}{S} \sum_{s=1}^{S} \kappa_h(z - z_s), \tag{10.95}$$

where $\kappa_h(z)$ is a **non-negative kernel function** with "bandwidth" h, such as the Gaussian kernel $\kappa_h(z) = \mathcal{N}(z|0, h)$.

Particle-Based Variational Distributions. The KDE-based KDE approximation (10.95) suggests the idea of defining the set $\mathcal{Q}_x$ of variational posteriors to include all distributions that are identified by a set $\varphi = \{(z_s)_{s=1}^S\}$ of particles via the KDE

$$q(z|x, \varphi) = \frac{1}{S} \sum_{s=1}^{S} \kappa_h(z - z_s). \tag{10.96}$$

Note that, in a manner similar to parametric VI, we do not make explicit the dependence of the particles on the fixed input x.

We will refer to techniques that tackle the VI problem (10.1) within the class of variational distributions of the form (10.96) as **particle-based VI**. Note that, strictly speaking, particle-based VI methods are also "parametric" in the sense that they are defined by a set of parameters, namely the particles. However, it is common to reserve the term "parametric VI" to the types of methods presented in the previous sections of this chapter, in which the variational parameter vector φ determines global properties of the variational posterior.

Unlike the parametric VI approach studied so far, by (10.96) particle-based VI does not impose a specific structure on the variational posteriors apart from the smoothness properties implied by the choice of a particular kernel, and apart from the number of particles, S. Therefore, the particle-based parametrization (10.96) of the variational posterior can more flexibly represent any posterior distribution as long as the number of particles is large enough and the posterior distribution is sufficiently smooth.

On the flip side, in a manner similar to KDE, this approach suffers from the **curse of dimensionality**: As the dimensionality of the latent space increases, the number of required particles grows, at least in principle, exponentially. That said, in the case of sufficiently low-dimensional latent vectors z, the flexibility of the particle-based model can yield more accurate approximations of the posterior that are not limited by the bias of parametric variational families.

Particle-based VI techniques can be interpreted as "**quantizers**" of the true posterior distribution: Like actual quantizers such as K-means (see Sec. 7.9), they optimize over a set of **prototype vectors** – the particles – with the goal of providing a useful representation of the input – the joint distribution with a fixed value x.

10.11.2 Stein Variational Gradient Descent

In this subsection, we review a specific, popular, particle-based VI scheme known as **Stein variational gradient descent (SVGD)**. SVGD is an iterative method that updates the set of particles φ in the direction that locally minimizes the free energy cost function of the VI problem (10.1). How should this update be designed?

If we were to simply update all particles by following the gradient of the free energy, all particles would end up converging to some local maximum of the posterior distribution – ideally to the MAP solution (10.27). This is not desirable, however, since we would like the particles to offer an estimate of the posterior distribution through the KDE (10.95), which requires the particles to be sufficiently **diverse**. To this end, the updates must include some *repulsive "force"* that preserves the diversity of the particles. Accordingly, as we discuss next, the definition of SVGD involves the selection of a positive semi-definite kernel function to quantify the similarity of the particles in the set φ. Importantly, this kernel function is distinct from the non-negative kernel function used by the KDE (10.96).

SVGD. SVGD is an iterative scheme that updates a set of S particles $\varphi^{(i)} = \{z_s^{(i)}\}_{s=1}^S$ along iteration index i with the goal of improving at each iteration the quality of the approximation provided by the particle-based estimate (10.96) of the posterior distribution. The scheme is derived within the framework of **reproducing kernel Hilbert spaces (RKHS)** by setting up the problem of maximizing the rate of local descent of the free energy between the particle-based approximation and the joint distribution $p(x, z)$.

As discussed in Sec. 4.11, an RKHS is determined by a **positive semi-definite kernel function** $\kappa(z, z')$, which can be thought of as defining a measure of similarity between two vectors z and z'. A typical example of a positive semi-definite kernel function is the "Gaussian" kernel $\kappa(z, z') \propto \exp(-||z - z'||^2)$ (note that the kernel need not be normalized).

Referring to Recommended Resources, Sec. 10.15, for a detailed derivation, SVGD operates as summarized in Algorithm 10.8. Accordingly, by (10.97) and (10.98), each particle z_s is updated in a direction that depends on all other particles $z_{s'}$ with $s' \neq s$. In particular, the first term in (10.98) moves each particle z_s in the direction that minimizes the supervised loss $-\log p(x, z)$, by weighting the contribution of each other particle $z_{s'}$ by its similarity $\kappa(z_s, z_{s'})$. Interestingly, this implies that, in some sense, particles "communicate" with each other to identify parts of the model space with low supervised loss $-\log p(x, z)$. By optimizing the joint distribution $p(x, z)$ through its log-loss, this first term equivalently seeks the MAP solution (10.27). In the absence of the second term in (10.98), all particles would hence move towards the MAP solution.

Algorithm 10.8: Stein variational gradient descent (SVGD)

initialize particles $\varphi^{(1)} = \{(z_s^{(1)})_{s=1}^S\}$ and $i = 1$
while *stopping criterion not satisfied* **do**
   for all particles $s = 1, \ldots, S$
   given a learning rate $\gamma^{(i)} > 0$, obtain the next iterate as

$$z_s^{(i+1)} \leftarrow z_s^{(i)} - \gamma^{(i)} r_s^{(i)} \tag{10.97}$$

   with

$$r_s^{(i)} = \sum_{s'=1}^S \left\{ \kappa(z_s^{(i-1)}, z_{s'}^{(i-1)}) \left[\underbrace{-\nabla_{z_{s'}} \log p(x, z_{s'}^{(i-1)})}_{\text{supervised log-loss gradient}} \right] - \underbrace{\nabla_{z_{s'}} \kappa(z_s^{(i-1)}, z_{s'}^{(i-1)})}_{\text{repulsive "force"}} \right\} \tag{10.98}$$

   end for
   set $i \leftarrow i + 1$
end
return $\varphi^{(i)} = \{(z_s^{(i)})_{s=1}^S\}$

The second term prevents this collapse of particle diversity from happening by causing a repulsion between particles that are "too similar". To see this, consider the use of the kernel $\kappa(z_s, z_{s'}) = \exp(-||z_s - z_{s'}||^2)$, for which we have the gradient

$$\nabla_{z_{s'}} \kappa(z_s, z_{s'}) = 2(z_s - z_{s'})\kappa(z_s, z_{s'}). \tag{10.99}$$

Using (10.99), for each $s' \neq s$ the second term in (10.98) drives particle z_s away from particle $z_{s'}$, i.e., in the direction $(z_s - z_{s'})$, to a degree that is proportional to the similarity $\kappa(z_s, z_{s'})$: Similar particles z_s and z'_s repel one another.

Example 10.12

Consider a joint distribution $p(x, z)$ given, for a fixed x, by an unnormalized mixture of two Gaussians with weights 0.7 and 0.3, and means 1 and -1, respectively, and a standard deviation for both Gaussians of 0.5. In Fig. 10.10, the dashed lines represent the true posterior $p(z|x)$ obtained by normalizing $p(x, z)$, while the solid lines represent the KDE, with Gaussian kernel and bandwidth $h = 0.1$ obtained from $S = 20$ particles as a function of the SVGD iteration number i. The particles are shown on the horizontal axis as circles. A constant learning rate is selected of $\gamma^{(i)} = 0.01$. In line with the discussed behavior of the SVGD update (10.98), the particles first move towards the nearest peak of the posterior driven by the first term in the update, and are then repulsed to cover the support of the distribution by the second term.

10.12 Amortized Variational Inference

In the VI problem, one obtains an approximation of the posterior $p(z|x)$ for a given value x. Therefore, the VI problem must be solved anew for every value x of interest. Amortized VI aims to "amortize" the complexity of VI by sharing the variational parameters φ of the variational

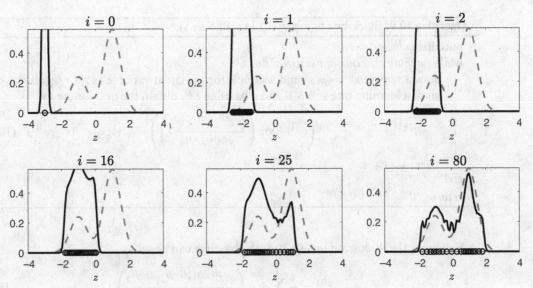

Figure 10.10 KDE (solid line) obtained with $S = 20$ particles updated via SVGD (circles) across the SVGD iteration index i, with dashed lines representing the true posterior.

posterior $q(z|x, \varphi)$ across all values x. As a result, amortized VI requires a single optimization to obtain a variational posterior $q(z|x, \varphi)$ applicable to all values x.

Formally, given a family of parametric variational posteriors $\mathcal{Q}$ of the form $q(\cdot| \cdot, \varphi)$, the amortized VI problem (10.3) is formulated as the minimization of the **average free energy**

$$\min_{\varphi} \mathrm{E}_{\mathrm{x} \sim p(x)}[\mathrm{F}(q(z|x, \varphi)||p(x, z))]. \qquad (10.100)$$

As mentioned, the key difference with respect to the VI problem is that the optimization is not tackled for a fixed value x; instead, the cost function in (10.100) is averaged over the marginal distribution $p(x)$ of the input. Therefore, the variational parameter φ obtained by solving the amortized VI problem provides a variational posterior $q(z|x, \varphi)$ that "performs well" on average over all values x – and not just for a single value.

Both black-box VI (Sec. 10.8) and reparametrization-based VI (Sec. 10.9) can be applied to the amortized VI problem. As described in the rest of this section, their respective derivations follow along the same lines as for the VI problem.

10.12.1 Black-Box Amortized Variational Inference

Black-box amortized VI addresses problem (10.100) by generalizing the black-box VI algorithm in Algorithm 10.3. As a result, it applies to all variational posteriors $q(z|x, \varphi)$ that enable sampling and whose log-loss can be differentiated. The derivation follows directly from Sec. 10.8 with a single caveat: The cost function (10.100), and its gradient, include an additional average over the input $\mathrm{x} \sim p(x)$. This additional expectation can be approximated by drawing one or more samples from the marginal $p(x)$.

The resulting algorithm, summarized in Algorithm 10.9, applies a **doubly stochastic gradient** estimate: The gradient is estimated by drawing samples for both input rv x and latent rv z.

Algorithm 10.9: Black-box amortized VI (with $S = 1$)

---

initialize $\varphi^{(1)}$ and $i = 1$
while *stopping criterion not satisfied* **do**
  draw a sample $\mathrm{x}^{(i)} \sim p(x)$ and a sample for the latent variable as $\mathrm{z}^{(i)} \sim q(z|\mathrm{x}^{(i)}, \varphi^{(i)})$
  given a learning rate $\gamma^{(i)} > 0$ and a baseline $c^{(i)}$, obtain the next iterate as

$$\varphi^{(i+1)} \leftarrow \varphi^{(i)} + \gamma^{(i)} \left(c^{(i)} - \log \left(\frac{q(\mathrm{z}^{(i)}|\mathrm{x}^{(i)}, \varphi^{(i)})}{p(\mathrm{x}^{(i)}, \mathrm{z}^{(i)})} \right) \right) \cdot \nabla_\varphi \log q(\mathrm{z}^{(i)}|\mathrm{x}^{(i)}, \varphi^{(i)}) \quad (10.101)$$

  set $i \leftarrow i + 1$
end
return $\varphi^{(i)}$

Following the discussion in Sec. 10.8, the baseline can be set as

$$c^{(i)} = \frac{1}{L} \sum_{j=1}^{L} \log \left(\frac{q(\mathrm{z}^{(i-j)}|\mathrm{x}^{(i-j)}, \varphi^{(i-j)})}{p(\mathrm{x}^{(i-j)}, \mathrm{z}^{(i-j)})} \right). \quad (10.102)$$

Black-box amortized VI in Algorithm 10.9 is a generalization of the black-box VI algorithm in Algorithm 10.3, which is obtained as a special case of black-box amortized VI by choosing the marginal $p(x)$ as a distribution concentrated on the value x of interest.

10.12.2 Reparametrization-Based Amortized Variational Inference

Black-box amortized VI is a broadly applicable tool that only requires the feasibility of sampling from and differentiating the variational posterior. When the variational posterior class $\mathcal{Q}$ is reparametrizable, one can obtain a more effective algorithm, characterized by lower-variance updates, via the reparametrization trick. In this subsection, we focus on the most common case in which prior and variational posterior are Gaussian.

Reparametrizable Gaussian Amortized Variational Posterior. Generalizing the discussion in Sec. 10.9, a typical choice is the **Gaussian amortized variational posterior**

$$q(z|x, \varphi) = \mathcal{N} \left(z \left| \begin{bmatrix} \nu_1(x|\varphi) \\ \nu_2(x|\varphi) \\ \vdots \\ \nu_M(x|\varphi) \end{bmatrix}, \begin{bmatrix} \sigma_1(x|\varphi)^2 & 0 & \cdots & 0 \\ 0 & \sigma_2(x|\varphi)^2 & & \vdots \\ \vdots & 0 & \ddots & 0 \\ 0 & \cdots & 0 & \sigma_M(x|\varphi)^2 \end{bmatrix} \right. \right)$$

$$= \mathcal{N} \left(z \left| \nu(x|\varphi), \mathrm{Diag}(\sigma(x|\varphi)^2) \right. \right). \quad (10.103)$$

Unlike its non-amortized counterpart (10.66), the amortized posterior (10.103) is characterized by an $M \times 1$ mean vector $\nu(x|\varphi)$ and by an $M \times 1$ vector of variances $\sigma(x|\varphi)^2$ that are functions of the input x, as well as of the variational parameters φ. Therefore, given the variational parameters φ, the variational posterior (10.103) provides a soft prediction for every value x.

The functions $\nu(x|\varphi)$ and $\sigma(x|\varphi)^2$ are typically selected as neural networks with input x and respective weights being subsets of the entries of vector φ. More precisely, following the parametrization (10.72), one typically writes the vector of standard deviations as $\sigma(x|\varphi) = \exp(\varrho(x|\varphi))$, where function $\varrho(x|\varphi)$ is modeled as a neural network. As mentioned, this allows the output of the function $\varrho(x|\varphi)$ to be unconstrained, while still guaranteeing the positivity of the entries of vector $\sigma(x|\varphi)$.

Reparametrization-Based Amortized VI. Adopting (10.103) as the amortized variational posterior and assuming a Gaussian prior $p(z) = \mathcal{N}(z|0_M, I_M)$, we now apply the reparametrization trick (Algorithm 10.6) to obtain a reparametrization-based amortized VI scheme. There are two differences as compared to the standard reparametrization-based VI strategy introduced in Sec. 10.9. The first is that, as for black-box amortized VI, the estimated gradient is **doubly stochastic**, since it needs to account for the expectations over the input x, as well as over the latent rv z. The second difference is that, by (10.103), the reparametrization transformation depends on the input x through the following procedure:

- **draw** a sample from the auxiliary distribution $\mathcal{N}(0_M, I_M)$ as

$$\mathrm{e} \sim q(e) = \mathcal{N}(e|0_M, I_M); \text{ and} \qquad (10.104)$$

- **compute**

$$z = r(\mathrm{e}|x, \varphi) = v(x|\varphi) + \exp(\varrho(x|\varphi)) \odot \mathrm{e}, \qquad (10.105)$$

where, unlike (10.68), we have highlighted the dependence of transformation $r(\mathrm{e}|x, \varphi)$ on the input x.

Therefore, the application of the reparametrization trick via (10.70) requires the Jacobian

$$\nabla_\varphi r(e|x, \varphi) = \nabla_\varphi v(x|\varphi) + \nabla_\varphi \varrho(x|\varphi) \mathrm{Diag}(\exp(\varrho(x|\varphi)) \odot e), \qquad (10.106)$$

where the $M \times M$ Jacobians $\nabla_\varphi v(x|\varphi)$ and $\nabla_\varphi \varrho(x|\varphi)$ can be obtained via automatic differentiation if the functions $v(x|\varphi)$ and $\varrho(x|\varphi)$ are implemented as neural networks (see Sec. 5.15). Note that the Jacobian (10.106) reduces to (10.74) when $v(x|\varphi) = v$ and $\varrho(x|\varphi) = \varrho$, with vector φ concatenating vectors v and ϱ. Overall, we have derived the algorithm detailed in Algorithm 10.10.

10.13 Variational Expectation Maximization

As seen in Sec. 7.8, optimal soft prediction plays a key role in the design of training algorithms for models with latent variables. In this section, we elaborate on the use of VI as a means to make the implementation of training algorithms for latent-variable models more efficient and scalable when exact optimal soft prediction is intractable. To this end, we will address the VEM problem (10.4).

10.13.1 A Brief Recap of the EM Algorithm

In order to put the VEM problem in context, let us start by connecting it with the EM algorithm introduced in Sec. 7.8. The EM algorithm tackles the special case of the VEM problem (10.4) in which the set of variational distributions $q(z|x)$ in $\mathcal{Q}$ is unconstrained. As detailed in Sec. 7.8, this unconstrained problem coincides with the ML training of a latent-variable generative model $p(x, z|\theta)$, encompassing observed vector x and latent vector z. Applications include the

---

Algorithm 10.10: Reparametrization-based amortized VI with Gaussian prior and variational posterior (with $S = 1$)

---

initialize $\varphi^{(1)}$ and $i = 1$
while *stopping criterion not satisfied* **do**

> draw a sample $\mathrm{x}^{(i)} \sim p(x)$ and an auxiliary sample $\mathrm{e}^{(i)} \sim \mathcal{N}(0, I)$
> compute $\mathrm{z}^{(i)} = \nu(\mathrm{x}^{(i)}|\varphi^{(i)}) + \exp(\varrho(\mathrm{x}^{(i)}|\varphi^{(i)})) \odot \mathrm{e}^{(i)}$
> estimate the gradient as
>
> $$\begin{aligned} \hat{\nabla}_\varphi^{(i)} =& \nabla_\varphi r(\mathrm{e}^{(i)}|\mathrm{x}^{(i)}, \varphi^{(i)}) \cdot \nabla_z(-\log p(\mathrm{x}^{(i)}|\mathrm{z}^{(i)})) \\ &+ \nabla_\varphi \nu(\mathrm{x}^{(i)}|\varphi^{(i)}) \cdot \nu(\mathrm{x}^{(i)}|\varphi^{(i)}) \\ &+ \nabla_\varphi \exp(\varrho(\mathrm{x}^{(i)}|\varphi^{(i)})) \cdot (\exp(\varrho(\mathrm{x}^{(i)}|\varphi^{(i)})) - \exp(-\varrho(\mathrm{x}^{(i)}|\varphi^{(i)}))) \end{aligned} \qquad (10.107)$$
>
> given a learning rate $\gamma^{(i)}$, obtain the next iterate as
>
> $$\varphi^{(i+1)} \leftarrow \varphi^{(i)} - \gamma^{(i)} \hat{\nabla}_\varphi^{(i)} \qquad (10.108)$$
>
> set $i \leftarrow i + 1$

end
return $\varphi^{(i)}$

---

unsupervised learning of directed and undirected generative models (Secs. 7.7 and 7.8), as well as the supervised learning of mixture models (see Sec. 6.9).

As also discussed in Sec. 7.8, the EM algorithm tackles problem (10.4), within an unconstrained set of distributions $\mathcal{Q}$, by alternating between optimization over the variational posteriors $q(z|x)$ in the E step and optimization over model parameter θ in the M step. Importantly, the EM algorithm assumes that both steps can be carried out *exactly*:

- given the current iterate $\theta^{(i)}$, the **E step** returns the posterior distribution $p(z|x, \theta^{(i)})$ evaluated at the training data points, i.e., $p(z_n|x_n, \theta^{(i)})$ for $n = 1, \ldots, N$; and
- the **M step** solves the problem of minimizing the training free energy

$$\min_\theta \left\{ \frac{1}{N} \sum_{n=1}^{N} \mathrm{F}(p(z_n|x_n, \theta^{(i)}) || p(x_n, z_n|\theta)) \right\}. \qquad (10.109)$$

VEM is useful when either or both of these steps are intractable.

10.13.2 Variational EM

In order to reduce the complexity of EM, VEM generalizes EM in the following two ways:

- In the **E step**, VEM replaces the exact calculation of the posterior distribution $p(z|x, \theta^{(i)})$ carried out by EM with **amortized VI** over a given set $\mathcal{Q}$ of parametric variational posteriors $q(z|x, \varphi)$. The variational parameter φ is shared across all data points, and hence the variational model is amortized across the entire data set.
- The optimization problems defining **E step** and **M step** are carried out approximately via **local optimization**.

Figure 10.11 Illustration of the distributions involved in the VEM problem (10.110).

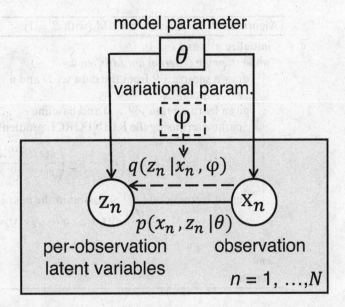

Overall, (parametric) VEM tackles the problem

$$\min_{\theta,\varphi} \frac{1}{N} \sum_{n=1}^{N} F(q(z_n|x_n,\varphi)||p(x_n,z_n|\theta)) \qquad (10.110)$$

of minimizing the **training free energy** simultaneously over the model parameters θ and the variational parameters φ. An illustration of the distributions involved in this problem can be found in Fig. 10.11. Depending of the form of the variational posterior, optimization over φ can be done by using either the REINFORCE gradient or the reparametrization trick. Furthermore, there are several ways to carry out the optimization over model parameters θ and variational parameters φ, with the most common being to perform simultaneous doubly stochastic gradient updates.

10.13.3 Black-Box VEM

For reference, in this subsection we develop a VEM algorithm based on the REINFORCE gradient, which we term **black-box VEM**. Black-box VEM carries out optimization steps for E and M steps simultaneously by leveraging **doubly stochastic gradient** estimators based on samples from both observed and latent variables. In particular, the gradient over the variational parameters is estimated via black-box amortized VI, as studied in the preceding section. The gradient of the free energy with respect to the model parameter θ can be directly estimated by noting that we have the equality

$$\nabla_\theta F(q(z|x,\varphi)||p(x,z|\theta)) = E_{z \sim q(z|x,\varphi)}[\nabla_\theta(-\log p(x,z|\theta))]. \qquad (10.111)$$

Algorithm 10.11 details the resulting procedure, in which the baseline can be set as (10.102).

Algorithm 10.11: Black-box VEM (with $S = 1$)

---

initialize $\varphi^{(1)}$ and $i = 1$

while *stopping criterion not satisfied* **do**

    draw a sample $x^{(i)}$ from the data set $\mathcal{D}$ and a sample for the latent variable as $z^{(i)} \sim q(z|x^{(i)}, \varphi^{(i)})$

    given learning rate $\gamma^{(i)} > 0$ and baseline $c^{(i)}$, obtain the next iterate for the variational parameters using the REINFORCE gradient (black-box VI) as

$$\varphi^{(i+1)} \leftarrow \varphi^{(i)} + \gamma^{(i)} \left(c^{(i)} - \log \left(\frac{q(z^{(i)}|x^{(i)}, \varphi^{(i)})}{p(x^{(i)}, z^{(i)}|\theta^{(i)})} \right) \right) \cdot \nabla_\varphi \log q(z^{(i)}|x^{(i)}, \varphi^{(i)}) \quad (10.112)$$

    given learning rate $\xi^{(i)} > 0$, obtain the next iterate for the model parameters as

$$\theta^{(i+1)} \leftarrow \theta^{(i)} + \xi^{(i)} \nabla_\theta \log p(x^{(i)}, z^{(i)}|\theta^{(i)}) \quad (10.113)$$

    set $i \leftarrow i + 1$

end

return $\varphi^{(i)}$

Example 10.13

In this example, we revisit the beta–exponential model (10.31) by assuming that the likelihood contains a trainable parameter θ. Mathematically, we have the following joint distribution $p(x, z|\theta)$:

$$z \sim p(z) = \text{Beta}(z|a, b)$$

$$(x|z = z) \sim p(x|z, \theta) = \text{Exp}(x|\theta z) = \frac{1}{\theta z} \exp\left(-\frac{x}{\theta z}\right) \mathbb{1}(x \geq 0), \quad (10.114)$$

where the prior hyperparameters ($a = 3, b = 3$) are fixed. Unlike the previous examples, in which we focused on approximating the posterior $p(z|x)$, we are now interested in training the generative model $p(x, z|\theta) = p(z)p(x|z, \theta)$ via ML based on a data set of observations $\mathcal{D} = \{(x_n)_{n=1}^N\}$. Specifically, we simplify ML training by addressing the VEM problem (10.110) under a beta variational distribution class $\text{Beta}(z|\varphi_1, \varphi_2)$ as in Example 10.9.

For reference, we first evaluate a **non-amortized solution** to the VEM problem in which, at each iteration i, we run several (here, 100) iterations of black-box VI (Algorithm 10.3) in order to obtain a variational distribution $q(z|x^{(i)}, \varphi^{(i)}) = \text{Beta}(z|\varphi_1^{(i)}, \varphi_2^{(i)})$ that approximates the posterior $p(z|x^{(i)}, \theta^{(i)})$. The model parameters are then updated via the doubly stochastic gradient (10.113). This approach requires running a separate black-box VI process for each iteration.

We then consider the black-box VEM algorithm detailed in Algorithm 10.11, which amortizes VI across the entire data set. For this purpose, we set the amortized posterior as $q(z|x, \varphi) = \text{Beta}(z|\varphi_1(1 + x), \varphi_2)$, with variational parameter vector $\varphi = [\varphi_1, \varphi_2]^T \in \mathbb{R}^2$. Note that, unlike the non-amortized reference solution, here the variational parameters evolve with the model parameters along the same number of iterations as per Algorithm 10.11.

We generate $N = 100$ data samples from the model (10.114), and implement both algorithms with constant learning rates $\gamma^{(i)} = 0.02$ and $\xi^{(i)} = 0.01$. Fig. 10.12 shows a number of random runs for both algorithms by plotting the evolution of the (scalar) iterate $\theta^{(i)}$ (solid lines) along with the ground-truth value $\theta = 1$ (dashed line). Despite being much simpler to implement, avoiding the multiple (here, 100) VI updates per iterate i of the non-amortized strategy, the amortized scheme is seen to have comparable performance, although its convergence may not always be guaranteed.

Figure 10.12 Evolution of the iterate $\theta^{(i)}$ (solid lines) along with the ground-truth value $\theta = 1$ (dashed line) for both non-amortized and amortized black-box VEM solutions. The non-amortized solution carries out 100 VI update steps per iteration i.

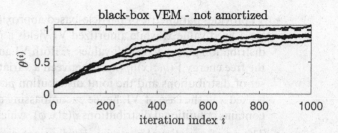

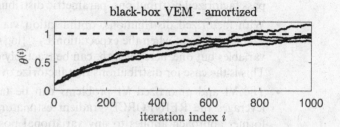

10.13.4 Variational Autoencoders

As a direct application of VEM, we now briefly introduce **variational autoencoders (VAEs)**. A VAE is a generative model $p(x, z|\theta)$ with Gaussian latent prior $p(z) = \mathcal{N}(z|0_M, I_M)$ and conditional distribution $p(x|z, \theta)$ defined by a neural network model. With reference to the terminology for autoencoders introduced in Sec. 7.5, VAEs interpret the amortized variational posterior $q(z|x, \varphi)$ as a **probabilistic encoder** mapping observation x to latent rv z, and the model $p(x|z, \theta)$ as a **probabilistic decoder** mapping latent rv z back to observation x. VAEs tackle the ML training of the generative model $p(x, z|\theta)$ by addressing the VEM problem. Specifically, by assuming the Gaussian variational posterior (10.103), VAEs leverage the reparametrization trick in order to estimate the gradient over the variational parameters φ. Details can be worked out directly, as in the preceding subsection, by replacing the REINFORCE gradient estimator with the reparametrization-based estimator of the gradient over φ. The reader is referred to Recommended Resources, Sec. 10.15, for more information on VAEs.

10.14 Summary

- Bayesian inference – also known as optimal soft prediction – of a latent rv z given a jointly distributed observation x = x requires the calculation of the posterior distribution $p(z|x)$. The optimal soft predictor $p(z|x)$ may be of interest for a given fixed value x or for all possible values x. Exact Bayesian inference is tractable only for specific models, most notably for joint distributions in the conjugate exponential family. In this case, the posterior $p(z|x)$ is in the same class of distributions as the prior $p(z)$.

- As a step towards the definition of tractable Bayesian inference procedures, the Laplace approximation fits a Gaussian distribution to the MAP solution by evaluating the Hessian of the log-loss $-\log p(x, z)$ with respect to z for a given fixed value x.

- VI provides parametric or particle-based approximations of the posterior distribution $p(z|x)$ for a fixed value x, while amortized VI yields a parametric approximation of the posterior distribution $p(z|x)$ across all values x. Both VI and amortized VI target the minimization of the free energy $F(q(z|x)||p(x, z))$ between a variational posterior distribution $q(z|x)$ in some set of distributions and the joint distribution $p(x, z)$. The set of distributions is defined for a fixed x in the case of VI, hence encompassing distributions in the domain of rv z; while it contains conditional distributions $q(z|x, \varphi)$, which are functions of x, for amortized VI.

- The set of variational posterior distributions assumed by VI and amortized VI can encompass factorized distributions, parametric distributions, or a combination of both.

- With factorized distributions, optimization via coordinate descent yields mean-field VI, which is efficient when the expectation $E_{z_{-m}}[\log p(x, z_m, z_{-m})]$ of the log-loss over all latent variables but one, here the mth, can be tractably computed or estimated for all values of m. This is the case for distributions that factorize over graphs, such as the Ising model.

- The VI and amortized VI problems can be tackled by means of black-box VI, which leverages the REINFORCE gradient estimator, or via the reparametrization trick. The former approach applies to any variational posterior from which samples can be drawn efficiently and whose log-loss can be differentiated with respect to the variational parameters; while the latter is only applicable to choices for the variational posteriors that enable reparametrization.

- Reparametrization requires that the rv z, distributed according to the variational posterior $q(z|x, \varphi)$, can be generated by first drawing an auxiliary variable $e \sim q(e)$ with distribution $q(e)$ not dependent on φ, and then setting $z = r(e|\varphi)$ for some function $r(e|\varphi)$ that is differentiable in the variational parameters φ for every fixed value of e. A typical example is given by Gaussian variational posteriors with variational parameters given by mean vector and covariance matrix, or functions thereof.

- Particle-based VI approximates the posterior distribution with a set of particles. As a notable example, SVGD updates the particles jointly in an SGD-like fashion to maximize the rate of decrease of the free energy, maintaining the diversity of the particle set. Particle-based strategies can be particularly advantageous when the posterior is multimodal and hard to fit with pre-defined parametric models.

- VEM provides an approximate solution for the ML problem of maximizing the likelihood in generative models by replacing the E step in the EM algorithm with amortized VI, and carrying out optimization of the free energy via SGD. VAEs are a popular application of the VEM framework.

10.15   Recommended Resources

Conjugate exponential-family models are covered in a number of textbooks including [1]. For the Laplace approximation, an excellent reference is [2]. Recommended reviews on VI and amortized VI, also covering the case with discrete latent rvs, are [3, 4, 5, 6]. A useful short summary of reparametrization-based VI methods can be found in [7]. Connections to probabilistic graphical models are extensively described in [8]. VEM is discussed in [9], and a more modern presentation focusing on VAE can be found in [10]. For SVGD, the reader is referred to [11], and the more general problem of quantization for a probability distribution is covered in [12].

Problems

10.1 Consider the generative model defined by prior $z \sim \mathcal{N}(0,1)$ and likelihood $(x|z = z) \sim \mathcal{N}(\mu(z), 1)$. (a) For $\mu(z) = z + a$ for an arbitrary constant a, evaluate the exact posterior $p(z|x = 1)$. (b*) For $\mu(z) = \mathbb{1}(z \geq 0)$, plot the posterior $p(z|x = 1)$.

10.2 You are managing a social media platform, and are interested in predicting the "likes" of two users for the updates from some politician. The first user is connected to other users that tend to like similar content, and she is hence a priori expected to "like" updates from the politician, while you have no prior information about the second user. You observe 20 updates, and record 6 "likes" from the first user and 10 for the second. (a) Explain how to use a beta–Bernoulli model to study this setting. (b) Justify the choice of prior Beta$(9,1)$ for the first user and Beta$(5,5)$ for the second. (c) What are the MAP estimates of the probabilities that the two users will each "like" the next update?

10.3 Consider the joint distribution defined as $z \sim \mathcal{N}(0,1)$ and $(x|z = z, \theta) = \text{Bern}(\sigma(\theta z))$ and fix $x = 1$. (a*) Evaluate the MAP solution z^{MAP} when $\theta = 10$. [You can approximate the value of z^{MAP} using numerical optimization.] (b) Compute the second derivative of the log-loss $-\log p(x = 1, z|\theta)$ with respect to z. (c*) Use the result from point (b) to evaluate and plot the Laplace approximation of the posterior $p(z|x = 1, \theta)$ for $\theta = 10$. On the same plot, draw the true posterior distribution $p(z|x = 1, \theta)$ for $\theta = 10$. (d*) Repeat point (c*) for $\theta = 1$.

10.4 Load the data set penguin from the book's website. It contains a single noisy binary image x. (a*) Display the image. (b*) Implement mean-field VI based on the Ising model. To this end, note that each pixel will need to be converted to the alphabet $\{-1,1\}$; initialize the variational parameters μ_m to equal the corresponding pixel in the image x (after conversion to the $\{-1,1\}$ alphabet); and set $\eta_1 = 1$ and $\eta_2 = 0.1$. (c*) Display the estimated image after each iteration. To estimate the image, select 1 for a pixel if the corresponding probability is larger than $1/2$, and 0 otherwise. (d*) What happens if η_2 is increased? (e*) What happens if we make η_2 negative?

10.5 We study the stochastic optimization problem $\min_\eta \mathrm{E}_{z \sim \text{Bern}(z|\sigma(\eta))}[z^2 + 2z]$. (a) Compute the derivative of the cost function with respect to η by direct calculation. (b) Compute the derivative of the cost function with respect to η by using the formula for the REINFORCE gradient, and show that it coincides with the derivative computed in point (a).

10.6 Consider the stochastic optimization problem $\min_\varphi \mathrm{E}_{z \sim \mathcal{N}(z|\varphi, 1)}[z^2 + 2z]$. Compute the derivative of the cost function with respect to φ by (a) direct calculation; (b) using the formula for the REINFORCE gradient; and (c) using the reparametrization trick. [Hint: $\mathrm{E}_{z \sim \mathcal{N}(z|\varphi, 1)}[z^3] = \varphi^3 + 3\varphi$.]

10.7 We study the stochastic optimization problem $\min_\varphi \mathrm{E}_{z \sim \mathcal{N}(z|\varphi, I_2)}[z_1 \exp(-3z_1 - 2z_2)]$. (a*) Using empirical averages, plot the contour lines of the objective in the plane defined by coordinates $\varphi = [\varphi_1, \varphi_2]^T$. (b*) Implement the REINFORCE gradient algorithm and plot a number of iterates superimposed on the contour lines obtained in point (a*). Compare the performance with and without a baseline. (c*) Implement the reparametrization-based gradient algorithm, and display a number of iterates as in point (b*). (d*) Repeat the previous point by considering a variational distribution $q(z|\varphi) = \mathcal{N}(z|\nu, \text{Diag}(\sigma^2))$ with $\sigma = \exp(\varrho)$ and variational parameters $\varphi = [\nu^T, \varrho^T]^T$.

10.8 Consider the model given by $z \sim \mathcal{N}(0,1)$ and $(x|z = z) = \text{Bern}(\sigma(10z))$ (see Problem 10.3). (a) Assume that the variational posterior is $\mathcal{N}(\nu, \exp(2\varrho))$, and write the

REINFORCE gradient estimator for the gradient of the free energy. (b*) Implement black-box VI by using the gradient derived in the preceding point and assuming x = 1. Plot the variational posterior obtained for several iterations until convergence.

10.9   For the same model and variational posterior considered in Problem 10.8: (a) Assume that the variational posterior is $\mathcal{N}(\nu, \exp(2\varrho))$, and write the reparametrization-based gradient estimator for the gradient of the free energy. (b*) Implement reparametrization-based VI assuming x = 1, and plot the variational posterior obtained for several iterations until convergence.

10.10 Consider again the model $z \sim \mathcal{N}(0, 1)$ and $(x|z = z) = \mathrm{Bern}(\sigma(10z))$, but assume an amortized variational posterior $q(z|x, \varphi) = \mathcal{N}((2x - 1)\nu, \exp(2\varrho))$. (a) Write the REINFORCE gradient estimator for the gradient of the free energy. (b*) Implement black-box amortized VI, and plot the amortized variational posterior $q(z|x, \varphi)$ obtained for several iterations until convergence for $x = 1$ and $x = 0$.

10.11 Repeat Problem 10.10 by applying reparametrization-based amortized VI in lieu of black-box amortized VI.

10.12 Consider the joint distribution $p(x, z) \propto 0.5\mathcal{N}(-1_2, I_2) + 0.5\mathcal{N}(1_2, I_2)$. (a) Write the SVGD update equation assuming the Gaussian kernel $\kappa(z, z')$. (b*) Implement SVGD, and plot the contour lines of the variational posterior estimated by using KDE for several iterations until convergence. For the KDE, use the Gaussian kernel $\kappa_h(z)$ and try different values of the bandwidth h. (c*) Vary the number of particles and observe the effect of this choice on the quality of the variational posterior.

10.13 Consider the generative model $z \sim \mathcal{N}(0, 1)$ and $(x|z = z, \theta) = \mathrm{Bern}(\sigma(\theta z))$, where θ is a trainable parameter. (a*) Generate a data set $\mathcal{D}$ with $N = 100$ samples x_n drawn from the model when assuming the ground-truth parameter $\theta_0 = 2$. (b*) Implement black-box VEM and apply it to the data set $\mathcal{D}$ to train parameter θ. To this end, use the amortized variational posterior $q(z|x, \varphi) = \mathcal{N}((2x - 1)\varphi, 1)$. Show the evolution of the model parameter θ along the iterations. (c*) Repeat point (b*) with a non-amortized posterior $\mathcal{N}(\varphi, 1)$ by optimizing a different value of φ for each iteration via black-box VI.

Appendices

Appendix 10.A: Dirichlet–Categorical Model

The Dirichlet–categorical model is another example of a conjugate exponential-family distribution. The likelihood for a single observation x is given by the categorical distribution

$$p(x|z) = \mathrm{Cat}(x|z) = \prod_{k=0}^{C-1} z_k^{\mathbb{1}(x=k)}, \tag{10.115}$$

and the conjugate prior is the Dirichlet distribution

$$p(z_0, \ldots, z_{C-1}|\alpha_0, \ldots, \alpha_{C-1}) = \mathrm{Dir}(z_0, \ldots, z_{C-1}|\alpha_0, \ldots, \alpha_{C-1}) \propto \prod_{k=0}^{C-1} z_k^{\alpha_k-1}, \tag{10.116}$$

with parameter vector $\alpha = [\alpha_0, \ldots, \alpha_{C-1}]^T$. The Dirichlet distribution has mean vector and mode vector with kth entries, for $k = 0, 1, \ldots, C - 1$, given as

$$E_{z_k \sim \text{Dir}(z_0,\ldots,z_{C-1}|\alpha_0,\ldots,\alpha_{C-1})}[z_k] = \frac{\alpha_k}{\sum_{j=0}^{C-1}\alpha_j}$$

$$\text{mode}_{z_k \sim \text{Dir}(z_0,\ldots,z_{C-1}|\alpha_0,\ldots,\alpha_{C-1})}[z_k] = \frac{\alpha_k - 1}{\left(\sum_{j=0}^{C-1}\alpha_j\right) - C}. \tag{10.117}$$

The posterior distribution given L i.i.d. observations $x_1, \ldots, x_L \sim \text{Cat}(x|z)$ is the Dirichlet distribution

$$p(z|x_1, \ldots, x_L) = \text{Dir}(z_0, \ldots, z_{C-1}|\alpha_0 + L[0], \ldots, \alpha_{C-1} + L[C-1])$$

$$\propto \prod_{k=0}^{C-1} z_k^{\alpha_k + L[k] - 1}, \tag{10.118}$$

where $L[k]$ represents the number of observations equal to k, i.e., $L[k] = \sum_{j=1}^{L} \mathbb{1}(x_j = k)$.

Appendix 10.B: Derivation of the Update Equation for Mean-Field VI

In this appendix, we show that the solution of problem (10.36) is given by (10.38). Under the mean-field factorization, the entropy $H(q(z|x))$ for any fixed x equals the sum

$$H(q(z|x)) = \sum_{m=1}^{M} H(q_m(z_m|x)), \tag{10.119}$$

since the rvs $\{z_m\}_{m=1}^{M}$ are modeled as being conditionally independent given $x = x$. Furthermore, by using the law of iterated expectations, we can write the average supervised log-loss as

$$E_{z \sim q(z|x)}\left[-\log p(x, z)\right] = E_{z_m}\left[E_{z_{-m}}[-\log p(x, z_m, z_{-m})]\right], \tag{10.120}$$

where $z_m \sim q_m(z_m|x)$. Therefore, problem (10.36) can be equivalently expressed as

$$\min_{q_m(z_m|x)} E_{z_m \sim q_m(z_m|x)}\left[E_{z_{-m}}[-\log p(x, z_m, z_{-m})]\right] - H(q_m(z_m|x)), \tag{10.121}$$

where the cost function equals the free energy $F(q(z_m|x)\| \exp(E_{z_{-m}}[\log p(x, z_m, z_{-m})]))$. (Recall that the second term in the free energy can be unnormalized as discussed in Sec. 3.9). Hence, by (3.102), the optimal solution $q_m^*(z_m|x)$ is given by the (normalized) distribution (10.38).

Appendix 10.C: Proof of the REINFORCE Gradient Formula (10.51)

By direct calculation, assuming first a discrete rv z, we have the following equalities:

$$\nabla_\varphi E_{z \sim q(z|x,\varphi)}[g(z|\varphi)] = \sum_z \nabla_\varphi\left(q(z|x,\varphi)g(z|\varphi)\right)$$

$$= \sum_z \left(\nabla_\varphi q(z|x,\varphi) \cdot g(z|\varphi)\right) + \sum_z \left(q(z|x,\varphi) \cdot \nabla_\varphi g(z|\varphi)\right)$$

$$= \sum_z \left(q(z|x,\varphi) \cdot \nabla_\varphi \log q(z|x,\varphi) \cdot g(z|\varphi)\right) + E_{z \sim q(z|x,\varphi)}[\nabla_\varphi g(z|\varphi)]$$

$$= E_{z \sim q(z|x,\varphi)}[g(z|\varphi) \cdot \nabla_\varphi \log q(z|x,\varphi)], \tag{10.122}$$

where for the third equality we have used the identity $\nabla_\varphi \log q(z|x,\varphi) = \nabla_\varphi q(z|x,\varphi)/q(z|x,\varphi)$, while for the fourth equality we have used the definition (10.60) and the equality

$$E_{z \sim q(z|x,\varphi)}[\nabla_\varphi \log q(z|x,\varphi)] = 0, \tag{10.123}$$

which follows from the zero-mean of the score vector (see Sec. 9.7). The proof follows in a similar way for continuous rvs z by replacing sums by suitable integrals.

Appendix 10.D: Optimal Selection of the Baseline for the REINFORCE Gradient

In this appendix, we elaborate on the design of an optimal baseline for the REINFORCE gradient that minimizes the variance of the gradient estimate. To start, let us introduce the simplifying notation $\hat{\nabla}_\varphi(z|c) = (g(z)-c) \cdot \nabla_\varphi \log q(z|\varphi)$, where $z \sim q(z|\varphi)$, for the REINFORCE estimator (10.54) with $S = 1$. By (10.55), this estimator is unbiased, and for simplicity we denote the true gradient as $\nabla_\varphi = \nabla_\varphi E_{z \sim q(z|\varphi)}[g(z)] = E_{z \sim q(z|\varphi)}[\hat{\nabla}_\varphi(z|c)]$. Since the estimator is unbiased, its **variance** can easily be seen to equal

$$\begin{aligned}
\text{Var}(\hat{\nabla}_\varphi(z|c)) &= E_{z \sim q(z|\varphi)}[||\hat{\nabla}_\varphi(z|c) - \nabla_\varphi||^2] \\
&= E_{z \sim q(z|\varphi)}[||\hat{\nabla}_\varphi(z|c)||^2] - ||\nabla_\varphi||^2, \tag{10.124}
\end{aligned}$$

where

$$E_{z \sim q(z|\varphi)}[||\hat{\nabla}_\varphi(z|c)||^2] = E_{z \sim q(z|\varphi)}[(g(z) - c)^2 ||\nabla_\varphi \log q(z|\varphi)||^2]. \tag{10.125}$$

What is the value of the baseline c that minimizes the variance (10.124) of the estimator? Differentiating the variance with respect to c yields

$$\frac{d}{dc} \text{Var}(\hat{\nabla}_\varphi(z|c)) = -2E_{z \sim q(z|\varphi)}[(g(z) - c)||\nabla_\varphi \log q(z|\varphi)||^2], \tag{10.126}$$

and solving the resulting first-order optimality condition $d\text{Var}(\hat{\nabla}_\varphi(z|c))/dc = 0$ yields

$$c^* = \frac{E_{z \sim q(z|\varphi)}[g(z)||\nabla_\varphi \log q(z|\varphi)||^2]}{E_{z \sim q(z|\varphi)}[||\nabla_\varphi \log q(z|\varphi)||^2]}. \tag{10.127}$$

Since the function $\text{Var}(\hat{\nabla}_\varphi(z|c))$ is strictly convex in c, the baseline (10.127) is the only global minimum of the variance. The optimal baseline c^* in (10.127) is a weighted average of the values taken by function $g(z)$ that depends on the norm of the score function of the variational posterior. Intuitively, this baseline centers the estimator at this weighted average so as to reduce the variance.

The solution c^* is difficult to evaluate since it requires computing expectations over the variational posterior. If we approximate the expectations in (10.127) using the past iterates, we get the approximate optimal baseline

$$c^{(i)} = \frac{\frac{1}{L}\sum_{j=1}^{L} g(z^{(i-j)})||\nabla_\varphi \log q(z^{(i-j)}|\varphi^{(i-j)})||^2}{\frac{1}{L}\sum_{j=1}^{L} ||\nabla_\varphi \log q(z^{(i-j)}|\varphi^{(i-j)})||^2}. \tag{10.128}$$

Appendix 10.E: Derivation of the Gradient of the KL Divergence (10.82)

In this appendix, we derive the expression (10.82) for the gradient of the KL divergence between the variational distribution $q(z|\mu(\varphi))$ and prior $q(z|\eta_p)$ with respect to φ. To start, consider two distributions in the same class within the exponential family, i.e.,

$$q(z|\eta) = \exp\left(\eta^T s(z) + M(z) - A(\eta)\right) \tag{10.129}$$

and

$$p(z|\eta_p) = \exp\left(\eta_p^T s(z) + M(z) - A(\eta_p)\right), \tag{10.130}$$

with respective natural parameters η and η_p. From the general formula (9.59) for the KL divergence $\mathrm{KL}(q(z|\eta)||p(z|\eta_p))$, we can compute the gradient with respect to the first natural parameter as

$$\nabla_\eta \mathrm{KL}(q(z|\eta)||p(z|\eta_p)) = \mathrm{FIM}(\eta)(\eta - \eta_p), \tag{10.131}$$

where we recall that $\mathrm{FIM}(\eta)$ is the FIM with respect to the natural parameter η. This expression follows from (9.59) by using the mapping between natural and mean parameters $\nabla_\eta A(\eta) = \mu$ (see (9.44)) and the characterization $\mathrm{FIM}(\eta) = \nabla_\eta^2 A(\eta)$ of the FIM in terms of the Hessian of the log-partition function (see (9.70)).

Let us now assume the parametrizations to be minimal. By (9.110) and (10.131), we can obtain the gradient with respect to the mean parameters μ corresponding to the natural parameters η as

$$\nabla_\mu \mathrm{KL}(q(z|\eta)||p(z|\eta_p)) = \eta - \eta_p. \tag{10.132}$$

For example, for Bernoulli distributions, we have

$$\frac{\mathrm{d}}{\mathrm{d}\mu}\mathrm{KL}(\mathrm{Bern}(z|\mu)||\mathrm{Bern}(z|\mu_p)) = \log\left(\frac{\mu}{\mu_p}\right) + \log\left(\frac{1-\mu_p}{1-\mu}\right). \tag{10.133}$$

Using (10.132), the gradient (10.82) is obtained via the chain rule.

Appendix 10.F: Gradient of the Entropy between Exponential-Family Distributions

Following Appendix 10.E, it is also useful to derive the gradient of the entropy for exponential-family distributions. We focus on distributions with $M(z) = 0$, i.e., of the form

$$q(z|\eta) = \exp\left(\eta^T s(z) - A(\eta)\right). \tag{10.134}$$

Using the general expression (9.58) for the entropy, we obtain

$$\nabla_\eta \mathrm{H}(q(z|\eta)) = \nabla_\eta(-\eta^T \mu + A(\eta)) = -\mu - (\nabla_\eta \mu)\eta + \nabla_\eta A(\eta)$$
$$= -\mathrm{FIM}(\eta)\eta. \tag{10.135}$$

Therefore, for minimal families we have the gradient

$$\nabla_\mu \mathrm{H}(q(z|\eta)) = -\eta. \tag{10.136}$$

For example, for Bernoulli distributions, we can write

$$\frac{\mathrm{d}}{\mathrm{d}\mu}\mathrm{H}(\mathrm{Bern}(\mu)) = -\log\left(\frac{\mu}{1-\mu}\right). \tag{10.137}$$

Bibliography

[1] C. M. Bishop, *Pattern Recognition and Machine Learning*. Springer, 2006.

[2] D. J. MacKay, *Information Theory, Inference and Learning Algorithms*. Cambridge University Press, 2003.

[3] D. M. Blei, A. Kucukelbir, and J. D. McAuliffe, "Variational inference: A review for statisticians," *Journal of the American Statistical Association*, vol. 112, no. 518, pp. 859–877, 2017.

[4] E. Angelino, M. J. Johnson, and R. P. Adams, "Patterns of scalable Bayesian inference," *Foundations and Trends in Machine Learning*, vol. 9, no. 2–3, pp. 119–247, 2016.

[5] C. Zhang, J. Bütepage, H. Kjellström, and S. Mandt, "Advances in variational inference," *IEEE Transactions on Pattern Analysis and Machine Intelligence*, vol. 41, no. 8, pp. 2008–2026, 2018.

[6] S. Mohamed, M. Rosca, M. Figurnov, and A. Mnih, "Monte Carlo gradient estimation in machine learning," arXiv preprint arXiv:1906.10652, 2019.

[7] T. Geffner and J. Domke, "On the difficulty of unbiased alpha divergence minimization," arXiv preprint arXiv:2010.09541, 2020.

[8] D. Koller and N. Friedman, *Probabilistic Graphical Models: Principles and Techniques*. The MIT Press, 2009.

[9] R. M. Neal and G. E. Hinton, "A view of the EM algorithm that justifies incremental, sparse, and other variants," in *Learning in Graphical Models*. Springer, 1998, pp. 355–368.

[10] D. P. Kingma and M. Welling, "An introduction to variational autoencoders," *Foundations and Trends in Machine Learning*, vol. 12, no. 4, pp. 307–392, 2019.

[11] Q. Liu and D. Wang, "Stein variational gradient descent: A general purpose Bayesian inference algorithm," arXiv preprint arXiv:1608.04471, 2016.

[12] S. Graf and H. Luschgy, *Foundations of Quantization for Probability Distributions*. Springer, 2007.

11 Information-Theoretic Inference and Learning

11.1 Overview

In this chapter, we consider inference and learning problems that can be formulated as the optimization of an **information-theoretic measure (ITM)**. Somewhat more formally, we are interested in problems of the form

$$\underset{q \in \mathcal{Q}}{\text{optimize}} \; \text{ITM}(q,p), \tag{11.1}$$

where "optimize" refers to a minimization or maximization; and $\text{ITM}(q,p)$ is an ITM, such as the KL divergence or free energy, that involves a **true distribution** p and a **model distribution** q. The model distribution q is constrained to lie in a given set of distributions $\mathcal{Q}$. Note that we are dropping the argument of the distributions p and q to simplify the notation and keep the discussion at a high level.

As summarized in Table 11.1, problems of the form (11.1) can be categorized as:

- **inference**, when the true distribution p is known, possibly in the form of an **unnormalized** version $\tilde{p}$ (see Sec. 3.9); or
- **learning**, when the true distribution p is unknown and only accessible via data samples.

Furthermore, the model class $\mathcal{Q}$ may contain:

- **likelihood-based models**, that is, distributions q with a differentiable, **analytical** form as a function of the model parameters, which defines the likelihood of the model parameters; or
- **likelihood-free models**, that is, distributions q that are defined in terms of **simulators** that produce samples drawn from q.

As notable examples, likelihood-based inference includes the VI and amortized VI problems, while likelihood-based learning includes the VEM problem, all of which were studied in Chapter 10. Likelihood-free inference and learning will be studied for the first time in this chapter. To compare likelihood-based and likelihood-free methods, note that typical likelihood-based models q can be used as simulators; while, generally, likelihood-free simulators do not provide analytical expressions for the distribution q. For instance, we have seen in Chapter 10 that VI and VEM solutions based on the REINFORCE gradient or the reparametrization trick leverage samples from a likelihood-based model q in order to obtain efficient optimization schemes for problem (11.1).

As we will detail in this chapter, the choice of the ITM, as well as the method used to compute it or estimate it, must be adapted to the specific inference/ learning and likelihood-based/ likelihood-free setting. The following are some key observations in this regard:

- If the true distribution p is only known in an unnormalized form $\tilde{p}$, an ITM such as the free energy $\text{F}(q||\tilde{p})$ should be used that does not require access to the corresponding normalized distribution p. For instance, this is the approach taken by VI and amortized VI.

419

- If the true distribution p is only accessible through data samples, i.e., in a learning problem, it is necessary to *estimate* the ITM using only these samples and not the actual distribution p. VEM can be interpreted as an example of this methodology.
- If the distributions p and q are both only accessible through samples, i.e., in a likelihood-free learning problem, it is necessary to estimate the ITM using *two* sets of samples, one for each distribution. This situation has not yet been encountered in this book.

Learning Objectives and Organization of the Chapter. By the end of this chapter, the reader should be able to:

- understand the impact of the choice of the ITM through the example of I-projections and M-projections (Sec. 11.2);
- understand the generalized VI (GVI) and generalized VEM (GVEM) problems, which provide broad frameworks to formulate problems of the form (11.1) for likelihood-based inference and learning, respectively (Sec. 11.3);
- formulate and address maximum-entropy learning (Sec. 11.4), InfoMax learning (Sec. 11.5), information bottleneck (Sec. 11.6), and rate-distortion encoding (Sec. 11.7) as special cases of the GVI and GVEM problems;
- understand and implement two-sample estimators of the KL divergence (Sec. 11.8) and of the more general class of f-divergences (Sec. 11.9);
- use two-sample estimators to address likelihood-free learning via generative adversarial networks (GANs) (Sec. 11.10); and
- understand the formulation of distributionally robust learning via ITMs (Sec. 11.11).

11.2 I-Projection and M-Projection

An important special case of the problem (11.1) concerns the approximation of a given target, true, distribution p within a class $\mathcal{Q}$ of variational distributions q via the minimization of an ITM such as the KL divergence. All inference and learning problems seen so far in this book that involve probabilistic models can be formulated in this way, including the VI, amortized VI, and VEM problems studied in Chapter 10. To start, in this section, we discuss the impact of the choice of the ITM on the properties of the approximation q obtained as a result of the optimization (11.1). The general considerations given in this section apply broadly to all all four settings listed in Table 11.1, as they pertain to the role of the ITM in determining the properties of the solution of problem (11.1). In this regard, we note that, for learning problems, the "true" distribution p should be replaced by the empirical distribution $p_{\mathcal{D}}$ obtained from the available data (see, e.g., Appendix 4.A).

Table 11.1 A taxonomy of information-theoretic inference and learning problems

	p (true)	q (model)
likelihood-based inference	analytical, possibly unnormalized	analytical
likelihood-based learning	data	analytical
likelihood-free inference	analytical, possibly unnormalized	simulator
likelihood-free learning	data	simulator

11.2.1 Defining I-Projection and M-Projection

In this section, we compare two standard metrics, namely the "direct" KL divergence $\text{ITM}(p,q) = \text{KL}(p||q)$ and the "reverse" KL divergence $\text{ITM}(p,q) = \text{KL}(q||p)$. To see the importance of both ITMs for inference and learning problems, note that the direct KL divergence is minimized over q by ML learning (see Sec. 4.9); while minimizing the reverse KL divergence over q is equivalent to the minimization of the free energy, which is adopted by VI and VEM (see Sec. 3.9).

Given a target distribution $p(x)$ and a family $\mathcal{Q}$ of parametric distributions $q(x|\phi)$, we accordingly consider the following two problems:

- **Information (I)-projection:**

$$\min_{\phi} \text{KL}(q(x|\phi)||p(x)); \tag{11.2}$$

- **Moment (M)-projection:**

$$\min_{\phi} \text{KL}(p(x)||q(x|\phi)). \tag{11.3}$$

The terminology "projection" suggests that, through problems (11.2) and (11.3), the true distribution $p(x)$ is "projected" into the space $\mathcal{Q}$ of distributions that can be parametrized in the form $q(x|\phi)$.

11.2.2 Mode-Seeking vs. Inclusive Approximations

As we discuss next, of the two ITMs defining I- and M-projections, one choice yields an "inclusive" approximation $q(x|\phi)$ that covers the entire domain of the target distribution $p(x)$, while the other produces an "exclusive" approximation that accounts only for the main mode of the target distribution (see also Sec. 3.6). We start with an example.

Example 11.1

To illustrate the key ideas, consider the "true" mixture-of-Gaussians distribution

$$p(x) = 0.3\mathcal{N}(x|-1, 0.3) + 0.7\mathcal{N}(x|1, 0.3), \tag{11.4}$$

and assume the approximating distribution $q(x|\phi) = \mathcal{N}(x|\nu, \sigma^2)$ with parameter vector $\phi = [\nu, \sigma^2]^T$ to be optimized. The M-projection amounts to an ML problem (with $p(x)$ representing the empirical distribution), and hence, as seen in Sec. 9.5, the optimal solution is given by the mean-matching parameters $\nu = \text{E}_{x \sim p(x)}[\text{x}] = 0.3 \cdot (-1) + 0.7 \cdot (1) = 0.4$ and $\sigma^2 = \text{E}_{x \sim p(x)}[\text{x}^2] - \nu^2 = 0.3 \cdot (1 + 0.3) + 0.7 \cdot (1 + 0.3) - (0.4^2) = 1.14$. In contrast, as discussed in Chapter 10, the I-projection can be computed numerically, e.g., via the reparametrization trick (see Sec. 10.9). The resulting approximations are illustrated in Fig. 11.1. The M-projection is seen to be "inclusive" of the entire support of the target distribution, while the I-projection is "exclusive" or mode-seeking.

The conclusions in Example 11.1 apply more generally. I-projections tend to be **mode-seeking** and **exclusive**. Mathematically, this is because the distribution $q(x|\phi)$ determines the support over which the distributions $p(x)$ and $q(x|\phi)$ are compared by the KL divergence

Figure 11.1 Ground-truth distribution $p(x)$ (dashed line) and approximations $q(x|\phi) = \mathcal{N}(x|\nu,\sigma^2)$ obtained via I- and M-projections (solid lines).

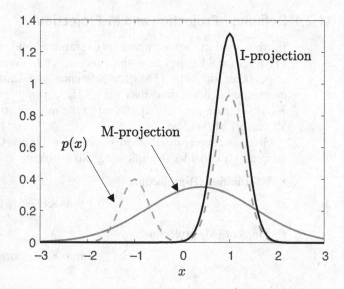

$\mathrm{KL}(q(x|\phi)||p(x))$. Therefore, an optimal solution can focus on a single mode of the distribution $p(x)$ without incurring a penalty in terms of KL divergence $\mathrm{KL}(q(x|\phi)||p(x))$. In contrast, M-projections tend to be **inclusive** and to span the entire support of $p(x)$. This is because the M-projection must avoid assigning zero values to distribution $q(x|\phi)$ at points x at which $p(x)$ is positive, so as to ensure a finite KL divergence $\mathrm{KL}(p(x)||q(x|\phi))$.

These conclusions have important implications for inference and learning. As an example, for ML learning, which is based on an M-projection, the trained model may overestimate the variance and produce distorted approximations of the population distribution. In contrast, VI and VEM, which are based on an I-projection, may underestimate the variance of the posterior distribution, and produce an approximation that does not properly account for the support of the posterior.

11.2.3 Bridging the Gap between I-Projection and M-Projection

An ITM that bridges the gap between the "reverse" and "direct" KL divergences used by the I- and M-projections, is the α-**divergence**. The α-divergence between two distributions $p(x)$ and $q(x)$ is defined as

$$\mathrm{D}_\alpha(p||q) = \frac{1}{\alpha(\alpha-1)} \mathrm{E}_{\mathrm{x}\sim q(x)}\left[\left(\frac{p(\mathrm{x})}{q(\mathrm{x})}\right)^\alpha - 1\right], \qquad (11.5)$$

where $\alpha \in \mathbb{R}$, with $\alpha \notin \{0,1\}$, is a parameter. It can be proved that, as $\alpha \to 0$, we obtain $\mathrm{D}_\alpha(p||q) = \mathrm{KL}(q||p)$, and, when $\alpha \to 1$, we have $\mathrm{D}_\alpha(p||q) = \mathrm{KL}(p||q)$. Consistently with the properties of I- and M-projections, performing projections of the type $\min_\phi \mathrm{D}_\alpha(p(x)||q(x|\phi))$ with $\alpha \leq 0$ yields mode-seeking, or exclusive, solutions, which become more so as α is decreased. Furthermore, for $\alpha \geq 1$, one obtains inclusive solutions, which become increasing so as α grows larger. A related definition of "α-divergence" with similar properties is given by Rényi's divergence (cf. (7.122) in Appendix 7.D with $S = 1$). We refer to Recommended Resources, Sec. 11.13, for further discussion.

11.3   Generalized Variational Inference and Generalized Variational Expectation Maximization

In this section, we introduce two broad frameworks for the definition of **likelihood-based** inference and learning problems of the form (11.1) that generalize the (amortized) VI and VEM problems studied in Chapter 10. The next four sections will then describe specific instantiations of the problem (11.1) that can be addressed within the two formulations presented here.

11.3.1   Generalized Variational Inference

As detailed in Chapter 10, VI provides a framework for the definition of **likelihood-based** **inference** problems when exact Bayesian inference is intractable. Given a joint distribution $p(x, z)$, for a fixed input value x, VI tackles the problem (10.1) of minimizing the free energy $F(q(z|x)||p(x, z))$ with respect to the soft predictor $q(\cdot|x)$, also known as **variational posterior**, in a set Q_x of distributions. When the set Q_x in unconstrained, the exact minimization of the free energy yields the posterior $p(z|x)$. VI allows for more general choices of the set Q_x of variational posteriors and of optimization algorithms, so as to trade accuracy for tractability. In this section, we introduce an extension of the VI framework that encompasses a larger set of ITMs.

To elaborate, we start by rewriting the VI problem (10.1) equivalently as (10.79), i.e., as

$$\min_{q(\cdot|x) \in Q_x} \left\{ \underbrace{E_{z \sim q(z|x)} \left[-\log p(x|z) \right]}_{\text{average log-loss in reconstructing } x \text{ from } z} + \underbrace{KL(q(z|x)||p(z))}_{\text{divergence between variational posterior and prior}} \right\},$$

(11.6)

where we recall that the observation x is fixed. In (11.6), the first term is a measure of loss, while the second is one of "complexity" of the variational posterior $q(z|x)$ as measured by the KL divergence to the prior distribution $p(z)$ of the latent rv.

Generalized VI (GVI) allows for more general choices of loss function and complexity measure than the log-loss and KL divergence, respectively, that are adopted by the VI formulation (11.6). To define a GVI problem, one chooses a loss function $\ell(x, z)$, accounting for how well the observation x is represented by z, as well as a regularizing complexity measure $C(q(z|x))$ of the variational posterior $q(z|x)$. With these two choices, the **GVI problem** is defined as

$$\min_{q(\cdot|x) \in Q_x} \left\{ \underbrace{E_{z \sim q(z|x)} \left[\ell(x, z) \right]}_{\text{average loss in reconstructing } x \text{ from } z} + \alpha \cdot \underbrace{C(q(z|x))}_{\text{regularizing complexity measure for the variational posterior}} \right\},$$

(11.7)

where $\alpha > 0$ is a **"temperature" parameter** (not to be confused with the α parameter in the α-divergence). The temperature parameter dictates the trade-off assumed by the objective (11.7) between loss and regularization, and is to be considered as a hyperparameter. We will refer to the objective of problem (11.7) as a **generalized free energy**.

The VI problem (11.6) is obtained as a special case of the GVI problem (11.7) by setting $\alpha = 1$, loss function as the log loss $\ell(x,z) = -\log p(x|z)$, and complexity measure as the KL divergence $C(q(z|x)) = \mathrm{KL}(q(z|x)||p(z))$.

Complexity measures $C(q(z|x))$ for the variational posterior $q(z|x)$ are typically chosen to be **convex** functions of $q(\cdot|x)$ (again, recall that x is fixed). Examples of convex regularizing functions include:

- **divergence**, or **distance**, measures between $q(z|x)$ and a prior distribution $p(z)$, such as the KL divergence or, more generally, one of the f-divergences to be reviewed in Sec. 11.9 or one of the integral probability metrics (IPMs) to be described in Sec. 11.10;
- the negative **entropy** $-\mathrm{H}(q(z|x))$.

Accordingly, a variational posterior $q(z|x)$ may be considered more "complex" if it is further from a reference prior, or if it has a small entropy. In the latter case, the soft predictor becomes excessively confident, possibly failing to account for the existing predictive uncertainty.

When a divergence metric to a prior is used as the complexity measure, the temperature parameter α can be used as a means to quantify how reliable the prior is as compared to the data. For instance, if the available data are noisy – perhaps because the labels were annotated by an unreliable crowd of volunteers or low-paid workers – one may wish to increase α.

Amortized VI can be similarly generalized. We will discuss this generalization in the context of VEM in the next subsection.

11.3.2 Generalized Variational Expectation Maximization

The VEM problem introduced in Chapter 10 provides a framework for the definition of **likelihood-based learning** problems in the presence of latent variables. In such problems, one has access to a data set $\mathcal{D} = \{(x_n)_{n=1}^N\}$ of N training points, and the goal is to optimize over a model parameter vector θ defining a joint distribution $p(x,z|\theta)$. The loss function is the log-loss $-\log p(x,z|\theta)$, which depends not only on the observation x but also on a latent variable z. Accordingly, VEM learning involves the inference of the latent rvs via VI as a key subtask. In this subsection, we extend the VEM framework in a manner analogous to GVI by allowing for general loss functions and regularizing complexity measures.

To elaborate, let us fix a parametric family $\mathcal{Q}$ of variational posterior distributions $q(z|x,\varphi)$, which depend on the variational parameter vector φ. Note that, unlike VI, the variational parameters φ are "amortized" among all values x. The VEM problem is then formulated as (10.4), which is restated here for convenience as

$$\min_{\theta} \min_{\varphi} \frac{1}{N} \sum_{n=1}^{N} \left\{ \mathrm{E}_{z_n \sim q(z_n|x_n,\varphi)} \left[-\log p(x_n|z_n,\theta) \right] + \mathrm{KL}(q(z_n|x_n,\varphi)||p(z)) \right\}, \qquad (11.8)$$

for a prior distribution $p(z)$. As in the case of VI, the objective is the sum of a loss term and a regularizer that gauges the complexity of the variational posterior.

In a manner similar to the generalization of VI to GVI, we can extend the VEM problem (11.8) by allowing for any loss functions $\ell(x,z|\theta)$ – now dependent on the model parameter vector θ – and for any complexity measure on the variational posterior $q(z|x,\varphi)$. This yields the **generalized VEM (GVEM)** problem

$$\min_{\theta} \min_{\varphi} \frac{1}{N} \sum_{n=1}^{N} \left\{ E_{z_n \sim q(z_n|x_n,\varphi)} \left[\ell(x_n, z_n|\theta) \right] + \alpha C(q(z_n|x_n,\varphi)) \right\}, \tag{11.9}$$

where $\alpha > 0$ is the temperature parameter. We refer to the previous subsection for a discussion of complexity measures $C(q(z|x,\varphi))$.

The VEM problem (11.8) is obtained as a special case of the GVEM problem (11.9) by setting $\alpha = 1$, loss function as the log loss $\ell(x, z|\theta) = -\log p(x|z,\theta)$, and complexity measure as the KL divergence $C(q(z|x,\varphi)) = KL(q(z|x,\varphi)||p(z))$. The generalized amortized VI problem can also be obtained as a special case of (11.9) when the model parameter θ is fixed and not subject to optimization.

11.3.3 Solving the GVI and GVEM Problems: Generalized Posterior Distributions

As in the case of VI, the GVI problem (11.7) can in principle be solved exactly when no constraints are imposed on the set Q_x, although the solutions are only computationally tractable for small-dimensional problems. For VI, the optimal unconstrained solution is the posterior distribution $p(z|x)$, while the optimal unconstrained solution of the GVI problem is referred to as **generalized posterior distribution**. In this subsection, we will discuss generalized posterior distributions for some instances of the GVI problem. These solutions can in principle also be used to address the corresponding unconstrained GVEM problem via an iterative two-step approach like the EM algorithm.

To start, consider the most common choice for the complexity measure, namely the KL divergence $C(q(z|x)) = KL(q(z|x)||p(z))$. Unlike the VI problem, we allow for a general loss function, and the resulting optimal solution of the GVI problem (11.7) is proved in Appendix 11.A to be the generalized posterior

$$q^*(z|x) \propto p(z) \exp\left(-\frac{1}{\alpha} \ell(x, z) \right), \tag{11.10}$$

which is also known as the **Gibbs posterior**. The Gibbs posterior (11.10) corresponds to the standard posterior distribution $p(z|x)$ if $\alpha = 1$ when the loss function is the log-loss, i.e., when $\ell(x, z) = -\log p(x|z)$. By (11.10), the Gibbs posterior "**tilts**" the prior $p(z)$ with a term that depends, exponentially, on the loss $\ell(x, z)$.

For other complexity measures, the **generalized posteriors** can be computed by leveraging connections to **Fenchel duality**, as detailed in Appendix 11.B. Examples of generalized posteriors can be found in Table 11.2. In the table, we have defined the squared ℓ_2 distance

Table 11.2 Some generalized posterior distributions (solutions of the GVI problem)

$C(q(z\|x))$	generalized posterior $q^*(z\|x)$
$KL(q(z\|x)\|\|p(z))$	$\propto p(z) \exp\left(-\frac{1}{\alpha} \ell(x, z) \right)$
$-H(q(z\|x))$	$\propto \exp\left(-\frac{1}{\alpha} \ell(x, z) \right)$
$\frac{1}{2}\|\|p(z) - q(z\|x)\|\|^2$	$\left(p(z) - \frac{1}{\alpha} \ell(x, z) - \tau \right)^{+}$

$$C(q(z|x)) = \frac{1}{2}||p(z) - q(z|x)||^2 = \frac{1}{2}\sum_z (p(z) - q(z|x))^2, \qquad (11.11)$$

where the sum is replaced by an integral for pdfs; and parameter τ satisfies the equality $\sum_z \left(p(z) - \frac{1}{\alpha}\ell(x, z) - \tau\right)^+ = 1$, with $(x)^+ = \max(x, 0)$, where, again, an integral should replace the sum for continuous rvs.

Example 11.2

Figure 11.2 shows the generalized posterior for prior $p(z) = \mathcal{N}(z|0, 1)$ and for the loss function plotted in the figure as a dashed black line. When $\alpha = 10$, the generalized posterior puts more weight on minimizing the complexity term $C(q(z|x))$. As a result, for $C(q(z|x)) = -H(q(z|x))$, the generalized posterior is close to a uniform distribution, whereas it approximates the prior $p(z)$ for $C(q(z|x)) = KL(q(z|x)||p(z))$ and $C(q(z|x)) = \frac{1}{2}||p(z) - q(z|x)||^2$. For smaller temperature values α, the emphasis shifts on minimizing the average loss, and the generalized posterior increasingly concentrates on the minimizer of the loss function $\ell(x, z)$.

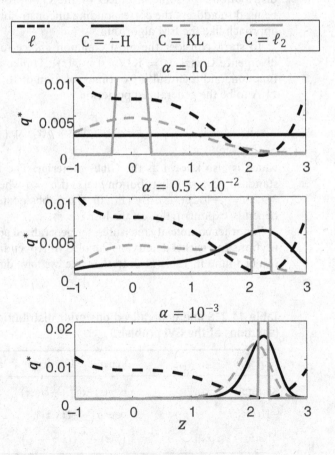

Figure 11.2 Illustration of generalized posteriors for different complexity measures $C(q(z|x))$. The dashed black line represents the loss function $\ell(x, z)$ (for a fixed value of x); the solid black line is the generalized posteriors for the KL measure; the solid gray line is for the quadratic distance; and the dashed gray line is for the negative entropy.

11.3.4 Approximate Solution of the GVI Problem

The generalized posteriors are difficult to compute explicitly for problems with a large latent space. In such cases, gradient-based solutions based on the REINFORCE gradient or on the reparametrization trick, detailed in Chapter 10, can be applied to tackle the GVI and GVEM problems. In this chapter, we will tacitly assume that any of the encountered GVI or GVEM problems can be addressed either exactly for small-dimensional problems or, more generally, via one of the scalable optimization studied in Chapter 10.

11.4 Maximum-Entropy Learning

As seen in Sec. 4.9, ML learning aims to find a model that maximizes the probability of the training set. In this section, we review an alternative **likelihood-based learning** paradigm, whereby data are used to define constraints on the model, and the model is optimized within such constraints to maximize its entropy. In this way, the model preserves the largest degree of uncertainty, while satisfying the constraints imposed by the data.

11.4.1 Defining Maximum-Entropy Learning

We consider the standard frequentist learning formulation in which we are given a data set $\mathcal{D} = \{(x_n)_{n=1}^N\}$ drawn i.i.d. from an unknown population distribution $p(x)$. Furthermore, we have defined, as part of the inductive bias, a family $\mathcal{H}$ of parametric distributions $p(x|\theta)$. We are interested in selecting a model $p(x|\theta)$ that approximates the unknown population distribution $p(x)$. Hence, the problem can be described as **density learning** (see Sec. 7.3).

Maximum-entropy learning assumes that there are some specific functions, or statistics, $s_k(x)$ of the data, whose average we wish the trained model $p(x|\theta)$ to preserve. This imposes the following **mean-matching constraints** on the model $p(x|\theta)$:

$$E_{x \sim p(x|\theta)}[s_k(x)] = \underbrace{\frac{1}{N}\sum_{n=1}^N s_k(x_n)}_{s_k(\mathcal{D})} \tag{11.12}$$

for $k = 1, \ldots, K$. In words, the expectation under the model $p(x|\theta)$ of each statistic $s_k(x)$ should equal the corresponding empirical average $s_k(\mathcal{D})$ observed in the data.

The **maximum-entropy learning problem** is accordingly defined as the maximization of the entropy of the model distribution $p(x|\theta)$ subject to the moment constraints (11.12), i.e., as

$$\max_{\theta} \text{H}(p(x|\theta))$$

$$\text{s.t. } E_{x \sim p(x|\theta)}[s_k(x)] = s_k(\mathcal{D}) \text{ for } k = 1, \ldots, K. \tag{11.13}$$

As anticipated, through this maximization, one searches for the distribution $p(x|\theta)$ that preserves the largest degree of uncertainty, while being consistent with the empirical averages of the sufficient statistics.

11.4.2 Addressing Maximum-Entropy Learning

Maximum-entropy learning is typically addressed indirectly by considering the *unconstrained* problem

$$\max_{\theta} \left\{ H(p(x|\theta)) + \sum_{k=1}^{K} \lambda_k \left(E_{x \sim p(x|\theta)} [s_k(x)] - s_k(\mathcal{D}) \right) \right\} \tag{11.14}$$

for some *Lagrange multipliers* $\{\lambda_k\}_{k=1}^{K}$ to be optimized as hyperparameters. The Lagrange multipliers are fixed constants that determine the relative importance of the entropy criterion and the satisfaction of the constraints in (11.13) in the optimization (11.14).

We now show that this optimization belongs to the class of GVI problems (11.7). To see this, we can write (11.14) in the equivalent form

$$\min_{\theta} \left\{ \underbrace{E_{x \sim p(x|\theta)} \left[\sum_{k=1}^{K} \lambda_k \left(s_k(\mathcal{D}) - s_k(x) \right) \right]}_{\text{average loss}} + \underbrace{(-H(p(x|\theta)))}_{\text{regularizing complexity measure}} \right\}. \tag{11.15}$$

Problem (11.15) is obtained as a special case of the GVI problem (11.7) by setting $\alpha = 1$, the loss function as $\ell(x) = \sum_{k=1}^{K} \lambda_k (s_k(\mathcal{D}) - s_k(x))$, and the complexity measure as the negative entropy. Note that rv x in (11.15) plays the role of rv z in the GVI problem (11.7).

If the class of model distributions $p(x|\theta)$ is unconstrained, the optimal solution follows from Table 11.2 as

$$p(x) \propto \exp \left(- \sum_{k=1}^{K} \lambda_k \left(s_k(\mathcal{D}) - s_k(x) \right) \right) \propto \exp \left(\sum_{k=1}^{K} \lambda_k s_k(x) \right). \tag{11.16}$$

This confirms the conclusion reached in Sec. 9.3 that, if the family of parametric distributions is large enough, the optimal solution of the entropy maximization problem under constraints on the expectations of the statistics $\{s_k(\cdot)\}_{k=1}^{K}$ is given by the exponential family distribution with sufficient statistics given by the $K \times 1$ vector $s(x) = [s_1(x), \ldots, s_K(x)]^T$.

11.5 InfoMax

In this section, we move from entropy maximization for density estimation to mutual information maximization for the **likelihood-based learning** problem of training discriminative latent-variable models. As introduced in Sec. 7.6, discriminative latent-variable models are of the form $p(z|x, \theta)$, in which x is the observation and rv z is a hidden representation of the input. As such, model $p(z|x, \theta)$ plays the role of a **stochastic encoder** from observation to representation.

11.5.1 Defining InfoMax

As in the usual formulation of unsupervised learning, we are given data set $\mathcal{D} = \{(x_n)_{n=1}^{N}\}$ of values x_n, drawn i.i.d. from the unknown population distribution $p(x)$. Combining the unknown population distribution with the stochastic encoder, we can define the joint distribution

$$p(x, z|\theta) = p(x)p(z|x, \theta). \tag{11.17}$$

We will write $p(x|z, \theta)$ for the conditional distribution of rv x given $z = z$ under this joint distribution. We are interested in the problem of finding a stochastic encoder $p(z|x, \theta)$, within the given set of parametric conditional distributions $p(z|x, \theta)$, that maximizes the **mutual information** between input and representation, i.e.,

$$\max_{\theta} \mathrm{I}(x; z|\theta). \tag{11.18}$$

The notation $\mathrm{I}(x; z|\theta)$ emphasizes that the mutual information is computed with respect to the joint distribution (11.17), which depends on the model parameter vector θ.

Unlike the maximum-entropy problem studied in the preceding section, if we do not impose any constraint on the model class for $p(z|x, \theta)$, the solution of the maximum mutual information problem (11.18) is trivial: The maximum mutual information is in fact obtained by simply setting $z = x$, i.e., by using the input itself as the representation. Therefore, we henceforth assume that we have some meaningful constraint on the model class $p(z|x, \theta)$ that limits its capacity. For instance, the dimension of the representation z may be constrained to live in a space of smaller dimension than the input.

In general, the mutual information can be expressed as

$$\mathrm{I}(x; z|\theta) = \mathrm{E}_{(x,z) \sim p(x)p(z|x,\theta)} \left[\log \left(\frac{p(x|z,\theta)}{p(x)} \right) \right]. \tag{11.19}$$

Hence, since the population distribution $p(x)$ does not depend on the model parameter vector θ, the optimization problem (11.18) can be written as the minimization

$$\min_{\theta} \mathrm{E}_{(x,z) \sim p(x)p(z|x,\theta)} \left[-\log p(x|z,\theta) \right]. \tag{11.20}$$

In problem (11.20), the model parameters θ determine both the averaging distribution and the loss function being averaged. This is akin to the GVI problem. However, unlike the GVI problem, the loss function $-\log p(x|z, \theta)$ depends on the unknown data distribution $p(x)$, since we have $p(x|z, \theta) = p(x)p(z|x, \theta)/p(z|\theta)$ with $p(z|\theta) = \mathrm{E}_{x \sim p(x)}[p(z|x, \theta)]$. As we see next, this dependence can be addressed by optimizing a bound on the cost function in (11.20).

11.5.2 Addressing InfoMax

To proceed, we apply the variational bound (3.90) to the mutual information. To this end, we introduce a parametric **stochastic decoder** $q(x|z, \varphi)$, which is to be considered as a variational distribution, and obtain from (3.90) the inequality

$$\mathrm{E}_{(x,z) \sim p(x)p(z|x,\theta)} \left[-\log p(x|z,\theta) \right] \le \mathrm{E}_{(x,z) \sim p(x)p(z|x,\theta)} \left[-\log q(x|z,\varphi) \right]. \tag{11.21}$$

We recall from Chapter 3 that this bound follows directly from Gibbs' inequality. The bound is clearly tight if the stochastic decoder $q(x|z, \varphi)$ is identical to the conditional distribution $p(x|z, \theta)$.

We can now consider the problem of minimizing the right-hand side of (11.21) over model parameters θ and variational parameters φ, i.e.,

$$\min_{\theta} \min_{\varphi} \mathrm{E}_{(x,z) \sim p(x)p(z|x,\theta)} \left[\underbrace{-\log q(x|z,\varphi)}_{\text{decoder's log-loss}} \right]. \tag{11.22}$$

The rationale for this choice of problem formulation is that, if the class of probabilistic decoders is large enough, the inner minimization returns the cost function of the original problem (11.20).

The problem (11.22) has an interesting interpretation in its own right. For a fixed vector φ, it seeks to find an **encoder** $p(z|x,\theta)$ that produces representations z from which the original data point x can be reconstructed (decoded) with a small loss via the decoder $q(z|x,\varphi)$. Intuitively, if, through the outer minimization, we can find a variational parameter φ that yields a small average decoder's loss, then the output z of the encoder must be a good representation of x.

A key computational advantage of the formulation (11.22) is that the log-loss being averaged does not depend on the unknown population distribution. Thanks to this, the cost function in (11.22) can be estimated using the available data set via an empirical average, yielding the problem

$$\min_{\theta} \min_{\varphi} \frac{1}{N} \sum_{n=1}^{N} E_{z_n \sim p(z|x_n,\theta)}[-\log q(x_n|z_n,\varphi)]. \tag{11.23}$$

This formulation is referred to as **variational InfoMax**.

To connect this optimization to the GVEM (11.9), let us swap the roles of the parameter vectors θ and φ, in the sense that model and variational distributions in (11.9) take the place of the variational and model distributions in (11.23), respectively. Then, the minimization in (11.23) is in the form of the GVEM problem with $\alpha = 0$ and log-loss $\ell(x,z|\varphi) = -\log q(x|z,\varphi)$. Accordingly, there are two key differences with respect to the standard VEM formulation, and hence to VAEs (see Sec. 10.13): First, no regularization term is included; and, second, the roles of model parameter θ and variational parameters φ are switched.

11.6   Information Bottleneck

In this section, we study another **likelihood-based** unsupervised **learning** problem that is based on the mutual information metric, namely the **information bottleneck**. As InfoMax, the information bottleneck aims at learning representations z from an input x via a **stochastic encoder** $p(z|x,\theta)$. Unlike InfoMax, the training data includes not only the input x, but also a target variable t. That is, the training set is given as $\mathcal{D} = \{(x_n, t_n)_{n=1}^{N}\}$. The variables x and t are jointly distributed according to an unknown population distribution $p(x,t)$ in a manner similar to supervised learning. For example, the pair (x, t) may include input x and a classifying label t. Unlike supervised learning, however, the goal is not to train a predictor of t from x, but instead to train a stochastic encoder $p(z|x,\theta)$ that extracts a representation z from x that is informative about the target variable t.

While the InfoMax approach studied in the preceding section aims to find a representation z that is maximally informative about the input x, the information bottleneck method seeks a representation z that is maximally informative about the target t, while discarding any information about x that is not relevant for t. Mathematically, we would like to simultaneously maximize the mutual information $I(t;z|\theta)$ between target t and representation z and minimize the mutual information $I(x;z|\theta)$ between input x and representation z. As in the preceding section, the notation emphasizes the dependence of the mutual information terms on the encoder's parameters θ.

Accordingly, the information bottleneck problem is formulated as

$$\max_{\theta} \{I(t;z|\theta) - \lambda I(x;z|\theta)\}, \tag{11.24}$$

where the mutual information is evaluated with respect to the joint distribution

$$p(x, t, z|\theta) = p(x, t)p(z|x, \theta), \tag{11.25}$$

combining population distribution $p(x, t)$ and stochastic encoder $p(z|x, \theta)$, and the maximization is over the parameters θ of the encoder $p(z|x, \theta)$. The hyperparameter $\lambda > 0$ determines the relative importance of the two mutual information terms in the problem (11.24).

As a first observation, unlike the InfoMax problem, the encoder $p(z|x, \theta)$ obtained by the information bottleneck is not trivial even when no constraints are imposed on it. This is because the term $I(x; z|\theta)$ in (11.24) penalizes the trivial solution $z = x$. We will describe a procedure to optimize an unconstrained encoder in the context of a related problem in the next section.

From a computational perspective, like InfoMax, the information bottleneck problem (11.24) is made complicated by the dependence of the joint distribution (11.25) on the unknown population distribution $p(x, t)$. To address this problem, we follow a similar approach to that in the preceding section, and leverage the variational bounds (3.89) and (3.90) on the mutual information.

To elaborate, let us introduce two variational distributions, namely a parametric **stochastic decoder** $q(t|z, \varphi)$ that predicts the target t given the representation z; and a marginal distribution $q(z)$ on the space of the representations. We take the distribution $q(z)$ to be fixed, although it could more generally depend also on the variational parameters φ. The maximization of the resulting upper bound on the information bottleneck objective can be formulated as the problem (see Appendix 11.C)

$$\min_{\theta} \min_{\varphi} \left\{ E_{(x,t,z)\sim p(x,t)p(z|x,\theta)} \underbrace{[- \log q(t|z, \varphi)]}_{\text{decoder's log-loss}} + \lambda E_{(x,z)\sim p(x)p(z|x,\theta)} \left[\log \left(\frac{p(z|x, \theta)}{q(z)} \right) \right] \right\}. \tag{11.26}$$

As in the variational InfoMax, a key advantage of the formulation (11.26) is that the objective can be estimated by using the data set via the problem

$$\min_{\theta} \min_{\varphi} \left\{ \frac{1}{N} \sum_{n=1}^{N} E_{z_n \sim p(z|x_n, \theta)}[- \log q(t_n|z_n, \varphi)] + \lambda E_{z_n \sim p(z|x_n, \theta)} \left[\log \left(\frac{p(z_n|x_n, \theta)}{q(z_n)} \right) \right] \right\}. \tag{11.27}$$

This optimization is typically referred to as the **variational information bottleneck problem**.

Interpreting $q(z)$ as a prior and swapping the roles of the parameter vectors θ and φ, as in the preceding section, the minimization (11.26) is again in the form of the GVEM problem (11.9). This can be seen by choosing the log-loss $\ell(x, t, z|\varphi) = - \log q(t|z, \varphi)$ and complexity measure $C(p(z|x, \theta)) = E_{z\sim p(z|x,\theta)} \left[\log \left(\frac{p(z|x,\theta)}{q(z)} \right) \right]$.

11.7 Rate-Distortion Encoding

Related to the information bottleneck is the standard setting of **rate-distortion theory**, whereby one seeks a **stochastic encoder** $p(z|x, \theta)$ that minimizes the mutual information $I(x; z|\theta)$ while guaranteeing some level of average distortion with respect to a target variable t. In the rate-distortion problem, we specifically choose the target variable as the input itself, i.e., $t = x$, and we allow for general distortion metrics.

To elaborate, define a **distortion function** $d(x, z)$, which can be interpreted as the loss of using the representation z to estimate x. Correspondingly, the average distortion is given as $E_{(x,z) \sim p(x,z|\theta)}[d(x, z)]$, where the joint distribution is as in (11.17), combining population distribution $p(x)$ and stochastic encoder $p(z|x, \theta)$. The **rate-distortion problem** of interest is stated as the minimization

$$\min_{\theta} \left\{ I(x; z|\theta) + \lambda E_{(x,z) \sim p(x,z|\theta)}[d(x, z)] \right\}. \tag{11.28}$$

In a manner similar to the information bottleneck problem (11.24), the cost function in (11.28) describes a trade-off between "complexity" of the representation – measured by the mutual information $I(x; z|\theta)$ – and accuracy of the representation – as accounted for by the average distortion. The hyperparameter $\lambda > 0$ determines the relative weights of complexity and accuracy.

As in the information bottleneck problem, the rate-distortion problem (11.28) is not trivial even with an unconstrained encoder. We now elaborate on the case in which problem (11.28) is addressed with an unconstrained encoder. To this end, we assume a stochastic encoder $q(z|x)$ that can be any arbitrary conditional distribution. Accordingly, we drop the dependence on the model parameters θ, and use the notation $q(z|x)$, in lieu of $p(z|x, \theta)$, in lieu of $p(z|x, \theta)$, to indicate that the distribution is under optimization. Furthermore, in order to use the variational bound (3.89), we introduce a variational marginal distribution $q(z)$.

The problem (11.28) can now be equivalently stated as

$$\min_{q(\cdot)} E_{x \sim p(x)} \left[\min_{q(\cdot|x)} \left\{ \underbrace{E_{z \sim q(z|x)} \left[\log \left(\frac{q(z|x)}{q(z)} \right) \right]}_{KL(q(z|x)||q(z))} + \lambda E_{z \sim q(z|x)}[d(x, z)] \right\} \right]. \tag{11.29}$$

The equivalence follows from (3.89), since, for a fixed encoder $q(z|x)$, the solution $q^*(z)$ of the outer minimization gives the marginal distribution $p(z)$ of the joint distribution $p(x)q(z|x)$, i.e.,

$$q^*(z) = p(z) = \sum_x p(x)q(z|x), \tag{11.30}$$

with the sum replaced by an integral for continuous rvs x.

The inner optimization over the encoder $q(z|x)$ in (11.29) can be carried out separately for each value x without loss of optimality. The resulting problem can be expressed as a GVI by writing the objective, scaled by $1/\lambda$, as

$$E_{z \sim q(z|x)}[d(x, z)] + \frac{1}{\lambda} KL(q(z|x)||q(z)). \tag{11.31}$$

Therefore, by Table 11.2, for a fixed variational marginal $q(z)$, the optimal solution of the inner optimization of problem (11.29) is given by the stochastic encoder

$$q^*(z|x) \propto q(z) \exp(-\lambda d(x, z)), \tag{11.32}$$

which "tilts" the prior as an exponential function of the distortion metric. Furthermore, as mentioned, for a fixed encoder $q(z|x)$, the optimal solution of the outer minimization is given by the marginal (11.30).

Overall, to tackle problem (11.26) we have obtained an **alternating optimization** procedure that is known as the **Blahut–Arimoto algorithm** in information theory. The Blahut–Arimoto algorithm alternates between computing the stochastic encoder (11.32) for a fixed $q(z)$ and evaluating the marginal $p(z)$ in (11.30) to update $q(z)$.

In this section, we have focused on the case of an unconstrained encoder. Extensions using parametric encoders follow in a manner analogous to the information bottleneck problem studied in the preceding section.

As a final remark, it is possible to extend the rate-distortion encoding problem by including so-called **perceptual distortion** constraints. These penalize encoders that yield large values for some measure of divergence between the marginal distribution of the reconstruction x and the population distribution $p(x)$.

11.8   Two-Sample Estimation of Information-Theoretic Metrics

So far, we have studied likelihood-based learning problems, in which the model distribution under optimization, q, depends on model parameters through an explicit, differentiable, analytical expression. Accordingly, the learning methods studied up to now are based on the minimization of empirical estimates of an ITM that leverage the known analytical expression for the model q as well as samples from the population distribution p. In **likelihood-free learning** problems, the analytical form of the model distribution q is not known, but we have access to a *simulator* that can produce samples from it. Likelihood-free models are more flexible, and may be preferable in applications that require complex generative mechanisms, e.g., for image generation. For likelihood-free learning, we hence need to estimate ITMs based on samples from *both* population distribution p and model distribution q.

In order to build the necessary tools to tackle the problem of likelihood-free learning, in this section and the next we investigate the problem of estimating ITMs based on data sets of samples drawn from the two distributions p and q. We refer to estimators of this type as **two-sample estimators**. Specifically, in this section, we derive two-sample estimators for the KL divergence and for the mutual information. In the next section, the KL divergence is generalized, along with the corresponding two-sample estimators, to a broader class of ITMs known as f-divergences.

11.8.1   Two-Sample Estimation of the KL Divergence

Consider the problem of estimating the KL divergence $\mathrm{KL}(p||q)$ for two unknown distributions $p(x)$ and $q(x)$. Assume that we are given the following two data sets:

$$\mathcal{D}_1 = \{(\mathrm{x}_n)_{n=1}^{N_1}\} \underset{\text{i.i.d.}}{\sim} p(x)$$

$$\mathcal{D}_2 = \{(\mathrm{x}'_n)_{n=1}^{N_2}\} \underset{\text{i.i.d.}}{\sim} q(x), \tag{11.33}$$

with data set $\mathcal{D}_1$ consisting of N_1 i.i.d. samples drawn from $p(x)$ and data set $\mathcal{D}_2$ consisting of N_2 i.i.d. samples drawn from $q(x)$. We would like to use these two data sets to estimate $\mathrm{KL}(p||q)$. Recall that the KL divergence $\mathrm{KL}(p||q)$ depends on the LDR $\log(p(x)/q(x))$ through an average over distribution $p(x)$ as per the definition

$$\mathrm{KL}(p||q) = \mathrm{E}_{\mathrm{x} \sim p(x)}\left[\log\left(\frac{p(\mathrm{x})}{p(\mathrm{x})}\right)\right]. \tag{11.34}$$

A simple approach would be to use a **plug-in estimator**: We first evaluate an estimate of the distribution $p(x)$ from data set $\mathcal{D}_1$ as well as an estimate of $q(x)$ from $\mathcal{D}_2$, and then we plug these estimates into the KL divergence definition (11.34). This approach tends to be inefficient, as it is unnecessary to estimate the two distributions $p(x)$ and $q(x)$ (two functions) when the goal is to estimate their KL divergence (a single scalar). This becomes increasingly more problematic as the dimension of the vectors x increases.

Many modern machine learning techniques leverage not the original definition (11.34) of the KL divergence, but an alternative, equivalent formulation known as the **Donsker–Varadhan (DV) variational representation**. The DV representation turns out, in fact, to be more amenable to the derivation of empirical estimators. The DV representation of the KL divergence is given as the maximization

$$\mathrm{KL}(p||q) = \max_{T(\cdot) \in \mathcal{T}} \left\{ \mathrm{E}_{\mathrm{x} \sim p(x)}[T(\mathrm{x})] - \log \left(\mathrm{E}_{\mathrm{x} \sim q(x)}[\exp(T(\mathrm{x}))] \right) \right\}, \tag{11.35}$$

where $\mathcal{T}$ is the space of (bounded) real-valued functions $T(\cdot)$. The advantage of this formulation is that, for any fixed function $T(\cdot)$, the expression on the right-hand side of (11.35) can be directly estimated by means of empirical averages that use the available data sets $\mathcal{D}_1$ and $\mathcal{D}_2$.

That said, in order to obtain the KL divergence, by (11.35), one needs to address the technical challenge of optimizing (11.35) over function $T(\cdot) \in \mathcal{T}$. There are two main ways to do this, which are discussed next.

11.8.2 Discriminator-Based Two-Sample Estimators of the KL Divergence

A natural idea is to constrain the function $T(\cdot)$ to lie within a class of parametric functions $T(\cdot|\varphi)$ with parameter vector φ in order to facilitate the optimization in (11.35). The function $T(\cdot|\varphi)$ is referred to as the **discriminator**. **Discriminator-based two-sample estimators** of the KL divergence are hence defined by the optimization

$$\mathrm{KL}(p||q) \simeq \max_{\varphi} \left\{ \frac{1}{N_1} \sum_{n=1}^{N_1} T(x_n|\varphi) - \log \left(\frac{1}{N_2} \sum_{n=1}^{N_2} \exp(T(x'_n|\varphi)) \right) \right\} \tag{11.36}$$

over the discriminator model parameter φ. In (11.36), we have estimated the two expectations in (11.35) through empirical averages based on the two available data sets (11.33). In the standard implementation, the discriminator is modeled as a neural network, and the optimization (11.36) is done via GD and backprop.

11.8.3 Classifier-Based Two-Sample Estimators of the KL Divergence

As a more flexible solution to the problem of optimizing (11.35) over function $T(\cdot) \in \mathcal{T}$, we now introduce classifier-based two-sample estimators of the KL divergence.

To start, note that the exact optimal solutions $T^*(\cdot)$ of problem (11.35) include functions of the general form (see Appendix 11.D)

$$T^*(x) = \log \left(\frac{p(x)}{q(x)} \right) + c, \tag{11.37}$$

where c is an arbitrary constant. That is, an optimal function $T^*(\cdot)$ can be obtained from the LDR $\log(p(x)/q(x))$.

It follows that, if we have an estimate $T(x|\varphi)$ of the LDR, typically dependent on some trainable parameter φ, we can plug it into the DV representation (11.35) to obtain an estimate of the KL divergence as in (11.36), i.e.,

$$\text{KL}(p||q) \simeq \frac{1}{N_1} \sum_{n=1}^{N_1} T(x_n|\varphi) - \log\left(\frac{1}{N_2} \sum_{n=1}^{N_2} \exp(T(x'_n|\varphi))\right). \tag{11.38}$$

Note that, while in (11.36) the parameter vector φ is optimized by directly maximizing the expression for the estimator, in (11.38) vector φ is obtained via the preliminary step of estimating the LDR. As such, the quality of the estimate of the KL divergence (11.38) depends both on the amount of data available and on the accuracy of the approximation $T(x|\varphi)$ of the LDR.

As we have studied in Sec. 7.3.3, an estimate of the LDR can be obtained via **contrastive ratio estimation** by training a binary **classifier** $p(t|x,\theta)$ to distinguish between samples from distribution $p(x)$ and samples from distribution $q(x)$. In particular, an estimate $T(x|\varphi)$ of the LDR can be obtained as

$$\log\left(\frac{p(x)}{q(x)}\right) \simeq \log\left(\frac{p(t=1|x,\varphi)}{p(t=0|x,\varphi)}\right) = T(x|\varphi) \tag{11.39}$$

when $N_1 = N_2$. Using (11.39) into (11.38) yields a **classifier-based two-sample estimator** of the KL divergence. Note that it is also possible to use a different set of samples for training of the binary classifier and for estimation of the DV representation in (11.38).

11.8.4 Properties of Two-Sample Estimators of the KL Divergence

Both discriminator and classifier-based two-sample estimators of the KL divergence are **biased** because of the presence of the logarithm in the second term of the DV representation. On the positive side, the estimates are **consistent**: As the number of data points increases, as long as the discriminator or the classifier has enough capacity, the estimate will tend to the ground-truth value. That said, the choice of large-capacity models yields a larger variance in the non-asymptotic regime, establishing a bias-variance, or bias-estimation error, trade-off akin to that studied in Sec. 4.5. See Recommended Resources, Sec. 11.13, for further details.

Example 11.3

Consider two-dimensional multivariate Gaussian distributions $\mathcal{N}(-1_2, I_2)$ and $\mathcal{N}(1_2, 0.5I_2)$. From Table 9.3, the KL divergence can be expressed in closed form as

$$\text{KL}(\mathcal{N}(-1_2, I_2)||\mathcal{N}(1_2, 0.5I_2)) = \log(0.5) + 8. \tag{11.40}$$

Assume now that we have $N_1 = N_2 = N$ samples from each distribution. To implement a classifier-based two-sample estimator of the KL divergence, we train a logistic regression binary classifier $p(t=1|x,\varphi) = \sigma(\varphi^T u(x))$ with 3×1 feature vector $u(x) = [x^T, 1]^T$. The top part of Fig. 11.3 shows $N_1 = N_2 = 100$ samples from the two distributions, along with the decision line of the trained logistic regression classifier. The bottom figure plots the evolution of the KL estimate (11.38) across the iterations used to train the classifier with a learning rate $\gamma = 0.1$ (solid line). Note that the estimate of the KL divergence can be negative. Furthermore, validating its consistency property, it converges to the ground-truth value (11.40) (dashed line).

Figure 11.3 Top: samples from two multivariate Gaussian distributions, along with the decision line of a trained logistic regression classifier; bottom: evolution of the KL estimate, obtained with a classifier-based two-sample estimator, across the iterations used to train the classifier with a learning rate $\gamma = 0.1$ (solid line), along with the ground-truth value (dashed line).

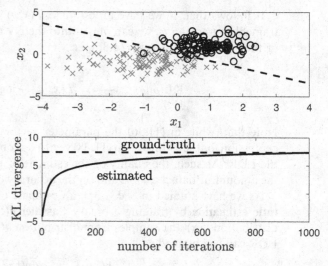

11.8.5 Estimation of the Mutual Information

As an application of the estimators for the KL divergence developed in this section, we now derive an estimator of the mutual information. The mutual information $\mathrm{I}(\mathrm{x};\mathrm{y})$ between two random variables (x,y) jointly distributed according to $p(x,y)$ can be written as the KL divergence

$$\mathrm{I}(\mathrm{x};\mathrm{y}) = \mathrm{KL}(p(x,y)\|p(x)p(y)) \tag{11.41}$$

between the joint distribution and the product of its marginals. Assume that the joint distribution $p(x,y)$ is unknown, but a data set

$$\{(\mathrm{x}_n, \mathrm{y}_n)_{n=1}^N\} \underset{\mathrm{i.i.d.}}{\sim} p(x,y), \tag{11.42}$$

with N i.i.d. samples from the joint distribution $p(x,t)$, is given. Can we leverage the estimators developed above for the KL divergence in order to estimate the mutual information based on these samples?

To do this, we can use the available $N_1 = N$ samples from the joint distribution $p(x,y)$. But we also need N_2 samples from the product-of-marginals distribution $p(x)p(y)$. One way to obtain $N_2 = N$ such samples is to randomly permute the samples y_n, so as to create new pairs $(x_n, y_{\pi(n)})$, with $n = 1, \ldots, N$, where $\pi(n)$ is a permutation of the integers $\{1, 2, \ldots, N\}$. The resulting approach is typically referred to as **mutual information neural estimator (MINE)** when a neural network is used as function $T(\cdot|\varphi)$ for the estimation of the KL divergence. Note that the label "MINE" is sometimes applied only to schemes that use a discriminator-based estimator.

Example 11.4

Consider two-dimensional multivariate Gaussian variables with zero mean, variance equal to 1, and covariance coefficient $\rho = 0.7$. The top part of Fig. 11.4 shows samples from the

Figure 11.4 Top: samples from the joint distribution $p(x, t)$ as crosses, and samples from the product-of-marginals distribution obtained by randomly permuting the samples $\{y_n\}_{n=1}^{N}$ as circles; bottom: evolution of the mutual information estimate, obtained with a classifier-based two-sample estimator, as a function of the number of data points, N, along with its ground-truth value.

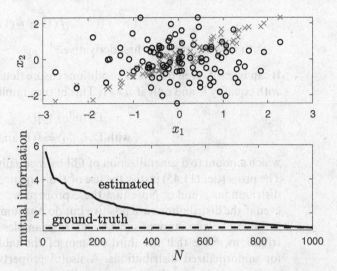

joint distribution as crosses, and samples from the marginal distribution obtained by randomly permuting the samples $\{y_n\}$ as circles. As can be inferred by inspection, a linear classifier would not be a good choice to distinguish between the two classes of samples. Therefore, we adopt QDA as described in Sec. 6.5 to implement a classifier-based two-sample estimator of the mutual information. The bottom part of the figure shows the mutual information estimate as a function of the number of data points, N. The trend of the estimated mutual information towards its ground-truth value is in line with the consistency of the KL estimator.

11.9 Beyond the KL Divergence: *f*-Divergences and Two-Sample Estimators

In this section, we move beyond the KL divergence, defining a broader class of information-theoretic divergences and developing corresponding two-sample estimators.

11.9.1 *f*-Divergences

The KL divergence is part of the larger class of *f*-divergences between two distributions $p(x)$ and $q(x)$. An **f-divergence** has the form

$$D_f(p\|q) = E_{x \sim q(x)}\left[f\left(\frac{p(x)}{q(x)} \right) \right],$$ (11.43)

where the **defining function** $f(\cdot)$, with domain $\mathbb{R}^+$ and output in $\mathbb{R}$, satisfies the following conditions:

- it is strictly convex; and
- it is such that

$$f(1) = f'(1) = 0, \tag{11.44}$$

with $f'(\cdot)$ denoting the first derivative.

It can be easily seen that these conditions ensure that the function satisfies the inequality $f(u) \geq 0$ with equality if and only if $u = 1$. This in turn implies the useful conditions

$$D_f(p||q) \geq 0$$
$$\text{with } D_f(p||q) = 0 \text{ if and only if } p = q, \tag{11.45}$$

which amount to a generalization of Gibbs' inequality from the KL divergence to f-divergences. The properties (11.45) justify the use of f-divergences as metrics for the "distance" between two distributions p and q. Note that these properties apply even if p and q are *unnormalized*, i.e., even if the distributions are positive but do not sum or integrate to 1.

Table 11.3 summarizes some standard examples of f-divergences between normalized distributions. Note that the third column of the table can be easily modified to account also for unnormalized distributions. A useful property is that all f-divergences can be locally approximated by χ^2-divergences (see Recommended Resources, Sec. 11.13).

To avoid possible confusion, note that it is also common not to impose the condition $f'(1) = 0$ in (11.44) when defining f-divergences, hence assuming only that the function $f(\cdot)$ is strictly convex and that it satisfies the equality $f(1) = 0$. In fact, even under this smaller set of assumptions, the Gibbs-like inequality (11.45) can be guaranteed to hold for *normalized* distributions by Jensen's inequality (5.114). Furthermore, removing the condition $f'(1) = 0$ allows one to also include among f-divergences the α-**divergence** (11.5) with $f(u) = 1/(\alpha(1-\alpha))(u^\alpha - u)$; as well as the **total variation (TV) distance**, which is defined as

$$D_{TV}(p(x)||q(x)) = \frac{1}{2}\sum_x |q(x) - p(x)|, \tag{11.46}$$

with the sum replaced by an integral for continuous latent rvs, by choosing $f(u) = 0.5|u - 1|$.

11.9.2 Two-Sample Estimation of f-Divergences

In a manner analogous – but not equivalent – to the DV representation (11.35) of the KL divergence, one can obtain the variational representation of f-divergences as (see Appendix 11.E)

Table 11.3 Examples of f-divergences

divergence	$f(u)$	formula
KL	$u \log u - u + 1$	$\mathrm{KL}(p||q) = \mathrm{E}_{x \sim p(x)}\left[\log\left(\frac{p(x)}{q(x)}\right)\right]$
reverse KL	$-\log u + u - 1$	$\mathrm{KL}(q||p) = \mathrm{E}_{x \sim q(x)}\left[\log\left(\frac{q(x)}{p(x)}\right)\right]$
χ^2	$(u-1)^2$	$\chi^2(p||q) = \mathrm{E}_{x \sim p(x)}\left[\frac{1}{p(x)q(x)}(p(x) - q(x))^2\right]$
$(2\times)$ JS	$u \log(u) - (u+1)\log(u+1) + u\log(2) + \log(2)$	$2\mathrm{JS}(q||p) = \mathrm{KL}\left(p \left\| \frac{p+q}{2}\right.\right) + \mathrm{KL}\left(q \left\| \frac{p+q}{2}\right.\right)$

$$D_f(p||q) = \max_{T(\cdot)\in\mathcal{T}}\{E_{x\sim p(x)}[a_f(T(x))] - E_{x\sim q(x)}[b_f(T(x))]\}, \tag{11.47}$$

where $\mathcal{T}$ is the space of (bounded) real-valued functions, and we define functions

$$a_f(T) = f'(\exp(T))$$
$$b_f(T) = \exp(T)f'(\exp(T)) - f(\exp(T)), \tag{11.48}$$

where we recall that $f'(\cdot)$ is the first derivative of function $f(\cdot)$.

Example 11.5

For the KL divergence, using Table 11.3, the variational representation (11.47) yields

$$KL(p||q) = \max_{T(\cdot)\in\mathcal{T}}\{E_{x\sim p(x)}[T(x)] - E_{x\sim q(x)}[\exp(T(x))] + 1\}, \tag{11.49}$$

where we can verify that we have the functions $a_f(T) = T$ and $b_f(T) = \exp(T) - 1$. The expression (11.49) differs from the DV representation (11.35). In fact, for any fixed function $T(\cdot)$, the right-hand side of (11.49) is smaller than that of (11.35), indicating that the DV representation provides stronger bounds on the KL divergence. (This can be seen by using the inequality $x - 1 \geq \log(x)$.)

Example 11.6

For the JS divergence, using Table 11.3, the variational representation (11.47) amounts to

$$JS(p||q) = \max_{T(\cdot)\in\mathcal{T}}\left\{\frac{1}{2}E_{x\sim p(x)}[\log(\sigma(T(x)))] + \frac{1}{2}E_{x\sim q(x)}[\log(\sigma(-T(x)))] + \log(2)\right\}. \tag{11.50}$$

This problem corresponds to the minimization of the population cross-entropy loss for a discriminative model $p(t = 1|x) = \tilde{\sigma}(T(x))$ when the label t is a Bern(0.5) rv and the two conditional distributions are given as $p(x|t = 1) = p(x)$ and $p(x|t = 0) = q(x)$. Appendix 3.B offers additional details on this interpretation.

Discriminator and Classifier-Based Two-Sample Estimators of *f*-Divergences. Following the same steps as for the KL divergence in the preceding section, one can develop discriminator- and classifier-based two-sample estimators for *f*-divergences based on the variational representation (11.47). Since discriminator-based solutions are straightforward to derive, we now briefly discuss classifier-based estimators.

As in the DV representation of the KL divergence, the minimizer of the variational representation (11.47) is given by the LDR $T^*(x) = \log(p(x)/q(x))$ (see Appendix 11.E). Therefore, as in the preceding section, we can estimate the optimal discriminator $T^*(x)$ by means of **contrastive ratio estimation**. Based on such an estimate $T(x|\varphi)$ of the LDR, one can then derive an estimator of any *f*-divergence based on data sets (11.33) as

$$D_f(p\|q) \simeq \frac{1}{N_1} \sum_{n=1}^{N_1} a_f(T(x_n|\varphi)) - \frac{1}{N_2} \sum_{n=1}^{N_2} b_f(T(x'_n|\varphi)). \tag{11.51}$$

11.10 Generative Adversarial Networks

With the background on two-sample estimators in the last two sections under our belts, we are now ready to tackle **likelihood-free learning** problems. In such problems, the model distribution, defined by model parameters θ, is only accessible through a simulator that produces samples drawn from it.

There are several ways to address this class of learning problems. For instance, one could estimate a "surrogate" likelihood on the basis of the samples produced by the simulator, and then use one of the methods developed for likelihood-based learning. In this section, we will focus on a state-of-the-art approach based on two-sample estimators of ITMs – **generative adversarial networks (GANs)**.

11.10.1 GAN Model

GANs are likelihood-free directed generative models of the form

$$z \sim \underbrace{p(z)}_{\text{latent prior}}$$

$$x = \underbrace{G(z|\theta)}_{\text{decoder}}, \tag{11.52}$$

where $G(z|\theta)$ is a deterministic decoding function parametrized by a vector θ, which is also referred to as **generator**. In most implementations, the generator is defined by a neural network with weights θ. The distribution $p(x|\theta)$ of the observation rv x obtained from the model (11.52) is known as the **pushforward distribution** of the latent distribution $p(z)$ through the generator $G(z|\theta)$. This distribution is not available in an explicit analytical form unless the generator is invertible (see Appendix 11.F) – which is not the case for most neural-network generators. Instead, the model (11.52) can be used as a *simulator* to produce samples of rv x by first sampling the latent rv z and then applying the generator as x = $G(z|\theta)$.

11.10.2 GAN Learning Problem

GANs aim to minimize an f-divergence between the unknown population distribution $p(x)$ and the pushforward distribution $p(x|\theta)$ entailed by the generative model (11.52). Mathematically, the problem is formulated as the optimization

$$\min_{\theta} D_f(p(x)\|p(x|\theta)). \tag{11.53}$$

Note that the original GAN scheme used a specific f-divergence, namely the JS divergence, while the more general formulation (11.53) is sometimes referred to as f-**GAN**. We will not make this distinction here.

The use of an f-divergence other than the KL divergence can have important practical advantages for the generation of natural signals, such as images. In fact, as seen in Sec. 11.2, using the "direct" KL divergence, one obtains M-projections, which tend to be inclusive, and hence to produce distributions that smooth out the modes of the data distribution; while adopting the "reverse" KL divergence yields I-projections, which tend to be exclusive, hence discarding some of the modes in the data distribution. These two approaches have been demonstrated to yield unrealistic or insufficiently diverse outputs in applications such as image synthesis. Through experiments, using alternative f-divergences such as the JS divergence has been seen to provide more realistic outputs.

GANs leverage the variational representation (11.47) of the f-divergence, which yields the following problem, equivalent to (11.53):

$$\min_{\theta} \max_{T(\cdot) \in \mathcal{T}} \{E_{x \sim p(x)}[a_f(T(x))] - E_{z \sim p(z)}[b_f(T(G(z|\theta)))]\}, \tag{11.54}$$

where we have used the GAN model (11.52) to describe samples from distribution $p(x|\theta)$. To tackle problem (11.54), GANs optimize a two-sample estimate of the f-divergence.

To elaborate, assume, as usual, that we are given a data set $\mathcal{D} = \{x_n\}_{n=1}^{N}$. In order to estimate the learning criterion in (11.54), for any fixed value θ of the generator's parameters, we can generate N samples $\{z_n\}_{n=1}^{N}$ from the latent prior $q(z)$, as well as the corresponding samples $\{(x_n' = G(z_n|\theta))_{n=1}^{N}\}$ from the model (pushforward) distribution $p(x|\theta)$. As a result, we have two data sets, one consisting of data points and one of **"fake" data** that are produced by the generator.

Based on these two data sets, we can estimate the f-divergence in (11.54) by using a two-sample estimator for the given fixed model parameter vector θ, and then using this estimate to update the model parameters θ. The process can be repeated by alternating between estimating the f-divergence through the generation of "fake" samples and updating the model parameters θ.

Example 11.7

Consider a population distribution $p(x)$ given by $\mathcal{N}(\mu, I_2)$ with $\mu = [3, 3]^T$, for which we have available a data set $\mathcal{D}$ of $N = 1000$ samples. Let us assume a simple generative model defined by latent prior $p(z) = \mathcal{N}(z|0_2, I_2)$ and generator $G(z|\theta) = z + \theta$ with model parameter $\theta \in \mathbb{R}^2$. Note that, in this toy example, the likelihood $p(x|\theta)$ can be computed explicitly. (Furthermore, the model is well specified.) As in the original GAN scheme, we use the JS divergence as the f-divergence in problem (11.53).

To train the generator's parameters θ, for each update of θ we apply a classification-based two-sample estimator of the JS divergence. For this estimator, we adopt a logistic regression classifier with feature vector $u(x) = [1, x^T]^T$. Note that the LDR obtained from this classifier is given as $T(x|\varphi) = \varphi^T u(x)$, where φ is the classifier's weight vector. We perform alternate optimization over the classifier's weights φ and over the generator's parameters θ: For a fixed θ, we carry out multiple SGD steps over φ by using standard logistic regression updates (see Sec. 6.3); and then we carry out one SGD step for θ with learning rate 0.1.

Fig. 11.5 shows samples from the data as gray circles, and samples drawn from the current model (11.52) at iteration i as black crosses. The number of iterations, i, refers to the number of updates of the generator's parameters θ. The current logistic regression decision line is shown as a dashed line. The figure shows that the outlined alternating optimization procedure can produce a pushforward distribution that matches the original data distribution well. When this happens, the classifier is unable to distinguish real samples (circles) from "fake" samples drawn from the model (crosses).

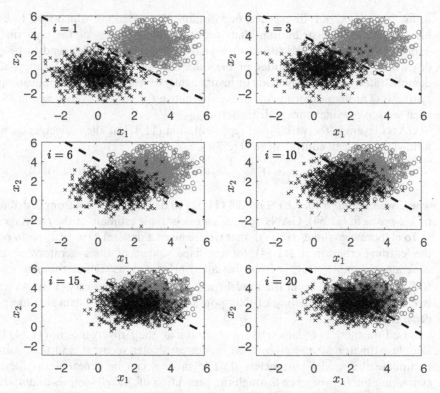

Figure 11.5 Training data as gray circles, as well as samples from the GAN model at training iteration i as black crosses. The current logistic regression decision line used to estimate the JS-divergence is shown as a dashed line.

11.10.3 GAN As a Min-Max Optimization Problem

In the most common implementation, unlike Example 11.7, GANs leverage a *discriminator-based* two-sample estimator of the f-divergence. From (11.54), this yields the problem

$$\min_{\theta} \max_{\varphi} \left\{ \frac{1}{N} \sum_{n=1}^{N} a_f(T(x_n|\varphi)) - \frac{1}{N} \sum_{n=1}^{N} b_f(T(G(z_n|\theta)|\varphi)) \right\}. \tag{11.55}$$

Problem (11.55) has a different form as compared to the learning formulations encountered so far: It does not address the *minimization* of a training loss; instead, it poses a **min-max problem**. While the goal of minimization is to obtain stationary points or, ideally, local or global minima of the training loss, min-max problems seek a **saddle point** of the training criterion.

An optimal solution (θ^*, φ^*) of problem (11.55) should be such that: (*i*) θ^* minimizes the training loss within the curly brackets with respect to θ for fixed $\varphi = \varphi^*$; while (*ii*) φ^* maximizes the same criterion over φ for a fixed $\theta = \theta^*$. Therefore, at an optimal point (θ^*, φ^*), the curvature of the function along any direction obtained by varying θ for fixed φ^* must be non-negative, while the curvature of the function along any direction obtained by varying φ for a fixed θ^* cannot be positive (see Sec. 5.7). This implies that (θ^*, φ^*) must be a saddle point for the learning criterion.

Optimizing problem (11.55) requires a careful handling of minimization and maximization steps, and is typically carried out by alternating GD steps over θ and gradient ascent steps over φ. In practice, various tricks are required to avoid the instability of the resulting dynamics in the space of parameters (θ, φ). The choice of the f-divergence is also important in facilitating convergence. See Recommended Resources, Sec. 11.13, for further discussion.

11.10.4 GAN Training As a Game between Generator and Discriminator

The min-max optimization in (11.55) can be interpreted as a **zero-sum game** between generator $G(z|\theta)$ and discriminator $T(x|\varphi)$. This is in the sense that both generator and discriminator aim to optimize the same learning criterion, with the generator wishing to minimize it and the discriminator to maximize it. Intuitively, the discriminator aims to distinguish between samples from the data set and samples produced by the generator. Conversely, the objective of the discriminator is to "fool" the discriminator into classifying generated samples as being real samples from the data set.

11.10.5 Integral Probability Metrics-Based GANs

An alternative to f-divergences that has also proved useful for the implementation of GANs is given by **integral probability metrics (IPMs)**. In this subsection, we provide a brief introduction.

An IPM between distributions $p(x)$ and $q(x)$ is defined by the variational representation

$$\mathrm{IPM}(p||q) = \max_{T(\cdot) \in \mathcal{T}} \{ \mathrm{E}_{x \sim p(x)}[T(x)] - \mathrm{E}_{x \sim q(x)}[T(x)] \}, \tag{11.56}$$

where different IPMs are obtained by choosing distinct optimization domains $\mathcal{T}$. Intuitively, for a fixed **test function** $T(\cdot)$, the metric (11.56) measures how distinguishable the averages of rv $T(x)$ are when evaluated under the two distributions $p(x)$ and $q(x)$. The maximization in (11.56) selects the test function that yields maximally distinct averages within the class of functions $\mathcal{T}$. We elaborate on two examples.

A first notable example is given by the **Wasserstein distance**, for which the optimization domain $\mathcal{T}$ is the set of all functions that are 1-Lipschitz. (This is also known as 1-Wasserstein distance: The Wasserstein distance family includes other distance metrics that are not IPMs.) A function $T(\cdot)$ is 1-Lipschitz if it satisfies the inequality $||T(x) - T(x')|| \leq ||x - x'||$ for all x, x' in its domain. This implies that the function $T(x)$ is continuous, and does not change too quickly. For differentiable functions, this condition is equivalent to the inequality $||\nabla T(x)|| \leq 1$ on the gradient.

Another important example is given by the **maximum mean discrepancy (MMD)**, for which the optimization domain $\mathcal{T}$ is the set of all functions that belong to an RKHS defined by some positive semi-definite kernel function $\kappa(x, x')$ (see Sec. 4.11) and have unitary norm. In this case, the optimization in (11.56) can be obtained in closed form, yielding the simple expression

$$\mathrm{MMD}(p||q) = \sqrt{\mathrm{E}_{x_1, x_2 \sim p(x), x_1', x_2' \sim q(x)}[\kappa(x_1, x_2) + \kappa(x_1', x_2') - 2\kappa(x_1, x_1')]}, \tag{11.57}$$

where all rvs x_1, x_2, x_1', and x_2' are mutually independent. Note that we have $\mathrm{MMD}(p||q) = 0$ if the two distributions p and q are identical.

Other IPMs include **Stein discrepancies** and the TV (11.46), which correspond to different choices of the set of functions $\mathcal{T}$. Stein discrepancies have the interesting property that they can be estimated using samples from one distribution, say p, and the score vector $\nabla_x \log q(x)$ from the other, making it ideal for evaluating and training energy-based models (see Sec. 7.7.3). Note that IPMs are distinct from f-divergences, and that they have in common with f-divergences (when removing the constraint $f'(1) = 0$ in (11.44)) only the TV distance.

To see the potential benefits of IPMs as compared to f-divergences such as the KL divergence, consider two pdfs $p(x) = \delta(x - x_0)$ and $q(x) = \delta(x - x_0')$ with $x \in \mathbb{R}$, where $\delta(\cdot)$ is the Dirac delta function. The two distributions are concentrated at two different values $x_0 \neq x_0'$ on the real line $\mathbb{R}$. The KL divergences $\mathrm{KL}(p\|q)$ and $\mathrm{KL}(q\|p)$ are both equal to infinity, and the JS divergence takes as its maximal value $\log(2)$. In contrast, with the kernel $\kappa(x, x') = xx'$, the MMD yields $\mathrm{MMD}(p\|q) = |x_0 - x_0'|$. The MMD, and more generally other IPMs such as the Wasserstein distance, can thus provide a more informative assessment of the difference between the two distributions, which may be more useful for guiding a training algorithm.

In terms of estimation, when the optimization in (11.56) can be computed explicitly, as for the MMD, a two-sample estimate can be directly obtained by averaging over the two data sets. More generally, as long as the functions in the optimization domain $\mathcal{T}$ can be parametrized in the form of a discriminator $T(x|\varphi)$, IPMs can be estimated in a manner similar to f-divergences via discriminator-based two-sample estimators. We refer to Recommended Resources, Sec. 11.13, for details.

11.11   Distributionally Robust Learning

In this chapter, we have discussed the use of several information-theoretic divergence measures involving two distributions as learning criteria. An alternative formulation of the learning problem uses divergence measures, not a learning criteria, but as a means to ensure the robustness of the trained model against the imperfect knowledge of the population distribution that arises from the availability of limited data. This section provides an introduction to this topic.

To start, let us define as $p_{\mathcal{D}}$ the empirical distribution of the observable variables given the training set $\mathcal{D}$. The empirical distribution $p_{\mathcal{D}}$ is an estimate of the true, unknown population distribution p. Distributionally robust learning caters for the discrepancy between the population distribution and the empirical distribution. Accordingly, it aims to find a model parameter θ that performs well for *all* possible population distributions $\tilde{p}$ that are "close" to the empirical distribution $p_{\mathcal{D}}$. This condition is made explicit by using some divergence measure $\mathrm{D}(\tilde{p}\|p_{\mathcal{D}})$, e.g., the KL divergence, through the inequality $\mathrm{D}(\tilde{p}\|p_{\mathcal{D}}) \leq \epsilon$ for some hyperparameter threshold value $\epsilon > 0$.

This leads to the **min-max** learning problem

$$\min_{\theta} \max_{\tilde{p}:\, \mathrm{D}(\tilde{p}\|p_{\mathcal{D}}) \leq \epsilon} L_{\tilde{p}}(\theta), \tag{11.58}$$

where we have defined $L_{\tilde{p}}(\theta)$ as the population loss under the candidate population distribution $\tilde{p}$. In words, distributionally robust learning aims to find the model parameter θ that is guaranteed to yield the minimum population loss when the population distribution is chosen adversarially to maximize the population loss within the set of plausible distributions defined by the inequality $\mathrm{D}(\tilde{p}\|p_{\mathcal{D}}) \leq \epsilon$. Note that setting $\epsilon = 0$ recovers the standard ERM approach. Solving problem (11.58) is generally more complex as it requires us to alternate between minimization and maximization steps in a manner similar to GAN training.

11.12   Summary

- Most inference and learning frameworks can be described as prescribing the optimization of an information-theoretic metric $\text{ITM}(q, p)$ over a model distribution q for a ground-truth, population distribution p.
- In inference problems, the distribution p is known, while in learning problems it is only available through data samples.
- When the distribution q is defined in terms of a differentiable analytical model, we have likelihood-based inference or learning problems. Generalized VI (GVI) and generalized VEM (GVEM) provide broad frameworks for the definition of these problems that recover as special cases maximum entropy, InfoMax, information bottleneck, and rate-distortion encoding, as well as the VI and VEM problems studied in Chapter 10.
- Models that are only accessible through a simulator are referred to as being likelihood-free. For likelihood-free models, two-sample estimators of ITMs are typically used in order to formulate the learning problem.
- Two-sample estimators can be derived for the KL divergence, as well as for the more general class of f-divergences and for the alternative class of integral probability metrics (IPMs). These estimators are obtained via variational representations that require optimization over auxiliary functions taking the form of a discriminator or of a (binary) classifier.
- With two-sample estimators, the corresponding learning problems result in min-max optimizations. An important instance is given by GANs, which are likelihood-free directed generative models.
- ITMs can also be useful in defining distributionally robust learning problems in which training is done by targeting the worst-case distribution within some information-theoretic "distance" from the empirical distribution.

11.13   Recommended Resources

I-projections and M-projections are detailed in [1], and the α-divergence is presented in [2]. The GVI and GVEM problems are motivated and detailed in [3, 4], while, to the best of my knowledge, their connections to ITM-based learning problems are explicitly elucidated in this chapter for the first time. The presentation of entropy maximization, InfoMax, information bottleneck, and rate-distortion encoding are in line with recent literature on variational learning [5, 6]. Rate-distortion encoding builds on the classical Blahut–Arimoto algorithm in rate-distortion theory [7]. Useful references on two-sample estimators and corresponding training algorithms, include GANs, are [8, 9, 10, 11, 12]. Specifically, the presentation of f-divergences in this chapter follows the review paper [8]. Finally, it is mentioned that the label "information-theoretic learning" is also applied to training strategies that optimize an information-theoretic metric, such as Rényi's entropy, of the empirical distribution of the error $t - \hat{t}(x|\theta)$ estimated via kernel methods [13].

Problems

11.1 For a two dimensional mixture-of-Gaussians distribution

$$p(x) = 0.4\mathcal{N}(-1_2, 0.3I_2) + 0.6\mathcal{N}(1_2, 0.3I_2),$$

address the following points. (a) Obtain the M-projection in the family of Gaussian distributions $p(x|\phi)$ with parameters ϕ including mean vector and covariance matrix. (b*) For the same family of distributions, develop a reparametrization-trick approach to tackle the I-projection problem. (c*) Illustrate contour lines of the true distribution $p(x)$ and of the M- and I-projections. Describe the two projections as mode-seeking or inclusive approximations of the original distribution $p(x)$.

11.2 Consider the GVI problem with a Gaussian two-dimensional prior $p(z) = \mathcal{N}(0_2, I_2)$ and the loss function $\ell(x, z) = -\log(z_1) + z_1^2 + z_2^2$ (note that x is fixed and implicit in this expression). (a) Derive the generalized posteriors obtained as the exact solutions of the respective GVI problems under KL, negative entropy, and quadratic distance for temperature $\alpha = 10$. (b*) Plot the contour lines of the generalized posteriors computed in point (a). (c*) Repeat the previous two points for a smaller value of α and comment on your results.

11.3 Assume that we are given the empirical mean $s_1(\mathcal{D}) = 1/N \sum_{n=1}^{N} x_n = 1$ for some data set $\mathcal{D}$ and that the rv x, which is to be modeled is known to be non-negative. (a) If no constraints are imposed on the distribution $p(x)$ of rv x, what is its maximum-entropy distribution? (b) Assume now a variational distribution $p(x|\phi)$ modeled as a Gaussian $\mathcal{N}(\mu, \sigma^2)$ with parameter vector $\phi = [\nu, \sigma^2]^T$. Develop a gradient descent-based algorithm that leverages the reparametrization trick to tackle the maximum-entropy learning problem for a fixed Lagrangian multiplier λ_1. (c*) Implement the algorithm, and plot the resulting distribution $p(x|\phi)$ superimposed on the optimal unconstrained distribution obtained in point (a). (d*) Comment on the role of the Lagrange multiplier λ_1.

11.4 (a) For the variational InfoMax problem, develop the updates of a gradient-based solution via the REINFORCE gradient. (b) Comment on the differences between the scheme developed in point (a) and black-box VI introduced in Chapter 10. (c) Then, assume a Gaussian encoder with parametrized mean and covariance matrix, and develop a reparametrization-based strategy. (d) Comment on the differences between the solution developed in point (c) and reparametrization-based VI discussed in Chapter 10.

11.5 Develop an iterative algorithm for the solution of the information bottleneck problem when no constraints are imposed on the stochastic encoder. [Hint: Try to follow the same steps as detailed in the text for rate-distortion encoding.]

11.6 For rate-distortion encoding, assume binary rvs x and z and the detection-error loss as distortion function. Detail the two steps of the Blahut–Arimoto algorithm as a function of λ.

11.7 Consider the problem of two-sample estimation of the KL divergence. (a*) Implement and verify Example 11.3 by reproducing Fig. 11.3. (b*) Propose and implement an alternative solution based on a two-layer neural network classifier, producing an illustration similar to Fig. 11.3. (c*) Repeat point (b*) for a discriminator-based two-sample estimator.

11.8 Consider the estimation of the mutual information based on MINE. (a*) Implement and validate Example 11.4, reproducing Fig. 11.4. (b*) Propose and test an alternative based on a discriminator-based two-sample estimator.

11.9 Implement and validate Example 11.7, reproducing Fig. 11.5.

Appendices

Appendix 11.A: Proof of the Optimality of the Gibbs Posterior Distribution (11.10)

Let us write the generalized free energy in the GVI problem (11.7) with complexity measure $C(q(z|x)) = \text{KL}(q(z|x)||p(z))$ as

$$E_{z \sim q(z|x)}\left[\ell(x,z)\right] + \alpha \text{KL}(q(z|x)||p(z)) = E_{z \sim q(z|x)}\left[\ell(x,z)\right] + \alpha E_{z \sim q(z|x)}\left[\log\left(\frac{q(z|x)}{p(z)}\right)\right]$$

$$= \alpha E_{z \sim q(z|x)}\left[\log\left(\frac{q(z|x)}{p(z)\exp\left(-\frac{1}{\alpha}\ell(x,z)\right)}\right)\right]$$

$$= \alpha \text{F}(q(z|x)||p(z)\exp(-\alpha^{-1}\ell(x,z))). \tag{11.59}$$

It follows that the GVI problem (11.7) with an unconstrained variational posterior can be reformulated as the minimization

$$\min_{q(\cdot|x)} \text{F}(q(z|x)||p(z)\exp(-\alpha^{-1}\ell(x,z))). \tag{11.60}$$

By (3.102), the solution is given by the normalized distribution (11.10), which concludes the proof.

Appendix 11.B: Minimization of the Free Energy and Fenchel Duality

In this appendix, we show that the GVI problem (11.7) with an unconstrained set Q_x of variational distributions is equivalent to the problem of computing the convex dual, or Fenchel dual, function of the regularizing complexity measure $C(q(z|x))$ (see Appendix 9.A for an introduction to convex duality). This result is useful since the convex dual of most common complexity measures is known, and hence the expressions of the generalized posteriors can be obtained directly from convex duality theory. (That said, these expressions may be intractable to compute as they involve hard-to-evaluate normalization constants.) The generalized posteriors in Table 11.2 can be derived through this correspondence between GVI and Fenchel duality.

To start, we recall from Appendix 9.A that, given a function $f(u)$, with u being a finite-dimensional real vector, its convex dual function $f^*(v)$ is given as the solution of the problem

$$f^*(v) = \max_u \left\{ u^T v - f(u) \right\}, \tag{11.61}$$

where vector v has the same dimension as u. How is the GVI problem related to the computation of the Fenchel dual?

Assume that the alphabet $\mathcal{Z}$ of rv z is discrete and finite. We will focus on this case, although the main conclusion applies more generally. To fix some notation, we set $\mathcal{Z} = \{1,\ldots,N_z\}$, although the elements of the alphabet are arbitrary. We can then write the generalized free energy – i.e., the objective of the GVI problem (11.7) – scaled by the temperature α and with a sign change as

$$-E_{z \sim q(z|x)}\left[\frac{\ell(x,z)}{\alpha}\right] - C(q(z|x)) = q^T\left(-\frac{l}{\alpha}\right) - C(q), \tag{11.62}$$

where vectors q and l are defined in Table 11.4. In the table, we have introduced the function $I_S(q) = 0$ if $||q||_1 = 1$ and $I_S(q) = \infty$ otherwise. This represents the support function for the set of all possible distributions q, i.e., for the probability simplex. We have also written $C(q)$ for $C(q(z|x))$.

With the correspondence established by Table 11.4, minimizing the generalized free energy – that is, minimizing (11.62) – is an instance of the Fenchel duality problem (11.61). Specifically, minimizing (11.62) produces the convex dual function of the complexity metric $C(q)$ evaluated at the vector $-l/\alpha$.

Table 11.4 Correspondence between free energy minimization and Fenchel duality

Fenchel duality	free energy minimization
u	$q = [q(1|x), \ldots, q(N_z|x)]^T$
v	$-l/\alpha = -[\ell(1,x), \ldots, \ell(N_z,x)]^T/\alpha$
$f(u)$	$C(q) + I_S(q)$

Appendix 11.C: Derivation of the Variational Information Bottleneck Criterion (11.26)

Let us introduce two variational distributions $q(t|z, \varphi)$ and $q(z)$. The objective in the information bottleneck criterion (11.24) can then be bounded by applying the variational bounds (3.89) and (3.90) as

$$
\begin{aligned}
I(t;z) - \lambda I(x;z) &\geq E_{(x,t,z)\sim p(x,t)p(z|x,\theta)}\left[\log\left(\frac{q(t|z,\varphi)}{p(t)}\right)\right] - \lambda E_{(x,z)\sim p(x)p(z|x,\theta)}\left[\log\left(\frac{p(z|x,\theta)}{q(z)}\right)\right] \\
&= H(p(t)) + E_{(x,t,z)\sim p(x,t)p(z|x,\theta)}[\log q(t|z,\varphi)] - \lambda E_{(x,z)\sim p(x)p(z|x,\theta)}\left[\log\left(\frac{p(z|x,\theta)}{q(z)}\right)\right].
\end{aligned}
\tag{11.63}
$$

Neglecting the entropy $H(p(t))$, which does not depend on the parameters θ and φ, yields the variational information bottleneck problem (11.26).

Appendix 11.D: Proof of the Optimality of Function (11.37)

Consider for simplicity the case of a rv x taking values in a discrete and finite set. Problem (11.35) can be seen to be convex in the values of the function $T(\cdot)$ under optimization. Therefore, the first-order optimality condition to problem (11.35) is necessary and sufficient to obtain globally optimal solutions. Addressing the resulting equation gives solutions of the form

$$
T^*(x) = \log\left(\frac{p(x)}{q(x)}\right) + \log\left(E_{x\sim q(x)}[\exp(T^*(x))]\right).
\tag{11.64}
$$

Finally, it is easy to check that functions of the form (11.37) satisfy condition (11.64).

Appendix 11.E: Proof of the Variational Inequality (11.47) for f-Divergences

Since the defining function $f(u)$ is strictly convex, by definition, it is strictly larger than its first-order Taylor approximation evaluated at any point u_0 except for $u = u_0$ (see Sec. 5.6). More precisely, we have the inequality

$$
\begin{aligned}
f(u) &\geq f(u_0) + f'(u_0)(u - u_0) \\
&= uf'(u_0) - u_0 f'(u_0) + f(u_0),
\end{aligned}
\tag{11.65}
$$

with equality holding if and only if $u = u_0$. Setting $u_0(x) = \exp(T(x))$ and $u(x) = p(x)/q(x)$ and applying (11.65) to the definition (11.43) of f-divergences, we get the desired inequality as follows:

$$D_f(p||q) = E_{x \sim q(x)} \left[f\left(\frac{p(x)}{q(x)} \right) \right]$$

$$\geq E_{x \sim q(x)} \left[\frac{p(x)}{q(x)} f'(\exp(T(x))) - \exp(T(x)) f'(\exp(T(x))) + f(\exp(T(x))) \right]$$

$$= E_{x \sim p(x)}[a_f(T(x))] - E_{x \sim q(x)}[b_f(T(x))]. \tag{11.66}$$

Furthermore, equality holds if and only if we set $\exp(T(x)) = p(x)/q(x)$, i.e., $T(x) = T^*(x) = \log(p(x)/q(x))$.

Appendix 11.F: Change-of-Variables Formula and Likelihood-Free Models

Consider a continuous rv generated as

$$z \sim p(z)$$
$$x = G(z), \tag{11.67}$$

where $p(z)$ is a joint pdf and $G(\cdot)$ is a function. Note that this is the type of directed generative model adopted by GANs (cf. (11.52)). When the function $G(\cdot)$ is invertible, we can explicitly compute the distribution of x using the **change-of-variable formula**

$$p(x) = p(z = G^{-1}(x))|\det(\nabla G^{-1}(x))|, \tag{11.68}$$

where $\nabla G^{-1}(x)$ is the Jacobian of the inverse $G^{-1}(x)$ of transformation $G(z)$.

Considering the case of scalar functions to build an intuition, the determinant of the Jacobian $\nabla G^{-1}(x)$ accounts for the fact that the probability density $p(z = G^{-1}(x))$ is "diluted" over a larger interval of values x if the transformation $G(z)$ has a large slope at $z = G^{-1}(x)$, and hence if the inverse transformation has a smaller slope at the value x of interest.

If the transformation is not invertible, the formula (11.68) is not applicable, and one needs to integrate over all values z that yield the equality $G(z) = x$ in order to compute $p(x)$. This is not feasible for complex transformations $G(x)$ such as those implemented by neural networks.

Bibliography

[1] I. Csiszár and P. C. Shields, *Information Theory and Statistics: A Tutorial*. Now Publishers, 2004.

[2] T. Minka, "Divergence measures and message passing," Microsoft Research, Tech. Rep., 2005.

[3] P. G. Bissiri, C. C. Holmes, and S. G. Walker, "A general framework for updating belief distributions," *Journal of the Royal Statistical Society. Series B, Statistical Methodology*, vol. 78, no. 5, pp. 1103–1130, 2016.

[4] J. Knoblauch, J. Jewson, and T. Damoulas, "Generalized variational inference: Three arguments for deriving new posteriors," arXiv preprint arXiv:1904.02063, 2019.

[5] B. Poole, S. Ozair, A. Van Den Oord, A. Alemi, and G. Tucker, "On variational bounds of mutual information," in *The 36th International Conference on Machine Learning*. ICML. 2019, pp. 5171–5180.

[6] A. Zaidi and I. Estella-Aguerri, "On the information bottleneck problems: Models, connections, applications and information theoretic views," *Entropy*, vol. 22, no. 2, p. 151, 2020.

[7] T. Berger, *Rate Distortion Theory: A Mathematical Basis for Data Compression*. Prentice-Hall, 1971.

[8] M. Shannon, "Properties of f-divergences and f-GAN training," arXiv preprint arXiv:2009.00757, 2020.

[9] A. D. Joseph, B. Nelson, B. I. Rubinstein, and J. Tygar, *Adversarial Machine Learning*. Cambridge University Press, 2018.

[10] J. Song and S. Ermon, "Understanding the limitations of variational mutual information estimators," in *International Conference on Learning Representations*. ICLR, 2020.

[11] I. J. Goodfellow, J. Pouget-Abadie, M. Mirza, B. Xu, D. Warde-Farley, S. Ozair, A. Courville, and Y. Bengio, "Generative adversarial networks," arXiv preprint arXiv:1406.2661, 2014.

[12] I. Gulrajani, F. Ahmed, M. Arjovsky, V. Dumoulin, and A. Courville, "Improved training of Wasserstein GANs," arXiv preprint arXiv:1704.00028, 2017.

[13] J. C. Principe, *Information Theoretic Learning: Renyi's Entropy and Kernel Perspectives*. Springer Science+Business Media, 2010.

12 Bayesian Learning

12.1 Overview

Previous chapters have formulated learning problems within a **frequentist** framework. Frequentist learning aims to determine a value of the model parameter θ that approximately minimizes the population loss. Since the population loss is not known, this is in practice done by minimizing an estimate of the population loss $L_p(\theta)$ based on training data – the training loss $L_{\mathcal{D}}(\theta)$. It is often the case that there are multiple model parameter vectors θ that provide similar performance in terms of the training loss, i.e., *within the training data*, but may provide significantly different decisions *outside it*. With frequentist learning, such *disagreements* among models that are almost equally plausible given the available data are discarded, as the training procedure produces a single model parameter vector. As a consequence, frequentist learning tends to produce decisions that are *badly calibrated* – often *overconfident* – when data are limited, as they fail to account for the residual uncertainty in the choice of the population-optimal within-class model.

Accounting for the residual uncertainty in the population-optimal within-class model, known as **epistemic uncertainty** (see Sec. 4.5), requires the learning algorithm to move away from the frequentist goal of producing a *point estimate* θ of the model parameter vector. Instead, the learning algorithm should produce a *distribution* $q(\theta)$ over the model parameter space. This way, each model parameter θ can be assigned a weight $q(\theta)$ that quantifies its compatibility with the available evidence, as well as prior knowledge. The final prediction of output t given input x can be obtained either by drawing a random sample $\theta \sim q(\theta)$ for use in the soft predictor $p(t|x,\theta)$ assumed by the given model class – an approach known as **Gibbs prediction** – or else by averaging the soft predictions over model parameter vectors $\theta \sim q(\theta)$ to obtain the **ensemble predictor** $E_{\theta \sim q(\theta)} \left[p(t|x,\theta) \right]$ (see Sec. 6.9). In either case, the contributions of various model parameters θ are determined by their weights $q(\theta)$.

When the number of data samples is limited, the distribution $q(\theta)$ should rely not only on training data but also on any available **prior** information available on the population distribution (see Sec. 4.10). To this end, Bayesian learning optimizes Gibbs or ensemble predictors by incorporating an information-theoretic regularization penalty that accounts for the "distance" of distribution $q(\theta)$ to a prior distribution $p(\theta)$ over the model parameters θ. As we will see in this chapter, the upshot of this approach is that Bayesian learning conceptually turns the learning problem into a Bayesian inference problem, whereby the model parameter vector θ is treated as a latent variable. A key consequence of this reformulation of learning is that the scalable solutions introduced in Chapter 10 for variational inference (VI), as well as for Monte Carlo (MC) sampling introduced in Sec. 7.7.3, become directly applicable as tools for training.

Learning Objectives and Organization of the Chapter. By the end of this chapter, the reader should be able to:

- understand the differences between frequentist learning and Bayesian learning (Secs. 12.2 and 12.3);

451

- understand (and maybe agree with) the main reasons why Bayesian learning may provide a more useful, and general, learning framework as compared to frequentist learning for some applications, particularly when training data are scarce (Sec. 12.4);
- address exact Bayesian learning for one of the most important models under which Bayesian learning is tractable, namely the Gaussian–Gaussian GLM (Sec. 12.5);
- approximate the posterior distribution in the model parameter space via Laplace approximation (Sec. 12.6);
- develop the MC sampling-based methods of rejection sampling, Metropolis–Hastings, and stochastic gradient Markov chain MC (MCMC) (Sec. 12.7);
- develop parametric and particle-based VI solutions for Bayesian learning (Sec. 12.8);
- understand model selection in Bayesian models via empirical Bayesian, also known as type-II ML, learning (Sec. 12.9);
- understand and implement non-parametric Bayesian learning through Gaussian processes (GPs) (Sec. 12.10);
- understand Bayesian models with local latent variables (Sec. 12.11); and
- understand generalized Bayesian learning (Sec. 12.12).

12.2 Frequentist Learning and Calibration

In this section and the next, we present a gentle introduction to Bayesian learning by contrasting it to frequentist learning. Here, and throughout most of this chapter, we will focus on parametric discriminative models $p(t|x,\theta)$ for supervised learning, but the discussion applies more broadly to generative models, as well as to unsupervised learning.

A Note on Terminology. In this chapter, in order to avoid any confusion that may arise from the inference-centric view of training adopted by Bayesian learning, we will refer to rv t as **"label"** – although the presentation applies also to regression – and to the observations x = x as **covariates**.

In this section, we start by reviewing frequentist learning via maximum likelihood (ML), and by introducing the concept of calibration. This discussion will eventually reveal Bayesian learning as a generalization of ML that encompasses an information-theoretic regularizer based on a prior distribution in the model parameter space.

12.2.1 Frequentist Learning via ML

Given a training set $\mathcal{D} = \{(x_n, t_n)_{n=1}^N\}$, with data points drawn i.i.d. from an unknown population distribution $p(x,t)$, as introduced in Chapter 4, ML learning tackles the problem

$$\theta_{\mathcal{D}}^{ML} = \arg\min_{\theta} L_{\mathcal{D}}(\theta) \tag{12.1}$$

of minimizing the training log-loss

$$L_{\mathcal{D}}(\theta) = \frac{1}{N}\sum_{n=1}^{N}(-\log p(t_n|x_n,\theta)) \tag{12.2}$$

over model parameter vector θ. As such, ML aims to find a *single* model parameter vector θ that maximizes the probability of the observed labels for the given covariates in the training set (see Sec. 4.9).

12.2.2 Calibration

As discussed, ML collapses any uncertainty on the population-optimal model parameter vector θ arising from the use of the training loss as a proxy for the population loss, by selecting a single value $\theta_{\mathcal{D}}^{ML}$. Because of the resulting inability to quantify epistemic uncertainty, frequentist learning can provide *badly calibrated* decisions, whose confidence levels do not reflect their actual accuracy.

Broadly speaking, a soft predictor $q(t|x)$ is said to be **well calibrated** if it assigns scores $q(t|x)$ to all possible values of the label t in a manner that is faithful to the population distribution. Somewhat more formally, a soft predictor $q(t|x)$ is well calibrated if it reflects the ground-truth **conditional population distribution** $p(t|x) = p(x,t)/p(x)$, with $p(x)$ being the marginal of rv x under $p(x,t)$, i.e., if we have the approximate equality

$$q(t|x) \simeq p(t|x). \tag{12.3}$$

Let us interpret, and elaborate on, this condition by focusing on classification problems. To this end, consider a hard predictor $\hat{t}(x)$, such as the optimal (MAP) hard predictor $\hat{t}^{MAP}(x) = \arg\max_t q(t|x)$ obtained from the soft predictor $q(t|x)$. The probability $p(\hat{t}(x)|x)$ represents the actual (unknown) probability that the decision $\hat{t}(x)$ is correct given input x. In contrast, the probability $q(\hat{t}(x)|x)$ represents the level of **confidence** of the predictor in its hard decision $\hat{t}(x)$.

Therefore, when the inequality

$$q(\hat{t}(x)|x) > p(\hat{t}(x)|x) \tag{12.4}$$

holds, the soft predictor $q(t|x)$ produces an **overconfident** assessment of the accuracy of the hard prediction $\hat{t}(x)$. This is in the sense that, under the soft predictor $q(t|x)$, the prediction $\hat{t}(x)$ is assigned a probability of error, $1 - q(\hat{t}(x)|x)$, that is smaller than the actual conditional probability of error $1 - p(\hat{t}(x)|x)$. Conversely, when we have $q(\hat{t}(x)|x) < p(\hat{t}(x)|x)$, the soft predictor is **underconfident**, as it underestimates the accuracy of the predictions $\hat{t}(x)$.

More generally, the calibration condition (12.3) requires the soft predictor to provide accurate probability estimates for *all* label values t given the covariate x. This can be particularly useful if the predictor is tasked to provide a *set-valued* prediction in the form of a high-probability set of values t for a covariate x. For a formal introduction to calibration and uncertainty quantification measures, please consult Recommended Resources, Sec. 12.14.

Example 12.1

Consider a binary classification problem with label $t \in \{0, 1\}$ and covariate $x \in \mathbb{R}$. Assume that the model class $\mathcal{H}$ includes two models identified with the binary parameter $\theta \in \{-1, +1\}$. The models define the likelihoods

$$(t|x = x, \theta) \sim p(t|x, \theta) = \text{Bern}(t|\sigma(10(\theta x + 0.5))), \tag{12.5}$$

which are illustrated in Fig. 12.1 for both values of the model parameter θ. Note that the model with $\theta = 1$ tends to classify positive values of x as $t = 1$, while the model with $\theta = -1$ tends to classify negative values of x as $t = 1$.

We are given the training set $\mathcal{D} = \{(-0.79, 0), (0.1, 1), (-0.1, 1), (0.8, 0)\}$ with $N = 4$ data points. On this data set, both models give a small log-loss for data points $(0.1, 1)$ and $(-0.1, 1)$, since they provide similar soft predictions for covariates around $x = 0$. However, model $\theta = 1$ yields a small log-loss on $(-0.79, 0)$ and a large log-loss on $(0.8, 0)$, while model $\theta = -1$ has a small log-loss on $(0.8, 0)$ and a large loss on $(-0.79, 0)$. Therefore, all in all, the two models fit the training data set $\mathcal{D}$ almost equally well. However, their predictions are very different for covariates further from $x = 0$:

Figure 12.1 The two soft predictors $p(t|x,\theta)$ with $\theta = \{-1, +1\}$ used in Examples 12.1 and 12.2 that illustrates the differences between frequentist and Bayesian learning in the presence of limited data. While frequentist learning must always select one or the other of the two predictors, a Bayesian predictor can optimize a distribution $q(\theta)$ on $\theta = \{-1, +1\}$. After training, one can either draw one of the two predictors at random from $q(\theta)$ for each test covariate x, or else use the ensemble predictor $p(t = 1|x, \mathcal{D})$. The latter, obtained with data and prior detailed in the text, is also shown.

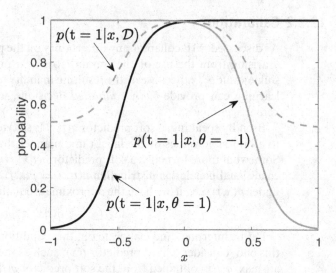

model $\theta = 1$ gives a higher score to label $t = 1$ for positive covariates x, and to label $t = 0$ for sufficiently negative covariates x; while the reverse is true for model $\theta = -1$.

ML learning computes the training log-loss

$$L_{\mathcal{D}}(\theta) = \frac{1}{4}\sum_{n=1}^{4}\log(1 + \exp(-t_n^{\pm}(10(\theta x_n + 0.5)))) \tag{12.6}$$

for the two models, obtaining

$$L_{\mathcal{D}}(1) = 3.27 \text{ and } L_{\mathcal{D}}(-1) = 3.24. \tag{12.7}$$

The training log-loss values under the two models are very similar, since the two models fit the training data almost equally well. However, ML chooses $\theta_{\mathcal{D}}^{ML} = -1$ because it has a slightly smaller training loss. In making this choice, ML learning disregards the epistemic uncertainty that exists in the selection of the optimal model as a result of the availability of a limited data set. As we argue next, this can cause the ML predictor $p(t|x,\theta)$ to be badly calibrated.

To this end, assume that the model is **well specified**, so that there exists a ground-truth value θ_0 of the model parameter such that the conditional population distribution coincides with model $p(t|x,\theta_0)$, i.e., $p(t|x) = p(t|x,\theta_0)$. We will see in Sec. 12.3 that Bayesian learning implicitly posits that the model is indeed well specified. For this example, in the worst case in which $\theta_0 = 1$, the predictor obtained by ML provides badly calibrated decisions.

To see this, let us take a covariate, say $x = -0.8$, and evaluate the soft prediction produced by the ML predictor:

$$p(t = 1|x = -0.8, \theta_{\mathcal{D}}^{ML} = -1) \simeq 1. \tag{12.8}$$

This prediction confidently favours label $t = 1$, while the actual conditional population distribution $p(t = 1|x = -0.8) = p(t = 1|x = -0.8, \theta_0 = 1) = 0.05$ gives close-to-zero probability to the event $t = 1$ when $x = -0.8$. Therefore, the calibration condition (12.3) is far from being satisfied, and the predictor is (extremely) overconfident in evaluating the accuracy of the MAP prediction $\hat{t}^{MAP}(-0.8) = 1$.

12.3 Basics of Bayesian Learning

In a nutshell, Bayesian learning views the problem of learning as one of inference, or soft prediction, of the model parameter θ given training data $\mathcal{D}$. While frequentist learning seeks a single value of the model parameter θ based on the training data $\mathcal{D}$, Bayesian learning optimizes over the space of *distributions* $q(\theta|\mathcal{D})$ on the model parameter vector θ. The distribution $q(\theta|\mathcal{D})$ can account for epistemic uncertainty by assigning to each model parameter θ a value $q(\theta|\mathcal{D})$ of the distribution that depends on the degree to which the model parameter θ is consistent with prior information and with the data set $\mathcal{D}$ (see Sec. 4.5). Having obtained distribution $q(\theta|\mathcal{D})$ based on the training set $\mathcal{D}$, Bayesian learning uses it for inference of the label t by implementing either a Gibbs or an ensemble predictor.

12.3.1 Bayesian Learning As Inference

As a brief recap from Chapter 3, an inference problem involves an input and a target variable that are jointly distributed according to a known joint distribution. Optimal soft prediction can be formulated as the minimization of the free energy between the soft predictor and the joint distribution, which returns the posterior distribution of the target variables given the observation of the input variables (see (3.97)).

To apply the framework of optimal inference to learning, we therefore need to specify input, target, and their joint distribution:

- The **input** is given by the training set $\mathcal{D} = \{(x_n, t_n)_{n=1}^N\}$. More precisely, assuming a discriminative model, we view the labels $t_{\mathcal{D}} = \{t_1, \ldots, t_N\}$ as rvs, while the covariates $x_{\mathcal{D}} = \{x_1, \ldots, x_N\}$ are taken to be fixed constants. Generative models could also be treated in a similar fashion by assuming the covariates to be jointly distributed with the labels.
- The **target** is the model parameter vector θ.
- The **joint distribution** of input and output is

$$p(\theta, \mathcal{D}) = \underbrace{p(\theta)}_{\text{prior}} \times \underbrace{p(\mathcal{D}|\theta)}_{\text{likelihood}}, \tag{12.9}$$

which depends on the **prior distribution** $p(\theta)$ of the model parameters, as well as on the data **likelihood** $p(\mathcal{D}|\theta)$. Given our focus on discriminative models, the shorthand notation $p(\theta, \mathcal{D})$ more precisely denotes the joint distribution of model parameter vector θ and labels $t_{\mathcal{D}}$, conditioned on the covariates $x_{\mathcal{D}}$:

$$p(\theta, \mathcal{D}) = p(\theta, t_{\mathcal{D}}|x_{\mathcal{D}}). \tag{12.10}$$

Using the standard assumption that the labels t_n are conditionally independent given the covariates $x_{\mathcal{D}} = x_{\mathcal{D}}$ and the model parameter vector θ (cf. (4.56)), we can write the likelihood as

$$p(\mathcal{D}|\theta) = p(t_{\mathcal{D}}|x_{\mathcal{D}}, \theta) = \prod_{n=1}^N p(t_n|x_n, \theta), \tag{12.11}$$

where again we have used a shorthand notation, $p(\mathcal{D}|\theta)$, which makes dependence on the covariates $x_{\mathcal{D}}$ implicit.

The choices of prior and likelihood define the **inductive bias** of Bayesian learning. The selection of the prior is particularly sensitive in the regime in which data availability is limited and hence epistemic uncertainty is high. In fact, the prior distribution describes the state of knowledge, and the corresponding level of uncertainty, of the Bayesian learner before any data are observed. As we will see, the relative contribution to the learning process of the prior as compared to the data likelihood decreases as more data are collected.

With prior and likelihood fixed as part of the inductive bias, the problem of optimal soft prediction of the model parameter vector θ can be formulated as the **minimization of the free energy**,

$$\min_{q(\cdot|\mathcal{D})} \mathrm{F}(q(\theta|\mathcal{D})||p(\theta,\mathcal{D})). \tag{12.12}$$

By (3.97), the optimal solution is given by the posterior distribution

$$p(\theta|\mathcal{D}) = \frac{p(\theta,\mathcal{D})}{p(\mathcal{D})}, \tag{12.13}$$

which can be written explicitly in terms of labels and covariates as

$$p(\theta|\mathcal{D}) = p(\theta|x_\mathcal{D},t_\mathcal{D}) = \frac{p(\theta,t_\mathcal{D}|x_\mathcal{D})}{p(t_\mathcal{D}|x_\mathcal{D})} = \frac{\underbrace{p(\theta)}_{\text{prior}}\underbrace{\prod_{n=1}^{N}p(t_n|x_n,\theta)}_{\text{likelihood}}}{\underbrace{p(t_\mathcal{D}|x_\mathcal{D})}_{\text{marginal likelihood}}}. \tag{12.14}$$

In (12.14), the normalizing constant, known as the **marginal likelihood**, is given as

$$p(t_\mathcal{D}|x_\mathcal{D}) = \mathrm{E}_{\theta\sim p(\theta)}\left[\prod_{n=1}^{N}p(t_n|x_n,\theta)\right]. \tag{12.15}$$

More discussion on the marginal likelihood can be found later in Sec. 12.3.4.

12.3.2 Bayesian vs. Frequentist Learning

As reviewed in the preceding section, in the frequentist framework we have *two distinct distributions*: the population distribution $p(x,t)$, which is unknown; and the model distribution $p(t|x,\theta)$, which depends on a model parameter θ to be trained. Furthermore, in a frequentist framework, inference and learning are conceptually and algorithmically distinct tasks.

In contrast, as detailed in the preceding subsection, in the Bayesian learning framework there is a *single joint distribution* of training data and model parameters, $p(\theta,\mathcal{D})$ in (12.10), that is given by the product of prior and likelihood as in (12.9). Conceptually, in Bayesian learning, one assumes that this joint distribution is **well specified**, that is, that it correctly reflects the mechanism underlying the generation of the data. Note that, as compared to the frequentist definition of a well-specified model class given in Sec. 9.7, here this assumption also entails the correctness of the prior distribution in representing the learner's state of knowledge before data are collected.

Under the assumption of a well-specified model, Bayesian learning applies the standard rules of probability to the joint distribution (12.10) in order to infer the model parameters θ using Bayesian inference. This yields the posterior distribution $p(\theta|\mathcal{D})$ in (12.14). Therefore, within a Bayesian framework, learning amounts to Bayesian inference; and learning and inference do not require different formulations.

This discussion provides another perspective on the importance of the selection of a realistic prior. As Bayesian learning "believes" in the correctness of the joint distribution $p(\theta, \mathcal{D})$ under which the posterior $p(\theta|\mathcal{D})$ is computed, a severely misspecified prior would be problematic, particularly in the regime of limited availability of data. We refer to Sec. 12.12 for a discussion of generalized Bayesian learning, which addresses the issue of misspecification.

In order to elaborate further on the relationship between frequentist and Bayesian learning, let us now observe more closely the free energy criterion minimized in (12.12) by Bayesian learning. By (3.99), the free energy criterion in (12.12) can be written as

$$\text{F}(q(\theta|\mathcal{D})||p(\theta, \mathcal{D})) = N \underbrace{\text{E}_{\theta \sim q(\theta|\mathcal{D})}\left[L_{\mathcal{D}}(\theta)\right]}_{\text{average training loss}} + \underbrace{\text{KL}(q(\theta|\mathcal{D})||p(\theta))}_{\text{information-theoretic regularizer}}, \qquad (12.16)$$

where the training loss is defined in (12.2) as for frequentist learning. Accordingly, the first term in (12.16) is the (scaled) *average* training loss, and the second is an information-theoretic regularizer that penalizes deviations of the soft predictor $q(\theta|\mathcal{D})$ with respect to the prior $p(\theta)$. Note that the average of the training loss is taken with respect to model parameters drawn from the distribution $q(\theta|\mathcal{D})$ under optimization.

As can be easily seen, in the absence of the regularizer, the distribution $q(\theta|\mathcal{D})$ obtained by minimizing the first term in (12.16) would be concentrated at the ML solution (12.1), i.e., we would obtain the solution $q(\theta|\mathcal{D}) = \delta(\theta - \theta_{\mathcal{D}}^{ML})$, where $\delta(\cdot)$ denotes the Dirac delta function (in the standard case of continuous-valued model parameters). Therefore, the inclusion of the regularizer is of critical importance in order to differentiate frequentist and Bayesian learning. The regularizer prevents the minimizer of the free energy (12.16) – that is, the posterior distribution $p(\theta|\mathcal{D})$ – from collapsing to the single value $\theta_{\mathcal{D}}^{ML}$ produced by ML. In this way, the posterior $p(\theta|\mathcal{D})$ can retain information about the prior uncertainty encoded by distribution $p(\theta)$.

The relative weights of the evidence brought by data and of prior knowledge encoded by the prior distribution in the free energy (12.16) are dictated by the number of data points, N. As N increases, the first term in (12.16) tends to dominate, and, in the asymptotic regime $N \to \infty$, the posterior distribution $p(\theta|\mathcal{D})$ collapses to the ML solution.

12.3.3 Gibbs Sampling and Bayesian Ensembling

Once a distribution $q(\theta|\mathcal{D})$ is obtained as a result of Bayesian learning – ideally providing a close approximation of the posterior $p(\theta|\mathcal{D})$ in (12.14) – how should it be used for inference on a new, test, covariate vector x? This can be done in one of two main ways: Gibbs prediction or ensemble prediction.

- **Gibbs prediction**. For any test covariate x, the Gibbs predictor draws a random sample of model parameter vector θ from the distribution $q(\theta|\mathcal{D})$ to produce the soft predictor $p(t|x, \theta)$. A different vector $\theta \sim q(\theta|\mathcal{D})$ is sampled for each new covariate x. In this regard, note that the average loss term in the free energy criterion (12.16) can be interpreted as evaluating the expected performance of the Gibbs predictor over random draws of the model parameter vector $\theta \sim q(\theta|\mathcal{D})$.

- **Ensemble prediction**. The ensemble soft predictor is defined as the average predictive distribution over the distribution $q(\theta|\mathcal{D})$ of the model parameters, i.e.,

$$q(t|x, \mathcal{D}) = \mathrm{E}_{\theta \sim q(\theta|\mathcal{D})}\left[p(t|x, \theta)\right]. \tag{12.17}$$

Accordingly, the ensemble predictor averages the predictions obtained under all model parameters θ, each weighted by the corresponding score $q(\theta|\mathcal{D})$. Unlike for the Gibbs predictor, the ensemble soft predictor $q(t|x, \mathcal{D})$ in (12.17) generally does not belong to the class $\mathcal{H}$ of soft predictors of the form $p(t|x, \theta)$, but is instead a **mixture** of such predictors.

Let us now specialize the ensemble predictor (12.17) to the ideal case in which the distribution $q(\theta|\mathcal{D})$ equals the posterior $p(\theta|\mathcal{D})$ in (12.14), yielding the mixture predictor

$$q(t|x, \mathcal{D}) = \mathrm{E}_{\theta \sim p(\theta|\mathcal{D})}\left[p(t|x, \theta)\right] = p(t|x, \mathcal{D}). \tag{12.18}$$

As suggested by the notation $p(t|x, \mathcal{D})$ in (12.18), the ensemble soft predictor (12.18) coincides with the **posterior distribution of the test label** t given the training set $\mathcal{D}$ and the test covariates x.

To make this last statement precise, consider the *extended* joint distribution that also includes the test example (x, t):

$$p(\theta, t_{\mathcal{D}}, t|x_{\mathcal{D}}, x) = p(\theta)\left(\prod_{n=1}^{N} p(t_n|x_n, \theta)\right)p(t|x, \theta). \tag{12.19}$$

In words, in (12.19) the test example is assumed to be conditionally independent given the model parameter θ, following the same model as for the training samples. Under the joint distribution (12.19), it can be directly checked that the ensemble soft predictor in (12.17) coincides with the posterior distribution $p(t|x, x_{\mathcal{D}}, t_{\mathcal{D}}) = p(t|x, \mathcal{D})$.

We can conclude that the Bayesian ensemble predictive distribution $p(t|x, \mathcal{D})$ in (12.18) can be equivalently viewed as optimally solving the Bayesian inference problem of predicting the output t for a test covariate x, given the training set $\mathcal{D}$, under the joint distribution in (12.19). Note again the key assumption made by Bayesian learning, that the model distribution (12.19) is a **well-specified** representation of the data generation process.

Example 12.2

Let us return to Example 12.1, which was studied in the preceding section under ML learning. To apply Bayesian learning, we choose the prior probability $p(\theta = 1) = p(\theta = -1) = 0.5$, which makes explicit a prior belief in the equivalence of the two models before the observation of any data. The posterior probability over the parameters can be directly computed as

$$p(\theta = -1|\mathcal{D}) = \frac{\overbrace{0.5}^{\text{prior}} \overbrace{\exp(-12.97)}^{\text{likelihood}}}{\underbrace{0.5\exp(-12.97) + 0.5\exp(-13.07)}_{\text{marginal likelihood}}} = 0.52. \tag{12.20}$$

Therefore, Bayesian learning retains information about the existing epistemic uncertainty (caused by the fact that both models have very similar training losses) by assigning almost equal scores to both models $\theta = 1$ and $\theta = -1$.

For example, when applied to the covariate $x = -0.8$, the ensemble soft predictor (12.18) yields

$$p(t = 1|x = -0.8, \mathcal{D}) = 0.52 \cdot p(t = 1|x = -0.8, \theta = -1) + 0.48 \cdot p(t = 1|x = -0.8, \theta = 1)$$
$$= 0.52 \cdot 0.99 + 0.48 \cdot 0.05 = 0.54. \qquad (12.21)$$

Unlike the ML soft prediction (12.8), this predictive probability reflects the uncertainty caused by disagreement outside the training set of the two models that fit the training data almost equally well.

The practical upshot of this discussion is illustrated in Fig. 12.1, which shows the Bayesian predictive distribution $p(t = 1|x, \mathcal{D})$ (dashed gray line) based on the given training set. For inputs x over which the two predictors are in disagreement, that is, for sufficiently large absolute value $|x|$, the Bayesian predictor assigns approximately equal probabilities to both values $t \in \{0, 1\}$.

This probability assignment retains information about the predictive uncertainty caused by limited availability of data. As a result, the Bayesian ensemble predictor is better calibrated than its frequentist counterpart.

For instance, following Example 12.1, the probability assigned to label $t = 1$ for $x = -0.8$ by the ensemble predictor is approximately 0.5, which is closer to the ground-truth value 0.05 than the probability of 1 assigned by the ML solution as in (12.8). The soft prediction provided by Bayesian learning is hence more "cautious" as it accounts for the existing epistemic uncertainty.

Example 12.3

To build some analytical intuition about the ensemble predictor (12.17), consider logistic regression under the assumption that the posterior is the Gaussian distribution $p(\theta|\mathcal{D}) = \mathcal{N}(\theta|v, \sigma^2 I)$ for some mean v and variance σ^2. Recall that logistic regression sets the likelihood as $p(t = 1|x, \theta) = \sigma(\theta^T u(x))$ for a vector of features $u(x)$. For this model, it turns out that the ensemble predictive distribution (12.18) can be well approximated as

$$p(t|x, \mathcal{D}) = \mathrm{E}_{\theta \sim \mathcal{N}(\theta|v, \sigma^2 I)}[\sigma(\theta^T u(x))]$$
$$\simeq \sigma\left(\frac{v^T u(x)}{\sqrt{1 + \pi \sigma^2 / 8}}\right). \qquad (12.22)$$

By (12.22), a larger variance σ^2 of the posterior decreases the confidence of the prediction by reducing the absolute value of the logit (i.e., of the argument of the sigmoid function).

12.3.4 Marginal Likelihood and Marginal Training Log-Loss

A final remark is in order regarding the marginal likelihood $p(t_{\mathcal{D}}|x_{\mathcal{D}})$ in (12.15), i.e., about the normalizing term in the posterior distribution (12.14). This is the probability – more precisely, the marginal distribution – assigned by the joint distribution (12.10) to the data set $\mathcal{D}$. In line with the notation (12.10), we can hence write the marginal likelihood as $p(\mathcal{D}) = p(t_{\mathcal{D}}|x_{\mathcal{D}})$. A useful perspective is to view the negative logarithm of the marginal likelihood,

$$-\log p(\mathcal{D}) = -\log p(t_{\mathcal{D}}|x_{\mathcal{D}}), \qquad (12.23)$$

as the training log-loss under the inductive bias defined by the joint distribution (12.10). Accordingly, we will refer to (12.23) as the **marginal training log-loss**, and to $\log p(t_{\mathcal{D}}|x_{\mathcal{D}})$ as the

marginal training log-likelihood, of the data $\mathcal{D}$ under the joint distribution (12.10). The qualifier "marginal" serves as a reminder that the distribution $p(\mathcal{D})$ is obtained by marginalizing over the model parameters θ. This viewpoint will be useful when discussing model selection in Bayesian learning in Sec. 12.9.

12.4   Why Bayesian Learning?

Having reviewed the basic differences between frequentist and Bayesian learning, this section lists a number of reasons why Bayesian learning may provide a preferable alternative to conventional frequentist techniques. This discussion leaves aside the computational challenges of Bayesian inference, which, as seen in Chapter 10, motivate the use of approximate inference techniques.

12.4.1   Bayesian Learning Quantifies Epistemic Uncertainty and Can Detect "Out-of-Distribution" Data

As we have seen in the preceding section, Bayesian learning can quantify **epistemic uncertainty** by converting the learning problem into one of soft prediction of the model parameters. In this way, Bayesian learning accounts for disagreements among models that, while fitting the training set well, may yield different predictions outside it. This disagreement exists when the training data set is not large enough to unambiguously pin down the optimal within-class model parameter, yielding a non-negligible epistemic uncertainty.

By providing a quantification of epistemic uncertainty, Bayesian learning can also be useful in identifying **"out-of-distribution" data**, that is, data whose generation does not follow the same distribution as the training data. In fact, for test covariates x that are not well supported by the training data, the epistemic uncertainty level should ideally be large, since the models competing to explain the training data should generally disagree outside it (assuming that the model class is sufficiently rich). When the level of epistemic uncertainty is large, one may conclude that input x is an outlier with respect to the training data distribution.

12.4.2   Bayesian Learning is Robust to Overfitting and Enables Model Selection Without Validation

As seen in Sec. 4.4, in frequentist learning, a model class with a larger capacity yields a lower training loss and may incur overfitting. This motivates the introduction of validation as a means of estimating the population loss and carrying out model selection. In stark contrast, in Bayesian learning, a larger model capacity does not generally cause overfitting, making it possible to select a model class solely on the basis of training data. Let us see why.

Example 12.4

Consider the example in Fig. 12.2, which is characterized by a scalar model parameter θ. The top figure depicts the case of a model class with small capacity, in which the model parameter θ can only be selected in the interval $\Theta = [-1, 1]$; while the bottom figure corresponds to a larger-capacity model class, in which θ can range in the interval $\Theta = [-3, 3]$. The training likelihood (dashed black line) measures how well a model parameter θ fits the given training data. The best-fitting model

Figure 12.2 Illustration of the robustness to overfitting of Bayesian learning: When data are limited, larger model classes yield posterior distributions that are more spread out. Therefore, the ensemble predictive distribution (12.17) and the corresponding Gibbs predictor weigh down the contributions of model parameters that may overfit the training data.

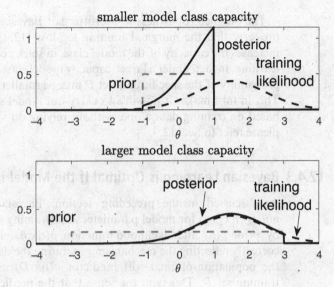

parameter is seen to be $\theta = 1.5$, with model parameters θ further apart from it yielding increasingly worse fits (i.e., lower training likelihood).

In frequentist learning, the choice of the model class with larger capacity would yield a lower training loss – or a higher training likelihood – upon training. In fact, the larger model class includes the best-fitting model parameter $\theta = 1.5$, while that is not the case for the smaller model class. But, as we discussed, this model parameter may overfit the data and yield a high population loss (not shown in the figure).

In Bayesian learning, the prior distribution (dashed gray line) for the smaller-capacity model class is spread on the smaller interval $\Theta = [-1, 1]$, while the larger-capacity model class has a broader prior covering the interval $\Theta = [-3, 3]$. Accordingly, while the posterior distribution for the larger-capacity model class covers better-fitting models, it is also spread out on a larger domain. As a result, although the larger-capacity model includes choices for θ that fit the data more closely and are hence more prone to overfitting, the contributions of these model parameters are weighted down by a small posterior distribution $p(\theta|\mathcal{D})$ in the ensemble and Gibbs predictors.

The upshot is that Bayesian learning is in principle more robust to overfitting, and an increase in model capacity does not necessarily yield a lower overall training loss.

The key observation in the preceding example is that a larger model class requires the prior $p(\theta)$ to cover a larger portion of the model parameter space, which in turn implies that the posterior distribution $p(\theta|\mathcal{D}) \propto p(\theta)p(\mathcal{D}|\theta)$ is more spread out. As a result, the optimal ensemble soft predictor (12.18), or the corresponding Gibbs predictor, must account for the contribution of a larger range of model parameters through the posterior distribution, and this helps prevent overfitting. In fact, even when the larger-capacity model class contains models that fit the data better, the contributions of these models are weighted down by the smaller value of the posterior distribution.

The outlined robustness to overfitting of Bayesian learning translates into the mathematical property that the marginal training log-loss (12.23) is not monotonically decreasing as one increases the capacity of the model class, in stark contrast to the training log-loss in frequentist learning. In fact, under a larger-capacity model class, due to the broader support of the prior, the probability of the specific data set $\mathcal{D}$ may be smaller than under the lower-capacity model class. This in turn makes it possible to carry out model selection and hyperparameter optimization based on training data alone, without relying on validation. For further detail on this point, please refer to Sec. 12.9.

12.4.3 Bayesian Learning is Optimal If the Model is Well Specified

As discussed in the preceding section, Bayesian learning assumes a joint distribution $p(\theta, t_{\mathcal{D}}, t | x_{\mathcal{D}}, x)$ for model parameter and (training and test) data that are defined as in (12.19) by prior $p(\theta)$ and likelihood function $p(t | x, \theta)$. If this joint distribution is **well specified**, correctly reflecting the mechanism generating the data, the ensemble prediction (12.18) returns the population-optimal soft predictor $q(t | x, \mathcal{D})$ as a function of the test covariate x and training set $\mathcal{D}$. This is in the sense that the predictor (12.18) minimizes the population log-loss $\mathrm{E}[-\log q(t | x, \mathcal{D})]$, where the expectation is taken over distribution (12.19). Therefore, the Bayesian ensemble predictor (12.18) is optimal under a well-specified model.

Under the assumption of a well-specified model, it is also possible to evaluate analytically the optimality error of the Bayesian ensemble soft predictor (12.18) with respect to the ideal predictor that is aware of the true model parameter θ by using information-theoretic metrics. The analysis also provides a neat way to quantify aleatoric and epistemic uncertainty, and is detailed in Appendix 12.A.

12.4.4 Bayesian Learning Minimizes a Bound on the Population Loss (Even for Misspecified Models)

A model such as the joint distribution (12.19) assumed by Bayesian learning is practically always misspecified, as both the prior and the likelihood may be incorrect ("All models are wrong but some are useful"). Bayesian learning is known to also have optimality guarantees for misspecified models when Gibbs predictors are used. Specifically, using generalization analysis via **PAC Bayes theory** (see Sec. 8.6), it can be shown that a Gibbs predictor with model parameter vector θ drawn from the posterior $p(\theta | \mathcal{D})$ optimizes an upper bound on the population loss, defined in the classical, frequentist, sense, irrespective of whether the model is well specified or not. We now briefly elaborate on this result.

Consider a general Gibbs predictor that selects a random parameter $\theta \sim q(\theta | \mathcal{D})$ based on an arbitrary distribution $q(\theta | \mathcal{D})$, which may depend on training set $\mathcal{D}$, and then applies the resulting (random) soft predictor $p(t | x, \theta)$ on a test covariate x. Suppose that, as in Sec. 8.6, we are interested in the population log-loss averaged over the choice of the predictor, i.e.,

$$L_p(q(\theta | \mathcal{D})) = \mathrm{E}_{\theta \sim q(\theta | \mathcal{D})} \left[\mathrm{E}_{(x,t) \sim p(x,t)} \left[-\log p(t | x, \theta) \right] \right]. \tag{12.24}$$

In this formulation, as in standard frequentist learning, the population distribution $p(x, t)$ is unknown and arbitrary, and there is generally a mismatch between the population distribution $p(x, t)$ and the model (12.19). By PAC Bayes theory (see Sec. 8.6), under suitable technical assumptions the average population loss $L_p(q(\theta | \mathcal{D}))$ can be upper bounded by the free energy

$\mathrm{F}(q(\theta|\mathcal{D})||p(\theta,\mathcal{D}))$ in (12.16), where $p(\theta)$ is an arbitrary prior distribution independent of the training set. More precisely, the bound includes constants independent of the soft predictor $q(\theta|\mathcal{D})$; it applies with high probability on the choice of the training set $\mathcal{D}$; and it holds *uniformly* for all population distributions $p(x,t)$ generating the set $\mathcal{D}$. The bound can be obtained by further elaborating on the results presented in Sec. 8.6 (see Recommended Resources, Sec. 12.14, for details).

Since, as discussed in the preceding section, the posterior $p(\theta|\mathcal{D})$ used by Bayesian learning minimizes the free energy in (12.16), it also minimizes the mentioned PAC Bayes upper bound on the population loss, no matter what the true population loss is. Therefore, Bayesian learning, via Gibbs prediction, provides a principled way to also account for epistemic uncertainty in the case of misspecified models.

It should be noted again that the PAC Bayes upper bound applies to Gibbs predictors, and hence the approach sketched in this subsection does not account for the performance of the ensemble predictor (12.17). That said, by Jensen's inequality (Appendix 3.A), the log-loss of the ensemble predictor is no larger than the average log-loss of the Gibbs predictor, making the PAC bound applicable, but generally loose, for ensemble predictors. This points to the fact that ensemble predictors can outperform Gibbs predictors in the presence of misspecification even when abundant data are available. This advantage stems from the capacity of ensemble predictors to combine the predictions of multiple models specialized to different parts of the problem domain (see Sec. 6.9).

Another, somewhat related, theoretical justification for Bayesian learning, which centers on the **minimum description length** principle, can be found in Appendix 12.B.

12.4.5 Bayesian Learning Facilitates Online Learning

Bayesian learning provides a principled framework for the definition of **online learning** schemes that operate on **streaming data**. To elaborate, assume that, after having observed a training set $\mathcal{D} = \{(x_n, t_n)_{n=1}^N\}$ and having computed the posterior (12.14), i.e.,

$$p(\theta|\mathcal{D}) = p(\theta|t_\mathcal{D}, x_\mathcal{D}) \propto p(\theta) \prod_{n=1}^N p(t_n|x_n, \theta), \tag{12.25}$$

a new observation (x_{N+1}, t_{N+1}) is made. How should one update the Bayesian predictor?

Using the Bayesian approach, we can obtain the updated model posterior as

$$p(\theta|\mathcal{D}, x_{N+1}, t_{N+1}) \propto p(\theta) \left(\prod_{n=1}^N p(t_n|x_n, \theta) \right) p(t_{N+1}|x_{N+1}, \theta)$$
$$\propto p(\theta|\mathcal{D}) p(t_{N+1}|x_{N+1}, \theta). \tag{12.26}$$

Therefore, we can update the posterior $p(\theta|\mathcal{D})$ by multiplying it by the likelihood $p(t_{N+1}|x_{N+1}, \theta)$ of the new observation and by renormalizing. Accordingly, all the information from the observation of the training set $\mathcal{D}$ that one needs to store in order to be able to compute the updated posterior $p(\theta|\mathcal{D}, x_{N+1}, t_{N+1})$ is given by the current posterior $p(\theta|\mathcal{D})$. The procedure in (12.26) can repeated be for successive observations, and it can be also applied to observations of batches of new data.

12.4.6 Bayesian Learning Facilitates Active Learning

In **active learning**, a learner selects data points adaptively: Given the available data points $\mathcal{D} = \{(x_n, t_n)_{n=1}^N\}$, the learner chooses the next input x_{N+1} and observes the corresponding (random) target value t_{N+1}. How should we select x_{N+1} so as to maximize the amount of information that we can extract from the new target value t_{N+1}, given what we already know from the training set $\mathcal{D}$? Bayesian learning provides a principled mechanism for this choice.

To see why, recall that a key property of Bayesian learning is that the ensemble predictive distribution $p(t|x, \mathcal{D})$ provides a well-calibrated measure of uncertainty. Therefore, through this distribution, the learner can gauge the current predictive uncertainty associated with any given covariate vector x, e.g., through the entropy $H(p(t|x, \mathcal{D}))$ of the label t given x and $\mathcal{D}$. The learner can now select the input x_{N+1} that yields the largest predictive uncertainty, e.g., the largest value of $H(p(t|x_{N+1}, \mathcal{D}))$. The rationale is that, in this way, the learner collects data so as to adaptively, and maximally, reduce uncertainty in the label.

Metrics such as the entropy of the predictive distribution capture both epistemic and aleatoric uncertainty, with the latter ideally accounting for the randomness in the ground-truth conditional distribution $p(t|x)$. More effectively, active learning may try to reduce **epistemic uncertainty** alone, rather than the overall predictive uncertainty. This approach is justified by the fact that the aleatoric contribution to the predictive uncertainty cannot be reduced by collecting more data. Appendix 12.A provides some detail on this point.

12.5   Exact Bayesian Learning

As we have discussed in Sec. 12.3, Bayesian learning solves an optimal inference problem by computing the posterior distribution $p(\theta|\mathcal{D})$ of the model parameters given the training set. Therefore, exact Bayesian learning is possible only for a limited set of models, most notably when the prior of the model parameters is the conjugate prior for the given likelihood (see Sec. 10.3). More commonly, exact Bayesian inference is not feasible, and approximate solutions based on VI or MC sampling are called for. In this section, we elaborate on a standard case in which exact Bayesian learning is possible, namely the Gaussian–Gaussian GLM.

12.5.1   When x and t are Independent

Before we start, let us briefly consider the simpler scenario in which the likelihood $p(t|x, \theta)$ does not depend on covariate x, so that we can write the likelihood function as $p(t|\theta)$. This setting is of interest, for instance, for density estimation problems (see Sec. 7.3). (The choice of the notation "t" is arbitrary here, and one can obtain the setting studied in Chapter 7 by replacing it with the notation "x".) If the prior $p(\theta)$ and the likelihood $p(t|\theta)$ form a conjugate exponential family, the posterior distribution for θ can be computed directly by adding to the natural parameters of the prior a vector that includes the sufficient statistics of the likelihood. The reader is referred to Sec. 10.3 for details.

12.5.2 Gaussian–Gaussian GLM

For the general case in which the likelihood $p(t|x,\theta)$ depends on input x, there is no general class of likelihood functions that admit a conjugate prior. For instance, there is no general prior (independent of the data) for GLM likelihoods (see Sec. 9.8). However, exact Bayesian learning is possible in special cases, including the classical **Gaussian–Gaussian GLM** example, which we study in the rest of this section.

Consider again the regression problem studied in Chapter 4 (Example 4.2), which is defined as follows. The likelihood is given by the GLM

$$p(t|x,\theta) = \mathcal{N}(t|\theta^T u(x), \beta^{-1}) \tag{12.27}$$

with fixed precision β and given feature vector $u(x)$; and the prior is given as

$$p(\theta) = \mathcal{N}(\theta|0, \alpha^{-1}I) \tag{12.28}$$

with a fixed precision α. It is recalled that the vector of features can contain non-linear functions of the input, such as the monomials considered in Chapter 4 for polynomial regression.

Following Sec. 3.4, the posterior over the model parameters can be computed directly as the Gaussian distribution

$$(\theta|\mathcal{D}) \sim \mathcal{N}(\theta|\theta_\mathcal{D}^{MAP}, \Sigma_\mathcal{D}^{MAP}), \tag{12.29}$$

with mean given by the MAP estimate of the model parameters

$$\begin{aligned}\theta_\mathcal{D}^{MAP} &= \beta \left(\alpha I + \beta X_\mathcal{D}^T X_\mathcal{D}\right)^{-1} X_\mathcal{D}^T t_\mathcal{D}\\ &= \left(\lambda I + X_\mathcal{D}^T X_\mathcal{D}\right)^{-1} X_\mathcal{D}^T t_\mathcal{D},\end{aligned} \tag{12.30}$$

where $\lambda = \alpha/\beta$, and covariance matrix

$$\Sigma_\mathcal{D}^{MAP} = \left(\alpha I + \beta X_\mathcal{D}^T X_\mathcal{D}\right)^{-1}. \tag{12.31}$$

The definition of data matrix $X_\mathcal{D}$ and label vector $t_\mathcal{D}$ are as in Sec. 4.3.

To remember this result, it is helpful to note that the natural parameters of the posterior $p(\theta|\mathcal{D}) = \mathcal{N}(\theta|v, \Theta^{-1})$ in (12.29) – namely, the scaled mean vector $\Theta^{-1}v$ and the precision matrix Θ (see Table 9.1) – are obtained as the sum of the natural parameters of the prior – namely, 0 and αI – and the corresponding parameters of the training data likelihood, which turn out to be $\beta X_\mathcal{D}^T t_\mathcal{D}$ and $\beta X_\mathcal{D}^T X_\mathcal{D}$, respectively (see Appendix 12.C for details). Hence, intuitively, the matrix $\beta X_\mathcal{D}^T X_\mathcal{D}$ quantifies the precision of the measurements.

The model in (12.27) and (12.28) can be equivalently understood as assuming, a priori, the relationship

$$t = \theta^T u(x) + \mathcal{N}(0, \beta^{-1}) \tag{12.32}$$

between covariates and labels, where the model parameter vector $\theta \sim \mathcal{N}(0, \alpha^{-1}I)$ is random and the notation $\mathcal{N}(0, \beta^{-1})$ represents a noise variable with the given Gaussian distribution. So, the input–output relationship is modeled as a **random linear function**.

A posteriori, having observed the training data $\mathcal{D}$, the distribution of the model parameter vector is changed to the posterior $p(\theta|\mathcal{D})$ in (12.29). Accordingly, from (12.32), the predictive distribution (12.18) can be directly obtained as

$$(t|x, \mathcal{D}) \sim \mathcal{N}(t|(\theta_\mathcal{D}^{MAP})^T u(x), \sigma^2(x)), \tag{12.33}$$

where

$$\sigma^2(x) = \beta^{-1} + u(x)^T \Sigma_{\mathcal{D}}^{MAP} u(x). \tag{12.34}$$

Therefore, in a manner similar to (12.32), the predictive distribution can be expressed as the random linear function

$$t = (\theta_{\mathcal{D}}^{MAP})^T u(x) + \mathcal{N}(0, \sigma^2(x)), \tag{12.35}$$

where, importantly, the predictive variance $\sigma^2(x)$ depends on the covariate x.

To elaborate further on this point, by (12.33) the optimal hard predictor under the ℓ_2 loss is the MAP predictor $\hat{t}_{\mathcal{D}}^{MAP}(x) = (\theta_{\mathcal{D}}^{MAP})^T u(x)$, and the predictive distribution (12.33) provides information about the uncertainty of this prediction through this covariate-dependent variance $\sigma^2(x)$ (see Sec. 4.5). This dependence on x of the variance in (12.33) is in contrast to the predictive distribution obtained by the MAP (frequentist) approach, i.e.,

$$p(t|x, \theta_{\mathcal{D}}^{MAP}) = \mathcal{N}(t|(\theta_{\mathcal{D}}^{MAP})^T u(x), \beta^{-1}), \tag{12.36}$$

which has a fixed variance. While the variance β^{-1} quantifies the aleatoric uncertainty associated with the likelihood (12.32), the variance $\sigma^2(x)$ produced by Bayesian learning accounts also for epistemic uncertainty.

Example 12.5

As a numerical example, consider the vector of $M + 1$ monomial features, $u(x) = [1, x, \ldots, x^M]^T$ in Example 4.2, and set $\beta = 10$ and $\alpha = 0.5 \times 10^{-4}$. Figure 12.3 shows the MAP predictor $\hat{t}_{\mathcal{D}}^{MAP}(x) = (\theta_{\mathcal{D}}^{MAP})^T u(x)$ as a dashed line, while the shaded area represents the **credible interval** $\hat{t}_{\mathcal{D}}^{MAP}(x) \pm \sigma(x)$ under the Bayesian predictive distribution (12.33). The credible interval indicates a range of values over which the label t is likely to fall under the predictive distribution $p(t|x, \mathcal{D})$.

Figure 12.3 MAP hard predictor (dashed line) and Bayesian credible interval – covering one standard deviation on either side of the mean of the predictive distribution (12.33) – for the Gaussian–Gaussian GLM with polynomial feature vector of degree M and data points illustrated as circles. The top and bottom figures differ in the capacity M (degree of the polynomial) of the assumed model class.

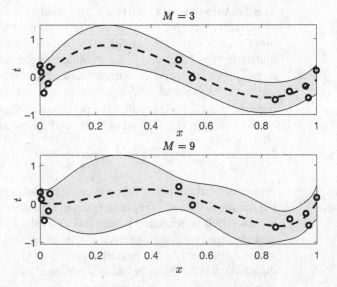

Specifically, since the predictive distribution is Gaussian, the label t, given the observation x = x, falls in the interval $\hat{t}_{\mathcal{D}}^{MAP}(x) \pm \sigma(x)$ with probability 68% (see Fig. 2.2).

Because of the uneven distribution of the observed values of x (circles), the accuracy of the prediction of the output t depends on the value of x: Covariates x closer to points in the training sets yield a smaller variance $\sigma(x)$, while covariates x in intervals of the input variable with fewer data points have a larger variance $\sigma(x)$. Furthermore, comparing the top figure, which corresponds to a smaller model class capacity $M = 3$, with the lower figure, which is obtained with $M = 9$, reveals the different impact of an increase in model class capacity on Bayesian learning as compared to frequentist learning. In Bayesian learning, increasing the model class capacity M does not cause overfitting. Instead, it yields a larger estimate of the epistemic uncertainty outside the training set – as evidenced by the size of the credible interval – owing to the larger number of model parameters θ that can represent the available data (and whose predictions generally disagree outside the training set).

12.6 Laplace Approximation

As introduced in Sec. 10.4 in the context of Bayesian inference, the Laplace approximation provides an efficient way to approximate the posterior distribution via a Gaussian distribution with mean given by the MAP solution. The Laplace approximation can be thought of as implementing *post-hoc* Bayesian inference in the sense that it is applied *after* conventional frequentist MAP training is completed. In this section, we will develop this approach, and detail a simplification based on the FIM that yields a more tractable, albeit approximate, implementation.

12.6.1 Laplace Approximation Fits a Gaussian pdf at the MAP Solution

The Laplace approximation proceeds in two steps:

1. First, it tackles the **MAP problem**

$$\theta_{\mathcal{D}}^{MAP} = \arg\min_{\theta}\{-\log p(\theta|\mathcal{D})\}$$

$$= \arg\min_{\theta}\{-\log p(\theta, \mathcal{D})\}$$

$$= \arg\min_{\theta}\left\{-\log p(\theta) - \sum_{n=1}^{N}\log p(t_n|x_n, \theta)\right\}. \tag{12.37}$$

2. Then, it **fits a Gaussian distribution**

$$q(\theta|\mathcal{D}) = \mathcal{N}(\theta|\theta_{\mathcal{D}}^{MAP}, \Theta^{-1}) \tag{12.38}$$

around the MAP solution $\theta_{\mathcal{D}}^{MAP}$, for an optimized precision matrix Θ. The precision matrix is specifically chosen so as to match the curvature of the log-loss under the posterior $p(\theta|\mathcal{D})$ around $\theta_{\mathcal{D}}^{MAP}$.

For the second step, the Laplace approximation matches the second-order Taylor expansion of the negative logarithm $-\log q(\theta|\mathcal{D})$ of the approximant (12.38) around $\theta_{\mathcal{D}}^{MAP}$ with the second-order expansion of the negative log-posterior $-\log p(\theta|\mathcal{D})$ around the same point.

Since the gradient of both log-loss functions is zero at $\theta_{\mathcal{D}}^{MAP}$ by construction, this amounts to matching the Hessian matrices of the two functions.

The Hessian of the negative logarithm $-\log q(\theta|\mathcal{D}) = -\log \mathcal{N}(\theta|\theta_{\mathcal{D}}^{MAP}, \Theta^{-1})$ with respect to θ is equal to Θ for any value of θ. Furthermore, the Hessian of the negative log-posterior $-\log p(\theta|\mathcal{D})$ with respect to θ – which is unknown – equals the Hessian of the negative log-joint distribution $-\log p(\theta, \mathcal{D})$, which can be computed from the model as

$$\nabla_\theta^2 (-\log p(\theta|\mathcal{D})) = \nabla_\theta^2 (-\log p(\theta, \mathcal{D})). \tag{12.39}$$

This yields

$$\nabla_\theta^2 (-\log p(\theta_{\mathcal{D}}^{MAP}|\mathcal{D})) = \nabla_\theta^2 (-\log p(\theta_{\mathcal{D}}^{MAP})) + \nabla_\theta^2 \left(\sum_{n=1}^{N} (-\log p(t_n|x_n, \theta_{\mathcal{D}}^{MAP})) \right)$$

$$= \nabla_\theta^2 (-\log p(\theta_{\mathcal{D}}^{MAP})) + N \cdot \nabla_\theta^2 L_{\mathcal{D}}(\theta_{\mathcal{D}}^{MAP}). \tag{12.40}$$

Finally, imposing the equality $\Theta = \nabla_\theta^2 (-\log p(\theta_{\mathcal{D}}^{MAP}|\mathcal{D}))$ returns the Laplace approximation

$$p(\theta|\mathcal{D}) \simeq q(\theta|\mathcal{D}) = \mathcal{N}(\theta|\theta_{\mathcal{D}}^{MAP}, (\nabla_\theta^2(-\log p(\theta_{\mathcal{D}}^{MAP}, \mathcal{D})))^{-1}). \tag{12.41}$$

Note that this approximation assumes the invertibility of the Hessian (12.40) of the log-posterior at the MAP solution. Following the example in Sec. 10.4, the Laplace approximation can be accurate for posterior distributions that have a single mode.

12.6.2 Simplification of Laplace Approximation via the Empirical FIM

Computing the Hessian (12.40) in the Laplace approximation (12.41) may be prohibitively complex, since it requires the evaluation of the second derivatives of the prior, as well as of the log-losses evaluated at each training point, for each pair of entries in the model parameter vector. As we briefly discuss in this subsection, a simplification of the Laplace approximation is sometimes used that replaces the Hessian with a matrix – known as the empirical FIM – which does not involve any second derivative but only gradient information.

As we have detailed in Sec. 9.7, when the model is well-specified and hence there is a ground-truth value for the model parameter θ, the Hessian of the *population* log-loss evaluated at the ground-truth vector θ equals the FIM of the underlying probabilistic model. The simplification at hand uses this observation to make the following approximations: (*i*) assuming that N is large enough, we can approximate the training log-loss with the population log-loss in (12.40); and (*ii*) assuming that the MAP solution $\theta_{\mathcal{D}}^{MAP}$ is sufficiently close to the ground-truth model parameter (and that the latter exists), we can replace the Hessian of the population loss with a FIM.

Developing this idea and using the expression (9.68) for the FIM, this simplified approach prescribes the use of the **empirical FIM**

$$\hat{\text{FIM}}(\theta_{\mathcal{D}}^{MAP}) = \frac{1}{N} \sum_{n=1}^{N} (\nabla_\theta \log p(t_n|x_n, \theta_{\mathcal{D}}^{MAP}))(\nabla_\theta \log p(t_n|x_n, \theta_{\mathcal{D}}^{MAP}))^T \tag{12.42}$$

in lieu of the Hessian of the training log-loss in (12.40). The empirical FIM corresponds to the correlation matrix of the gradients $\{(\nabla_\theta \log p(t_n|x_n, \theta_{\mathcal{D}}^{MAP}))_{n=1}^{N}\}$, and it can be computed directly from the gradient vectors without having to evaluate second derivatives.

It should be emphasized that this simplification of the Laplace approximation via the empirical FIM (12.42) is well justified under the mentioned strong assumptions concerning the availability of a large data set and the closeness of the MAP solution $\theta_{\mathcal{D}}^{MAP}$ to the ground-truth value of the model parameter vector. We refer to Recommended Resources, Sec. 12.14, for more discussion on this point.

12.6.3 Simplification of Laplace Approximation via the Generalized Gauss–Newton Method

When neural network models are used to define the likelihood $p(t|x,\theta)$, an alternative way to simplify the Laplace approximation is by means of the **generalized Gauss–Newton (GGN)** method. The GGN approximation applies the chain rule to the direct computation of Hessian $\nabla_\theta^2(-\log p(\mathcal{D}|\theta))$ by writing the likelihood $p(t|x,\theta)$ as a function of the feature vector output by the neural networks. By omitting a term in the resulting expression that depend on the Hessian of the vector of features, one obtains a simpler expression that only requires the computation of the Jacobian of the feature vector with respect to the model parameters.

As an example, consider an L-layer neural network for binary classification (see Table 6.3), for which the log-loss for a data point (x_n, t_n) can be written as $\ell(a_n^L) = \log(1 + \exp(-t_n^\pm a_n^L))$, where $a_n^L = (w^L)^T u(x_n|\theta)$ is the (scalar) pre-activation of the classification layer (cf. (6.52)). The GGN method leverages the following approximation:

$$\nabla^2(-\log p(t_n|x_n,\theta)) \simeq \left.\frac{\partial^2 \ell(a)}{\partial a^2}\right|_{a=a_n^L} \cdot \nabla a_n^L \cdot \left(\nabla a_n^L\right)^T \tag{12.43}$$

$$= \left(\sigma\left(a_n^L\right)\left(1 - \sigma\left(a_n^L\right)\right)\right) \cdot \nabla a_n^L \cdot \left(\nabla a_n^L\right)^T, \tag{12.44}$$

where the gradient $\nabla_\theta a_n^L$ can be obtained directly via backprop, and the second derivative is computed from (6.33). See Recommended Resources, Sec. 12.14, for further details.

12.7 MC Sampling-Based Bayesian Learning

What to do when it is not computationally feasible to obtain the posterior $p(\theta|\mathcal{D})$, and the Laplace approximation is not sufficiently accurate? In such cases, there are two main categories of approximate Bayesian methods one can resort to, namely VI algorithms and MC sampling techniques. As studied in Chapter 10, VI algorithms minimize the free energy by restricting the optimization domain to a tractable set of variational posteriors. In contrast, MC sampling techniques aim to produce samples from the posterior distribution, so as to enable the estimate of posterior averages such as the predictive distribution (12.18). An illustration of the difference between the two methodologies can be found in Fig. 12.4.

In this section, we focus on MC sampling, while the next section concentrates on VI. Specifically, in the rest of this section, we first describe the goals of MC sampling, and then cover a number of standard MC sampling algorithms in order of scalability, from rejection sampling to stochastic gradient MCMC.

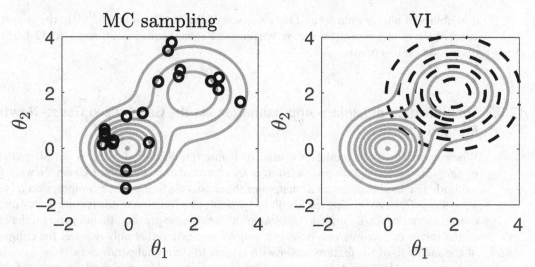

Figure 12.4 Contour lines of a ground-truth posterior distribution $p(\theta|\mathcal{D})$ (solid gray lines), along with (left) samples generated by an MC sampling technique with the aim of approximating averages with respect to the posterior; and (right) contour lines (dashed lines) of a Gaussian variational posterior obtained by a VI algorithm.

12.7.1 Goal of MC Sampling

MC sampling aims to draw S samples $\theta_1, \ldots, \theta_S$ i.i.d. from the posterior distribution $p(\theta|\mathcal{D})$. Once these samples are available, one can estimate posterior averages of the form $E_{\theta \sim p(\theta|\mathcal{D})}[f(\theta)]$ for some function $f(\cdot)$ via the empirical average

$$E_{\theta \sim p(\theta|\mathcal{D})}[f(\theta)] \simeq \frac{1}{S} \sum_{s=1}^{S} f(\theta_s). \tag{12.45}$$

An important example of posterior average is the ensemble predictive distribution (12.18), which can be estimated via (12.45) as

$$p(t|x, \mathcal{D}) \simeq \frac{1}{S} \sum_{s=1}^{S} p(t|x, \theta_s). \tag{12.46}$$

A key challenge in MC sampling is that the posterior distribution $p(\theta|\mathcal{D})$ is not known, and one has access only to the joint distribution $p(\theta, \mathcal{D}) = p(\theta)p(\mathcal{D}|\theta)$.

12.7.2 Rejection Sampling

Suppose that we can draw i.i.d. S' samples $\theta'_1, \ldots, \theta'_{S'}$ i.i.d. from the prior $p(\theta)$. Rejection sampling selects each **candidate sample** θ'_s independently with some **acceptance probability** $p(\text{acc}|\theta'_s)$, where we used "acc" to denote the event that a candidate sample is accepted. As we will see, it is possible to design the acceptance probability $p(\text{acc}|\theta')$ as a function of the candidate sample θ' and the training set $\mathcal{D}$ so that the distribution of an accepted sample equals the posterior $p(\theta|\mathcal{D})$. Specifically, we will show that, with a proper choice of the acceptance probability, rejection sampling leads to the selection of a subset of $S \leq S'$ samples $\theta_1, \ldots, \theta_S$ that have the same distribution of i.i.d. samples drawn from the posterior distribution $p(\theta|\mathcal{D})$. We emphasize that the acceptance probability should not depend on the true posterior distribution but only on the joint distribution $p(\theta, \mathcal{D})$ in order for this approach to be useful. Furthermore, ideally, the acceptance probability should be large enough to prevent most of the candidate samples from being rejected.

Deriving Rejection Sampling. For any acceptance probability $p(\mathrm{acc}|\theta)$, the distribution of an accepted sample is given as the conditional distribution

$$p(\theta|\mathrm{acc}) = \frac{p(\theta)p(\mathrm{acc}|\theta)}{p(\mathrm{acc})} \propto p(\theta)p(\mathrm{acc}|\theta), \qquad (12.47)$$

since an accepted sample is first drawn from the prior $p(\theta)$ and then selected with probability $p(\mathrm{acc}|\theta)$. Note that the probability of acceptance, $p(\mathrm{acc}) = \int p(\theta')p(\mathrm{acc}|\theta')\mathrm{d}\theta'$, is a normalizing constant in (12.47) that does not depend on θ. We wish the distribution (12.47) of an accepted sample to match the posterior, i.e., to satisfy

$$p(\theta|\mathrm{acc}) = p(\theta|\mathcal{D}) \propto p(\theta)p(\mathcal{D}|\theta). \qquad (12.48)$$

Combining (12.47) and (12.48) yields the condition

$$p(\mathrm{acc}|\theta) = \frac{p(\mathcal{D}|\theta)}{B}, \qquad (12.49)$$

where the constant B should satisfy the following conditions: (i) it should not depend on θ; and (ii) it should ensure the inequality $p(\mathrm{acc}|\theta) \leq 1$ for all possible values of the candidate sample θ.

Conditions (i) and (ii) are satisfied if we set the normalizing constant as $B = p(\mathcal{D}|\theta_{\mathcal{D}}^{ML})$, where we recall that the ML parameter is defined as $\theta_{\mathcal{D}}^{ML} = \arg\max_\theta p(\mathcal{D}|\theta)$. (If there are multiple ML solutions, one can choose any of them since they all yield the same value for B.) In fact, (i) the constant $B = p(\mathcal{D}|\theta_{\mathcal{D}}^{ML})$ does not depend on the candidate sample θ; and (ii) it guarantees that the acceptance probability

$$p(\mathrm{acc}|\theta) = \frac{p(\mathcal{D}|\theta)}{p(\mathcal{D}|\theta_{\mathcal{D}}^{ML})} \qquad (12.50)$$

is no larger than 1 for any θ. The latter condition follows from the inequality $p(\mathcal{D}|\theta) \leq p(\mathcal{D}|\theta_{\mathcal{D}}^{ML})$ that holds for all possible values of θ by the definition of the ML solution. Furthermore, the acceptance probability equals 1, i.e., $p(\mathrm{acc}|\theta) = 1$, only if θ coincides with an ML solution $\theta_{\mathcal{D}}^{ML}$.

Intuitively, with the choice (12.50), rejection sampling has a larger probability of accepting a candidate θ_s' if the sample has a larger likelihood $p(\mathcal{D}|\theta_s')$, that is, informally, if it better fits the data $\mathcal{D}$. The resulting algorithm is summarized in Algorithm 12.1.

Limitations of Rejection Sampling. Rejection sampling is very inefficient when the acceptance probability (12.50) tends to be small for samples drawn from the prior $p(\theta)$. In fact, in this case, in order to obtain a reasonably large number S of samples, one must start with a very large number S' of candidate samples.

By (12.50), the acceptance probability is small when the candidate sample θ_s' has a poor fit of the data set $\mathcal{D}$ in terms of the likelihood $p(\mathcal{D}|\theta)$ as compared to the ML solution. Therefore, if the prior $p(\theta)$ is too broad, or if it only covers model parameters θ with a low likelihood $p(\mathcal{D}|\theta)$, rejection sampling does not provide a satisfactory solution to the problem of MC sampling. The upshot of this discussion is that rejection sampling is typically only used for small-dimensional problems.

Algorithm 12.1: Rejection sampling

---

initialize $s = 0$

draw S' candidate samples $\theta'_1, \ldots, \theta'_{S'}$ i.i.d. from the prior $p(\theta)$

for $s' = 1, \ldots, S'$ **do**

  select each candidate sample $\theta'_{s'}$ with probability

$$p(\text{acc}|\theta'_{s'}) = \frac{p(\mathcal{D}|\theta'_{s'})}{p(\mathcal{D}|\theta_{\mathcal{D}}^{ML})}$$

  if accepted, set $s = s + 1$ and $\theta_s = \theta'_{s'}$ **end**

end

return accepted samples $\{\theta_s\}_{s=1}^{S}$ with $S = s$

12.7.3 Markov Chain Monte Carlo: Metropolis–Hastings Algorithm

As we have seen in the preceding subsection, rejection sampling produces a sequence of samples $\theta_1, \theta_2, \ldots, \theta_S$ that are equivalent to *i.i.d.* samples drawn from the posterior $p(\theta|\mathcal{D})$, hence providing an *exact* solution to the MC sampling problem. However, the approach is inefficient when the candidate samples produced by the prior provide low values for the training likelihood, yielding a small acceptance probability. In order to enhance the efficiency of sample generation, MCMC schemes introduce two key *approximations* in the goals of MC sampling explained in Sec. 12.7.1:

- The produced samples $\theta_1, \theta_2, \ldots, \theta_S$ are generally **correlated**, and not i.i.d.
- Their respective marginal distributions equal the posterior only **asymptotically**, i.e., only as $S \to \infty$.

In this subsection, we present a general template for the definition of MCMC methods, namely the **Metropolis–Hastings algorithm**. As we will see, the approach follows a sequential procedure whereby the next sample is generated using a random transformation of the previous accepted sample.

Specifically, in order to enhance the probability of accepting a sample, MCMC methods generate candidate samples successively, in an *adaptive* way, with the next sample θ_{s+1} being generated "in the neighborhood" of the last accepted sample θ_s. The rationale is that, if we do not stray too far from an accepted sample, we are likely to hit on another sample that provides a reasonable fit of the data and may be accepted. This local sample generation mechanism can potentially address the inefficiency of rejection sampling, which uses the prior distribution to generate all samples. On the flip side, by design, the samples produced by MCMC are correlated, and the method may require a long time to explore the domain of the posterior distribution.

Formally, MCMC methods define a **transition distribution** $p(\theta'|\theta)$ that produces the next candidate sample θ'_{s+1} given the previous accepted sample θ_s as $\theta'_{s+1} \sim p(\theta'|\theta_s)$; then, they accept this sample with a probability that ensures that the asymptotic distribution of the accepted samples coincides with the posterior. As in rejection sampling, the acceptance probability should not depend on the actual posterior $p(\theta|\mathcal{D})$, but only on the joint $p(\theta, \mathcal{D}) = p(\theta)p(\mathcal{D}|\theta)$, in order for the approach to be viable. The general form of MCMC methods is defined by the Metropolis–Hastings algorithm, which is described in Algorithm 12.2.

How can we interpret the acceptance probability (12.51)? First, it should be noted that, unlike rejection sampling, the acceptance probability $p(\text{acc}|\theta', \theta_s)$ depends also on the last accepted

---

Algorithm 12.2: Metropolis–Hastings (MH) MCMC sampling

---

initialize sample θ_0 and $s = 0$
while $s < S$ **do**

    generate a candidate sample $\theta' \sim p(\theta'|\theta_s)$
    accept the candidate sample with probability

$$p(\text{acc}|\theta', \theta_s) = \min\left(1, \frac{p(\theta')p(\mathcal{D}|\theta')}{p(\theta_s)p(\mathcal{D}|\theta_s)} \cdot \frac{p(\theta_s|\theta')}{p(\theta'|\theta_s)}\right) \qquad (12.51)$$

    if accepted, set $s = s + 1$ and $\theta_s = \theta'$ **end**
end
return accepted samples $\{\theta_s\}_{s=S_{\min}+1}^{S}$

---

sample θ_s, reflecting the local nature of sample generation in MCMC. Second, the acceptance probability (12.51) increases with the product of two ratios:

- the **posterior ratio**

$$\frac{p(\theta'|\mathcal{D})}{p(\theta|\mathcal{D})} = \frac{p(\theta')p(\mathcal{D}|\theta')}{p(\theta_s)p(\mathcal{D}|\theta_s)} \qquad (12.52)$$

between the posterior distributions evaluated at the candidate sample θ' and at the latest accepted sample θ_s; and

- the **transition ratio** $p(\theta_s|\theta')/p(\theta'|\theta_s)$ between the likelihood of generating θ_s from θ' and vice versa for the given transition distribution.

The first ratio is quite intuitive: If the posterior distribution at the candidate sample is larger than that of the last accepted sample, we should tend to accept the candidate sample. Importantly, from the right-hand side of (12.52), this ratio does not depend on the individual posterior distributions for the two samples. As for the second ratio, in many practical implementations it is eliminated by choosing a **symmetric transition distribution**, which satisfies the condition

$$p(\theta|\theta') = p(\theta'|\theta) \qquad (12.53)$$

for all θ and θ'. In fact, with a symmetric transition distribution, the transition ratio equals $p(\theta_s|\theta')/p(\theta'|\theta_s) = 1$. An example of a symmetric transition distribution is the Gaussian conditional distribution

$$p(\theta'|\theta) = \mathcal{N}(\theta'|\theta, \beta^{-1}). \qquad (12.54)$$

The use of this transition distribution amounts to adding Gaussian noise with zero mean and variance β^{-1} to the latest accepted sample θ. The MH algorithm with the transition distribution (12.54) is known as **random walk MH**.

To sum up, with a symmetric transition distribution such as (12.54), the probability of acceptance in MCMC sampling is proportional to the improvement in the posterior obtained by using candidate sample θ' as opposed to the last accepted sample θ_s.

It can be proved that MCMC sampling produces a sequence of accepted samples whose marginal distributions converge asymptotically to the posterior $p(\theta|\mathcal{D})$. That said, it is generally hard to quantify the time needed to obtain a desirable approximation (also known as the **mixing time**). For a formal statement and a proof of this claim, see Recommended Resources, Sec. 12.14.

In light of this asymptotic property, posterior expectations, such as the ensemble predictive distribution, are estimated, not by directly using (12.45) and (12.46), but by discarding the first $S_{\min}$ samples, which constitute the transient, or **burn-in**, period. This yields the estimate of the ensemble predictive distribution

$$p(t|x, \mathcal{D}) \simeq \frac{1}{S - S_{\min}} \sum_{s=S_{\min}+1}^{S} p(t|x, \theta_s), \tag{12.55}$$

and similarly for any other posterior expectations, where we recall that S is the total number of generated samples.

In practice, the **initialization** θ_0 in the MH MCMC sampling algorithm is important to ensure that the distribution of the samples converges quickly to the desired posterior. A useful idea is to use as initialization a model parameter vector θ_0 obtained following a standard frequentist learning approach, e.g., SGD for the ML problem.

12.7.4 Stochastic Gradient Markov Chain Monte Carlo

Stochastic gradient MCMC (SG-MCMC) methods provide an efficient implementation of MCMC that requires minor modifications as compared to the standard SGD algorithm introduced in Chapter 5. To improve over random walk MH methods, SG-MCMC techniques incorporate gradient information about the objective function (12.37) in the transition distribution. In this subsection, we describe a popular representative of SG-MCMC schemes, namely the **stochastic gradient Langevin dynamics (SGLD)** algorithm, which we previously encountered in Sec. 7.7 in the context of the problem of inference for energy-based models.

To start, let us review the operation of SGD when applied to the problem (12.37) that defines MAP learning for a given training set $\mathcal{D} = \{(x_n, t_n)_{n=1}^{N}\}$. We do so by modifying the notation in Chapter 5 in order to suit the discussion in this section by referring to the current iteration with subscript s. At each iteration s, SGD picks a mini-batch $\mathcal{S}_s$ of S_s indices from the set $\{1, \ldots, N\}$ uniformly at random (following the random uniform mini-batching assumption in Chapter 5); and then it updates the model parameter vector as

$$\theta_{s+1} \leftarrow \theta_s + \gamma_s \left(\frac{1}{S_s} \sum_{n \in \mathcal{S}_s} \nabla_\theta \log p(t_n|x_n, \theta_s) + \frac{1}{N} \nabla_\theta \log p(\theta_s) \right). \tag{12.56}$$

Following the analysis in Sec. 5.11, the term in parentheses in the update (12.56) is an unbiased estimate of the gradient of the objective function in problem (12.37) (divided by $1/N$ for consistency with the standard definition of training loss).

Stochastic Gradient Langevin Dynamics. SGLD modifies the SGD update (12.56) by adding Gaussian noise with zero mean and power equal to γ_s. The algorithm is detailed in Algorithm 12.3.

SGLD as MCMC Sampling. SGLD can be thought of as an MCMC sampling scheme in which the transition distribution is defined by the update (12.57) and no acceptance mechanism is used, or, equivalently, the acceptance probability equals 1. Note from Algorithm 12.3 that, by (12.57), the transition distribution varies across the iterations. It can be proved that, under the

---

Algorithm 12.3: Stochastic gradient Langevin dynamics (SGLD)

---

initialize sample θ_0 and $s = 0$
while $s < S$ **do**

  pick a mini-batch $\mathcal{S}_s \subseteq \{1, \ldots, N\}$ of S_s samples from the data set $\mathcal{D}$
  produce the next sample of the model parameter vector as

$$\theta_{s+1} \leftarrow \theta_s + \gamma_s \left(\frac{1}{S_s} \sum_{n \in \mathcal{S}_s} \nabla_\theta \log p(t_n | x_n, \theta_s) + \frac{1}{N} \nabla_\theta \log p(\theta_s) \right) + v_s, \qquad (12.57)$$

  where the added noise $v_s \sim \mathcal{N}(0, 2\gamma_s I)$ is independent across the iterations
  set $s \leftarrow s + 1$

end
return $\{\theta_s\}_{s=S_{\min}+1}^{S}$

Munro–Robbins conditions $\sum_{s=1}^{\infty} \gamma_s = \infty$ and $\sum_{s=1}^{\infty} (\gamma_s)^2 < \infty$ (see Chapter 5) and other suitable technical assumptions, the marginals of the samples generated using SGLD tend to the posterior distribution $p(\theta|\mathcal{D})$.

Operationally, in earlier iterations, when the learning rate is large, the SGLD iterates tend to move towards a MAP solution; while, in later iterations, as the learning rate decreases, the Gaussian noise contribution tends to dominate the stochastic gradient update in (12.57), since the standard deviation of the noise, $(2\gamma_s)^{1/2}$, becomes larger than the learning rate γ_s.

Extensions and Other SG-MCMC Methods. When the mini-batch includes the entire data set, i.e., when SGD is replaced by GD, SGLD is referred to as **Langevin MC (LMC)** or **unadjusted Langevin algorithm (ULA)**. Furthermore, when an MH acceptance step is included at each iteration, the resulting algorithm is known as **Metropolis adjusted Langevin algorithm (MALA)**.

More complex SG-MCMC methods have been introduced that improve on SGLD, as well as on its corresponding full-gradient variants, in terms of speed of convergence and accuracy of estimation of the posterior distribution. A notable example is **stochastic gradient Hamiltonian MC (SG-HMC)**. SG-HMC keeps track of an additional momentum variable with the goal of counteracting the impact of noise introduced by mini-batching in the SGLD updates, enabling larger update steps. More broadly, the general form for SG-MCMC schemes involves auxiliary variables, and is specified by diffusion matrices that control the level of noise introduced by the updates and by curl matrices that determine the pre-conditioning of the gradient updates (see Recommended Resources, Sec. 12.14).

12.7.5 Monte Carlo Methods for Likelihood-Free Models

While this chapter focuses on likelihood-based models, it is worth noting that there exist MC methods that operate in the likelihood-free setting, in which the likelihood is only available through a simulator and not as an analytical parametric expression (see Chapter 11). These are known as **approximate Bayesian computation (ABC)** methods. A popular ABC scheme applies a form of rejection sampling based on the comparison of sufficient statistics between data and samples generated from the model.

12.7.6 Importance Sampling

An alternative to MC sampling-based techniques introduced in Sec. 12.7.1 is offered by schemes that produce not only a sequence of samples $\theta_1, \ldots, \theta_S$ but also corresponding non-negative weights $w_1, \ldots, w_S$, satisfying the condition $\sum_{s=1}^{S} w_s = 1$, in such a way that posterior averages of a function $f(\cdot)$ can be estimated via the *weighted* empirical average

$$E_{\theta \sim p(\theta|\mathcal{D})}[f(\theta)] \simeq \sum_{s=1}^{S} w_s f(\theta_s) \tag{12.58}$$

in lieu of (12.45). In particular, the ensemble predictive distribution (12.18) can be estimated as

$$p(t|x, \mathcal{D}) \simeq \sum_{s=1}^{S} w_s p(t|x, \theta_s). \tag{12.59}$$

Importantly, in order for the right-hand sides of (12.58) and (12.59) to be "useful" estimates, there is no need for the samples $\{\theta_s\}_{s=1}^{S}$ to be (approximately) distributed according to the posterior $p(\theta|\mathcal{D})$. In fact, if properly chosen, the weights $\{w_s\}_{s=1}^{S}$ can compensate for the mismatch between the distribution of the samples and the posterior. The resulting approach is known as **importance sampling**.

To briefly elaborate on importance sampling, assume that we can draw samples θ_s i.i.d. from some distribution $q(\theta)$. Ideally, this distribution would be "closer" to the posterior than the prior $p(\theta)$ is, addressing some of the limitations of rejection sampling. Then, we can associate each sample θ_s with an intermediate weight

$$v_s = \frac{p(\theta_s, \mathcal{D})}{q(\theta_s)}, \tag{12.60}$$

which equals the ratio between joint distribution $p(\theta_s, \mathcal{D})$ and sampling distribution $q(\theta_s)$. The intermediate weights (12.60) are then **self-normalized** in order to obtain the final weights

$$w_s = \frac{v_s}{\sum_{s'=1}^{S} v_{s'}} \tag{12.61}$$

to be used in the empirical averages in (12.58) and (12.59). With this choice, it can indeed be proven that, as $S \to \infty$, the importance sampling estimates (12.58) and (12.59) are consistent, as they tend, with high probability, to the corresponding ensemble averages.

Particle-based importance sampling strategies that propagate a set of samples along with their weights are known as **sequential MC**. Techniques that combine MCMC with importance sampling include, notably, **annealed importance sampling** (see Recommended Resources, Sec. 12.14).

12.8 Variational Bayesian Learning

In VI-based Bayesian learning, the minimization of the free energy (12.12) is addressed directly by restricting the optimization domain to a set of **variational posteriors** $q(\theta|\varphi)$ identified by a **variational parameter vector** φ. The resulting problem can accordingly be expressed as

$$\min_{\varphi} F(q(\theta|\varphi)\|p(\theta, \mathcal{D})). \tag{12.62}$$

The optimization (12.62) defines a VI problem, and it can be solved using the techniques introduced in Chapter 10, namely the REINFORCE gradient and the reparametrization trick. It is also possible to use particle-based methods such as SVGD. In this section, we first develop the common reparametrization-based approach based on a Gaussian variational posterior. Then, we present particle-based VI via SVGD, and comment on the relationship of SVGD with the MCMC methods studied in the preceding section.

12.8.1   Reparametrization-Based Variational Bayesian Learning

In this subsection, we briefly discuss how the reparametrization trick can be used to develop a VI-based method for Bayesian learning under the assumption of a Gaussian variational posterior. When applied to neural networks, the approach is also known as **Bayes-by-backprop**.

Deriving Reparametrization-Based Variational Bayesian Learning. To start, following Sec. 10.9.7, let us assume the Gaussian prior $p(\theta) = \mathcal{N}(\theta|0, I)$, as well as the Gaussian variational posterior

$$q(\theta|\varphi) = \mathcal{N}(z|\nu, \mathrm{Diag}(\exp(2\varrho))), \tag{12.63}$$

where the variational parameter vector is set as $\varphi = [\nu^T, \varrho^T]^T$. Vectors ν and ϱ both have the same dimension as the model parameter θ, and they are unconstrained. To develop the method, one now only needs to specify how to estimate the gradient of the free energy in (12.62).

Algorithm 12.4: Reparametrization-based variational Bayesian learning (with $S = 1$)

initialize $\varphi^{(1)} = [(\nu^{(1)})^T, (\varrho^{(1)})^T]^T$ and $i = 1$
while *stopping criterion not satisfied* **do**
  draw a sample $e^{(i)} \sim \mathcal{N}(0, I)$
  draw a sample $(x^{(i)}, t^{(i)})$ from the data set $\mathcal{D}$
  compute $\theta^{(i)} = \nu^{(i)} + \exp(\varrho^{(i)}) \odot e^{(i)}$
  estimate the gradient of the free energy as

$$\hat{\nabla}_{\varphi}^{(i)} = \begin{bmatrix} \hat{\nabla}_{\nu}^{(i)} \\ \hat{\nabla}_{\varrho}^{(i)} \end{bmatrix} \tag{12.64}$$

  with

$$\hat{\nabla}_{\nu}^{(i)} = \nabla_{\theta}(-\log p(t^{(i)}|x^{(i)}, \theta^{(i)})) + \nu^{(i)}$$
$$\hat{\nabla}_{\varrho}^{(i)} = \exp(\varrho^{(i)}) \odot e^{(i)} \odot \nabla_{\theta}(-\log p(t^{(i)}|x^{(i)}, \theta^{(i)}))$$
$$+ \exp(\varrho^{(i)}) \odot (\exp(\varrho^{(i)}) - \exp(-\varrho^{(i)})) \,. \tag{12.65}$$

  given a learning rate $\gamma^{(i)} > 0$, obtain the next iterate as

$$\varphi^{(i+1)} \leftarrow \varphi^{(i)} - \gamma^{(i)} \hat{\nabla}_{\varphi}^{(i)} \tag{12.66}$$

  set $i \leftarrow i + 1$
end
return $\varphi^{(i)}$

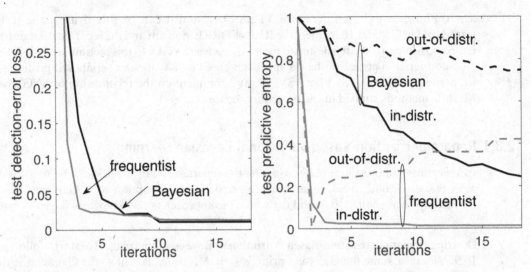

Figure 12.5 Test detection-error loss (left) and test predictive entropy for in-distribution and out-of-distribution samples (right) under frequentist (MAP) and Bayesian VI-based learning.

As detailed in Sec. 10.9, this can be done via the reparametrization trick. Accordingly, at each iteration, we apply a **doubly stochastic gradient estimator**, which relies on one or more samples from the training set to approximate the training loss, as well as one or more auxiliary samples $e \sim \mathcal{N}(0, I)$ in order to apply the reparametrization trick. The resulting algorithm, detailed in Algorithm 12.4 for the case of a single auxiliary sample ($S = 1$), can be obtained directly from Algorithm 10.7. In it, the gradient $\nabla_\theta(-\log p(t|x, \theta))$ can be computed using the methods discussed in Chapter 5, notably backprop in the case of neural network models.

Example 12.6

In this example, we compare frequentist and Bayesian learning for the problem of training a logistic regression classifier $p(t|x, \theta)$ on the task of distinguishing handwritten digits 0 and 1 from the USPS image data set (see Sec. 6.5.6). The set of training data images of digits 0 and 1 consists of around 1000 examples per class. We compare the standard MAP frequentist approach (see (12.37)) with Bayesian learning via the reparametrization-based scheme in Algorithm 12.4. For both methods, we set mini-batch size 20 and constant learning rate $\gamma = 0.5$.

The left-hand side of Fig. 12.5 shows the test detection-error loss vs. the number of iterations i for one run of the algorithm. It is observed that the accuracy levels of both frequentist and Bayesian schemes are comparable after a sufficiently large number of iterations. The benefits of Bayesian learning become apparent, however, when one considers the problem of **uncertainty quantification**.

In the right-hand side of Fig. 12.5, we plot the entropy of the predictive distribution, i.e., the **test predictive entropy**

$$H(E_{\theta \sim q(\theta|\varphi^{(i)})}[p(t|x, \theta)]), \tag{12.67}$$

across the iteration index i. This entropy is approximated by using 1000 samples from the variational posterior $q(\theta|\varphi^{(i)})$. For frequentist learning, we accordingly evaluate the predictive entropy $H(p(t|x, \theta))$. We first evaluate the test predictive entropy **in-distribution**, i.e., by averaging

(12.67) over images x of 0 and 1 digits in the test set. The resulting average test predictive entropy for in-distribution data (i.e., for digits 0 and 1) is a measure of the level of uncertainty of the predictor on the type of data it has been trained on. More precisely, the predictive uncertainty metric (12.67) captures both epistemic and aleatoric uncertainty contributions (see Appendix 12.A).

Ideally, the in-distribution test predictive entropy should decrease across the iterations as more and more data are collected, and it should be large for earlier iterations. This is indeed seen to be the case in Fig. 12.5 for both frequentist and Bayesian learning, although frequentist learning becomes quickly very confident, even while the test error is still relatively large.

For further insight into the uncertainty quantification properties of Bayesian learning, we finally consider the test predictive entropy (12.67) averaged over **out-of-distribution data** samples x. Out-of-distribution data are generated via a different mechanism then the training set. For this example, we specifically choose images corresponding to digits 2 and 9 from the USPS data set. The resulting average test predictive entropy for out-of-distribution samples (i.e., for digits 2 and 9) is a measure of the level of (combined epistemic and aleatoric) uncertainty of the predictor on data it has received no information about.

Ideally, the out-of-distribution test predictive entropy should be constant during the iterations, as no information is collected about out-of-distribution data. From Fig. 12.5 it is observed that Bayesian learning provides a more accurate quantification of uncertainty, as its out-of-distribution test predictive entropy decreases only slightly across the global iterations. In contrast, frequentist learning offers an unreliable estimate of out-of-distribution uncertainty.

Sparsity and Parametric Variational Posteriors. The choice of the set of variational posteriors $q(\theta|\varphi)$ may be guided by specific properties one would like to enforce on the output of the learning process. One of the most useful such properties is **sparsity**. Sparse model parameter vectors have few non-zero inputs, and hence "sparse" variational posteriors $q(\theta|\varphi)$ are such that, with high probability, a number of the entries of vector θ are zero or much smaller than the rest of the entries. This implies that, when the model parameters θ are drawn from $q(\theta|\varphi)$, the resulting vector is likely to be sparse. Sparse variational posteriors are typically accompanied by sparse priors in the model definition, encoding a prior belief in the sparsity of the model.

Sparse priors are typically defined by **spike-and-slab** distributions, which consist of a mixture of a delta function at the all-zero vector and a proper distribution such as a Gaussian pdf. Variational posteriors can be defined in various ways, yielding different instantiations of **variational dropout** schemes. One approach is to model the variational posterior $q(\theta|\varphi)$ so that each entry θ_i is the product $\theta_i = \mathrm{b}_i \cdot \bar{\theta}_i$ of a Bernoulli rv $\mathrm{b}_i \sim \mathrm{Bern}(p_i)$ and a (deterministic) parameter $\bar{\theta}_i$. With this choice, the variational parameter vector φ includes the probabilities p_i and the parameters $\bar{\theta}_i$. The probability p_i determines how likely it is for the entry θ_i to be non-zero, with $1 - p_i$ being the probability of "dropping" the ith weight. In a neural network implementation, the variables b_i may be reused across all weights applied to the output of a neuron, so that, when $\mathrm{b}_i = 0$, the neuron is effectively "dropped" by the network.

12.8.2 Adversarial Variational Bayesian Learning

Reparametrization-based variational Bayesian learning assumes that one has access to an analytical, reparametrizable, variational posterior $q(\theta|\varphi)$. In order to increase the flexibility of the approximation of the posterior distribution, it is possible to generalize variational Bayesian learning to the case in which the variational posterior is defined by a **pushforward distribution** of a (global) latent rv θ' (see Sec. 11.10). Accordingly, the variational distribution is only defined **implicitly** as the output of the parametrized simulator

$$\theta' \sim p(\theta')$$
$$\theta = G(\theta'|\varphi), \tag{12.68}$$

where the prior $p(\theta')$ is fixed and the variational parameters φ define the operation of the function $G(\cdot|\varphi)$. In order to minimize the free energy over φ, one can leverage the reparametrization trick to compute the gradient of the first term in (12.16). For the second term, one can estimate the KL divergence using the two-sample estimators introduced in Sec. 11.8. This approach is known as **adversarial variational Bayesian learning** (see Recommended Resources, Sec. 12.14, for details).

12.8.3 Particle-Based Variational Bayesian Learning

Particle-based VI methods such as SVGD (see Sec. 10.11) describe the variational posterior $q(\theta|\varphi)$ distribution in terms of a set of particles $\varphi = \{\theta_s\}_{s=1}^{S}$. As in MCMC, the particles can be used to approximate posterior expectations such as the predictive distribution through the empirical averages (12.45) and (12.46). However, unlike the samples produced by MCMC methods, the particles produced by VI are *deterministic*, *interacting*, and generated *jointly* so as to approximately minimize the free energy function. We start by elaborating on the relationship between MCMC and particle-based VI, and then detail the application of SVGD to Bayesian learning.

MCMC vs. Particle-Based Variational Learning. In MCMC, samples are generated sequentially, and they have the property that, asymptotically, the marginal distribution of each sample coincides with the desired posterior $p(\theta|\mathcal{D})$. However, MCMC methods do not provide any guarantee that, for a finite value of the number S of samples, the samples produced by this process are optimal in terms of offering the best estimate for empirical averages such as (12.45) and (12.46). In particular, successive samples produced by MCMC tend to be correlated, which makes them generally less useful in estimating empirical averages.

Algorithm 12.5: Particle-based variational Bayesian learning via SVGD (with full minibatches)

---

initialize particles $\varphi^{(1)} = \{(\theta_s^{(1)})_{s=1}^{S}\}$ and $i = 1$
while *stopping criterion not satisfied* **do**
    for all particles $s = 1, \ldots, S$
    given a learning rate $\gamma^{(i)} > 0$, obtain the next iterate as

$$\theta_s^{(i+1)} \leftarrow \theta_s^{(i)} - \gamma^{(i)} r_s^{(i)} \tag{12.69}$$

    with

$$r_s^{(i)} = \sum_{s'=1}^{S} \kappa(\theta_s^{(i-1)}, \theta_{s'}^{(i-1)}) \underbrace{\left[-\nabla_{\theta_{s'}} \log p(\theta_{s'}^{(i-1)}, \mathcal{D}) \right]}_{\text{log-loss gradient}} - \underbrace{\nabla_{\theta_{s'}} \kappa(\theta_s^{(i-1)}, \theta_{s'}^{(i-1)})}_{\text{repulsive "force"}} \tag{12.70}$$

    end for
    set $i \leftarrow i + 1$
end
return $\varphi^{(i)} = \{(\theta_s^{(i)})_{s=1}^{S}\}$

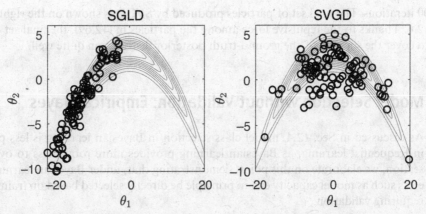

Figure 12.6 Ground-truth posterior (gray contour lines) along with $S = 100$ samples produced by SGLD (left) and with the $S = 100$ particles output by SVGD after 200 iterations (right).

In contrast, non-parametric VI methods like SVGD optimize directly and jointly over the set of S particles in order to minimize some measure of discrepancy between the distribution obtained from the particles (e.g., using KDE) and the desired distribution. As such, they can be more effective in producing accurate estimates of ensemble quantities when the number S of particles is limited.

Particle-Based Variational Bayesian Learning via SVGD. Referring to Sec. 10.11 for details (see Algorithm 10.8), when applied to Bayesian learning, SVGD prescribe the joint update of the set of particles described in Algorithm 12.5. The algorithm is stated for convenience with full mini-batches, but it can easily be generalized to account for smaller mini-batch sizes.

The interpretation of the SVGD update (12.69) follows the discussion in Sec. 10.11. In particular, the first term in (12.69) pushes the particles towards the MAP solution (12.37), while the second tries to ensure diversity by causing a repulsion between close particles.

Example 12.7

Consider a banana-shaped ground-truth posterior distribution of the form

$$p(\theta|\mathcal{D}) \propto \exp\left(-\frac{1}{2}\frac{\theta_1^2}{100} - \frac{1}{2}(\theta_2 + 0.03\theta_1^2 - 3)^2\right), \tag{12.71}$$

whose contour lines are shown in Fig. 12.6. In this example, we implement and compare SGLD (Algorithm 12.3) and SVGD (Algorithm 12.5). Both algorithms only require an unnormalized version of the posterior, and hence they can be derived directly from the right-hand side of (12.71) without requiring any normalization. In particular, the gradient $\nabla_\theta \log p(\theta, \mathcal{D})$ can be directly computed by differentiating the right-hand side of (12.71).

First, SGLD is implemented starting from initial point $\theta = [-20, -10]^T$ with a constant learning rate $\gamma^{(i)} = 0.1$ for $S = 100$ iterations. The samples produced by one run of the algorithm are shown on the left-hand side of Fig. 12.6. The samples tend to move towards the peak of the distribution (at $[0, 3]^T$), while exploration of other parts of the distribution is limited to the local perturbations caused by the addition of Gaussian noise in (12.57).

Then, SVGD is implemented with $S = 100$ particles, which are initialized to equal the 100 samples produced by SGLD. The learning rate is selected as $\gamma^{(i)} = 0.2$, and the algorithm is run for 200 iterations. The final set of particles produced by SVGD is shown on the right-hand side of Fig. 12.6. Thanks to the repulsive force among the particles in (12.69), the final set of particles is seen to cover the support of the ground-truth posterior distribution quite well.

12.9   Model Selection without Validation: Empirical Bayes

As discussed in Sec. 12.4, model class selection in Bayesian learning is less problematic than in frequentist learning, as Bayesian learning provides more robustness to overfitting. In this section, we elaborate on this point, demonstrating that, under Bayesian learning, hyperparameters such as model capacity can in principle be directly selected based on training data without requiring validation.

12.9.1   Empirical Bayes, aka Type-II ML

As detailed in Sec. 4.4, in a frequentist framework, the (scaled) training log-loss

$$-\log p(\mathcal{D}|\theta_{\mathcal{D}}^{ML}) = -\log p(t_{\mathcal{D}}|x_{\mathcal{D}}, \theta_{\mathcal{D}}^{ML}) = -\sum_{n=1}^{N} \log p(t_n|x_n, \theta_{\mathcal{D}}^{ML}) \quad (12.72)$$

evaluated at the ML trained model $\theta_{\mathcal{D}}^{ML}$ generally decreases – or is anyway non-increasing – with the model class capacity. In fact, a larger-capacity model class cannot worsen the fit of the training data. Therefore, model capacity selection requires the evaluation of the log-loss $-\log p(\mathcal{D}^v|\theta^{ML})$ on a separate validation set $\mathcal{D}^v$ in order to enable the choice of a model class that approximately minimizes the population loss.

In Bayesian learning, the assumed joint distribution (12.9), specified by prior and likelihood, defines a distribution across data sets $\mathcal{D}$ via the marginal training likelihood

$$p(\mathcal{D}) = p(t_{\mathcal{D}}|x_{\mathcal{D}}) = \int p(\theta) \prod_{n=1}^{N} p(t_n|x_n, \theta)\mathrm{d}\theta. \quad (12.73)$$

Unlike the training log-loss (12.72), the marginal training log-loss $-\log p(\mathcal{D})$ does not necessarily decrease with the model capacity. In fact, as reviewed in Sec. 12.4, a larger model capacity necessarily implies a more spread-out prior over the model parameter vectors, which prevents the best-fitting models for the data set $\mathcal{D}$ from dominating the expectation in (12.73). As a result, a more spread-out prior must necessarily assign a smaller probability to at least some of the data sets $\mathcal{D}$, potentially decreasing the marginal likelihood and increasing the marginal training log-loss $-\log p(\mathcal{D})$.

Overall, a model capacity that is too small "covers" too few data sets, yielding a high value of the marginal training log-loss; while a model capacity that is too large "covers" too many data sets, diluting the marginal training likelihood across a larger domain and again producing a high value of the marginal training log-loss. The model class that minimizes the marginal training log-loss is hence one that has enough capacity, but not more than necessary – an embodiment of the **Occam's razor** principle.

The problem of maximizing the marginal training likelihood over the model capacity, or more generally any other hyperparameters that determine prior and likelihood, is known as **empirical Bayes**, or **type-II ML**.

   The discussion in this subsection suggests that the marginal likelihood is a useful measure of the generalization capacity of a model. As such, it can be adopted as a metric for model selection, without requiring validation. An alternative way to understand this property of the marginal likelihood is to note that the distribution (12.73) can be expressed by the chain rule of probability as

$$p(\mathcal{D}) = \prod_{n=1}^{n} p(t_n|x_{\mathcal{D}}, t^{n-1}) = \prod_{n=1}^{n} p(t_n|x_n, \mathcal{D}^{n-1}), \tag{12.74}$$

where $t^{n-1} = \{t_1, ..., t_{n-1}\}$ for $n > 1$ and t^0 is an empty set, and we have $\mathcal{D}^{n-1} = \{(x_{n'} t_{n'})_{n'=1}^{n-1}\}$, with $x^{n-1} = \{x_1, ..., x_{n-1}\}$ for $n > 1$ and x^0 is an empty set. Each term $p(t_n|x_n, \mathcal{D}^{n-1})$ is a soft predictor of the nth response variable t_n given the data set $\mathcal{D}^{n-1}$, and therefore the marginal training log-loss

$$-\log p(\mathcal{D}) = -\sum_{n=1}^{n} \log p(t_n|x_n, \mathcal{D}^{n-1}) \tag{12.75}$$

carries out a form of cross-validation (see Sec. 4.4). In it, a model trained on the first $n - 1$ samples of the training set, i.e., on data set $\mathcal{D}^{n-1}$, is tested on the nth training data (x_n, t_n). This perspective also highlights a limitation of the marginal training log-loss as a measure of generalization, in that only when n is large enough does the soft predictor

$$p(t_n|x_n, \mathcal{D}^{n-1}) = \mathrm{E}_{\theta \sim p(\theta|\mathcal{D}^{n-1})}[p(t|x, \theta)] \tag{12.76}$$

reflect well the performance of the soft predictor (12.18) trained on the entire data set $\mathcal{D}$.

12.9.2 Gaussian–Gaussian GLM

In this subsection, we develop empirical Bayes for the Gaussian–Gaussian GLM prior and likelihood in (12.27) and (12.28). Under this model, the marginal training likelihood for a given data set $\mathcal{D}$ can be computed explicitly. Denoting as D the dimension of model parameter vector θ, the model in (12.27) and (12.28) can be equivalently stated as the linear relationship

$$t_{\mathcal{D}} = X_{\mathcal{D}}\theta + \mathcal{N}(0, \beta^{-1}I_N) \tag{12.77}$$

between the $D \times 1$ model parameter vector θ, distributed according to the prior as $\theta \sim \mathcal{N}(0_D, \alpha^{-1}I_D)$, and the $N \times 1$ label vector $t_{\mathcal{D}}$. This yields the marginal training likelihood

$$p(\mathcal{D}) = p(t_{\mathcal{D}}|x_{\mathcal{D}}) = \mathcal{N}(t_{\mathcal{D}}|0_N, C_{\mathcal{D}}), \tag{12.78}$$

with $N \times N$ covariance matrix

$$C_{\mathcal{D}} = \alpha^{-1}X_{\mathcal{D}}X_{\mathcal{D}}^T + \beta^{-1}I_N. \tag{12.79}$$

Recall that the data matrix $X_{\mathcal{D}}$ is of dimension $N \times D$, collecting by rows the feature vectors corresponding to the N training points.

   Therefore, the marginal training log-loss is

$$-\log p(\mathcal{D}) = -\log p(t_{\mathcal{D}}|x_{\mathcal{D}}) = \frac{1}{2}t_{\mathcal{D}}^T C_{\mathcal{D}}^{-1} t_{\mathcal{D}} + \frac{1}{2}\log\det(2\pi C_{\mathcal{D}}). \tag{12.80}$$

The two terms in the marginal training log-loss (12.80) depend in opposite ways on the model capacity, which is defined by the size, D, of the model parameter θ. A larger model capacity generally increases the eigenvalues of the covariance matrix $C_{\mathcal{D}}$. Therefore, the first term in

Figure 12.7 Marginal training likelihood as a function of the model capacity M for the Gaussian–Gaussian GLM applied to polynomial regression (Example 12.8).

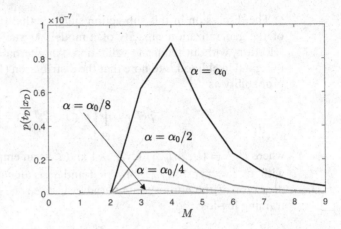

(12.80) tends to decrease with the model capacity, and is related to the training error; while the second term tends to increase with the model capacity and is a measure of the model complexity.

Example 12.8

Figure 12.7 shows the marginal training likelihood (12.78) as a function of the degree M for the polynomial regression problem studied in Example 12.5 ($D = M + 1$). We also set $\beta = 10$ and $\alpha_0 = 10^{-3}$. The figure confirms that, unlike in frequentist learning, one can identify an optimal value for the model class capacity M based solely on training data by maximizing the marginal training likelihood. It is also observed that a less precise prior, i.e., a smaller value of α, yields a smaller marginal training likelihood because of the larger spread of the prior distribution.

12.9.3 Laplace Approximation for Empirical Bayes

For more general models, the marginal training log-likelihood is not available in closed form because of the need to evaluate the integral in (12.73) in order to marginalize out the model parameter vector θ. In these cases, the Laplace approximation described in Sec. 12.6 can be conveniently applied to obtain an estimate of the marginal training log-loss, as detailed in this subsection.

To this end, we expand the log-loss $-\log p(\theta, \mathcal{D})$ around the MAP solution $\theta_{\mathcal{D}}^{MAP}$ in (12.37) using a second-order Taylor approximation to obtain

$$p(\theta, \mathcal{D}) \simeq p(\theta_{\mathcal{D}}^{MAP}, \mathcal{D}) \cdot \exp\left(\frac{1}{2}(\theta - \theta_{\mathcal{D}}^{MAP})^T \nabla_\theta^2(-\log p(\theta_{\mathcal{D}}^{MAP}, \mathcal{D}))(\theta - \theta_{\mathcal{D}}^{MAP})\right). \quad (12.81)$$

Plugging this approximation into the integral (12.73) yields the **Laplace approximation of the marginal training log-likelihood**,

$$-\log p(\mathcal{D}) \simeq -\log p(\theta_{\mathcal{D}}^{MAP}, \mathcal{D}) + \frac{1}{2}\log\det\left(2\pi\nabla_\theta^2(-\log p(\theta_{\mathcal{D}}^{MAP}, \mathcal{D}))\right), \quad (12.82)$$

where the second term in (12.82) is computed from the normalization constant of the unnormalized Gaussian pdf that appears in (12.81). This approximation can be further simplified by estimating the Hessian, as discussed in Sec. 12.6.

The approximation (12.82) has the same general form as the marginal training log-loss (12.80) for the Gaussian–Gaussian GLM. The first term in (12.82) is a measure of training loss, and it tends to decrease with the model capacity. In contrast, the second term acts as a regularizer that penalizes the complexity of the model and accounts for the generalization error, with the Hessian of the log-loss playing the role of the covariance matrix in (12.79).

Another way to see the connection between the second term in (12.82) and generalization is through the discussion in Sec. 5.13. The second term is in fact minimized for models with Hessians of the training log-loss $-\log p(\theta, \mathcal{D})$ that have small eigenvalues, and hence a small determinant, at the MAP solution $\theta_{\mathcal{D}}^{MAP}$. Small eigenvalues imply a small curvature, which in turn corresponds to a "wide" minimum $\theta_{\mathcal{D}}^{MAP}$ of the log-loss $-\log p(\theta, \mathcal{D})$. As discussed in Sec. 5.13, "wide" minima have been demonstrated empirically, as well as in some analytical studies, to provide solutions in the parameter space that generalize better outside the training set.

12.10 Bayesian Non-parametric Learning

Parametric models provide strong inductive biases under which a predictor is identified via a finite-dimensional vector θ, which in turn determines the "capacity" of the model. After training, prediction on a test input x can be carried out based solely on the trained parameter $\theta_{\mathcal{D}}$, without requiring access to the training set $\mathcal{D}$. In Sec. 4.11, we introduced the alternative framework of non-parametric models, in which the predictor on a new covariate vector x depends directly on the training samples $\{(x_n, t_n)_{n=1}^{N}\}$ in the data set $\mathcal{D}$ through the similarity of the test covariate x with each of the covariates $\{(x_n)_{n=1}^{N}\}$. Non-parametric models define lighter inductive biases that only specify a measure of similarity between covariates via a (positive semi-definite) kernel function $\kappa(x, x')$. In this section, we present non-parametric models from a Bayesian perspective.

The prior distribution $p(\theta)$ assumed by the parametric Bayesian models studied so far in this chapter indirectly defines a distribution over the class of soft predictors $p(t|x, \theta)$ determined by the likelihood. In contrast, **Bayesian non-parametric models** assume a prior distribution directly **on the space of predictive functions**, without having to explicitly specify a parametric model class. As we will see in this section, when the prior is Gaussian, this approach yields a Bayesian counterpart of the frequentist kernel techniques studied in Sec. 4.11 that are known as Gaussian processes (GPs).

12.10.1 Gaussian–Gaussian GLM Revisited

To start, let us reconsider the Gaussian–Gaussian GLM defined by (12.28) and (12.32) for the special case in which the precision of the likelihood, β, is infinite. The main goal of this subsection is to reinterpret the Gaussian–Gaussian GLM as a random predictor. In the next subsection, we will see how GPs leverage this viewpoint to define a non-parametric prior on the space of predictors.

By (12.28) and (12.32), the Gaussian–Gaussian GLM fixes a $D \times 1$ vector of features $u(x) = [u_1(x), \ldots, u_D(x)]^T$, and assumes the random input–output relationship

$$\mathrm{t} = t(x|\theta) = \theta^T u(x) = \sum_{d=1}^{D} u_d(x), \text{with } \theta \sim \mathcal{N}(0_D, \alpha^{-1} I_D). \tag{12.83}$$

Accordingly, the Gaussian–Gaussian GLM posits a stochastic relationship between covariates and labels that is defined by the random linear combination $t(x|\theta) = \theta^T u(x)$ of a fixed vector of features $u(x)$ through the random vector of coefficients θ.

Consider a set $\mathcal{D}$ of N labels $t_{\mathcal{D}}$ given the corresponding covariates $x_{\mathcal{D}}$. By (12.83), the Gaussian–Gaussian GLM views the outputs $t_{\mathcal{D}}$ as being generated by first drawing the model parameter vector $\theta \sim \mathcal{N}(0_D, \alpha^{-1} I_D)$ and then computing $t_n = t(x_n|\theta)$ for $n = 1, \ldots, N$. The resulting marginal joint distribution of the labels is hence given as

$$p(t_{\mathcal{D}} = t_{\mathcal{D}}|x_{\mathcal{D}} = x_{\mathcal{D}}) = \int p(\theta) \prod_{n=1}^{N} p(t_n|x_n, \theta) d\theta$$

$$= \mathcal{N}(t_{\mathcal{D}}|0, K_{\mathcal{D}}), \tag{12.84}$$

where the $N \times N$ covariance matrix $K_{\mathcal{D}}$ has entries

$$[K_{\mathcal{D}}]_{m,n} = u(x_m)^T u(x_n) = \kappa(x_m, x_n) \tag{12.85}$$

for $m, n = 1, \ldots, N$. In words, under the model (12.83), any two observations t_m and t_n are correlated, with a covariance (12.85) that equals the inner product $\kappa(x_m, x_n) = u(x_m)^T u(x_n)$ between the corresponding feature vectors $u(x_m)$ and $u(x_n)$. As introduced in Sec. 4.11, function $\kappa(x_m, x_n)$ in (12.85) is an example of a positive semi-definite kernel, and the covariance matrix (12.85) is the **kernel matrix** for the given data set $\mathcal{D}$.

It is important to stress the nature of the correlation among observations described by the marginal (12.84). Had the model parameter θ been known, by (12.83), any two observations t_m and t_n would have been deterministic and hence independent. But the model parameter θ is unknown and assumed to be distributed according to the Gaussian prior (12.28). For this reason, observing an output t_m generally provides information about another output t_n: The observation of t_m is informative about the unknown model parameter θ, which, in turn, brings evidence about t_n.

12.10.2 Gaussian Processes

We have seen in the preceding subsection that the Gaussian–Gaussian GLM, with $\beta \to \infty$, assumes a linear random predictive function of the form (12.83), under which the labels $t_{\mathcal{D}}$ produced by a vector of covariates $x_{\mathcal{D}}$ are correlated as per the joint distribution (12.84). A **Gaussian process (GP)** models a random predictive function $t(x)$ by directly specifying the joint distribution of any subset of its output values $t_{\mathcal{D}}$, conditioned on $x_{\mathcal{D}}$, as a jointly Gaussian vector. In a manner similar to (12.84), the joint distribution is fully specified by a positive semi-definite kernel function (as well as, possibly, a mean function). Therefore, following the discussion on kernel methods in Sec. 4.11, one can think of a GP as assuming a random predictive function of the form (12.83) that is not constrained to use *finite-dimensional* feature vectors $u(x)$. Instead, depending on the kernel function, the random predictor can effectively implement even infinite-dimensional vectors of features $u(x)$.

Inductive Bias. To elaborate, generalizing (12.83), a GP assumes, as its inductive bias, that covariates and labels are related via a random function $t(x)$ as

$$t = t(x). \tag{12.86}$$

The prior over the set of all functions $t(x)$ between input x and output t is defined by a mean function $\mu(x)$ which returns a scalar for each covariate vector x, and by a positive semi-definite kernel function $\kappa(x, x')$ which returns a scalar for each pair of covariate vectors x and x'. We refer to Sec. 4.11 for a definition of positive semi-definite kernel function.

The **mean function** $\mu(x)$ models the mean of the random function (12.86) as

$$E[t(x)] = \mu(x), \tag{12.87}$$

where the expectation is taken over the prior distribution of the random function $t(x)$. The mean function is typically set to be a constant, e.g., zero, unless one has information about the likely shape of function $t(x)$.

The **kernel function** $\kappa(x, x')$ models the covariance of any two outputs $t(x)$ and $t(x')$ of the random function as

$$E[(t(x) - \mu(x))(t(x') - \mu(x'))] = \kappa(x, x'), \tag{12.88}$$

where again the expectation is over the prior distribution of the random function $t(x)$. A common choice is the **Gaussian kernel**

$$\kappa(x, x') = a \exp\left(-\frac{1}{2h}||x - x'||^2\right) \tag{12.89}$$

for some parameter $a > 0$. With the Gaussian kernel, a larger "bandwidth" h gives a higher prior chance of generating smoother predictive functions $t(x)$ that vary more slowly over x. This is because the input values x and x' need to be further from one another, say at a distance equal to 3 to 4 times $\sqrt{h}$, in order for the covariance $\kappa(x, x')$ of the corresponding outputs to be small. Conversely, a smaller h is more likely to generate less smooth functions with faster variations.

As a result of the described choice of prior over the space of random functions $t(x)$ from x to t, GPs model the joint distribution of any subset of outputs as

$$p(t_{\mathcal{D}} = t_{\mathcal{D}}|x_{\mathcal{D}} = x_{\mathcal{D}}) = \mathcal{N}(t_{\mathcal{D}}|\mu_{\mathcal{D}}, K_{\mathcal{D}}), \tag{12.90}$$

where we have defined the $N \times 1$ mean vector $\mu_{\mathcal{D}} = [\mu(x_1), \dots, \mu(x_N)]^T$, and introduced the $N \times N$ kernel matrix

$$[K_{\mathcal{D}}]_{m,n} = \kappa(x_m, x_n), \tag{12.91}$$

which generalizes (12.85) to any positive semi-definite kernel function.

Gaussian "Processes". The terminology "GP" reflects the interpretation of the random mapping $t(x)$ in (12.86) as a stochastic process defined on the space of the covariates x. For instance, if the covariate x is a scalar real number, one can think of $t(x)$ as a standard stochastic process varying over the "time" variable x. For translation-invariant kernels such as (12.89), the process can be seen to be stationary.

Predictive Distribution with Noiseless Observations. Using the expression for the conditional distribution of jointly Gaussian rvs (see Chapter 2), the posterior distribution of the random function $t(x)$ for a new input x can be directly obtained as

$$p(t(x) = t|\mathcal{D}) = \mathcal{N}(t|\hat{t}^{MAP}(x), \sigma^2(x)), \tag{12.92}$$

where the MAP estimate, i.e., the mean of the posterior, is given as

$$\hat{t}^{MAP}(x) = \mu(x) + \kappa_{\mathcal{D}}(x)^T K_{\mathcal{D}}^{-1}(t_{\mathcal{D}} - \mu(x_{\mathcal{D}})) \tag{12.93}$$

and the variance is

$$\sigma^2(x) = \kappa(x, x) - \kappa_{\mathcal{D}}(x)^T K_{\mathcal{D}}^{-1} \kappa_{\mathcal{D}}(x), \tag{12.94}$$

with the $N \times 1$ vector

$$\kappa_{\mathcal{D}}(x) = \begin{bmatrix} \kappa(x, x_1) \\ \kappa(x, x_2) \\ \vdots \\ \kappa(x, x_N) \end{bmatrix}. \tag{12.95}$$

The complexity of computing the predictor (12.93) scales with the cube of the number of data points, N, due to the need to invert the $N \times N$ kernel matrix $K_{\mathcal{D}}$. Several schemes have been proposed to alleviate this computational burden, including the idea of "summarizing" the data set $\mathcal{D}$ based on a subset of optimized virtual data points, known as **inducing points**. When the summary is computed by neural networks, one obtains the class of **conditional neural processes**.

Predictive Distribution with Noisy Observations. Let us now assume that the observations t are obtained as the result of a random function modeled via a GP with the addition of observation noise $\mathcal{N}(0, \beta^{-1})$, generalizing the likelihood model (12.32). This modifies the mean and variance of the predictive distribution as

$$\hat{t}^{MAP}(x) = \mu(x) + \kappa_{\mathcal{D}}(x)^T (K_{\mathcal{D}} + \beta^{-1} I_N)^{-1} (t_{\mathcal{D}} - \mu(x_{\mathcal{D}})) \tag{12.96}$$

and

$$\sigma^2(x) = \beta^{-1} + \kappa(x, x) - \kappa_{\mathcal{D}}(x)^T (K_{\mathcal{D}} + \beta^{-1} I_N)^{-1} \kappa_{\mathcal{D}}(x), \tag{12.97}$$

which reduce to (12.93) and (12.94), respectively, when $\beta \to \infty$.

Example 12.9

We return to Example 12.5, and apply GP regression with zero mean function, $\mu(x) = 0$, and the Gaussian kernel (12.89) with $a = 1$. Unlike Example 12.5, here we do not need to select a priori the capacity of the model via the degree M. In fact, the only assumption made by the inductive bias of GPs is encoded in the bandwidth h of the Gaussian kernel function (12.89). In Fig. 12.8, we draw the credible interval $\hat{t}^{MAP}(x) \pm \sigma(x)$, with MAP predictor $\hat{t}^{MAP}(x)$ represented as a dashed line. We set $\beta = 10$ and $\alpha = 0.5 \times 10^{-4}$, and explore two different values of the kernel bandwidth h.

A large bandwidth h (top figure) yields a prior that is supported only over smooth input–output functions. This is akin to choosing a parametric model with a small capacity, although no assumption is made here on the parametric form of the regression function. As seen in Fig. 12.8 (top), this model tends to underfit the data, and yields lower estimates of the predictive uncertainty. These estimates are low because of the smaller set of functions that are a priori under consideration by the model with a large h: With a small set of functions, the "disagreement" among models is also less pronounced.

Conversely, a smaller value of h (bottom figure) provides an accurate match of the data in regions of the covariate space that contain a sufficient number of data points. It also produces a well-calibrated estimate of the uncertainty where data availability is sparse. Importantly, choosing a smaller value of h, and hence effectively increasing the capacity of the model, does not cause overfitting.

Figure 12.8 Predictive (MAP) mean and credible interval (covering one standard deviation on either side of the mean) for GP regression with different values of the bandwidth h (Example 12.9).

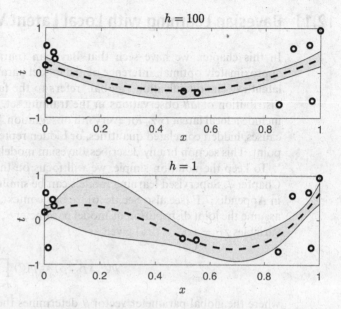

Empirical Bayes for GPs. Since the marginal likelihood can be written in closed form in a manner akin to (12.80), empirical Bayes can be easily applied in GPs to select hyperparameters such as the kernel function. This approach leads to popular methods such as **deep kernels**.

GP Classification. While the application to regression outlined above is more natural and common, GPs can also be used for classification. For instance, for binary classification, one could use a GP to model the logit, to which a "squashing" non-linearity such as a sigmoid is applied in order to obtain a predictive probability. Unlike the regression setting, the posterior over the label for a new covariate vector x is not available in closed form, and is often estimated using methods such as Laplace approximation. For multi-class classification, one can use multiple GPs to model the vector of logits. See Recommended Resources, Sec. 12.14, for details.

12.10.3 Dirichlet Process Mixture Models

While GPs define priors over the space of *deterministic* mappings between inputs and outputs, it is also possible to define non-parametric prior distributions over *probabilistic* predictive models. This is done by assuming mixture predictive models of the form (see Sec. 6.9)

$$\sum_{k=1}^{K} \pi_k p(t|x, \theta_k) \tag{12.98}$$

in which the number of mixture components K is a rv. The prior over the mixture probabilities π_k is supported over the infinite-dimensional space of all possible distributions of categorical rvs. Specifically, the prior is typically defined by a **Dirichlet process**, whose realizations, having a random number K of components, can be generated through a procedure known as *stick-breaking sampling*. An alternative is to use **determinantal point processes** that directly model the distribution of the model parameters $\{\theta_k\}$. For details, please see Recommended Resources, Sec. 12.14.

12.11 Bayesian Learning with Local Latent Variables

In this chapter, we have seen that Bayesian (parametric) learning carries out optimal, or approximately optimal, inference of the model parameter vector θ, which is viewed as a **global latent rv**. The qualification "global" refers to the fact that the parameter vector θ affects the distribution of *all* observations in the training set. Chapters 7 and 10 described models that include a **local latent rv** z_n for every nth observation. The local latent rv z_n may represent hidden causes, hidden correlated quantities, or hidden representations, that are specific to the nth data point. This section briefly describes Bayesian models that include both global and local rvs.

To keep the notation simple, we will focus on the unsupervised learning setting studied in Chapter 7. Supervised learning models can be similarly considered by following the formalism in Appendix 7.F (see also Sec. 13.6). In this context, **Bayesian models with local latent variables** assume the joint distribution of model parameter θ, training data $x_{\mathcal{D}} = \{x_1, \ldots, x_N\}$ and latent variables $z_{\mathcal{D}} = \{z_1, \ldots, z_N\}$ given as

$$p(\theta, x_{\mathcal{D}}, z_{\mathcal{D}}) = p(\theta) \prod_{n=1}^{N} p(x_n, z_n | \theta), \tag{12.99}$$

where the global parameter vector θ determines the joint distribution $p(x, z | \theta)$, which may be modeled using directed or undirected generative models as discussed in Chapter 7. For example, in a directed generative model $p(x, z | \theta) = p(z)p(x | z, \theta)$, the prior $p(z)$ may be fixed and only the conditional distribution $p(x | z, \theta)$ may depend on θ.

The ideal goal of Bayesian learning with local latent variables is to compute the posterior distribution $p(\theta, z_{\mathcal{D}} | x_{\mathcal{D}})$ of all the latent variables, both global (θ) and local ($z_{\mathcal{D}}$), along with the corresponding marginals. For example, when directed generative models are used for clustering, one is interested in the marginal $p(z_n | x_{\mathcal{D}})$ to carry out the soft imputation of each nth input x_n to a cluster identified by the latent rv z_n (see Secs. 7.9 and 7.10). For supervised learning, as we have studied in this chapter, one needs the marginal $p(\theta | x_{\mathcal{D}})$ in order to obtain a Gibbs predictor or an ensemble predictive distribution for new inputs.

To design scalable solutions, one can develop approximate methods based on VI or MC sampling. Let us briefly elaborate on VI. The problem of computing the joint posterior $p(\theta, z_{\mathcal{D}} | x_{\mathcal{D}})$ can be formulated as the minimization of the free energy

$$\min_{q(\cdot, \cdot | x_{\mathcal{D}})} \mathrm{F}(q(\theta, z_{\mathcal{D}} | x_{\mathcal{D}}) || p(\theta, x_{\mathcal{D}}, z_{\mathcal{D}})) \tag{12.100}$$

with respect to the **variational joint posterior** $q(\theta, z_{\mathcal{D}} | x_{\mathcal{D}})$ of global and local latent rvs. To directly derive all the marginals, a natural approach is to assume a **mean-field** form for the variational joint posterior that factorizes into separate terms for global and local latent rvs (see Sec. 10.6), i.e.,

$$q(\theta, z_{\mathcal{D}} | x_{\mathcal{D}}) = q(\theta | x_{\mathcal{D}}) \prod_{n=1}^{N} q_n(z_n | x_{\mathcal{D}}), \tag{12.101}$$

where we have introduced the **global variational posterior** $q(\theta | x_{\mathcal{D}})$ and **local variational posteriors** $q_n(z_n | x_{\mathcal{D}})$ for $n = 1, \ldots, N$. It is emphasized that an optimizer of the free energy within this set of variational posteriors is only an approximation of the true posterior $p(\theta, z_{\mathcal{D}} | x_{\mathcal{D}})$. In fact, given the observations $x_{\mathcal{D}}$, global and latent rvs are generally not independent.

With the set of mean-field variational posteriors in (12.101), the free energy in (12.100) can be written as

$$F(q(\theta, z_{\mathcal{D}}|x_{\mathcal{D}})||p(\theta, x_{\mathcal{D}}, z_{\mathcal{D}})) = \underbrace{\frac{1}{N}\sum_{n=1}^{N}E_{\theta \sim q(\theta|x_{\mathcal{D}})}\left[F(q_n(z_n|x_{\mathcal{D}})||p(x_n, z_n|\theta))\right]}_{\text{average training local free energy}}$$

$$+ \underbrace{KL(q(\theta|x_{\mathcal{D}})||p(\theta))}_{\text{global information-theoretic regularizer}} . \qquad (12.102)$$

The first term in (12.102) is the empirical average of the expected local free energies, where the expectation is taken with respect to the global variational posterior $q(\theta|x_{\mathcal{D}})$. Each nth local free energy term depends on the local variational posterior $q_n(z_n|x_{\mathcal{D}})$; while the second term is the global information-theoretic penalty accounting for the prior $p(\theta)$ on the global rv. Assuming unconstrained global and local variational posteriors, the problem of minimizing the free energy (12.102) can in principle be addressed via a message-passing scheme akin to that described in Sec. 10.6 for mean-field VI.

Applications of Bayesian models with local variables will be covered in Sec. 13.6.

12.12 Generalized Bayesian Learning

Throughout this chapter, we have seen that Bayesian learning formulates training as an inference problem at the level of the model parameter vector θ (or of the predictive function for non-parametric models). As discussed in Sec. 11.3, Bayesian inference can be generalized by allowing for arbitrary loss functions and more general complexity measures for the variational posterior. The same approach can also be naturally applied to Bayesian learning, yielding the framework of **generalized Bayesian learning**.

Generalized Bayesian learning revolves around the optimization of the **generalized free energy** as per the problem

$$\min_{q(\cdot|\mathcal{D})}\left\{\underbrace{N E_{\theta \sim q(\theta|\mathcal{D})}\left[L_{\mathcal{D}}(\theta)\right]}_{\text{average training loss}} + \alpha \cdot \underbrace{C(q(\theta|\mathcal{D}))}_{\text{complexity measure}}\right\}, \qquad (12.103)$$

where $\alpha \geq 0$ is a temperature parameter; the training loss $L_{\mathcal{D}}(\theta)$ may depend on a general loss function (not necessarily the log-loss); and $C(q(\theta|\mathcal{D}))$ is a complexity metric for the distribution $q(\theta|\mathcal{D})$, such as a divergence measure with respect to a prior distribution on the model parameter θ. With $\alpha = 0$, this formulation reduces to frequentist learning via ML, while, with $\alpha = 1$, the log-loss as loss function, and the KL divergence as complexity measure, we have the standard Bayesian learning problem (12.12) with (12.16). Following the discussion in Sec. 12.4.4, the choice of different complexity measures can be justified via the derivation of bounds on the population loss (see Recommended Resources, Sec. 12.14). Solutions to problem (12.103) are generally referred to as **generalized posteriors**.

Generalized Posteriors. Assuming an unrestricted set of distributions $q(\theta|\mathcal{D})$, the generalized posteriors can be computed using Table 11.2 for different complexity measures. As a notable example, when the KL divergence is used as complexity measure, the generalized posterior is given as

$$q(\theta|\mathcal{D}) \propto p(\theta) \left(\prod_{n=1}^{N} p(t_n|x_n,\theta) \right)^{\frac{1}{\alpha}}, \tag{12.104}$$

and is referred to also as $(1/\alpha)$-**posterior**. Other names used for it include **fractional posterior**, **tempered posterior**, and **power posterior**. In (12.104), the contribution of the likelihood is "tempered" by taking its $(1/\alpha)$th power. Therefore, as compared to the standard posterior distribution (obtained with $\alpha = 1$), if $\alpha > 1$ (large temperature), the generalized posterior downweighs the contribution of the data over that of the prior, while if $0 < \alpha < 1$ (small temperature), the contribution of the data is emphasized over that of the prior. For instance, if the likelihood $p(t|x,\theta)$ is Gaussian, the effective variance of the observations t_n in (12.104) is multiplied by α, and hence a large temperature indicates an increase in the noise affecting the observations.

While generalized posteriors, such as (12.104), may not be tractable owing to the need to obtain normalization constants, approximate generalized Bayes solutions can be devised based on VI, by restricting the optimization domain in (12.103); or on MC, by approximately drawing samples from the generalized posterior, e.g., via SGLD.

Robustness to Misspecified Model. The generalized Bayesian learning formulation has been proven to be particularly advantageous in cases where the model is not well specified, thanks to the flexibility it allows in choosing loss function, temperature parameter, and complexity measure. To elaborate on this point, consider a situation in which data are abundant, i.e., N is large, so that the standard posterior distribution concentrates around the MAP solution. This is desirable if the model is well specified, and hence there is a single ground-truth value for the model parameter; but it is generally not a satisfactory solution in models with misspecification in which multiple models may be needed to properly "cover" the problem space. In contrast, by choosing a value of the temperature $\alpha > 1$ in the generalized posterior (12.104), one can control the level of concentration of the generalized posterior around its mode.

Broader Applicability. Another advantage of the generalized Bayesian problem (12.103) is that it may be formulated even when the likelihood is not computable in normalized form. This is the case, for instance, in energy models, such as Boltzmann machines for classification (see Appendix 7.F), in which the dependence of the likelihood (x,t) and model parameters θ is through an energy function.

Gibbs vs. Ensemble Predictors. Following the discussion in Sec. 12.4.4, the generalized free energy performance criterion (12.103) implicitly assumes Gibbs predictors. This is because the first term in (12.103) evaluates the average training loss with respect to random draws of the model parameter vector θ. In order to ensure that the learning criterion takes full advantage of ensembling, particularly in the presence of misspecification, the first term in (12.103) should be modified by considering a loss function defined directly on the ensemble predictor.

An Alternative Approach to Defining Generalized Bayesian Learning. Before concluding this section, we point to an alternative direction one may follow to define a generalized form of Bayesian learning. By minimizing the free energy $\mathrm{F}(q(\theta|\mathcal{D})||p(\theta,\mathcal{D}))$, Bayesian learning tackles an I-projection problem whereby the unnormalized distribution $p(\theta,\mathcal{D})$ is projected into a class of variational distributions $q(\theta|\mathcal{D})$. As detailed in Sec. 11.2, the I-projection tends to be exclusive and to underestimate the spread of the target distribution, here $p(\theta, \mathcal{D})$. Therefore, one may prefer to address a generalized version of the problem in which the free energy is replaced by a different measure of distance, or divergence, between the (normalized) distribution $q(\theta|\mathcal{D})$ and the (unnormalized) distribution $p(\theta, \mathcal{D})$. Such optimization criteria, like the free energy, may in some cases be obtained as upper or lower bounds on convex or concave functions of (estimates of) the marginal distribution $p(\mathcal{D})$ (see Appendix 7.D), although the tightness of such bounds does not necessarily translate into more effective generalized Bayesian criteria. The resulting problem is generally distinct from (12.103), and the choice between the two approaches depends on the specific application, although the generalized free energy metric in (12.103) appears to be more flexible and may yield gradients that are easier to estimate.

12.13 Summary

- Frequentist learning relies on the choice of a single model parameter θ, disregarding any disagreement outside the training set among models $p(t|x, \theta)$ that fit the training set almost equally well. As a result, frequentist learning fails to properly quantify epistemic uncertainty, that is, the uncertainty in the selection of the optimal model that arises from the limited availability of data.

- Bayesian learning focuses on inferring a distribution $q(\theta|\mathcal{D})$ on the space of model parameters that captures the relative ability of model parameters θ to explain the data $\mathcal{D}$, as well as the model's consistency with the prior belief $p(\theta)$. This distribution is then used to make a prediction through an ensemble of models, each weighted by the corresponding value of $q(\theta|\mathcal{D})$, or through a (random) Gibbs predictor.

- Bayesian learning offers a number of unique benefits over frequentist learning, including its robustness to overfitting and its ability to facilitate online and active learning.

- Bayesian learning can be formulated as the problem of minimizing a free energy functional. The problem can only be solved exactly in special cases such as conjugate families of distributions. Otherwise, it requires the use of MC-based or VI-based approximate solution methods.

- MC-based algorithms generate samples approximately drawn from the posterior of the model parameters in order to enable the approximation of posterior averages such as the ensemble predictive distribution. In contrast, parametric VI-based methods optimize over a class of parametric distributions. Particle-based VI methods such as SVGD provide a hybrid approach that describes the distribution $q(\theta|\mathcal{D})$ through particles (similar to MC) that are deterministic parameters updated via SGD (like VI).

- Bayesian learning can alternatively be formulated as the problem of Bayesian inference of a random function $t(x)$ mapping input x to output t. This yields Bayesian non-parametric models, with the most notable example being GP regression.

- Bayesian learning can be generalized by allowing for arbitrary loss functions and complexity regularizers so as to enhance robustness to model misspecification.

12.14 Recommended Resources

Excellent textbooks on Bayesian learning include [1, 2, 3, 4], with theoretical aspects covered in [5]. Connections with PAC Bayes theory are elaborated on in [6, 7, 8]. Discussions on calibration and uncertainty quantification metrics can be found in [9]. An extensive review of formal methods for selecting prior distributions is offered by [10]. A useful recent paper on the Laplace method and empirical Bayes is [11]. Monte Carlo methods are detailed in [12], and reference [13] introduces annealed importance sampling. An early, and still very relevant, in-depth discussion of Bayesian neural networks is given in [14]. Bayesian model selection and the Laplace approximation are particularly well explained in [15]. More on the empirical FIM can be found in [16]. SG-MCMC is covered in recent references such as [17, 18, 19]. Useful references on VI-based learning include [17, 20, 21], and adversarial variational Bayesian learning is discussed in [22]. For pointers to the literature on SVGD, please see Chapter 10. GPs are discussed in depth in the book [23], and Bayesian non-parametrics in [24]. For Dirichlet processes, readers are referred to [25]. A statistician's view on the interplay between frequentist and Bayesian data analysis is reported in [26]. A useful introduction to methods, known as **conformal prediction**, that obtain credible intervals, or credible sets for discrete rvs, with exact (frequentist) coverage probability is [27]. Generalized Bayesian learning is elaborated on in [28, 29, 30, 31], and the definition of a generalized posterior based on the α-divergence can be found in [32].

Problems

12.1 For the variational posterior $q(\theta|\mathcal{D}) = \mathcal{N}(\theta|0, \gamma^2)$ and the likelihood given by the logistic regression model $p(t = 1|x, \theta) = \sigma(\theta x)$, (a*) evaluate and plot the Bayesian ensemble predictor $q(t|x, \mathcal{D}) = \mathrm{E}_{\theta \sim q(\theta|\mathcal{D})}[p(t = 1|x, \theta)]$ for $\gamma^2 = 1$ as a function of x. (b*) Repeat for $\gamma^2 = 10$, and comment on the comparison with point (a*). (c*) For both cases $\gamma^2 = 1$ and $\gamma^2 = 10$, plot the approximation (12.22).

12.2 (a) Prove that the optimal soft predictor $p(t|x, \mathcal{D}) = \mathrm{E}_{\theta \sim p(\theta|\mathcal{D})}[p(t|x, \theta)]$ is the posterior distribution of the test label t given the training data $\mathcal{D}$ and the test input x. (b) Write the free energy criterion minimized by the optimal soft predictor $p(t|x, \mathcal{D})$.

12.3 We have a regression problem with input x constrained in the interval $[0, 1]$ and model class $\mathcal{H}$ consisting of soft predictors $p(t|x, \theta) = \mathcal{N}(t|\theta \cos(2\pi x), 0.1)$ with $\theta \in \{0, 1\}$. Note that there are only two models in the class. Assume that the prior puts equal probability on both models. A limited data set $\mathcal{D} = \{(0, 0.9), (0.5, -0.08)\}$ is available. (a) Obtain the ML and MAP solutions $\theta_{\mathcal{D}}^{ML}$ and $\theta_{\mathcal{D}}^{MAP}$. (b) Compute the posterior distribution $p(\theta|\mathcal{D})$. (c*) For $x = 1$, plot the ensemble predictive distribution $p(t|x, \mathcal{D})$ as a function of t. In the same plot, also show the corresponding frequentist predictor $p(t|x, \theta_{\mathcal{D}}^{ML})$.

12.4 For the regression example in the preceding problem, assume that the population distribution $p(x, t)$ is such that $p(x)$ is uniformly distributed in the interval $[0, 1]$ and the likelihood is given by $p(t|x) = p(t|x, \theta = 1) = \mathcal{N}(t|\cos(2\pi x), 0.1)$. (a*) Generate a data set $\mathcal{D}$ of $N = 2$ data points from the population distribution. Based on this data set, obtain both ML and Bayesian ensemble predictive distributions. (b*) Repeat for $N = 30$ and comment on your results.

12.5 (a*) Implement the Gaussian–Gaussian GLM example in Example 12.5, reproducing Fig. 12.3. (b*) In the resulting figure, identify epistemic and aleatoric uncertainty, and comment on the dependence of these two types of uncertainty on the model class M. (c) What is the Laplace approximation for this example?

12.6 For the prior $p(\theta) = \mathcal{N}(0_2, I_2)$ and the logistic regression likelihood $p(t = 1|x, \theta) = \sigma(\theta^T x)$, (a*) generate a data set of $N = 10$ points assuming that the model is well specified. (b*) For this data set, evaluate the Laplace approximation, and plot the corresponding contour lines. (c*) Evaluate the simplified Laplace approximation based on the empirical FIM, and compare it with the distribution obtained in point (b*).

12.7 Implement and validate Example 12.7 comparing SGLD and SVGD, reproducing Fig. 12.6.

12.8 Implement and validate Example 12.9 involving GP regression, reproducing Fig. 12.8.

12.9 To evaluate the impact of a model mismatch, consider a population distribution $p(t) = 0.2\mathcal{N}(t|0, 1) + 0.8\mathcal{N}(t|3, 1)$, while the model class is defined by models of the form $p(t|\theta) = \mathcal{N}(t|\theta, 1)$. (a*) Generate a data set $\mathcal{D}$ of $N = 50$ examples from the population distribution. (b*) For a Gaussian prior $p(\theta) = \mathcal{N}(\theta|0, 1)$, evaluate and plot the posterior distribution $p(\theta|\mathcal{D})$. (c*) Consider a generalized Bayesian formulation with KL divergence as the complexity measure (using prior $p(\theta) = \mathcal{N}(\theta|0, 1)$), and evaluate the generalized posterior for different values of the temperature α. Compare the generalized posteriors with the posterior obtained in point (b*).

Appendices

Appendix 12.A: Quantifying Epistemic and Aleatoric Uncertainty in Bayesian Learning

In this appendix, we show how Bayesian learning can neatly quantify the contributions of epistemic uncertainty and aleatoric uncertainty to the overall predictive uncertainty (see Sec. 4.5) in terms of information-theoretic metrics. To this end, let us assume that the joint distribution (12.19) is **well specified**, and that exact Bayesian learning is implemented. Under these assumptions, the Bayesian ensemble predictor is the posterior of the label of the test example, i.e., $p(t|x, \mathcal{D})$ in (12.18), and the corresponding average log-loss is the conditional entropy

$$H(t|x = x, \mathcal{D}) = E_{\mathcal{D} \sim p(\mathcal{D})} E_{t \sim p(t|x, \mathcal{D})} [-\log p(t|x, \mathcal{D})], \qquad (12.105)$$

where $p(\mathcal{D})$ represents the distribution of the data set. This conditional entropy captures the **overall predictive uncertainty** via the population log-loss of the predictor. In the reference ideal case in which the ground-truth model parameter vector θ is known, the minimum predictive uncertainty, i.e., the minimum population log-loss, would be given by the conditional entropy

$$H(t|x = x, \theta) = E_{(\theta, t) \sim p(\theta)p(t|x, \theta)} [-\log p(t|x, \theta)]. \qquad (12.106)$$

This is a measure of **aleatoric uncertainty**, since it corresponds to the irreducible part of the population log-loss that cannot be decreased by collecting more data.

The difference between these two average log-losses measures the **"regret"** of Bayesian learning for not having used the true model parameter θ. This difference is evaluated as

$$H(t|x = x, \mathcal{D}) - H(t|x = x, \theta) = H(t|x = x, \mathcal{D}) - H(t|x = x, \theta, \mathcal{D})$$
$$= I(t; \theta|x, \mathcal{D}), \qquad (12.107)$$

where, by (12.107), the *conditional* mutual information is defined in an analogous manner to the standard mutual information (see Sec. 3.7). In the first equality, we have used the fact that, under the model (12.19), the label t is independent of $\mathcal{D}$ once θ is fixed, i.e., we have the equality $p(t|x, \theta) = p(t|x, \theta, \mathcal{D})$.

Being the difference between overall uncertainty and aleatoric uncertainty, the conditional mutual information $I(t; \theta|x, \mathcal{D})$ quantifies **epistemic uncertainty**, that is, the part of the overall predictive uncertainty that can be reduced by observing more data. Intuitively, as the training set $\mathcal{D}$ grows in size, the residual information that can be extracted about the model parameter θ from the observation of the test data point (x, t) – which is quantified by $I(t; \theta|x, \mathcal{D})$ – decreases.

The metrics (12.105)–(12.107) are averaged over the data set $\mathcal{D}$. For a fixed data set $\mathcal{D}'$, the mutual information in (12.107) can be written as

$$I(t; \theta|x, \mathcal{D} = \mathcal{D}') = H(t|x = x, \mathcal{D} = \mathcal{D}') - H(t|x = x, \theta, \mathcal{D} = \mathcal{D}')$$
$$= E_{t \sim p(t|x, \mathcal{D}')}[-\log p(t|x, \mathcal{D}')] - E_{(\theta, t) \sim p(\theta|\mathcal{D}')p(t|x, \theta)}[-\log p(t|x, \theta)]. \quad (12.108)$$

Note that the average of the mutual information (12.108) over the data set $\mathcal{D} \sim p(\mathcal{D})$ yields (12.107). Interestingly, the first entropy in (12.108) can be interpreted as the average log-loss of the ensemble predictor $p(t|x, \mathcal{D}')$ in (12.18), while the second entropy is the average log-loss of the Gibbs predictor obtained by sampling the model parameter vector from the posterior $p(\theta|\mathcal{D})$.

(More precisely, the second term measures the average log-loss of the Gibbs predictor in the ideal case in which the model parameter vector θ drawn from the posterior $p(\theta|\mathcal{D})$ underlies the generation of the label via the conditional distribution $p(t|x, \theta)$. The average log-loss of the Gibbs predictor is actually no smaller than that of the ensemble predictor, by Jensen's inequality.)

Therefore, the epistemic uncertainty measured by (12.108) at an input x is larger when ensemble and Gibbs predictors obtain significantly different performance levels in terms of log-loss. In this case, the predictors in the support of the ensemble defined by the posterior $p(\theta|\mathcal{D})$ **"disagree"** significantly on their individual soft predictions.

The criterion (12.108) is used as an **acquisition function** for active learning by an approach known as **Bayesian active learning by disagreement (BALD)**. As introduced in Sec. 12.4.6, given the current training set $\mathcal{D}'$, the problem of active learning is to select the next input x so as to add the resulting labelled sample (x, t) to the data set $\mathcal{D}'$. BALD chooses the input x that maximizes the epistemic uncertainty measure (12.108), obtaining an input–output pair (x, t) that is maximally informative about the model parameter vector θ given $\mathcal{D}'$.

Appendix 12.B: Bayesian Learning Implements Minimum Description Length Inference

We have seen in Sec. 4.10 that (frequentist) MAP learning can be interpreted as a form of MDL inference. In this appendix, we ask whether a similar interpretation can be given of Bayesian learning. Throughout this appendix, we assume discrete rvs for both model parameter θ and label t. Furthermore, for concreteness, we will take all logarithms to be in base 2, so as to be able to talk about bits rather than nats as the unit of measure of information.

To start, let us review the MDL inference interpretation of MAP learning. The key observation underlying it is that the regularized log-loss $(-\log_2 p(\theta) - \log_2 p(t_\mathcal{D}|x_\mathcal{D},\theta))$ minimized by MAP learning can be interpreted as the description length (in bits) of a **two-step compression code**. The code first describes the model parameter θ using a compressor designed using the prior distribution $p(\theta)$, and then encodes the labels $t_\mathcal{D}$ using a compressor designed using the likelihood $p(t|x,\theta)$. (The decoder is assumed to know the distributions $p(\theta)$ and $p(t|x,\theta)$, as well as the inputs $x_\mathcal{D}$.) Therefore, the model parameter vector $\theta_\mathcal{D}^{MAP}$ selected by MAP learning minimizes the description length of this code, and hence MAP implements MDL inference.

We will now see that an analogous interpretation is possible for Bayesian learning, by invoking a different form of compression known as **bits-back coding**. Note that a different form of compression is needed, given the reliance of Bayesian learning on a distribution $q(\theta|\mathcal{D})$ in the model parameter space rather than on a point estimate like $\theta_\mathcal{D}^{MAP}$.

A first, natural, attempt in this direction would be to draw a model parameter vector as $\theta \sim q(\theta|\mathcal{D})$ from the data-dependent distribution $q(\theta|\mathcal{D})$, and then to apply a two-step compression code. The average description length of this approach would be

$$\mathrm{E}_{\theta \sim q(\theta|\mathcal{D})}[-\log_2 p(\theta) - \log_2 p(t_\mathcal{D}|x_\mathcal{D},\theta)] = N\mathrm{E}_{\theta \sim q(\theta|\mathcal{D})}[L_\mathcal{D}(\theta)] + \mathrm{H}(q(\theta|\mathcal{D})||p(\theta)), \quad (12.109)$$

where $L_\mathcal{D}(\theta) = 1/N(-\log_2 p(t_\mathcal{D}|x_\mathcal{D},\theta))$ is the training log-loss (in bits). The free energy (12.16) optimized by Bayesian learning is closely related, but not equivalent, to (12.109).

In fact, the free energy (12.16) can be obtained from (12.109) by replacing the cross entropy $\mathrm{H}(q(\theta|\mathcal{D})||p(\theta))$ with the KL divergence $\mathrm{KL}(q(\theta|\mathcal{D})||p(\theta))$. Since we have the equality

$$\mathrm{H}(q(\theta|\mathcal{D})||p(\theta)) = \mathrm{H}(q(\theta|\mathcal{D})) + \mathrm{KL}(q(\theta|\mathcal{D})||p(\theta)), \quad (12.110)$$

by the non-negativity of the entropy (for discrete rvs), the criterion (12.16) optimized by Bayesian learning is smaller than the average description length (12.109) by a term equal to the entropy $\mathrm{H}(q(\theta|\mathcal{D}))$.

Based on this observation, it would seem that the distribution $q(\theta|\mathcal{D})$ chosen by Bayesian learning does not minimize a meaningful measure of description length. We will now see that this is not the case: The free energy minimized by Bayesian learning is the average description length of a more sophisticated coding scheme known as bits-back coding.

The key idea in bits-back coding is that the encoder can embed an information message of size $\mathrm{H}(q(\theta|\mathcal{D}))$ bits in the selection of rv $\theta \sim q(\theta|\mathcal{D})$. This way, when the decoder recovers vector θ, these bits can also be recovered, and the *effective* communication overhead is reduced accordingly. The resulting description length, discounted by $\mathrm{H}(q(\theta|\mathcal{D}))$, is exactly the free energy optimized by Bayesian learning.

Measuring entropies in bits, bits-back coding operates as follows:

1. The encoder observes a message from a source of information with entropy $\mathrm{H}(q(\theta|\mathcal{D}))$.
2. The resulting $\mathrm{H}(q(\theta|\mathcal{D}))$ bits are used to select model parameter θ in such a way that the distribution of the model parameter vector θ is approximately $q(\theta|\mathcal{D})$ (information theory guarantees that this is possible).
3. A two-step code with description length $(-\log_2 p(\theta) - \log_2 p(t_\mathcal{D}|x_\mathcal{D},\theta))$ bits is used to describe θ and $t_\mathcal{D}$.
4. The decoder recovers θ and $t_\mathcal{D}$.
5. From θ, the decoder recovers also the information message of $\mathrm{H}(q(\theta|\mathcal{D}))$ bits.

Appendix 12.C: "Natural Parameters" of the Gaussian Likelihood

The likelihood for the Gaussian GLM can be written as

$$p(t_\mathcal{D}|x_\mathcal{D},\theta) \propto \exp\left(-\frac{\beta}{2}(t_\mathcal{D} - X_\mathcal{D}\theta)^T(t_\mathcal{D} - X_\mathcal{D}\theta)\right)$$

$$\propto \exp\left(-\frac{\beta}{2}\theta^T X_\mathcal{D}^T X_\mathcal{D}\theta + \beta\theta^T X_\mathcal{D}^T t_\mathcal{D}\right). \tag{12.111}$$

By comparison with the multivariate Gaussian distribution in Table 9.1, this shows that the terms $-\frac{\beta}{2}X_\mathcal{D}^T X_\mathcal{D}$ and $\beta X_\mathcal{D}^T t_\mathcal{D}$ play the same role as the natural parameters in multiplying the sufficient statistics θ and $\theta\theta^T$ of the multivariate Gaussian vector.

Bibliography

[1] D. Sivia and J. Skilling, *Data Analysis: A Bayesian Tutorial*. Oxford University Press, 2006.

[2] S. Theodoridis, *Machine Learning: A Bayesian and Optimization Perspective*. Academic Press, 2015.

[3] D. Barber, *Bayesian Reasoning and Machine Learning*. Cambridge University Press, 2012.

[4] A. Gelman, J. B. Carlin, H. S. Stern, D. B. Dunson, A. Vehtari, and D. B. Rubin, *Bayesian Data Analysis*. CRC Press, 2013.

[5] J. M. Bernardo and A. F. Smith, *Bayesian Theory*. Wiley Series in Probability and Statistics, vol. 405. John Wiley & Sons, 2009.

[6] P. Alquier, "User-friendly introduction to PAC-Bayes bounds," arXiv preprint arXiv:2110.11216, 2021.

[7] T. Zhang, "Information-theoretic upper and lower bounds for statistical estimation," *IEEE Transactions on Information Theory*, vol. 52, no. 4, pp. 1307–1321, 2006.

[8] S. T. Jose and O. Simeone, "Free energy minimization: A unified framework for modeling, inference, learning, and optimization [lecture notes]," *IEEE Signal Processing Magazine*, vol. 38, no. 2, pp. 120–125, 2021.

[9] C. Guo, G. Pleiss, Y. Sun, and K. Q. Weinberger, "On calibration of modern neural networks," in *The 34th International Conference on Machine Learning*. ICML, 2017, pp. 1321–1330.

[10] R. E. Kass and L. Wasserman, "The selection of prior distributions by formal rules," *Journal of the American Statistical Association*, vol. 91, no. 435, pp. 1343–1370, 1996.

[11] A. Immer, M. Bauer, V. Fortuin, G. Rätsch, and M. E. Khan, "Scalable marginal likelihood estimation for model selection in deep learning," arXiv preprint arXiv:2104.04975, 2021.

[12] C. Robert and G. Casella, *Monte Carlo Statistical Methods*. Springer Science+Business Media, 2013.

[13] R.M. Neal, "Annealed importance sampling," *Statistics and Computing*, vol. 11, pp. 125–139, 2001.

[14] — Neal, *Bayesian Learning for Neural Networks*, Lecture Notes in Statistics, vol. 118. Springer Science+Business Media, 2012.

[15] D. J. MacKay, *Information Theory, Inference and Learning Algorithms*. Cambridge University Press, 2003.

[16] F. Kunstner, L. Balles, and P. Hennig, "Limitations of the empirical Fisher approximation for natural gradient descent," arXiv preprint arXiv:1905.12558, 2019.

[17] E. Angelino, M. J. Johnson, and R. P. Adams, "Patterns of scalable Bayesian inference," *Foundations and Trends in Machine Learning*, vol. 9, no. 2–3, pp. 119–247, 2016.

[18] Y.-A. Ma, T. Chen, and E. B. Fox, "A complete recipe for stochastic gradient MCMC," in *NIP'15: Proceedings of the 28th International Conference on Neural Information Processing Systems - Volume 2*. Curran Associates, 2015, pp. 2917–2925.

[19] C. Nemeth and P. Fearnhead, "Stochastic gradient Markov chain Monte Carlo,"

Journal of the American Statistical Association, vol. 116, no. 533, pp. 433–450, 2021.

[20] C. Blundell, J. Cornebise, K. Kavukcuoglu, and D. Wierstra, "Weight uncertainty in neural network," in *The 32nd International Conference on Machine Learning*. ICML, 2015, pp. 1613–1622.

[21] Y. Gal and Z. Ghahramani, "Dropout as a Bayesian approximation: Representing model uncertainty in deep learning," in *The 33rd International Conference on Machine Learning*. ICML, 2016, pp. 1050–1059.

[22] S. Rodriguéz-Santana and D. Hernández-Lobato, "Adversarial α-divergence minimization for Bayesian approximate inference," *Neurocomputing*, vol. 471, pp. 260–274, 2020.

[23] C. E. Rasmussen, "Gaussian Processes in Machine Learning," in *Summer school on machine learning*. Springer, 2003, pp. 63–71.

[24] S. Ghosal and A. Van der Vaart, *Fundamentals of Nonparametric Bayesian Inference*. Cambridge University Press, 2017.

[25] D. M. Blei and M. I. Jordan, "Variational inference for Dirichlet process mixtures," *Bayesian Analysis*, vol. 1, no. 1, pp. 121–143, 2006.

[26] M. J. Bayarri and J. O. Berger, "The interplay of Bayesian and frequentist analysis," *Statistical Science*, vol. 9, no. 1, pp. 58–80, 2004.

[27] A. N. Angelopoulos and S. Bates, "A gentle introduction to conformal prediction and distribution-free uncertainty quantification," arXiv preprint arXiv:2107.07511, 2021.

[28] P. G. Bissiri, C. C. Holmes, and S. G. Walker, "A general framework for updating belief distributions," *Journal of the Royal Statistical Society. Series B, Statistical Methodology*, vol. 78, no. 5, pp. 1103–1130, 2016.

[29] J. Knoblauch, J. Jewson, and T. Damoulas, "Generalized variational inference: Three arguments for deriving new posteriors," arXiv preprint arXiv:1904.02063, 2019.

[30] T. Matsubara, J. Knoblauch, F.-X. Briol, and C. J. Oates, "Robust generalised Bayesian inference for intractable likelihoods," arXiv preprint arxiv:2104.07359, 2021.

[31] M. A. Medina, J. L. M. Olea, C. Rush, and A. Velez, "On the robustness to misspecification of α-posteriors and their variational approximations," arXiv preprint arXiv:2104.07359, 2021.

[32] S.-I. Amari, "Integration of stochastic models by minimizing α-divergence," *Neural Computation*, vol. 19, no. 10, pp. 2780–2796, 2007.

Part IV

Beyond Centralized Single-Task Learning

13 Transfer Learning, Multi-task Learning, Continual Learning, and Meta-learning

13.1 Overview

As discussed so far in this book, the standard formulation of machine learning makes the following two basic assumptions:

1. **Statistical equivalence of training and testing**. The statistical properties of the data observed during training match those to be experienced during testing – i.e., the population distribution underlying the generation of the data is the same during both training and testing.
2. **Separation of learning tasks**. Training is carried out separately for each separate learning task – i.e., for any new data set and/or loss function, training is viewed as a new problem to be addressed from scratch.

These assumptions are not necessarily valid or desirable in practical applications:

1. **Distributional shift between training and testing**. The distribution of the training data may be different from the distribution of the test data. As an example, one may train a recommendation algorithm by using data from a number of individuals with certain shared characteristics, and then deploy it to serve a different community with distinct preferences.
2. **Shared information across learning tasks**. Different learning tasks can be "similar" in the sense that there may be useful information that can be **transferred** from data collected for one task to another. Transferring information across tasks is especially relevant in situations in which data from each task are limited, and/or when the time available to train for any given task is limited. As an example, in the case of a recommendation algorithm, a model to be trained using data from one community may benefit from consideration of data collected from another set of individuals, particularly when the two groups have similar preferences.

In some applications, multiple learning tasks are naturally encountered by an agent over time. In such settings, the agent must often leverage data from previously observed tasks in order to adapt to a new task based on limited data, while also retaining the ability to perform well on previously encountered tasks. As an example, consider a probe sent to explore a far-off planet. The probe needs to adjust to a priori unknown and changing conditions on the ground as it operates. To this end, it should be able to quickly modify its operation based on data from a new environment, without forgetting how to deal with previously seen ground conditions.

In this chapter, we review four basic frameworks that move beyond the standard formulation of machine learning based on the statistical equivalence of training and testing and on the separation of learning tasks: transfer learning, multi-task learning, continual learning, and meta-learning. We will discuss principles, along with some representative techniques and simple examples. A summary illustration of the settings to be covered in this chapter can be found in Fig. 13.1.

Learning Objectives and Organization of the Chapter. By the end of this chapter, the reader should be able to:

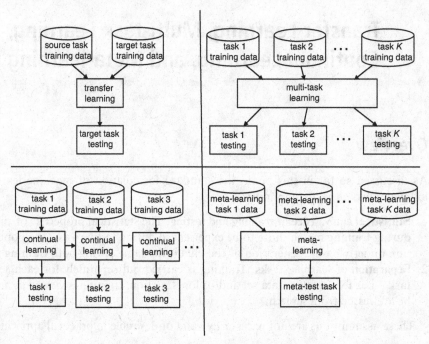

Figure 13.1 Illustration of transfer learning, multi-task learning, continual learning, and meta-learning.

- understand the principles of transfer learning, and implement transfer learning methods based on the likelihood ratio-based approach (Sec. 13.2);
- understand principles and algorithms for multi-task learning (Sec. 13.3);
- understand and implement coreset-based and regularization-based methods for continual learning (Sec. 13.4);
- understand the basics of meta-learning, and of gradient-based meta-learning solutions (Sec. 13.5); and
- revisit and formulate transfer learning, multi-task learning, continual learning, and meta-learning from a Bayesian perspective (Sec. 13.6).

13.2 Transfer Learning

As we have seen in the previous chapters, a **learning task** is defined by the combination of a population distribution, which underlies the generation of data, and a loss function, which is used to gauge the accuracy of a model. In **transfer learning**, we have two different learning tasks, known as **source task** and **target task**. Training data are generally available to the learner for both tasks. However, the data set from the target task is typically small, and insufficient to obtain desirable performance levels on test data from the target task. Transfer learning seeks ways to use both data sets – from source and target tasks – in order to enhance the generalization performance on the target task. The key design problem for the learner is thus to define which aspects of the data or of the model can be transferred from the source task to the target task with the aim of improving the population loss for the latter. In this section, we elaborate on this problem, starting with a taxonomy of the different types of "shifts" that can occur between training and testing phases.

13.2.1 Defining Transfer Learning

In transfer learning, there are *two unknown population distributions*, one for the source task – denoted as $p^S(x, t)$ – and one for the target task – denoted as $p^T(x, t)$. The only loss function that matters is the one for the target task, since the performance is measured in terms of population loss for the target task. A labeled data set $\mathcal{D}^S = \{(x_n^S, t_n^S)_{n=1}^{N^S}\}$ with N^S examples is assumed to be available for the source task. The outputs t_n^S can be either discrete – in which case we have a classification problem – or continuous – in which case we have a regression problem. Following the standard frequentist formulation, the N^S training examples are assumed to be produced by the underlying unknown population distribution $p^S(x, t)$ in an i.i.d. manner. For the target-task data, we can distinguish two scenarios depending on whether the target-task data are labeled or not:

- **Unlabeled target-task data**. In the first case, only unlabeled data are available for the target task data. Specifically, the learner has access to a data set

$$\mathcal{D}^T = \{(x_n^T)_{n=1}^{N^T}\} \tag{13.1}$$

of N^T covariate vectors for the test task. (The superscript "T" here identifies quantities of interest for the target task, and does not denote transposition.) Each example x_n^T is assumed to be generated i.i.d. from the marginal $p^T(x)$ of the joint distribution $p^T(x, t)$. One may also refer to this setting as **"transfer unsupervised learning"** due to the lack of labels for the target task. However, we will not use this terminology here to avoid possible confusion stemming from the fact that we do have access to labeled data for the source domain.

- **Labeled target-task data**. In the second setting, labeled data are available for the target task in the form of a data set

$$\mathcal{D}^T = \{(x_n^T, t_n^T)_{n=1}^{N^T}\}, \tag{13.2}$$

where every example (x_n^T, t_n^T) is assumed to be generated i.i.d. from the target-task population distribution $p^T(x, t)$. This case may be referred to as **"transfer supervised learning"**.

It is also possible to consider **"semi-supervised"** settings in which target-task data are only partially labeled.

Along a different direction, we can distinguish transfer learning problems on the basis of the type of change occurring between source-task and target-task population distributions, which can take the form of a covariate shift or a general shift.

13.2.2 Covariate Shift

A **covariate shift** indicates that the marginals of the covariates are different between source and target tasks, i.e., $p^S(x) \neq p^T(x)$, but the conditional distributions of the output given the input are the same, i.e., $p^S(t|x) = p^T(t|x) = p(t|x)$. Assuming a covariate shift thus amounts to positing that there exists a common discriminative distribution $p(t|x)$ that provides accurate predictions for both tasks. Importantly, under a covariate shift, it may not be necessary to have labeled data from the target task, since data from the source task are informative about the desired predictive distribution $p(t|x)$.

As an example of covariate shift, consider the problem of designing a diagnostic tool that identifies the presence or absence of a certain disease, defined by binary variable t, given the outcomes x of some observations made on a patient, such as symptoms and test results. The distribution of the observations x may be different in distinct geographic regions, yielding different marginals $p(x)$. In contrast, the predictive distribution $p(t|x)$ may be assumed to be the same, since equal diagnoses should be offered for the same symptoms and test results.

A covariate shift may also result from a **sampling bias**, which occurs when data are collected in a way that is not representative of the true underlying population distribution that will be encountered at test time, causing a covariate shift.

Example 13.1

Consider a regression problem with one-dimensional covariates x. In Fig. 13.2, data samples for the source task are represented as crosses, while those for the target task are represented as circles. The covariates for the source task (horizontal coordinate of the crosses) have a different distribution than the covariates for the target task (horizontal coordinate of the circles). In contrast, the conditional distribution $p(t|x)$, represented here by the population-optimal MAP predictor (dashed line), is the same for both tasks.

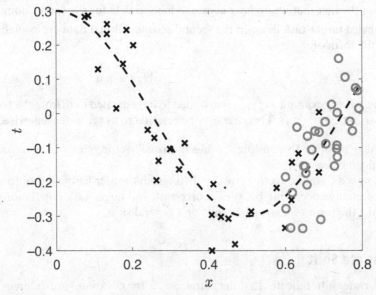

Figure 13.2 An example of covariate shift between source task, whose training samples are represented as crosses, and target task, whose training samples are shown as circles. The common population-optimal MAP predictor is shown as a dashed line.

13.2.3 General Shift

In other settings, the source and target population distributions may have distinct conditional target distributions, i.e., $p^S(t|x) \neq p^T(t|x)$, implying that a common discriminative model that applies to both tasks does not exist. Under such a **general shift**, labeled data from the target domain are typically required in order to enable transfer of information between the source and target tasks regarding the desired predictive distribution $p^T(t|x)$. This is because the labels from the source task are, by themselves, not informative about the predictive distribution $p^T(t|x)$ for the target task.

Under a general shift, a positive transfer of information may take place from source task to target task if the distributions $p^S(t|x)$ and $p^T(t|x)$ are sufficiently similar. As an example, consider the diagnosis of two distinct, but related, diseases, such as different forms of the flu. The two conditional distributions relating diagnosis t to a patient's symptoms and test results x are generally different, but they may share common aspects. For instance, the two conditional probabilities may rely on common features of the input x, such as a combination of symptoms, that can be reused to obtain accurate predictions for both diseases.

To formalize this idea, we can assume that there exists a **common set of features** $u(x)$ such that the conditional distributions of the target rv t given the observation of vector $u(x)$ under the two tasks are equal, i.e.,

$$p^S(t|u(x)) = p^T(t|u(x)). \tag{13.3}$$

In words, there exist common features that are *invariant* with respect to the given general shift. We can think of these shared features as being, in some sense, the true **"cause"** of the target variable t, irrespective of the specific characteristics of the two tasks. This idea yields methods such as **invariant risk minimization**, for which we refer to Recommended Resources, Sec. 13.8 (see also Chapter 15 for more discussion on causality).

A special case of general shift occurs when the covariate distribution is the same, i.e., $p^S(x) = p^T(x)$, but the conditional distributions $p^S(t|x)$ and $p^T(t|x)$ are different. This setting is known as a **concept shift**. Under a concept shift, transfer of information from source to target task can occur through the common marginal of the input.

Example 13.2

Consider again a regression problem with one-dimensional covariates x, as in Example 13.1. In Fig. 13.3, the covariates for the source task – given by the horizontal coordinate of the crosses – have the same distribution as the target task – given by the horizontal coordinate of the gray circles. In contrast, the conditional distributions $p^S(t|x)$ and $p^T(t|x)$, which are represented here by the corresponding population-optimal MAP predictors – dashed black and dashed gray lines, respectively – are different. This is an example of concept shift.

13.2.4 Likelihood Ratio-Based Transfer Learning

As a notable example of transfer learning algorithms, we now describe **likelihood ratio-based transfer learning**. We will focus on a model class $\mathcal{H}$ consisting of hard predictors $\hat{t}(x|\theta)$, although extensions to probabilistic models are direct. Accordingly, we fix a loss function $\ell(t, \hat{t})$ that

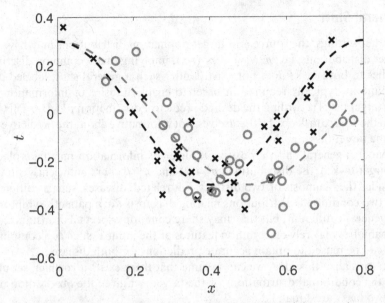

Figure 13.3 An example of concept shift between source task, whose training samples are represented as crosses, and target task, whose training samples are shown as circles. The corresponding conditional distributions $p^S(t|x)$ and $p^T(t|x)$ are shown as dashed black and dashed gray lines, respectively, by drawing the corresponding population-optimal MAP predictors.

is relevant for the target task, and the population losses under target-task and source-task population distributions are defined as

$$L_p^T(\theta) = E_{(x,t) \sim p^T(x,t)}[\ell(t, \hat{t}(x|\theta))] \tag{13.4}$$

and

$$L_p^S(\theta) = E_{(x,t) \sim p^S(x,t)}[\ell(t, \hat{t}(x|\theta))], \tag{13.5}$$

respectively. Note that the two population losses differ only in the distinct averaging distributions.

Likelihood ratio-based transfer learning leverages the following key observation: One can move from the average with respect to the source-task population distribution to the desired average with respect to the target-task population distribution through the **importance sampling trick**

$$\begin{aligned} L_p^T(\theta) &= E_{(x,t) \sim p^T(x,t)} \left[\ell(t, \hat{t}(x|\theta)) \right] \\ &= E_{(x,t) \sim p^S(x,t)} \left[r^{T/S}(x,t) \ell(t, \hat{t}(x|\theta)) \right], \end{aligned} \tag{13.6}$$

where

$$r^{T/S}(x,t) = \frac{p^T(x,t)}{p^S(x,t)} \tag{13.7}$$

is the ratio between the two population distributions, to be referred to as the **population likelihood ratio**. Therefore, by (13.6), the desired target-task population loss can be obtained by averaging over the source-task population distribution, as long as the contribution of each sample (x, t) is weighted by the ratio $r^{T/S}(x, t)$. The population likelihood ratio indicates how important – more precisely, how frequent – sample (x, t) is for the target task relatively to the source task.

Covariate Shift. Let us now develop likelihood ratio-based transfer learning under a covariate shift. In this case, the population likelihood ratio is

$$r^{T/S}(x, t) = \frac{p^T(x, t)}{p^S(x, t)} = \frac{p^T(x)}{p^S(x)} = r^{T/S}(x), \tag{13.8}$$

where we have introduced the notation $r^{T/S}(x)$ to indicate that the population likelihood ratio depends only on the ratio of the covariate distributions. To proceed, assume that we have access to unlabeled data $\mathcal{D}^T = \{(x_n^T)_{n=1}^{N^T}\}$ from the target task. Using data set $\mathcal{D}^T$ along with the covariates $\{(x_n^S)_{n=1}^{N^S}\}$ from the source-task data set $\mathcal{D}^S = \{(x_n^S, t_n^S)_{n=1}^{N^S}\}$, we can obtain an estimate $\hat{r}^{T/S}(x)$ of the population likelihood ratio. An effective way to obtain an estimate $\hat{r}^{T/S}(x)$ is to use contrastive density ratio learning based on the training of binary classifiers, as described in Sec. 7.3.5.

Once an estimate $\hat{r}^{T/S}(x)$ of the population likelihood ratio is available, one can tackle the following ERM problem by using data from the source task:

$$\min_\theta L_{\mathcal{D}^S}^T(\theta), \tag{13.9}$$

where

$$L_{\mathcal{D}^S}^T(\theta) = \frac{1}{N^S} \sum_{n=1}^{N_S} \hat{r}^{T/S}(x_n^S) \ell(t_n^S, \hat{t}(x_n^S|\theta)) \tag{13.10}$$

is, by the importance sampling trick, an empirical estimate of the target-task population loss. Accordingly, each training example (x_n^S, t_n^S) from the source task is weighted based on the relevance $\hat{r}^{T/S}(x_n^S)$ of the corresponding covariate x_n^S for the target task.

Example 13.3

For the data shown in Fig. 13.2, consider the problem of training a linear predictor $t(x|\theta) = \theta^T u(x)$ with feature vector $u(x) = [1, x]^T$ under a quadratic loss function. Figure 13.4 shows as a dashed black line the predictor obtained by solving the standard ERM problem using source-task data (which is an LS problem). This predictor is clearly not a good fit for the target-task data (circles in Fig. 13.2).

Let us now apply likelihood ratio-based transfer learning. To this end, we train the weight function $\hat{r}^{T/S}(\cdot)$ in (13.10) by using contrastive density ratio learning with a logistic regression classifier. For the classifier, we again use the feature vector $u(x) = [1, x]^T$, and obtain the weight as (cf. (6.30))

$$\hat{r}^{T/S}(x) = \exp(\varphi^T u(x)), \tag{13.11}$$

where φ is the model parameter vector of the classifier.

Figure 13.4 shows each sample (x_n^S, t_n^S) from the source domain (crosses) with size proportional to its weight $\hat{r}^{T/S}(x_n^S)$. Furthermore, the solid line is the predictor obtained via likelihood-ratio transfer learning. (Note that this is also an LS problem for this example.) The resulting predictor clearly reflects much more accurately the distribution of the target task.

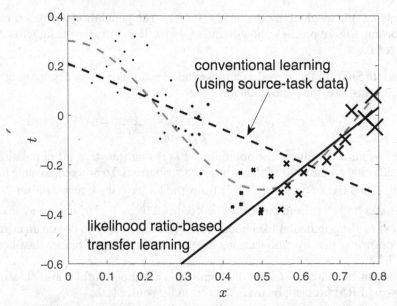

Figure 13.4 Likelihood ratio-based transfer learning weighs each data point from the source domain (crosses) with a weight represented in the figure by the size of the corresponding marker. The figure compares conventional learning, which applies the same weight to each sample, with likelihood ratio-based transfer learning, for the covariate shift example in Fig. 13.2.

General Shift. Consider now the case of a general shift. We assume that a labeled data set $\mathcal{D}^T$ from the target task is available, alongside the labeled data set $\mathcal{D}^S$ from the source task. First, note that, by (13.6), we can write the target-task population loss (13.4) as

$$L_p^T(\theta) = \alpha \mathrm{E}_{(\mathrm{x,t}) \sim p^S(x,t)}[r^{T/S}(\mathrm{x,t})\ell(\mathrm{t},\hat{t}(\mathrm{x}|\theta))] + (1-\alpha)\mathrm{E}_{(\mathrm{x,t}) \sim p^T(x,t)}[\ell(\mathrm{t},\hat{t}(\mathrm{x}|\theta))] \qquad (13.12)$$

for any constant $0 \leq \alpha \leq 1$, where we have used the population likelihood ratio (13.7). From data sets $\mathcal{D}^T$ and $\mathcal{D}^S$ we can obtain an estimate of the population likelihood ratio $\hat{r}^{T/S}(x,t)$ using, e.g., contrastive density ratio learning (see Sec. 7.3.5). Unlike the case of a covariate shift, under a general shift, estimating the population likelihood ratio requires labels from the target task.

Once an estimate $\hat{r}^{T/S}(x,t)$ of the likelihood ratio is available, one can tackle the following ERM problem by using data from both source and target tasks:

$$\min_\theta L_{\mathcal{D}^S \cup \mathcal{D}^T}^T(\theta), \qquad (13.13)$$

where

$$L_{\mathcal{D}^S \cup \mathcal{D}^T}^T(\theta) = \frac{\alpha}{N^S} \sum_{n=1}^{N^S} \hat{r}^{T/S}(x_n^S, t_n^S)\ell(t_n^S, \hat{t}(x_n^S|\theta)) + \frac{(1-\alpha)}{N^T} \sum_{n=1}^{N^T} \ell(t_n^T, \hat{t}(x_n^T|\theta)) \qquad (13.14)$$

is an empirical estimate of the target-task population loss (13.12). In (13.14), the contribution of the source-task data is collectively weighted by factor α, while each sample (x_n^S, t_n^S) is further weighted by the likelihood ratio $\hat{r}^{T/S}(x_n^S, t_n^S)$. The hyperparameter α can be selected based on domain knowledge or validation.

Example 13.4

Assume a concept shift setting in which the covariate distribution is the same for both source and target tasks and given by $p^S(x) = p^T(x) = \mathcal{U}(x|0,1)$. The target variables for the source task are generated from the population distribution $p^S(t|x) = \mathcal{N}(0.3\cos(2\pi x), 0.1^2)$, while the corresponding target distribution is $p^T(t|x) = \mathcal{N}(0.3\cos(2\pi x + \Delta\pi), 0.1^2)$ for some shift parameter Δ. Note that parameter Δ provides a measure of the difference between the two tasks. For polynomial regression with degree $M = 4$ and quadratic loss function (see Chapter 4), the approach (13.13) is applied with the trivial choice of the weights $\hat{r}^{T/S}(x_n^S, t_n^S) = 1$ and $\alpha = N^S/(N^S + N^T)$, so that all samples are equally weighted. The resulting optimization amounts to a standard LS problem (see Sec. 4.3). We compare the test loss obtained on the target task when we use only the N^T training samples from the target task with those resulting from the use of both source-task and target-task data.

Figure 13.5 shows the test loss for the target task as a function of the number of training points, N^S, in the source-task training set. If the target data set is small, e.g., if $N^T = 5$, transfer learning is seen, in the left part of the figure, to improve the test loss even for large values of the shift parameter Δ. This is an example of **positive transfer**: Using data from the source domain improves the performance on the target domain. Note also that the test loss saturates when N^S is large enough, at which point additional target data would be needed in order to improve the estimate of population loss under the target-task population distribution $p^T(x,t)$.

In contrast, when target data are more abundant, here with $N^T = 10$ data points, adding source-task data is seen in the right-hand part of the figure to be deleterious when the shift parameter Δ is sufficiently large. This is an example of **negative transfer**: Using source-domain data increases the test loss for the target task.

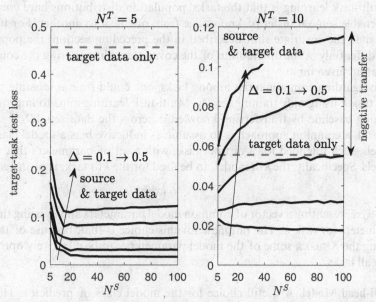

Figure 13.5 Test loss on the target task vs. number of examples in the source domain, N^S, for transfer learning with different numbers N^T of examples from the target domain. The difference between source and task domains is measured by a shift parameter Δ. For reference, the dashed line represents the performance obtained by using only target data.

Multi-source Transfer Learning. An extension of the transfer learning setting studied in this section is multi-source transfer learning. In this case, one has data from multiple source tasks that can be used to improve training on a target task. An important design question for this problem is how to combine data from multiple source tasks in a way that reflects their relevance for the target task.

13.3   Multi-task Learning

As studied in the preceding section, in transfer learning the goal is to identify a model that generalizes well on a target task by also using data from a related source task. Two tasks are hence involved in the problem, with data from the source task serving as auxiliary information used to (possibly) enhance the performance of the model on test data from the target task. In multi-task learning, one is interested in the generalization performance on K tasks, rather than on a single target task. For each task, we are given labeled training data, and the goal is to identify K models, each performing well when applied to test data of one of the K tasks. In this section, we provide a brief introduction to multi-task learning.

13.3.1   Inductive Bias Assumed by Multi-task Learning

To formulate multi-task learning, we focus on model classes of hard predictors, and assume that each task k is characterized by a population distribution $p_k(x, t)$ and by a loss function $\ell_k(t, \hat{t})$. For each task k, we are given a data set $\mathcal{D}_k$ of N_k training examples $\{(x_{k,n}, t_{k,n})_{n=1}^{N_k}\}$, which are assumed to be generated i.i.d. from the population distribution $p_k(x, t)$. The driving assumption in multi-task learning is that the tasks' population distributions have common characteristics that enable some transfer of knowledge from one task to another. For instance, in a manner similar to the covariate shift described in the preceding section, the population distributions may differ only in the distribution of the covariates, while sharing the conditional distribution of output given input.

Disregarding commonalities among tasks, one could train a separate predictor $\hat{t}_k(x|\theta_k)$ for each task k using only training set $\mathcal{D}_k$. Multi-task learning aims to improve on this single-task learning baseline by transferring knowledge across the data sets $\{\mathcal{D}_k\}_{k=1}^K$ of the K tasks. To this end, a common approach is to assume as inductive bias a model class $\mathcal{H}$ that includes K models, each intended for a distinct task, with a set of parameters that are shared across all models. Specifically, the kth model, to be used for the kth task, is of the general form

$$\hat{t}_k(x|\theta_0, \theta_k), \tag{13.15}$$

with θ_0 representing a vector of common model parameters and θ_k being the task-specific model parameters for task k. The rationale for this choice is that, because of the possible similarity among the K tasks, some of the model parameters can be effectively optimized based on data from all tasks.

Multi-head Model. A useful choice for the model class of predictors (13.15) is given by the multi-head model illustrated in Fig. 13.6. This predictor contains a common feature extraction function, which is parametrized by the shared vector θ_0, followed by separate "heads", one for each task k, which are instead each parametrized by the per-task vector θ_k and produce the final predictions $\hat{t}_k(x|\theta_0, \theta_k)$.

Figure 13.6 A multi-head architecture for multi-task learning.

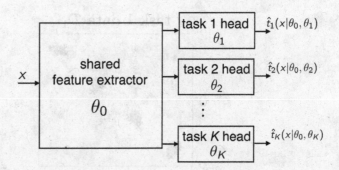

Attention-Based Model. Another idea is to use the per-task parameters θ_k to select how the shared parameters θ_0 should be combined to produce a prediction for task k. To illustrate this idea, consider a **recommendation system** in which we have a certain finite number of items indexed by an integer x. We would like to produce a prediction $\hat{t}_k(x)$ of the rating for an item with index x by each user k, based on a subset of ratings from all K users. A useful model class would assume as part of the inductive bias predictors of the form

$$\hat{t}_k(x|\theta_0,\theta_k) = [\theta_0]_{x:}\theta_k, \tag{13.16}$$

where θ_0 is a matrix containing, by row, feature vectors representing each of the items; $[\theta_0]_{x:}$ represents the xth row of matrix θ_0; and θ_k is a column feature vector representing user k. Accordingly, the rating produced by model (13.16) is obtained via the inner product between item x's feature vector $[\theta_0]_{x:}$ and user k's feature vector θ_k. The feature matrix θ_0 representing the items is naturally shared, while the user-specific vector θ_k is a local parameter. In (13.16), the per-user parameter θ_k defines which aspects of the item vector $[\theta_0]_{x:}$ the model should pay **"attention"** to when predicting the rating for user k.

13.3.2 Two-Level ERM Formulation

Since we have training data $\{\mathcal{D}_k\}_{k=1}^K$ from all tasks, a standard formulation of multi-task learning aims to minimize the **sum-training loss** $\sum_{k=1}^K L_{\mathcal{D}_k}(\theta_0,\theta_k)$ over shared parameters θ_0 and per-task parameters $\{\theta_k\}_{k=1}^K$. This optimization can be formulated as the **two-level ERM problem**

$$\min_{\theta_0} \sum_{k=1}^K \min_{\theta_k} L_{\mathcal{D}_k}(\theta_0,\theta_k), \tag{13.17}$$

where the empirical training loss for task k is given as

$$L_{\mathcal{D}_k}(\theta_0,\theta_k) = \frac{1}{N_k} \sum_{n=1}^{N_k} \ell_k(t_{k,n}, \hat{t}_k(x_{k,n}|\theta_0,\theta_k)). \tag{13.18}$$

In the two-level problem (13.17), the outer minimization is over the shared parameters θ_0, while there are K inner minimizations, each with respect to the respective per-task parameter vector θ_k. Note that the two optimization levels are mutually dependent, and hence, generally, they

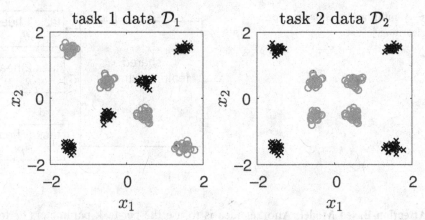

Figure 13.7 Data for Example 13.5, which focuses on multi-task learning for binary classification. The gray circles and black crosses represent data from two different classes.

cannot be tackled separately. Problem (13.17) should be addressed by jointly optimizing over θ_0 and $\{\theta_k\}_{k=1}^K$, e.g., by using SGD.

Example 13.5

Consider the training set shown in Fig. 13.7 for two binary classification tasks. Examples belonging to either class are identified by (gray) circles and (black) crosses, respectively. The distribution of the covariates is the same for both tasks, while the label assignment is different. To tackle this problem, a two-layer two-head network (see Fig. 13.6) is adopted, in which the first layer is shared between the two tasks, and so are its weights θ_0; while the last layer consists of two separate heads implementing logistic classifiers with respective weights θ_1 and θ_2. The D^1 hidden neurons use ReLU activation functions. Training is done by using standard GD with decaying learning rate for the joint optimization of the parameters $(\theta_0, \theta_1, \theta_2)$. In Fig. 13.8, the performance of single-task training (dashed lines), whereby only data from the respective task are used, is compared with multi-task learning (solid lines). Note that, for single-task training, the number of iterations indicated in the figure applies separately to each task. Therefore, multi-task training uses half the number of iterations as the single-task training of both tasks.

In this problem, the label assignment of task 1 is akin to the XOR problem (cf. Fig. 6.4). As such, it follows from Sec. 6.4 (see Example 6.10) that $D^1 = 3$ hidden units are necessary to classify exactly all examples in the training set. Furthermore, it is easy to see that, for task 2, a single quadratic feature would be sufficient to perfectly separate the two classes. Confirming this observation, Fig. 13.8 shows that, with $D^1 = 3$ hidden neurons, as well as with $D^1 = 4$, single-task learning obtains zero training error at convergence for both tasks. In contrast, with $D^1 = 3$, multi-task learning yields a residual error for task 2. Increasing the number of hidden units to $D^1 = 4$ allows perfect classification to be obtained for both tasks using multi-task learning.

This example suggests that the capacity of the shared feature extractor in the multi-head architecture in Fig. 13.6 should be selected as a function of the compatibility of the different tasks. A smaller-dimensional feature space is sufficient for tasks that share significant commonalities in terms of input–output mapping, while a larger common feature vector may otherwise be needed.

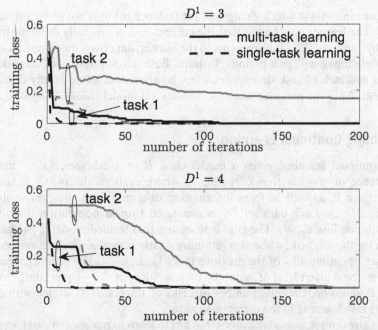

Figure 13.8 Training loss vs. the number of iterations for conventional single-task learning, as well as for multi-task learning with a multi-head model, for the binary classification tasks in Fig. 13.7. Top and bottom figures correspond to different values of the number of hidden neurons D^1 for the neural networks used by both single-task and multi-task learning.

13.3.3 Non-parametric Multi-task Learning

It is also possible to define non-parametric multi-task methods. A particularly useful idea is to consider **kernel methods** (see Sec. 4.11), in which the kernel function $\kappa(x,k,x',k')$ depends not only on the pair of inputs x and x', but also on the respective task identifiers k and $k' \in \{1,\dots,K\}$. Accordingly, the kernel function $\kappa(x,k,x',k')$ is meant to capture the "similarity" between input x for task k and input x' for task k'. Following the discussion in Sec. 4.11, the kernel $\kappa(x,k,x',k')$ is used to weight the available prediction for example x' in task k' when designing the predictor for input x in task k.

A typical choice is to partition the kernel as the product $\kappa(x,k,x',k') = \kappa_X(x,x')\kappa_T(k,k')$, where $\kappa_X(x,x')$ corresponds to the usual kernel applied on the inputs, while $k_T(k,k')$ is a measure of the similarity between tasks. The similarity between tasks can be encoded, for instance, via a graph with nodes representing tasks and edges included only between "similar" tasks.

13.4 Continual Learning

As discussed in the preceding section, in multi-task learning, data from all tasks are fully available for training on all tasks at once. In contrast, in continual learning, tasks are encountered sequentially, with data set $\mathcal{D}_k$ for task k observed after data set $\mathcal{D}_{k-1}$ for task $k-1$. We hence have a sequence of tasks with corresponding sequence of data sets $\mathcal{D}_1, \mathcal{D}_2, \dots$. This section focuses mostly on the standard formulation of continual learning in which the learner is informed about the time instants at which changes in the task occur but is not given a descriptor of the current task. The learner only knows that the current task is generally different from the previous, but not by "how much". Unbeknownst to the learner, data from the same task may even be observed multiple times, e.g., in a periodic fashion. Both assumptions – of clear demarcations between tasks and lack of task descriptors – can be alleviated, as we briefly discuss at the end of this section. As in the previous sections, we focus on model classes of hard predictors.

13.4.1 Defining Continual Learning

In continual learning, given a model class $\mathcal{H}$ of predictors. $\hat{t}(x|\theta)$, the learner produces a sequence of models $\hat{t}(\cdot|\theta_1)$, $\hat{t}(\cdot|\theta_2),\ldots$, where each model $\hat{t}(\cdot|\theta_k)$ is trained using data $\mathcal{D}_k$ from task k, as well as from information obtained from the previously observed data sets $\mathcal{D}_1,\ldots,\mathcal{D}_{k-1}$. Each data set $\mathcal{D}_k$ is generated from a population distribution p_k, yielding a population loss $L_{p_k}(\theta)$. The goal is to ensure that trained model $\hat{t}(\cdot|\theta_k)$ generalizes well on test data for the kth task, while also retaining a satisfactory generalization performance on a given subset – possibly all – of the previous tasks $1,\ldots,k-1$. For instance, one may be interested in preserving a given level of accuracy on a window of $K-1$ preceding tasks. Note that, under this assumption, the total number of tasks of interest is K, in line with the notation used for multi-task learning in the previous section.

As an example, consider a robot that has to learn to navigate through various types of terrain. Each type of terrain corresponds to a different task. We generally wish the robot to "transfer" knowledge of the previously encountered terrain to new terrain in order to limit the amount of data it neededs to adapt to its changed circumstances. Furthermore, we would like the robot to "remember" how to navigate previously encountered terrain, which it could be faced with in the future.

13.4.2 Stability vs. Plasticity

In training model $\hat{t}(\cdot|\theta_k)$ for the current task k, transferring information from previous tasks can be beneficial if the learner can leverage commonalities between the tasks in order to reduce data requirements and training complexity. On the flip side, the constraint that model θ_k maintain a suitable performance level on at least some of the previous tasks may impose limitations on the performance achievable on the current task k.

Catastrophic Forgetting and Backward Knowledge Transfer. In light of this, a key problem in continual learning is to ensure that the update of the model θ_k made for the purpose of guaranteeing generalization for the kth task does not yield a "catastrophic forgetting" of previous tasks. Catastrophic forgetting refers to a situation in which the updated model θ_k does not maintain suitable generalization performance levels on some of the previous tasks that are still of interest.

More generally, considering a past task $k-l$ for $l \geq 1$, we can distinguish among the following effects upon training for task k:

- **Backward positive transfer**. Adding data from a new task k yields a model θ_k that reduces the test loss for the previously observed task $k-l$, i.e., $L_{p_{k-l}}(\theta_k) < L_{p_{k-l}}(\theta_{k-1})$.
- **Backward negative transfer**. Adding data from a new task k yields a model θ_k that increases the test loss for the previously observed task $k-l$, i.e., $L_{p_{k-l}}(\theta_k) > L_{p_{k-l}}(\theta_{k-1})$. A significant backward negative transfer is referred to as **catastrophic forgetting**.

Overall, the problem of designing continual learning is one of **stability vs. plasticity**: We would like to adapt to new tasks (plasticity), while retaining the capacity to operate on previously observed tasks (stability).

Forward Knowledge Transfer. One can also consider the effect of an update of the model for the current task k on a future task $k + l$ with $l \geq 1$:

- **Forward positive transfer**. Adding data from a new task k yields a model θ_k that reduces the test loss for the future task $k + l$ as compared to the case in which only data from task $k + l$ are used.

- **Forward negative transfer**. Adding data from a new task k yields a model θ_k that increases the test loss for the future task $k + l$ as compared to the case in which only data from task $k + l$ are used.

It should be noted that the specific definitions given here of positive/ negative forward/ backward transfer are somewhat arbitrary, and several variants and specific metrics are proposed in the literature.

13.4.3 Coreset-Based Continual Learning

Information about previous tasks can be retained for future use in various ways, yielding the two main classes of continual learning algorithms, namely **coreset-based** and **regularization-based**. In this subsection, we study coreset-based techniques, while the next focuses on regularization-based methods. In Sec. 13.6, we will see that, from a Bayesian perspective, these two classes of techniques correspond to **likelihood-based** and **prior-based** methods, respectively.

Coreset-Based Training. Coreset-based strategies maintain a **coreset** of samples from prior tasks. At each step k, the coreset includes a (possibly empty) subset $\mathcal{C}_{k-l} \subseteq \mathcal{D}_{k-l}$ of examples from each of the previous tasks $k - l$ with $l \geq 1$. Training at each step relies not only on the data set $\mathcal{D}_k$ but also on the current coresets $\mathcal{C}_{k-l}$ for $l \geq 1$. Note that the coresets are generally updated from one iteration to the next, as we discuss next.

At each step k, an ERM problem is addressed by averaging over all the examples in the sets $\mathcal{D}_k$ and $\mathcal{C}_{k-l}$, with $l \geq 1$. This problem can be written as

$$\theta_k = \arg\min_{\theta} \left\{ L_{\mathcal{D}_k}(\theta) + \sum_{l \geq 1} L_{\mathcal{C}_{k-l}}(\theta) \right\}, \tag{13.19}$$

where $L_{\mathcal{D}_k}(\theta) = \frac{1}{N_k} \sum_{n=1}^{N_k} \ell_k(t_{k,n}, \hat{t}_k(x_{k,n}|\theta))$ is the training loss for the kth task, and the training loss functions $L_{\mathcal{C}_{k-l}}(\theta)$ for the coresets are defined in a similar fashion by limiting the empirical average on the samples in each coreset $\mathcal{C}_{k-l}$.

One way to think about coreset-based methods is in terms of **"experience replay"**: The coresets stored in the memory allow the learner to "replay", or **rehearse**, previous experiences with the goal of retaining previously learned skills while honing new ones. Accordingly, there is generally a trade-off between the size of the memory and the capacity of the continual learner to maintain desirable performance levels on previous tasks.

Managing the Coresets. A key issue in coreset-based schemes is the management of the coresets. The simplest approach is to store an equal number of samples from each task within a window of $K - 1$ previous tasks, and to discard samples from older tasks. To this end, one can keep a first-in-first-out queue of data points from $K-1$ tasks: When an equal number of training points from the new, kth, task is added to the queue, the samples from the oldest task are removed from the queue. This approach relies implicitly on the assumption that only the most recent $K - 1$ tasks should be catered for by the current trained model.

Beyond Coresets. While the strategies discussed so far in this subsection store data points from previous tasks, there are other ways to manage the memory at the learner. Here we briefly review three approaches.

- Instead of storing randomly selected samples, one can optimize the information retained in the buffer by using various forms of clustering, e.g., by memorizing suitable **"cluster prototypes"** that describe the data distributions under previous tasks (see Sec. 7.9).
- In addition to samples from previous tasks, one may also store **gradient vectors** so as to enable an assessment of how "aligned" the loss functions of different tasks are. This information can be used to ensure that current update directions are not likely to degrade the training loss of previous tasks.
- An alternative to storing data is to train a **generative model** (see Sec. 7.8) that is capable of generating samples that are "similar" to those observed in prior tasks. This approach may be particularly useful in the presence of **privacy constraints**. In fact, in some applications, one may wish not to store any data from previous tasks in order to avoid possible leaks of sensitive information.

13.4.4 Regularization-Based Continual Learning

In contrast to coreset-based methods, regularization-based continual learning techniques do not store data from previous tasks. Instead, they summarize information about previous tasks in the form of a **regularization** term that is applied to the learning problem for the current task. Recalling the relationship between regularization and MAP (Sec. 4.10), these methods can be thought of as maintaining a distribution in the model parameter space to serve as a **prior** for the current task. The prior at each step k is meant to summarize relevant information from a window of previous tasks. This idea can be made formal by adopting a Bayesian setting, as discussed in Sec. 13.6.

Elastic Weight Consolidation. As a notable example of regularization-based continual learning, we now describe elastic weight consolidation (EWC). Let us assume, to fix the ideas, that we wish not to "forget" tasks encountered within a window of $K - 1$ previous tasks. Note that other formulations are also possible. In its simplest form, for each of the previous tasks $k - l$ with $l = 1, \ldots, K - 1$, EWC stores:

- the $M \times 1$ model parameter vector θ_{k-l} updated after observing data from task $k - l$; and
- an auxiliary $M \times 1$ vector $F_{k-l} = [F_{1,k-l}, \ldots, F_{M,k-l}]^T$, with entry $F_{m,k-l}$ representing the "importance" of the mth entry $\theta_{m,k-l}$ of the model parameter vector $\theta_{k-l} = [\theta_{1,k-l}, \ldots, \theta_{M,k-l}]^T$ for task $k - l$.

We will discuss how to compute these vectors below. Note that this approach requires storing $2M$ numbers, which may be smaller than the amount of memory required by coreset-based schemes. There are also several ways to further reduce the memory overhead, such as storing suitable linear combinations of these parameters.

At the kth step, the learner obtains the updated model θ_k by tackling the regularized ERM problem

$$\min_{\theta_k} \left\{ L_{\mathcal{D}_k}(\theta_k) + \lambda \sum_{l=1}^{K-1} ||\text{Diag}(F_{k-l})(\theta_k - \theta_{k-l})||^2 \right\}, \tag{13.20}$$

where hyperparameter $\lambda > 0$ accounts for the relative relevance of the previous tasks with respect to the current task. Intuitively, the regularization term in (13.20) penalizes deviations of the updated model parameter θ_k with respect to the previous model parameters $\{\theta_{k-l}\}_{l=1}^{K-1}$. Writing $\theta_k = [\theta_{1,k}, \ldots, \theta_{M,k}]^T$, each lth regularization term in problem (13.20) can be expressed as

$$||\text{Diag}(F_{k-l})(\theta_k - \theta_{k-l})||^2 = \sum_{m=1}^{M} F_{m,k-l}^2 (\theta_{m,k} - \theta_{m,k-l})^2. \tag{13.21}$$

This shows that the term $F_{m,k-l}^2$ acts as a weight for the mth entry of the model parameter vector as it pertains to task $k-l$: If $F_{m,k-l}^2$ is large, the regularization term in problem (13.20) penalizes large deviations of the mth parameter $\theta_{m,k}$ from the corresponding entry $\theta_{m,k-l}$ obtained when training for task $k-l$. If the loss is quadratic, the EWC problem (13.20) is easily seen to be a LS problem that can be solved in closed form (see Sec. 4.3).

How to obtain the vectors F_{k-l} of relevance weights? EWC chooses each mth entry of vector F_{k-l} as

$$F_{m,k-l} = \left| \frac{\partial L_{\mathcal{D}_{k-l}}(\theta_{k-l})}{\partial \theta_{m,k-l}} \right|, \tag{13.22}$$

which is meant to measure the sensitivity of the training loss for task $k-l$ to changes in the mth entry of the model parameter vector. The rationale for this choice is that, if the partial derivative $\partial L_{\mathcal{D}_{k-l}}(\theta_{k-l})/\partial\theta_{m,k-l}$ is large in absolute value, one expects that changing the mth entry $\theta_{m,k-l}$ would significantly affect the performance of task $k-l$, potentially causing catastrophic forgetting.

Importantly, the partial derivative $\partial L_{\mathcal{D}_{k-l}}(\theta_{k-l})/\partial\theta_{m,k-l}$ is computed at the previous model parameter θ_{k-l}. This is done in order to avoid having to recompute gradients on previous tasks at every update of the model parameters – an operation that is in fact generally impossible since regularization-based methods do not store prior data. This also implies that the information encoded by (13.22) about the sensitivity of the training loss for task $k-l$ may in fact be outdated, potentially limiting the performance of EWC.

More technically, when the log-loss is used, the term $F_{m,k-l}^2$ in (13.22) can be interpreted as the mth diagonal element of the empirical FIM for the loss function of task $k-l$ evaluated as θ_{k-l} (see Sec. 9.7). Accordingly, it estimates the amount of information that the data for task $k-l$, when generated with model parameter θ_{k-l}, carries about the mth element of the model parameter vector. This quantity can conversely be taken to measure the relevance of the mth parameter θ_m for the loss function of task $k-l$.

Overall, intuitively, the regularization term in the EWC objective (13.20) prevents the model from moving too far away from the parameter vectors that are important for the given window of $K - 1$ prior tasks, while leaving less crucial parameters as degrees of freedom to optimize the model for the kth task.

Example 13.6

Consider a binary image classification task in which the inputs for task 1 are grayscale images representing handwritten digits 0 for one class and digits 1 for the other; while for tasks 2 and 3, the inputs are given by the same images as for task 1 but the pixels are randomly permuted, with the same permutation applied to all examples in the same task. The permutations are independent for the two tasks. A two-layer neural network is assumed with a logistic classifier for the second layer. Training and testing are based on the standard log-loss.

As a first benchmark, we assume that, at each iteration, an SGD update is carried out with a mini-batch of size 10 for the current task. The task is changed every 20 SGD iterations. Figure 13.9 show the test loss for the three tasks across the SGD iterations, with bold lines representing the loss of the current task. Note that, in this experiment, data from the same task are presented periodically to the learner. The results in the figure demonstrate that, when training on a given task, the test loss for the other tasks can increase, in which case we have a **negative transfer**; but it can also decrease, in which case we have **positive transfer**. For this simple example in which tasks are periodically presented to the learner, no catastrophic forgetting occurs since the test losses of all tasks demonstrate an overall decreasing trend.

Assume now that data from task 3 are presented only for the first 20 iterations. After that, only samples from the first two tasks are observed. We again implement SGD using samples from the current task. Figure 13.10 shows that in this case we have catastrophic forgetting for task 3.

Finally, let us implement a simplified version of EWC in which we only penalize deviations from the model parameter θ_3 of task 3 (with $\lambda = 3$). The results in Fig. 13.11 shows that the approach is indeed useful in mitigating catastrophic forgetting, and also illustrates the plasticity-stability trade-off: In order to "stabilize" the solution and avoid forgetting for task 3, adaptation, i.e., "plasticity", for the other two tasks is reduced.

Figure 13.9 Test loss vs. the number of SGD iterations for continual learning under periodic task presentation. Tasks change every 20 SGD iterations and conventional SGD is applied on the training loss of the current task. The test loss for the three tasks is shown with bold lines representing the loss of the current task. The figure illustrates examples of both negative and positive transfer.

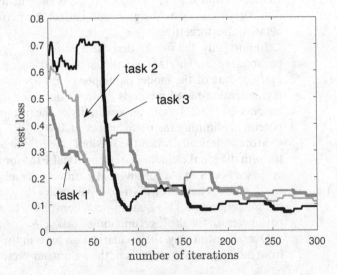

Figure 13.10 Test loss vs. the number of SGD iterations for continual learning with a disappearing task. Task 3 is presented only for the first 20 iterations, and then tasks 1 and 2 are alternated for the remaining iterations, with tasks changing every 20 iterations. The test loss for the three tasks is shown with bold lines representing the loss of the current task. The figure illustrates catastrophic forgetting for task 3.

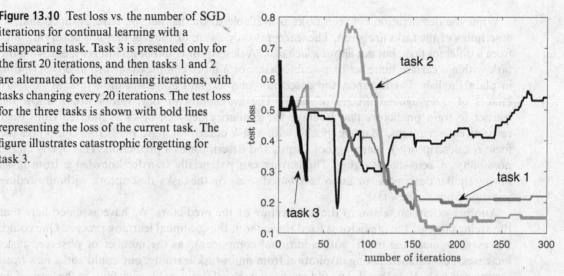

Figure 13.11 Continual learning with a disappearing task as in Fig. 13.10 under a simplified version of EWC with a quadratic regularization term that penalizes deviations from the learned model for task 3 after the first 20 SGD iterations. The figure shows that regularization can mitigate catastrophic forgetting, while inducing a trade-off between stability and plasticity.

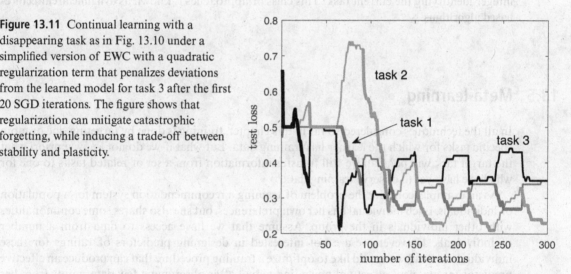

13.4.5 Extensions

We conclude this section by briefly discussing several extensions that alleviate some of the limitations of the setting studied in this section. First, we have assumed that the **demarcation between successive tasks** is known to the learner. This allows the learner to manage its data buffer for coreset-based strategies, as well as to maintain the data structures necessary to compute the regularization penalty for regularization-based algorithms. If the demarcation between tasks were not known, the learner could run a *change detection algorithm* to automatically detect changes in the statistics of the data.

While the demarcation between tasks is available to the learner, we have assumed that no **descriptors of the tasks** are given. The learner is only aware of the fact that it is observing data from a different task, but not about *which* task. A descriptor could be an integer identifying the task within a certain finite set of possible tasks, or a more complex tag such as a description in plain English. For instance, the descriptor may explain a classification task as "classify images of dogs against images of cats". The availability of descriptors would allow the learner to train predictors that leverage the similarity between tasks as inferred from their respective descriptions. Mathematically, with a task descriptor, hard predictors could take the form $\hat{t}(x, \text{descriptor}|\theta)$, mapping both input and descriptor to a prediction. This opens up the possibility of **zero-shot learning**: The learner can potentially transfer knowledge from tasks having similar descriptors to a new task based solely on the tasks' descriptors, without needing any data from the new task.

Another extension relates to the **architecture** of the predictors. We have assumed here that the architecture of the predictor is fixed throughout the continual learning process. One could, however, augment the model with additional components as the number of observed tasks increases. For instance, taking inspiration from multi-task learning, one could add a new head for any new task. Note that this would require having a descriptor for each task, in the form of an integer identifying the current task. This class of approaches is known as **dynamic architecture-based algorithms**.

13.5   Meta-learning

In all the techniques considered so far in this chapter, training is done by targeting one or more specific tasks for which the learner has training data. But what if we do not know a priori what the target task will be? Can we still transfer information from a set of related tasks to one for which we have yet to observe training data?

As an example, consider the problem of training a recommendation system for a population of individuals. Each individual has her own preferences, but she also shares some commonalities with other individuals in the group. Assume that we have access to data from a number of individuals. However, we are not interested in designing predictors of ratings for these individuals. Instead, we would like to optimize a training procedure that can produce an effective predictor for any new, as yet unknown, individual after observing a few data points from her activity. Since a priori we do not yet have data from the new individual, the techniques developed so far in this chapter are not applicable. This is the domain of meta-learning.

In meta-learning, we target an entire **class of tasks**, also known as a **task environment**, and we wish to "prepare" for any new task that may be encountered from this class. To formalize this point, we assume that the task environment includes tasks with "similar" population distributions and loss functions. Data for any task in the task environment can be modeled as being generated by first picking a task – and hence a population distribution – at random from the task environment, and then drawing data from the selected population distribution.

By focusing on the problem of adaptation to new tasks, meta-learning operates at a higher level of abstraction than conventional learning. As we will see in this section, while conventional learning techniques focus on optimizing model parameters, the goal of meta-learning is to optimize **hyperparameters**, such as learning rate or initialization of an SGD-based training algorithm.

13.5.1 Meta-training and Meta-testing

Meta-learning applies to scenarios in which, prior to observing the (small) data set to be used for training on a new task, one has access to a larger data set of examples from related tasks, which is known as the **meta-training data set**. All tasks are assumed to be part of the same task environment. Meta-learning consists of two distinct phases:

- **Meta-training**. Given the meta-training data set for multiple related tasks from the task environment, meta-learning optimizes a set of hyperparameters.
- **Meta-testing**. After the meta-learning phase is completed, data for a target task, known as **meta-test task**, in the task environment are revealed, and model parameters are optimized using the "meta-trained" hyperparameters.

As such, the meta-training phase aims to optimize hyperparameters that enable efficient training on a new, a priori unknown, target task in the meta-testing phase.

Few-Shot Classification. A classical application of meta-learning is the problem of few-shot classification. To describe it, consider a general multi-class classification task with C classes, which is also referred to as *C-way classification*. For each class, we observe L examples as training data. The resulting problem is known as *C*-**way** *L*-**shot classification**. Note that the total number of training examples per task is $N = C \cdot L$.

To fix the ideas, think of the task of classifying images of animal breeds of a certain family based on a small number L of images for each breed. For instance, one task could involve classifying breeds of cats; another breeds of dogs; yet another breeds of horses; and so on. A priori, we do not know which family of animals we will have to classify at run time, and we would like to use meta-learning to prepare for any family of animals that may come our way.

Using conventional machine learning, given a family of animals of interest, say dogs, one should collect a sufficiently large number L of examples of each dog breed, and then train a machine learning model, e.g., softmax regression, from scratch. But what if L is small? Conventional methods would typically fail as a result of underfitting.

Assume now that we can collect a meta-training data set including separate data sets for the related tasks of classifying images of different cat breeds, bird breeds, horse breeds, and so on. While these tasks are all distinct, we may consider them to be part of the same task environment. The goal of meta-learning is to ensure that, when faced with an a priori unknown, and an as yet unobserved, category of animals, the hyperparameter vector optimized based on the meta-training data facilitates fast and effective adaptation of a classifier based on a small number L of examples per breed.

13.5.2 Reviewing Conventional Learning

In order to introduce the notation necessary to describe meta-learning, let us briefly review the operation of conventional machine learning.

Training and Testing. In conventional machine learning, one starts by fixing an inductive bias, namely a model class $\mathcal{H}$ and a training algorithm. The model class $\mathcal{H}$ contains models

Figure 13.12 Illustration of conventional learning.

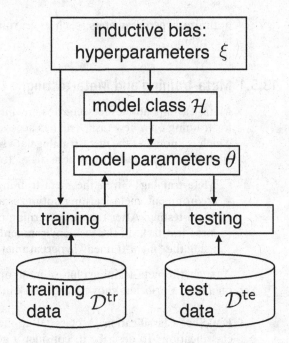

parametrized by a vector θ. Model class and training algorithm depend on a *fixed* vector of hyperparameters, ξ. Hyperparameters specify, for instance, the vector of features to be used in a linear model, or the initialization and learning rate of an SGD-based optimizer. The training algorithm is applied to a training set $\mathcal{D}^{tr}$, which may also include a separate validation set. Note that we have added a superscript "tr" in order to simplify the discussion to follow. The training algorithm produces a model parameter vector θ, which is finally tested on a separate test data set $\mathcal{D}^{te}$. The overall process is summarized in Fig. 13.12.

Drawbacks of Conventional Learning. Conventional machine learning suffers from two main potential shortcomings that meta-learning can help address, namely large **sample complexity** and large **iteration complexity**. As seen in Chapter 8, sample complexity refers to the number of training samples, N, required to obtain a suitable test performance. Iteration complexity is used loosely here to define the number of iterations, and hence the complexity, of iterative training algorithms such as SGD (see Appendix 5.C). Importantly, both issues can potentially be mitigated if the inductive bias is optimized based on domain knowledge. For instance, if, as part of the inductive bias, we choose informative features or a "good" initialization point for the model parameters θ, we can generally reduce both sample and iteration complexities. But what if we do not have the domain knowledge necessary to optimize the inductive bias, at least not to the desired extent?

13.5.3 Joint Learning

Suppose now that we have access to training data sets $\mathcal{D}^{tr}_k$ for a number of distinct learning tasks in the same task environment that are indexed by the integer $k = 1, \dots, K$. Each data set $\mathcal{D}^{tr}_k$ contains N training examples. How should we use these data?

Figure 13.13 Illustration of joint learning.

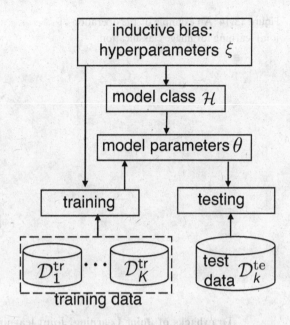

Training and Testing. A first idea would be to adopt **joint learning**: Pool together all the training sets $\{\mathcal{D}_k^{\mathrm{tr}}\}_{k=1}^K$, and use the resulting aggregate training loss

$$L_{\{\mathcal{D}_k^{\mathrm{tr}}\}_{k=1}^K}(\theta) = \frac{1}{K} \sum_{k=1}^K L_{\mathcal{D}_k^{\mathrm{tr}}}(\theta) \tag{13.23}$$

as the learning criterion to train a shared model parameter θ. By (13.23), joint learning corresponds to a form of multi-task learning (cf. (13.17)) when only shared parameters across tasks are assumed. As such, as illustrated in Fig. 13.13, joint learning inherently caters only to the K tasks in the original pool, and expects to be tested on data on any one of these tasks. As we further discuss in the next subsection, this differs from the goal of meta-learning, which prepares for new, as yet unknown, tasks.

Example 13.7

Consider a 2-way L-shot classification problem in which inputs are images representing different categories of objects. As an example, in Fig. 13.14 we have data from two tasks ($K = 2$): The first data set includes images of flowers, labeled as $t = 0$, and of backpacks, labeled as $t = 1$; while the second contains images of mugs, labeled as $t = 0$, and of crayons, labelled as $t = 1$. Note that, in the pooled data set, images that are assigned the same label are heterogeneous. For instance, the class $t = 0$ includes images of flowers and mugs. The dashed line in Fig. 13.14 represents the decision region of the single classifier trained based on the available pooled training data via joint learning. Testing should be done on test data from one of the $K = 2$ tasks, since joint learning does not prepare for the possibility of new tasks.

Figure 13.14 An example of the operation of joint learning for image classification.

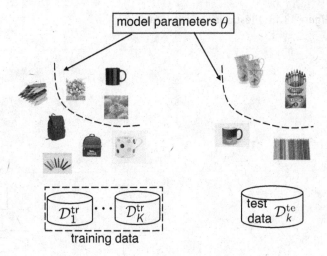

Drawbacks of Joint Learning. Joint learning has some evident appeal as a means to mitigate sample and iteration complexities of conventional learning. First, by pooling together data from K tasks, the overall size of the training set is $K \cdot N$, which may be large even when the available data per task is limited, i.e., when N is small. Second, training only once for K tasks amortizes the iteration complexity across the tasks, yielding a potential reduction of the number of iterations by a factor of K.

But joint learning has two potentially critical shortcomings. First, the jointly trained model only works if there is a single model parameter θ that "works well" for all tasks. As suggested by Fig. 13.14, this may often not be the case. Second, even if there is a single model parameter θ that yields desirable test results on all K tasks, this does not guarantee that the same is true for a new task. In fact, by focusing on training a common model, joint learning is not designed to enable adaptation based on training data for a new task.

As a remedy for this second shortcoming, one could use the jointly trained model parameter θ to initialize the training process on a new task – a process known as **fine-tuning**. However, there is generally no guarantee that this would yield a desirable outcome, since the training process used by joint learning does not optimize for the subsequent step of adaptation to a new task.

13.5.4 Introducing Meta-learning

As in joint learning, in meta-learning we assume the availability of data from K related tasks from the same task environment, which are referred to as **meta-training tasks**. However, unlike joint learning, data from these tasks are kept separate, and a distinct model parameter θ_k is trained for each kth task. As illustrated in Fig. 13.15, what *is* shared across tasks is a **common hyperparameter vector** ξ that is optimized based on meta-training data, and is applied across all tasks to define a **shared inductive bias**.

In this subsection, we introduce meta-learning by emphasizing the differences with respect to joint learning and by detailing the meta-training and meta-testing phases. This will eventually lead to the description of a state-of-the-art meta-learning algorithm, known as MAML.

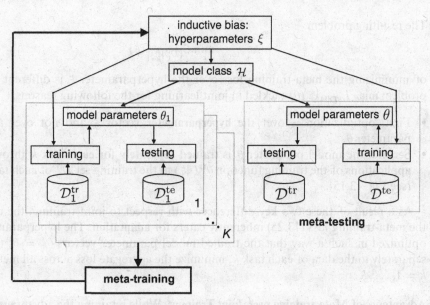

Figure 13.15 Illustration of meta-learning.

Inductive Bias and Hyperparameters. As discussed, the goal of meta-learning is to optimize the hyperparameter vector ξ and, through it, the inductive bias that is applied for the training of each task – both observed, in the meta-training set, and new, for meta-testing. To simplify the discussion and focus on the most common setting, let us assume that the model class $\mathcal{H}$ is fixed, while the training algorithm is a mapping $\theta^{\text{tr}}(\mathcal{D}|\xi)$ between a training set $\mathcal{D}$ and a model parameter vector θ that depends on the hyperparameter vector ξ, i.e.,

$$\theta = \theta^{\text{tr}}(\mathcal{D}|\xi). \tag{13.24}$$

As an example, the training algorithm $\theta^{\text{tr}}(\mathcal{D}|\xi)$ could output the last iterate of a GD-based optimizer. The hyperparameter ξ can affect the output $\theta^{\text{tr}}(\mathcal{D}|\xi)$ of the training procedure in different ways. For instance, it can determine the regularization constant; the learning rate and/or the initialization of an iterative training procedure such as GD or SGD; the mini-batch size in SGD; a subset of the parameters in vector θ, e.g., used to define a shared feature extractor; the parameters of a prior distribution; and so on.

The output $\theta^{\text{tr}}(\mathcal{D}|\xi)$ of a training algorithm is generally random. This is the case, for instance, if the algorithm relies on SGD because of its use of stochastic gradients. In the following discussion, we will assume for simplicity a deterministic training algorithm, but the approach described here carries over directly to the more general case of a random training procedure by adding an average over the randomness of the trained model $\theta^{\text{tr}}(\mathcal{D}|\xi)$.

Meta-training. To formulate meta-training, a natural idea is to use as the optimization criterion the aggregate training loss

$$\mathcal{L}_{\{\mathcal{D}_k^{\text{tr}}\}_{k=1}^K}(\xi) = \sum_{k=1}^K L_{\mathcal{D}_k^{\text{tr}}}(\theta^{\text{tr}}(\mathcal{D}_k^{\text{tr}}|\xi)), \tag{13.25}$$

which is a function of the hyperparameter ξ. This quantity is known as the **meta-training loss**.

The resulting problem

$$\min_{\xi} \mathcal{L}_{\{\mathcal{D}_k^{\mathrm{tr}}\}_{k=1}^{K}}(\xi) \tag{13.26}$$

of minimizing the meta-training loss over the hyperparameter ξ is different from the ERM problem $\min_{\theta} L_{\{\mathcal{D}_k^{\mathrm{tr}}\}_{k=1}^{K}}(\theta)$ tackled in joint learning for the following reasons:

- First, optimization is over the **hyperparameter** vector ξ and not over a shared model parameter θ.
- Second, the model parameter θ is trained **separately** for each task k through the parallel applications of the training function $\theta^{\mathrm{tr}}(\cdot|\xi)$ to the training set $\mathcal{D}_k^{\mathrm{tr}}$ of each task $k = 1, \ldots, K$ (see Fig. 13.15).

As a result of these two key differences with respect to joint training, the minimization of the meta-training loss (13.25) inherently caters for **adaptation**: The hyperparameter vector ξ is optimized in such a way that the trained model parameter vectors $\theta_k = \theta^{\mathrm{tr}}(\mathcal{D}_k^{\mathrm{tr}}|\xi)$, adapted separately to the data of each task k, minimize the aggregate loss across all meta-training tasks $k = 1, \ldots, K$.

Advantages of Meta-training over Joint Training. While retaining the advantages of joint learning in terms of sample and iteration complexity, meta-learning addresses the two highlighted shortcomings of joint learning:

- First, meta-learning does not assume that there is a single model parameter θ that "works well" for all tasks. It only assumes that there exists a common model class and a common training algorithm, as specified by **hyperparameters** ξ, that can be effectively applied across the class of tasks of interest.
- Second, meta-learning prepares the training algorithm $\theta^{\mathrm{tr}}(\mathcal{D}|\xi)$ to **adapt** to potentially new tasks through the selection of the hyperparameters ξ. This is because the model parameter vector θ is left free by design to be adapted to the training data $\mathcal{D}_k^{\mathrm{tr}}$ of each task k.

Meta-testing. As mentioned, the goal of meta-learning is to ensure generalization to any new task that is drawn at random from the same task environment. For any new task, during the meta-testing phase, we have access to training set $\mathcal{D}^{\mathrm{tr}}$ and test set $\mathcal{D}^{\mathrm{te}}$. The new task are referred to as the **meta-test task**, and is illustrated in Fig. 13.15 along with the meta-training tasks.

The training data $\mathcal{D}^{\mathrm{tr}}$ of the meta-test task are used to adapt the model parameter vector to the meta-test task, obtaining $\theta^{\mathrm{tr}}(\mathcal{D}^{\mathrm{tr}}|\xi)$. Importantly, the training algorithm depends on the hyperparameter vector ξ. The performance metric of interest for a given hyperparameter vector ξ is the test loss for the meta-test task, or **meta-test loss**, which can be written as

$$L_{\mathcal{D}^{\mathrm{te}}}(\theta^{\mathrm{tr}}(\mathcal{D}^{\mathrm{tr}}|\xi)). \tag{13.27}$$

In (13.27), as usual, the population loss of the trained model is estimated via the test loss evaluated with the test set $\mathcal{D}^{\mathrm{te}}$.

Example 13.8

Consider again the 2-way L-shot image classification problem described in Example 13.7. Fig. 13.16 illustrates two meta-training tasks – distinguishing backpacks against flowers and distinguishing crayons against mugs – along with a meta-training task – classifying images of furniture

Figure 13.16 An example of the operation of meta-learning for image classification. The meta-training data are used to identify a hyperparameter vector ξ, which is in turn leveraged to adapt a model parameter θ to the training data of a meta-test task.

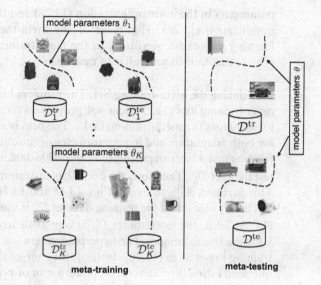

meta-training

meta-testing

against images of cameras. Unlike joint learning, in which a single classifier is trained for the overall meta-training data set (see Fig. 13.14), meta-learning treats the two meta-training tasks separately, obtaining a different model parameter θ_k for each task k. The meta-training data are used to identify a hyperparameter vector ξ, which is in turn leveraged to adapt a model parameter θ to the training data of the meta-test task. The figure also hints at an important extension of the meta-training approach considered in this subsection, whereby meta-training data are also split into training and test subsets.

13.5.5 An Improved Meta-learning Criterion: Splitting the Meta-training Set

We have just seen that meta-testing requires a split of the data for the new task into a training part, used for adaptation, and a test part, used to estimate the population loss (13.27) (see Fig. 13.16). In this subsection, we discuss how the idea of splitting per-task data sets into training and testing parts can also be useful during the meta-training phase.

Meta-overfitting. As explained in the preceding subsection, the training algorithm $\theta(\mathcal{D}^{\mathrm{tr}}|\xi)$ is defined by an optimization procedure for the problem of minimizing the training loss on the training set $\mathcal{D}^{\mathrm{tr}}$ (or a regularized version thereof). We can write this informally as

$$\theta^{\mathrm{tr}}(\mathcal{D}^{\mathrm{tr}}|\xi) \leftarrow \min_{\xi} \min_{\theta} L_{\mathcal{D}^{\mathrm{tr}}}(\theta) \tag{13.28}$$

to highlight the dependence of the training algorithm on the training loss $L_{\mathcal{D}^{\mathrm{tr}}}(\theta)$ and on the hyperparameter ξ.

In light of this, let us take another look at the definition of the meta-training loss (13.25) and at the optimization (13.26). Because of (13.28), in problem (13.26) one is effectively optimizing the training losses $L_{\mathcal{D}_k^{\mathrm{tr}}}(\theta)$ for the meta-training tasks $k = 1, \ldots, K$ twice, first over the model parameters in the inner optimization (13.28) and then over the hyperparameters ξ in the outer optimization (13.26). This reuse of the meta-training data for both adaptation and meta-learning may cause overfitting to the meta-training data, and results in a training algorithm $\theta^{\text{tr}}(\cdot|\xi)$ that fails to generalize to new tasks.

Partitioning the Meta-training Set. The problem highlighted above is caused by the fact that the meta-training loss (13.25) does not provide an unbiased estimate of the sum of the population losses across the meta-training tasks. The bias is a consequence of the reuse of the same data for both adaptation and hyperparameter optimization. To address this problem, for each meta-training task k, we can partition the available data into two data sets, a training data set $\mathcal{D}_k^{\text{tr}}$ and a test data set $\mathcal{D}_k^{\text{te}}$. Therefore, the overall meta-training data set is given as $\mathcal{D}^{\text{mtr}} = \{(\mathcal{D}_k^{\text{tr}}, \mathcal{D}_k^{\text{te}})_{k=1}^K\}$. This partition is illustrated in Figs. 13.15 and 13.16.

The key idea is that the training data set $\mathcal{D}_k^{\text{tr}}$ is used for adaptation using the training algorithm (13.28), while the test data set $\mathcal{D}_k^{\text{te}}$ is kept aside to estimate the population distribution of task k for the trained model. The hyperparameters ξ are not optimized to minimize the sum of the training losses as in (13.26). Instead, they target the sum of the test losses, which provides an unbiased estimate of the corresponding sum of population losses.

Meta-learning as Nested Optimization. To summarize, the general procedure followed by many meta-learning algorithms consists of a nested optimization of the following form:

- **Inner loop.** For a fixed hyperparameter vector ξ, training on each task k is done separately, producing per-task model parameters

$$\theta_k = \theta^{\text{tr}}(\mathcal{D}_k^{\text{tr}}|\xi) \leftarrow \min_{\xi} \min_{\theta} L_{\mathcal{D}_k^{\text{tr}}}(\theta) \tag{13.29}$$

for $k = 1, \ldots, K$.
- **Outer loop.** The hyperparameter vector ξ is optimized as

$$\xi_{\mathcal{D}^{\text{mtr}}} = \arg\min_{\xi} \mathcal{L}_{\mathcal{D}^{\text{mtr}}}(\xi), \tag{13.30}$$

where the **meta-training loss** is (re-)defined as

$$\mathcal{L}_{\mathcal{D}^{\text{mtr}}}(\xi) = \sum_{k=1}^K L_{\mathcal{D}_k^{\text{te}}}(\theta^{\text{tr}}(\mathcal{D}_k^{\text{tr}}|\xi)). \tag{13.31}$$

The specific implementation of a meta-learning algorithm depends on the selection of the training algorithm $\theta^{\text{tr}}(\mathcal{D}|\xi)$ and on the method used to solve the outer optimization. In the next subsection we will provide the details of a reference scheme that optimizes over the initialization of SGD as an inner loop solver.

13.5.6 MAML: Meta-learning the Initialization

In its most basic form, **model-agnostic meta-learning (MAML)** assumes a training algorithm (13.29) of the form

$$\theta^{\text{tr}}(\mathcal{D}^{\text{tr}}|\xi) = \xi - \gamma \nabla_\theta L_{\mathcal{D}^{\text{tr}}}(\theta)|_{\theta=\xi}$$
$$= \xi - \gamma \nabla_\xi L_{\mathcal{D}^{\text{tr}}}(\xi). \tag{13.32}$$

In words, MAML assumes that a GD step is made to adapt the model parameter vector based on the training data $\mathcal{D}^{\mathrm{tr}}$, starting from an initial point given by the hyperparameter vector ξ. Note that the hyperparameter vector ξ is of the same size as the model vector θ. In practice, one can extend the approach by allowing for multiple GD steps, as well as by replacing GD with SGD. These extensions will be discussed later in this section.

MAML uses GD – or, more generally, SGD – also for the outer optimization (13.30) over the hyperparameters ξ. To this end, one needs to compute the gradient

$$\nabla_\xi \mathcal{L}_{\mathcal{D}^{\mathrm{mtr}}}(\xi) = \sum_{k=1}^{K} \nabla_\xi L_{\mathcal{D}_k^{\mathrm{te}}}(\theta^{\mathrm{tr}}(\mathcal{D}_k^{\mathrm{tr}}|\xi)). \tag{13.33}$$

Note that a stochastic estimate of this gradient can be easily obtained by sampling a subset of tasks and averaging only over the selected tasks.

Computing the MAML Gradient (13.33). To proceed, define as

$$\theta_k = \xi - \gamma \nabla_\xi L_{\mathcal{D}_k^{\mathrm{tr}}}(\xi) \tag{13.34}$$

the trained model parameter for task k after one GD iteration in (13.32), and M for the dimension of model parameter θ and hyperparameter ξ. Using the chain rule of differentiation, we have the per-task gradient

$$\begin{aligned}
\nabla_\xi L_{\mathcal{D}_k^{\mathrm{te}}}(\theta^{\mathrm{tr}}(\mathcal{D}_k^{\mathrm{tr}}|\xi)) &= \nabla_\xi L_{\mathcal{D}_k^{\mathrm{te}}}(\xi - \gamma \nabla_\xi L_{\mathcal{D}^{\mathrm{tr}}}(\xi)) \\
&= \nabla_\xi(\xi - \gamma \nabla_\xi L_{\mathcal{D}^{\mathrm{tr}}}(\xi)) \nabla_{\theta_k} L_{\mathcal{D}_k^{\mathrm{te}}}(\theta_k) \\
&= (I_M - \gamma \nabla_\xi^2 L_{\mathcal{D}^{\mathrm{tr}}}(\xi)) \nabla_{\theta_k} L_{\mathcal{D}_k^{\mathrm{te}}}(\theta_k),
\end{aligned} \tag{13.35}$$

where we write $\nabla_\theta L_{\mathcal{D}_k^{\mathrm{te}}}(\theta_k)$ for the gradient $\nabla_\theta L_{\mathcal{D}_k^{\mathrm{te}}}(\theta)|_{\theta=\theta_k}$. The gradient applied by MAML is hence the product of two terms:

- The second term is the gradient of the test loss $\nabla_{\theta_k} L_{\mathcal{D}_k^{\mathrm{te}}}(\theta_k)$ evaluated at the adapted parameter θ_k. This term provides the direction of local maximal change of the test performance at the adapted parameters θ_k. It can be computed by using standard backprop if the model is a neural network.
- The first term is the Jacobian $\nabla_\xi \theta_k = (I_M - \gamma \nabla_\xi^2 L_{\mathcal{D}^{\mathrm{tr}}}(\xi))$ of the updated model parameter θ_k with respect to the hyperparameters ξ. This term quantifies the direction of maximal local change of the updated model parameters θ_k with respect to the hyperparameter vector at its current value ξ.

An important observation from (13.35) is that, since the derivatives with respect to the hyperparameters need to "flow through" the per-task update operation (13.32), MAML is a **second-order scheme**, requiring the Hessian $\nabla_\xi^2 L_{\mathcal{D}^{\mathrm{tr}}}(\xi)$ of the training loss functions. More specifically, for the update (13.35), one needs to compute the matrix–vector product $\nabla_\xi^2 L_{\mathcal{D}^{\mathrm{tr}}}(\xi) \nabla_{\theta_k} L_{\mathcal{D}_k^{\mathrm{te}}}(\theta_k)$. Appendix 13.A describes how to approximate the product of the Hessian and a vector without computing the Hessian. This approximation has a computational complexity of the order of the computation of two gradients.

MAML Gradient with Multiple Per-task Iterations. We now generalize the MAML update (13.35) by allowing for SGD-based training algorithms carrying out multiple iterations in lieu of the single GD step in (13.32). Accordingly, $I \geq 1$ SGD iterations are carried out for per-task training, with each iteration using a stochastic estimate $\nabla_\theta \hat{L}_{\mathcal{D}^{\mathrm{tr}}}(\theta)$ of the gradient $\nabla_\theta L_{\mathcal{D}^{\mathrm{tr}}}(\theta)$, where we denote as $\hat{L}_{\mathcal{D}^{\mathrm{tr}}}(\theta)$ the training loss estimated on the basis of an arbitrary mini-batch of examples from data set $\mathcal{D}^{\mathrm{tr}}$.

The resulting training algorithm (13.29) can be expressed as the sequence of operations

$$\theta^{(2)} = \xi - \gamma^{(1)} \nabla_\theta \hat{L}_{\mathcal{D}^{\mathrm{tr}}}(\theta)|_{\theta=\xi}$$

$$\theta^{(3)} = \theta^{(2)} - \gamma^{(2)} \nabla_\theta \hat{L}_{\mathcal{D}^{\mathrm{tr}}}(\theta^{(1)})$$

$$\vdots$$

$$\theta^{\mathrm{tr}}(\mathcal{D}^{\mathrm{tr}}|\xi) = \theta^{(I+1)} = \theta^{(I)} - \gamma^{(I)} \nabla_\theta \hat{L}_{\mathcal{D}^{\mathrm{tr}}}(\theta^{(I-1)}), \tag{13.36}$$

where the initialization is given by the hyperparameter vector, i.e., $\theta^{(1)} = \xi$ and we have allowed for varying learning rates $\gamma^{(i)}$ across the iterations $i = 1, \ldots, I$.

In order to apply SGD-based meta-learning for the outer problem (13.30), we need the gradient $\nabla_\xi L_{\mathcal{D}_k^{\mathrm{te}}}(\theta^{\mathrm{tr}}(\mathcal{D}_k^{\mathrm{tr}}|\xi))$ for all meta-training tasks $k = 1, 2, \ldots, K$. Using the chain rule of differentiation, this is given as

$$\nabla_\xi L_{\mathcal{D}_k^{\mathrm{te}}}(\theta^{\mathrm{tr}}(\mathcal{D}_k^{\mathrm{tr}}|\xi)) = \nabla_\xi \theta^{(2)} \cdots \nabla_{\theta^{(I-1)}} \theta^{(I)} \cdot \nabla_{\theta^{(I)}} \theta^{(I+1)} \cdot \nabla_\theta L_{\mathcal{D}_k^{\mathrm{te}}}(\theta^{(I+1)}), \tag{13.37}$$

where the ith Jacobian is given as

$$\nabla_{\theta^{(i)}} \theta^{(i+1)} = I_M - \gamma^{(i+1)} \nabla_\theta^2 \hat{L}_{\mathcal{D}^{\mathrm{tr}}}(\theta^{(i)}) \tag{13.38}$$

for $i = 1, 2, \ldots, I$, with $\theta^{(1)} = \xi$. Note that with $I = 1$, one recovers the gradient (13.35). By (13.37), this generalization to multiple SGD iterations per task requires I successive Hessian matrix–vector multiplications.

Simplified Versions of MAML. Various less principled simplifications of MAML have been proposed to reduce complexity. As an example, a popular approach, referred to as **first-order MAML**, approximates the MAML gradient (13.35) by the gradient of the test loss evaluated at the adapted parameters, as in

$$\nabla_\xi L_{\mathcal{D}_k^{\mathrm{te}}}(\theta^{\mathrm{tr}}(\mathcal{D}_k^{\mathrm{tr}}|\xi)) \simeq \nabla_\theta L_{\mathcal{D}_k^{\mathrm{te}}}(\theta_k). \tag{13.39}$$

This amounts to discarding the Jacobians in the MAML gradient (13.37).

Example 13.9

Consider a learning setting involving multiple regression tasks. In each task, the input x is sampled from a distribution $p(x)$ that is uniform in the interval $[-5, 5]$. The tasks are distinguished by the respective conditional distributions $p(t|x) = \delta(t - t^*(x))$ of the target variable, which is concentrated at the value $t^*(x) = a\sin(2\pi x + b)$ for some constants a and b. For each task, parameter a is drawn uniformly in the interval $[-5, 5]$, while the phase is uniform in the interval $[0, \pi]$. As an example, Fig. 13.17 shows the ground-truth sinusoidal relationship $t^*(x)$ between input x and output t for a given meta-test task as a dashed gray line. The ℓ_2 loss is adopted, and the model class contains hard predictors $\hat{t}(x|\theta)$ implemented as three-layer neural networks ($L = 3$) with 40 neurons in each hidden layer, ReLU activation functions, and a linear last layer. For all schemes, we use per-task SGD updates with constant learning rate $\gamma = 0.01$, while the outer optimization in meta-learning uses a learning rate 0.001 with the Adam optimizer (see Appendix 5.D).

Figure 13.17 Hard predictors produced by conventional learning, joint learning, and MAML for a regression problem. The dashed gray lines represent the ground-truth sinusoidal relationship between x and t for the meta-test task at hand; the dashed black lines are initializations for the hard predictor; the solid gray line is the adapted hard predictor after 1 per-task SGD step; and the solid black line is the adapted hard predictor after 10 per-task SGD steps.

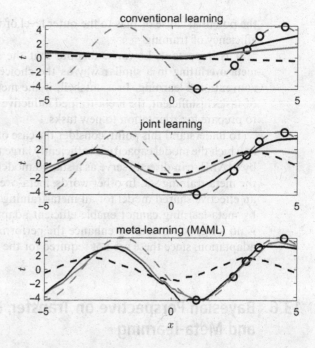

For each task, we have access to only $N = 5$ training samples, which are shown in Fig. 13.17 as circles for the given meta-test task. As seen in Fig. 13.17 (top), the hard predictor $\hat{t}(x|\theta)$ trained via conventional learning, starting from the initialization represented as a dashed black line, fails to adapt after 1 SGD step (solid gray line) or 10 SGD steps (solid black line). The available data are insufficient for conventional learning to perform well.

For joint learning and meta-learning, $K = 25$ tasks are generated at random at each outer iteration. Joint learning obtains the predictor represented as the dashed black line in the central panel of Fig. 13.17. From this predictor, fine-tuning using the $N = 5$ available data points improves the predictor after 1 and 10 SGD steps (solid gray and black lines, respectively), but the latter is still quite far from the ground-truth.

For MAML, the per-task optimization carried out during meta-learning assumes a single SGD step. As illustrated by the bottom panel of Fig. 13.17, MAML obtains an initialization (dashed black line), from which adaptation using the few data points can be effectively accomplished even after a single SGD step (solid gray line).

13.5.7 Meta-inductive Bias

While the inductive bias underlying the training algorithm used in the inner loop is optimized by means of meta-learning, the meta-learning process itself assumes a **meta-inductive bias**. The meta-inductive bias encompasses the choices of the hyperparameters to optimize in the outer loop – e.g., the initialization of an SGD training algorithm – as well as the optimization algorithm used in the outer loop. There is of course no end to this nesting of inductive biases: Any new learning level brings its own assumptions and biases. Meta-learning moves the potential cause of bias to the outer level of the meta-learning loop, which may improve the efficiency of training.

It is important, however, to note that the selection of a meta-inductive bias may cause **meta-overfitting** in a similar way as the choice of an inductive bias can cause overfitting in conventional learning. In a nutshell, if the meta-inductive bias is too broad and the number of tasks insufficient, the meta-trained inductive bias may overfit the meta-training data and fail to prepare for adaptation to new tasks.

To understand this point, consider the case of MAML. Meta-overfitting refers to a situation in which the model capacity is sufficiently large to guarantee that the initialization meta-trained by MAML can directly serve as a shared model parameter, with no adaptation, for all tasks in the meta-training set. In other words, meta-overfitting occurs when joint training can produce an effective shared model for all meta-training tasks. In this case, the inductive bias produced by meta-learning cannot enable efficient adaptation for new tasks. In fact, the initialization is no longer optimized to enhance the performance of the trained model after a few steps of adaptation, since these are not required for the meta-training tasks.

13.6 Bayesian Perspective on Transfer, Multi-task, Continual, and Meta-learning

Transfer, multi-task, continual, and meta-learning – for short, TMCM learning methods – involve data from multiple tasks, and differ in the way these data sets are presented to the learner and in the goals of the learning process. In this section, we will discuss how all TMCM methods can be described within the formalism of Bayesian learning with local rvs introduced in Sec. 12.11. This is done by viewing the per-task model parameters θ_k as local latent rvs, and by interpreting shared model parameters or hyperparameters as global rvs, denoted here as vector ϕ. For example, the global parameter vector ϕ can be used to represent the shared parameter vector θ_0 in multi-task learning (see Fig. 13.6), or the hyperparameter vector ξ in meta-learning (see Fig. 13.15). A mapping between local and global rvs, on the one hand, and TMCM quantities, on the other, can be found in Table 13.1.

General Formulation. To define a general formulation that applies, *mutatis mutandis*, to all cases, consider the presence of K tasks, with each task k being associated with training data $\mathcal{D}_k^{\mathrm{tr}}$. We focus on supervised learning, and write the training data sets as $\mathcal{D}_k^{\mathrm{tr}} = \{(x_{k,n}, t_{k,n})_{n=1}^{N_k}\}$. Following the hierarchical Bayesian model introduced in Sec. 12.11 and using the correspondence in

Table 13.1 Correspondence between Bayesian learning with local latent variables (Sec. 12.11) and Bayesian formulations of transfer, multi-task, continual, and meta-learning

Bayesian learning with local latent variables	TMCM Bayesian learning
observation x_n	per-task data $\mathcal{D}_k^{\mathrm{tr}}$ (and $\mathcal{D}_k^{\mathrm{te}}$ for meta-learning)
local latent rv z_n	per-task parameters θ_k
global latent rv θ	shared (hyper)parameters ϕ

Table 13.1, we consider the joint distribution of global parameters ϕ, per-task model parameters $\{\theta_k\}_{k=1}^K$, and training data sets $\{\mathcal{D}_k^{\text{tr}}\}_{k=1}^K$ as

$$p(\phi, \{\theta_k\}_{k=1}^K, \{\mathcal{D}_k^{\text{tr}}\}_{k=1}^K) = p(\phi) \prod_{k=1}^K p(\theta_k, \mathcal{D}_k^{\text{tr}} | \phi), \tag{13.40}$$

where $p(\phi)$ is the prior distribution of the global parameter vector, and the per-task joint distribution factorizes as

$$p(\theta_k, \mathcal{D}_k^{\text{tr}} | \phi) = p(\theta_k | \phi) p(\mathcal{D}_k^{\text{tr}} | \theta_k, \phi). \tag{13.41}$$

In (13.41), we have assumed that the global parameter vector ϕ generally affects both the per-task prior distribution $p(\theta_k | \phi)$ and the per-task likelihood $p(\mathcal{D}_k^{\text{tr}} | \theta_k, \phi)$. The latter, in turn, factorizes across training points as

$$p(\mathcal{D}_k^{\text{tr}} | \theta_k, \phi) = \prod_{n=1}^{N_k} p(t_{k,n} | x_{k,n}, \theta_k, \phi). \tag{13.42}$$

13.6.1 Bayesian Multi-task Learning

The most direct application of the outlined hierarchical Bayesian model (13.40) is to multi-task learning. In multi-task learning, following the notation in Sec. 13.3, the global parameter vector ϕ corresponds to the shared model parameter vector θ_0 and the local rv $\{\theta_k\}_{k=1}^K$ are the per-task model parameter vectors. (As compared to the notation used in Sec. 13.3, in (13.40) we have added the superscript "tr" for consistency with the notation adopted for meta-learning in the preceding section.) As discussed in Sec. 12.11, the goal is to infer the posterior distribution $p(\theta_0, \{\theta_k\}_{k=1}^K | \{\mathcal{D}_k^{\text{tr}}\}_{k=1}^K)$ of all model parameters, along with the corresponding marginals $p(\theta_0, \theta_k | \{\mathcal{D}_{k'}^{\text{tr}}\}_{k'=1}^K)$ for all tasks $k = 1, \ldots, K$. These posteriors can then be used to obtain the ensemble predictive distributions

$$p(t_k | x_k, \{\mathcal{D}_{k'}^{\text{tr}}\}_{k'=1}^K) = \mathrm{E}_{\theta_0, \theta_k \sim p(\theta_0, \theta_k | \{\mathcal{D}_{k'}^{\text{tr}}\}_{k'=1}^K)}[p(t_k | x_k, \theta_0, \theta_k)] \tag{13.43}$$

for all tasks $k = 1, \ldots, K$. In practice, calculation of the posteriors and the predictive distributions is often of prohibitive complexity, and one must resort to approximate inference Bayesian methods via VI or MC sampling as discussed in Sec. 12.11.

13.6.2 Bayesian Transfer Learning

Bayesian transfer learning can be formulated within a Bayesian framework in a similar way as multi-task learning by considering $K = 2$ tasks. The specific formulation depends on whether one assumes the presence of labeled or unlabeled data sets from the target domain, and details can be derived as discussed in Recommended Resources, Sec. 13.8.

13.6.3 Bayesian Continual Learning: Exact Solutions

As seen in Sec. 12.4, Bayesian learning provides a principled way to update the posterior of the model parameters as more data are processed in an online fashion. In this subsection and in the next, we apply the arguments of Sec. 12.4 at the level of per-task data sets instead of data points (see Table 13.1) to develop Bayesian continual learning. We start by considering in this subsection the ideal case in which posteriors can be computed exactly, while the more general case is covered in the next subsection.

Updating the Exact Posterior Distribution. Following the formulation of continual learning described in Sec. 13.4, the global parameters ϕ in the joint distribution (13.40) correspond to the model parameters θ, and no task-specific parameters are instantiated. Therefore, the goal of Bayesian learning is to update the posterior distribution of the model parameters θ. To this end, assume that, after observation of the training set $\mathcal{D}_1^{\text{tr}}, \ldots, \mathcal{D}_{k-1}^{\text{tr}}$ and computation of the exact posterior

$$p(\theta | \{\mathcal{D}_{k'}^{\text{tr}}\}_{k'=1}^{k-1}) \propto p(\theta) \prod_{k'=1}^{k-1} p(\mathcal{D}_{k'}^{\text{tr}} | \theta), \tag{13.44}$$

data set $\mathcal{D}_k^{\text{tr}}$ for the kth task is collected. The updated posterior including the new data set $\mathcal{D}_k^{\text{tr}}$ is then given as

$$p(\theta | \{\mathcal{D}_{k'}^{\text{tr}}\}_{k'=1}^{k}) \propto p(\theta) \left(\prod_{k'=1}^{k-1} p(\mathcal{D}_{k'}^{\text{tr}} | \theta) \right) p(\mathcal{D}_k^{\text{tr}} | \theta)$$

$$\propto p(\theta | \{\mathcal{D}_{k'}^{\text{tr}}\}_{k'=1}^{k-1}) p(\mathcal{D}_k^{\text{tr}} | \theta). \tag{13.45}$$

Therefore, we can update the posterior $p(\theta | \{\mathcal{D}_{k'}^{\text{tr}}\}_{k'=1}^{k-1})$ by multiplying it by the likelihood $p(\mathcal{D}_k^{\text{tr}} | \theta)$ of the new data set and normalizing. This update procedure can be repeated for successive tasks in a continual fashion.

Accordingly, all the necessary information from the observation of previous tasks is summarized in the posterior $p(\theta | \{\mathcal{D}_{k'}^{\text{tr}}\}_{k'=1}^{k-1})$. This implies that there is no need to store all previous per-task data sets $\{\mathcal{D}_{k'}^{\text{tr}}\}_{k'=1}^{k-1}$, since recalling only the current posterior $p(\theta | \{\mathcal{D}_{k'}^{\text{tr}}\}_{k'=1}^{k-1})$ is sufficient to evaluate the exact next posterior given the overall data set $\{\mathcal{D}_{k'}^{\text{tr}}\}_{k'=1}^{k}$. Note also that the exact posterior $p(\theta | \{\mathcal{D}_{k'}^{\text{tr}}\}_{k'=1}^{k}) \propto p(\theta) \prod_{k'=1}^{k} p(\mathcal{D}_k^{\text{tr}} | \theta)$ does not depend on the order in which the tasks are presented to the learner.

Updating the Exact Posterior Distribution as Free Energy Minimization. The exact posterior (13.45) can be viewed as the solution of two conceptually different problems. We will see in the next subsection that these two formulations yield likelihood-based and prior-based continual learning methods when one applies approximate Bayesian inference.

1. **Bayesian continual learning by minimization of the global free energy.** In the first formulation, the posterior $p(\theta | \{\mathcal{D}_{k'}^{\text{tr}}\}_{k'=1}^{k})$ is obtained as the minimum of the **global free energy**

$$\text{F}\left(q(\theta) \,\middle\|\, p(\theta) \prod_{k'=1}^{k} p(\mathcal{D}_{k'}^{\text{tr}} | \theta) \right) = \text{E}_{\theta \sim q(\theta)} \left[\sum_{k'=1}^{k} (-\log p(\mathcal{D}_{k'}^{\text{tr}} | \theta)) \right] + \text{KL}(q(\theta) \| p(\theta)), \tag{13.46}$$

which sums the likelihoods of all previous tasks. That the solution of this problem yields the posterior (13.45) follows directly from (3.102).

2. **Bayesian continual learning by minimization of the local free energy.** In contrast, the second formulation views the posterior (13.45) as the minimum of the **current local free energy**

$$F\Big(q(\theta) \,\Big\|\, p(\theta|\{\mathcal{D}_{k'}^{\text{tr}}\}_{k'=1}^{k-1})p(\mathcal{D}_k^{\text{tr}}|\theta)\Big) = \mathrm{E}_{\theta \sim q(\theta)}\big[-\log p(\mathcal{D}_k^{\text{tr}}|\theta)\big] + \mathrm{KL}(q(\theta)\|p(\theta|\{\mathcal{D}_{k'}^{\text{tr}}\}_{k'=1}^{k-1})), \tag{13.47}$$

in which the role of the prior is played by the current posterior $p(\theta|\{\mathcal{D}_{k'}^{\text{tr}}\}_{k'=1}^{k-1})$. The fact that the posterior (13.45) is the minimizer follows again from (3.102). Unlike the previous formulation, this second criterion accounts for the likelihood of only the current task. Therefore, in line with the discussion in this subsection, this second formulation shows that, if one can compute exact posteriors, previous data sets $\{\mathcal{D}_{k'}^{\text{tr}}\}_{k'=1}^{k-1}$ can be discarded and one only needs to maintain the posterior $p(\theta|\{\mathcal{D}_{k'}^{\text{tr}}\}_{k'=1}^{k-1})$.

13.6.4 Bayesian Continual Learning: Approximate Solutions

The discussion in the preceding subsection assumes that one can compute the exact posterior $p(\theta|\{\mathcal{D}_{k'}^{\text{tr}}\}_{k'=1}^{k})$ in (13.45), which is, as we know, often impractical. In this subsection, we discuss more efficient and scalable solutions based on approximate Bayesian inference.

We start by observing that, when solved exactly, the global formulation (13.46) and the local formulation (13.47) yield the same solution, namely the exact posterior (13.45). However, this equivalence no longer holds when approximate solutions are used. In particular, following the VI principle, the first formulation yields likelihood-based Bayesian learning methods, while the second leads to prior-based methods. These are the counterparts, respectively, of the coreset-based and regularization-based techniques studied in Sec. 13.4, and they are the focus of this subsection.

To develop VI-based techniques, we introduce a class $\mathcal{Q}$ of variational posteriors $q(\theta|\{\mathcal{D}_{k'}^{\text{tr}}\}_{k'=1}^{k})$, which may be parametric or particle-based, to approximate the true posterior $p(\theta|\{\mathcal{D}_{k'}^{\text{tr}}\}_{k'=1}^{k})$. To simplify the notation, we will write $q_k(\theta)$ for $q(\theta|\{\mathcal{D}_{k'}^{\text{tr}}\}_{k'=1}^{k})$ in this subsection. As a typical example, we may choose the Gaussian class of posteriors $q_k(\theta) = \mathcal{N}(\theta|\mu_k, \Sigma_k)$ with variational parameters (μ_k, Σ_k).

Likelihood-Based Continual Learning. The global free energy in (13.46) can be approximated by creating a (possibly empty) coreset $\mathcal{C}_{k'}^{\text{tr}} \subseteq \mathcal{D}_{k'}^{\text{tr}}$ of samples from each prior task $k' = 1, \ldots, k-1$, yielding the learning criterion

$$\min_{q_k(\theta) \in \mathcal{Q}} \left\{ \mathrm{E}_{\theta \sim q_k(\theta)} \left[-\log p(\mathcal{D}_k^{\text{tr}}|\theta) - \sum_{k'=1}^{k-1} \log p(\mathcal{C}_{k'}^{\text{tr}}|\theta) \right] + \mathrm{KL}(q_k(\theta)\|p(\theta)) \right\}. \tag{13.48}$$

Note that the first, loss-related, term in the objective (13.48) is equivalent to the cost function (13.19) assumed by the coreset-based approach. The optimization (13.48) can be addressed by using the VI techniques discussed in Chapter 10.

Prior-Based Continual Learning. When VI is applied, the local formulation (13.47) yields the class of prior-based strategies known as **variational continual learning** (VCL). At each step k, given the current approximation $q_{k-1}(\theta) \in \mathcal{Q}$, VCL tackles the problem

$$\min_{q_k(\theta) \in \mathcal{Q}} \left\{ \mathrm{E}_{\theta \sim q_k(\theta)} \big[-\log p(\mathcal{D}_k^{\text{tr}}|\theta) \big] + \mathrm{KL}(q_k(\theta)\|q_{k-1}(\theta)) \right\}, \tag{13.49}$$

in which the posterior $p(\theta|\{\mathcal{D}_k^{\text{tr}}\}_{k'=1}^{k-1})$ in the original formulation (13.47) has been replaced by the variational distribution $q_{k-1}(\theta)$. Accordingly, VCL introduces the information-theoretic regularization term $\text{KL}(q_k(\theta)||q_{k-1}(\theta))$, which penalizes deviations of the updated variational posterior $q_k(\theta)$ from the current variational posterior $q_{k-1}(\theta)$.

For an example of VCL, consider the class $\mathcal{Q}$ of variational distributions of the form $q(\theta) = \mathcal{N}(\theta|\mu,\sigma^2 I)$ with a fixed variance σ^2 and variational parameter μ. Writing $q_{k-1}(\theta) = \mathcal{N}(\theta|\mu_{k-1},\sigma^2 I)$, the optimization (13.49) of the variational parameter vector μ_k can be formulated as

$$\min_{\mu_k} \left\{ \mathrm{E}_{\theta\sim\mathcal{N}(\theta|\mu_k,\sigma^2 I)} \left[-\log p(\mathcal{D}_k^{\text{tr}}|\theta) \right] + \frac{1}{2}\frac{||\mu-\mu_{k-1}||^2}{\sigma^2} \right\}. \tag{13.50}$$

This minimization is reminiscent of the problem (13.20) tackled by EWC with $K = 2$, with the caveats that the training loss $L_{\mathcal{D}_k}(\theta)$ in EWC is replaced by the average training log-loss and that the regularization term does not include the relevance weight vector applied by EWC.

When substituting the free energy in (13.49) with other divergence measures (see Chapter 11), one obtains the more general class of **assumed density filtering (ADF)** strategies. As an example, replacing the standard I-projection carried out by VCL with an M-projection (see Sec. 11.2) yields a form of **expectation propagation (EP)**. See Recommended Resources, Sec. 13.8, for details.

13.6.5 Bayesian Meta-learning

The Bayesian formulation of meta-learning includes not only the training sets $\{\mathcal{D}_k^{\text{tr}}\}_{k=1}^K$ but also the test sets $\{\mathcal{D}_k^{\text{te}}\}_{k=1}^K$ for the K tasks comprising the meta-training data. Furthermore, the hyperparameter vector ξ plays the role of the global parameters ϕ, and the rvs $\{\theta_k\}_{k=1}^K$ represent the per-task model parameters. Overall, from (13.40), the joint distribution of hyperparameters, per-task parameters, and meta-training data can be expressed as

$$p(\xi,\{\theta_k\}_{k=1}^K,\{\mathcal{D}_k^{\text{tr}}\}_{k=1}^K) = p(\xi)\prod_{k=1}^K p(\theta_k|\xi)p(\mathcal{D}_k^{\text{tr}}|\theta_k,\xi)p(\mathcal{D}_k^{\text{te}}|\theta_k,\xi). \tag{13.51}$$

Accordingly, the hyperparameter vector ξ is assumed to generally affect both model parameters' prior $p(\theta|\xi)$ and likelihood function $p(t|x,\theta,\xi)$, with the latter in turn determining the conditional probabilities $p(\mathcal{D}_k^{\text{tr}}|\theta_k,\xi)$ and $p(\mathcal{D}_k^{\text{te}}|\theta_k,\xi)$.

If we were to model the hyperparameter as deterministic, the optimization of the resulting log-loss given the training data would amount to an **empirical Bayes** problem (see Sec. 12.9). In contrast, under the joint distribution (13.51), solutions aiming at obtaining the joint or marginal posteriors of hyperparameter and model parameter vectors can be devised based on VI or MC sampling methods by following the discussion in Sec. 12.11. See Recommended Resources, Sec. 13.8, for details.

13.7 Summary

- This chapter has studied learning settings that go beyond the standard learning problem by accounting for multiple learning tasks at once. A common thread among all the problems reviewed in this chapter is the goal of transferring useful knowledge from one set of tasks to another.

- In transfer learning, this goal is most explicit, since data from the source task are used only to improve the generalization performance for the target task. Differences between source and target tasks may account for domain and/or concept shifts, and likelihood ratio-based solutions can compensate for these shifts if data are available from both domains.
- In multi-task learning, tasks share part of a model. Transfer of information across tasks is positive if the resulting, partially shared, model is more effective on each task than one that is trained based solely on the available task's data. A standard approach is to train a shared feature extractor across different tasks.
- In continual learning, transfer occurs among multiple tasks that are observed sequentially over time. The update of a model for the current task may positively or negatively affect the performance for previously observed tasks, as well as for tasks that are yet to be encountered. Continual learning techniques manage memory and training objective in different ways by encoding information about previous tasks via coreset-based or regularization-based criteria.
- In meta-learning, transfer occurs between a set of observed tasks and a task that is a priori unknown and possibly yet to be encountered. This is done by sharing hyperparameters across tasks, while allowing for separate adaptation of models for each new task.
- Bayesian models with latent variables provide a unifying framework to study transfer learning, multi-task learning, continual learning, and meta-learning.

13.8 Recommended Resources

Transfer and multi-task learning are covered in the book [1]. Invariant risk minimization is described in [2]. Useful review papers on continual learning include [3]. For meta-learning, the reader is referred to the edited book [4] and to the more recent landmark paper [5] and review paper [6]. Bayesian models are studied in [7, 8].

Appendices

Appendix 13.A: Computing the Product of the Hessian and a Vector

In order to compute the MAML gradient (13.35), one needs to evaluate the product of the Hessian matrix of a loss function and a vector. This appendix describes an efficient numerical approach that obtains a finite-difference approximation for this product. The complexity of the finite-difference method is double that of computing a gradient vector for the same loss function.

Given a loss function $L(\theta)$ that is doubly continuously differentiable over a local neighborhood of the vector θ of interest, the finite-difference method approximates the Hessian–vector product Hg, where $H = \nabla_\theta^2 L(\theta)$ is the Hessian matrix and g is any vector. Specifically, the Hessian–vector product Hg is approximately computed as

$$Hg \simeq \frac{1}{\alpha}(\nabla_\theta L(\theta + \alpha g) - \nabla_\theta L(\theta)), \tag{13.52}$$

where $\alpha > 0$ is a sufficiently small constant value. A particular choice that has been proposed for α is

$$\alpha = \frac{2\sqrt{\epsilon}(1 + \|\theta\|)}{\|g\|}, \tag{13.53}$$

where $\epsilon = 1.19 \cdot 10^{-7}$ is an upper bound on the relative error due to rounding in single-precision floating-point arithmetic.

Bibliography

[1] Q. Yang, Y. Zhang, W. Dai, and S. J. Pan, *Transfer Learning*. Cambridge University Press, 2020.

[2] M. Arjovsky, L. Bottou, I. Gulrajani, and D. Lopez-Paz, "Invariant risk minimization," arXiv preprint arXiv:1907.02893, 2019.

[3] G. I. Parisi, R. Kemker, J. L. Part, C. Kanan, and S. Wermter, "Continual lifelong learning with neural networks: A review," *Neural Networks*, vol. 113, pp. 54–71, 2019.

[4] S. Thrun and L. Pratt, *Learning to Learn*. Springer Science+Business Media, 2012.

[5] C. Finn, P. Abbeel, and S. Levine, "Model-agnostic meta-learning for fast adaptation of deep networks," in *The 34th International Conference on Machine Learning*. ICML, 2017, pp. 1126–1135.

[6] T. Chen, X. Chen, W. Chen, H. Heaton, J. Liu, Z. Wang, and W. Yin, "Learning to optimize: A primer and a benchmark," arXiv preprint arXiv:2103.12828, 2021.

[7] S. Farquhar and Y. Gal, "A unifying Bayesian view of continual learning," arXiv preprint arXiv:1902.06494, 2019.

[8] J. Yoon, T. Kim, O. Dia, S. Kim, Y. Bengio, and S. Ahn, "Bayesian model-agnostic meta-learning," in *NIPS'18: Proceedings of the 32nd International Conference on Neural Information Processing Systems. Curran Associates*, 2018, pp. 7343–7353.

14 Federated Learning

14.1 Overview

So far, this book has focused on conventional *centralized* learning settings in which data are collected at a central server, which carries out training. When data originate at distributed agents, such as personal devices, organizations, or factories run by different companies, this approach has two clear drawbacks:

- First, it requires transferring data from the agents to the server, which may incur a prohibitive **communication load**.
- Second, in the process of transferring, storing, and processing the agents' data, **sensitive information** may be exposed or exploited.

Federated learning (FL) refers to distributed learning protocols involving agents that collaborate towards a *common* training task *without the direct exchange of data*. By transferring limited information about the local data sets, FL can reduce the communication load, as well as potentially decreasing the leakage of information between agents, while still enabling collaborative training across agents.

Why would agents risk disclosing information about private data to collaborate on a training task? Consider the example of a hospital interested in designing a medical diagnostic tool that maps images to diagnoses of some condition. Each hospital may have a relatively small data set, which covers only certain types of patients. Through FL, the hospital can potentially benefit from data and experience accrued at other hospitals in order to obtain a more accurate diagnostic tool, without having to acquire larger data sets. At the same time, the participating hospitals are mindful of the need to keep patients' information private. Therefore, the benefits of collaborative learning via FL must be weighed against any potential leakage of information, which should be kept to a minimum.

Given its broad scope, FL can apply at various scales, ranging from silos of different enterprises or organizations, such as hospitals, to mobile devices owned by different individuals. As such, it has a large number of applications, running the gamut from industrial process optimization to personal healthcare.

Learning Objectives and Organization of the Chapter. By the end of this chapter, the reader should be able to:

- understand and implement one-shot FL and federated SGD (FedSGD), as well as local SGD schemes, including federated averaging (FedAvg), that bridge the gap between one-shot FL and FedSGD (Sec. 14.2);
- understand how to quantify and ensure the privacy of FL protocols by means of differential privacy (DP)-based metrics and mechanisms (Sec. 14.3); and
- understand and design Bayesian FL protocols (Sec. 14.4).

14.2 Frequentist Federated Learning

In this section, we introduce FL within a frequentist framework for the standard **parameter server setting** illustrated in Fig. 14.1, in which a number of agents communicate through a **central server** (or server for short). To this end, we first describe the parameter server setting, and present the reference approaches of single-agent learning and global learning. Then, we discuss one-shot FL and FedSGD algorithms, and, finally, we introduce local SGD solutions, with a focus on FedAvg. Convergence and limitations of FedAvg are also discussed. Throughout this section, we focus on the problem of supervised learning of hard predictors, although extensions to soft predictors (see Sec. 14.4) and unsupervised learning are possible.

14.2.1 Parameter Server Setting

As illustrated in Fig. 14.1, the parameter server setting assumes the presence of K agents and of a server, which is connected to all agents. All communications between agents occur through the server. Each agent $k = 1, \ldots, K$, stores its own training data set $\mathcal{D}_k = \{(x_{k,n}, t_{k,n})\}_{n=1}^{N_k}$ of N_k labeled examples that are assumed to be generated i.i.d. from a population distribution $p_k(x, t)$. We fix, as part of the inductive bias, a model class $\mathcal{H}$ of parametric hard predictor $\hat{t}(x|\theta)$ with model parameter vector θ. All agents share the inductive bias, and aim to collaboratively optimize the model parameter vector θ. Variants of the setting involving FL or soft predictors can be similarly considered (see Sec. 14.4).

Importantly, the population distributions $p_k(x, t)$ are generally different across the agents indexed by $k = 1, \ldots, K$. For instance, distinct hospitals may "draw" their respective patients from different (literal) populations of individuals with distinct characteristics. As another example, the smart watches of distinct individuals may collect data that are distributed according to the unique properties of the wearer. We can concisely describe the FL setting as being **locally i.i.d.**, as the data are generated i.i.d. from the same population distribution locally at each agent; but **globally non-i.i.d.**, since the population distributions across different agents are generally distinct.

14.2.2 Single-agent Learning

As a benchmark, let us first consider the standard **single-agent learning** approach in which agents do not communicate, and each agent k carries out training based solely on its local data set $\mathcal{D}_k$.

Figure 14.1 Illustration of the parameter server setting assumed by FL.

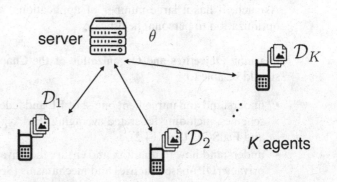

To this end, agent k can compute its **local training loss** $L_{\mathcal{D}_k}(\theta)$ based on its data set $\mathcal{D}_k$, and tackle the **local ERM** problem (or a regularized version thereof)

$$\min_{\theta} \left\{ L_{\mathcal{D}_k}(\theta) = \frac{1}{N_k} \sum_{n=1}^{N_k} \ell_k(t_{k,n}, \hat{t}(x_{k,n}|\theta)) \right\}. \tag{14.1}$$

Single-agent learning can yield desirable performance if the agent has a sufficiently large data set. FL is of interest when this is not the case and the data available at each agent k are inadequate to meet the application's requirements.

14.2.3 Global Learning

Now consider the other benchmark extreme situation in which the **global training data set** $\mathcal{D} = \cup_{k=1}^{K} \mathcal{D}_k$, which pools together all the local data sets $\mathcal{D}_k$, is available at the server. Note that the global data set $\mathcal{D}$ contains $N = \sum_{k=1}^{K} N_k$ data points. In this case, the server could directly tackle the **global ERM problem** (or a regularized version thereof) of minimizing the **global training loss** across all agents, i.e.,

$$\min_{\theta} \left\{ L_{\mathcal{D}}(\theta) = \frac{1}{N} \sum_{k=1}^{K} N_k L_{\mathcal{D}_k}(\theta) \right\}. \tag{14.2}$$

Note that the multiplication of the local training loss $L_{\mathcal{D}_k}(\theta)$ by N_k is necessary for (14.2) to weigh every sample in the global training set equally. As we discuss next, FL aims to match the performance of this centralized training strategy *without requiring that the local data leave the respective agents*.

Before we discuss how this can be done, it is useful to ask: Are we guaranteed that a global model trained via the centralized strategy (14.2) would outperform, in terms of local population losses, the local solutions obtained via (14.1)? In short, the answer is no. Global training, and hence also FL, can only be advantageous if **positive transfer** of information is possible across the data sets and population distributions of different users. As discussed in the preceding chapter, positive transfer implies that one can extract useful information from the data of one agent about the learning task of interest for another. Whether positive transfer is possible or not generally depends on the amount of data available at each agent, as well as on the level of *global "non-i.i.d.-ness"* across the agents.

14.2.4 One-shot, or "Embarrassingly Parallel", FL

FL aspires to match the performance of the centralized learner (14.2) by means of local computations at the agents and limited inter-agent communications. In its simplest embodiment, FL is implemented using a **one-shot** protocol, which is an example of **"embarrassingly parallel"** decentralized computing. Accordingly, each agent first carries out separate training by tackling the local ERM problem (14.1). Then, the resulting locally trained model parameter vector θ_k at each agent k is transmitted to the server, which combines all **local models** θ_k, with $k = 1, \ldots, K$, to produce a **global model** θ. A typical way to do this (but not the only one) is by computing the average

$$\theta = \frac{1}{K} \sum_{k=1}^{K} \theta_k. \tag{14.3}$$

The global model parameter vector (14.3) is sent back to the agents, who can use it to carry out prediction by using the globally trained model $\hat{t}(x|\theta)$.

One-shot FL has the advantage of requiring a *single communication round*, but this comes at the cost of *limited cooperation* among agents. For instance, it can be proved that only under restrictive conditions does the combined model (14.3) improve the local training loss of at least some agents as compared to single-agent learning. (This is the case, for instance, if the loss functions are convex in θ.) In general, the benefits accrued by cooperative training via one-shot FL may be modest.

14.2.5 FedSGD

While one-shot FL requires a single communication round, **FedSGD** goes to the other extreme of requiring one communication round for each update of the model parameter vector. To describe it, let us fix some initialization $\theta^{(1)}$ of the **global iterate** of the model parameter vector, which is maintained by the server and updated across the **global iteration** index $i = 1, 2, \ldots$. At each global iteration i, each agent k downloads the current iterate $\theta^{(i)}$ and computes a stochastic estimate of its local gradient $\nabla_\theta L_{\mathcal{D}_k}(\theta^{(i)})$ (see Chapter 5). The server collects all the stochastic gradients to obtain a stochastic estimate of the global gradient $\nabla_\theta L_{\mathcal{D}}(\theta^{(i)})$, and uses this global stochastic gradient to update the model parameter and obtain the next iterate $\theta^{(i+1)}$.

FedSGD requires one communication round for each SGD update. While this comes at a potentially high price in terms of *communication overhead*, FedSGD can be interpreted as implementing *centralized* SGD optimization for the global ERM problem (14.2). In fact, the global stochastic gradient obtained by FedSGD corresponds to the stochastic gradient that a centralized implementation of SGD would produce for problem (14.2). (Technically, this is true if one assumes that the mini-batch is formed in centralized SGD by picking samples from all agents in the same way as FedSGD.)

The upshot of this discussion is that FedSGD is essentially equivalent to centralized SGD for the global ERM problem (14.2), but this desirable property comes at the cost of a large communication load, with one communication round per SGD update.

14.2.6 Local SGD-Based FL

Local SGD-based FL refers to a broad class of FL protocols that bridge the gap between one-shot FL and FedSGD, trading communication load for adherence to the global ERM problem (14.2). The idea is to generalize FedSGD by carrying out *multiple local SGD steps* before communicating and aggregating information from the agents at the server. As a result, local SGD-based FL protocols perform multiple local SGD steps for each communication round.

More precisely, as detailed in Algorithm 14.1, at each global iteration i, a subset of agents participates in the global update of the model parameter. Each active agent downloads the current model $\theta^{(i)}$ from the server; obtains an updated model parameter $\theta_k^{(i+1)}$ through several local SGD updates based on the local data set $\mathcal{D}_k$; and forwards $\theta_k^{(i+1)}$ to the server. Having collected the updated models from all the agents that are active at iteration i, the server combines the model parameters to obtain the next global iterate $\theta^{(i+1)}$. Note that each global iteration, encompassing multiple local SGD steps, requires one communication round.

The general template established by Algorithm 14.1 can recover one-shot FL and FedSGD as special cases when all agents are active at each iteration. Specifically, if each agent carries out

---

Algorithm 14.1: General template for local SGD-based FL protocols

---

initialize $\theta^{(1)}$ and $i = 1$

while *stopping criterion not satisfied* **do**

    a subset $\mathcal{K}^{(i)}$ of $K^{(i)} \leq K$ agents is active

    each active agent $k \in \mathcal{K}^{(i)}$ downloads the current model $\theta^{(i)}$ from the server

    each active agent $k \in \mathcal{K}^{(i)}$ carries out a number of local SGD steps starting from the
       initial value $\theta^{(i)}$ using its local data set $\mathcal{D}_k$, and returns the updated model
       parameters $\theta_k^{(i+1)}$ to the server

    the server combines the received updated model parameters $\{\theta_k^{(i+1)}\}_{k \in \mathcal{K}^{(i)}}$ and
       (possibly) the previous model parameter $\theta^{(i)}$ to obtain the next iterate $\theta^{(i+1)}$

    set $i \leftarrow i + 1$

end

return $\theta^{(i)}$

---

a single SGD iteration at each global iteration i, Algorithm 14.1 obtains FedSGD, whereby the server aggregates the local stochastic gradients computed at the current global iterate. If a single global iteration – and hence a single communication round – is performed, we recover one-shot FL, whereby communication only takes place once, at the end of the local computations.

For reference and completeness, Algorithm 14.2 describes how the local SGD updates are implemented by an active agent k at each global iteration i of Algorithm 14.1. In this algorithm, we let the number of **local SGD iterations** carried out by agent k at global iteration i equal some integer $J_k^{(i)}$, which may vary across agents k and iterations i. For instance, some agents k may have more computing power available, allowing the selection of a larger value $J_k^{(i)}$.

Algorithm 14.2: Local SGD updates at each active agent k at global iteration i

---

initialize $\theta_k^{[1]} = \theta^{(i)}$ (model parameters downloaded from the server) and $j = 1$

while $j \leq J_k^{(i)}$ **do**

    pick a mini-batch $\mathcal{S}_k^{[j]} \subseteq \{1, \ldots, N_k\}$ of $S_k^{[j]}$ samples from the data set uniformly at
    random

    given a local learning rates $\gamma_k^{[j]}$, obtain next local iterate via mini-batch SGD as

$$\theta_k^{[j+1]} \leftarrow \theta_k^{[j]} - \frac{\gamma_k^{[j]}}{S^{[j]}} \sum_{n \in \mathcal{S}^{[j]}} \nabla \ell_k(t_{k,n}, \hat{t}(x_{k,n}|\theta_k^{[j]})) \tag{14.4}$$

    set $j \leftarrow i + 1$

end

return $\theta_k^{(i+1)} = \theta_k^{[J_k^{(i)}+1]}$

The number $J_k^{(i)}$ of local SGD updates at each global iteration dictates the trade-off between communication and accuracy: A larger $J_k^{(i)}$ decreases the communication load, but potentially it obtains a global trained model that is significantly different from the solution of the global

ERM problem (14.2); while a smaller $J_k^{(i)}$ approximates the FedSGD protocol more closely, but at the cost of a larger communication overhead.

The general template of local SGD-based FL schemes described by Algorithms 14.1 and 14.2 leaves open a number of questions, most notably (a) how the updated models are to be aggregated at the server, and (b) how the subset of active agents is to be selected at each global iteration. We address these issues next.

14.2.7 Aggregation Function at the Server

In Algorithm 14.1, the aggregation function combines the local updates $\{\theta_k^{(i+1)}\}_{k \in \mathcal{K}^{(i)}}$ of the set $\mathcal{K}^{(i)}$ of $K^{(i)}$ active devices and (possibly) the previous model parameter $\theta^{(i)}$ to obtain the next iterate $\theta^{(i+1)}$. One can instantiate the aggregation function at the server in different ways. The most common choice amounts to the average of the local updates

$$\theta^{(i+1)} = \frac{1}{K^{(i)}} \sum_{k \in \mathcal{K}^{(i)}} \theta_k^{(i+1)}, \tag{14.5}$$

or more generally to the weighted average with the previous iterate

$$\theta^{(i+1)} \leftarrow (1 - \gamma_g^{(i)})\theta^{(i)} + \frac{\gamma_g^{(i)}}{K^{(i)}} \sum_{k \in \mathcal{K}^{(i)}} \theta_k^{(i+1)} \tag{14.6}$$

for some global learning rate $0 < \gamma_g^{(i)} \leq 1$. With one of these choices for the aggregation function, the local SGD-based protocol in Algorithm 14.1 is typically referred to as **federated averaging (FedAvg)**.

In a manner similar to the discussion around the aggregation (14.3) for one-shot FL, the functions (14.5) and (14.6) come with no guarantee of optimality in general. They also implicitly assume that all agents are well behaved. To elaborate on this point, assume that an agent is either **malfunctioning or malicious**. A malicious agent k can try to **"poison"** the trained model by sending an adversarially crafted vector $\theta_k^{(i+1)}$ back to the server with the goal of impairing the performance of the aggregated global model. As a notable example, **backdoor attacks** aim to ensure that the global model returns an incorrect output when fed with inputs affected by a "triggering" signal.

Indeed, the averages in (14.5) and (14.6) are not robust to updates that do not follow the protocol. An instance of a more robust aggregation function would be the element-wise median of the model parameters $\{(\theta_k^{(i+1)})_{k \in \mathcal{K}^{(i)}}\}$, instead of the element-wise mean in (14.5) and (14.6) used by FedAvg. With this choice, the server can effectively discard possible outliers, which may be produced by malfunctioning or malicious agents.

Example 14.1

Consider a setting with $K = 2$ agents that collaboratively train a logistic binary classifier via FedAvg for the problem of distinguishing handwritten digits 0 and 1 from the USPS data set (see Fig. 6.23). The training data set of images of digits 0 and 1 consists of around 1000 examples per class. These are partitioned between the two agents as follows. Agent 1 has a fraction f, with $0 \leq f \leq 1$, of examples for digit 0, i.e., around $1000f$ samples; while agent 2 has the remaining fraction $1 - f$ of examples for digit 0, i.e., around $1000(1 - f)$ samples. Conversely, agent 2 has a fraction f of examples for digit 1; while agent 1 has the remaining fraction $1 - f$ of examples for digit 1. Note that, with $f = 1$, agent 1 only has examples for digit 0, and agent 2 only has examples for digit 1. We implement SGD with mini-batch size 20 and constant learning rate $\gamma = 0.5$. Agents carry out two local SGD iterations, i.e., $J_k^{(i)} = 2$, per communication round.

As the performance metric of interest we choose the population detection-error loss evaluated on a test set that includes an equal number of images of digits 0 and 1. Therefore, the population distribution assumed during testing is distinct from the population distributions underlying the data available at each agent. In fact, the local population distributions at the agent imply an unbalanced generation of images of 0s and 1s, while the test distribution is balanced. See Sec. 14.2.10 for further discussion on this point.

We plot the test loss in Fig. 14.2 as a function of the total number of local iterations. (The zeroth iteration indicates the initialization in the plot.) We compare the performance of FL with that obtained via single-agent training, whereby each agent trains separately on its local data. Single-agent training is also tested under the mentioned balanced population distribution. We show the performance of agent 1 with dashed lines and of agent 2 with dashed-dotted lines. We specifically plot one representative run for all schemes.

The loss of FL (via FedAvg) is constant across two iterations, since the shared model parameter vector is updated only once every two iterations (i.e., $J_k^{(i)} = 2$). When $f = 1$, single-agent training is seen to fail, since each agent has examples of only one of the two classes. In contrast, at convergence, FL produces a shared classifier that is able to obtain a very small test loss.

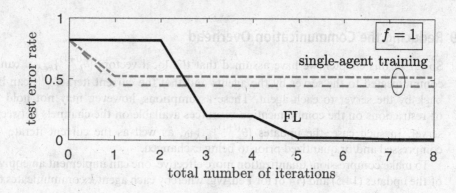

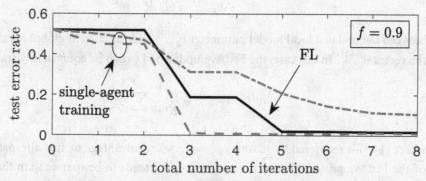

Figure 14.2 Test detection-error loss (error rate) vs. the number of global iterations (or communication rounds) for FL and single-agent training under different levels of data heterogeneity across the agents, with a larger f denoting a larger level of heterogeneity.

When $f = 0.9$, and hence each agent has examples from both classes, FL may not necessarily improve the performance for both agents. In the example in the figure, we specifically see that agent 1 can reduce the test loss more quickly based solely on local data, whereas agent 2 benefits from FL.

This example points to the fact that the advantages of FL may not be equally distributed across agents. In this particular setting, for instance, the few images of handwritten digits 1 at agent 1 are sufficient to obtain a well-performing classifier, and thus agent 1 does not need to collaborate with agent 2. In contrast, the few samples of images of digits 0 available at agent 2 are not enough to train an accurate classifier, and agent 2's classifier can be improved via FL.

14.2.8 Selecting the Set of Active Agents

Depending on the specific setting, the set of agents $\mathcal{K}^{(i)}$ that are active at each global iteration i may be the result of local decisions at the agents and/or of centralized decisions at the server. For example, in a common implementation of FL, mobile phones may only participate in FL when plugged into an electrical outlet so as to avoid running down the battery. Agents may also be unavailable when **"straggling"**, that is, when lagging behind because of malfunctioning or large computational loads. In other implementations, the set $\mathcal{K}^{(i)}$ may be centrally optimized based on agent availability, quality of the link between device and server, "informativeness" of the agents' updates, or timeliness of the agents' updates. All these criteria can take different forms, and yield distinct FL protocols. See Recommended Resources, Sec. 14.6, for details.

14.2.9 Reducing the Communication Overhead

So far in this chapter, we have assumed that the local vectors $\{\theta_k^{(i+1)}\}_{k \in \mathcal{K}^{(i)}}$ can be perfectly communicated to the server by the agents and that the current iterate $\theta^{(i)}$ can be ideally fed back by the server to each agent. These assumptions, however, may not hold true because of restrictions on the communication resources available on the channels between agents and server. In such case, the updates $\{\theta_k^{(i+1)}\}_{k \in \mathcal{K}^{(i)}}$, as well as the current iterate $\theta^{(i)}$, must be compressed and/or quantized prior to being exchanged.

To make compression/quantization more effective, one can implement an equivalent variant of the updates (14.5) and (14.6) for FedAvg, whereby each agent k communicates the difference

$$\Delta\theta_k^{(i+1)} = \theta_k^{(i+1)} - \theta^{(i)} \tag{14.7}$$

between the updated local model parameter $\theta_k^{(i+1)}$ and the current global iterate $\theta^{(i)}$, in lieu of the vector $\theta_k^{(i+1)}$. In this case, the FedAvg update (14.6) can be equivalently stated as

$$\theta^{(i+1)} \leftarrow \theta^{(i)} + \frac{\gamma_g^{(i)}}{K^{(i)}} \sum_{k \in \mathcal{K}^{(i)}} \Delta\theta_k^{(i+1)}, \tag{14.8}$$

where (14.5) is recovered by setting $\gamma_g^{(i)} = 1$. A key advantage of this alternative formulation of the FedAvg protocol is that the update $\Delta\theta_k^{(i+1)}$ tends to be sparser than the corresponding vector of model parameter vector $\theta_k^{(i+1)}$. This is because significant updates may only be made to a subset of the entries of the model parameter vector during the local SGD iterations, especially if the local learning rate and/or the number of local SGD updates is small.

14.2.10 Testing the Performance of FL

The way in which the performance of FL is tested depends on the specific setting and on the requirements of the individual agents. In some cases, such as in Example 14.1, there may be a natural "ground-truth" population distribution $p(x, t)$ that defines the population loss $L_p(\theta)$ that one should test for. In other cases, when a ground-truth common population distribution $p(x, t)$ cannot be identified, the performance of each agent can be measured by its local population loss $L_{p_k}(\theta)$. In order to obtain a single performance metric, one could then opt for the **average** population loss $1/K \sum_{k=1}^{K} L_{p_k}(\theta)$ across the agents; or, with an eye to ensuring "fairness" across the agents, one could adopt the **worst-case** population loss $\max_k L_{p_k}(\theta)$, where the maximum is taken over the agents' index $k = 1, \ldots, K$.

14.2.11 Convergence Properties of FedAvg

A basic question about the performance of FedAvg concerns its convergence to a local minimum, or at least to a stationary point, of the global training loss $L_{\mathcal{D}}(\theta)$ in the global ERM problem (14.2). This subsection elaborates on the additional challenges of ensuring the convergence of FL algorithms as compared to conventional local training.

To this end, we focus on the standard FedAvg update (14.5), and assume the simplifying conditions that (i) full gradients are computed by the agents at each local iteration, i.e., each mini-batch includes the full data set, $S_k^{[j]} = \{1, \ldots, N_k\}$; and that (ii) all agents are active at each global iteration, i.e., $K^{(i)} = K$. Furthermore, we assume that the losses $\ell_k(t_k, \hat{t}(x_k|\theta))$ are convex functions of the model parameter vector θ for all agents $k = 1, \ldots, K$. This condition holds, notably, for linear models under quadratic and logistic losses, and, more generally, for the log-loss of all GLMs (see Sec. 9.8). If the loss functions $\{\ell_k(t_k, \hat{t}(x_k|\theta))\}_{k=1}^{K}$ are convex, so are the per-task training losses $\{L_{\mathcal{D}_k}(\theta)\}_{k=1}^{K}$. Finally, we set $N_k = N/K$, so that all agents have the same number of data points, and the global training loss (14.2) can be expressed as

$$L_{\mathcal{D}}(\theta) = \frac{1}{K} \sum_{k=1}^{K} L_{\mathcal{D}_k}(\theta). \tag{14.9}$$

In order to address the convergence properties of FedAvg, we study the change in the global training loss (14.9) from one global iteration i to the next, $i + 1$. To start, let us note that, with a single local iteration, i.e., with $J_k^{(i)} = 1$ for all agents k and global iterations i, FedAvg corresponds to FedSGD, which in turn, under the given assumptions, is equivalent to conventional GD on the global training set. Therefore, the analysis of convergence follows from the standard theory developed for GD in Chapter 5. This suggests that, if the local updates do not deviate *too much* from a single step of GD, the convergence of FL will enjoy similar guarantees.

To elaborate further on this point, let us now consider the more general case with $J_k^{(i)} \geq 1$ local iterations. By (14.5), the sum of the local training losses at global iteration $i + 1$ can be written as

$$\sum_{k=1}^{K} L_{\mathcal{D}_k}(\theta^{(i+1)}) = \sum_{k=1}^{K} L_{\mathcal{D}_k}\left(\frac{1}{K} \sum_{k'=1}^{K} \theta_{k'}^{(i+1)}\right). \tag{14.10}$$

Furthermore, by Jensen's inequality, because of the assumed convexity of the training loss, we have the upper bound

$$\sum_{k=1}^{K} L_{\mathcal{D}_k}\left(\frac{1}{K}\sum_{k'=1}^{K}\theta_{k'}^{(i+1)}\right) \leq \frac{1}{K}\sum_{k=1}^{K}\sum_{k'=1}^{K} L_{\mathcal{D}_k}(\theta_{k'}^{(i+1)})$$

$$= \frac{1}{K}\sum_{k=1}^{K} L_{\mathcal{D}_k}(\theta_{k}^{(i+1)}) + \frac{1}{K}\sum_{k=1}^{K}\sum_{k'=1, k'\neq k}^{K} L_{\mathcal{D}_k}(\theta_{k'}^{(i+1)}). \qquad (14.11)$$

As we know from Chapter 5, if the learning rates are properly chosen, the training loss $L_{\mathcal{D}_k}(\theta_k^{(i+1)})$ of each agent k evaluated at the updated model parameter $\theta_k^{(i+1)}$ satisfies the inequality

$$L_{\mathcal{D}_k}(\theta_k^{(i+1)}) \leq L_{\mathcal{D}_k}(\theta_k^{(i)}), \qquad (14.12)$$

guaranteeing an improvement across successive iterations. Therefore, the first term in (14.11) decreases across the iterations.

This is not the case, however, for the second term in (14.11). In fact, each summand $L_{\mathcal{D}_k}(\theta_{k'}^{(i+1)})$ with $k \neq k'$ is the training loss for an agent k evaluated at the model parameter $\theta_{k'}^{(i+1)}$ updated by a different agent k'. This suggests that, in order to study convergence, one needs to impose some conditions on the heterogeneity of the learning tasks at the different agents. For instance, one could bound the differences between the gradients of the training losses at different agents. Various convergence results can be developed under such assumptions (see Recommended Resources, Sec. 14.6).

14.2.12 Beyond FedAvg

There are many ways to modify, and possibly improve, FedAvg. Specific weak points of FedAvg include its performance deterioration under scenarios in which the population distributions $p_k(x, t)$ are significantly different across agents $k = 1, \ldots, K$; its sensitivity to malicious or incorrect behavior on the part of the agents; its lack of privacy guarantees; and its requirement that a single model be trained across all agents. At the time of writing, all of these points, and more, are being addressed in the technical literature.

An approach to tackling settings with markedly heterogeneous statistics across the agents is to let agents **"personalize"** the local models by allowing each model to depend on both shared parameters and local parameters. This follows the principles of multi-task learning or meta-learning, both covered in Chapter 13.

14.3 Private Federated Learning

One of the main motivations for the introduction of FL is its limited leakage of information about the data stored at each agent during the communication rounds. The expectation of privacy arises from the constraint imposed in FL that agents only exchange model information, and not data samples directly. However, it has been shown that it is possible to "reverse engineer" the FL training process in order to obtain information about the training set directly from the trained model parameter vector. Therefore, FL, by itself, does not offer any formal guarantee of privacy, and additional mechanisms must be put in place for this purpose. Towards the goal of designing such mechanisms, the first question is how to *measure* privacy – a topic we turn to next before returning to the problem of designing private FL protocols.

14.3.1 Differential Privacy

As is the case for FL, the golden standard for privacy in settings where agents disclose statistics of local data sets is **differential privacy (DP)**.

Setting. To define DP, consider a given agent and the server. Unless stated otherwise, we drop the subscript k identifying the agent in order to simplify the notation. The agent has a local data set $\mathcal{D}$, and generates from $\mathcal{D}$ a statistic, that is, a function of $\mathcal{D}$. In the case of FL, the statistic consists of the sequence of model updates sent to the server. As we will see, in order to ensure privacy, it will be important for the statistic s to be a *random* function of the data set $\mathcal{D}$, which we model via a conditional distribution $p(s|\mathcal{D})$. Accordingly, the agent discloses a statistic

$$s \sim p(s|\mathcal{D}). \tag{14.13}$$

The server receives the statistic s for further processing. We assume there are aspects of the data $\mathcal{D}$ that the agent would like to hide from the server. Specifically, in the DP formulation, we would like to hide the presence or absence of any given data point in $\mathcal{D}$. That is, based on the observation of s, the server should not be able to reliably determine whether any specific data point was used to compute s. This type of constraint is known as **instance privacy**.

Modeling the Server. DP takes a *worst-case* detection-based approach to formulate this constraint. To describe it, consider two **"neighboring" data sets** $\mathcal{D}'$ and $\mathcal{D}''$ that are identical except for a single data point. The server knows that one of the two data sets was used to generate s via (14.13), but not which one. As a result, effectively, the server knows all data points in the true data set except for one. DP requires that the server should not be able to reliably detect whether $\mathcal{D}'$ or $\mathcal{D}''$ was used in generating s for *any* two neighboring data sets $\mathcal{D}'$ and $\mathcal{D}''$. The rationale for this constraint is that, if the server is unable to distinguish between any two such data sets, it cannot also possibly obtain information about any specific data point in the data set used to produce s.

To sum up, DP makes the following worst-case assumptions about the server: (*i*) the server knows the mechanism $p(s|\mathcal{D})$ used to generate the statistic s from data set $\mathcal{D}$; and (*ii*) it also knows the entire data sets $\mathcal{D}'$ and $\mathcal{D}''$, which only differ by one data point. Based on observation of the statistic $s \sim p(s|\mathcal{D})$, the server would like to determine which data set, $\mathcal{D}'$ or $\mathcal{D}''$, was used to generate the statistic s. This is a detection problem, and thus, as studied in Chapter 3, the optimal detector under the detection-error loss is given by the MAP predictor.

Assuming that the prior on the two data sets is uniform, the MAP predictor simply chooses data set $\hat{\mathcal{D}} = \mathcal{D}'$ if $p(s|\mathcal{D}') > p(s|\mathcal{D}'')$, or $\hat{\mathcal{D}} = \mathcal{D}''$ if $p(s|\mathcal{D}') < p(s|\mathcal{D}'')$. (In the case of a tie, the two choices are equivalent.) Therefore, the MAP detector can be implemented as follows: The server computes the LLR

$$\text{LLR}(s|\mathcal{D}', \mathcal{D}'') = \log\left(\frac{p(s|\mathcal{D}')}{p(s|\mathcal{D}'')}\right), \tag{14.14}$$

and then it choose $\hat{\mathcal{D}} = \mathcal{D}'$ if $\text{LLR}(s|\mathcal{D}', \mathcal{D}'') > 0$ or $\hat{\mathcal{D}} = \mathcal{D}''$ if $\text{LLR}(s|\mathcal{D}', \mathcal{D}'') < 0$.

ϵ**-DP Requirement.** The ϵ**-DP requirement**, for some $\epsilon > 0$, imposes that the inequality

$$\mathrm{LLR}(s|\mathcal{D}', \mathcal{D}'') = \log\left(\frac{p(s|\mathcal{D}')}{p(s|\mathcal{D}'')}\right) \leq \epsilon \tag{14.15}$$

hold for all values s and for all neighboring data sets $\mathcal{D}'$ and $\mathcal{D}''$. The parameter ϵ determines how strict the privacy requirement imposed by ϵ-DP is, with stronger privacy guarantees obtained as ϵ decreases. Intuitively, as we will detail next, the ϵ-DP requirement ensures that the probability that the server can decide correctly between the two data sets $\mathcal{D}'$ and $\mathcal{D}''$ – for any such pair of data sets – is small, since the observation of the statistic s does not provide sufficient evidence in support of either data set. Note that, since the inequality (14.15) must hold for all pairs of neighboring data sets $\mathcal{D}'$ and $\mathcal{D}''$, the ϵ-DP requirement effectively requires the LLR to be smaller than ϵ in absolute value (since we can swap the roles of the two data sets in the inequality (14.15)).

To understand the implications of the definition of ϵ-DP, assume that, for the given value $\mathsf{s} = s$, we have the inequality $p(s|\mathcal{D}') > p(s|\mathcal{D}'')$ and hence the MAP predictor is $\hat{\mathcal{D}} = \mathcal{D}'$. A similar calculation applies to the complementary case in which $p(s|\mathcal{D}') < p(s|\mathcal{D}'')$. Under this assumption, given the observation $\mathsf{s} = s$, the probability of a correct guess at the server is the probability that the true data set used to generate data is $\mathcal{D}'$. This probability is given by the posterior

$$p(\mathcal{D}'|s) = \frac{p(s|\mathcal{D}')}{p(s|\mathcal{D}') + p(s|\mathcal{D}'')}, \tag{14.16}$$

where we have used the assumption above that both data sets are a priori equally likely.

Under the ϵ-DP condition (14.15), we have the inequality

$$p(s|\mathcal{D}'') \geq \exp(-\epsilon)p(s|\mathcal{D}'), \tag{14.17}$$

which implies

$$p(\mathcal{D}'|s) \leq \frac{1}{1 + \exp(-\epsilon)} = \sigma(\epsilon). \tag{14.18}$$

It follows that, if ϵ is small, the probability of correct detection is close to 0.5. This corresponds to the same probability of correct detection that the server would have had it not observed the statistic s.

(ϵ, δ)**-DP Requirement.** Requesting that the ϵ-DP condition (14.15) hold for all possible values of the statistic s may be too demanding and in practice unnecessary: After all, not all values s are equally likely to be produced by the agent via the mechanism (14.13). Therefore, it is common to impose the relaxed condition

$$\Pr[\mathrm{LLR}(\mathsf{s}, \mathcal{D}', \mathcal{D}'') \leq \epsilon] \geq 1 - \delta \tag{14.19}$$

for all neighboring data sets $\mathcal{D}'$ and $\mathcal{D}''$, where the probability is over the random mechanism $\mathsf{s} \sim p(s|\mathcal{D}')$. This is known as the (ϵ, δ)**-DP requirement**. Accordingly, the (ϵ, δ)-DP condition only requires that the LLR be smaller than ϵ with a sufficiently large probability. As ϵ and δ decrease, the (ϵ, δ)-DP requirement becomes more stringent.

Beyond DP. Before we discuss how to ensure DP in FL, it is useful to remark that DP is not the only privacy criterion, and that it may have drawbacks in some applications. Notably, DP is agnostic to the distribution of the training sets, while one may wish to obtain specialized guarantees for the data distributions of interest in any given learning tasks. Furthermore, given its strong worst-case assumptions, DP may yield overconservative design solutions.

14.3.2   Gaussian Mechanism

How should we choose the mechanism $p(s|\mathcal{D})$ in (14.13) in order to satisfy the (ϵ,δ)-DP requirement? To start, let us try to choose a deterministic function of the data $\mathcal{D}$, say $s = f(\mathcal{D})$, as the statistic to be disclosed. For instance, the statistic could be obtained as the sequence of model parameters produced by deterministic optimization strategies such as GD (with fixed initialization). The choice of a deterministic statistic can be seen to violate the ϵ-DP constraint (14.15) for any positive privacy parameter $\epsilon > 0$. In fact, unless the function $f(\cdot)$ is constant, it is always possible to find two neighboring data sets $\mathcal{D}'$ and $\mathcal{D}''$ that produce different values s of the statistic. To address this issue, we can add add noise to the function output $f(\mathcal{D})$ in the hope that this randomization will sufficiently mask the presence of any specific data point.

To this end, a standard approach is the **Gaussian mechanism** $p(s|\mathcal{D}) = \mathcal{N}(s|f(\mathcal{D}),\sigma^2)$, i.e.,

$$s = f(\mathcal{D}) + \mathcal{N}(0,\sigma^2), \tag{14.20}$$

where we have denoted as $\mathcal{N}(0,\sigma^2)$ an additive Gaussian noise rv with zero mean and power σ^2. Can we choose the variance σ^2 of the additive noise to guarantee the (ϵ,δ)-DP requirement? Intuitively, we should add more noise if function $f(\cdot)$ is more "sensitive" to the input, that is, if its output reveals more information about the input. We define the **sensitivity** of the given function $f(\cdot)$ as

$$\Delta = \max_{\mathcal{D}',\mathcal{D}''} \|f(\mathcal{D}') - f(\mathcal{D}'')\|, \tag{14.21}$$

where the maximization is over any two neighboring data sets $\mathcal{D}'$ and $\mathcal{D}''$.

It can be proved that the Gaussian mechanism guarantees the (ϵ,δ)-DP condition if the inequality

$$\sigma^2 \geq \frac{\Delta^2}{2R_{DP}(\epsilon,\delta)}, \tag{14.22}$$

holds, where

$$R_{DP}(\epsilon,\delta) = \left(\sqrt{\epsilon + \left[C^{-1}\left(\delta^{-1}\right)\right]^2} - C^{-1}\left(\delta^{-1}\right) \right)^2 \tag{14.23}$$

and $C^{-1}(x)$ is the inverse function of $C(x) = \sqrt{\pi}xe^{x^2}$. The inequality (14.22) states that (ϵ,δ)-DP can be guaranteed if the variance of the added noise is larger than a threshold $\Delta^2/(2R_{DP}(\epsilon,\delta))$ that depends on the sensitivity of the function $f(\cdot)$. Note that the function $R_{DP}(\epsilon,\delta)$ decreases with ϵ and δ. Therefore, as the DP constraint becomes more stringent, the inequality (14.22) implies that a larger noise power σ^2 is required.

A useful way to think about this result is that the amount of information **leaked** by the disclosure of statistic s is quantified by the ratio $\Delta^2/2\sigma^2$, which must satisfy the inequality

$$\frac{\Delta^2}{2\sigma^2} \leq R_{DP}(\epsilon,\delta). \tag{14.24}$$

DP and Generalization. Before moving on to the application of the Gaussian mechanism to FL, it is useful to remark that the goals of privacy are not in disagreement with the general aim of learning algorithms to generalize outside the training set. In fact, as seen, DP imposes the constraint that the output of the learning process should not depend too much on any individual data point – which is also a requirement for generalization, since a strong dependence between data and training output is a hallmark of overfitting. Overall, this suggests that, for sufficiently large values of the DP parameters (ϵ, δ), imposing privacy constraints should not impair the generalization performance of a training algorithm. A formal framework to study generalization – **algorithmic stability** – is indeed based on gauging the impact on the output of a training algorithm of changing a single data point in a training set.

14.3.3 Gaussian Mechanism for Federated Learning

How do we apply the Gaussian mechanism (14.20) to FL? Let us impose that the (ϵ, δ)-DP constraint be satisfied for all agents across a number I of global iterations. Across these iterations, each agent k releases a sequence of model parameters $\theta_k^{(1)}, \ldots, \theta_k^{(I)}$, or the corresponding updates $\Delta\theta_k^{(1)}, \ldots, \Delta\theta_k^{(I)}$. We now need to quantify the sensitivity of these vectors with respect to changes in a single data point in the underlying local data set. The standard approach is to assume that each communicated vector, $\theta_k^{(i)}$ or $\Delta\theta_k^{(i)}$, is bounded in norm, i.e., $\|\theta_k^{(i)}\| \leq \Delta_{\max}$ or $\|\Delta\theta_k^{(i)}\| \leq \Delta_{\max}$. This can be ensured by **clipping** the vector before releasing it to the server. Under this boundedness assumption, the sensitivity Δ for each iteration i can be directly upper bounded as $2\Delta_{\max}$. This is because the norm of the difference between two vectors with norms at most $\Delta_{\max}$ cannot be larger than $2\Delta_{\max}$.

Following the discussion in the preceding subsection, at each global iteration i, the agent leaks an amount of information that can be upper bounded by

$$\frac{\Delta^2}{2\sigma^2} \leq \frac{\Delta_{\max}^2}{\sigma^2}. \tag{14.25}$$

By leveraging suitable **composition theorems**, one can prove that the overall information leaked across I iteration can in turn be upper bounded by the sum of the per-iteration terms in (14.25) across the iterations, i.e., by $I\Delta_{\max}^2/\sigma^2$. Choosing a noise power σ^2 that satisfies the inequality

$$I\frac{\Delta_{\max}^2}{2\sigma^2} \leq R_{DP}(\epsilon, \delta), \tag{14.26}$$

one can, in a manner similar to (14.24), ensure that the (ϵ, δ)-DP constraint is satisfied.

Despite the alignment between the goals of privacy and generalization discussed above, if the DP requirements are sufficiently strict, there is generally a trade-off between accuracy and privacy. In fact, in order to increase privacy via the Gaussian mechanism (i.e., to decrease ϵ and/or δ), one needs to carry out clipping and add noise, which may affect the overall generalization performance.

14.4 Bayesian Federated Learning

As discussed in Sec. 12.4, one of the key benefits of Bayesian learning over frequentist learning is the capacity of the Bayesian framework to quantify epistemic uncertainty, which is caused by the limited availability of data. In FL deployments involving mobile or embedded devices, the availability of small data sets is expected to be the rule rather than an exception. Furthermore, in uses cases involving sensitive applications, such as in personal health monitoring, quantifying the uncertainty with which predictions or recommendations are offered is critical to avoid making overconfident, and possibly hazardous, decisions. Finally, by accounting for the respective states of epistemic uncertainty at the agents, the server can make a better informed decision in aggregating the outcomes of the agents' local training processes. All in all, extending Bayesian methods to FL is appealing for many important scenarios in which FL is envisaged to be a suitable solution. In this section, we elaborate on **Bayesian FL (BFL)** through the lens of VI methods (see Chapter 12).

14.4.1   Defining (Generalized) Bayesian Federated Learning

In this section, we introduce BFL in the parameter server setting with K agents described in Sec. 14.2 and illustrated in Fig. 14.1. To keep the presentation as broad as possible, we adopt the generalized Bayesian learning framework detailed in Sec. 12.12.

To start, let us introduce a set $\mathcal{Q}$ of variational distributions $q(\cdot|\mathcal{D})$. The **global variational posterior distribution** $q(\cdot|\mathcal{D})$ is meant to capture the (epistemic) uncertainty that exists about model parameter θ upon the observation of the global data set $\mathcal{D}$. Ideally, **generalized BFL** obtains the global variational posterior distribution by minimizing the **global generalized free energy**

$$\min_{q(\cdot|\mathcal{D})\in\mathcal{Q}}\left\{\sum_{k=1}^{K}N_k\underbrace{\mathrm{E}_{\theta\sim q(\theta|\mathcal{D})}\left[L_{\mathcal{D}_k}(\theta)\right]}_{\text{average training loss for agent }k}+\alpha\cdot\underbrace{\mathrm{KL}(q(\theta|\mathcal{D})\|p(\theta))}_{\text{complexity measure}}\right\}. \qquad (14.27)$$

In (14.27), the first term is the average global training loss (14.2), with expectation taken over the global variational posterior distribution $q(\cdot|\mathcal{D})$; and the second term is a measure of complexity for the global variational posterior $q(\cdot|\mathcal{D})$, which consists of the KL divergence with respect to a prior distribution $p(\theta)$ (see Sec. 12.12). The temperature parameter $\alpha>0$ determines the relative weight of the two terms in the objective in (14.27). Note the key difference with respect to the frequentist FL algorithms discussed so far, which aim to optimize a shared model parameter θ, rather than a distribution in the model parameter space.

Under no constraints on the set $\mathcal{Q}$ of variational posteriors, it follows from Sec. 12.12 that the optimal solution of problem (14.27) is given by the **generalized global posterior** (cf. (12.104))

$$q^*(\theta|\mathcal{D})=\frac{1}{Z(\mathcal{D})}\underbrace{p(\theta)}_{\text{prior}}\prod_{k=1}^{K}\underbrace{\exp\left(-\frac{N_k}{\alpha}L_{\mathcal{D}_k}(\theta)\right)}_{\text{generalized likelihood for agent }k}$$

$$=\frac{1}{Z(\mathcal{D})}\underbrace{p(\theta)}_{\text{prior}}\underbrace{\exp\left(-\frac{1}{\alpha}\sum_{k=1}^{K}N_kL_{\mathcal{D}_k}(\theta)\right)}_{\text{global general likelihood}}, \qquad (14.28)$$

where we have made explicit the normalization constant $Z(\mathcal{D})$. The generalized global posterior reduces to the conventional **global posterior distribution** $p(\theta|\mathcal{D})$ if we select the log-loss as loss function and the temperature parameter as $\alpha=1$.

Crucially, the normalizing constant $Z(\mathcal{D})$ in the generalized global posterior (14.28) depends on the data of all agents, since we have

$$Z(\mathcal{D})=\int p(\theta)\exp\left(-\frac{1}{\alpha}\sum_{k=1}^{K}N_kL_{\mathcal{D}_k}(\theta)\right)\mathrm{d}\theta. \qquad (14.29)$$

Therefore, the generalized global posterior (14.28) at a model parameter vector θ cannot be obtained by multiplying the prior by the corresponding generalized likelihood for each agent. Instead, more extensive communication is needed in order to estimate also the constant $Z(\mathcal{D})$, whose calculation requires an integration over all values of the model parameter vector θ.

Example 14.2

In order to illustrate some of the differences between frequentist FL and BFL, consider a simple setting with two agents ($K = 2$) in which the **generalized local subposterior**

$$\tilde{q}_k(\theta) \propto p(\theta)^{1/K} \exp\left(-\frac{N_k}{\alpha} L_{\mathcal{D}_k}(\theta)\right) \tag{14.30}$$

of each agent k is given as shown in Fig. 14.3. The rationale behind the definition of the generalized local subposterior is that the product of all generalized subposteriors, upon normalization, yields the generalized global posterior (14.28). Note that this requires taking the prior to the power of $1/K$ in (14.30) in order to avoid counting the prior multiple times in the product of local subposteriors.

In a one-shot frequentist implementation of FL (Sec. 14.2.4), each agent k would solve the problem of maximizing the respective log-losses $-\log \tilde{q}_k(\theta)$ (treating $-1/K \log p(\theta)$ as a regularizing term). This would result in the local solutions $\theta_1 = -1$ and $\theta_2 = 1$, illustrated as black circles in Fig. 14.3. When combined using the aggregation function (14.3), this frequentist approach would yield the solution $\theta = 0$ shown as a black rectangle in Fig. 14.3. The obtained model parameter is clearly not useful for either of the two agents, since both agents assign a close-to-zero subposterior value to $\theta = 0$. Therefore, in this example, the summary of the local data sets given by the frequentist

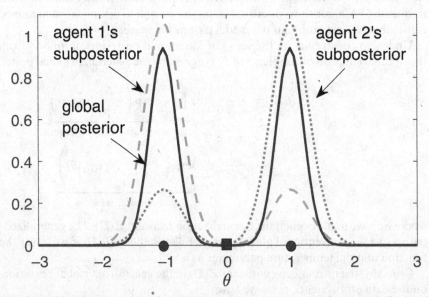

Figure 14.3 Generalized local subposteriors (14.30) at two agents ($K = 2$) shown as dashed and dotted gray lines, along with the generalized global posterior (14.28), illustrated as a black line. The optimal frequentist local solutions θ_k are shown as black circles, while the resulting aggregated solution (14.3) obtained via a one-shot frequentist protocol is displayed as a black rectangle.

solutions θ_k does not provide sufficient information to produce a model that caters to both agents (with a one-shot communication protocol).

In contrast, if each agent k solves the local Bayesian learning problem of computing the sub-posterior (14.30) and communicates the entire *function* $\tilde{q}_k(\theta)$ to the server, the server can multiply and normalize the functions $\tilde{q}_1(\theta)$ and $\tilde{q}_2(\theta)$ to obtain the generalized global posterior (14.28). As a result, ideally, BFL can solve the global learning problem of computing the generalized global posterior via a one-shot protocol based on the outcomes of local Bayesian learning processes. The problem is, of course, that this would require communicating an entire function of θ from the agents to the server, which is generally impractical. We will see next how to address this issue.

14.4.2 Federated Variational Inference: Model and Definitions

As in centralized Bayesian learning, approximate solutions to the global free energy minimization problem (14.27) can be devised based on VI by restricting the class $\mathcal{Q}$ of distributions in problem (14.27) to parametric distributions or to particle-based representations. It is also possible to design MC sampling-based methods, particularly for one-shot protocols, and we point to Recommended Resources, Sec. 14.6, for details. To simplify the terminology, we will henceforth refer to (14.28) as the **global posterior** even when we have $\alpha \neq 1$.

The general goal of **federated VI (FVI)** is to approximate the global posterior (14.28) by means of communications between agents and server, along with local computations involving each agent's local data set. The design of FVI starts from the observation that the global posterior (14.28) can be expressed as

$$q^*(\theta|\mathcal{D}) = p(\theta) \prod_{k=1}^{K} l_k^*(\theta|\mathcal{D}), \tag{14.31}$$

where we have defined the optimal **scaled likelihoods**

$$l_k^*(\theta|\mathcal{D}) = \frac{1}{Z_k(\mathcal{D})} \exp\left(-\frac{N_k}{\alpha} L_{\mathcal{D}_k}(\theta)\right) \tag{14.32}$$

for some constants $\{Z_k(\mathcal{D})\}_{k=1}^{K}$ such that the equality $\prod_{k=1}^{K} Z_k(\mathcal{D}) = Z(\mathcal{D})$ holds. For example, we can set $Z_k(\mathcal{D}) = (Z(\mathcal{D}))^{1/K}$. The scaled likelihoods (14.32) are not normalized in general (i.e., they do not integrate to 1), although the variational posterior $q^*(\theta|\mathcal{D})$ in (14.31) is normalized.

Importantly, as indicated by the notation, each optimal scaled likelihood (14.32) depends on both the training loss $L_{\mathcal{D}_k}(\theta)$ of agent k and the global normalizing constant $Z(\mathcal{D})$. While the first is a function only of the local data set $\mathcal{D}_k$, the latter depends on the global data set $\mathcal{D}$. Therefore, computing the optimal scaled likelihood $l_k^*(\theta|\mathcal{D})$ at agent k requires communicating with the other agents.

Based on the factorization (14.31), FVI constrains the set of the **global variational posteriors** under optimization, $\mathcal{Q}$, to have the form

$$q(\theta) = p(\theta) \prod_{k=1}^{K} l_k(\theta), \tag{14.33}$$

where each function $l_k(\theta)$ is meant to be an approximation of the optimal scaled likelihood (14.32), and is referred to as approximate scaled likelihood, or **approximate likelihood** for short.

Note that we do not indicate the dependence of the variational global posterior (14.33) on the global data set $\mathcal{D}$, since this distribution is obtained in FVI by using local computing and communications (and not by directly using the global data set $\mathcal{D}$).

As we discuss next, FVI addresses the problem (14.27) of minimizing the global free energy over the approximate likelihoods $\{l_k(\theta)\}_{k=1}^{K}$ in a distributed way, with each agent k updating its own approximate likelihood $l_k(\theta)$.

14.4.3 Federated Variational Inference with Unconstrained Approximate Likelihoods

For each global iteration i, let us define the current global variational posterior as

$$q^{(i)}(\theta) = p(\theta) \prod_{k=1}^{K} l_k^{(i)}(\theta), \tag{14.34}$$

where $l_k^{(i)}(\theta)$ is the current approximate likelihood for agent k. This distribution is maintained at the server. In this subsection, we assume that no constraint is imposed on the approximate likelihoods $l_k^{(i)}(\theta)$. Furthermore, we study the case in which a single agent k is scheduled at each iteration. Generalizations to multiple scheduled agents are direct (see Recommended Resources, Sec. 14.6), and the case with constrained approximate likelihoods will be discussed in the next subsection.

Under these simplified assumptions, initializing $l_k^{(1)}(\theta) = 1$ for all θ and $q^{(1)}(\theta) = p(\theta)$, FVI carries out the following steps at each global iteration i:

1. **Download of the current global variational posterior**. The scheduled agent k obtains the current global variational posterior $q^{(i)}(\theta)$ from the server.
2. **"Removal" of the current local approximate likelihood**. The scheduled agent k removes its current approximate likelihood $l_k^{(i)}(\theta)$ from the current global variational posterior $q^{(i)}(\theta)$ by computing the **cavity distribution**

$$q_{-k}^{(i)}(\theta) = \frac{q^{(i)}(\theta)}{l_k^{(i)}(\theta)}. \tag{14.35}$$

Note that, by (14.34), the cavity distribution is equal to

$$q_{-k}^{(i)}(\theta) = p(\theta) \prod_{k'=1, k' \neq k}^{K} l_{k'}^{(i)}(\theta), \tag{14.36}$$

that is, it includes the approximate likelihoods of all agents except agent k. Note also that the cavity distribution is unnormalized.
3. **"Inclusion" of the local likelihood**. The scheduled agent k treats the (normalized version of the) cavity distribution as prior, and updates it to obtain the next iterate of the global variational posterior as

$$q^{(i+1)}(\theta) \propto q_{-k}^{(i)}(\theta) \exp\left(-\frac{N_k}{\alpha} L_{\mathcal{D}_k}(\theta)\right). \tag{14.37}$$

Intuitively, this second step follows from the fact that the true global posterior (14.28) can be obtained by using the update (14.37) *if* the cavity distribution satisfies

$$q_{-k}^{(i)}(\theta) = p(\theta) \prod_{k'=1, k' \neq k}^{K} l_{k'}^{*}(\theta|\mathcal{D}), \tag{14.38}$$

that is, if the approximate likelihoods $l_{k'}^{(i)}(\theta)$ for the other agents have converged to their optimal values in (14.32). More formally, the updated variational posterior (14.37) can be obtained as the solution of the problem of minimizing the **local free energy** (see Sec. 12.12),

$$\min_{q(\cdot)} \left\{ N_k \mathrm{E}_{\theta \sim q(\theta)} \left[L_{\mathcal{D}_k}(\theta) \right] + \alpha \mathrm{KL}(q(\theta) \| q_{-k}^{(i)}(\theta)) \right\}. \qquad (14.39)$$

4. **Update of the local approximate likelihood**. Agent k updates its approximate likelihood as

$$l_k^{(i+1)}(\theta) = \frac{q^{(i+1)}(\theta)}{q_{-k}^{(i)}(\theta)}. \qquad (14.40)$$

This step follows from the fact that, by (14.34), we assume the updated global posterior to factorize as

$$q^{(i+1)}(\theta) = \left(\underbrace{p(\theta) \prod_{k'=1, k' \neq k}^{K} l_{k'}^{(i)}(\theta)}_{q_{-k}^{(i)}(\theta)} \right) l_k^{(i+1)}(\theta), \qquad (14.41)$$

that is, we assume that only the kth approximate likelihood is updated in going from iterate $q^{(i)}(\theta)$ to the next iterate $q^{(i+1)}(\theta)$. Accordingly, we also set

$$l_{k'}^{(i+1)}(\theta) = l_{k'}^{(i)}(\theta) \qquad (14.42)$$

for all $k' \neq k$, so that we also have the desired factorization (14.34), i.e.,

$$q^{(i+1)}(\theta) = p(\theta) \prod_{k=1}^{K} l_k^{(i+1)}(\theta). \qquad (14.43)$$

5. **Upload of the updated global variational posterior**. Agent k transmits the updated global variational posterior $q^{(i+1)}(\theta)$ to the server.

Convergence of FVI. Following arguments similar to those in Example 14.2, if each agent could solve problem (14.39) exactly, it would be sufficient to schedule each agent once in order to have the global variational posterior equal the global generalized posterior. When this is not the case, or when constraints are imposed on the approximate likelihoods, the convergence properties of FVI depend on how the agents are scheduled, and general theoretical results are scarce. What can be easily proved is that, assuming all agents are periodically scheduled, the global posterior $q^*(\theta|\mathcal{D})$ in (14.28) is the unique fixed point of the FVI procedure. This means that, if the current iterate is given by the variational posterior $q^{(i)}(\theta) = q^*(\theta|\mathcal{D})$ and by the approximate likelihoods $l_k^{(i)}(\theta) = l_k^*(\theta|\mathcal{D})$ for all $k = 1, \ldots, K$, then the next iterate $q^{(i+1)}(\theta)$ is also equal to the global posterior $q^*(\theta|\mathcal{D})$. Furthermore, no other distribution has this property.

14.4.4 Federated Variational Inference with Exponential-Family Models

In the preceding subsection, we introduced an idealized FVI protocol in which the updated variational posterior (14.37) can be computed exactly and communicated losslessly between agents and server. To limit computation complexity and communication load, in this subsection we address the more practical case in which the variational posterior $q(\theta)$ is constrained to lie in a parametric set $\mathcal{Q}$ of functions. Specifically, we will consider the case in which the approximate likelihoods are restricted to belong to an exponential-family distribution (see Chapter 9).

To proceed, let us now assume that the class $\mathcal{Q}$ of variational distributions is restricted to variational posteriors of the form

$$q(\theta|\eta) = p(\theta|\eta_0) \prod_{k=1}^{K} l_k(\theta|\eta_k), \tag{14.44}$$

where $p(\theta|\eta_0)$ is the (fixed) prior distribution and $l_k(\theta|\eta_k)$ is the approximate likelihood for agent k. Prior and approximate likelihoods are chosen as members of the same exponential-family distribution, i.e.,

$$p(\theta|\eta_0) = \text{ExpFam}(\theta|\eta_0)$$
$$\text{and } l_k(\theta|\eta_k) \propto \text{ExpFam}(\theta|\eta_k), \tag{14.45}$$

with respective natural parameters η_0 and $\{\eta_1, \eta_2, \ldots, \eta_K\}$. Note that the prior is normalized, while the approximate likelihoods are unnormalized.

Recall from Chapter 9 that a distribution in the exponential family is of the form

$$\text{ExpFam}(\theta|\eta) = \exp\left(\tilde{\eta}^T \tilde{s}(\theta) - A(\eta)\right)$$
$$\propto \exp\left(\tilde{\eta}^T \tilde{s}(\theta)\right), \tag{14.46}$$

where the augmented vector of natural parameter is $\tilde{\eta} = [\eta^T, 1]^T$ and the augmented vector of sufficient statistics is $\tilde{s}(x) = [s(x)^T, M(x)]^T$, with $s(x)$ being the vector of sufficient statistics and $M(x)$ the log-base measure. In this subsection, we drop the tilde to simplify the notation, and write $s(x)$ and η for the *augmented* vectors of sufficient statistics and of natural parameters. Note that if $M(x) = 0$, this redefinition is not necessary, and we can take η to represent the vector of natural parameters. This is the case for Bernoulli rvs, for categorical rvs, and for Gaussian rvs with parameters given by both mean and covariance (see Chapter 9).

With the choice (14.45), the global variational distribution (14.44) is given as

$$q(\theta|\eta) = p(\theta|\eta_0) \prod_{k=1}^{K} l_k(\theta|\eta_k)$$
$$= \text{ExpFam}\left(\theta\bigg|\eta = \eta_0 + \sum_{k=1}^{K} \eta_k\right). \tag{14.47}$$

Therefore, the global variational posterior belongs to the same exponential-family distribution as prior and approximate likelihoods, and the global variational (natural) parameter η is given by the sum $\eta = \eta_0 + \sum_{k=1}^{K} \eta_k$ of the natural parameters of prior and approximate likelihoods.

Let us now specialize the steps of the FVI algorithm introduced in the preceding subsection to this setting.

1. **Download of the current global variational posterior**. The scheduled agent obtains the current global variational parameter $\eta^{(i)}$ from the server.
2. **"Removal" of the current local approximate likelihood**. The scheduled agent k removes its current approximate likelihood $t_k^{(i)}(\theta)$ from the current variational posterior by computing the **cavity distribution**

$$q_{-k}^{(i)}(\theta) = \frac{q^{(i)}(\theta|\eta^{(i)})}{l_k^{(i)}(\theta|\eta_k^{(i)})} \propto \mathrm{ExpFam}(\theta|\eta^{(i)} - \eta_k^{(i)}). \tag{14.48}$$

Removing the contribution of agent k from the current global variational posterior $q^{(i)}(\theta|\eta^{(i)})$ thus only requires subtracting the corresponding local natural parameter $\eta_k^{(i)}$ from the global parameter $\eta^{(i)}$. Note that this relationship is valid if the difference $\eta^{(i)} - \eta_k^{(i)}$ is a feasible natural parameter. For instance, for a Gaussian distribution $\mathcal{N}(\nu, \beta^{-1})$, the second natural parameter is $-\beta/2$, and hence only vectors with negative second entry are feasible natural parameters.

3. **"Inclusion" of the local likelihood.** The scheduled agent k treats the (normalized version of the) cavity distribution as prior, and updates it to obtain the updated global variational parameter $\eta^{(i+1)}$ by minimizing the **local free energy** as per the problem

$$\min_{\eta} \left\{ N_k \mathrm{E}_{\theta \sim \mathrm{ExpFam}(\theta|\eta)} \left[L_{\mathcal{D}_k}(\theta) \right] + \alpha \mathrm{KL}(\mathrm{ExpFam}(\theta|\eta) || \mathrm{ExpFam}(\theta|\eta^{(i)} - \eta_k^{(i)})) \right\}. \tag{14.49}$$

This problem can be tackled using the techniques discussed in Chapter 10, such as reparametrization-based VI. Furthermore, in the special case of conjugate exponential families, the problem can be solved exactly (see Sec. 12.5).

4. **Update of the local approximate likelihood.** Agent k updates its approximate likelihood as

$$t_k^{(i+1)}(\theta) = \frac{q^{(i+1)}(\theta|\eta^{(i+1)})}{q_{-k}^{(i)}(\theta)} = \frac{\mathrm{ExpFam}(\theta|\eta^{(i+1)})}{\mathrm{ExpFam}(\theta|\eta^{(i)} - \eta_k^{(i)})}$$

$$= \mathrm{ExpFam}(\theta|\eta^{(i+1)} - (\eta^{(i)} - \eta_k^{(i)})). \tag{14.50}$$

Therefore, storing the approximate likelihood only requires keeping track of the natural parameter $\eta^{(i+1)} - (\eta^{(i)} - \eta_k^{(i)})$, assuming again that this is a feasible solution.

5. **Upload of the updated global variational posterior.** Agent k transmits the updated model parameter $\eta^{(i+1)}$ to the server.

14.5 Summary

- This chapter has provided a brief introduction to the subject of FL by covering standard frequentist solutions, as well as Bayesian approaches based on VI.

- FL protocols alternate local training steps, communication rounds, and aggregation operations. Inter-agent communication is used to share information about model parameters, such as the model parameter vector, the gradient, or probability distributions in the model parameter space.

- The design of FL protocols entails the consideration of trade-offs among accuracy, convergence speed, latency, and communication load. For instance, local SGD-based FL protocols depend on the selection of the number of local iterations per communication round: A larger number of local iterations can decrease the communication load and improve latency, but it may also impair the accuracy of the shared model.

- While data are not directly exchanged among the agents, privacy is not guaranteed, unless additional mechanisms are put in place to meet formal privacy requirements such as DP. A standard approach is to add noise via the Gaussian mechanism.

- Bayesian FL inherits the potential advantages of Bayesian learning summarized in Chapter 12, by keeping track of epistemic uncertainty through the optimization of a shared posterior distribution in the model parameter space.
- As hinted at throughout this chapter, at the time of writing the field of FL is the subject of active research, and it is expected that significant breakthroughs will build on the basic solutions described here.

14.6 Recommended Resources

In-depth reviews about the current state of the art on FL include [1, 2]. A general introduction to DP is provided by [3]. Discussions on distributed Bayesian learning, including MC sampling-based methods, can be found in [4, 5, 6].

Bibliography

[1] P. Kairouz, H. B. McMahan, B. Avent, A. Bellet, M. Bennis, A. N. Bhagoji, K. Bonawitz, Z. Charles, G. Cormode, and R. Cummings, "Advances and open problems in federated learning," arXiv preprint arXiv:1912.04977, 2019.

[2] T. Li, A. K. Sahu, A. Talwalkar, and V. Smith, "Federated learning: Challenges, methods, and future directions," *IEEE Signal Processing Magazine*, vol. 37, no. 3, pp. 50–60, 2020.

[3] C. Dwork and A. Roth, "The algorithmic foundations of differential privacy." *Foundations and Trends in Theoretical*

Computer Science, vol. 9, no. 3–4, pp. 211–407, 2014.

[4] E. Angelino, M. J. Johnson, and R. P. Adams, "Patterns of scalable Bayesian inference," *Foundations and Trends in Machine Learning*, vol. 9, no. 2–3, pp. 119–247, 2016.

[5] T. D. Bui, C. V. Nguyen, S. Swaroop, and R. E. Turner, "Partitioned variational inference: A unified framework encompassing federated and continual learning," arXiv preprint arXiv:1811.11206, 2018.

[6] R. Kassab and O. Simeone, "Federated generalized Bayesian learning via distributed Stein variational gradient descent," arXiv preprint arXiv:2009.06419, 2020.

Part V

Epilogue

15 Beyond This Book

15.1 Overview

This final chapter covers topics that build on the material discussed in the book, with the aim of pointing to avenues for further study and research. The selection of topics is clearly a matter of personal choice, but care has been taken to present both well-established topics, such as probabilistic graphical models, and emerging ones, such as causality and quantum machine learning. The topics are distinct, and each section can be read separately. The presentation is brief, and only meant as a launching pad for exploration.

Learning Objectives and Organization of the Chapter. By the end of this chapter, the reader should be able to:

- understand probabilistic graphical models, and specifically Bayesian networks and Markov random fields (Sec. 15.2);
- understand adversarial learning (Sec. 15.3);
- appreciate the novel challenges that arise when one is interested in drawing conclusions about causality relationships from data (Sec. 15.4);
- appreciate some of the new opportunities and challenges that are introduced by the use of quantum computing in machine learning (Sec. 15.5);
- understand the general concept of machine unlearning (Sec. 15.6); and
- discuss critically – possibly disagreeing with the author – the prospect of developing general AI (Sec. 15.7).

15.2 Probabilistic Graphical Models

In this book, we have mostly assumed that the joint distribution of the rvs describing each observation – input x, target t, and latent vector z – is *unstructured*. Two exceptions stand out, namely the RBM studied in Sec. 7.7 and the Ising model introduced in Sec. 10.6. In both cases, the rvs of interest are modeled as having a specific structure in terms of local statistical dependencies. This property was found to be instrumental in encoding inductive biases that are tailored to given classes of signals. For instance, the Ising model was used to capture the assumption that neighboring pixels in an image tend to be more correlated than pixels that are further apart. RBMs and Ising models are just two examples of **probabilistic graphical models (PGMs)**, a general formalism that enables the definition of structural properties in joint distributions.

"Structure" in probabilistic models refers more formally to **conditional independence properties**: The probabilistic model imposes that some variables in the input vector x, target vector t, and latent vector z are conditionally independent given others. For instance, in the mentioned

Ising model, a pixel is assumed to be independent of all pixels that are not immediate neighbors when conditioning on the neighboring pixels.

As with any other choice pertaining to inductive biases, the selection of a PGM carries the potential benefit of reducing overfitting, along with the possible downside of increasing the bias (see Chapter 4). A suitable selection should, therefore, be based on either domain knowledge, as in the image processing example, or on validation. There are two main types of PGMs, namely Bayesian networks and Markov random fields, which we briefly review in the rest of this section.

15.2.1 Bayesian Networks

To keep the discussion general, let us lump together all variables in rvs x, t, and z in a single collection of rvs $\{y_1, \ldots, y_L\}$. **Bayesian networks (BNs)** are defined by **directed acyclic graphs (DAGs)** that model conditional independence properties among the individual rvs in the set y.

Each rv y_i is assigned to a vertex of the DAG. A directed edge from one rv y_i to another rv y_j indicates, in a specific sense to be formalized shortly, that rv y_i is directly responsible for the distribution of rv y_j. In a DAG, there are no directed cycles, i.e., there are no closed paths following the direction of the arrows.

Denote as $\mathcal{P}(i)$ the set of "parents" of node y_i in the DAG, that is, the subset of rvs $\mathcal{P}(i) \subseteq \{y_1, \ldots, y_L\}$ that "emit" an edge ending at node y_i. We also define as $y_{\mathcal{P}(i)}$ the corresponding subset of "parent" rvs.

Owing to the acyclic structure of the DAG, it is always possible to order the rvs $\{y_1, \ldots, y_L\}$ such that the parents of any rv y_i precede it. Relabeling the indices of the variables to conform to this ordering, this means that for any rv y_j in $y_{\mathcal{P}(i)}$ we have the inequality $j < i$. The BN, through its associated DAG, imposes that the joint distribution of the rv y factorizes as

$$p(y_1, \ldots, y_L) = \prod_{i=1}^{L} p(y_i | y_{\mathcal{P}(i)}). \tag{15.1}$$

Recall from the chain rule for probabilities from Chapter 2 that a general, unstructured distribution satisfies the factorization

$$p(y_1, \ldots, y_L) = \prod_{i=1}^{L} p(y_i | y_1, \ldots, y_{i-1}). \tag{15.2}$$

Therefore, the BN factorization (15.1) imposes the following local structural property on the joint distribution: When conditioning on the parents $y_{\mathcal{P}(i)}$ of a rv y_i, the rv y_i is statistically independent of all the preceding variables $y_1, \ldots, y_{i-1}$ not included in $y_{\mathcal{P}(i)}$. Mathematically, the BN encodes the local conditional independence relationships

$$y_i \perp \{y_1, \ldots, y_{i-1}\} \backslash y_{\mathcal{P}(i)} | y_{\mathcal{P}(i)} \tag{15.3}$$

where the "perp" notation $\perp$ indicates statistical independence.

BNs are suitable models when one can identify "causality" relationships among the variables. In such cases, there exists a natural order of the variables, so that rvs that appear later in the order are "caused" by a subset of the preceding variables. The "causing" rvs for each rv y_i are included in the parent set $\mathcal{P}(i)$, and are such that, when conditioning on rvs $y_{\mathcal{P}(i)}$, rv y_i is independent of all preceding rvs $\{y_1, \ldots, y_{i-1}\}$.

A typical example is the case of sequential data, such as for speech processing, in which the next phoneme can be modeled as being "caused" by a limited subset of previous ones.

The "causality" relationship among the ordered variables $\{y_1, \ldots, y_L\}$ encoded by a BN makes it easy, at least in principle, to draw samples from a BN. In fact, this can be done by using **ancestral sampling**: Generate rv $y_1 \sim p(y_1)$; then, $y_2 \sim p(y_2|y_{\mathcal{P}(2)})$; and so on, with rv y_i being generated as $y_i \sim p(y_i|y_{\mathcal{P}(i)})$. For more discussion about the need for quotation marks when discussing "causality", please see Sec. 15.4.

15.2.2 Markov Random Fields

Like BNs, **Markov random fields (MRFs)** encode a probability factorization or, equivalently, a set of conditional independence relationships. They do so, however, through an **undirected graph**. As with BNs, each vertex in the graph is associated with one of the rvs in the set $\{y_1, \ldots, y_L\}$.

Given the underlying undirected graph, let us index the maximal cliques in the graph using an integer c. A **clique** is a fully connected subgraph, and a **maximal clique** is one that is not fully included in a larger clique. With this definition, an MRF imposes that the joint distribution factorizes as

$$p(y_1, \ldots, y_L) = \frac{1}{Z}\prod_c \psi_c(y_c),\qquad(15.4)$$

where c is the index of a clique in the graph; y_c is the subset of rvs associated with the vertices in clique c; $\psi_c(y_c) \geq 0$ is the **factor** or **potential** for clique c; and Z is the partition function. The partition function equals $Z = \sum_x \prod_c \psi_c(y_c)$ for discrete rvs, and $Z = \int \prod_c \psi_c(y_c)dy$ for continuous rvs. Examples of MRFs include both RBMs and Ising models. For instance, in the case of RBMs, the maximal cliques are given by all pairs of rvs consisting of one rv from the visible vector x and the other from the latent vector z.

Each factor $\psi_c(y_c)$ encodes the **compatibility** of the values y_c in each clique, with larger values of $\psi_c(y_c)$ corresponding to configurations y_c that are more likely to occur. Factors are generally not probability distributions, i.e., they are not normalized to sum or integrate to 1. Furthermore, unlike BNs, each factor does not distinguish between conditioning and conditioned rvs. Instead, all variables in the clique y_c can play the same role in defining the value of the potential $\psi_c(y_c)$.

Potentials are typically parametrized using the **energy-based form**

$$\psi_c(y_c) = \exp(-\mathcal{E}_c(y_c|\eta_c)),\qquad(15.5)$$

for some energy function $\mathcal{E}_c(y_c|\eta_c)$ with parameter vector η_c. This form ensures that the factors are strictly positive as long as the energy is upper bounded. A special class of such models is given by **log-linear models**, such as RBMs, in which the energy is a linear function of the parameters (see Sec. 7.7.3).

MRFs are especially well suited for modeling mutual relationships of compatibility, or lack thereof, among variables, rather than "causality" effects. This distinguishing feature of a MRF, while potentially yielding modeling advantages in some applications, comes with the added difficulties of *evaluating* the joint probability distribution (15.4) and *sampling* from it. In fact, the computation of the probability (15.4) requires the partition function Z, which is generally intractable when the alphabet of the rvs at hand is large enough. This is typically not the case for BNs, in which each conditional distribution can be selected from a known (normalized) distribution. Furthermore, unlike BNs, MRFs do not enable ancestral sampling, as all the conditional distributions of the individual rvs in x are tied together via the partition function.

As suggested by the discussion in this section, BNs and MRFs encode different types of statistical dependencies, with the former capturing "causality" while the latter accounting for mutual compatibility. Accordingly, there are conditional independence properties that can be expressed by a BN or a MRF but not by both. An example is the "V-structure" BN x → y ← z, defined by a joint distribution of the form $p(x, y, z) = p(x)p(z)p(y|x, z)$, whose independencies cannot be captured by an MRF.

That said, given a BN, one can always obtain an MRF by defining the potential functions

$$\psi_i(y, y_{\mathcal{P}(i)}) = p(y_i | y_{\mathcal{P}(i)}) \tag{15.6}$$

to yield the MRF factorization

$$p(y_1, \ldots, y_L) = \prod_{i=1}^{L} \psi_i(y_i, y_{\mathcal{P}(i)}) \tag{15.7}$$

with partition function $Z = 1$. This factorization defines an MRF in which each maximal clique contains a rv y_i and its parents $y_{\mathcal{P}(i)}$. As per the discussion in this subsection, the resulting MRF may not account for all the independencies encoded in the original graph of the BN.

15.3 Adversarial Attacks

A predictor $\hat{t}(x|\theta)$ has been trained, and hence the model parameter θ is fixed, and it is now deployed in the real world. How robust are its decisions? To stress the possible weak points of the system, suppose that an attacker wishes the predictor to malfunction. Needless to say, this may cause significant real-world problems if the predictor is used in sensitive applications, such as face recognition or autonomous driving.

Defining the Attacker. As a baseline scenario, assume that the attacker can perturb an input x fed to the predictor by producing a modified input

$$x^{\text{adv}} = x + \Delta x. \tag{15.8}$$

Without the perturbation vector Δx, the prediction of the target rv t is $t = \hat{t}(x|\theta)$, and the attacker wishes the predictor to output a different value for the target variable in the presence of the perturbation Δx.

More precisely, the attacker aims to satisfy two competing requirements:

1. **Successful attack**. First, the attacker wishes that the prediction $t^{\text{adv}} = \hat{t}(x^{\text{adv}}|\theta)$ be different from the correct prediction $t = \hat{t}(x|\theta)$ on the unperturbed input, i.e., $t^{\text{adv}} \neq t$. The attacker may have a specific output $t^{\text{adv}} \neq t$ that it wishes the predictor to produce, or it may be satisfied with any output t^{adv} as long as it is different from t.
2. **Hard-to-detect attack**. Second, the attacker wishes to keep the perturbation hard to detect in order to foil attempts by the predictor to spot the anomalous input x^{adv}. This requirement is typically imposed by limiting the ℓ_k norm $||\Delta x||_k$, for some integer k, of the perturbation Δx. A particularly useful norm is the ℓ_∞ norm, i.e., $||\Delta x||_\infty = \max_d\{|\Delta x_d|\}$, which corresponds to the maximum absolute value of the entries of the perturbation vector. With this choice, no element of the input vector x can be modified, in absolute value, by a term larger than $||\Delta x||_\infty$.

Another important aspect that defines the attacker is the type of information that is available to it for the purpose of designing the perturbation Δx. There are two main cases to consider:

- **White-box attacker**. The attacker has access to the predictive function $\hat{t}(\cdot|\theta)$ (where the model parameter θ is fixed).
- **Black-box attacker**. The attacker only has access to a limited set of input–output pairs (x, t) produced by the predictor $\hat{t}(\cdot|\theta)$.

To consider the worst case, let us assume a white-box attacker. We also assume that the attacker is satisfied with the predictor producing any output t^{adv} as long as it is different from t, and that the "size" of the perturbation is measured by the ℓ_k norm for some integer k, possibly equal to ∞.

Attack Optimization. Under these assumptions, the problem of designing the perturbation Δx at the attacker may be formulated as the maximization

$$\max_{\Delta x: \, ||\Delta x||_k \leq \Delta_{\max}} \ell(t, \hat{t}(x + \Delta x|\theta)), \tag{15.9}$$

where $\Delta_{\max} > 0$ is an upper bound on the norm of the input perturbation. Accordingly, the attacker maximizes the loss of the predictor when measured on the correct output t for the given input x. This makes output t less likely to be produced by the predictor. Alternatively, if the attacker had a specific desired output t^{adv} that it wished to enforce on the predictor, the design problem could be changed to the minimization

$$\min_{\Delta x: \, ||\Delta x||_k \leq \Delta_{\max}} \ell(t^{\text{adv}}, \hat{t}(x + \Delta x|\theta)). \tag{15.10}$$

Under the ℓ_∞-norm for the constraint, the first formulation (15.9) – the second can be tackled in an analogous way – can be addressed by the **fast gradient sign method**. This applies gradient descent followed by binary quantization, yielding the perturbation

$$\Delta x = \Delta_{\max} \text{sign}\{\nabla_x \ell(t, \hat{t}(x|\theta))\}, \tag{15.11}$$

where $\text{sign}\{x\} = 1$ if $x > 0$ and $\text{sign}\{x\} = -1$ if $x < 0$ and the operation is applied element-wise. Note that the sign operation ensures that the ℓ_∞ norm constraint in (15.9) is satisfied. Furthermore, a single gradient step is taken in accordance with the requirement that the perturbation be small.

Adversarial Training. So far, we have assumed that the predictor $\hat{t}(x|\theta)$, and hence the model parameter θ, are fixed. Is it possible to modify the training procedure so as to mitigate the effect of possible attacks? This generally requires some knowledge of the types of attack that are expected. As an example, assume that the designer knows that an attacker may try perturbing the input by tackling problem (15.9). Then, the learner may opt to minimize a "robustified" version of ERM, which is defined by the problem

$$\min_\theta \left\{ \frac{1}{N} \sum_{n=1}^{N} \left[\max_{\Delta x: \, ||\Delta x||_k \leq \Delta_{\max}} \ell(t_n, \hat{t}(x_n + \Delta x|\theta)) \right] \right\}. \tag{15.12}$$

In practice, this problem can be tackled by augmenting the data set $\mathcal{D}$ with **adversarial examples** obtained by addressing the inner problem in (15.12). For example, one could use the fast gradient sign method (15.11), and augment the data set $\mathcal{D}$ with examples of the form $(x_n + \Delta_{\max}\text{sign}\{\nabla_x \ell(t_n, \hat{t}(x_n|\theta))\}, t_n)$. Apart from making the trained model $\hat{t}(x|\theta)$ potentially less susceptible to adversarial attacks, this data augmentation-based approach may improve the generalization performance by reducing overfitting to the training set $\mathcal{D}$.

Underspecification. The discussion around adversarial learning raises a more general issue: There may be multiple models that can provide approximately the same population loss while having significantly different performance in terms of auxiliary measures such as adversarial robustness. So, models that are equivalent in terms of population loss can behave quite differently under **"stress tests"** that may represent deployment conditions of interest, such as the presence of adversaries. Note that this phenomenon, known as underspecification, is distinct from the presence of multiple models that fit the training data almost equally well – a hallmark of epistemic uncertainty and a rationale for using Bayesian learning (see Chapter 12). Underspecification may be addressed by adding further requirements to the design specification in the form of "stress test" criteria such as adversarial robustness, or by incorporating more detailed prior knowledge in the inductive bias.

15.4   Causality

Machine learning leverages statistical dependencies and regularities in the data in order to extrapolate relationships among variables of interest. As such, the outputs of a machine learning model capture correlations, and they cannot generally be used to draw any conclusions about **causality**, unless additional assumptions are made about the data generation process.

To elaborate on the possible pitfalls of causal interpretations of machine learning-based predictions, consider the top part of Fig. 15.1, which shows a distribution of data points on the plane defined by coordinates $x =$ exercise and $t =$ cholesterol (the numerical values are arbitrary). Training a model that relates t as the dependent variable to the variable x would clearly identify an upward trend – an individual that exercises more would be predicted to have a higher cholesterol level. This prediction is legitimate and supported by the available data, but can we also conclude that exercising less would reduce one's cholesterol? In other words, can we conclude that there exists a causal relationships between x and t? We know the answer to be no, but this cannot be ascertained from the data in the figure.

Following the Israeli-American computer scientist Judea Pearl's suggestion (see Recommended Resources, Sec. 15.8), a way out of this conundrum is to leverage prior information we may have about the problem domain. This information is used to define a class of models that encode, in an explicit way, our understanding of the causal relationships among the variables of interest. For the example at hand, based on our knowledge of the problem, we may decide that it is necessary to include another measurable variable in the model, namely age. In this model, we view age as a possible cause of both exercise and cholesterol.

Based on this model, in the bottom part of Fig. 15.1 we redraw the data by highlighting the age of the individual corresponding to each data point. The figure shows that we can **explain away** the negative correlation between exercise and cholesterol: Older people – within the observed bracket – tend to have higher cholesterol as well as to exercise more. Therefore, age is a common cause of both exercise and cholesterol level. We conclude that, in order to capture the causality relationship between exercise and cholesterol, we need to **adjust** for age. This requires considering the trend within each age bracket separately, recovering the expected conclusion that exercising is useful to lowering one's cholesterol. This example is an instance of the so-called **Simpson's paradox**: Patterns visible in the data disappear, or are even reversed, when controlling from some variable.

Figure 15.1 Illustration of Simpson's paradox.

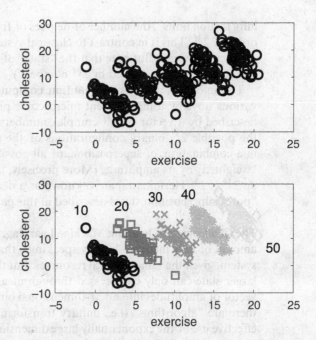

The correlation between x and t that is apparent in Fig. 15.1, while useful for prediction, should not be a basis for decision-making. When assessing the causal relationship between x and t, we should first understand which other variables may explain the observations and then discount any spurious correlations.

The discussion in this section reveals an important limitation of most existing machine learning algorithms when it comes to identifying causality relationships, or, more generally, to answering **counterfactual** ("what if?") queries. The study of causality can be carried out within the elegant framework of **interventions** on PGMs developed by Pearl. Within this framework, the inductive bias encodes assumptions about the impact of interventions on the variables of interest. For instance, in the example above, the model class we implicitly considered posited that any intervention on either exercise or cholesterol level does not directly affect the other variable, while "intervening" on the age variable may affect both exercise and cholesterol levels. Causality remains a subject of intense scientific discussion at the time of writing.

15.5   Quantum Machine Learning

At the time of writing, recent progress on the implementation of **quantum computers** has led to renewed efforts to understand the power of "quantum" in solving specific problems, including for data-driven algorithms.

The Promise of Quantum Machine Learning. Quantum computers leverage the microscopic behavior of the computational substrate, which is described by quantum physics. At a high level, the power of quantum computers stems from the exponentially larger space in which they operate as compared to classical computers. In fact, given a quantum computer with M "binary information units", the number of degrees of freedom that can potentially be explored is of the order of 2^M. This is in contrast to classical systems, which can only store and process M bits in M binary devices. (But note that the "state" of a vector of M *random* bits is also described by a probability vector defined by 2^M numbers.)

The information units in quantum computers, known as **qubits**, can be implemented in various ways by using different microscopic physical systems. The joint state of M qubits is described by a vector of 2^M complex numbers, each representing the "amplitude" of one of the possible 2^M binary configurations of the qubits. In fact, the system of M qubits exists in a combination, or **superposition**, of all possible 2^M configurations, with each configuration "weighted" by its amplitude. (More precisely, the joint state of M qubits can be described by an $2^M \times 2^M$ Hermitian matrix, known as a density matrix, which represents a mixture of the "pure" superposition states described in this paragraph.)

The Challenges of Quantum Machine Learning. So, a quantum system can store and process an amount of information – loosely speaking – that is exponentially larger than that of a classical system, given the same physical resources. But there is a fundamental caveat: The exponentially larger state can only be accessed through measurements that "collapse" the 2^M dimensional vector of amplitudes into an M-dimensional binary vector. The key design issue is then whether there are "algorithms" (i.e., unitary transformations in quantum computing) that can make effective use of the exponentially large dimensional space to produce the desired configurations before a measurement is made. It is well known that this is the case in important problems, such as factoring, but it is currently unclear if it is also true for broad classes of machine learning applications.

The Operation of Quantum Machine Learning. Quantum computing can be applied to **classical data**, i.e., to standard numerical vectors, as long as these are first encoded in the state of microscopic physical elements. Alternatively, one can leverage quantum computing to process the **state of quantum systems**, e.g., to implement quantum simulators replicating the behavior of quantum systems (in the words of Richard Feynman, "Nature isn't classical, dammit, and if you want to make a simulation of nature, you'd better make it quantum mechanical, and by golly it's a wonderful problem, because it doesn't look so easy."[1]).

Furthermore, there are two general design approaches for machine learning algorithms. The first is to transform existing classical algorithms for deployment on a quantum computer. The aim of these methods is to leverage **quantum speed-ups** for routines involving optimization and linear algebra. The second approach aims to identify **"native" quantum algorithms** that directly leverage the properties of quantum physics.

Let us briefly elaborate on the latter class. In this regard, an important observation is that the probabilities of the different configurations produced by a measurement are linear functions of the complex state of the system. (More precisely, they are linear in the density matrix.) The situation is reminiscent of **kernel methods** (see Sec. 4.11). In kernel methods, inputs are implicitly mapped to a large feature space, which is generally inaccessible and may even be infinite-dimensional. The final result is obtained through a linear combination of correlations between the input and the training data, where the correlations are computed in the feature space. Importantly, the number of degrees of freedom, i.e., the number of trainable parameters, does not scale with the size of the feature space, but rather with the number of training data points.

[1] See J. Preskill, "Quantum computing 40 years later," arXiv:2106.10522, 2021.

In an analogous way, a quantum computer can be used to map a classical input into the state of a quantum system. This state is inaccessible, and operations on it are only consequential insofar as they affect the output of the final measurement. One can thus think of quantum computers as providing an opportunity to operate with effective vectors of features that live in an exponentially large space. Whether this choice of inductive bias can be leveraged to produce machine learning algorithms with unique properties in terms of accuracy, as well as computational and sample efficiency, for specific problems remains to be seen.

15.6 Machine Unlearning

Suppose that one has trained a model parameter θ based on training data $\mathcal{D}$, which may have been collected from different individuals or organizations. Later on, one or more of these individuals or organizations may request that the contribution of their data be "deleted" from the trained model θ. This is indeed a legal right, known as the **right to be forgotten** or **right to erasure**, in several countries at the time of writing.

The process of "deleting" a subset of the training data, $\mathcal{D}_u \subset \mathcal{D}$, from a trained model θ is referred to as **machine unlearning**. Machine unlearning is successful if one cannot infer any information about the data in $\mathcal{D}_u$ by observing the "unlearned" model θ_u.

Unlearning can be trivially accomplished by retraining from scratch using the data set $\mathcal{D} \setminus \mathcal{D}_u$, i.e., by excluding the data set $\mathcal{D}_u$ to be deleted. However, this may be too computationally intensive, and it would be preferable to define more efficient procedures that start with model θ trained from the entire data set $\mathcal{D}$, and produce an approximation of the model that would have been produced by training from scratch using the data set $\mathcal{D} \setminus \mathcal{D}_u$.

A simple approach to achieving this goal is to train a number of separate models, each from a distinct subset of data points from $\mathcal{D}$. These models can then be combined using a mixture model to obtain a final prediction. When one wishes to delete a subset of data points, only the component models that depend on the data points to be unlearned need to be retrained. This method enhances the efficiency of unlearning at the cost of relying on mixture models that may be not as effective as a single model trained on the entire data set.

15.7 General AI?

To conclude this book, I would like to return to the relationship between machine learning and general AI that was mentioned in Chapter 1. Broadly speaking, general AI refers to the capability of hypothetical agents to carry out cognitive tasks at the level of human beings. There is a sense among a number of researchers that we may be on the verge of developing such agents. As may be clear from this book, I side with the skeptics on this matter. Following American scientist Melanie Mitchell (see Recommended Resources, Sec. 15.8), here are five ideas underpinning the view that general AI is within reach, and why they may be flawed.

- The first idea is that "narrow AI", i.e., machine learning, is on a continuum with general AI. Should this be true, solving narrow tasks such as playing chess would entail progress towards the goal of developing an intelligent being.
- The second is our tendency to deem tasks that are difficult for us – such as playing chess – as *hard*, and to consider tasks that come naturally to us – such as common-sense reasoning – as

easy. Should this be the case, chess-playing software should be able to exhibit the cognitive capabilities of a toddler.

- Third, there is our propensity to describe the operation of artificial agents in terms of human capabilities. Should this type of language be accurate, a classifier may actually "observe" and "understand" the composition of images.
- Fourth, intelligence is often seen as residing inside the brain, whereas human intelligence appears to be inherently embodied and defined in relation to interactions with the world.
- Finally, we tend to view intelligence as "pure" and external to biases, goals, and limitations of a human being. But intelligence may actually be inextricable from such factors.

Ultimately, general AI appears to be outside the reach of machine learning and current AI systems based on induction, as well as of GOFAI solutions implementing deductive inference (see Chapter 1). Rather than chasing the ghost of general AI, a more useful driving objective for machine learning research seem to be that of developing methods that can augment human capacity to think, communicate, and create, within an agreed-upon set of ethical and societal principles, while being mindful of the limitations of the technology. In Norbert Wiener's words, written in 1954,[2]

The present desire for the mechanical replacement of the human mind has its sharp limits. When the task done by an individual is narrowly and sharply understood, it is not too difficult to find a fairly adequate replacement either by a purely mechanical device or by an organization in which human minds are put together as if they were cogs in such a device. ... To expect to obtain new ideas of real significance ... without the leadership of a first-rate mind is another form of the fallacy of the monkeys and the typewriter.

15.8 Recommended Resources

A comprehensive source of information on probabilistic graphical model is [1], and other useful references include [2] and [3]. The use of probabilistic graphical models for risk management is specifically discussed in [4]. Adversarial attacks and adversarial training are reviewed in [5, 6]. Discussion of model underspecification can be found in [7] and in the many follow-up comments available in the literature. A useful technical introduction to causality is provided by the book [8], and a higher-level discussion is offered by Judea Pearl in [9]. Quantum machine learning is the subject of [10, 11, 12] and references therein, and machine unlearning is studied in [13]. For a poetic and philosophical take on information and quantum theory by a physicist, the reader is referred to [14]. A cautionary view on the complexity of developing general AI, which inspired Sec. 15.7, is offered by Melanie Mitchell in [15]; and the dangers of the "myth of AI" are expounded in [16]. Finally, the book [17] elaborates on a useful analogy between the use human beings have made of animals for millennia and the integration of AI and robotics into our societies and workplaces.

---

[2] See N. Wiener, *Invention: The Care and Feeding of Ideas*. The MIT Press, 1993.

Bibliography

[1] D. Koller and N. Friedman, *Probabilistic Graphical Models: Principles and Techniques*. The MIT Press, 2009.

[2] R. M. Neal, "Connectionist learning of belief networks," *Artificial Intelligence*, vol. 56, no. 1, pp. 71–113, 1992.

[3] K. P. Murphy, *Machine Learning: A Probabilistic Perspective*. The MIT Press, 2012.

[4] N. Fenton and M. Neil, *Risk Assessment and Decision Analysis with Bayesian Networks*. CRC Press, 2018. .

[5] B. Biggio and F. Roli, "Wild patterns: Ten years after the rise of adversarial machine learning," *Pattern Recognition*, vol. 84, pp. 317–331, 2018.

[6] I. J. Goodfellow, J. Shlens, and C. Szegedy, "Explaining and harnessing adversarial examples," arXiv preprint arXiv:1412.6572, 2014.

[7] L. Breiman, "Statistical modeling: The two cultures (with comments and a rejoinder by the author)," *Statistical Science*, vol. 16, no. 3, pp. 199–231, 2001.

[8] J. Peters, D. Janzing, and B. Schölkopf, *Elements of Causal Inference: Foundations and Learning Algorithms*. The MIT Press, 2017.

[9] J. Pearl and D. Mackenzie, *The Book of Why: The New Science of Cause and Effect*. Basic Books, 2018.

[10] M. Schuld, *Supervised Learning with Quantum Computers*. Springer, 2018.

[11] M. Schuld, "Quantum machine learning models are kernel methods," arXiv preprint arXiv:2101.11020, 2021.

[12] S. Mangini, F. Tacchino, D. Gerace, D. Bajoni, and C. Macchiavello, "Quantum computing models for artificial neural networks," *EPL (Europhysics Letters)*, vol. 134, no. 1, p. 10002, 2021.

[13] L. Bourtoule, V. Chandrasekaran, C. A. Choquette-Choo, H. Jia, A. Travers, B. Zhang, D. Lie, and N. Papernot, "Machine unlearning," arXiv preprint arXiv:1912.03817, 2019.

[14] C. Rovelli, *Helgoland*. Prometheus, 2021.

[15] M. Mitchell, "Why AI is harder than we think," arXiv preprint arXiv:2104.12871, 2021.

[16] E. J. Larson, *The Myth of Artificial Intelligence*. Harvard University Press, 2021.

[17] K. Darling, *The New Breed*. Penguin Books, 2021.

Index